PROGRESS(ES) – THEORIES AND PRACTICES

PROCEEDINGS OF THE 3RD INTERNATIONAL MULTIDISCIPLINARY CONGRESS (PHI 2017 – PROGRESS(ES) – THEORIES AND PRACTICES), BARI, ITALY, OCTOBER 4–7, 2017

Progress(es) – Theories and Practices

Editors

Mário S. Ming Kong & Maria do Rosário Monteiro
CIAUD-FA ULisboa/CHAM, FCSH, Universidade NOVA de Lisboa, Universidade dos Açores, Portugal

Co-editor

Maria João Pereira Neto
CIAUD- FA ULisboa/CHAM, FCSH, Universidade NOVA de Lisboa, Universidade dos Açores, Portugal

CRC Press
Taylor & Francis Group
Boca Raton London New York Leiden

CRC Press is an imprint of the
Taylor & Francis Group, an informa business
A BALKEMA BOOK

CRC Press/Balkema is an imprint of the Taylor & Francis Group, an informa business

Typeset by MPS Limited, Chennai, India
Printed and bound in Great Britain by CPI Group (UK) Ltd, Croydon, CR0 4YY

Published by: CRC Press/Balkema
Schipholweg 107C, 2316 XC Leiden, The Netherlands
e-mail: Pub.NL@taylorandfrancis.com
www.crcpress.com – www.taylorandfrancis.com

ISBN: 978-0-8153-7415-2 (Set of Hbk and CDROM)
ISBN: 978-1-351-24269-1 (eBook)

Table of contents

Editorial foreword

It is our pleasure to present the third volume of Proportion, Harmony, and Identities (PHI). It follows PHI 2016, held in Lisbon and entitled Utopia(s); Worlds and Frontiers of the Imaginary, also published by CRC Press/Balkema[1]. The present volume is the result of the 3rd INTERNATIONAL MULTIDISCIPLINARY CONGRESS PHI 2017, held in Bari-Taranto-Italy, at the Swabian Castle and Sala Murat in Bari and the University of Trento in Taranto between the second and seventh of October, 2017.

The congress was designed as a platform for researchers, academics, and students to present, share and exchange ideas, visions of the past and the future and research results applicable to Architecture, Urban Planning, Landscape, Industrial Design, Design, Arts and Humanities, Engineering, Social and Natural Sciences and Technology, amongst others, on the importance of harmony and proportion as subjects that define, differentiate, and unite identities. Scholars were invited, through an international CFP, to submit research on theoretical and methodological aspects related to Progress in the aforementioned scientific fields and present their technical and theoretical research results carried out on this year's Congress core theme: "PROGRESS(ES) – THEORIES AND PRACTICES.

We received seventy-five papers from nine countries, all of which were subjected to double blind peer review, distributed according to each paper's scientific area to senior researchers for evaluation. All members of the scientific committee were also peer reviewers of papers within their field of expertise. After an intense process of scrutiny and after a rigorous linguistic review sixty-eight papers were selected for oral presentation and publication as chapters of this volume. In this sense, this book represents the combined effort of scholars from Brazil, Finland, Italy, Poland, Portugal, Romania, Sweden, the United Kingdom and the United States of America.

We decided to organize the publication in five major sections. The first, bearing the same title as the book itself, is formed by our guest speakers' articles. They serve, in our opinion, as important introductions to the diversity and complexity of themes involving Progress(es) – Theories and Practices but also to the articles of the following sections, with contributions on sciences and social implications of the recent discoveries in the realm of High Energy Particle Physics; the physical destruction of architectural heritage as a result of war and the experience of political and racial discrimination, during the transition between the first and second halves of the 20th century; the concept of progress in relation to urban and territorial planning and how debates in planning theory and practice can illuminate perspectives for creating a more progressive, participative and equalitarian urban society; a study focused on the physical, physiological and psychological colour effects and perception, in its interaction with human beings through objects and environment, underlining the importance of colour ergonomics.

The second section approaches Progress, Theories and Practices from the point of view of Architecture, Urban Planning and Design, or putting it in a different formulation, this section deals with the city, and its evolution based on the concept from theory to practice.

The third section assembles texts dealing with Progress, Theories and Practices in Arts, approaching different perspectives on their importance and influence on the development of society, knowledge, times and spaces.

The fourth section assembles the articles one can unite under the vast umbrella of Humanities: culture, literature, philosophy, cinema, science fiction, music and poetry.

The fifth and final section is dedicated to Science and Technology. It is organized chronologically and features articles on different fields of human realization and scientific advances.

Apart from our intervention in the organization of the volume, all individual chapters are the sole responsibility of their respective authors.

We thank all members of the scientific committee, partner universities, and their research centres, organizing committee members, Fondazione Puglia, Regione Puglia – Dipartimento Turismo, Economia della Cultura e Valorizzazione del Territorio, Regione Puglia – Assessorato alla Formazione e Lavoro – Politiche per il lavoro, Diritto allo studio, Scuola, Università, Formazione Professionale, Polo Museale della Puglia – MiBACT, Comune di Bari – Assessorato alle "Culture, Turismo, Partecipazione e attuazione del programma", Confindustria Giovani

[1] It came to our attention too late that in the proceedings of PHI 2016, in the chapter written by V. Mariz, Restoration of historic monuments as utopia: The bell tower(s) of Santa Maria de Belém church, there is a mistake in identifying a source (in page 207, lines 32 and 36; where one reads "Neto, Maria João Pereira" one should read "Neto, Maria João Baptista". We apologise to the author for this unfortunate mistake.

Imprenditori, and especially all the congress participants for making this congress and this book of proceedings possible.

Editors-in-Chief, PHI 2017 Congress Organizing Chairpersons

Mário S. Ming Kong
(Professor/Researcher, CIAUD-FA ULisboa/CHAM, FCSH, Universidade NOVA de Lisboa, Universidade dos Açores)

Maria do Rosário Monteiro
(Professor/Researcher, CHAM, FCSH, Universidade NOVA de Lisboa, Universidade dos Açores)

Co-Editor, PHI 2017 Congress Co-Organizing Chairperson

Maria João Pereira Neto
(Professor/Researcher, CIAUD-FA ULisboa/CHAM, FCSH, Universidade NOVA de Lisboa, Universidade dos Açores)

Committees

ORGANIZING COMMITTEES

Chief-Coordinators

Mário S. Ming Kong
Senior Researcher, (CIAUD) Centro de Investigação em Arquitetura, Urbanismo e Design – Faculdade de Arquitetura – Universidade de Lisboa/CHAM, FCSH, Universidade NOVA de Lisboa, Universidade dos Açores

Maria do Rosário Monteiro
Senior Researcher, CHAM, FCSH, Universidade NOVA de Lisboa, Universidade dos Açores

Co-Coordinator

Maria João Pereira Neto
Senior Researcher, (CIAUD) Centro de Investigação em Arquitetura, Urbanismo e Design – Faculdade de Arquitetura – Universidade de Lisboa/CHAM, FCSH, Universidade NOVA de Lisboa, Universidade dos Açores

SCIENTIFIC COMMITTEE

Aleksander Olszewski
Faculty of Art, Kazimierz Pułaski University of Technology and Humanities, Radom, Poland

He studied at the Academy of Fine Arts in Lódź. He is a Professor at the UTH in Radom, a Former Dean of the Faculty of Arts (2009–2016), Head of the Department of Digital Media and Photography. He works in the field of painting and computer graphics. Author of 40 solo exhibitions. Editor of a yearly "Arteria" published by the Art Department of the UTH in Radom. He was awarded the Cross of the Knight to the Order of Poland Restored.

Ana Cardoso de Matos
Member of Research Centre CIDEHUS/UE) and Professor at the Évora University, Department of History

Professor at the Évora University (Department of History) and a member of Research Centre CIDEHUS/UE. She is the responsible at the University of Évora of the Erasmus Mundus Master TPTI-Techniques, patrimoines, territoires de l'industrie. Her research interests are focussed on urban history and on the history and heritage of technology, engineering and industry. She has been Visiting Professor at École des Hautes Études en Science Sociales, Paris (2010, 2012). She publishes regularly both in national and international journals.

Ana Cristina Guerreiro
Senior Researcher, (CIAUD) Centro de Investigação em Arquitetura, Urbanismo e Design – Faculdade de Arquitetura – Universidade de Lisboa

Professor at the Faculty of Architecture, Technical University of Lisbon. She holds a PhD in Visual Communication from the same Faculty and a Post-degree in History of Art by the FCSH, Nova University of Lisbon. She is graduated in Fine Arts, by the Faculty of Fine Arts in Lisbon. She works and researches in Fine Arts and Theory of Art, including the relation between Arts and Architecture.

Ana Isabel Buescu
History Department/CHAM, FCSH, Universidade NOVA de Lisboa, Universidade dos Açores

Position Associate Professor History Department Faculdade de Ciências Sociais e Humanas Universidade Nova de Lisboa. anabuescu@netcabo.pt
www.fcsh.unl.pt/deps/historia
http://www.cham.fcsh.unl.pt/files/file_000992.pdf
Scientific domains & interests (15th–16th centuries) Portuguese History. Education of princes. Court culture. Royal ceremonies. Royal and aristocratic libraries. Biographical History.
Books (since 2005) Author:
2011 – *Na Corte dos Reis de Portugal. Saberes, ritos e memórias. Estudos sobre o Século XVI*, Lisboa, Colibri, 266 pp., 1st ed., 2010
2008 – *D. João III (1502–1557)*, 2nd ed. Lisboa, Temas e Debates, 416 p.

2007 – *Catarina de Áustria (1507-1578) Infanta de Tordesilhas, Rainha de Portugal*, Lisboa, A Esfera dos Livros, 491 p.
2005 – *D. João III (1502-1557)*, Lisboa, Círculo de Leitores, 344 p.Indexed Articles (since 2013)
– "Os Santos na Corte de D. João III e de D. Catarina", *Lusitania Sacra*, 2a série, 28 (Jul.-Dez.), 2013, pp. 49-72. ISSN: 0076-1508 | e ISSN: 2182-8822 ISBN: 978-972-8361-55-6, http://www.ft.lisboa.ucp.pt/resources/Documentos/CEHR/Pub/LS/2/Lusitania_Tomo28_indice.pdf
– "Utopia e profetismo no século das Luzes. D. José, Príncipe do Brasil, Imperador do Mundo", *Revista de História da Sociedade e da Cultura* (Coimbra), 13, 2013, pp. 283-310. ISSN: 1645-2259
– "Dimensão política e de poder da comida régia e do corpo do rei", Librosdelacorte.es (publicação electrónica www.librosdelacorte.es), 7, año 5, Otoño-invierno 2013 ISSN 1989-6425. http://www.librosdelacorte.es
– "Livros em castelhano na livraria de D. Teodósio I (c.1510-1563), 5° duque de Bragança", *Revista de Estudios Humanísticos*. Historia (Univ. Léon), 12, 2013, pp. 105-126. ISSN: 1696-0300
– "A livraria de D. Teodósio (1510–1563), duque de Bragança. A sua dimensão numa perspectiva comparada", *Ler História*, vol. 65, 2013, pp.59-73. ISSN: 0870-6182
Current Research (Projects)
– De Todas as Partes do Mundo, O património do 5° Duque de Bragança, D. Teodósio I. (PTDC/EAT- HAH/098461/2008), coord. Jessica Hallett/CHAM, FCSH/NOVA-UAc
http://www.cham.fcsh.unl.pt/teodosio/ (researcher)

Ana Leonor Madeira Rodrigues
Senior Researcher, (CIAUD) Centro de Investigação em Arquitetura, Urbanismo e Design – Faculdade de Arquitetura – Universidade de Lisboa

Visual Artist Full Professor at University of Lisbon – Faculty of Architecture
https://analeonorrodrigues. myportfolio.com
pt.linkedin.com/pub/ana-leonor-madeira-rodrigues/11/a99/a43
She graduated in Fine Arts at ESBAL (Lisbon). Between 1989 and 1992 she lived in Berlin where she specialized in Aesthetics and Artistic Communication, at Hochschule der Kunst TU.
She holds a PhD in Architecture (Visual Communication) and is Head of the Department for Drawing and Visual Communication and is full Professor at the Faculty of Architecture – University of Lisbon.
As an academic researcher, she is interested in the cognitive processes of the act of drawing and in its specificity as a mode of non-verbal communication.
Published several books. Ana Leonor M. Madeira Rodrigues is also an artist who exhibits regularly and since the 1990s is investigating and recording a self-created/invented bacteriological accident, that alters the relationship between closed inhabited spaces (kitchens) and their inhabitants: the Burned By Blue collapse.
She works on this accident since then. Two recent exhibitions: My Laboratory – Luanda, Angola. Compulsive Acts of Communication between Bacteria and Humans, Library of the College of Science
Member of CIAUD
Member of CAA – College of Art Association Member of ASCI – Art and Science Collaborations
Responsible for the Journal for Drawing *Linha do Horizonte*
http://linhadohorizonte.fa.utl.pt/

Ana Maria Martinho
Senior Researcher, CHAM, FCSH, Universidade NOVA de Lisboa, Universidade dos Açores

She has supervised postgraduate students across the world (including in the United States, Angola, Portugal, Cabo Verde and Mozambique). A number of her postgraduate students have attained positions in academia and politics.
Having been involved in over fifty international missions, she visited multiple times all five Lusophone African countries. She also has established academic links with Brazil, South Africa, Argentina,

Venezuela, Botswana, Canada, and the United States and, has lectured extensively in many of these locations.
She serves on the editorial board of 4 international academic publications. Since 1990, she has authored 71 publications; books, editions, book chapters, international papers.
Her current research focuses on one monograph project: Material Culture and Literature in Sub-Saharan Africa, as well as on 2 papers for peer-reviewed publications.

Ana Marta Feliciano
Senior Researcher, (CIAUD) Centro de Investigação em Arquitetura, Urbanismo e Design – Faculdade de Arquitetura – Universidade de Lisboa

Born in Torres Vedras, Degree in Architecture from the Faculty of Architecture of the Technical University of Lisbon in 1995, worked in the City Hall of Torres Vedras between 1996 and 1997, also exercising professional activity in Architecture and Urbanism.
In 2001 obtained the Master Degree in Housing Architecture from the Faculty of Architecture of the Technical University of Lisbon, with the dissertation "Habitação e Utopia nos Anos Sessenta; As Propostas do Grupo Archigram no Contexto de uma Década de Rupturas".
Between 2003 and 2007 held a scholarship from the Foundation for Science and Technology and attended the PhD course Teoría y Practica del Proyecto in the Polytechnic University de Madrid. Obtained the European Doctorate in Architecture in 2008, with the thesis "La Metáfora del Organismo en las Arquitecturas Visionarias de los Años Sessenta; La Obra del Grupo Archigram como Reinvención de un Nuevo Habitar".
Since 1997, is professor and researcher at the Faculty of Architecture of the University of Lisbon, teaching theoretical and practical disciplines of Architectural Project.
Writes on Art and Architecture. Books:
– *A Metáfora do organism nas Arquitecturas dos Anos Sessenta; a Obra dos Archigram como manifesto de um novo habitar* (ed. 2014; ISBN 978-989-658-264-7)
– *A Casa Senhorial como Matriz da Territorialidade; a Região de Torres Vedras entre o Tempo Medieval e o final do Antigo Regime* (ed. 2015; ISBN 978-989-658-335-4).

Annalisa di Roma
Associate Professor in Industrial Design, Polytechnic University of Bari

Since December 2016 PhD Associate Professor in Industrial Design, since December 2008 PhD Assistant professor in Industrial Design (S.S.D. Icar-13) at the Polytechnic of Bari. Enrolled at Architects Council of Taranto, n. 604.
In 2005-2007 – Post PhD Fellowship; in 2004 – PhD in Architectural Design; 2000 – Graduated in Architecture. She has obtained many acknowledgments, prizes and mentions in the field of applied activity in the architectural project and industrial design for international and national competitions.
Main research interests, since her PhD, are focused on updating the language of the artifacts, from the artistic industry issues until the digital industry and craft; particularly she has focused on the cultural heritage. Actively involved in the Apulia Regional industrial and creative districts.

António Leite
Senior Researcher, (CIAUD) Centro de Investigação em Arquitetura, Urbanismo e Design – Faculdade de Arquitetura – Universidade de Lisboa

Degree in Architecture from the Faculty of Architecture of the Technical University of Lisbon in 1995, professional exercise since 1995 is responsible for several projects of Architecture and Urbanism.
In 2001 received the degree of Master of Housing Architecture in the Architecture Faculty of the Technical University of Lisbon. Between 2003 and 2007, was research fellow of Foundation for Science and Technology, obtaining in 2008 the degree of 'European Doctor' conferred by Polytechnic University de Madrid, with the thesis "La Casa Romántica; de la Matriz Romántica a un Concepto Acrónico y Operativo en la Contemporaneidad".
Is professor and researcher at the Faculty of Architecture of the University of Lisbon since 1996/1997, teaching theoretical and practical disciplines of Architecture Project, taught project workshops of Urban Planning and Architecture in seminars in Portugal and abroad.
Writes for Art and Architecture publications, of which stand out the books *A Casa Romântica; uma Matriz para a Contemporaneidade*

(ed. 2014; ISBN 978-989-658-263-0)) and *A Casa Senhorial como Matriz da Territorialidade; a Região de Torres Vedras entre o Tempo Medieval e o final do Antigo Regime* (ed. 2015; ISBN 978-989-658-335-4).

Antonio Monte
Ricercatore del Consiglio Nazionale delle Ricerche-Istituto per i Beni Archeologici e Monumentali (CNR-IBAM))

Architect, Ph.D.
He is a researcher at the National Research Council for Archaeological and Monumental Heritage (CNR-IBAM) in Lecce. Since 2002, he has been teaching at the Master in Conservation, Management and Valorization of Industrial Heritage, University of Padua.
From 2005 to 2015 he taught industrial archeology, history of craftsmanship, Architecture and Restoration for Historical Built, History of Industrial Settlements and Industrial Heritage and Landscape Architecture and Industrial Heritage, University of Salento and Basilicata Studies. For more than 20 years he has been involved in knowledge, capitalization, conservation and enhancement of industrial heritage, business history and corporate museums and has been the author of numerous publications. In 2013 with the volume "Salento, Art of Producing Artisans, factories and captains of business between the nineteenth and twentieth century", editions Grifo, Lecce 2012, won the "Terni City of Archeology" Engineer-humanist Gino Papuli, father of Italian industrial archeology. On 22 January 2016 in Milan, under the Fifth Edition of OlioOfficinaFestival, he was awarded the Oil Culture Award. He is Vice President of the Italian Association for Industrial Archaeological Heritage (AIPAI). É vice presidente nazionale dell'Associazione Italiana per il Patrimonio Archeologico Industriale (AIPAI).

Barbara vaz Massapina
Senior Researcher, (CIAUD) Centro de Investigação em Arquitetura, Urbanismo e Design – Faculdade de Arquitetura – Universidade de Lisboa

Received her Architect's degree in 1994, Master's degree in Housing Architecture 2002 entitled "TA: Shelter Typologies. Homés proposal for the Homeless" and her PhD in Architecture in 2011 entitled "Brise-soleil in Architecture", all from the Faculty of Architecture, Technical University of Lisbon. The PhD thesis was awarded the SIL Prize for Development and Investigation.
Started her professional activity in 1989 while still a student in a private architectural office, working as a designer under the supervision of a senior architect.
In 1994 stayed in this office as a trainee architect, and worked as a coauthor on different projects in Architecture and Urban Design, some of them awarded with 1° Prize in the International Competition
for Cacilhas downtown planning, the 1° Prize for Unit 5 in the Vila Expo in Lisbon and the 1° Torres Novas City Prize for Rehabilitation.
In 1994 starts to teach 1st and 2nd year of architecture at a private University.
In 1999 started an academic career at the Faculty of Architecture, University of Lisbon, having taught Project to all the five years, as well as Conservation and Rehabilitation studies, Urban Planning and Design, and Interior Architecture.
Full Professor for the Department of Architecture, and effective member of CIAUD (Research Centre for Architecture, Urbanism and Design, Faculty of Architecture, University of Lisbon).
Author or coauthor of several papers presented at national and international congresses and seminars.
Coauthor of seven published books on the subject of Architecture and/or Conservation Studies.

Calogero Montalbano
Professor at the Polytechnic of Bari in Architectural and Urban Design

Architect, received his PhD in in Architectural Design in 2002. He is since 2008 an aggregate professor at the Polytechnic of Bari in Architectural and Urban Design. Member of Retevitruvio, Italian inter-university network of ssd icar/14/15/16 (since 2010). Member of the Academic Board of the PhD in "Architectural design for the Mediterranean countries" (2010-12). Elected member of the Academic Senate at the Polytechnic of Bari (2012-15). Elected member of the "Strategic Committee" of the Polytechnic of Bari (2013-15). Member of the Academic Board of the PhD "Architecture: Innovation And Heritage" (since 2013). Member of the Italian National Committee of ICOMOS – International Council on Monuments and Sites, Italian National Council of Monuments and Sites (since 2014). Member international Committee of ENHR – European Network for Housing Research (since 2015). Reviewer for the International Journal "Journal of Civil Engineering and Architecture Research" (since 2015). Deputy Rector of the Polytechnic of Bari for the new University Library System (since 2015). His research is based on the technical and technological updating of settlement and housing building techniques. He is author or co-author of 68 scientific papers in the context of Urban Enhancement of Cultural Heritage and Sustainable Development in Urban scale. He has also published 7 academic books and he took part in over 16 design competitions. He took part in over 7 international and 17 national programs. He is the inventor of one filed patent in the area of recycling and innovative sustainable buildings.

Carla Chiarantoni
Professor in Architectural Design at the School of Building Architecture of the Polytechnic of Bari

Engineer received her PhD in Building Engineering in 2005. Since 2011 she is an aggregate professor in Architectural Design at the School of Building Architecture of the Polytechnic of Bari. Her fields of research and teaching are focused on: building design and study of sustainable solutions in residential project of affordable housing and social housing; designing systems using renewable energy technologies in a residential project. Since 2011 she is Masters' Degree advisor for final studios on architectural and urban design at the Polytechnic of Bari.

Carlos Figueiredo
Senior Researcher, (CIAUD) Centro de Investigação em Arquitetura, Urbanismo e Design – Faculdade de Arquitetura – Universidade de Lisboa

Academic Position: Assistant Professor. Faculty of Architecture, Lisbon University
Main Research Fields:
– Visual Communication, Fictional Cinematic Visual Narratives in Documentary, Film and Animation
– Screenwriting and Conceptual Design, Media and Communication, Production Design, Scenography and Architecture
– Computer Graphics and Digital Tools of Project and Design, in its conceptual and representation components
– Material and Intangible Heritage, Identity, Memory and Society
Teaches in the 1st Cycle Degrees of Urban Planning, Architecture, Set Design, Design and Fashion Design and has a Degree in Architecture in the Faculty of Architecture of UTL (1986).
Short CV:
Post-Graduation:
– EU sponsored Course in "Informatics applied to Architecture and Urbanism", in FA-UTL, in 1987, being approved with a Distinctly Approved rating
– Post graduation in Architecture of Housing, in 1998-9, being approved with 18 values, in the Faculty of Architecture of UTL
PhD:
– In Communication Sciences, conferred by the University of Beira Interior, in 2008, under a dissertation entitled "Scenic and Film Space: Representation and Computer Graphics", approved unanimously by the Jury.
Main Contact (email): cfigpt@gmail.com/cfig@fa.ulisboa.pt

Clara Gonçalves
CITAD, Universidade Lusíada de Lisboa, Portugal/ISMAT, Portugal/Faculdade de Arquitectura da Universidade de Lisboa, Portugal

Degree in Architecture, 1993, FA/UTL | Master in Architecture, 1998, FA/UTL | Doctorate in Architecture, 2008, ETSA/US | Researcher, s. 2010, CITAD, FA/ULL | Member of ATINER, s. 2014 | Associate professor of Architecture and History and Theory of Architecture, s. 2008, ISMAT, Portimão | Author of published articles in several magazines on architecture | Copy-editor of Jornal Arquitectos, 2000-2003 and 2010-2012 | Practicing architect, s. 1995|

Cristina Figueiredo
Senior Researcher, (CIAUD) Centro de Investigação em Arquitetura, Urbanismo e Design – Faculdade de Arquitetura – Universidade de Lisboa

1995 to date – Assistant Professor of the Faculty of Architecture of ULisboa
2006 – PhD in Biotechnology at I.S.T./U.T.L., on topic: Development of a System of Biodegradation oxic – anoxic for Textile Wastewater, Lisbon/Portugal on 21/12/2006.
1999 – MSc in Biotechnology/Biochemical Engineering from IST/UTL, Lisbon/Portugal
1993 – Lic th in Textile Engineering (Production Branch) by U. Beira Interior, Covilhã/Portugal
1999-2006 – Member of ENVERG (Environment Research Group) IBB – IST/UTL
2007 to date – an Effective member of CIAUD (Research Centre for Architecture, Urbanism and Design), FA/ULisboa.
Research Areas of Interest: Textile Materials, Sustainable Textiles and Apparel, Biotechnology, Ecodesign.
2008-2012 – Coordinator of the Master in Fashion Design FA/UTL
2009 to date – Scientific Evaluator ANI (National Innovation Agency)
2007 to date – Scientific Advisor at the PhD in Design:
– The importance of biotechnology application in the textile industry: Natural textile dyes, public examination on 29/04/2016, Very Good with Praise
– New Materials and Techniques Applied to White Castle Embroidery Design in Textiles and Clothing, public examination on 12/06/2012, Very Good with Praise.
– Sustainability in Apparel Design, public evidence in 07/03/2013, Very Good with Praise.
Co -Advisor of the Scientific Research Project PhD in Design:
– Intelligent Clothing as a Tool to Design of Sports Performance, public evidence in 18/10/2012, Very Good with Praise.

Elisabete Rolo
Senior Researcher, (CIAUD) Centro de Investigação em Arquitetura, Urbanismo e Design – Faculdade de Arquitetura – Universidade de Lisboa

She is a graphic designer and teacher in the Design Department of the Faculty of Architecture at the University of Lisbon. As a graphic designer, she developed her career in a significant design studio (Atelier B2) for approximately nine years, and currently develops the activity as a freelancer.
Since 2010, she has been teaching and researching in graphic design, especially in the areas of editorial design, typography, brand design and graphic design history.
She developed her PhD thesis about Portuguese Graphic Design History, focusing on the work of Sebastião Rodrigues.

Fátima Vieira
Associate Professor (with "Agregação") at Faculdade de Letras da Universidade do Porto/CETAPS

She has been teaching since 1986. She is the Chairperson of the Utopian Studies Society/Europe since 2006. At ILCML – Instituto de Literatura Comparada Margarida Losa, she was the Coordinator for the three editions of the research project "Literary Utopias and Utopian Thought: Portuguese Culture and the Western Literary Tradition" (2001–2010) funded by FCT (Fundação para a Ciência e a Tecnologia). Now she is the Coordinator of the multidisciplinary research project (also funded by the FCT) "Utopia, Food and the Future: Utopian Thinking and the Construction of Inclusive Societies – A Contribution from the Humanities". Fátima Vieira is also the director of the collection "Nova Biblioteca das Utopias", of the Portuguese publishing house "Afrontamento", and the director of *E-topia*, an electronic journal on Portuguese utopianism since 2004. (http://ler.letras.up.pt/site/default.aspx?qry=id05id164&sum=sim), as well as of *Spaces of Utopia*, a transdisciplinary electronic journal on Utopia written in English (since 2006) – http://ler.letras.up.pt/site/default.aspx?qry=id05id174&sum=sim). She is also Book Review Editor for the North-American Journal *Utopian Studies* (edited by Penn State) since 2006.
In 2013, the American and Canadian association Society for Utopian Studies selected Fátima Vieira as a recipient of the Larry E. Hough Distinguished Service Award.

Fernando Moreira da Silva
Senior Researcher, Director of (CIAUD) Centro de Investigação em Arquitetura, Urbanismo e Design – Faculdade de Arquitetura – Universidade de Lisboa

Full Professor in Design at the Faculty of Architecture of the University of Lisbon (FA/ULisboa). MSc and PhD – University of Salford, UK; PhD – Technical University of Lisbon; Post-Doctorate from the University of Salford; President of CIAUD – Research Centre for Architecture, Urbanism and Design; Coordinator of the Design Scientific Area, Coordinator of the Masters and PhD degree in Design at FA/ULisboa; Member of the General Council of the University of Lisboa; Panel coordinator for the PhD and post-doc Scholarships in Design, Architecture and Urbanism at FCT; International consultant for CNPq, Brazil; Honorary Researcher at SURFACE, UK; Co-coordinator of the International PhD degree in Design and Innovation (general coordination of the University of Naples); Visiting Professor at several foreign universities in Brazil, Italy and Spain; Supervisor of 78 Master Dissertations and 48 PhD theses; Member of Scientific Committees of international scientific journals; Coordination and participation in funded scientific research projects; Publications in indexed scientific journals with peer review in databases such as Scopus and ISI; several book chapters and three books, published by editors such as Caleidoscopio or Taylor & Francis.

Giacomo Martines
Researcher in Architectonic Restoration – Polytechnic of Bari

He is University Researcher in Architectural Restoration at the Bari Polytechnic since 2009, PhD in Environmental Design.
Professor for the "Restoration" course in Architecture Degree Course and for the "Restoration Work-site" in the Specialization School. His research main theme is the valorization of the heritage. Was responsible for Bari Polytechnic in the project "Archeology of the Landscapes of Puglia Adriatic in Roman times". Is author of texts on landscaping and historical centers restoration.

Gianni Montagna
Senior Researcher, (CIAUD) Centro de Investigação em Arquitetura, Urbanismo e Design – Faculdade de Arquitetura – Universidade de Lisboa

PhD in Design by the Faculty of Architecture of the University of Lisbon since 2012, with a thesis on the application of smart systems to clothing, teaches Fashion Design at the first academic year of the Masters in Fashion Design and often teaches other disciplines at the same institution, as in the case of Accessory Design. In the textile development laboratory (D_TEX) of the Faculty of Architecture of Lisbon, he is responsible for the area of Textile Design and is responsible for the discipline of Laboratory III of the PhD in Design of the same Faculty. He is a guest lecturer at the Doctoral Program in Fashion Design at the University of Minho in collaboration with the Beira Interior University, where he collaborates in the teaching of "Interfaces Design Fashion and Technology".

Giovanni Leoni
Full Professor of History of Architecture at the University of Bologna

Leoni (b. 1958; Ma/PhD IUAV, Venice, Italy) is Full Professor in History of Architecture at the University of Bologna (www.unibo.it/sitoweb/giovanni.leoni). He coordinates the PhD Program in Architecture (phd.unibo.it/architettura/it). His main fields of research are: art and architecture theory in the 19th century, particularly John Ruskin; Post War architecture, particularly in Italy and in Portugal.

Iolanda Ramos
Senior Researcher, CETAPS, Universidade Nova de Lisboa – Faculdade de Ciências Sociais e Humanas

Professor at Universidade Nova de Lisboa, Portugal. She has published extensively within the field of Victorian Studies, mainly on political, economic and gender issues under the framework of Cultural Studies and Utopian Studies. Her doctoral thesis on Ruskin's social and political thought was published by the Gulbenkian Foundation in 2002. She is the author of *Matrizes Culturais: Notas para Um Estudo da Era Vitoriana* (Colibri, 2014) and co-editor of *Performing Identities and Utopias of Belonging* (Cambridge Scholars Publishing, 2013). Her essay "'Tell me what you like, and I'll tell you what you are' – An Overview of the Victorian Political Economy of Art" (2003) was published in *Cahiers Victoriens et Édouardiens*, a journal indexed in Scopus. She is a member of the Editorial Board of the online and double blind-refereed journals *The Eighth Lamp: Ruskin Studies Today* and *Spaces of Utopia: An Electronic Journal*. She has been carrying out research as part of the project “Mapping Dreams: British and North-American Utopianism” within the CETAPS (Centre for English, Translation and Anglo-Portuguese Studies).

Jorge Boueri
Universidade de São Paulo, Faculdade de Arquitetura e Urbanismo, Faculty Member

Architect, Urbanist and Designer. Full Professor at the University of São Paulo | Brazil.
Visiting Professor at Osaka City University, 2005, and the University of Lisbon, 2014-2015. He wrote four books on architecture dimension and housing.
He held several positions in the Public Administration of cities and produced several works of Urban Planning and Design.

João Cabral
Senior Researcher, (CIAUD) Centro de Investigação em Arquitetura, Urbanismo e Design – Faculdade de Arquitetura – Universidade de Lisboa

He is architect (Escola Superior de Belas Artes, Lisbon), Honours Diploma Urban and Regional Development Planning (Architectural Association, London), PhD (Urban and Regional Studies, University of Sussex, U.K.) is associate professor at the Faculty of Architecture (FA), University of Lisbon and researcher at the Research Centre for Architecture, Urbanism and Design. At FA he is currently president of the Department of Arts, Humanities and Social Sciences. He is also the coordinator of the “Colégio dos Urbanistas” at “Ordem dos Arquitectos”. His research interests and teaching activities range from urban planning, policies and methodologies to regional development, territorial governance, and spatial planning systems. He has been involved in research projects on governance and policies in urban regions. Recent relevant publications include Crespo, J. L. & Cabral J. (2012) “The institutional dimension to urban governance and territorial management in the Lisbon metropolitan area” in Seixas J. & Albet A. (ed.) *Urban Governance in Southern Europe*, Ashgate; and Cabral, J. (2015) Portugal, *disP – The Planning Review*, 51, Taylor & Francis.

João Luís Lisboa
Senior Researcher, Vice – Director of CHAM, FCSH, Universidade NOVA de Lisboa, Universidade dos Açores

PhD in History and Civilization, European University Institute, Florence, (*Mots (dits) écrits. Formes et valeurs de la diffusion des idées au 18ème siècle au Portugal*).
2001-2002, Director of the Portuguese Institute for the Book and Libraries, Ministry of Culture
2004-2014, director of the Centre for the History of Culture.
Fields of research: intellectual History, History of the book and reading, Modern European culture.
Some recent publications:
2016: “From publishing to the publisher – Portugal and the changes in the world of print in the 19th century” in Abreu and Silva (eds.), *The Cultural Revolution of the 19th century: Theatre, the Book-trade and Reading in the Transatlantic World*, London, I.B.Tauris Publ. (69-86 and 260-261).
2015: “Read, watch and laugh (with 18th century humorous books)” in Ferrão and Bernardo (eds.) *Views on Eighteenth Century Culture: Design, Books and Ideas*, Cambridge, CSP (346-357).
2011: “Ideas in(to) facts” in Irimia and Ivana (eds.), *Author(ity) and the Canon between institutionalization and questioning: Literature from high to late modernity*, Bucharest, ICR (144-152).
2009: “Facts being...” *Storia della storiografia*, 55, pp. 3-28.

João Paulo Oliveira E Costa
Senior Researcher, President Center of CHAM, FCSH, Universidade NOVA de Lisboa, Universidade dos Açores

1998 PhD in History, FCSH/NOVA University, Lisbon, Portugal.
1989 MSc in History
1984 Graduated in History, Faculty of Letters, University of Lisbon, Portugal.
Historian and novel writer, he is a professor and senior researcher with main scientific interests being the Renaissance and the European Maritime Discoveries.
1990- ...Professor, Department of History, FCSH/NOVA University, Lisbon, Portugal.
2002- Director of CHAM – Centre for the Humanities, (FCSH/NOVA-UAc), Lisbon, Portugal.
2013-...Coordinator of the E-learning Master Program in History of the Portuguese Empire, FCSH/NOVA University.

2013-...Member of the Scientific Council of FCSH/NOVA University
2016- ...Head of UNESCO Chair for Maritime Heritage
RESEARCH PROJECTS (selected)
2005-2015 Coordinator of the project EVE – Enciclopédia Virtual da Expansão Portuguesa (POS-C 7092/4.2/C/REG). http://www.fcsh.unl.pt/cham/eve/
2000-2015 Editor of *Bulletin of Portuguese Japanese Studies* (2000-2009) and Director of *Anais de História de Além-Mar* (2003-...).
2000-2008 Coordinator of the FCT funded research projects "Nobility in the India State in the 16th century" (POCTI/36414/HAR/99 e POCTI/HAR/55685/2004 2000-2008) and "Portuguese Jesuits missions in the Far EasT" (POCTI/HAR/66684/2004 – 2004-2008).
PUBLICATIONS (selected)
2014 (book) (coordinator and co-author with José Damião Rodrigues and Pedro Aires de Oliveira), *História da Expansão e do Império Português*, Lisbon, Esfera dos Livros.
2013 (book) *Mare Nostrum, em busca de honra e riqueza*, Lisbon, Temas & Debates (includes a chapter on the 15th century donations of Atlantic Islands, which were not yet discovered).
2011 (book chapter), "The beginnings of Portuguese Overseas Expansion" in *The Historiography of Medieval Portugal c. 1950-2010* (ed. José Mattoso), Lisbon, Instituto de Estudos Medievais, pp. 591-606
2009 (book) *Henrique, o Infante*, Lisboa, Lisbon, Esfera dos Livros.
2006 (book) *D. Manuel I, um príncipe do Renascimento*, Lisbon, Temas & Debates.

João Seixas
Physics Department of Instituto Superior Técnico (Lisbon, Portugal)/Member of Laboratório de Instrumentação e Física Experimental de Partículas (LIP)/Member of Centro de Física Teórica de Partículas (CFTP)/Member of CMS Collaboration

He is a Professor in the Physics Department of Instituto Superior Técnico (Lisbon) and researcher at the Particle Physics Laboratory (LIP). He is the Leader of the Portuguese Team in the NA60 Collaboration of the SPS Program at CERN and Team Leader of the Heavy Ions and Quarkonia Group in the CMS Collaboration of the Large Ion Collider (LHC) Program also at CERN. He has been a resident researcher at CERN and professor and researcher at several other institutions in Europe and outside Europe. He has authored more than 400 scientific publications in high impact journals and a book. He has also an important scientific outreach activity both in Portugal and abroad with regular presentations on Particle Physics for the public and for schools.
Besides his research work in theoretical and experimental Particle Physics, he also works in new energy efficiency methods and implementation of renewable energies in urban areas. He is responsible for pioneering studies in this scientific domain in Portugal.
More recently he has participated in the creation of an innovative didactic product (Euclides Box for the company Geometric World) aiming at radically changing the way Geometry and Mathematics are seen by students, parents and professors.

Jorge de Novaes Bastos
Senior Researcher, (CIAUD) Centro de Investigação em Arquitetura, Urbanismo e Design – Faculdade de

Full Professor at Faculdade de Arquitectura, Universidade de Lisboa
1. Academic background –
(a) Ph.D. (1987) – The University of Texas at Austin, Austin, TX, USA;
(b) M.Sc. (1983) – The University of Texas at Austin, Austin, TX, USA;
(c) Diplome (1978) – Universidade do Porto, Oporto, Portugal;
2. Professional activities – Rehabilitation projects (some studies)-
A.1. PUBLIC SPACES –
A.1./01 – Palácio dos Marqueses de Fronteira e de Alorna, Benfica – Lisboa, 1990-97
A.1./02 – Palácio de Pancas – Palha, Sta. Apolónia – Lisboa, 1991-94
A.1./03 – Adega dos Frades, Mosteiro de Santa Maria da Vitória – Batalha, 1994
A.1./04 – Picadeiro d' El – Rei, Praça – Forte de Almeida – Almeida, 1994
A.1./05 – Centro de Interpretação da Estação Arqueológica de Milreu – Algarve, 1996

A.1./06 – Arco Romântico do Parque de Monserrate – Parque Natural de Sintra – Cascais, 1998
A.1./07 – Centro de Interpretação do Parque Ecológico, CIPE, Monsanto – Lisboa, 1999-00
A.1./08 – Casa Rural do Séc. XVI, Estação Arqueológica de Milreu – Algarve, 1999-00
A.1./08 – Termas da Ferraria, São Miguel – Região Autónoma dos Açores, 2000
A.1./09 – Termas do Carapacho, Graciosa – Região Autónoma dos Açores, 2001
A.1./10 – Residência Nuno Vitorino, Quinta do Texugo – Sampaio, 2001
A.1./11 – Residência de Estudantes, Fábrica da Pólvora, Barcarena – Oeiras, 2003
A.1./12 – Recuperação da Envolvente do Jardim da Trv. do Hospital, Lisboa, 2004
A.1./13 – Fortaleza de Sagres – Túnel de Acesso e Torreão Central – Algarve, 2004-05
A.1./14 – Torre de Água, Jardim da Alameda Roentgen – Lisboa, 2004-05
A.2. ? PRIVATE SPACES –
A.2./01 – Casal da Beloura, São Pedro de Penaferrim – Sintra, 1989
A.2./02 – Edifício de Habitação, Rua Ribeiro Sanches, Lapa – Lisboa, 1990-97
A.2./03 – Refeitório da Fábrica de Linhas COATS & CLARK – Vila Nova de Gaia, 1993
A.2./04 – Casa da Várzea, Trescova – Felgueiras, 1996
A.2./05 – Casa de Santa Cruz em Felgar – Moncorvo, 1998
A.2./06 – Prédio na Rua de São Mamede ao Caldas – Lisboa, 1999
A.2./07 – Residência na Quinta de São João, Fontaínhas – Cascais, 2000-04

Jorge Firmino Nunes
Researcher, (CIAUD) Centro de Investigação em Arquitetura, Urbanismo e Design – Faculdade de Arquitetura – Universidade de Lisboa

He graduated from the Faculty of Architecture of Lisbon Technical University in 1993, where he obtained his PhD in Architecture in 2012, with the thesis *A Razão da Arquitectura: A Ideia de Metrópole nos Pensamentos de Manfredo Tafuri e Rem Koolhaas*. He teaches at the Department of Arts, Humanities, and Social Sciences of the Faculty of Architecture of Lisbon University. Projects Editor of Journal *Arquitectos* (2002-2005 e 2009-2012). He was part of the editorial team of Ordem dos Arquitectos between 2005 and 2007. Coordinator, with Ana Vaz Milheiro and João Afonso, for the exhibition Habitar Portugal 2003-2005 (2006 Architecture Venice Biennale). Researcher of the project Arquitecturas do Mar at the Faculty of Architecture of Lisbon University, supported by the Foundation for Science and Technology.

Julian Sobrino Simal
Profesor Titular de Universidad Universidad de Sevilla (Escuela Técnica Superior de Arquitectura)

Doctor in History of Art and Full Professor of the Technical School of Architecture of the University of Seville. He teaches in the Master of Architecture and Heritage and is part of the Doctorate of Architecture Program. He is a specialist in the History of Industrial Architecture and Systems of Interpretation of Industrial Heritage. Among his publications, we can emphasize "Arquitectura industrial en España, 1830-1990", "Arquitectura de la industria en Andalucía" y "Luces sobre la memoria. La Real Fábrica de Artillería de Sevilla". He is Vice-President of TICCIH-Spain.

Luís Crespo Andrade
Senior Researcher, CHAM, FCSH, Universidade NOVA de Lisboa, Universidade dos Açores

Professor in the Department of Philosophy of the Faculdade de Ciências Sociais e Humanas of the Universidade Nova de Lisboa.
Coordinator of the Early-Modern and Modern Thought Group of CHAM – Centre for the Humanities (FCSH/NOVA, UAc).
Coordinator of the Seminário Livre de História das Ideias.
Author of numerous studies on utopian thought and intellectual history, namely *Intellectuals, Utopia e Comunismo. A inscrição do marxismo na cultura portuguesa* (Lisbon, Fundação Calouste Gulbenkian – Fundação para a Ciência e a Tecnologia, 2010).

Maria de Fátima Nunes

President of the Advanced Research Scientific Board (IIFA) – évora University. Full Profofessor and Teacher of History, History of Cutlure and History, Science

Full Professor. Teacher of History, History of Cutlure and History of Science (18th-21st centuries) – 1st, 2nd and 3rd Bologna cycles.
Supervises PhD students, Master Students;
Scientific Coordination of 1st cycle: history and sources methodologies work; Director and scientific coordination of the Unit Research (R&D) History and Philosophy of Science Unit (CEHFCi) www.chefci.org (www.officinalis.org)
President of the Advanced Research Scientific Board (IIFA) – Évora University (2010_1013)
Co_coordination of the network History & Science with the IHC (Institute of Contemporary History-UNL)

Maria João Durão

Researcher, (CIAUD) Centro de Investigação em Arquitetura, Urbanismo e Design – Faculdade de Arquitetura – Universidade de Lisboa/Laboratório da Cor/ColourLab FAUL Coordinator

Prof. Maria João Durão, has a Diploma in Fine Arts, Faculty of Fine Arts, Lisbon University; obtained Ph.D. degree from The University of Salford, Manchester, UK (2000) funded by PRODEP; was granted equivalence to PhD in Scientific Area of Desenho e Comunicação Visual, Faculdade de Arquitectura de Lisboa, Universidade Técnica de Lisboa. Her post-Doctoral project funded by FCT- Foundation for Science and Technology: "Art, Science and Technology -Drawings of Santiago Calatrava, hosted at University of Salford Research Centre BUHU, Manchester, UK and Department of Architecture "Technion-Israel Institute of Technology Haifa", Israel (2009). She lectures courses of Drawing, and Architectural Drawing at Faculty of Architecture Lisbon since 2001, where she holds a wide range of scientific responsibilities within postgraduate courses: Member of Scientific Board of the Doctoral Course of Design, since 2006; Master and Doctoral Programme Modules: "Anthropology of Vision" (2003-present), and the nuclear module "Seminar of Thesis Project". She set up 'Master Degree in Colour for Architecture, Faculty of Architecture-Lisbon Technical University (2003) holding its scientific coordination. In 2004, she founded Colour Lab-Faculty of Architecture where she supervises doctoral and post-doctoral research projects. In 2016, she became coordinator of Colour and Light Research Group –CIAUD – Centre of Research of Architecture, Urbanism and Design. Co-founder of the Portuguese Colour Association (2003): elected President (re-elected 2009) and presently its Honorary President. Internationally elected twice for Colour Association (re-elected 2015). Vice-President/Portugal- International Association of Colour Consultants/Designers, presided by Frank Mahnke, Geneve (2003).
Maria João Durão has a wide range of international conference participations and scientific committee/peer review invitations.
Her research interests in Space Architecture were established in 2002: Member of Team 11 and co-author of *The Millennium Charter – Fundamental Principles for Space Architecture* Houston, Texas, USA (2002). She is a senior member of AIAA- American Institute of Aeronautics and Astronautics and its Aerospace Architecture Subcommittee (Design Engineering); and invited member "Human Factors Subcommittee", Space Architecture Committee, AIAA-Aerospace Sciences Meeting, Nevada (2007). Presently her research is linked to the Education Subcommittee of the Space Architecture Technical Committee -American Institute of Aeronautics and Astronautics.
She publishes mainly internationally and serves as editor or scientific coordinator in *The Portuguese Colour Association Newsletter* (2003-2010); *FABRIKART*, Universidad del País Vasco (2003-present); Scientific coordinator *Cor, Espaço Urbano, Arquitectura e Design*, Ed. Apcor & Robbialac (2013); ARCHINEWS -02, *Revista de Arquitectura, Urbanismo e Interiores*. Coordinator of *Light and Colour-Interdisciplinary and Transversal Phenomenon* (2012). Lately, she was nominated Editor-in-Chief of the *AEROSPACE AMERICA Journal* for the yearly article of the Space Architecture Technical Committee. ISSN: 0740-722X (2014; 2015; 2016).

Maria João Pereira Neto
Researcher, (CIAUD) Centro de Investigação em Arquitetura, Urbanismo e Design – Faculdade de Arquitetura – Universidade de Lisboa/CHAM, FCSH, Universidade NOVA de Lisboa, Universidade dos Açores

PhD History, Master Sociology.
Professor at Faculty of Architecture University of Lisbon. Member of the Department of Arts, Humanities, and Social Sciences. Main domains of teaching and research: applied Social Sciences, Humanities, Art and Architecture History, Design, Scenography, Heritage. Since 1978 scientific research experience and professional consulting in the domains of Social Sciences and Humanities. Member of the Scientific Board Faculty of Architecture (2008-2011; 2014…). Senator of The University of Lisbon, elected for the Scientific Board (2013-2017). Effective member CIAUD – Research Centre for Architecture, Urbanism and Design faculty of Architecture, University of Lisbon, associated member of CHAM – Centre for Humanities (FCSH/NOVA, UAc). Member of the Geographical Society – Lisbon, president of the Arts and Literature Section, and in 2016 also the Section of Heritage Studies, voter of the sections: Tourism, Communication Studies, Education and Ethnology. Author of several indexed texts in ISI, Scopus, Web of Science.

Maria da Luz Sampaio
Research Member of CIDEHUS – PHD in History of Science specializations in Museology in University of Évora

Graduated in History in the Faculty of Humanities from the University of Porto 1997; post-graduation in Social Museology at the University of Lusófona, Lisbon. Later on, in 2009 I have finished my Master in Local and Regional Studies in the Faculty of Arts and Humanities of University of Porto. Recently, in 6 july 2015, I have acomplish my PHD in History of Science specializations in Museology in University of Évora, with the title "From the factory to the Museum: identification, heritization and diffusion of technical-industrial culture.
From 1992 I was involved in the research project of the Inventory of Industrial Heritage of the city of Porto. In 1996 I was involved in the plan and opening of the Museum of Science and Industry of Porto, and later on, from 2000 to 2011 I was its Director. I have a long-standing interest and much experience in cultural management (specially museums), and also in the studies of industrial material culture. I have started in 2014 a training Course "Introduction to Industrial Tourism" in the Faculty of Arts and Humanities of Porto. Since 2013 Research Member of CIDEHUS – University of Évora. Since November 2016 Post-Doc CIDEHUS/University of Évora. Author of books and papers about local history, conversion of Industrial buildings, museum plan, industrial heritage.

Maria Leonor Garcia da Cruz
Professor and Researcher at Faculty of Arts and Humanities of the University of Lisbon

Professor and Researcher at Faculty of Arts and Humanities of the University of Lisbon-FLUL. PhD. in Early Modern History (UL 1999). Researcher in Early Modern History and European overseas Expansion, Maghreb, Asia and America, supervising master thesis, PhD. and postdoctoral dissertations in societies, mind-sets, institutions, political and financial management, representations.
Member of the Research Units Building and Connecting Empires and Cultural Encounters and Intersecting Societies at the History Centre UL, she supervises research projects in IMAGETICS: interdisciplinary studies on representations and identitarian constructions, and TREASURY: history of economic thought and management, taxation, social networks, politics and ethics.
Publications in RCAAP and in international journals indexed as *Vozes dos Vales* (Qualis-Capes, Latindex) and *O Ideário Patrimonial* (DRJI, Latindex).
https://lisboa.academia.edu/MariaLeonorGarciaCruz/CurriculumVitae

Maria Leonor Sampaio da Silva
Senior Researcher, CHAM, FCSH, Universidade NOVA de Lisboa, Universidade dos Açores

She is a Professor at the Faculty of Social Sciences and Humanities at the University of the Azores, where she teaches courses related to Translation and Portuguese and English Contemporary Culture. She holds a Master's degree in Anglo-Portuguese Studies and a PhD in English Culture. She has dedicated her research to English authors such as H. G. Wells and Bertrand Russell, and to themes that bridge Literature, Culture, Translation and the Visual Arts. She is a Researcher at CHAM – Centre for the Humanities (FCSH/NOVA, UAc) and has been involved in research projects concerning the cultural relevance of photographic documents. Besides several articles on the main academic fields of her research (utopian and science fiction writing, translation studies and dialogues of literature with other arts) her publications include a book on the Arts Academy of the Azores, several annotated translations of travel books by foreign authors who visited the Azores in the 19th century, and a book of short stories for which she was awarded in 2014 the Humanities Prize Daniel de Sá.

Maria Luísa Malato
Senior Researcher, CHAM, FCSH, Universidade NOVA de Lisboa, Universidade dos Açores

Associate Professor with tenure at the University of Porto (Faculty of Literature and Arts, Porto, Portugal) for rhetoric, and aesthetics. MA and PhD in Comparative Literature and Aggregation in Romance Literatures (2008). Main fields of research: Utopian Studies, Rhetorical Studies, and Theatre (18th-19th centuries). Since 1990, member of the French Society for Eighteenth-Century Studies (SFEDS). Member of the Institute of Comparative Literature Margarida Losa (ILCML). Since 2013, Vice-President of the Portuguese Comparative Literature Association (APLC). Co-Editor of *Pontes de Vista*, online journal about Literature and Philosophy. From her bibliography, we could highlight the following books: *A Little anthology of Portuguese Poems*, Porto, Cena Poética/Poetria, 2016.
"O Belo e os Monstros…, in *O Teatro em Portugal nos séculos XVIII e XIX*, eds. Ch. Müller e M. Neumann, Berlim/Lisboa, Ibero-Amerikanisches Institut, 2015, pp. 39-61;
L'Exil et le Royaume : d'Albert Camus à Vergílio Ferreira, Paris, Le Manuscrit, 2014, 240 pp.. Co-editor;
Uma História da Literatura Europeia, Lisboa, Quid Juris, 2008;
Le XVIIIe Siècle: une Époque d'Ombres et de Lumières. Co-author, Mons/Belgium, Univ. Mons, 2009;
Manual Anti-Tiranos, Porto Alegre/Brazil, 2009.

Maria do Rosário Monteiro
CHAM, FCSH, Universidade NOVA de Lisboa, Universidade dos Açores

Professor of Comparative Literature at the FCSH/UNL (New University of Lisbon). Graduated in Modern Languages and Literature at the University of Lisbon (1983), Master in Comparative Literary Studies from the Universidade Nova de Lisboa (1987) and PhD in Literary Sciences, speciality of Comparative Literature, from Universidade Nova de Lisboa (1997). Lectures Comparative Culture and Literature at graduate and postgraduate levels. Senior Researcher at CHAM – Centre for the Humanities (FCSH/NOVA, UAc). Editor of several books, both in Portuguese and English, author of the first academic books on Tolkien published in Portugal, and author of several essays on Utopia, Fantasy, and Science Fiction. Further information: Orcid ID: orcid.org/0000-0001-6214-5975; academia.edu: https://fcsh-unl.academia.edu/MariadoRos%C3%A0rioMonteiro. Full CV: http://www.degois.pt/visualizador/curriculum.jsp?key=4279105572167475.

Maria do Rosário Pimentel
Researcher, CHAM, FCSH, Universidade NOVA de Lisboa, Universidade dos Açores

Maria do Rosário Pimentel graduated in History at Universidade de Coimbra and was awarded her PhD in Portuguese Studies at Universidade Nova de Lisboa. She is currently Associate Professor at Universidade Nova de Lisboa. She has carried out activities predominantly in the area of Cultural History and Society with a special focus on Cultural topics, Slavery and the Slave Trade. She has published, among other works and papers, *Viagem ao Fundo das Consciências. A Escravatura na Época Moderna*. Lisboa, Colibri, 1995 (338 pages), which is a version of her PhD thesis, and *Chão de Sombras. Estudos sobre Escravatura*, Lisboa, Colibri, 2010 (200 pages).
She is currently researcher in the fields of Portuguese Culture and History of Slavery and is senior researcher at CHAM – Centre for the Humanities (FCSH/NOVA, UAc). Member of, Sociedade Portuguesa de Estudos do Século XVIII, and its international branch, Sociedade Internacional de Estudos do séc. XVIII, as well as with the Portuguese Committee for the UNESCO Project Slave route. Author of two monographes on slavery, editor of 5 proceedings on Portuguese Culture and Literature, and more than 20 articles published in national and international journals.

Mário Say Ming Kong
Researcher, (CIAUD) Centro de Investigação em Arquitetura, Urbanismo e Design – Faculdade de Arquitetura – Universidade de Lisboa / CHAM, FCSH, Universidade NOVA de Lisboa, Universidade dos Açores

Degree in Architecture AT FAUTL, PhD in Architecture in the field of drawing and visual communication at Escuela Superior Technical Architecture Barcelona - Universidad Politécnica de Cataluña (UPC-ETSAB). He is presently Phd Professor at FAUTL, Lecturer in ESELx and Visiting Professor at the Master Course in arts at ESBAL. Previously taught at Lusophona University (ULHT) and Independent University, respectively in the Departments of Urban Planning and. Architecture.
In 2000 was Coordinator of the first year of the course in Urban Planning ULHT In 1998 Regent of the discipline Design/CAD/Geometry in ULHT. He participated in scientific research studies and consulting work for outside entities. He has also participated in several publications, communications Indexed to Taylor & Francis and ISI, and training courses in order to disseminate the results of his research activities at national and international universities.
His main research areas are: Harmony and proportion in representation between West and East and its application to Sustainable Architecture, in particular by applying concepts of origami and Kirigami to materials such as paper and bamboo.

Paulo Pereira
Researcher, (CIAUD) Centro de Investigação em Arquitetura, Urbanismo e Design – Faculdade de Arquitetura – Universidade de Lisboa

Art and architecture historian, Master in Art History (Univerdidade Nova de Lisboa), PhD in History of Architecture (Faculdade de Arquitectura da Universidade de Lisboa) with the thesis *A Fabrica Medieval. Concepção e construção na arquitectura portuguesa, 1150-1550* (2012)
Invited lecturer in several academic meetings and seminars in Portugal, Spain, France (CNRS), Germany, Italy, U.S.A (George Washington University) and Brazil (USP). Curator of several exhibitions in Portugal and abroad Vice-President of IPPAR (National Heritage) between 1995 and 2002. Editor and author of *História de Arte Portuguesa*, (3 vols, Círculo de Leitores). Author of *Lugares Mágicos de Portugal* (8 vols., Círculo de Leitores, 2004-2005. Co-curator of the international exhibition *Neue Welten* (Novos Mundos), held in the Deutsche Historisches Museum, Berlin (2007/2008). Author of *Arte Portuguesa. História Essencial* (Temas & Debates, 2012) and *Decifrar a Arte Portuguesa* (6 vols Círculo de Leitores, 2014). Teacher at Faculdade de Arquitectura da Universidade de Lisboa.

Raffaella Maddaluno
Researcher, (CIAUD) Centro de Investigação em Arquitetura, Urbanismo e Design – Faculdade de Arquitetura – Universidade de Lisboa

Graduated in Architecture at the Faculty of Architecture at the Polytechnic of Bari in 2000. Gained a PhD in Architectural History during the course of Architecture and Construction at the Faculty of Engineering of the University of Tor Vergata in Rome in 2007 with a degree in architectural history entitled: *The housing estate of Inacasa Mario and Ridolfi Wolfgang Frankl in cerignola: Analysis and intervention strategies*. Obtained a specialization grant for study abroad in 2002 funded by the Polytechnic of Bari, at the Faculty of Architecture of the University of Porto (FAUP).
Attended a Master Second-level short specialisation degree in *Conservation, management and enhancement of industrial heritage* at The University of Padua with a thesis entitled *Valorization proposals of industrial heritage in Val Di Bisenzio: Isola wool mill case study. A Historical, architectural and territorial governance analysis*.
In 2012-2013 became a fellow of the Calouste Gulbenkian Foundation in Lisbon with a research project aimed at the publication of a book in Italian on *Fernando Távora: A Viagem do Gulbenkiam* 1960 and 2014 in the same Foundation with the theme of *Milling factories in portugal, the routes of wheat, flour and pasta in the transition to modernity 1860-1930*.
An Assistant Professor and Fellow at the Faculty of Architecture of the University of Matera in Basilicata from 2008 to 2011 in the courses of *History of the city and landscape, genealogy of the laboratory of architectural history, history of architecture* and an assistant on the course in the History of contemporary architecture at the Faculty of Architecture of the Polytechnic of Bari in 2003 and 2004.

Currently, she is Assistant Professor at the Faculty of Architecture of The University of Lisbon in the cousers of *History of the modern architecture and theory of design*.

Santi Centineo
Polytechnic University of Bari

Architect, graduated with full marks from Palermo University, specialized at Milan Teatro alla Scala, PhD in Interior and Exhibit Design at Politecnico di Milano under Andrea Branzí's tutorship.
He works as set-designer in many Italian theatres (such as Maggio Fiorentino, San Carlo di Napoli, La Fenice di Venezia, Carlo Felice di Genova, Comunale di Bologna, Verdi di Trieste, Petruzzelli di Bari, Regio di Torino) and abroad (Montpellier, Città del Mexico, Philadelphia, Tokyo, Avignone, Toulouse, Madrid, Burgos, Coruña, Bilbao, Santiago de Compostela, Tenerife, Salamanca).
In 2006 for Mozart's Don Giovanni at Genoa Carlo Felice, he was awarded "Pierluigi Samaritani Award" as best Italian young set-designer.
He is the author of many publications and essays on the interior, exhibit, furniture and set design.
He is an assistant professor of Interior, Furniture and Exhibit Design at Politecnico di Bari.

Teresa Botelho
Researcher, CETAPS, DLLM, FCSH, Universidade Nova de Lisboa

She is Associate Professor at The Faculty of Social and Human Sciences, Nova University of Lisbon, where she teaches American Studies. She has published extensively on African American, and Asian American culture and literature, theatre and drama. Her current interests include technological utopias/dystopias and the post-human, post-black literature, identity theory in its intersection with utopia, visual culture and cinema, the collaboration between sciences and literature, especially drama, and literary and visual representations of 9/11 with a special emphasis on Arab-American literature and televisual fiction.
Her most recent publications include:
– "Utopía y poder en Estados Unidos: Una unión más perfecta: utopías de pluralidad en los Estados Unidos." *Utopía y poder en Europa y América*. Ed. Moisés González García and Rafael Herrera Guillén. Madrid: Tecnos Editorial, 2015 pp. 251-268
– "Reimagining the Body in Post-Singularity Techno-Utopias" *Spaces of Utopia: An Electronic Journal*, 2nd series, no. 3, 2014 pp. 70-83- *Performing Identities and Utopias of Belonging*. Ed. Teresa Botelho/Iolanda Ramos. Newcastle: Cambridge Scholars Publishing, 2013
– "Dystopias of Belonging: Defamiliarizing Sameness in Danzy Senna's Symptomatic" *Performing Identities and Utopias of Belonging*. Ed. Teresa Botelho/Iolanda Ramos. Newcastle: Cambridge Scholars Publishing, 2013 pp. 52-68
– "Finding an Aesthetic of her Own: Partnering Identities in the Work of Faith Ringgold" *Women and the Arts: Dialogues in Female Creativity*. Ed. Diana V. Almeida. Bern: Peter Lang, 2013 pp.109-124
– "Becoming Tony Alfama: The American Education of a Portuguese Kid in Charles Reis Felix' Tony: A New England Boyhood." *Gávea-Brown: A Bilingual Journal of Portuguese-American Studies* Vol. XXXIV-XXXV, 2013 pp. 131-146.

Teresa Cabral
Researcher, (CIAUD) Centro de Investigação em Arquitetura, Urbanismo e Design – Faculdade de Arquitetura – Universidade de Lisboa

Professor and Coordinator of the master degree in Communication Design at the Faculty of Architecture at the University of Lisbon (FA-ULisboa) where she teaches Graphic Design, Communication Design, Typography and Infographics, since 2003.
Permanent Researcher at the CIAUD – Centro de Investigação em Arquitectura, Urbanismo e Design of the FA-ULisboa.
Professional practice in Graphic and Editorial Design since 1987.

Sponsors

This event and this book had the support of CHAM (FCSH/NOVA-UAc), through the strategic project sponsored by FCT (UID/HIS/04666/2013).

Preface

Mário S. Ming Kong
PHI 2017 Congress Organizing Chairperson
Professor/Researcher, CIAUD – FAUL, Faculty of Architecture – University of Lisbon, Lisbon, Portugal/CHAM, FCSH, Universidade NOVA de Lisboa, Universidade dos Açores
Secção de Desenho, Geometria e Computação, Departamento de Artes, Humanidades e Ciências Sociais, Faculdade de Arquitectura, Universidade de Lisboa, Portugal

We are pleased to welcome you to the book Proportion, Harmony and Identities "Progress(es) – Theories and Practices", the result of the the 3rd INTERNATIONAL MULTIDISCIPLINARY CONGRESS PHI held in Bari-Taranto, at Swabian Castle and Sala Murat in Bari and University of Trento in Taranto, between the second and the seventh of October 2017.

The congress Proportion, Harmony and Identities is an annual international event for the presentation, interaction and dissemination of research related to the topic of Harmony and Proportion relevant to Architecture, Arts and Humanities, Urban Planning, Design, Engineering and Social and Natural Sciences and aims to foster the awareness and discussion on the importance of this subject and its benefits for the community at large.

It is my pleasure to announce the participation of expert speakers from nine countries in this seven days event that, for the first time included workshops for the student community. We have received research papers from distinguished participating academics and researchers from countries accross three continents. Thus, this event revealed itself as being a platform for researchers in a vast variety of fields to discuss, share, and exchange experiences. The comprehensive content of the congress has attracted immense attention and the wealth of information shared through the papers is, in our point of view, extremely useful for professionals working in the related fields.

This book, divided into five parts, contains a collection of research papers, organized as chapters, presented at PHI 2017 in Bari and Taranto, Italy. This publication, containing the conference proceedings, documents and presentations made at PHI 2017, is the result of the creative work of their authors and of a highly selective peer-review process.

I would like to express my sincere thanks to all those who have contributed to the success of PHI 2017.

The 3rd INTERNATIONAL MULTIDISCIPLINARY CONGRESS PHI 2017: "PROGRESS(ES) – THEORIES AND PRACTICES" would not have been possible without the help of a group of people from the Faculty of Architecture of the University of Lisbon, the Faculty of Human and Social Sciences (FCSH) of the New University of Lisbon and the Department of Architectural and Urban Design from the Polytechnic University of Bari, (Poliba). Researchers affiliated to the research centres from these universities – CIAUD, CHAM, and the Polytechnic University of Bari, (Poliba) – selflessly and enthusiastically supported and helped me to overcome the many obstacles that arose during the preparation and organization of this event.

I want to thank all authors of submitted papers for their participation. All contributed a great deal of effort and creativity to produce this work, and I, in my quality of organizing chair, am especially happy that they chose PHI 2017, as the place to present it. Credit also goes to all collaborators, in particular, the Scientific Committee members and reviewers, who donated substantial time from their busy schedules to carefully read and conscientiously evaluate the submissions.

In addition, in the name of the Organizing Committee, I would like to take this opportunity to extend our sincere gratitude to all supporting Organizations, for their support and encouragement and for making the event a success.

Special thanks go to all our speakers, authors, and delegates for making PHI 2017 a fruitful platform for sharing, learning, networking, and inspiration.

We sincerely hope you find this publication enriching and thought provoking.

Bari, October 2017

Introduction

Mario S. Ming Kong
Professor/Researcher, CIAUD – FAUL, Faculty of Architecture – University of Lisbon, Lisbon, Portugal/CHAM, FCSH, Universidade NOVA de Lisboa, Universidade dos Açores
Secção de Desenho, Geometria e Computação, Departamento de Artes, Humanidades e Ciências Sociais, Faculdade de Arquitetura, Universidade de Lisboa, Portugal
mskong@fa.ulisboa.pt

"The greater part of progress is the desire to progress"[1].

Lúcio Aneu Sêneca (4 a.C–65 d.C)

"Every day you may make progress. Every step may be fruitful. Yet there will stretch out before you an ever-lengthening, ever-ascending, ever-improving path. You know you will never get to the end of the journey. But this, so far from discouraging, only adds to the joy and glory of the climb"[2].

Winston S. Churchill (1874–1965)

"We can only see a short distance ahead, but we can see plenty there that needs to be doné'[3].

Alan Turing (1912–1954)

PROGRESS(ES) – THEORIES AND PRACTICES": HUMANITY IN PROGRESS

The word progress derives from Latin *progressus* 'an advance', from the verb *progredi*, from *pro-* 'forward' plus *gradi* 'to walk'. The Cambridge dictionary defines progress as a movement to an improved or more developed state. So, when we speak of progress, we generally associate this concept with the idea that the world can become gradually better.

In this perspective, the concepts of progress and development combined form the essence of the significance of humanity's evolution.

Thus, progress is inextricably linked to man's need to create a better world, that is, progress is born of an ideal, of a utopia[4]. In the words of Oscar Wilde (1854–1900), "A map of the world that does not include Utopia is not worth even glancing at, for it leaves out the one country at which Humanity is always landing. And when Humanity lands there, it looks out, and, seeing a better country, sets sail. Progress is the realisation of Utopias"[5].

In this sense, progress is part of the traditional conception of history, which describes it as a univocal path for humanity: the path of development of science, technology, but also of social consciousness.

While progress is often associated with the notion of straight-line evolution, there are, however, alternative conceptions, such as the cyclical theory of the eternal return formulated by Friedrich Nietzsche (1844–1900); the dialectic theories, according to the design of Georg Hegel (1770–1831), or of Karl Marx (1818–1883); the theory that envisions a progress in the shape of a spiral, among others. Unlike the line of thought of linear progress, in these understandings, retrogression is part of the evolutionary process of development.

Rethinking paths, questioning dogmas, facing settled ideals, going against the tide is necessary to move forward and, sometimes, it is progressing by regressing.

If we look at history through this perspective, we can take as an example the development of the Renaissance period, undoubtedly a period in which great advances were made at all levels, but where traditional instituted

---

[1] Moral letters to Lucilius, Letter 71, 36, Translation Richard Mott Gummere.

[2] Cit. in Churchill by Himself (2008), ed. Langworth, PublicAffairs, p. 568.

[3] Computing Machinery and Intelligence (1950); Published in Mind – A Quarterly Review of Psychology and Philosophy, vol. 59, #236 (1950), p. 460.

[4] Cf, Maria do Rosário Monteiro; Mário S. Ming Kong (Chief-Editors); Maria João Pereira Neto (Co-Editor), (2017), *Utopia(s) – Worlds and Frontiers of the Imaginary*, CRC Press / Balkema, Taylor & Francis Group. London UK.

[5] Oscar Wilde, *The Soul of Man under Socialism 2004* by Kessinger Publishing (first published March 15th 1891).

and mythical dogmas had to be questioned, and a classical understanding of the world, more centred on man, had to be recovered, so to liberate minds and consciences and allow a new view of the world. It was looking back to the ideals of classical antiquity that allowed the philosophical, social, artistic and technological progress, so revolutionary, at that time.

Thus, we can assess that one of the main implications of the modern idea of progress lies in the significance of time at a given period, it is a cumulative vision of time that shifts the emphasis from cyclical time, hitherto predominant in traditional forms of thought and religiosity.

Progress refers, in this line of ideas, also to emancipation, to "human" liberation, to the evolution of knowledge and technique. This was the vision of progress that became a driving force in the eighteenth and nineteenth centuries as one of the main legacies of the age of Enlightenment[6].

The evolutionary-progressive view, established in the nineteenth century, inspired by Charles Darwin's theory of the evolution of species (1809–1882), was another striking moment in the history of humanity's progress. Again, a vision that promises to rupture with the established thought relative to the physical and spiritual nature of the Man, comes to give an impulse to the progress of the Human Sciences like Linguistics, Sociology, History and Philosophy.

The idea of progress is, therefore, a major matrix for development, its theoretical and practical foundations have become the working tools of scientists, philosophers, and artists, who seek strategies and policies to accelerate the development process in different contexts.

In fact, the last centuries were exceptional periods, where science and technology surprised and revolutionized social structures, accompanied by great contradictions. But technological and social scientific advances have always been accompanied by recess and social, ecological and ethical-moral crisis.

The more globalized world without frontiers is being built by building new, more subtle and efficient frontiers – microphysical frontiers, as the French philosopher Michel Foucault (1926–1984) would say[7].

We do not know what the Future holds for us, what progress humanity will face, and what setbacks it will take to reach them. But the process of progress itself is certain since the will to progress is part of human nature.

BIBLIOGRAPHY

Bernal, J. D. (1969) *Ciência na História*. v. VII. Lisboa: Livros Horizonte.

Echeverría, J. (2001) Tecnologias, espacios de interacion y valores. In: Cerezo, J. A. L.; Luján, J. L.; Palácios, E. M. G. (eds). *Filosofia de la tecnologia*. Madrid: Organização dos Estados Iberoamericanos para la Educación, la Ciência y la Cultura (OEI), pp. 15–29.

Foucault, Michel. (1982) *Microfísica do poder*. Rio de Janeiro: Graal.

Kong, Mário S. Ming. (2013) *Central Tejo. Uma abordagem da Arquitetura Industrial*. Lisboa: Editora Insidecity Lda.

Marx, Leo; Mazlish, Bruce. (1996) *Progress. Fact or illusion?*, Michigan: University of Michigan.

Misbett, R. (1985) *A história da idéia de progresso*. Brasília: Editora Universidade de Brasília.

Monteiro, Maria do Rosário; Kong, Mário S. Ming (Chirf-Editors); Neto, Maria João Pereira (Co-Editor). (2017). *Utopia(s) – Worlds and Frontiers of the Imaginary*. London: CRC Press Taylor & Francis Group.

Sack, Robert David. (2002) *Progress. Geographical Essays*. 3 ed. Baltimore, Maryland: The Johns Hopkins University.

Wilde, Oscar. (2014) *The Soul of man under socialism*. Londres: Kessinger Publishing. x, 99, p.

---

[6] Mário S. Ming Kong, (2013) – *Central Tejo. Uma abordagem da Arquitetura Industrial* Editora Insidecity Lda, Lisboa, pp. 185.

[7] Cf. FOUCAULT, Michel (1982). *Microfísica do poder*. Rio de Janeiro, Graal, 1982.

Part I

Progress(es) – theories and practices

Discovery, science and progress

João Seixas
Laboratório de Instrumentação e Física Experimental de Partículas, LIP-Lisboa, Lisboa, Portugal
Departamento de Física, Instituto Superior Técnico, Universidade de Lisboa, Lisboa, Portugal

ABSTRACT: I analyse the major scientific and societal implications of the recent discoveries in the realm of High Energy Particle Physics. These discoveries are considered in a broader context of a general shift in the way science is done by its practitioners and perceived by the public.

Keywords: Particle Physics, Higgs boson, Standard Model, Quantum Physics.

1 DISCOVERY

Since the early 20th century, we have witnessed a major revolution in the way we perceive and understand the Universe at its tiniest scales. Major breakthroughs in the realm of atomic, nuclear and subnuclear physics have brought with them a new vision of Nature and the role we, as humans, play in it. This revolution has one name: the quantum.

Until the end of the 19th century, Nature was assumed to exchange physical quantities continuously between its actors. Because this is so deeply rooted in our experience and education, it is normally forgotten that such quantities, namely, linear momentum, angular momentum, and energy, were actually fundamental discoveries of the Newtonian model of Nature. Contrarily to his predecessors, Newton imposed that it is not the velocity of a body, but the product of its mass by its velocity that is the fundamental dynamical quantity to be considered in any interaction. It is the product of mass and angular velocity that describes the kinematics and dynamics of rotating bodies. Energy, in its various forms, is the currency that Nature uses in all physical processes. The all-pervading concept of mass underlies the Newtonian Universe, defines the relation between motion and force (or between motion and energy in the Leibnitz formulation), and is the fundamental source for gravitation.

But what is mass? What is energy? These questions were left unanswered, for lack of appropriate data, during three centuries. Meanwhile, they have been appropriated by the "man in the street", especially at the turn of the 20th century, in many ways the original discoverers of the concepts would never abide by. This act of appropriation of words and concepts continues to this day mostly due to the general feeling that scientific concepts can be exported freely to other domains and used independently of the circumstances for which they were defined. This is, of course, a perilous act.

By the end of the 19th century, the three major branches of Classical Physics, namely Mechanics, Thermodynamics and Electromagnetism, were essentially complete, each in its own domain of relevance. Statistical Mechanics was put forth by Ludwig Boltzmann as an attempt to give a deeper microscopic meaning to the fundamental laws of Thermodynamics, but since the atomic structure of Nature was still somewhat foggy it did not enjoy the relevance it has today. James Clerk Maxwell had meanwhile shown that light corresponds to a vibration of the electromagnetic field and Heinrich Hertz produced electromagnetic waves for the first time in 1885. A question remained though: how does radiation interact with matter? How are the physical quantities described above exchanged in the process?

The answer to this question dwelled in the three known branches of Physics at the same time: it relied on a model for the motion of the atoms in a given material, its relation with temperature, and how the electromagnetic radiation was emitted and absorbed by them. As usual in Physics, a simple, manageable model encompassing all the necessary properties was necessary. Gustav Kirchhoff created this model in 1860: it consisted of a cavity with light absorbing inner walls and a very small hole such that any light entering the cavity through this hole would never leave it again.

One could, of course, also heat the cavity and radiation would then be emitted which could be observed and analysed through the hole. The spectrum of this radiation – that is, the relation between the intensity of the different light components and their colour– could then be observed with a spectrometer. Using an extremely beautiful and elegant argument based on the second Law of Thermodynamics Kirchhoff was able to show that the radiation spectrum was independent of the material constituting the cavity. This was, therefore, a universal model for the interaction of light

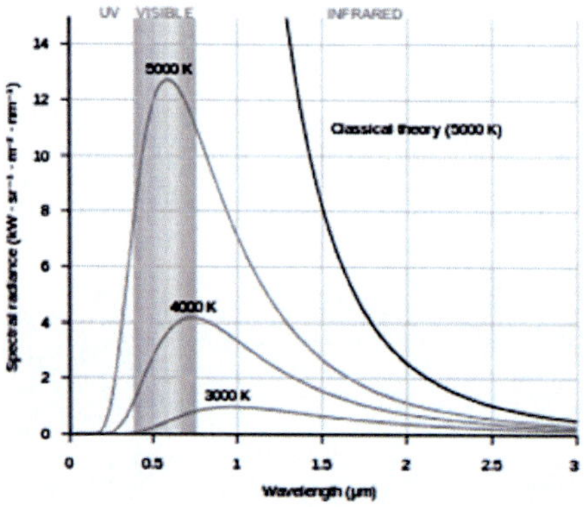

Figure 1. Blackbody radiation spectrum. The curve marked "Classical Theory" corresponds to the Raleigh-Jeans Law (source https://en.wikipedia.org/wiki/ Black-body radiation).

with matter. The name for this object was dubbed by Kirchhoff himself: the black body.

The experimentally observed blackbody radiation spectrum presented, however, a significant problem: for small wave-lengths it was in complete disagreement with the model for the interaction between radiation and matter stemming from the basic principles of Classical Physics (the so-called Raleigh-Jeans law) as can be seen in Fig. 1. Particularly worrisome was the fact that the Raleigh-Jeans Law formally implied that the total amount of energy in the cavity was infinite!

The solution to this problem came reluctantly from Max Planck in 1900. In order to explain the full spectrum of blackbody radiation, Planck proposed that energy was not exchanged continuously between radiation and the atoms contained in the cavity walls. The energy was exchanged instead in small parcels, which Planck named "quanta".

Despite their success in explaining the interaction between radiation and matter, quanta did not enjoy full recognition until Einstein showed how light quanta (or photons) could explain the recently discovered photoelectric effect.

Meanwhile, the wave nature of matter, first proposed by Luis de Broglie, found a confirmation in the Davisson–Germer experiment. The door to Quantum Theory was open.

2 SCIENCE

Science is the art of discovery. Building on the success of quanta in explaining the basic mechanism of interaction between light and matter, the first decades of the 20th century witnessed an epistemological revolution in the way the Universe was perceived. A systematic effort both on the experiment and theory fronts unravelled a new vision of reality hitherto hidden from humankind. Niels Bohr gave a first description of the atomic structure; Arnold Sommerfeld and his student Werner Heisenberg complemented this description with the first hints of spin, which were complemented by Wolfgang Pauli; the puzzling nature of spin was further disclosed by the Stern-Gerlach experiment. Photons were found to behave as particles in Compton processes. Wave-particle duality was recognised as a basic ingredient of Nature at the atomic scale (or smaller) and based on it a full quantum theory was first devised by Heisenberg and later by Erwin Schrödinger. Quantum Mechanics was born.

In the wake of this movement, important philosophical changes took place regarding the role of humans in Nature, both in its creation and evolution. Nowhere better is this change visible than in the discussions between Einstein and Bohr (BOHR 1949). Although somewhat dimmed by the 1960s-1980s motto of "calculate or perish", the quantum mechanical debate over the ultimate role of the observer (that is, the human) in quantum processes never fully left the stage as is demonstrated by the introduction of Bell's inequalities (BELL 1966). This discussion is presently revived by the recent developments in quantum computation and quantum optics.

Quantum Mechanics, as the name indicates, describes the evolution of quanta under the interactions that are present at a given instant. It is in this context that the first surprises appear in the quantum realm. There are only four basic interactions: electromagnetic and gravitational which have an infinite range; strong and weak interactions, which act only at the nuclear scale (that is within 1 femtometer or 10^{-15} m). There is no indication whatsoever that other interactions acting at intermediate scales exist.

The number of basic building blocks of Nature is equally limited and organised (by mass) in "families". In the so-called Standard Model (HERRERO 1998), they are three:

$$\begin{pmatrix} u \\ d \\ \cdots \\ e \\ \nu_e \end{pmatrix} \quad \begin{pmatrix} c \\ s \\ \cdots \\ \mu \\ \nu_\mu \end{pmatrix} \quad \begin{pmatrix} t \\ b \\ \cdots \\ \tau \\ \nu_\tau \end{pmatrix}$$

In each family, above the dotted line, lay the quarks with flavour (u)p, (d)own, (c)harm, (s)trange, (t)op, and (b)eauty (or (b)ottom) which participate in strong and weak interaction processes. Each of them can have one of three possible "strong charges" which are named "colours". They are the strong interaction equivalent of the electric charge for electromagnetic processes. The quarks also have an electric charge, albeit with quite unusual values of 1/3 or 2/3 of the electron charge.

Bellow the dotted line lay the "leptons" and their neutrino partners. They come in three (weak interaction) flavours: (e) electron, (μ) muon and (τ) tau, all of them having the (electric) charge of the electron (more about the electric charge of the leptons and quarks below). The neutrino partners have no electric charge, have very small or even zero mass, but they have a weak interaction charge or "weak isospin". They interact extremely weakly with matter. The role of neutrinos is presently under deep scrutiny since they can represent a window of observation of what lays beyond the present Standard Model.

The strong, electromagnetic and weak interactions are mediated by exchanged particles. The (massless) photon γ mediates the electromagnetic force; the weak interactions are mediated by three types of (quite massive) particles called W^+, W^- and Z^0 which, as the name indicates, have positive, negative and zero electric charge. The strong force is mediated by eight electrically neutral gluons that connect the colour charge of one quark and the colour charge of another quark within range.

The different interactions play very different roles. The strong interactions are responsible for the stability of the nucleus and thus the stability of matter. The electromagnetic interaction is responsible for the atomic and molecular structure. Finally, the weak interaction is responsible for the mixing between the families and thus ensures evolution at the microscopic level.

Each particle (except the neutral ones) has an electric counter partner or anti-particle. For instance the electron (e^-), of negative (electric) charge has a counter partner called positron (e^+) of opposite electric charge. When a particle and an anti-particle meet, they annihilate yielding a photon.

As mentioned before, particles also have an intrinsic (quantum) rotation or spin. The spin of a particle can only be either half-integer or integer. In the first case, the particle is called a "fermion"; in the second, it is called a "boson". Fermions and bosons have very different characteristics: two bosons can be at the same place at the same time, while fermions cannot (Pauli Exclusion Principle).

Since the 1980s "supersymmetry" (or "SUSY"), a new type of symmetry relating bosons and fermions, has been proposed as an extension of the Standard Model. However, to this day, no indication whatsoever has been found of its existence.

Underlying all the Standard Model structure described so far lays the ethereal Higgs field introduced in the 1960s by Peter Higgs, Robert Brout and François Englert and further exploited by Carl Hagen, Gerald Guralnik and Thomas Kibble (KIBBLE 2009). The Higgs field quantum, also called Higgs boson, is the only known fundamental particle with a spin consistent with zero (hence a boson). The precise nature of this particle is not yet fully clear. Recently discovered at the Large Hadron Collider at CERN (HIGGS 2012), the Higgs boson has been since then at the centre of an intense experimental and theoretical research effort. Up to this moment, it is unclear whether this particle is nothing but the keystone of the known subnuclear structure, or whether it is the herald of a new, yet to be discovered, the realm of new Physics beyond the Standard Model.

In this section we have described the ultimate "how things work", the ultimate material structure composing the Universe as we know it. A word of caution is in order, though, regarding the role of the Higgs boson. This particle is indeed a fundamental piece without which the whole construction of the Standard Model would fail. It does give mass to the elementary particles composing it but the vast majority of the mass in the Universe DOES NOT have its origin in the interaction with Higgs field. The (known) mass in our Universe comes mostly from the less known of the four interactions, namely the strong interaction, whose intensity (hence its associated energy) is orders of magnitude larger than the intensity of the other three. The well-known Einstein mass-energy formula (slightly tweaked) $m = E/c^2$ provides the necessary basis for this argument.

In addition, the attentive reader might have realised that we made no mention to gravity in the Standard Model. On one hand, it is true that at the scales at which we presently test the model gravity plays no role: its intensity, when compared to the other three interactions, is utterly irrelevant. However, this absence also reflects our inability so far to construct a quantum version of gravity.

3 A LIMITED UNIVERSE

Since humankind initiated its conscious journey towards the Universe the way we, as humans, perceive the ultimate reality evolves (or revolves) between an open Universe and a closed one. From Newton, we inherited an infinite and eternal Universe where (almost) everything was possible within the three basic Laws of Mechanics. However, this vision started to change in the late 19th century.

Following the null result of the Michelson-Morley experiment, Einstein showed that the speed of light in vacuum is an upper bound to all velocities. Communication between observers became thus severely limited.

Walther Nernst demonstrated that the absolute zero (−273.15 °C) could not be attained by any finite sequence of operations. There is, therefore, an inaccessible physical region.

In late 1920, Edwin Hubble showed that the Universe is expanding and George Lemaître demonstrated, consequently, that the Universe's history could be traced back to a single point (a singularity). The observable Universe itself thus became not only limited in its size (in particular because the speed of light is finite) but also in its age. One of the most striking demonstrations of this fact is provided, actually, by a graph akin to Fig. 2. In Fig. 2 we can see the spectrum of the Cosmic Microwave Background Radiation (CMBR), an electromagnetic remnant radiation from the Big Bang. This curve (the error bars in the points are smaller than the size of the symbols used to represent them) constitute the most precise measurement of a blackbody spectrum. Only tiny fluctuations in this background spectrum are responsible for the existing structure of the Universe. These fluctuations,

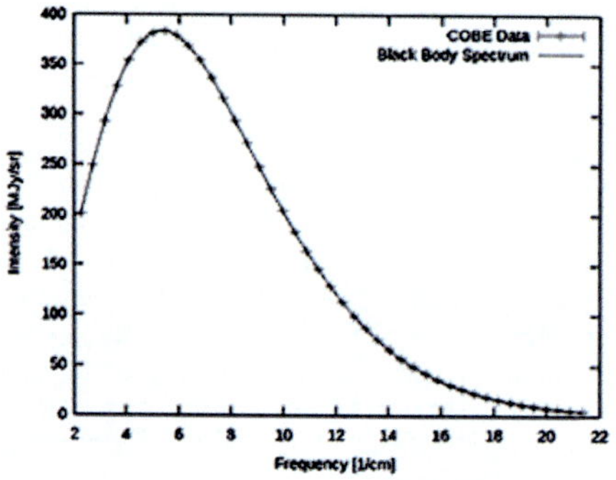

Figure 2. CMBR spectrum as measured by the FIRAS instrument in the COBE satellite. (SMOOT 2007).

Figure 3. Map of temperature fluctuations in the CMBR as obtained by the Planck satellite (PLANCK 2015). These tiny fluctuations are the result of quantum fluctuations when the Universe was a small fraction of a second old. These fluctuations ultimately led to the creation of stars and galaxies.

measured with the Planck Surveyor (PLANCK 2015) satellite, are shown in Fig. 2 and constitute the most precise data so far.

Last, but not least, Quantum Mechanics provides us with one of the most stringent bounds. Because of the wave nature of matter, linear momentum and position cannot precisely measure at the same time. Determinism is lost since position and velocity must be known precisely in order to predict the trajectory of a particle. Heisenberg's Uncertainty Principle tells us, consequently, that our knowledge of Nature cannot cross well-defined limits.

4 PROGRESS

Discovery, Science and Progress are three words with very different meanings and scope. A discovery changes the way we perceive the world around us while science systematises the construction procedure following discovery. Progress and innovation simply use the results of science to create economic value. These three pillars of human evolution complement each other but –it is important to stress it– are completely distinct. Confusing them leads, in fact, to the death of Discovery.

To understand the distinction let us take an example from Classical Physics. The relation between heat and work has been discovered (and rediscovered) many times by humankind. The science of Thermodynamics systematised a vast amount of experimental data, which led to the concept of energy relating heat and work: energy conservation became, in fact, the first law of Thermodynamics.

Energy conservation could then be used to build engines that, in turn, create economic value. Industry seeks progress and tries to create more and more efficient engines to increase the economic return.

Sadi Carnot finds, in the early 19th century, while questioning the limits of machine efficiency, an idealised model of a perfect machine. The concept of entropy is discovered and limits to energy conversion became unavoidable. Consequently, it was realised that evolution in Nature can only proceed in some specific ways. Science exploits this breakthrough in Physics and Chemistry setting limits for new heat engines and possible chemical reactions, thus creating new paths for economic value creation. This sequence of discovery, scientific evolution, and innovative progress continues to this day.

The founding fathers of Quantum Mechanics were acutely aware of the importance of their discovery but could not care less about its economic importance. This importance was actually only recognised decades after Quantum Mechanics was firmly established. Like great artists, these scientists only cared whether their inspired vision of Nature was perfectly depicted.

Discovery cannot proceed without freedom. Curiously, we are ready to accept this in Art but often fail to recognise how important this also is in Science. The reason is, of course, the confusion between Discovery, Science, and Progress. The situation is even more complex today, leading to an ever-deeper confusion between the three concepts. Large facilities such as the Large Hadron Collider at CERN create the set for discovery but are not a guarantee of discovery. As a facility, CERN provides an efficient stage for science and innovation to proceed and interact. It constitutes, therefore, a turning point for economic value creation, even if that is not its main goal. Its aim is Science, not economic progress.

In conclusion, the latest results on the ultimate structure of the Universe have provided a new vision of Nature. Science will continue to unfold the consequences of this new vision and as a side effect triggers progress for humankind. Discovery, however, is, by definition, unpredictable: it is precisely because of this that it always changes the way we perceive the Universe and ourselves.

BIBLIOGRAPHICAL REFERENCES

Bohr, N. (1949) "Discussions with Einstein on Epistemological Problems in Atomic Physics". The Value of Knowledge: A Miniature Library of Philosophy. Marxists Internet Archive. Retrieved 2010-08-30. From Albert Einstein: Philosopher-Scientist (1949), publ. Cambridge University

Press, 1949. Niels Bohr's report of conversations with Einstein.

Bell, J. S. (1966) "On the problem of hidden variables in quantum mechanics". *Rev. Mod. Phys*. 38 (1966) 447: J. S. Bell, *Speakable and Unspeakable in Quantum Mechanics* (Cambridge University Press 1987).

Herrero, M. J. (1998) Lectures presented at the NATO ASI 98 School, *Techniques and Concepts of High Energy Physics*. St. Croix, Virgin Islands, USA, June 18–29.

Kibble, T. W. B. (2009) "Englert–Brout–Higgs–Guralnik–Hagen–Kibble Mechanism". *Scholarpedia*. 4 (1): 6441. Bibcode:2009SchpJ...4.6441K. doi:10.4249/scholarpedia.6441. Retrieved 2012-11-23

Higgs G. Aad et al. (2012) (ATLAS Coll.) Physics Letters B 716 (2012) 1; S. Chatrchyan et al. (CMS Coll.) Physics Letters B 716 (2012) 30.

Planck. (2015) For an overview of Planck Satellite results see https://www.cos mos.esa.int/web/planck/publications#Planck2015.

Smoot, G. (2007) Rev. Mod. Phys 79, 1349.

Anonymous as a theme of discontinuity in the culture of Italian architecture between the first and second halves of the 20th century: E.N. Rogers and L. Ricci

Giovanni Leoni
Department of Architecture, University of Bologna, Bologna, Italy

ABSTRACT: The physical destruction of the architectural heritage as a result of the war and the experience of political and racial discrimination, then deportation, deeply marked the culture of Italian architecture during the transition between the first and second halves of the 20th century, changing its cultural sensitivity and paradigms. Among the themes that characterise the change, there is the reflection on the Anonymous not as an appeal to popular architectural cultures instead of designer-based, but rather as a change in the structure of the creative personality of the designer in the relationship between the individual conception of architectural form and the multi-personal dimension of architectural work.

Keywords: Anonymous/personality; cosmopolitanism/internationalism; Ernesto N. Rogers; Leonardo Ricci

1 THEMES OF DISCONTINUITY

The physical destruction of the architectural heritage as a result of the war and the experience of political and racial discrimination, then deportation, deeply marked the culture of Italian architecture during the transition between the first and second halves of the 20th century, changing its cultural sensitivity and paradigms. A study of discontinuity in Italian architectural culture in the late 20th century should start with a summary of the hypothesis of continuity. This would imply an extensive bibliographical essay and analysis of historiography – which we will undertake at another time – because the hypothesis of continuity between the first and second halves of the 20th century was, without a doubt, the prevailing one. The reasons are specific to the Italian architectural culture, but are part of a broader research into continuity between Modernism of the early 1900s and the second half of the century focused on the idea of a revision – or of a series of crises – of the Modern Movement rather than on the idea of an irreparable rift between the hypotheses that supported that project and the changed cultural conditions created by World War II. In particular, still today the theme of continuity stands on the repeated revisions of two key concepts of early 20th-century architectural culture: internationalism and the designer's personality. By internationalism I mean the hypothesis – that somehow supplants the search for "national styles" in the second half of the 1800s – that there are hegemonic centres that process cultural paradigms of international value, which thus define a periphery adhering more or less critically to this dominant culture, or refuses it or is refused, isolating themselves. It is impossible to summarise here, even briefly, the development of the internationalist perspective that infused the European architectural culture of the 1900s, from the cosmopolitan atmosphere of the artistic avant-garde of the early 20th century, which is its matrix, to the formulation of International Style refined by Henry-Russell Hitchcock and Philip Johnson in 1932, from the regionalist revisions of the Modern Movement immediately after World War II to relaunching the issue in a Postmodernist key and in response to Critical Regionalism. Even today there remains an "international" aspect of cultural architecture, supported both by trade and general media, which journalists have coined "starchitect" or "archistar" but that is based on a very precise organisation of the educational, cultural and professional world in the field of architecture and that still awaits an accurate historical analysis, especially today when its crisis seems to be definitive and to portend the transition to other cultural and economic models, even in Europe.

By architect personalities I mean the different weight that two different models of architectural creativity had during the 1800s and 1900s. On the one hand, the model of architect-artist driven by a strong individual personality of an intuitive nature, oriented towards the processing of prevailing, recognisable forms linked to him, with respect to the system of constraints (social, economic, constructive) that architecture faces in the process of becoming real. On the other hand, the model of an architect at the service of his project, whose job is to shape the process of realising the designed structure, subjecting conception

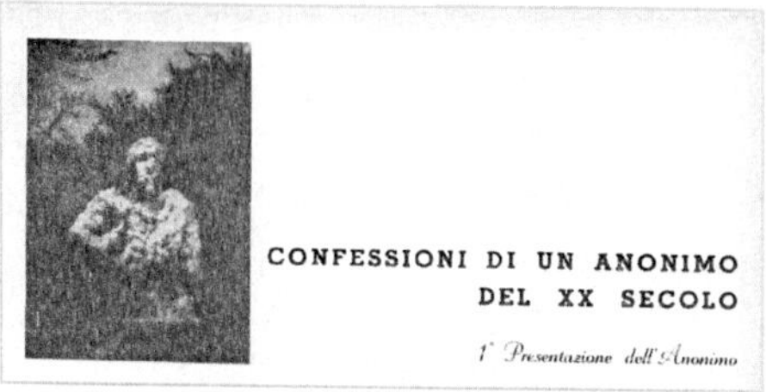

Figure 1. E.N. Rogers, Presentazione dell' Anonimo, *DOMUS*, 1941, 158: 45.

to the constraints of geography, collective nature and material existence. In spite of a 20th-century architectural historiography largely dominated by the cult of personality, a history of the models of creativity of 20th-century architects paradoxically remains to be written and the subject is extraordinarily subtle inevitably implying for any architectural work the coexistence of personality and anonymity. Focusing on the Italian situation, of particular interest are the years of the war and the immediate post-war period, a time of deep crisis followed by recovery in the transition between the fifties and sixties, where hypotheses of continuity marked both historiographical production regarding Italian architecture and its cultural identity.

Among the themes that characterise the change, there is the reflection on the Anonymous not as an appeal to popular architectural cultures instead of designer-based, but rather as a change in the structure of the creative personality of the designer in the relationship between the individual conception of architectural form and the multi-personal dimension of architectural work. Below we present two instances from a larger ongoing research on the topic of the Anonymous in Italian culture between World War II and the beginning of the sixties.

Between the beginning of the forties and the mid-sixties, two positions, among others, followed by a single formulation: 20th Century Anonymous. This is how Ernesto Nathan Rogers defined himself in his "Confessions" published in Domus between 1940 and 1941, and in 1962 it is the title that Leonardo Ricci gave to his book summarising the theoretical positions resulting from experimental designs started in 1949 with the project for Monterinaldi, and completed, together with the book, in the Monte degli Ulivi Village in Riesi.

1.1 *Ernesto Nathan Rogers: "Confessions of a 20th century anonymous"*

Rogers, outlining the characteristics of the Anonymous, even in the pain of a growing discriminatory climate, does not describe a defeat but rather portends a new cognitive and creative structure to be placed at architecture's foundation. The Anonymous is a designer who abandons the development of an individual artistic "personality" to become a means for conveying the expressions of others. However, it is not the disappearance of the personality that Rogers foresees, and in his pages lingers the figure of the genius architect, now with a dilated and "boundless" personality to the point of being suprapersonal. It is a mutation that entrusts to the Anonymous the new task of giving voice to "humanity" and that results from assuming as a field of action not the territories of a spirituality that is inaccessible to the common man, but the strange "place" in which Rogers declares to want to give "congress" to the reader, a place that lies "between mum" – the childhood memory of a caring gesture – "the cashier" – an erotic impulse that anyone can indulge – and "God" – taking design responsibility for common action, the only divine dimension remaining for architecture. A place where the simple act of common man encounters and blends into the simple act of the Anonymous designer, devoid of individual and subjective representation, intent on managing and bringing to form the common act as an act of design. It is the radical inversion of the functionalist principle and every associated legend, the abandonment of the task of interpreting classifiable human needs and offering them an architectural form that is viable for a community. There are only individual men, because:

> In space, some higher, some lower, to the left or to the right of the large cross, we left an empty tomb with underneath written 'place for Giovanni' or 'for Maria' or 'for Pietro' or 'for Ernesto' or 'for Natalina' or mine for Anonymous. ([Rogers, 1941, 159: 67)

The Anonymous therefore still finds himself in a special place, and the crowd – the destroying force of the 19th century intellectual thrust into the city, the term of comparison upon which the Vanguard built its poetics and a new prophet figure – still scares him, "pressing on every side; tearing clothes into tatters and risking the removal of limbs of life". But – and this is a tragic new development – the Anonymous finds the second solitude, even when thrust into the community and lost therein. [Ibid.] Lonely in the crowd, in the community, uncomfortable whenever "a function chains us to someone else performing or that has performed the same function", the Anonymous experienced a "dramatic conflict": "I am myself, but I am also one of you". (Rogers, 1941, 159: 67) Condemned to perceiving his unique personality not as exceptional but as similar to the common being, the Anonymous comes to terms with the "holy terror of one's corporeal existence".

> Every breath has a different cadence, yet you seek yours in others' breathless efforts; why not draw close to his for the suffering or joy he is interested in? (Rogers, 1941, 159: 67).

The role model, the guiding role of the artistic personality, is transformed into an effort to become part of the ordinary, requiring new extraordinary skills.

> We recognise that you shape a bit of my life, but I also do a bit of the same for yours. You change me, with your presence in my destiny,

> but I influence yours by giving back to you, in the arcane treasure trove of my works, your experiences that I have relived. Your solitude, my solitude that inhabits each other, because of anonymous love, loving your neighbour means populating one's loneliness with that of others. (Rogers, 1941, 159: 67)

It is still a dual figure, that cannot give up the modern condemnation of personalities but that understands the irrelevance of his personal, individual being. The Anonymous' s design task is not to outline what still is not, but rather to change the existing, letting himself be changed by it. There is no salvation, Rogers writes, "neither in the ivory tower of egotism nor in the dissolution of oneself in a common liquid mixture", only in the "acceptance of the battle" between personality and its disappearance. [Ibid.] The time of the Anonymous, in consequence of his destiny as a representative of the ordinary solitude of each, and the similarity with all, is a time that has nothing to do with the linear development from past to future that characterises the modern project. The Anonymous resists becoming part of a non-measurable time, one in which all "vision past and future" is dissolved, an instantaneous time, a constant present, the time of now, of the act, of living in the happening, "because I am my time and my time takes the appearance of my face" (Rogers, 1941, 162: 69).

It is interesting to note how Rogers still uses the term "style" in his pages about the Anonymous, though placed in quotation marks and defined as "a result of our lives" because:

> we are making a 'style' every day, as the river running over pebbles; it is a result of our lives, my office colleague, my fellow bus rider, my lady on the mezzanine. (Rogers, 1941, 158: 45)

The "style of the period" is, therefore, a result of the constant transcription in terms of the action of each person in his or her ordinary existence. The task, the "responsibility" of the Anonymous, far more onerous than the responsibility of recognising his isolated singularity outside of the crowd and outside of the community, is to give form to the simple act.

> Your toothache disturbs me, but – it depends on you – it could become a beautiful chair. Why don't we help each other to live? Who knows what huge buildings with so much pain. (Rogers, 1941, 158: 45)

Rogers defines his query of the Anonymous as an "open and cruel" confession because, without pretence, without taking refuge in specialised languages, using prose that is outrageously unscientific, he clearly anticipates the act of renouncing the structure of the architectural project based on artistic personality but also the abandonment of the idea of the project as work of a specialised nature intended to be studied in the laboratory, which would then produce an architecture in ordinary life. An argument against the specialism that had wide circulation in the Anonymous culture of the late 20th century, not only in Italy. The Anonymous does not deal with a special space, aesthetically founded and controlled, but rather

> a point in human space: you are just over there. You yourselves are well-defined points in this universal humanity. Just a moment of distraction and we will lose each other (Rogers, 1941, 159: 67)

For the Anonymous, pushing oneself to the limit of possibly losing every principle of personality corresponds to the identification of a new field of action for the project, which means probing the design theme in the moment and in the dimension in which it is still owned by all people, it is still a problem of life shared with those who have no responsibility to subject it to design hypotheses. In other words, it means tackling the project from the paradoxical prospect of an absence of a decision, of fully listening, of a balance between Name and Anonymous made even more onerous by comparison with absolute singularity, of the non-traceability of the individual choice, of the individual act with general parameters.

> The first law is to find the humanity in ourselves and ourselves in humanity. Even for Anonymous me, this is the first law. And so I can write it down like this: I have to be so deeply Anonymous that I arrive at a name, and if I had a name I would want it to be so vast that it became confused with the anonymous. Names and Anonymous derive from a common origin, like the axes of a Cartesian system forming a huge cross within whose infinite spaces are located all our points. (Rogers, 1941, 159: 67)

But the identity of the Anonymous, which is formed upon the renunciation of the Name and not of its affirmation, finds substance in the Work.

> Names drift through history, they are on the books, among the dates, on the streets signs, even far away from their works; the Anonymous, no, they cannot leave, and only when you get close to the remains of his work is life breathed into him on this side of dreams. Consider the importance of these facts, respect them: they are aspects of eternity. (Rogers, 1941, 158: 45)

1.2 *Leonardo Ricci: Anonymous 20th century*

In the early formulations of the Anonymous that accompany the presentation of Monterinaldi on the pages of *Domus* in 1957 (Ricci, 1957: 86–99), Leonardo Ricci seems to have already surpassed the heroic vision and commitment to the Work – and to the city, but the subject would open issues that we cannot address here – of the balance between Name and Anonymous discussed by Rogers in the journal more than 15 years earlier. However, there remain many similarities. Ricci also disputes the figure of an

Figure 2. L. Ricci, Casa Mann Borgese, Forte dei Marmi, 1957.

Figure 3. L. Ricci, Casa Balmain, Marciana, 1958.

Figure 4. L. Ricci, Casa Cardon, Castiglioncello, 1961.

architectural designer with an extraordinary personality anticipating the overcoming of the "loved and admired" masters in the 1962 volume – in favour of a new figure of a designer focused on sharing, according to Ricci a prelude to the definitive disappearance of architecture as specialised activity, as an activity separate from the ordinary and from everyone's everyday life. The second point of close contact with Rogers is the overcoming of functionalism to leave space for the figure of an entirely unique "client" having "infinite desires". As for Rogers, also for Ricci at this date, the issue is not to passively surrender to the difference and diversity of the subjects, but rather retain the educational role of the designer who, for Rogers had to expand his personality by offering a common voice, the Work, to the many names, while for Ricci he should instead interact with the client and teach him to distinguish between desires that are "fundamental" and "discretionary and even vain". That is why, as Ricci writes in *Domus* that one must not be "existentialist" but rather "existential", or recognise as "fundamental" only "acts that arise from existential truths of man and not from futile reasons of taste". The aesthetic criterion as a guide to the architectural project disappears, returning the focus of reflection on the theme of a "language of existence" also discussed by Rogers. And, as for Rogers in his pages on the Anonymous, the idea of "modernity" formulated by Ricci is not a progression of the new projected into the future but rather an "eternal present", a permanent modernity based on super historical constants and that sets aside the evolutionary idea of architectural languages processed on a formal basis. For Ricci, architectural forms are not "a priori inventions of architecture" but rather the a posteriori result of "life analysis" that generates "consequent spaces". In those same years Ricci performs experiments, designing and building works like Casa Mann Borgese, Casa Balmain and Casa Cardon – essentially contemporaneous – the practical results of the positions expressed in *Domus*.

An identical constructive grammar unites the three buildings: the stone for the substructural work that roots the building to the ground, the plasticity of concrete to shape the spatial fluidity that translates the vital movement. However, no formal personality is recognisable as a point of continuity in the three works, and the internationalist hypothesis is perfectly overturned in a heteronym that demonstrates, and not only theorises, the disappearance of the early 1900s architect artist's identity. The overturning returns in the full theoretical formulation of the work on the Anonymous in which Ricci describes a form-act that builds a non-mythical world, not absurd but "logical", "not in a rational way, but in the sense of logos". Architecture, "those stupid and still wrong things that are called houses", as Ricci describes them, must be the result of a query and transcription of "naked existence", and Ricci is fully aware that in this condition architecture is confined by the experience of the Holocaust – to which various explicit references are made in his book on the Anonymous – an experience that has cancelled every possible expectation of happiness bound to the community because "the unhappiness of others affects our happiness and cancels it" at the moment in which suffering is given in such a form devoid of any possible sense and perspective of redemption. Even "happiness" offered by architecture, the "world of form" as a promise or action of redemption, as a guide-world for the transformation of the real, is but a "drug", a fascinating but dangerous illusion. Ricci's interest in the subject of community is therefore not surprising, but his text on the Anonymous does not indulge in any illusion of a newfound original condition in which the relationship with the material is released from representation. The task Ricci sets for himself is not to draw on naked existence but rather to free architecture's content of existence from the constraints of form, transforming architecture from a representation of existence to a place where it freely occurs. The focus thus shifts to hosted life rather than represented life, in a radical surpassing of the prefiguration:

> There are no standards in the logical world, nothing fixed, nothing unchangeable. The

Figure 5. L. Ricci, A liveable space for two people, *DOMUS*, 1965, 426, pp. 130–31.

> standard is intrinsic in the act committed, not outside of it, in such a way that it can be used as a measuring rod. (Ricci, 1962: 26)

The "logical construction of architecture" is derived from the maintenance of the decision and of the gesture in a constant current condition and an interpretive attitude consisting of "two instances: receiving and returning". The instance of receiving is the non-specialised dimension of the architecture of which it was previously noted:

> This is the phase in which the architect is only a human being, not yet a specialised operator. The richer the humanity of this human being, the more complete will be the architect; and he will neglect none of the aspects of human life. This is the phase during which the architect must not use his pencil or specify anything. It is the phase of conception. This is the phase in which an architect is just a man, not yet a specific operator. The more this man will be full of humanity, the more the architect will be complete and will not overlook anything of life. It is the phase during which the architect must not take the pencil in hand, nor clarify anything. It is the stage of conception. (Ricci, 1962: 228)

A conception that does not represent the world but arises from a dispersion of individual personality among the things and the single individualities of the "clients", to establish a planning action based on query and not statement, an unattainable condition in the era of personality, but also a chance to "work in error" and attempt a "new world", the world of the Anonymous. Evidently, it is a project that leads to the disappearance of the discipline or if you prefer, its radical overhaul, a change not in linguistic codices but rather in cognitive and productive structure.

In the same years in which Ricci published his book on Anonymous, first in New York and then in Italy, projects such as "Living space for two people" or works such as the Monte degli Ulivi village bear witness of the design and plastic strength of the "new maîtrise" of the 20th Century Anonymous.

BIBLIOGRAPHICAL REFERENCES

Ricci L., (1957), Monterinaldi, *DOMUS*, 337, p. 87.

—. (1962), *Anonymous* (20 Century), New York, Braziller.

Rogers, Ernesto N., (1941), La presentazione dell'Anonimo, *DOMUS*, 158, p.45.

—. (1941), Le coordinate dell'Anonimo, *DOMUS*, 159, p.67.

—. (1941), L'Anonimo e la folla, *DOMUS*, 160, p.59.

—. (1941), I confini dell'Anonimo, *DOMUS*, 161, p. 69.

—. (1941), L'Anonimo nel tempo, *DOMUS*, 162, p. 69.

—. (1941), I sogni dell'Anonimo, *DOMUS*, 164, 31.

—. (1941), Responsabilità dell'Anonimo, *DOMUS*, 167, p. 17.

—. (1942), La personalità dell'Anonimo, *DOMUS*, 170, p. 94.

—. (1942), La casa dell'Anonimo, *DOMUS*, 176, p. 333.

Progress(es) in planning – theory and practice and the quest for activist modes

João Cabral
CIAUD, Faculdade de Arquitectura, Universidade de Lisboa, Lisboa, Portugal

ABSTRACT: The subject of this article is the concept of progress in relation to urban and territorial planning and how debates in planning theory and practice can illuminate perspectives for creating a more progressive, participative and equalitarian urban society. The article is divided into three sections. The first section elaborates on a concept of progress that is associated with the development and consolidation of a capitalist society geared by the accumulation of wealth. The concept of a model of development from the Regulation School is adopted to provide an understanding of the changing role of the state and civil society that is setting the conditions and practices for planning intervention. The second section focuses on current debates in planning theory, particularly the concept of public interest influencing decisions and processes for planning practice. The final section addresses the quest for an activist planning approach within a scenario dominated by the liberalisation of the economy and argues for reflexive and transdisciplinary approaches in planning research and practice.

Keywords: Progress, mode of regulation, planning, urban politics, activist planning

1 CAPITALISM AND THE IDEOLOGY OF PROGRESS

This article focuses on the role of planning as a human activity organising the allocation and redistribution of human and natural resources, with an emphasis on the role and the betterment of space and territory. In this sense, the concept of progress alludes to the history of mankind as a cyclical process, from the lessons learned and its contradictions and achievements as elaborated by Adorno (2003). The historical references are associated with the development of a western European progress-oriented culture with its roots in the secularisation of Christianity and the Enlightenment ideas of the 18th century. Thus, progress became associated with economic growth, scientific discovery, technological advance, and human development as part of the particular Occidental ideology (Atkinson, 2014; Vairley, 2013).

Progress follows changes and that can be understood by the different and successive revolutions (English, French, Russian...) which, from the seventeenth century, resulted in power changes and brought about different groups with varying interests (Vairley, 2013). The shift from absolutism to liberalism and from feudalism to capitalism reorganised the social and class hierarchy and widened the economy and the production and consumption basis. The progress fuelled by social struggle and by commercial interests became associated with the accumulation of wealth and a strong work ethic, as explained but also criticised by Max Weber (Weber, 1976).

This link between accumulation and its associated circuits of production and consumption, and the ethics, rules and conventions guaranteeing the coherence of these processes needs to be understood through adopting a thorough analytical manner. The approach and the key intermediate concepts of regime of accumulation and mode of regulation developed by the Regulation School in the early 1970s provide a good framework for analysis (Aglietta, 1979; Boyer, 1988; Cabral, 1998). A regime of accumulation expresses the particular organisation and set of regularities involving production, income distribution, exchange of the social product and consumption. The concept of mode of regulation considers the materialisation and coherence of the regime of accumulation, i.e., the norms, habits, laws, the institutional forms and structures which represent the socialisation of production, securing, therefore, capitalist reproduction. Thus, a model of development represents the combination of an accumulation regime and a mode of regulation. Utilising a Regulation School approach involves an identification of two other elements: a hegemonic structure and a technological or industrial paradigm. Both elements refer to the concrete historical character of regulation. Hegemonic structures follow Gramsci's development of the notion of hegemony and the formation of a social or 'historic bloc' (Gramsci, 1971: 366, 377). A technological or industrial paradigm reflects the close relation between technical advancements and long-term changes in patterns of accumulation as labour processes developed technically (see Boyer, 1988). The emergence of the regulation concept follows a recognition of the impact that advances in computer and communications technologies are having in the organisation of labour and in the participation and control of workers in the production process.

The model of development known as Fordism, an explanation of its crisis and the definition of a new coming model (post-Fordism) are the main study subject of regulationists. Fordism as a model of development was based on a direct relation between productivity gains and workers cost of living, integrated as consumers and guaranteeing the reproduction of the system. This model entered into a structural crisis during the mid-1970s because of the greater complexity of economic relations associated with the internationalisation of production and markets. The main features of this crisis are:

a) problems of consistency in the production system that is unable to diversify and restructure to ensure increases in productivity and capital efficiency in the face of changes in demand patterns;
b) problems of maintaining a stable wage relationship, insofar as the internationalisation of production has reduced the importance of the consumption of workers and the ability to demand and organise trade unions;
c) changes in the relationship between modes of regulation and accumulation regimes and the capacity to intervene at the level of national economies through the 'Keynesian' social, fiscal and monetary policies of the welfare state.

Thus, as the socio-economic limits of space are redefined and integrated into new social and economic contexts, the important question for understanding the new ideology of progress through accumulation and consumption and the resulting patterns of social and spatial development lies in identifying the new regulatory and institutional mechanisms. Neo-liberalism, as a concept, is used to characterise this new model of the political system and urban governance, which succeeds Fordism. As a concept of political economy, it is associated with a greater openness to market dynamics of forms of governance and control of the spheres of social and economic life. In this sense, the 'new economic policy' is expressed, among other components, through privatisation processes, the 'deregulation' of the control and management chains with decentralisation of functions and measures to make the labour market more flexible. It is also represented by a separation in state functions between the conditions for the reproduction of a system traditionally based in public investment in infrastructures and the conditions for mediation, control and promotion of private sector activities.

The emergence of hybrid situations unfolds, in which the role of public administration articulates and complements the interests of civil society and the market under practices and institutional forms known as systems of governance. These complementarities are functional insofar as the two spheres of action, the state and the market, represent different interests. The advantages of the two levels of interests are verified and evaluated in each concrete situation, depending on several factors, namely the conditions of representation and defence of the public interest and in the design of policies and in the role of planning, to which we now turn.

2 THE CASE OF PLANNING

We now focus on the place and role of planning (urban, territorial, spatial), as situated within the idea of progress as part of the development of productive relations and the western capitalist culture. In the introduction of his seminal book, *Urban and Regional Planning* Peter Hall (1992: 3) defined:

> planning as a general activity is the making of an orderly sequence of actions that will lead to the achievement of a stated goal or goals.

In the case of urban planning:

> it has a spatial, or geographical, component, in which the general objective is to provide for a spatial structure of activities (or of land uses) which in some way is better than the pattern that would exist without planning. (Hall 1992: 4)

Thus, planning is essentially an activity and a process, concerned with shaping and governing spaces, taking into account social, economic, and environmental matters. The way we address planning activities, formulate, and decide on planning actions is the result of contributions and lessons learned from history and practice. These form the body of theory which influence and frame decisions and processes.

This body of theory and knowledge and its usefulness for planning practice has been the subject of much debate and inquiry into the meaning of theory and the definition of practice (e.g. Alexander, 2016; Allmendinger, 2002; Campbell et al., 1996; Fainstein et al., 2016). Fainstein (2016) identified six sets of issues associated with:

(i) the history of planning,
(ii) its justification (means or ends, processes or outcomes),
(iii) its rational (why intervene? The role and justification for public sector intervention),
(iv) the role of values (ethics),
(v) how to deal with the constraints of the capitalist political economy (of democracy, of bureaucracy), and
(vi) the question of the public interest.

Although the debates concerning the six issues can be associated with different time-specific changes and historic moments, they can be seen alongside each other, raising new questions about today's role of theory and practice. As Fainstein and DeFilippis stated in the introduction to the 2016 edition of *Readings in Planning Theory*, there were important advances, particularly for an understanding of power relations in planning, the role of mediation, addressing complexity and establishing links between diversity, equity, democracy and community. In terms of the main concern of this paper – the quest for activist modes in planning practice – the debate around the concept of public interest is critical. The principle of public interest has been the historic justification for the existence of planners and plans but its rationale is also the result of the role of the state in the economy (and in the urbanisation process). Alford and Friedland (1985) explained that the role of the state has to be understood in terms of the contradictory relations among the prevailing logics for its functioning represented by the state capitalist, bureaucratic and democratic features. Interestingly enough, much of the debate around the role of public interest in planning does not explicitly address the theory of the state. The public interest is, however, at the forefront of the arguments put forward by professional planners, although it is also used to justify individual decisions and actions. In the introduction to a special issue of *Planning Theory* Malcolm Tait identified three critical questions on this matter.

> Firstly, is public interest still a relevant concept for planning and professional planners, and if so, what is the role of the professional planner in identifying the public interest? Secondly, how might communities identify and support the interests of those outside their own community, even if these clash with their own interests? And thirdly, what is the role of the state in promulgating wider public interests against the wishes of individual communities? (Tait, 2016: 336)

These are relevant issues. On the one hand, they raise the discussion of the role of the state in addressing the role of planners under neo-liberalism regulatory policies. On the other hand, they contribute to identifying the particular contexts and arenas where planners can have a more activist role in the urban scene. We will review this in the next section.

3 THE 'NEW URBAN QUESTION' AND THE QUEST FOR AN ACTIVIST PLANNING APPROACH – THE IMPERATIVE FOR TRANSDISCIPLINARITY AND REFLEXIVITY

In the first part of this article, neo-liberalism was referred to in connection with the ideology of progress associated with the drive for the accumulation of wealth and for consumerism facilitated by the liberalisation of regulatory processes and the financialization of the economy. However, the neo-liberal trend has also provoked forms of resistance and mobilisation, with cities and the urban becoming the arena for alternative movements and the place of social unrest. In contrast with the social movements of the 1970s, which reclaimed a new role for a state concerned with the reproduction of labour power, as described by Manuel Castells (1977) in *The Urban Question*, the centre of dispute became a capitalist state centred on the extended reproduction of capital and on 'productive' consumption, "through financialization and dispossession, through dispossession and reconfiguration of urban space" (Merrifield, 2014: 18). The notion of reproduction is used here in relation to Regulation Theory and the argument of a mutual reciprocity between a regime of accumulation and its mode of regulation. Merrifield also added that although regulationists did not discuss the urban question,

> It is perhaps not too hard to project their analytical devices onto the plane of urbanization itself, to frame our current global urban condition as a regime of accumulation and mode of (de)regulation that could be best described as 'parasitic'. (Merrifield, 2014: 118)

Thus, the 'new urban question' meant a different perspective and gave a central role to urban politics. It became associated with the claim for 'the right to the city' as enunciated by Lefebvre, for 'the inclusive city' and for a collaborative and an activist planning approach. The reference to collaborative planning as developed by Patsy Healey in her 1997 book *Collaborative Planning: Shaping Places in Fragmented Societies*. is prior to the reaction to the neo-liberal trend referred to above, but established the idea of planning as an interactive process, a governance activity, with a concern on the quality of places and territories and a "moral commitment to social justice" (Healey, 2003). Her ideas, together with the work of Forester and others, established a practice of planning with a focus on the role of agency and on a relational approach to understanding and addressing complexity and diversity in complex urban governance contexts. The reference here to an activist planning approach is based upon Tore Sager's defence of critical planning theory and more explicitly as a symbol of a more direct form of resistance to mainstream policies associated with neo-liberalism (Sager, 2016; 2011). In a recent paper Sager (2016: 1262) elaborates on modes of activist planning used for "organizing alternative thinking in planning contexts where policy change is higher on the activists' agenda than system change" with examples of recognised activist planning including insurgent and radical forms.

However, the challenge for a progressive and activist practice also means that processes and actions have to be adequate, with the capacity for continuous mobilisation. The conditions for this to happen, however, are not guaranteed from the outset. They depend on the concrete local conditions and have to be constructed over time, through consensus but also through dissent, assuming the confrontation of democratic political positions and the role of agonistic politics (Mouffe, 2013). The role of the process is, therefore, critical for guaranteeing community access to the relevant information and to decision making. The debate around what is a good planning process relates to the design of the planning methodology and the collection of the appropriate evidence necessary for understanding trends, defining options, generating alternatives and motivating and involving stakeholders.

Davoudi (2012: 436–427) refers to the use of 'evidence' as a key aspect which in interpretative planning is an iterative rather than linear process, as an ensemble of different knowledge informing policy to an 'informed society'. The imperative for an interpretative approach, in contrast to a positivist tradition, as been argued by Davoudi (2012) is associated with a conceptualization of space as relational and dependent on the social and cultural processes that produce it.

> Understanding of spatiality as matters of fact is combined with its understanding as matters of concern. ... The desire to connect, rather than to control, spaces and times is what drives interpretive planner. (Davoudi, 2012: 438–439)

Thus, the evidence is not the outcome of a value-free knowledge, as positivists believed, but the object of scientific investigation where reflection and action are thoroughly articulated (Beunen et al., 2013; Schön, 1983). By reflecting content and process, reflexivity produces action-oriented knowledge integrating new skills, theories and concepts from different disciplines (Beunen et al., 2013). The role of transdisciplinarity for research and planning is, therefore, paramount as the outcome of a process involving multidisciplinary and interdisciplinary contributions for addressing complex urban problems (Després et al., 2011).

> ... instead of reducing reality to the parts researchable at the intersection of multiple disciplinary perspectives, transdisciplinary research includes at once what stands between disciplines, across disciplines and beyond any discipline, thus combining all the processes of multidisciplinarity and interdisciplinarity. (Després et al., 2011: 35)

Achieving transdisciplinarity as a process and an outcome guaranteeing that local planning is a credible and transparent process, accountable to the democratic governance structure at the scales where cooperation and mobilisation are critical, is not an easy task. Corporative, class and group interests and divisions permeate all sectors of government, academic, public, and private life, and shape administrative procedures, research objectives and the conditions for community organisation and forms of action. The end results tend to morph into a deviation and sometimes the opposite of the initial goals, as corporative, class and group interests influence methodologies and strategies for their own benefit, from decision making to research, design and implementation. Thus, the struggle for 'the just city' for the activist planner is not only to achieve given objectives and results but an everyday assignment for guaranteeing reflexivity and transdisciplinarity in planning research and practice.

BIBLIOGRAPHICAL REFERENCES

Adorno, Theodor W. (2003) Progress *Can One Live After Auschwitz?* New York: Stanford University Press.

Aglietta, Michel. (1979) *A theory of capitalist regulation.* London: New Left Books.

Alexander, Ernest. (2016) There is no planning — only planning practices: Notes for spatial planning theories. *Planning Theory* 15 (1): 91–103.

Alford, Robert & Friedland, Roger. (1985) *Powers of Theory – capitalism, the state and democracy.* Cambridge: Cambridge University Press.

Allmendinger, Philip. (2002) *Planning Theory.* London: Palgrave.

Atkinson, Adrian. (2014) Urbanisation: A brief episode in history. *City* 18 (6): 609–632.

Barr, Alfred H. (1934) Foreword. *Museum of Modern Art. Machine Art.* New York: Museum of Modern Art. https://www.moma.org/documents/moma_catalogue_1784_300061872.pdf. (Accessed 8 June 2017).

Benevolo, Leonardo. (2002) *Historia de la arquitectura moderna.* Gustavo Gili.

Beunen, R, Van Assche, K & Duineveld, M. (2013) The importance of reflexivity in planning and design education. *Wageningen University Working Papers in Evolutionary Governance Theory* 14: 23.

Boyer, Robert H. (1988) Technical Change and the Theory of Regulation. In: Dosi, G (ed) *Technical change and Economic Theory.* London: Macmillan.

Bronowski, J & Mazlisch, Bruce. (1960) A tradição intelectual do Ocidente. Lisboa: Edições 70.

Bury, John Bagnell. (1920) *The Idea of Progress. An inquiry into its origin and growth.* London: Macmillan & Co.

Cabral, João. (1998) Spaces of regulation, the changing role of the state and the case for regional planning in Portugal. *XII Congress of the Association of European Schools of Planning.* University of Aveiro.

Campbell, Scott & Fainstein, Susan (eds). (1996) *Readings in Planning Theory.* Malden: Blackwell Publishers.

Castells, Manuel. (1977) *The Urban Question: A Marxist Approach.* London: Edward Arnold.

Colquhoun, Alan. (2009) Symbolic and Literal Aspects of Technology. *Collected Essays in Architectural Criticism.* London: Black Dog, 23–26.

Davoudi, Simi. (2012) The Legacy of Positivism and the Emergence of Interpretive Tradition in Spatial Planning. *Regional Studies* 46 (4): 429–441.

Després, Carole, Vachon, Geneviève & Fortin, Andrée. (2011) Implementing Transdisciplinarity: Architecture and Urban Planning at Work. In: Doucet, I and Janssens, N (eds). *Transdisciplinary Knowledge Production in Architecture and Urbanism, Urban and Landscape Perspectives.* 11. Dordrecht: Springer, 33–49.

Fainstein, Susan & DeFilippis, James. (2016) *Readings in Planning Theory.* 4th ed. New Jersey: Wiley & Blackwell Publishers

Gramsci, Antonio. (1971) *Selections from prison notebooks.* Hoare, Q and Smith, G. N. London: Lawrence and Wishart.

Griffiths, Paul. (1996) *Modern Music: A concise History.* London: Thames and Hudson.

Hall, A. Rupert. (2014) *The Revolution in Science.* Abingdon: Routledge.

Hall, Peter. (1992) *Urban and regional planning.* 3rd ed. London: Routledge.

Healey, Patsy. (2003) Collaborative Planning in Perspective. *Planning Theory* 2 (2): 101–123.

—. (1997) *Collaborative Planning, shaping places in fragmented societies.* Vancouver: UBC Press.
Johnson, Philip. (1994) Preface to Sixtieth-Anniversary Edition. *Museum of Modern Art. Machine Art.* New York: Museum of Modern Art.
Kultermann, Udo. (1993) *Architecture in the 20th Century.* New York: Van Nostrand Reinhold.
Masiero, Roberto. (2003) *La estética de la arquitectura.* Madrid: Machado Libros.
Merrifield, Andy. (2014) *The New Urban Question.* London: Pluto Press.
Mouffe, Chantal. (2013) *Agonistics – thinking the world politically.* London: Verso.
Nesbit, Robert A. (2009) *History of the Idea of Progress.* London: Transaction.
Pérez-Goméz, Albert. (1996) *Architecture and the Crisis of Modern Science.* Cambridge: MIT Press.
Sager, Tore. (2016) Activist planning: a response to the woes of neo-liberalism? *European Planning Studies* 24 (7): 1262–1280.
—. (2011) Neo-liberal urban planning policies: A literature survey 1990-2010. *Progress in Planning* 76 (4): 147–199.
Sant'Elia, Antonio & Marinetti, Filippo Tommaso. (1971) Futurist architecture. In: Conrads. In: Conrads, Ulrich (ed) *Programs and Manifestos on 20th Century Architecture.* Cambridge: MIT Press, 34–38.
Schön, Donald A. (1983) *The reflective practitioner: how professionals think in action.* New York: Basic Books.
Struik, Dirk J. (1992) *História concisa das matemáticas.* Lisboa: Gradiva.
Tait, Malcolm. (2016) Planning and the public interest: Still a relevant concept for planners. *Planning Theory* 15 (4): 335–343.
Tatarkiewicz, Wladyslaw. (2013) *A History of Six Ideas: An Essay in Aesthetics.* Berlin: Springer.
Tiekso, Tanja. (2016) Luigi Russolo's Art of Noises and the Aesthetics of Musical Experimentalism. In: Berghaus, Günter (ed) *International Yearbook of Futurism.* Berlin: De Gruyter, 71–93.
Vairley, Francis. (2013) *Reflections on the Fall of Modern Civilisation.* London: New Generation Publishing.
Weber, Max. (1976) *The Protestant Ethic and the Spirit of Capitalism.* London: George Allen and Unwin.
Wittkower, Rudolf. (1998) *Architectural principles in the age of humanism.* 5th ed. London: Academy Editions.

Colour effects and human perception: Contributions to architecture and design project

Fernando Moreira da Silva
University of Lisbon, Faculty of Architecture, CIAUD, Lisbon, Portugal

ABSTRACT: Every Human Being establishes a personal relationship with Colour. Often we give ourselves an instinctive colour treatment, just by choosing a particular and specific colour, or putting around us certain colours in our homes or gardens. Most of our reactions are, however, unaware and it is only when we begin to use colour in an informed way that we can take advantage of this extraordinary life force in order to improve the quality of life and our well-being, especially if we consider the preparation of architects and designers to develop projects.

This paper presents a research project which aims to study physical, physiological and psychological colour effects, and perception, in its interaction with human beings through objects and environment, underlining the importance of colour ergonomics. The main objectives focus on the acquisition of scientific knowledge in the area serving as a projective tool for architects and designers, as well as to contribute, through the dissemination of its results, as a reference to the use of colour to users in general.

Keywords: Colour, Colour Effects, Colour Perception, Colour Contrast.

1 INTRODUCTION

Nature provided us with colour to fuel our body and spirit. Colour enriches our entire system, giving us vital energy as a supplement which is an essential part of life.

According to Kress and Leeuwen (2002), colour communicates meanings at three levels: ideational, interpersonal and textual. Colour communicates ideational meanings because colour helps to identify the representations of the world, including specific people, places or ideas. Interpersonal meanings when it is used to do things. Colour can also be used to convey unity and cohesion.

By observing nature, colour often appears as a means of defence and conservation, as with plants and poisonous animals whose showy colour warns potential predators or even a function of attraction that allows plants reproduction and mating calls of certain animals. The human being, in its evolution, inherited psychological and physiological reactions to colour, although they cannot be controlled nor explained objectively make colour as a necessary medium for the information, communication and understanding of the environment. Colour has the function to attract attention, convey information add emotions and stimulate illusions. (Gamito and Silva, 2009)

As highly coloured beings, our forms are made of vibrant and ever-changing colours and as human beings, we answer colours actively or passively in everything we do. Light waves affect us in every minute of our lives and enter our energy system, whether we are awake, whether we are sleeping, visual or blind. Our growth, blood pressure, pulse, temperature, muscle activity, immune system, etc., they are all affected by light rays. The coloured rays affect not only our bodies but also our emotions, dispositions and mental faculties.

We all have a personal relationship with colour. Most of our reactions are, however, unaware and it is only when we begin to use colour in an informed way that we can take advantage of this extraordinary life force in order to improve the quality of life and our well-being. Humans are not the only ones to be affected by colour. In the animal and vegetable world, colour may mean survival or extinction. Colour is used to attract, camouflage, as a sign of danger or send sexual signals. Colour is instinctive to life and it is as important for our species as it is for the animal and plant kingdoms. There are many ways to introduce colour in our system. Through the understanding of the physiological and psychological colour effects, we can select the best colours for our clothes to our home and work environments.

Of course, one cannot forget colour symbolic, religious and cultural power. Always used in religious rituals, colour is the language of the soul. "Color is a property and the language of form". (Mahnke, 1993)

There is a colour language which we have forgotten and we have to learn it again, which brings benefits to our health, happiness, and perhaps even for the survival of our planet. There are very few studies on the subject, and even these are merely descriptive of a period of time. Until now there has never been an attempt to understand the global phenomenon. Do architects or designers have proper training on colour and its potential, to use it properly in their projects? If people possess more knowledge about the use of colour, and the effects it produces in every one of us, will use it differently?

The present research project aims to continue a former study on the physical, physiological and psychological colour effects, in their interaction with the human beings, in the formation of architects and designers. The main objective focus on the acquisition of scientific knowledge in the area that can serve as a tool to be used in the project development, as well as to contribute, through the dissemination of its results, as a reference to the use of colour to users in general. A user-centred research methodology has been used, with participatory design, survey, direct observation and active research, supported by mechanical means.

2 CONTEXT

To work in architecture or in design one needs to have knowledge of different correlated areas and how to manipulate them. However, there is a gap in colour knowledge. Gibson (1971) affirms that colour perception is "richer and more open to personal interpretation".

The visual and non-verbal systems operate relatively without a "tutor" in our society, at least in comparison with language. Visual perception is a primary system, which is located at the same expression level as is verbal language. In terms of human development, the visual system of signs occurs earlier than language itself. In terms of complexity, one can find visual interpretation more complex than verbal interpretation, because of the absence of a conventional sign system and the formalisation of a training protocol.

Visual perception and communication could also be considered primary because the observer has to learn how to control better, and independently, the visual interpretative function. So, this kind of perception is neither derivative nor peripheral to verbal language, and consequently, the designation of visual perception and communication as secondary, tertiary or of a "superstructure" built in verbal language is not right. In Paivio notion (1971; 1986), a more appropriated model would be built of a dual codification, saying that visual and verbal information are codified and uncodified by perceptive and cognitive separated systems. One of the systems is considered visual/pictorial and manipulates simultaneously the elements of imagination; the other one is linguistic and propositional and operates in sequence. Messaris explains that the two systems are structural and functionally distinct (1994:73). Although being independent, the two systems are also interdependent in a way that a certain visual concept may be converted in a verbal etiquette and vice versa. A new interpretation to explain the interaction between the two systems is the metaphor of processing interactively in parallel (Messaris, 1994). Redefining the notion of a "primary" system, and including visual communication and perception as well as the verbal one, one can achieve a complete analysis of the complexity of this process.

3 HUMAN STRIATE CORTEX AND COLOUR CONTRAST PROCESSING

According to Weiskrantz et al (2007), colour appearance is not simply determined by the composition of reflected light. Colour appearance is more closely related to a constant property of a surface, its spectral reflectance. The relative reflectance of adjacent surfaces at any specific wavelength of light determine the relative intensities of light reflected from them at that wavelength, regardless of the spectral composition of the light illuminating them.

Ratios of cone signals are approximately invariant under changes of illumination (Foster & Nascimento, 1994). If the reflectance properties of one surface are known, then the properties of all other surfaces seen under the same illuminant can be inferred on the basis of these cone ratios or colour contrasts (this is the essence of Land's "retinex" colour constancy algorithm) (Land & McCann, 1971).

Questions about the role striate cortex, the first visual area of the higher mammalian brain, plays in colour perception have been reignited in the last years (Johnson et al, 2004).

These cells in the striate cortex are excited by light of one distribution of wavelengths and inhibited by the light of another distribution falling in one small region of space. Adjacent or surrounding regions show the opposite pattern of responses. The cells, therefore, respond optimally to borders with a specific chromatic contrast that falls across the centre and surrounding regions of their spatial receptive fields. These findings suggest colour-specific processing starts earlier in the visual system than previously thought. The earliest mechanisms of colour constancy, our ability to recognise objects' colours regardless of the colour of light illuminating them, might occur in striate cortex. If so, people without striate cortex, like some older people, will lose colour contrast processing.

They should behave as if objects change colour as the light illuminating them changes, their apparent colour depending on the wavelengths of light they reflect, rather than the properties of their surface material (Weiskrantz et al, 2007).

In their book "Color contrast processing in human striate cortex", Weiskrantz et al (2007) have shown that our response to carefully tailored visual illusions can reveal whether we discriminate between stimuli on the basis of the physical properties of light reaching our eyes from those stimuli (wavelength and intensity) or the contrasts the stimuli make with their backgrounds. If the background luminance of an image varies gradually across it, then two stimuli presented against it at different positions but having identical luminances will be perceived by a normal observer as being different, presumably as a consequence of their different immediate backgrounds. Two stimuli that differ in brightness but have the same contrast relative to their background will, however, be perceived as similar. This illusion is also effective for colour using stimuli matched for chromatic contrast in terms of the changes they elicit in the excitation of different retinal cone types.

Many eye conditions that result in low vision also result in colour perception and vision deficiencies. While the most common congenital colour defects produce little in the way of performance limitations in and of themselves, the additional loss of sensory information that arises from a colour defect for an already compromised visual system could have greater consequences.

Conversely, the residual colour discrimination that some people with low vision show could provide a viable avenue along which information could be presented to enhance performance on certain tasks dependent upon visual input. This line of reasoning suggested that a useful approach would be: a) to evaluate what residual colour discrimination low vision people display and b) to evaluate what influence the introduction of chromatic contrast might have on specific tasks for normal and low vision people. (Arditi & Knoblauch, 1996)

With knowledge on colour perception, the degree to which colour cues influence performance on specific tasks can be evaluated.

4 COLOUR PERCEPTION

For a good comprehension of colour and an adequate colour application, in architecture or in the design area, it is essential to have knowledge of colour perception, which includes colour characteristics and proprieties, and the different colour effects.

The colour characteristics or attributes, which allow an objective description and contribute to a good perception and identification, are: hue, luminosity (value), and saturation (chroma). Hue is a universal variable present in all colours (Munsell 1976) and is defined by its wavelength that places it on the visible spectre. On common language, it is confused with the colour's name. To these three fundamental attributes, we may aid temperature. Luminosity or value is the lightness graduation of a hue: is the distinction between any colour and a lighter or darker one (Munsell 1976).

By the perception laws, when all the eye sensibilities are activated similarly and simultaneously, to the higher luminosity corresponds white, and to the less sensibility corresponds black. On the neutral luminosity axis, going from white to black through the greys, the three sensibilities are activated in various degrees, but no one is dominant.

The third attribute, chroma or saturation, defines the state of purity of a colour. All pure colours have their saturation at 100% corresponding to their wavelength. When saturation is non-existent, the colour is achromatic and belongs to the neutral luminosity axis. By the perception laws, to brighten up a colour it is necessary to activate partly the sensibilities that don't belong to this colour, although, when the colour becomes lighter, it becomes also less saturated. To darken a colour its sensibility must be less activated, and the colour loses luminosity and saturation.

Colour temperature may be considered in comparison with other colours or by its effects. However, it can be also considered a colour characteristic having a correspondence to its wavelength. Colour measurement by electronic sensors shows that surfaces covered with a range of reds, orange, or yellow are warmer than surfaces covered with blues or green colours. Dark colours are warmer than lighter ones, absorbing more incident light. So, within a hue, the temperature may change with its luminosity. In an architectonic, environmental or design composition, the human eye is spontaneously attracted by warm colours, and need more concentration to perceive cold colours. The relative colour temperature makes that a colour may seem more or less warm in comparison with other colours.

It is common to represent colour through a tri-dimensional form, known as "colour solid" and adopted by several authors, in chromatic systems used in arts and in sciences, following the fundamental relation between the three original dimensions. Judging colour in spatial issues and when comparing it with the perception of space regarding the light, one positions in first place the attribute of luminosity or value. Nevertheless this characteristic, or attribute, in terms of chromatic definition evolves into an integrated dimension to the attribute of hue, which in terms of an informative meaning of colour is the one that characterises it the most. In a former research study (Moreira da Silva, 1999), we proved that in the tri-dimensional conception, when getting together the three attributes, the following relationship is established:

$$\text{Hue}\ \frac{\text{Luminosity (or value)}}{\text{Saturation (or chroma)}}$$

It was also proved that there is a tri-dimensional relationship of colour/space, which forms a unity, and is represented by the following expression:

$$C/S = (hv/c)/S$$

Being: h = hue; v = value; c = chroma; C = colour; S = space

This relationship becomes the perceptive unity which is the basis for the other possible formations in colour/space relationship.

The perception of a colour hue and luminosity is activated by the reaction of the eye sensibilities to the emission of certain wavelengths. The wavelengths that would activate the eye sensibilities, not yet activated, constitute the complementary colour. Therefore, the complementary colour is the same as light rays absorbed, and the addition of the two colours rebuilds the white light. Complementarity to white is black, and primary colours are complementary to secondary colours. However, the totality of complementary colours is not limited to these colours and every colour from the visible spectre has its complementary colour, its pair to the white colour reconstruction.

After-image is a complementarity relationship provoked by eye tiredness and by visual memory. This phenomenon consists in, after observing one colour insistently, seeing the complementary colour over a white background. This reaction is more immediate when the observed colour has great intensity, and it happens because the cones sensible to this colour become saturated and, when the eye moves to the white background, are temporarily impeached and can only activate the complementary sensibilities. The complementary afterimage is usually called negative after image. Nevertheless, after-image can be positive when it happens under the stimuli of an intensive light and is similar to the observed colour. Every hue form an after image with its complementary hue, and the after image of white is black.

The after-image effect can be neutralised by a complementary background colour, it can also be diminished or neutralised, applying a white or black contour. Successive contrast is the reaction to after image provoked by colours, observed one after the other. When the eye is moving rapidly over a coloured surface, it observes the colour in which it focuses and also, the after-image from the previously observed colour. This phenomenon happens because the eye, trying to activate all the sensibilities, tints the adjacent colours with the observed complementary hue.

The successive contrast is provoked only by hues, independently of their luminosity. If two colours with a great luminosity contrast are placed side by side when they separate the lighter colour darkens and the darker one lightens.

Simultaneous contrast is the result of an after image which alters the appearance of a colour, by the influence of the adjacent colours, having the influenced colour a minor area than the influent colour. This reaction occurs on saturated, or less saturated hues, and neutral colours, and may affect an object shadow.

Neutral colours are the most affected by the simultaneous contrast, and in consequence, a grey background makes the other colours look more intense. Being a form/ground contrast, it can happen in relation to all colour characteristics.

The shadow colour of the object depends on the incident light. Every coloured light creates a shadow in its complementary colour, which is not the effect of simultaneous contrast, but real colour. White light, the daylight, activates every eye sensibilities and its shadow is black. However, the colours of the shadows provoked by various coloured lights correspond to additive colour mixtures. On the other hand, on daylight absence, the coloured light resultant shadows are different. One coloured light has a black shadow because it is the absence of light. When the object is illuminated by two complementary colours, each colour provokes a complementary shadow and where the shadows overlap the black colour is seen, while the mixed light is the result of the two colours additive synthesis. With analogue lights, each light projects a shadow in the other incident colour, the overlapped shadow still will be black, and the mixed light will be the two colour additive synthesis. The results of the three primary colours projection are three primary colours shadows, one overlapped black shadow, and a white light mixture because the additive synthesis of the three light beams reconstitutes the sunlight.

5 HOW DOES IMPAIRED VISION AFFECT COLOUR PERCEPTION?

Arditi (1999) designed three basic guidelines for making effective colour choices that work for nearly everyone. Following the guidelines are explanations of the three perceptual attributes of colour – hue, value (lightness) and saturation (chroma) – as they are used by vision scientists. This study, as most of the research implemented until now, was developed for digital displays projects, not for architecture or product design.

Partial sight, ageing and congenital colour deficits all produce changes in perception that reduce the visual effectiveness of certain colour combinations.

Two colours that contrast sharply to someone with normal vision may be far less distinguishable to someone with a visual disorder. It is important to appreciate that it is the contrast of colours one against another that makes them more or less discernible rather than the individual colours themselves.

Here are three simple rules for making effective colour choices:

1. Exaggerate lightness differences between foreground and background colours, and avoid using colours of similar lightness adjacent to one another, even if they differ in saturation or hue. Don't assume that the lightness you perceive will be the same as the lightness perceived by people with colour deficits. You can generally assume that they will see less contrast between colours than you will. If you lighten the light colours and darken the dark colours in your design, you will increase its visual accessibility;
2. Choose dark colours with hues from the bottom half of the hue circle against light colours from the top half of the circle. Avoid contrasting light colours from the bottom half against dark colours from the top half. For most people with partial sight and/or

congenital colour deficiencies, the lightness values of colours in the bottom half of the hue circle tend to be reduced.

3. Avoid contrasting hues from adjacent parts of the hue circle, especially if the colours do not contrast sharply in lightness.

Colour deficiencies associated with partial sight and congenital deficiencies make it difficult to discriminate between colours of similar hue.

Lightness, like hue, is a perceptual attribute that cannot be computed from physical measurements alone. It is the most important attribute in making contrast more effective.

With colour deficits, the ability to discriminate colours on the basis of lightness is reduced.

6 COLOUR PHYSIOLOGY

The context used herein this document to "colour physiology" is the following one: the effect and the extension the brain stimulation (involved the process described above) has on other human body systems.

According to Mahnke (1996), the environment may be responsible for the introduction of "nervousness, headaches, lack of concentration, inefficiency, poor provisions, visual disturbances, anxiety and stress". All of these are physiological responses caused by a large number of stimuli; the effect that colour, in particular, has in the human being has been investigated for many years. Freiling (1990) points out the largest comebacks in the light of a particular colour: red colour is devastating; blood pressure is inconsistent, evidencing a pulse increase, it induces dryness in the throat and headaches. The subjects tend to move away from the source. The yellow light leads the subject to the tendency to move away from the source, and a subjective sensation in hands. "Yellow causes tension but at the same time releases and activates motor activity"; violet-blue light (and blue) "...leads to calm and to concentration capacity"; the green light "...has a similar effect to the stimulus of light balancing heterogeneous trends." (Freiling, 1990: 23–26)

The following investigation was conducted by Birren (1982). Although not specifically related to the behaviour of humans, some of the findings are particularly relevant. Different coloured lights affect the rate of increase of various plants, albeit stimulating hue varies with the species. The behaviour of insects, fishes, reptiles and birds is affected by the variation of wavelengths, including the end of the spectrum, tending to ultraviolet light. The hue of light has a profound effect on birth ratio of various animals, e.g. mice living under a full light spectrum reproduce at sex ratio of 50:50; mice living under blue light produce 30% of males and 70% of females, while living under the pink colour light produce 70% of males and 30% females. Surprisingly these ratios also reflected in blind mice under similar lighting. In studies specifically directed to the approach of the physiological effects of direct exposure of coloured light, according to Birren (1982), the researcher Metzger (n.d.) used light sources directed only to one eye. The effect was that the muscles, only on the same side of the body, responded to the stimulus. Metzger also found changes "...in the superficial and deep sensations, both demonstrating a deep dependence on the optical stimulus." The subjects were asked to stay standing with arms stretched horizontally forward, placing the light source directed only to one eye. When the coloured light was used, the red led the arm of the illuminated side of the body moved out. The green light caused the arm movement inside. Ellinger (1953) developed similar experiments directing light to only one side of the face and neck of subjects. When the red light was used, the subjects with arms stretched likewise drove to the light source, verifying the removal with the use of blue light. According to Ellinger "This reaction also occurs when the eyes are closed."

Birren (1982) concluded: "It seems clear that humans, like all other living beings, react to radiation ... We can therefore generalized that colour affects muscle tension, cortical activation (brain waves), heart rate, breathing, and other functions of the autonomous nervous systems – and certainly this brings defined emotional and aesthetic reactions, tastes, pleasant and unpleasant associations" (...) "The skin sees in technicolour and it is also a good detector and seems to reflect the direction in which the brain neurons process colour information. The experiments demonstrating body reactions to colour argue that colours lead to emotional states, which are specific to different shades" (Birren, 1982: 56).

There are two phenomena that are fully recognised in the neurological world, which are "the constancy of Colour' and "the observation of the negative." The constancy of colour is the name given to the eye tendency in viewing objects with the same form under different light conditions. For example, a white sheet of paper appears white under a red light or a bright blue one (although different if viewed simultaneously under the two sources separately). The observation of the negative is the name given to the eye's ability to, after a prolonged exposure to a specific hue, create an image of its complementary colour (or opposite). For example, if the eye constantly observes a blue object, and suddenly the eye sees a white area, a Pink-orange version of the blue object on white surface will appear, for a while. Some surgeons have used this principle to prevent this green observation purposes after lengthy operations involving the sight of blood (red). All operating theatres are now painted green. Experiments involving hypnotised subjects have confirmed the existence of this phenomenon of observation of the negative.

Besides colour physiology, there are other colour dimensions which are also very important for the study, such as Colour Ergonomics, Colour Symbology, Colour Perception. However, we decided to focus this paper on Colour Psychological Associations.

7 COLOUR PSYCHOLOGICAL ASSOCIATIONS

Even without detailed knowledge about the psychological effects of colour, it is well known that it can affect our moods, our disposition. There are some colours that are exciting and inspiring, just as there are others that are depressed. How often we use expressions like "I feel blue", "green with envy", "red with anger", without thinking about the meaning that lies behind each of these words. Our feelings and emotions are directly affected by the hormonal balance or imbalance in our body. Once this is affected by the colours, they have, of course, an indelible influence on our feelings and dispositions.

"Research on the psychological aspects of colour is difficult for the mere reason that human emotions are none too stable and the psychic make-up of human beings varies from person to person" (Wright 1998: 28).

The psychological effects of colour have been extensively investigated through the use of colourful cards, instead of using painted compartments. As Sivik (1997) concluded, "... there is considerable disparity between the conclusions drawn by cards or buildings".

It is believed that most of the colour effect is based on associations of education from childhood, such as the sky and sea are blue, the grass is green, etc. The principles are applied in architecture due to taste associations, which can be translated into benefits to the overall effect of the building. The second way of the experiment involves the projection of light of a particular colour directly onto the retina of a subject so as to compare the physiological reactions (Ellinger, 1953).

Some colours can calm us, while others can stimulate our mental activity. Through the process of restoring the balance of colour energy directed to the pituitary, it is possible to restore the metabolic and emotional balance. This can relieve stress, tension, anxiety or depression. Certain colours can help us to deal with our feelings of loneliness, frustration or pain. The use of colour to modify the emotional energy also results in the change of perception of the world and our way of living. Once the colour is directly linked to the subconscious, we can use it to diagnose and treat a problem at a deep level (Moreira da Silva, 2013).

So far we have developed some studies based on relevant literature data and experiments with users, crossed with scientific research conducted in several countries, which have already conducted to some psychological associations to some of the most used colours. We also added a survey by inquiry, questionnaire-based, involving 623 people from Portugal and UK, both gender, with ages between 23 and 72.

Here we present some of the results, as an example:

RED – vital, powerful, ambitious, hot, intimate, sensual, determined, friendly, brave, antidepressant, furious, impatient, angry; ORANGE – hot, insurance, creative, stimulant, entertaining, cheerful, with humour, independent, antidepressant; YELLOW – cheerful, light, bright, mentally stimulating, logical, smart, orderly, optimistic, clear thinking, fearless; GREEN – harmonious, mental and physical relaxing, peaceful, natural, refreshing, calm, sincere, insurance, free, generous, restrained, personal domain; TURQUOISE – refreshing, cool, calm mentally, youth, power concentration, control, communication power, confidence; BLUE – cold, clear, relaxing, mentally calm, brings peace, tranquility, wisdom, spacious like the sea or the sky, sensitive, hopeful, faithful, believer, flexible, quiet; INDIGO/VIOLET/PURPLE – dramatic, spiritual, creative, intuitive, media, mystical, inspiring beauty and art, protective, clean; MAGENTA – support, natural, kind, considerate, condescending, compassionate; BROWN – support, mentally mean, land, seclusion; WHITE – peaceful, cathartic, cold, isolated; BLACK – comforting, mysterious, female, protective, restrictive; GREY – independent, secure, tab, lonely, self-critical; SILVER – changeable, harmonious, feminine, sensitive; GOLD – desire, understandable, powerful, high ideals, abundance.

8 CONCLUSION

Having in mind colour effects and perception when developing projects in architecture or design, there are some general principles to be present, such as:

Simplicity – simple, logical design of all layouts with the use of colour contrast and the right lighting will make the final project easy and simple;

Consistency – the project should aim to achieve consistency of form and colour use;

Providing contrasts – the project must provide bigger, bolder and brighter contrasts between forms and their background, creating differences in brightness/luminance or differences in colour/hue, or both;

Inclusivity – while giving due regard to the prevalence and symptoms of vision impairments, such as in older people, one must promote inclusive design based on the social model of disability – which focuses on the design of the built environment rather than on an individual's impairment.

Through recent researches, we know which areas of the brain are activated, as well as we can measure the behaviour of humans in what concerns colour issues. So, an experiment with users is under development in order to check brain reactions to the different colour dimensions, comparing the results with those obtained by the other previously used methods. The main objectives focus on the acquisition of scientific knowledge in the area that can serve as a projective tool for architects and designers, as well as to contribute, through the dissemination of its results, as a reference to the use of colour to users in general.

As expected, results, we intend to achieve a systematisation of scientific knowledge reusable by all within the interaction Colour/User and also guidelines for the use and application of colour in architecture and design projects.

ACKNOWLEDGMENTS

The author would like to acknowledge the support given by CIAUD, Research Centre for Architecture, Urbanism and Design, Faculty of Architecture, University of Lisbon and FCT, Foundation for the Science and Technology, Portugal.

BIBLIOGRAPHICAL REFERENCES

Arditi, A. & Knoblauch, K. (1996). Effective color contrast and low vision. In B. Rosenthal and R. Cole (Eds.) *Functional Assessment of Low Vision*. St. Louis, Mosby, pp. 129–135.

Arditi, A. (1999). *Effective color contrast: designing for people with partial sight and color deficiencies*. New York, Arlene R. Gordon Research Institute, Lighthouse International.

Birren, F. (1982). *Light, Colour and Environment*. New York: Van Nostrand Reinhold Co.

Ellinger, R.G. (1953). *The organization of colour*. New York: Edward Bros.

Foster, D.H. & Nascimento, S.M.C. (1994). Proc R Soc London Ser B, 257, pp. 115–121.

Frieling, H. (1990). *The Colour Mirror: The Quicktest for Character Diagnosis with the Colours of the "Frieling-Test"*. Berlin: Musterschmidt.

Gamito M. and Moreira da Silva, F. (2009). *Cor no Mobiliário Urbano: um factor de Inclusividade, Orientação e Identificação*. 5th CIPED, Brasil: Bauru.

Gibson, J. (1971). *The Information Available in Pictures*. Leonardo, 4 (Winter), pp. 27–35.

Johnson, E., Hawken, M. & Shapley, R. (2004). *Journal Neurophysiol*, 91, pp. 2501–2514.

Kress, G. and Leeuwen, T.V. (2002). Colour as a Semiotic Mode: Notes for a Grammar of Colour, *Visual Communication*, 1(3), pp. 343-368.

Land, E. & McCann, J. (1971). Journal Opt Soc Am, 61, pp. 1–11.

Mahnke, F. & Mahnke, R. (1993). *Color and Light in man-made environments*. New York: Van Nostrand Reinhold

Mahnke, F. (1996). *Colour, Environment, and Human Response*. USA: John Wiley & Sons, Inc.

Messaris, P. (1994). *Visual Literacy: Image, Mind, & Reality*. Boulder CO: Westview Press

Moreira da Silva, F. (1999). *Colour-Space: its qualitative management in architecture*. PhD's Thesis. The Salford University.

Moreira da Silva, F. (2013). *Colour and Inclusivity*. Lisboa: Editora Caleidoscópio.

Munsell, A.H. (1976). *Munsell book of color*. Baltimore: Macbeth Division of Kollmorgen.

Pavio, A. (1971). *Imagery and Verbal Processes*. NY: Holt, Rinehart and Winston.

Pavio, A. (1986). *Mental Representations: A Dual Coding Approach*. Oxford: Oxford University Press

Sivik, L. (1997). *Generality aspects of colour naming and colour meaning*. Department of Psychology, Göteborg University.

Weiskrantz, L. et al (2007). *Color contrast processing in human striate cortex*. USA: PNAS.

Wright, A. (1998). *The beginner's guide to Colour Psychology*. London: Colour Affects Lt

Part II

Architecture/urbanism/design; (theory to practice)

From potency in theory to act in practice

José Lopes Morgado
University of Lisbon, Faculty of Architecture, CIAUD, Lisbon, Portugal

ABSTRACT: Progress is a man's proper thing that reflects his rational ability to realise a cultural evolution autonomously from natural evolution and translates the Aristotelian passage from the potency (matter) to act (form). This paper poses the question of the Albertian relation of the theory of architecture to the practice of architecture, in the actual imagining and concretizing of a possible better world.

Over the centuries, the civilizational advancement has operated mainly in the field of science and technology; since always and in contrast, the cultural progress, especially in the ambit of the artistic dimension, has generated a triadic architecture, which has passed the test of time, with characteristics that favour an architectural aesthetic harmony that, when experienced, allows a human ethical harmony.

By imagining a desired horizon, with a sense of *right rule* directed at improving reality, the architectural design applies theoretical knowledge, the result of research, with the required purpose of developing a useful practical knowledge.

The communication argues that, as the transformation of the world is conceived and constructed, establishing a certain human order, more fair and appropriate, a progressive change is fulfilled.

In conclusion, it is the human desire and will that makes it possible, to achieve progress with a survival advantage, that is holistically sustainable by architecture.

Keywords: Progress, Architecture, Potency and Act, Theory and Practice, Aristotle and Alberti

1 FROM POTENCY TO ACT

Natural evolution is unconscious; but, in a different way, in not being genetic, cultural progress is conscious.

Sometimes recapitulating good ideas from the past, in reconciliation with nature and with human nature, already indicates some progression; but it is not enough, as our discontent suggests.

The passage from potency (*dunamis*) to act (*energeia*) shows the real power of progress; from an aetiology (*origin*) to a teleology (*end*), integrating an ontology (*nature*), this is the formula of progress; in developing the potential, the best possibility can be promoted; the human being is faced with inevitable, gradual and sometimes disconcerting progress, given the admirable mutation between the everlasting natural *will to live* and long-lasting cultural *will to live better*.

Because nature favours the selection of the fittest creature, in the *aesthetics of survival*, we can "find some fairly close matches between characteristics we like and characteristics that would have improved our chances of survival" (Hildebrand, 1999: 10).

Actually, as stated in the congress paper "*Harmony as a Tangible Utopia*" (cf. Morgado, 2016: 53), in the long process of biological survival, life regulation, genetic mutation and adaptation to the environment, in a natural selection and evolution of the species, the construction of the shelter played an important role; as a survival idea, architecture participates in the evolution of the human being, for a better existence. "In general, men seek what is good, not what is traditional" (Aristotle, *Pol*, II, 8, 1269a4).

2 FROM THEORY TO PRACTICE

According to with Birgit Cold, in order to develop consciousness and responsibility and to benefit, in future, aesthetically satisfying or supportive environments, we really need "to discover what is theoretically accepted and empirically known about aesthetics, well-being and health. [...] we realise that aesthetics is part of the total experience and utility of the environment, involving functional, ethical, social, technical and other factors" (Cold, 1998: 3).

In applying the theory to the practice, the architectural triad – necessity, convenience and pleasure of beauty – presents itself as a factor of quality and progress, contributing to the possibility of a better world; the adaptive force of harmonization of the triad rests on the absolutely necessary, the contingently useful and the spiritually beautiful; the Albertian triad, has the power to better configure architecture, implementing a way of sustained reform to "avoid the revolution" (Le Corbusier, 1923: 243).

The human triad of need, desire and aspiration (NDA), the DNA of human progress, is associated with another triad, of will, choice and decision; in reality, given the discontent of man, there has been a continuous progression of needs, desires and aspirations, which implies the architecture.

The overcoming of the natural state to the cultural state, reveals the passage from necessity to convenience; the evolution of the art of edification to the art of dwelling, generates a better built and acceptable living conditions, in the *art of place* for the proper *world of life* (cf. Norberg-Schulz, 2000: 56).

The potential for transformation enables evolution; a spiritual transformation is capable of magnifying the potential of the human being, making him virtuous and with a better destiny.

A worthy *double harmony*, aesthetic in architecture and ethics in man, allows the transforming effects that make possible the passage from human misery to human dignity, and the pragmatic approach to the *best of possible worlds*.

By building better, we can reform the human way of life; man is influenced by this power of architecture; therefore, given the recognised noble influence of architecture, and its effects and consequences, there is a real need for a change in our practice.

Just as natural selection enhances the cultural selection, the ontology of man is well favoured by the ontology of architecture; by improving the confining structure, we can modify the world.

In theory and practice considerations, under the mediation of the architecture and from an extraordinary dwelling, the dignity of architecture really favours the dignity of man (cf. Alberti, 2011: 431).

Good sense combined with common sense, admits that the old potentiate the new, that the tradition potentiate the innovation; the test of time and the test of the triad are the criteria for the consistency of a friendly architecture to the inhabitant, for the establishment of a best practice; is the pragmatic test that guarantees the path of progress; in developing and transforming the built environment, architecture's endeavour is in order to reconcile the improvement of human life; "the highest community [...] shall aim for the highest good of all" (Aristotle, *Pol*, I, 1, 1252a5).

Indeed, in architecture it is crucial the diligence of moving from theory to practice, the method of moving from design to production, the responsibility to move from promise to fulfilment; and the design is intended to transform into a preferred condition; but the difficulty lies in the good application of the idea of good configuration, and it is the architect's intention and responsibility to make achievable this purpose.

The quality of a triadic architecture is a strong contribution to the building of a better world and society; in practice, among other edificatory design criteria, for this fair purpose concurs certain essential characteristics made more evident when magnified in the perceived attributes; with a common and a good willingness, architecturally, the sublime causes aim the sublime effects.

History inclines to the prevalence of progress; and we aspire to a balanced progress, scientific and technological, but also ethical, social and political progress; both in theory and practice, we are longing for a harmony between the laws of nature and the laws of culture; survival depends on the positive direction of progression, and extinction derives from the negative direction of regression; as there is a good utopia and a bad utopia, there is a good progress and a bad progress; the movement from less to more, from worse to better, in itself, is good.

And theoretically and empirically, in the praxis, the "challenge is to understand the background of aesthetic preferences and human needs and within the available resource framework transform these into artistic and functional wholes" (Cold et al., 1998: 23).

3 ARCHITECTURAL PROGRESS

In a post-triad consciousness of progress, really, today's solutions must not be the problems of tomorrow; and with better procedures in the architectural praxis, either in the process of creation or in the process of fruition, happiness can be generated; an attitude of prevention, in the action of building well, precede an attitude of therapy, in the reaction of restore and repair; above all else, difficulties can be avoided.

Also in connection, in architecture the prior conception of the "design of the architect and the production of the building aspire to the coincidence"; indeed, Alberti's theoretical approach to architectural discipline is "always with an interventional and prepositional purpose" (Krüger, 2011: 72, 375).

There still is the perspective of a benign practice, and in which Alberti's window promises to be an *open window* to a better world; but it is not the lines that make the *lineamentum*; in reality, "it is an architect who, with a sure and perfect method, knows not only to design in theory but also to carry out in practice all works" (Alberti, 2011: 138).

By improving reality, progress can bring forth, hopefully, the superior in human beings, our best inclinations; moulding the circumstances for the best, in the better alternative of possibility, it is the superior contribution of the responsible architect, always concretize well the improvement of harmony in individual and of the order in society, convoking the psychology and sociology.

The sense of *right rule* (cf. Aristotle, *EN*, II, 2, 1103b32), confer the purpose of the theory and practice of Alberti, operating in the direction of progress; architecture is not neutral and, by influencing the experience of *vivential space*, enables the overcoming of the eternal human existential conflict; we can be focused on raising the issues without diminishing the alternatives; willing to set a place more prone to the dream is possible, and to make a better future is tangible, in this horizon of possibilities, to carrier good things.

Figure 1. *"Light remains through the Shadow"*. A hope metaphor for the human progress despite the real or imaginary obstacles; in the long run, it is the idea that survives. Drawing: José Lopes Morgado.

With a spiritual mission that aspires to greater things, the idea of progress prevails in the liberating capacity of the human condition, through an alternative world, always supported, assisted and oriented architecturally.

Alberti's lesson establishes the duty of *aedificatoria*, an architecture subject to an ethical program in a risk-free choice; informed by the knowledge of the theory, the wise architect accomplishes a balanced praxis and, in the magnified application of the triad, the progressive architecture goes beyond the triad, reaching the harmony of unity, with the purpose of making our daily life an extraordinary thing, of changing for the better the habitual indwelling, with an edificatory capacity in favour of the man; only this *metatriadic architecture* is worthy of its creator.

As we have seen, architecture can pass from the potency to the act, really concretizing the best possibility; the interest in potential is a certain capacity for the creative transformation, by the architectural act; surely "all excellence is capable of fully developing the potential" (Aristotle, *EN*, II, 6, 1106a16).

Possibly, the mechanism for creativity, innovation and change exists "because somewhere along the road of evolution we developed a divine discontent with the status quo" (Smith, 2003: 12).

In the great transfiguration of the real, architecture truly cooperates with man; for Battista Alberti, in Choay' s version, architecture seems to *promise singular fruits* (cf. Choay, 2004: 47).

Always denoting a care for individual, the potential of the architect's praxis can be well achieved, since by his word and action he transforms matter into form, revealing an order; and the potential of architecture mainly lies in responding creatively and "satisfactorily to the complex desire that defines humanity. As humans, our greatest gift is love [...]. Despite our suspicions, architecture has been and must continue to be built upon love" (Pérez-Gómez, 2006: 3).

The effective achievement of civilising potential and progress of architecture is promoted by man, the fundamental ally of architecture; but the limitations of architecture are associated with the limitations of man; with a power that brings responsibilities to the architect, a preventive architecture takes care of difficulties and of the normality of man.

The progress by architecture is a fact. "Buildings are the pressure cookers of cultural change. [...] as artefacts, they are large and unavoidable. They frequently challenge the accepted norms and as such stretch the mind and thereby improve what psychologists call the 'optimum perceptual rate'; in other words, our capacity to adjust to novelty and surprise" (Smith, 2003: 15).

As knowing the theory is surely useful to understand the practice, knowing the past and the present is essential to discern the direction of progress; that is why, we do not always need to know from effective experience about a real phenomenon; the actual idea of progress maybe is unacceptable, although "it is said of someone who undermined the foundations of his house that he could know *a priori* that it was to collapse" (Kant, 2013, Int, 1, B2: 37).

If progress is an alteration of the *status quo*, according to Kant, all changes must have a cause; and all causes contains a necessary bond with an effect; through careful investigation, the theoretical foundations are assured for, in the good praxis, to proceed to the construction of a building that does not collapse; and, in order to progress, necessarily, this must be solved as best as possible.

There is evidence that the preventive architectural solutions may promote the progressive human well-being, above all when the need of the visible outer reality coincides with the desire of the invisible inner reality, and with the will to find that correspondence.

For this purpose, it is fundamental to know when and why the possible gap arises and how to bridge it, in order to find the humanity's equilibrium and to achieve a harmonious progress.

4 HARMONIOUS PROGRESS

Thus, to conclude this assertion, in the real issue of *progress(es) – theories and practices*, with reciprocal consequences, surely the potential of architecture is proclaimed by the potential of man.

In conclusion, the greatest existential action is the satisfactory daily life in architecture; with an availability to intervene in the everyday reality, the good architecture has a guiding and regulatory action that provides the improvement in dwelling; the progress is well accepted when it comes from a harmonious development, capable of changing to a better condition and of generating a real transcendence of the human; this conscious awareness is a means of modifying the existence.

To develop the consciousness promotes the improvement of the architecture that shapes our future environment.

In addition, in a sublimated experience in our best architecture, we can flourish and bear fruit in the triadic harmony of the Albertian *res aedificatoria*, which propitiates and regulates the balance, and contributes to the elevation of the spiritual mood state of the human being; and, with a firm well-being in our soul, satisfaction is the effective and ultimate criterion of living successfully.

In the path of the potency in the theory to the act in the practice, and with an Aristotelian teleological sense, there is a real eudemonic potential of architecture that legitimates the individual and the public happiness of the human being.

According to with Alberti, in the existent potential of architecture is a perceived sense of harmony, a natural and cultural fittingness and fulfilment that bring a valuable and proper happiness.

In conclusion, with a co-feeling for the other, in the fair relationship between the aesthetic equity and ethical equity, harmonious development is essential for the invested in progress *aedificatoria*; with a consequent potential in effect, we are in presence of a certain power of a harmonious "cathartic architecture" (Morgado, 2016: 309).

Interpreting Aristotle and Alberti, in the *polis* and in the architecture, all the excellence is able to develop the full potential of the human life; restoring a triadic necessary function in a correct way, it achieves the beautiful harmony; that is progress indeed.

How to configure our progress is literally in our hands or, in other words, in our instinct and in our rationality; we don't know if the future of mankind will be on this wonderful planet; but, till that uncertain moment, we must configure ourselves, and surely an environmentally ecological progress.

Therefore, already aware of the harmful consequences of the truly unbalanced, incomplete and imperfect progress, thus resulting from a transitory, unstable, and labile human growth, our deep faith in progress is better today, for we only believe in our good progress.

Perhaps, in a certain and appropriate way, the progress is impossible without architecture, fortunately inducing that the progress of architecture only reveals the progress of man.

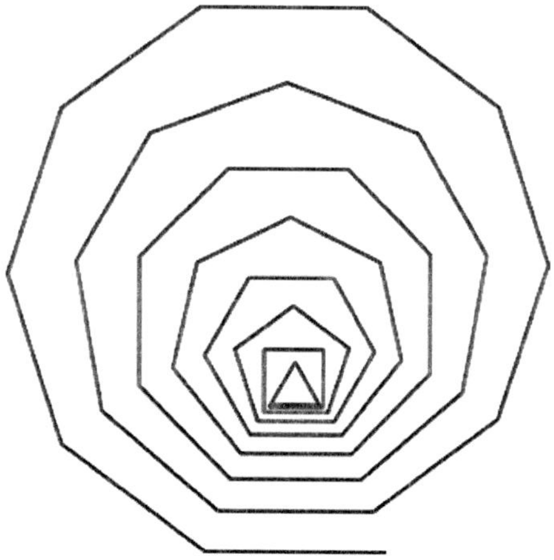

Figure 2. "*Spiral of Polygons*". It is generated by eight revolutions and with centres in ascension, developing from an equilateral triangle to the decagon; by an evolutive progression, geometrically embodies an Albertian complexity model of harmony, imbued with a Pythagorean spirit, and denotes an Aristotelian practical theory. Concept: José Lopes Morgado (1989). Drawing: Maria Cândida Morgado (1991).

BIBLIOGRAPHICAL REFERENCES

Alberti, Leon Battista. (2011). *Da Arte Edificatória (De Re Aedificatoria)*. Translated by Arnaldo Espírito Santo; rev. by Mário Krüger. Lisboa: Fundação Calouste Gulbenkian.

Alberti, Leon Battista, (2004). *L'Art d'Édifier (De Re Aedificatoria)*. Translated by Pierre Caye; Footnotes and Françoise Choay. Paris: Éditions du Seuil.

Aristotle. (2004) *Ética a Nicómaco*. Translated by António de Castro Caeiro. Lisboa: Quetzal Editores.

Aristotle. (2016) *Política*. (Translated by António Campelo Amaral and Carlos de Carvalho Gomes. Lisboa: Nova Vega, Lda.

Cold, B., et al. (1998) *Aesthetics, Well-being and Health – abstracts on theoretical and empirical research within environmental aesthetics*. Oslo: Norsk Form. NTNU Trondheim.

Hildebrand, G. (1999) *Origins of Architectural Pleasure*. Berkeley and Los Angeles: University of California Press.

Kant, I. (2013). *Crítica da Razão Pura*. (Translated by Manuela Pinto dos Santos). Lisboa: Fundação Calouste Gulbenkian.

Le Corbusier. (1977) *Vers une Architecture*. Paris: Éditions Arthaud.

Morgado, José Lopes. (2016). *Harmonia: o Potencial Catártico da Arquitectura*. Lisboa: Caleidoscópio-Edição e Artes Gráficas, S.A.

Morgado, José Lopes. (2016) Harmony as a tangible utopia. *Utopia(s) – Worlds and Frontiers of the Imaginary*. CRC Press, 53–58. https://doi.org/10.1201/9781315265322-10. (2017/07/09).

Norberg-Schulz, Christian. (2000) *Architecture: Presence, Language and Place*. Milan, Skira Editore.

Pérez-Gómez, A. (2006) *Built Upon Love. Architectural Longing after Ethics and Aesthetics*. Cambridge, Massachusett: Massachusetts Institute of Technology Press.

Smith, P. F. (2003) *The Dynamics of Delight. Architecture and Aesthetics*. London: Routledge, Taylor & Francis.

From divine to human: Architecture and music celebrating the machine, but still searching for Plato

Clara Germana Gonçalves
CITAD, Universidade Lusíada de Lisboa, Portugal/University of Lisbon, Faculty of Architecture Portugal/ISMAT, Portimão, Portugal

ABSTRACT: This paper reflects on why and how artists in general, and architects and musicians in particular, stopped following the laws of nature, through natural mimesis, and, after changes introduced in the course of the 17th and 18th centuries – with the Scientific Revolution –, switched to following the laws of the machine, the fruit and contribution of progress. A machine that was now human, in opposition to the previous world machine of antiquity.

It looks at how Renaissance Humanism was, on the one hand, a period that placed major focus on Platonic geometries and highly intellectualised thought, following doctrines based on ideals, and, on the other, was an epoch in which a new paradigm associated with the machine (human invention par excellence) was able to take form.

In addition, it draws a parallel between the emergence of fascination with the machine – fed by the notion of Progress – and a vision of the world as a gigantic machine devoid of Divine intervention.

In this shift towards positivism and functionalism, artists from many fields were also responding to an appeal and a need for an outlook that established a connection to an ideal, eternal world.

Keywords: the idea of progress, machine, platonic geometry, geometrical beauty, Futurism

1 "MACHINE ART AND GEOMETRICAL BEAUTY"

A text by Philip Johnson (1906–2005) is particularly insightful of this development and illustrates very well the idea that this paper seeks to throw light on.

In 1994, sixty years after the *Machine Art* exhibition he curated at the MoMA, Philip Johnson (1994) wrote:

> Sixty years later, the *Machine Art* exhibition naturally seems dated. In 1934 we were in our twenties. Adrenaline was running high; machine worship was running wild. After all, the influence of Marinetti's machine-crazed Futurist movement was not so long before.

Somewhat paradoxically he goes on to state: "Platonic dreams of perfection were the ideal" (Johnson, 1994).

This reference to Plato and antiquity had already been made in the catalogue to the exhibition, the foreword to which, written by the museum's director, Alfred H. Barr (1902–81), is preceded by passages from Plato (quoting the *Philebus*) and St. Thomas Aquinas (quoting *Summa Theologiae* – and its definition of beauty based on three characteristics: *integritas, consonantia* e *claritas* [perfection, harmony and clarity]. The preface begins with a section sub-titled "Machine Art and Geometrical Beauty" in reference to precisely the extract from the *Philebus,* that precedes it.

Moreover, in this context of fascination with the machine, it was seen as an instrument that made it possible, easier than before, to achieve forms that came closer to the Platonic ideals:

> The beauty of machine art is in part the abstract beauty of 'straight lines and circles' made into actual tangible 'surfaces and solids' by means of tools, 'lathes and rulers and squares.' In Plato's day the tools were simple handworker's implements but today, as a result of the perfection of modern materials and the precision of modern instruments, the modern machine-made object approaches far more closely and more frequently those pure shapes the contemplation of which Plato calls the first of the 'pure pleasures'. (Barr, 1934).

Philip Johnson and Alfred H. Barr were not alone: Le Corbusier (187–1965) was one of the outstanding examples of the taste for the new machines (manifested as early as 1923 in *Vers une architecture*), but also of a seemingly paradoxical nostalgia for an unaltered and unalterable Platonic interpretation of the geometry of the world and architecture (as evidenced in *Le Modulor* and *Modulor 2*)[1].

2 THE DIVINE MACHINE

2.1 *Mathematical beauty*

In addition to the due historical contextualization, in mirroring what was going on in the early decades of the 20th century, Johnson's text refers to the same ideals as Renaissance writings. Platonism was very much the dominant ideal during the formation of the Humanist movement. Being Mathematics the language of nature.

Mathematics was the foremost science during the Renaissance, also in the abstract and theoretical sense of metalogic (Masiero, 2003: 96). Mathematics was seen as a pure language, a human invention that could help making Divine thought clear. And mathematics was the first science – both for the followers of Plato, who believed in the harmony of the universe and for the Cartesians, who believed in reason-based methods (Struik, 1992: 165).

The way architecture and music present themselves and share these ideals can be referred to as an example of this paradigm. Though perhaps not on an equal basis. Architecture should follow music, for music reveals Divine harmony. This is the Pythagorean tradition that establishes the relationship between musical consonance and whole number quotients. Those quotients were to be applied in music. Alberti (1404–72) proposed:

> We shall therefore borrow all our rules for harmonic relations (*'finito'*) from the musicians to whom this kind of numbers is extremely well known, and from those particular things wherein Nature shows herself most excellent and completed. (apud. WITTKOWER, 1998: 109)

2.2 *The end of the divine machine*

The idea of Progress that emerged during the Enlightenment and flourished in the 19th century, anticipated the development of a mechanical view of the world that had no Divine intervention. This was also how the machine emerged in the social imaginary. As a metaphor: when the world forfeited its status of Divine machine – ruled by mathematics, the divine laws – the new human-machine replaced it as the centre of attention.

However, it was a long process only matured at the end of the 19th century.

> [I]t would certainly be erroneous to suppose that natural philosophers of around 1600 were less devoutly Christian than those of three centuries before. (Hall, 2014: 8)

Moreover, God remained in the picture, though in the background. Because Deism did not negate the existence of the Divine. As J.B. Bury (1920) argues in *The Idea of Progress*:

> The principle of unchangeable laws in nature is intimately connected with the growth of Deism which is a note of this period. The function of the Deity was virtually confined to originating the machine of nature, which, once regulated, was set beyond any further interference on His part, though His existence might be necessary for its conservation. (1920: 118)

Modern science, which "is highly mathematical in structure and argument", which "demands rigorous standards in observation and experiment", and which "excludes spiritual agencies from its province and accepts a pure materialism", was "finally established only in the late nineteenth century" (Hall, 2014: 4).

Or, in a more generic perspective: "It has only been during the last two centuries that the transcendental dimension of meaning has been questioned" (Pérez-Goméz, 1996: 5).

3 THE SCIENTIFIC REVOLUTION AND THE EMERGENCE OF FUNCTIONALISM

Interest in the machine can be seen as the culmination of an interest in science – more specifically, in applied science (Pérez-Goméz, 1996) – which began with the Scientific Revolution. As Masiero (2003: 95) argues, since the modern age there was a clear diminishing of the gap between theory and practice. Theory became, in this period, science, defining itself from the formal point of view and becoming verifiable.

During the 15th and 16th centuries, one witnessed, for the first time, a conscious literary interest in crafts and technologies. A huge body of literature was created on this topic, based on the idea that innovation and the advancement of technologies were both possible and necessary (Hall, 2014: 13).

There was a transition from a classical Renaissance to a popular Renaissance, from an idealistic Renaissance to an empirical one. And one moved from a veneration of ancient humanism to a strong belief in the human present (Bronowski et al., 1960: 30).

In the field of architecture, treatises attempted the resolution of concrete problems, distancing themselves from preceding theories in the manner of authors such as Alberti. We were rapidly approaching functionalism and, with that, the machine and

[1] It would seem, however, that Le Corbusier's interest in the mechanically sophisticated object did not weaken his Platonic side, where there was also a certain amount of nostalgia for a past that was more marked by cosmological symbolism. His interest in purer mathematics and geometry are well known; examples of this interest are the Mathematical Object in the Phillips Pavilion at the Brussels Exhibition of 1958, and his design system, the *Modulor*. The relationship Le Corbusier established between architecture and music was indicative of this.

the lack of rationale not intrinsic to the machine. To use the words of Pérez-Gómez (1996: 4): Durand's (1760–1834) theories already "avoid all reference to philosophy or cosmology".

The new applied science was instrumental not only as a source of "inspiration". Inherited from the 19th century – the century of dominance of engineers – this is a science that gave rise to technology, and no longer the theory that gave rise to the practice (Masiero, 2003: 95), removed from the preceding notion of science associated with holistic cosmology.

In harmony with the spirit that animated this new science, the new "machine" distanced itself even more from tradition and became a technological application devoid of symbolism and uncertainty. Science, in its technological aspect, and the Promethean idea of the "machine" thus became part of a new reality.

The machines featured numerous elements that fed the idea of Progress and that were to fit perfectly in the functionalist theory of architecture: faster, not susceptible to illness or any other human weaknesses, more easily controllable, lacking free will, more precise, specialised. Like gearing or cogwheels – an organised object with only necessary elements; a plastic object – aesthetically appealing. A symbol of individuals overcoming their own limitations. In his *Historia de la arquitectura moderna* [History of Modern Architecture], Leonardo Benevolo (2002: 17) lists and explains many of the factors that served as proof that mechanisation and industry brought only benefits to mankind: food, hygiene, salubrity, sanitation and particularly, in his opinion, health.

Despite the constant presence of the two views of progress (hopeful or suspicious), it was its apologia that won and dictated the rules. Accepted by the more sceptical as a necessary evil (not all succumbed to a fascination for the machine[2]), the apology of the machine was all driven and encouraged especially by the vanguards.

Although it was already in the embryonic stage before the World War, it should be pointed out that this development became most pronounced after the World War I. Despite the fact that the idea of progress had forfeited some of its leading roles. The development of the mode of transport engineering, which experienced a strong impulse with the war, was decisive. The new means of transport – automobiles, aeroplanes, ships – were "envied", if not as works of art, then at least as aesthetic objects. The notion that the machine was not adverse to humanity, as previously argued, but could be employed as useful instruments, was now becoming widespread. In addition, the conclusion was now common that making things by hand, as William Morris (1834–1896) idealised, did not make individuals any happier, but became an obstruction to the production of objects that would be accessible to all.

[2] See, for example, Van de Velde (1863–1957), who, in the context of his work with the Deutsche Werkbund, maintained a heated debate with Hermann Muthesius (1861–1927) on this issue.

4 THE HUMAN MACHINE

The machine that was celebrated essentially from the 19th century onwards, was a human machine, the fruit of Progress, and one that can be seen metaphorically as a counterpoint to the earlier Divine machine: as a metaphor for the once Divine world now devoid of Divine intervention.

Whereas it was God who controlled the world machine (even if the latter functioned without permanent Divine intervention), it was now humanity who controlled the machine. Progress had replaced God (or man?). We went from the mathematical God – the geometer of the universe – to the absent God, a notion that was contrary to the concept of the original machine, the model machine that was the world; from the omnipresent God to the absent God: from the World to the Machine. From the Divine and perfect machine to the human, super-human (enchanting?) machine: Progress was the new God or a demiurge that brought order to chaos.

The universe, now seen also as a machine, went from a Divine creation to a mere practical and functionalist universe, a practical and functionalistic machine that resolved concrete immediate problems.

If one can say that the Renaissance introduced the idea of overcoming Nature, then the machine gave rise to the desire to replace Nature.

4.1 *Another form of mimesis*

The idea of mimesis is a particularly interesting one, as it is closely linked to the topic of the relationship that Humanity in general, and art in particular, establish with Nature and, consequently or not, with God. That machine was to be imitated at the beginning following the idea of mimesis, and in that imitation, architecture and music (again as examples) drank from the same source.

As the Renaissance matured, the idea that the work of art overcame nature began to impose itself. Marsilio Ficino (1433–99) argued that art is "wiser than nature" (Tatarkiewicz, 2013: 278); and Giorgio Vasari (1511–1574) advanced that "natura vinta dall'arte" [nature is conquered by art] (Tatarkiewicz, 2013: 247).

Whilst traditional mimesis created a microcosm analogous to the macrocosm, in the new era of the machine one can almost say that it saw the macrocosms as a macro machine analogous to the micro-machine. Going from God to Human.

Placing the work of art and the machine – the great creations, because human – side by side, Gino Severini (1883–1966) (apud. Kultermann, 1993: 3) regarded the machine as an object which creation was identical to that of a work of art:

> All elements of matter, for example, the components of a machine, are assembled by one mind, the mind of the inventor (…). The process of developing a machine is analogous to the creation of a work of art.

4.2 *The new machine as an object without a past*

Despite their great heterogeneity, most of the experiments carried out in the early 20th century, in both architecture and music (and, more broadly speaking, in all artistic fields), revealed a general common attitude: a voluntary reaction against the qualities and values defined and established during the preceding century. That was the mentality installed in the various vanguard movements that emerged in the European context at the beginning of the new century.

The reaction against the past found in the machine – a human-made, new, artificial object par excellence – a new model. Since nature had already served as a model, through its laws and in its image, now laws and image would be those of the machine.

The fact that the "machine" presented itself as a topic without a past was fundamental for its choice as the generator of a new aesthetic. The new leading role was assigned to the machine by artists in general, and by architects and musicians in particular. One can observe how the imaginary associated with the machine – both in its mechanical/functional and its metaphorical aspects – played a decisive role in architecture and music universes. The shared fascination with the machine proved, therefore, to be a means of achieving the desired emancipation from the past and was the basis for the proximity between the architectural and musical discourses, giving formal expression to the desire for independence and overcoming. That desire was now realised. Deification of human creation took the place of fascination with Divine creation.

Theo Van Doesburg (1883–1931) (apud. Kultermann, 1993: 3), for example, spoke of the need to incorporate the machine into questions of style:

> If it is correct, that the culture means in the broadest sense independence from nature, then we should not be surprised, if the machine dominates with regard to cultural wishes concerning style.

J. J. P. Oud (1890–1963) (apud. Kultermann, 1993: 2) saw the machine – which, in contrast to craftwork production methods, facilitated high-precision labour – not as an auxiliary instrument that substituted what came before it, but as a new value. In addition to being a source of inspiration and information for architecture, that characteristic was to be divulged, optimised and applied.

> The products of the machine should be accepted on the basis of the uniqueness of the machine, and no longer as substitute for handcrafted work. The preciseness of the purely technical products which we admire (automobile, ship, instruments, etc.) should be an example for reorganization in architecture in the sense that the demands made on a structure should be as precisely fixed, and realized with the latest materials, construction, and work technology.

Claude Debussy (1862–1918) (apud. Griffiths, 1996: 98), showed, in 1913, his fascination with the "century of the aeroplane":

> Is it not our duty to find a symphonic means to express our time, one that evokes the progress, the daring and the victories of modern days? The century of the aeroplane deserves its music.

5 CELEBRATING THE MACHINE

In rejecting the past and celebrating the machine, Futurists and Constructivists drew attention to the new environment and new opportunities provided by the large city, the machine and dynamism.

The Futurists were to markedly leave their mark on the history of both architecture and music in this respect. In "The Foundation and Manifesto of Futurism" (1908), Filippo Tommaso Marinetti (1876–1944) (apud. Griffiths, 1996: 286) stated:

> We declare that the splendour of the world has been enriched with a new form of beauty, the beauty of speed. A race-automobile adorned with great pipes like serpents with explosive breath (...) a race-automobile which seems to rush over exploding powder is more beautiful than the Victory of Samothrace.[3]

While uttering, in "Futurist Architecture" (1914) that "[e]verything must be revolutionary" (Sant'Elia et al., 1971: 36), Antonio Sant'Elia (1888–1916) and Marinetti, proclaimed the Futurist city and (in accordance with the future statements of Le Corbusier) the house as a machine as well:

> We must invent and rebuild the *Futurist* city: it must be like an immense, tumultuous, likely, noble work site, dynamic in all its parts; and the *Futurist* house must be like an enormous machine. (Sant'Elia et al., 1971: 36)

Luigi Russolo (1845–1947), a leading Futurist composer, "called for a music that would relate to the sounds and rhythms of machines and of factories, an 'art of noises' which must be strident, dynamic and eagerly in tune with modern life." The composition method he developed, Bruitism, was based on the use of all kinds of noises produced by industrial society; any noise or sound could become musical material. He built a number of "intonarumori" [noise intoners]: mechanical contraptions designed to produce exploding, crackling, rubbing, buzzing noises. Russolo organised "concerts" with such instruments. Despite performances in Italy and Britain before WW I, it was only in 1921 that they received particular attention, in Paris. Russolo maintained his interest in such intoners up until roughly 1930, but he left them in

[3] In the same spirit, one can also point, in the Portuguese context, to a poem from 1917, "Elogio da Velocidade" ["Ode to Speed"], by Almada Negreiros, or the affirmation by Álvaro de Campos (one of Pessoa's heteronyms): "Newton's binomial formula is as beautiful as the Venus de Milo", published in 1944.

Paris, where they were destroyed during World War II (Griffiths, 1996: 98–100).

Freeing himself from the past, Russolo (apud. Tiekso, 2016: 77) celebrated the new age:

> We futurists have all deeply loved and enjoyed the harmonies of the great masters. Beethoven and Wagner have stirred our hearts and nerves for many years. But now we are satiated with them, and we derive far more pleasure from ideally combining the noises of trams, internal-combustion engines, carriages, and noisy crowds than from rehearing, for example, the 'Eroica' or the 'Pastorale'.

Another interesting aspect was that this period gave rise to many works of art whose titles reflected the fascination with the machine and industry. As examples: Paris witnessed the presentation of the composition *Ballet Mécanique* (1926) by George Antheil (1900–1959) – paradigmatically scored for eight pianos, eight xylophones, a pianola, two electric doorbells and an aeroplane propeller; or *Pacific 231* (1923) by Arthur Honegger (1892–1955) – an orchestral work depicting a locomotive in motion.

The design for the huge Monument to the Third International (1920) by Vladimir Tatlin (1885–1953) presented itself as the veritable idealisation of that new stance. According to the spirit of the times, Tatlin wanted to break with the past – with the realism of the tsars' epoch.

6 AFTER THE CELEBRATION: STILL SEARCHING FOR PLATO

The vision of the world governed by Divine laws, full of mystery and poetry; the vision of a world/stage generating mundane music – music muted and replaced by the festive noises of modernity – was supplanted by a vision of the world as a machine that is merely the sum of the gearing elements, now governed by the same physical laws that determined the construction of bridges, roads, skyscrapers, ships, trains, cars, planes and other constructions that pay homage to or celebrate the new construction techniques, and no longer by the laws of the old *mathesis universalis*.

However, the end of the century contrasts with its beginning, as Robert A. Nesbit points out:

> Everything now suggests, however, that Western faith in the dogma of progress is waning rapidly in all levels and spheres in this final part of the twentieth century. (2009: 9)

Alan Colquhoun (2009: 24) stresses that in the Modern Movement "the new technology was an idea rather than a fact" and that, although a lot of the materials used by architects of the Modern Movement "were conceived to be the product of machine techniques", those architects never saw said materials as "'ready-made' but adapted them to a preconceived plastic form".

In reality, what Philip Johnson wanted was some sort of a formal approximation to the Platonic ideas. Between the ideal and functionalism, neither was ever satisfactory. In the first case because the ideal was never realised; and in the second, because there was always the imponderable and human accident. The first choice: ultra-human and the second: infrahuman.

After all, Talin's tower would house each of the four agencies of the Comintern in four internal glass volumes – a cube, a tetrahedron, a cylinder and a semi-sphere. Each of these volumes would revolve according to astronomic time – once a year, once a month, once a day and once an hour, respectively.

In the end, the machine is only an instrument. The machine always ends up becoming obsolete: to will always be dated is a *machining* problem. Eternity is never dated.

BIBLIOGRAPHICAL REFERENCES

Barr, Alfred H. (1934) Foreword. *Museum of Modern Art. Machine Art.* New York: Museum of Modern Art. https://www.moma.org/documents/moma_catalogue_1784_300061872.pdf. (Accessed 8 June 2017).

Benevolo, Leonardo. (2002) *Historia de la arquitectura moderna.* Gustavo Gili.

Bronowski, J & Mazlisch, Bruce. (1960) A tradição intelectual do Ocidente. Lisboa: Edições 70.

Bury, John Bagnell. (1920) *The Idea of Progress. An inquiry into its origin and growth.* London: Macmillan & Co.

Colquhoun, Alan. (2009) Symbolic and Literal Aspects of Technology. *Collected Essays in Architectural Criticism.* London: Black Dog, 23–26.

Griffiths, Paul. (1996) *Modern Music: A Concise History.* London: Thames and Hudson.

Hall, A. Rupert. (2014) *The Revolution in Science.* Abingdon: Routledge.

Johnson, Philip. (1994) Preface to Sixtieth-Anniversary Edition. *Museum of Modern Art. Machine Art.* New York: Museum of Modern Art.

Kultermann, Udo. (1993) *Architecture in the 20th Century.* New York: Van Nostrand Reinhold.

Masiero, Roberto. (2003) *La estética de la arquitectura.* Madrid: Machado Libros.

Nesbit, Robert A. (2009) *History of the Idea of Progress.* London: Transaction.

Pérez-Goméz, Albert. (1996) *Architecture and the Crisis of Modern Science.* Cambridge: MIT Press.

Sant'Elia, Antonio & Marinetti, Filippo Tommaso. (1971) Futurist architecture. In: Conrads. In: Conrads, Ulrich (ed) *Programs and Manifestos on 20th Century Architecture.* Cambridge: MIT Press, 34–38.

Struik, Dirk J. (1992) *História concisa das matemáticas.* Lisboa: Gradiva.

Tatarkiewicz, Wladyslaw. (2013) *A History of Six Ideas: An Essay in Aesthetics.* Berlin: Springer.

Tiekso, Tanja. (2016) Luigi Russolo's Art of Noises and the Aesthetics of Musical Experimentalism. In: Berghaus, Günter (ed) *International Yearbook of Futurism.* Berlin: De Gruyter, 71–93.

Wittkower, Rudolf. (1998) *Architectural Principles in the Age of Humanism.* London: Academy Editions.

Note: This paper was funded by National Funds through FCT – Fundação para a Ciência e a Tecnologia under the Project UID/AUR/04026/2013.

Interstitial spatialities in progress: From identity to relationality

Ana Vasconcelos
Department of Architecture, Urbanism and Design, Faculty of Architecture, University of Lisbon, Lisbon, Portugal

ABSTRACT: Can it not be said that "interstitial spatialities" of architecture and contemporary life are in progress or development? There where science, culture, the arts, general thought and the forms and spaces in which we live interrelate with one another with increasing transparency and consciousness? There where background and figure, territory and construction, natural and artificial, interior and exterior, structure and detail, well-defined and ambiguous, space-form and function-use are intertwined, blend or mix together through a geometry that is designed between visibilities and invisibilities in search of greater fluidity, contiguity and relationality. This is the in-between or interstitial space that we provide, an actual space in lived experience, ever more present in today's architectural projects.

We reflect on the "in-between" condition, which is borderline, ambivalent, dynamic and imprecise, on its increasingly topological geometric characteristics and cultural references linked to the current paradigm of complexity, in order to understand its presence, relationship or involvement in significant works of contemporary architecture, revealing its sense and importance in the present day and in the discipline.

The value of the relationship and interaction is situated in contemporary architectural conception, where the form is more matrix and less configuration, where the in-between realm appears in conceptual, formal and experiential aspects, becoming increasingly more consciously a geometric place. It is an architecture that reflects (or senses) the premises of the current Zeitgeist (spirit of the time), based on a more holistic, transdisciplinary, systemic and complex thought, and a new collective consciousness regarding the reality that foretells a crisis of perception, a change in paradigms and new values. What emerges is a more interactive perspective that is increasingly more focused on the study of the relationships between things, from which another way of feeling and perceiving the world, places and spaces emerge, which little by little is changing the way we think about and draw architecture and, consequently, interact with and in it.

Keywords: in-between space, geometry, architecture, culture, contemporaneity

In the field of geometry, in its form, concept and symbolism, and inherently in the field of contemporary architecture, we consider the "in between" condition (borderline, ambivalent/multivalent, tensed and dynamic) that we are presently investigating is being manifested/materialized in the drawn form and in the built form, through a series of figures. The Vesica-Pisces, or other figures in which the conjunction or intersection are implicit, is one of them, as are non-orientable surfaces (the Möbius strip, Klein bottle, etc. (Math, topology: Poincaré, H. (1895) Analysis Situs), spirals, dynamic and intersecting folded, fractal or rhizomatic systems, where shapes often oscillate between angular and rounded forms. These geometries make it possible to build in-between or interstitial spaces, in their quality as dynamic systems of units or multiplicities. Or better yet, rather than spaces, said geometries constitute physical and perceptive spatialities that are energetically dynamic, capable of "activating" users, of "moving them" and transporting them to the internal or external movement, inviting them to immerse themselves in the "high game of architecture" as it is called by C. Jencks (Jencks, 2000), in reference to the architecture of Palladio and Eisenman.

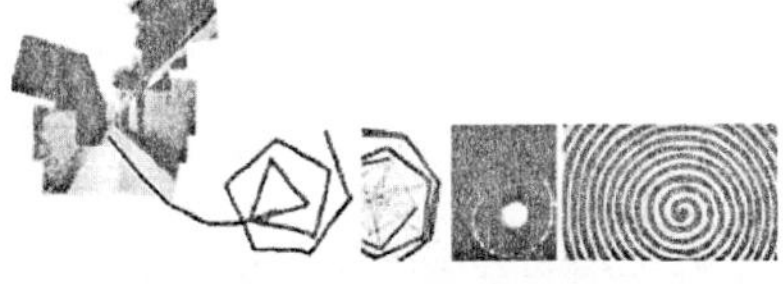

Figure 1. Continuity, discontinuity and repetition in difference: Spiral. Left: Verso il maré, E. Miralles. Middle: Fifteen variations on the same theme, M. Bill (Aavv, 2001). Middle: Golden Spiral's stair (photo), H. Vöth (Ruiz De La Puerta, Sanchez Gonzalez, 2007). Right: Spiral (http://pt.photaki.com/picture-espiral-hipnotica_412038.htm).

These figures, forms, structures and geometric matrices implicitly contain the in-between condition. In the Vesica-Pisces, among other regular and irregular shapes, regardless of whether they are similar, this condition is seen through the conjunction or intersection of polarities, entities or worlds. In the case of Möbius strip

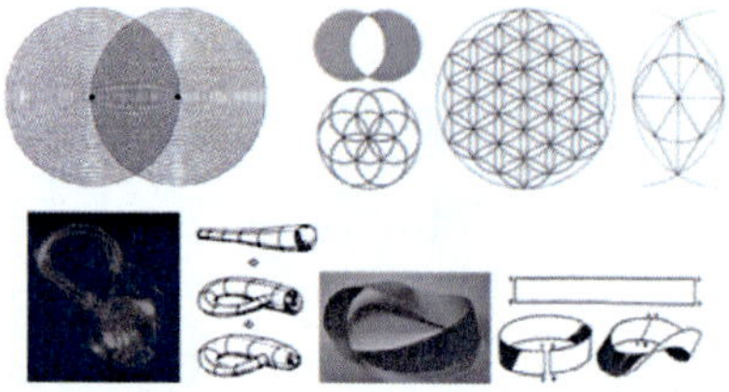

Figure 2. Above: Conjunction and intersection: Vesica-Piscis, Flower of Life symbol, Vesica-Piscis and The Tai Chi triangles (Freitas, 1984). Below: Continuity, fluidity and topological: Klein bottle (http://www.google.com/imgres?hl=pt-BR&tbm=isch&tbnid=57GpcpY_Yj67wM:&imgrefurl. (Freitas, 1984); Möbius strip (http://www.google.com/imgres?hl=pt-BR&tbm=isch&tbnid=5eUT6ZFjYxUFeM:&imgrefurl).

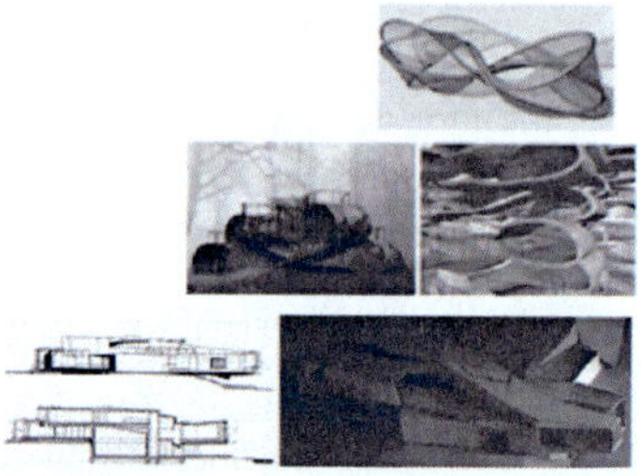

Figure 3. Above: Virtual House, USA 1997, FOA (http://www.google.com/imgres?hl=pt-BR&tbm=isch&tbnid=Zx0QgpUYAFXj0M:&imgrefurl). Below: Möbius House, 'T Gooi-Utrecht, 1993–1998, UN-Studio (http://bartlettyear1architecture.blogspot.pt/2010/03/mobius-house-in-amsterdam-by-ben-van.html).

and Klein bottle, among other non-orientable surfaces, the in-between condition manifests itself through its permanent continuities and topological contiguities. In spirals, the in-between condition emerges through the dynamic conciliation between continuity and discontinuity on their multiple levels. In terms of the formal systems consisting of folds and furrows (fissures, fractures and marks/traces) in a sequence of repetition in difference, the in-between condition emerges through the in-between or interstitial formal-spatial conditions that can be generated in architecture, constituted between continuities (connections, affinities) and discontinuities (changes, differences) in terms of position, direction, scale, configuration, texture, colour or brightness[1]. These formal systems are many times associated with fractals or rhizomes insofar as they are geometric matrices of contiguity, interrelationships and self-similarity.

Except for the Vesica-Pisces, all these geometric shapes are linked in some way to what is known as topological geometry (non-Euclidian), which studies and defines the structures of the dynamic transformation processes (change, deformation and evolution) and their invariant properties, and which are based on the concepts of proximity, continuity, contiguity, limits and self-similarity[2]. All these concepts are associated with a purely "in-between/intermediate" borderline condition based on the value of the "relation" (nexus, ligation, connection or interrelationship), and no longer just the mere value of the "configuration" (of the object, identity or the exact measurement/dimension) characteristic of Euclidian geometry.

The folds, fractals and rhizomes are figures/shapes from non-Euclidian geometry which, in the midst of the continuity and discontinuity of form, position or scale, allude to a simultaneous dynamic condition of repetition and difference. We can also include in this group non-orientable surfaces, such as the Möbius strip, the Klein bottle or Boy's surface, spiral systems, all types of repeated furrows (fissures, fractures, cracks, etc.), as well as any formal system resulting from formal operations of intersection, overlapping, cutting or folding. All these figures can constitute spatial-formal interstitial systems that place more value on "similarity" and "relationality" (relationships, interrelationships, interactions, etc.) than on equality, repetition or identity. In other words, these are systems whose formal condition is characterised not by equality or difference, but rather by the similarity of the parts of the whole and their inherent tension.

These figures can be used to generate complex and self-similar architectural spatial and formal systems that are conceptually and perceptively dynamic, where the iteration of the irregular, the broken, the interrupted, the similar and the idea of evolution, expansion and flow are inherent in the concept of repetition in difference that characterises them.

We have chosen these figures, forms, structures and sequences as key figures in the concept of the in-between because, besides all being borderline geometric (formal, conceptual or symbolic, with inherent tension or dynamism), they are increasingly more commonly used by architects today as a source of inspiration or a drawing tool in architectural design. In the formalisation process insofar as they form part of the space matrix (conceptual-formal diagrams,

[1] Expanding on the concept of repetition in difference, we can say that it also includes the underlying value of the "modalities", insofar as they are variations or gradations of the same thing, colour, light, etc.: modes that are both repetition and difference; modes that are repeated in different ways; modes that resemble one another. For example, the colour white (which, in all reality, is never only or purely white) has an infinite number of degrees of intensity, variables that are its modes of existence, which are what architecture reveals.

[2] However, both spirals and the fractal order can also be integrated within classic Euclidian geometry. With regard to rhizomes, so idolized by Deleuze and Guattari, we could say that they belong to neither topological geometry (due to their breaks) nor Euclidian geometry (as the result of their flows, expansions, agencies and becoming, and of their imperfections, imprecisions and inflections), but rather to another type of geometry, perhaps an "in-between geometry" that contemplates both Euclidian and topological geometries.

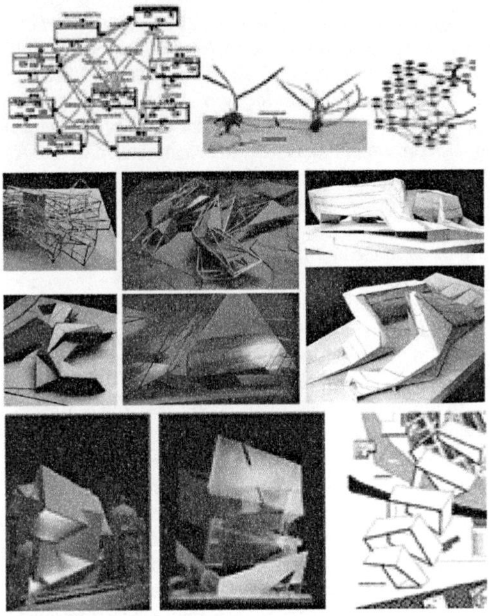

Figure 4. Above: In-between continuity and discontinuity, multiplicity of connections, flows and events (poliorganic, non-hierarchical, non-centered): Rhizomes (https://www.google.com/search?um=1&hl=pt-BR&biw=781&bih=655&tbm=isch&sa=1&q=organograma+informatico&oq. http://portfolioensinodaarte.blogspot.pt/2010/06/rizoma-o-caule-das-gramineas.html. https://www.google.com/search?um=1&hl=pt-BR&biw=781&bih=655&tbm=isch&sa=1&q=representa%C3%A7%C3%A3o+fractal&oq). Middle: L'huile Library, Switzerland, P. Eisenman. Church for the year 2000, Rome, 1996, P. Eisenman (El Croquis n°83 (1997): Peter Eisenman 1990–1997, El Croquis Editorial, Madrid). Below: Victoria & Albert Museum, London, 1996, D. Libeskind (http://forum.skyscraperpage.com/showthread.php?p=4789045, http://www.telegraph.co.uk/culture/3624000/The-Architectureof-Daniel-Libeskind.html?image=9).

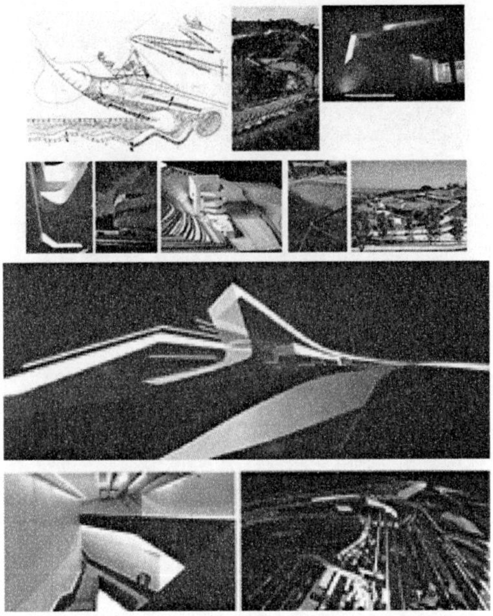

Figure 5. Above: Igualada Cemetery, 1986–90, E. Miralles, C. Pinós (http://www.google.pt/imgres?hl=ptPT&sa=X&tbo=d&rlz=1T4GZAZ_ptPTPT419PT420&biw=1920&bih=941&tbm=isch&tbnid=8A2cwFpazfj6bM:&imgrefurl). Middle, left: Iberê Camargo Foundation, Porto Alegre, 2000–08, A. Siza Vieira (En Blanco nº 1 (2008), General de Ediciones de Arquitectura). Middle, right: Botanical Garden, Barcelona, 1989–99, C. Ferrater (https://www.google.com/search?um=1&hl=pt-BR&biw=889&bih=655&tbm=isch&sa=1&q=jardin+botanico+carlos+ferrater&oq). Middle: LF One pavilion, Weil am Rhein, 1993, Z. Hadid (http://newsfekosmograd.com/kosmograd/graphic_design/page/2/). Below: Maxxi-Contemporary Art Museum, Rome, 1999/2003–10, Z. Hadid (https://www.google.com/search?um=1&hl=pt-BR&biw=897&bih=655&tbm=isch&sa=1&q=maxxi+zaha+hadid&oq).

drafts, diagrams, etc.)[3], or as a final formal-spatial proposal in the configuration (building or the building complex). All of them are geometries that can generate in an architectural site fluid and topological spatialities with intentionally undefined limits, and even more complex, fractalised and densified spatialities, through the use of breaks, furrows, angularities or autonomies, both topological and non-topological, which we refer to as "interstitial topographies". Both types of spatialities create formal, energetic and perceptive tensions among visibilities, invisibilities and resonances, proposing polytopic perspectives, in other words, multiple perspectives between spaces, forms, luminosities, virtualities, illusions, reversibility, etc. that oscillate in a continuous interrelationship, always between a here, a there and a further beyond.

These spatialities appear to be liquid, sometimes diaphanous in appearance and other times more constrained, where all the entities or components among the imperceptibilities (energy flows) and perceptibilities (constructed forms) are actively and simultaneously correlated without any of them losing its identity or inherent potential for contradiction. They are spatialities and poetic operations that accompany (or perhaps sense) the progress made in a science that in the early 20th century proposed that all matter, all system or event, are nothing more than mere "energy relationships" of greater or lesser complexity. They are ultimately "substance, space, time and information in permanent mutation" or rather, in "a perpetual

[3] We consider categories of "the in-between" (spatial or formal) to be: matrix and object. In this case, the diagram (conceptual,-formal, draft, diagram, etc.) is understood as the space-matrix, an in-between space of the matrix category in the architectural project. Cf. author's PhD: Vasconcelos, Ana. (2016).

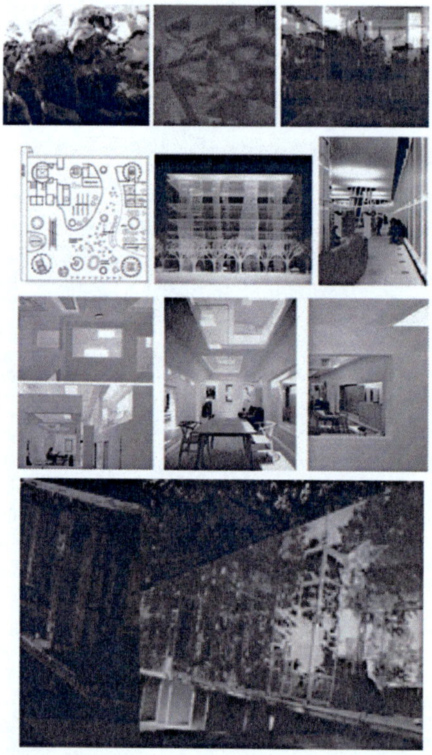

Figure 6. Above: Translucent: Photo ARCO 2006, Madrid (author's photo); Piranesian Matrix, 2009, P. Burgaleta (author's courtesy); Incipiente Proposal, video photogram, Iris Astor (student's work, ETSAM). Middle: Media library, Sendai, 1995–2000, T. Ito (El Croquis nº123 (2004): Toyo Ito 2001–2005 – más allá del movimiento moderno, El Croquis Editorial, Madrid). Middle: House N, Oita, Japan, 2006–08, S. Fujimoto (2G Revista internacional de arquitectura nº50 (2009): Sou Fugimoto, Gustavo Gili). Below: Glass Video Galery, 1990, Tschumi (A+U, Architecture and Urbanism (1994): Bernard Tschumi, 1983–1993).

oscillation between actualization and potentiation" (Nicolescu, 2005: 164)[4] , where the notion of an object is replaced by that of an event, occurrence, relationship, interconnection, movement and transition[5].

These are geometries that many architects today use, or draw inspiration from when formalising their poetic operations and inquiring into new formal, spatial and functional possibilities, as can be seen in these works shown here. Almost all of them are architectures whose geometry is no longer so specific, due to aspects of the two- or three-dimensional drawing or of the construction. Today more than ever, architecture, geometry and computer technology are linked together in a way that is freer to investigate, materialize or recreate ideas, many of which come from scientific discoveries, philosophic proposals or cinematic fictions of essays, movies or the Internet, with the aim of conquering new ways of thinking and creating places.

Currently, in both the artistic and the architectural realm, increasingly more common are poetic operations of deformation or transformation, which are based on the geometry of folds, fractalised forms or rhizomatic meshes, operations that build formal systems in which, ultimately, everything is similar. We are talking about systems whose order is based on repetition and difference, where there is formally something that always persists and something that always differs; where the difference, the equality, the continuity, the discontinuity and all the other polarities are dynamically and interstitially integrated to build the configuration or spatiality. They are poetic operations of deformation and transformation where both the intersection and the overlapping may be implicit, operations in which the primary objective is achieving spatialities with imprecise limits and polytopic perspectives (interstitial topographies), as opposed to spaces or places with rigid limits and autonomous configurations.

The concept of repetition in difference, alluding to the meanings given by Leibniz, Nietzsche and Deleuze, and inherent to topological geometry, rhizomatic matrices (even though these accept breaks), folded, fractal and self-similar shapes and non-Cartesian space, confirm randomness, the process or event (the occurrence, the action, the event), movement and the iteration in difference, as well as surprise or astonishment, the unexpected, continuous transformation and the tension of relationships or interactions. Repetitions in difference, interstitial topographies and polytopic perspectives, increasingly more common in current architecture, are conceptually and formally expressed in many works (such as those we present here) through folds, furrows, fissures, fractures and breaks, which are frequently intercepted, intertwined or overlapping with simple cubic geometries subject to a fractal and/or rhizomatic order.

We are talking about shapes/structures/geometries that, through operations of fragmentation, deformation, rotation, intersection, combination and overlapping, formalize spaces with an identity and difference at the same time, between continuities and discontinuities, in a myriad of connections, reverberations, flows and occurrences, in search of a dynamic appearance and greater interaction with the nearby context and its users. More than spaces, architectural spatialities are generated with complex or undefined limits and with a dynamic formal contiguity; spatialities where the repetition in difference is order and form, a space

[4] Translated by the author.

[5] These are all important concepts in current scientific thought, which evoke the "in-between condition". Energetic dynamism comes to govern the group of physical phenomena and announces a new concept of space and time, as shown by different elements that have already been proven by science, such as wave-particle concepts, the overlapping of states or the possibility of the existence of quantum bodies, Goethe's protophenomenon, Sheldrake's morphic field and Pischinger's extracellular matrix.

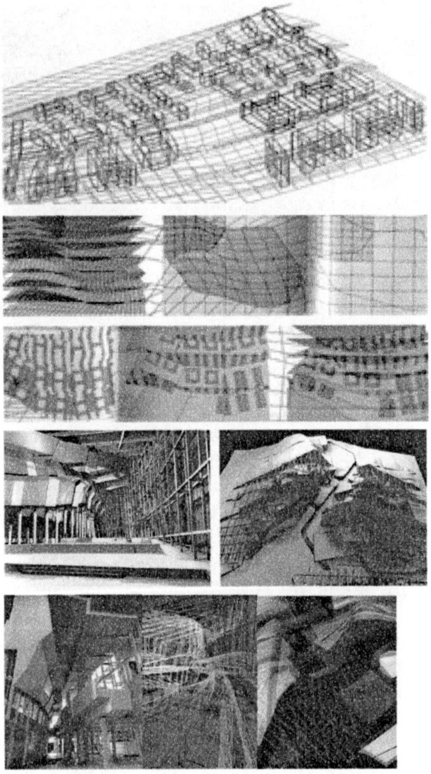

Figure 7. Above: Rebstockpark, Germany, 1991, P. Eisenman (https://www.google.com/search?um=1&hl=pt-BR&biw=889&bih=655&tbm=isch&sa=1&q=eisenman+rebstockpark&oq). Below: The City of Culture of Galicia, Santiago de Compostela, 1999, P. Eisenman (http://www.gocsa.es/Obr7.asp?id_Obra=42&Idioma=3&iframe=true&width=1074&height=600. https://www.google.com/search?um=1&hl=pt-BR&biw=985&bih=655&tbm=isch&sa=1&q=ciudad+cultura+eisenman&oq).

matrix[6] in which self-similarity, resemblance and proximity prevail over difference and equality, be it conceptual, formal or material. More than autonomous or closed-in buildings, topographies are generated that are interstitial, continuous and receptive to several types of occupancy, activities or programs.

In his book *The Architecture of the Jumping Universe*, Charles Jencks maintains that fractality is one of the five facets of the new cultural and architectural paradigm in which we live, which has emerged from the sciences of complexity and their theories of chaos, cosmology and philosophy and contemporary thought, with the other facets being organic technology, topographic forms, wave shapes and cosmogenic architecture. This new paradigm, as Jencks explains, exhorts the adoption of self-organizing systems and similar forms with a dynamic appearance, that boasts freedom, complexity, variety, fluidity polyvalence and emotion. The new paradigm began to take shape in architecture over the last three decades of the 20th century, through a quest for other spatialities, other possibilities for design, construction and the occupancy of space, and as a result, other forms of interpretation and rehabilitation of the cities that were built[7]. According to Jencks, all these facets have come to overthrow the concept of "repetition" (associated with equality, consistency, identity and monotony), translated into architecture according to simple, pure and reductionist forms that "release constant sensations, regardless of the origin of the observer" (Jencks, op. cit.), as occurs with spheres, cubes and other solid shapes considered to be pure. All these concepts and geometries, linked to the four basic aspects of modernity (mechanism, materialism, determinism and reductionism) were commonly used in modern architecture (Mies' paradigmatic 'less is more') and are equally common in classic architecture. They are concepts and geometries, however, which take on a new value in the formally changing and dynamic reality of current architecture. This value is notoriously recognised and sought after in the careers of some contemporary architects, such as Eisenman, Tschumi, Holl, Libeskind, Hadid, FOA, Van Berkel and Bos (UN-Studio), Ito, Fujimoto, Miralles, Ferrater and Siza, among others. All these presented architecture works (from fig. 3 to fig. 11) were studied in the scope of "interstitial spatialities" (Vasconcelos 2016)[8], however, are not specifically described in this paper.

The entire "in-between realm" of "figural" modality, because of its imprecision, ambivalence/multivalence and ambiguity, is conceptualised and formalised in a condition of identity and difference at the same time[9]. This is a condition that contemporary architecture investigates and experiments with through operations with digital diagrams, for example[10]. The conceptual-generative diagrams, far from being rigid or conclusive structuring systems, are geometric matrices (space-matrices), the form of which is governed by an order (flexible) that oscillates between repetition (identity) and variation (difference).

---

[6] The "space-matrix" is introduced into the overarching category of "the in-between" (form, formation, generative structure, "becoming"). Cf. author's PhD: Op. cit., chap. II, 5.4.

[7] Jencks considers two great changes in the paradigms of the history of culture: from the Christian to the modern paradigm and from the modern paradigm to that of complexity. For this reason, he refers to the current paradigm of complexity as post-modern (using "post-modern" in the broadest sense of the term).

[8] Cf. author's PhD: Vasconcelos, Ana. (2016) Lo intermedio como lugar. Lo intersticial, lo fronterizo y lo impreciso en la arquitectura contemporánea, Escuela Técnica Superior de Arquitectura – Universidad Politécnica de Madrid, chap.II, chap.IV and chap.V.

[9] We consider modalities of "the in-between" to be the figurative and the figural (between figure and background). Cf. author's PhD: Op. cit., chap. II, 5.3 and 5.5.

[10] The diagram is understood as the "space-matrix" in the architectural project. Cf. author's PhD: Op. cit., chap. IV, 1.

On the other hand, the trend of the current project is to conceive of spaces with an increasing degree of formal and functional ambiguity, given that it is more and more common to restrict areas in the main urban centre for economic reasons and the scarcity of lands, thus making greater spatial versatility and capacity for adaptation a necessity. The objective of everything constantly becoming more transformable, adaptable and flexible in response to every need is a premise/concern in current architectural work, which demands a greater compromise between formal, material and multi- or polyfunctional aspects, and between the qualities of being long-lasting, temporary, adjustable and creative.

As a result, contemporary architecture is dedicating more and more efforts to studying/experimenting with spatialities that are structured around an open formal/inconclusive matrix, the limits of which are more imprecise and with a greater possibility of transformation and change. The value of the relationship and interaction now lies in the contemporary architectural conception. This value is no longer strictly understood to be between form and function, rather in a broader and more integrated sense, among idea/concept, form, function, multiple possibilities for use, energy dynamics and those of perception, and between the truly inhabitable and the virtually liveable. We are talking about an architecture where the in-between realm appears in conceptual, formal and experiential aspects, ever more consciously becoming a geometric place. In other words, in this architecture, the form is increasingly more geometrically focused on the matrix and less on configuration. It is more structuring, transformable and adjustable, and less precise or specific, resulting in a compromise between the matrix and the object[11].

In some way, as held by Charles Jencks, in contemporary times, architectural works reflect the new paradigm of the complexity typical of the current Zeitgeist or spirit of the time. They do this through the manipulation of these complex and dynamic geometries, with the aid of computers and scientific topics and developments. This new Zeitgeist summons a change in model from objects to relationships, to a more holistic, systemic and complex thought (E. Morin), and a new collective conscience on the reality that foretells, as referred to by Fritjof Capra in his book The Web of Life (1996), the crisis of perception, the change of paradigm and new values.

The verification and consideration of reality by science, as an open unit and a complex, multi-level plurality, continues to inspire architecture and through in general, with the "in-between spaces" we are investigating as evidence of this. As previously mentioned, more than spaces, they are "spatialities" that somehow reflect the new perspective that has been established in the broad field of contemporary culture. It is a more interactive perspective that intends to be non-logocentric, systematic and transdisciplinary, increasingly more focused on the study of the relationships between things, from which another way of feeling and perceiving the world, places and spaces emerge, which little by little is changing the way we think about and draw architecture and, consequently, interact with and in it.

Each of the architectural works that we present in a peculiar manner presents us with the current trope of the interstitial, in which architecture, territory and topology, both already existing and new, collective and individual, public and private, interior and exterior, flow and configuration, colour and form, light, image and space, and other uses and functions are increasingly interwoven or blended under the rule of a hybrid geometry designed between a topological and a Euclidean condition. Jencks describes this geometry very well in his text "The New Paradigm: Non-Linear Architecture", addressing the following topics: From matter to mutter; the cinematic section; The fractal landform; Fluid fractals; Not inform but emergent form.

Are these "new" geometries and spatialities, currently so widely studied by architects (and increasingly so by mathematicians and physicists), not a specific testimony that questions the lack of contradiction of the linear Aristotelian logic and which extols another type of logic? vIs it, not a more complex, intricate, dynamic and mysterious logic, existing in both architecture and life itself?

An example of said logic is the "logic of energetic antagonism" described by Lupasco, which is established in a dynamic and coexisting ternary structure that oscillates unstable between la potentiation and actualization, or what is the same, between imperceptibility and perceptibility, preserving the antagonistic qualities of all the components. In other words, it is a formal logic that is met between the notion of the object/configuration and the notion of energy/flow, with this also being understood as an event, occurrence, relationship, interconnection, movement and transition[12].

---

[11] Categories of "the in-between": matrix (form, formation, generative structure, "becoming": the space-matrix) and object (figure, configuration, limit/contour, "being": the limit-realm). Cf. author's PhD: Op. cit., chap. II, 5.4.

[12] Lupasco' s logic of energetic antagonism is a formal tri-dialectic logic that oscillates from potential to actualization, and vice versa, from actualization to potential, implicating or formalizing a dynamic energetic unit in which the antagonistic qualities of all the elements/occurrences/events remain intact, even though they are united or interwoven. That is to say, it considers a dynamic energetic unit system in which there is an active and simultaneous correlation of polarities-identities, without any of them losing its potential for contradiction, contrary to the synthesis of the dialectic Hegelian logic, in which the thesis and the antithesis lose even their very potential for contradiction. As Nicolescu states, the tri-dialectic logic of Lupasco, insofar as it is a dynamic unit system, overthrows Hegel's synthetic dialectic logic in favour of the simultaneous coexistence of the three different and contradictory polarities in each manifestation, and by the fact that the third term (T-state) is a dynamic mechanism independent of the other two terms. (Cf. Nicolescu, 2005: 163–169).

We are talking about a logic that ousts the notion of the object from its traditional central role, raising the notion of energy to an essential spot and which understands reality as a perpetual oscillation between actualization and potentiation, where the energetic dynamism comes to dominate the set of physical phenomena, announcing a new concept of space-time.

It is a logic conquered by relativistic and quantum physics almost a century ago, and which has had repercussions (more or less conscious) in the field of architecture, especially in certain architectural experiences carried out as of the last two decades of the 20th century. They are experiences that are implemented under interstitial spatialities that are increasingly more polytopic, that appear more dynamic and interrelated than autonomous, between continuities and discontinuities, repetitions and differences, angularities and rounded forms, as we can see in the works we have selected.

They are all architectural designs whose geometries and poetic operations constitute interstitial spatialities in which all of the qualities (formal, luminous and functional) are interwoven. We are referring to geometries and poetic operations that are formalized in dynamic formal systems of folds, fractals or rhizomes, and architectures that, rather than considering form in an autonomous manner, where the "form follows the function" or "form contains the function", the proposal is to think about the relationality of form, space and function from a non-rigid structuring geometric matrix, in search of space-formal alternatives that can be more versatile, multivalent and interactive; in other words, more easily adjustable to the increasingly more complex and challenging demands of everyday life today.

BIBLIOGRAPHICAL REFERENCES

AAVV. (2001) *Max Bill*, DPA – Documents de Projectes d'Arquitectura. Barcelona: UPC.

Capra, Fritjof. (2003) *La trama de la vida, una nueva perspectiva de los sistemas vivos*. Barcelona: Anagrama.

Deleuze, Gilles. (2002) Diferencia y *repetición*. Madrid: Júcar Universidad.

—. (1994) *Lógica del sentido*. Barcelona: Paidos Iberica.

Deleuze, Gilles, Guattari, Félix. (1997) *Mil Mesetas – Capitalismo y esquizofrenia*. Valencia: Pre-textos.

Eisenman, Peter. (1997) "Procesos de lo Intersticial", *El Croquis* 83.

Esteban Penelas, José Luís. (2007) *Superlugares, los espacios inter-media*. Madrid: Rueda.

Freitas, Lima. (1984) *Pintar o Sete – Ensaios sobre Almada Negreiros, o Pitagorismo e a Geometria Sagrada*. Lisboa: INCM.

Jencks, Charles. (2000) *The Architecture of the Jumping Universe*. London: Academy Editions, Great Britain.

—. (1997) "Nonlinear Architecture, New Science = New Architecture? *Architectural Design* 129.

—. (1997) Landform Architecture, Emergent in the Nineties. *Architectural Design* 129.

Nicolescu, Basarab. (2005) *Nós, a Partícula e o Universo*. Lisboa: Esquilo.

Ruiz de la Puerta, Félix. Sánchez Gonzaléz, Juana. (2007) *La espiral en la arquitectura, espacios pictóricos y arquitectónicos*. Madrid: Mairea Libros.

Vasconcelos, Ana. (2016) Lo intermedio como lugar. Lo intersticial, lo fronterizo y lo impreciso en la arquitectura contemporánea. PhD Escuela Técnica Superior de Arquitectura – Universidad Politécnica de Madrid.

The bitter gifts of progress: Russian revolution, "barrackization" of living space and the poverty of experience

Irina Seits
Department of Aesthetics, Södertörn University, Stockholm, Sweden

ABSTRACT: The year 2017 marks the centenary of Russian Revolution that greatly affected the development of twentieth century's aesthetic and political thought both in Russia and abroad. Bolshevik Revolution of October 1917 ultimately transformed young state's living space with the help of architectural Avant-garde that passionately supported ideas of liberation from Imperial ruins in the name of Progress. Revolution, Progress and Avant-garde were synonymous then, calling to move forward towards future. That move meant fast and reckless drive away from the past, divorce with heritage, experience and history.

The land fertilised by Russian Revolution attracted many thinkers to observe it in action, to witness immediate achievements of Progress. One of those greatest minds was Walter Benjamin, who stayed in Moscow in the fall and winter of 1926–27 and whose later concept of new barbarism and poverty of experience put critique on modernity and progress.

Guided by Benjamin's observations of post-revolutionary Russia I think on the paths of Avant-Garde during reformation of the urban living space inspired by the faith in ideological and technological Progress. Post-revolutionary Russia is a great example to reflect on how Progress, when seen as a drive to the future and escape from the past meets its own end and finds itself stuck on the ruins of modernity.

Keywords: Russian Revolution, Avant-garde, progress, living space, W. Benjamin.

1 WALTER BENJAMIN'S STAY IN MOSCOW

Benjamin's stay in Moscow was described in the "Moscow Diary" and the "Moscow" essay. The latter is a text that is devoted completely to Benjamin's impressions from Russian capital as an urban and spatial phenomenon. I take it to the closer reading in order to reflect on the urban space of Moscow as on the ground where new Soviet reality was being constructed, and where Benjamin intended to observe Bolshevik Revolution in action.

His other works that I refer to as well, such as "Experience and Poverty" (1933), and "The Work of Art in the Age of Mechanical Reproduction" (1935) allow for deeper analyses of how ideology of unconditional devotion to technological and political progress affected people's lives, squeezing them into the new forms of living that were defined by the break with past and evolution of the new poverty of experience – the theory that reveals the early critique of modernity and progress per se. The chosen texts along with analyses of transformations in the urban space of the early Soviet state help reveal the origins of functionalism, which is regarded as the most modern and progressive method in the architecture of that time.

Benjamin had not found Revolution in Moscow. He had observed the city that was going through the transformation on a high level of speed and intensity. He had not described the total reorganisation of space either in terms of the new functionalist architecture that was being constructed in his presence nor through the notice of demolition of the old buildings that possessed reference to the ruined Empire. Benjamin turned to other features and spatial dimensions that he brought to light and that were not directly related to neither architecture nor construction. Nevertheless, he fully revealed the reformation of Moscow space that was initiated by constructivists and supported by the new regime. Though, as it turned out, they possessed rather different goals, both the state policy and constructivist practice prepared the space of the new Russian capital for the further appropriation, using the terms by Lefebvre.

Another concept that is introduced by Benjamin and that is highly important when approaching the issues on modernist architecture and modernity is the notion of aura. Though Benjamin does not use the term "aura" in his essay on Moscow, he actually talks about the aura of the city when he describes his stay in Russia.

1.1 *The failed Revolution and invisible Avant-garde in Moscow*

Benjamin came to Moscow to search for the Revolution and did not found it. Neither had he found

Figure 1. The Izvestiya Newspaper Building, arch. G.B. Barkhin, M.G. Barkhin, 1925–27. Photo: ITAR-TASS.

Avant-garde in the Soviet capital. Yet, the year 1927 was the period when many bright pieces of architectural functionalism were being widely discussed and constructed in Moscow, e.g. the Radio Tower by arch. V. G Shukhov (1919–1922), the building of the Izvestiya Newspaper by architect Barkhins (1925–27) along with the number of dom-kommunas (communal houses), factory-kitchens, garages, Metro stations and administrative buildings by prominent architects of Avant-garde such as M. Ginzburg, K. Melnikov, A. Shchusev and others.

Still, those works had not turned Moscow into the city of modernist architecture, as Bruno Taut's and Martin Wagner's Siedlungen had turned Berlin into the functionalist capital of Europe. Moscow was not the city of Revolution and not the proper ground for the Avant-garde architecture. Its nature was not progressive in a revolutionary sense, and Revolution itself had not happened there. Moscow became the city of power, but not of the Revolution. It did not allow experimenters of Avant-garde to progress into its urban landscape. Moscow carries the monuments of modernist architecture, but they are single islands, the sites that are disconnected from the city's absolute urban space, which is shaped to the higher extent by the architecture of ancient churches and Socialist realism. That architecture embodies establishment of traditional classical forms of spatial organisation.

The turn to the Socialist realism with its reference to the Roman Imperial symbolism after the decade of the modernist experiments in art and architecture is often seen as a turn back, as regression to the sustainable ideology of traditional methods and forms that were used in the accomplishment of totalitarisation of the state.

One of the major features of Avant-garde architecture is its immediacy and connection to the historical contemporaneity and chronology. The high speed and intensity of living that was noticed by Benjamin in Moscow ("For each citizen in Moscow the days are full to the brim") (Benjamin, 1999:31), was characteristic to the Art of Russian Avant-garde as well. The construction of the buildings had to be done fast, followed by intensive discussion of the projects in the architecture journals (e.g. SA – "Sovremennaya Arkhitektura") and mass media editions. The materials used for construction were cheap and of low quality, and thus they were inevitably meant for the short-term existence, but it was important for constructivists to give a visual presentation of the latest achievements of technological progress through their works. They had not worked for eternity but cared for the immediate realisation of their projects.

It was an internal contradiction within the working method as well as in the whole constructivist aesthetics that the commitment to the immediacy of the architectural experience was to comply with the ideas of the development of the reproducible standard as the guiding idea for the constructivist architectural practice. The goal to produce the algorithm that would lead to the creation of the reproducible standard, which was understood as an ideal form for the architectural materialisation of the function, was put at the core of the method. Moreover, it was not the invention by constructivists, but rather a declaration of their devotion to the technological progress and technological means of art production that had penetrated the very creative process from the end of the 19th century. Benjamin, in his essay "The Work of Art in the Age of Mechanical Reproduction" (1935) had summarized the idea of the reproducibility that was intentionally inscribed to the very aesthetics of functionalism: "Around 1900, technological reproduction not only had reached a standard that permitted it to reproduce all known works of art, profoundly modifying their effect, but it also had captured a place of its own among the artistic processes" (Benjamin, 1999:253, author's italics).

The functionalist method consciously sacrificed the authenticity of its produced works thus eliminating the possibility for aura to emerge. As Benjamin writes in the same essay, "in even the most perfect reproduction, one thing is lacking: the here and now of the work of art – its unique existence in a particular place" (Benjamin, 1999:253).

On the other hand, the immediacy of constructivist architecture and subjectivity of its expression had led to the creation of masterpieces – the authentic works of constructivist architecture that cannot be reproduced, as Benjamin writes that "the whole sphere of authenticity eludes technological – and, of course, not only technological – reproducibility" (Ibid). Those masterpieces, that were built in Moscow by Melnikov, Golosov, brothers Vesnin, Ladovsky, Ginsburg, Nickolaev and others developed the aura that had transmitted through their uniqueness and through the temporal distance that separates them from the era when they were created.

The nature of Avant-garde of the 1920s was hallmarked with immediacy, expressiveness and devotion to the technological progress. The storm of its liberated energy that was blowing from the throat of Revolution should have been enormous. Yet Moscow resisted Avant-garde and, according to Benjamin's observations, dismissed it even earlier than the Stalin's decree of 1932 that officially declared the turn from functionalism to the appropriation of classical heritage of Roman Empire, which led to the development of Socialist Realism as an artistic style and working method.

1.2 *The new optics for Moscow*

What Benjamin wins from his Moscow experience is "a new optics" that is "the most undoubted gain from a stay in Russia" (Benjamin, 1927:22). He sees Berlin as a deeply European city through the careful observation of what he sees in Moscow: "However little you know Russia, what you learn is to observe and judge Europe with the conscious knowledge of what is going on in Russia" (Benjamin, 1927:22). The political history of the 20th century is highlighted with the permanent resistance to the conscious comprehension of its former experience. It is in that divorce with the past and in the unwillingness to face the history, which is as strong today as it was in Benjamin's time, and where the poverty of experience fully reveals itself.

The divorce with experience after the WWI was unintentional and unconscious for most people; it was a matter of survival. People became "not richer but poorer in communicable experience" (Benjamin, 1999:731). The strings of communication through which experience had been transmitted for ages were torn and Benjamin was wandering through the jungles of their snatches in the streets of the frazzled out Moscow.

The loss of the art of the story-telling provoked by the muteness of communicable experience, and the everyday life full to the brim were revealed in the forms, materials (glass, reinforced concrete, iron) and content that were most characteristic of the modernist architecture of that time.

The Russian Avant-garde architecture was not the child of Bolshevik Revolution and either was German functionalism the child of Weimar Republic. They were born before the WWI when the technological progress of the 19th century prepared the fertile ground for its growth. Functionalism would have changed the face of Europe regardless of the dramatic political events. Both Russian constructivism and German functionalism were different in nature and origin from Communism and Nazism and thus were abandoned in both countries as alien to the newly empowered ideologies.

Post-Revolutionary architecture in Russia is the most immediate architecture; it requires not the contemplation but the immediate experience of its space, forms and rhythms. The inhumanly intensified mobility of life demanded the immense mobility of architecture and the living space that it provided. Benjamin notices in his essay the "unconditional readiness for mobilisation" (Benjamin, 1999:29) of Moscow's population. Everything moves and changes, everything and everybody is involved in the certain transition of practices and meanings. The mobility and intensity of life have the nature of laboratory experimentation "to the point of exhaustion", and "no organism, no organisation, can escape this process" (Benjamin, 1999:28). He continues that "employees in their factories, offices in buildings, pieces of furniture in apartments are rearranged, transferred, and shoved about" (Benjamin, 1927:28–29).

Mobility and transition capture all spheres of life. The living itself stopped being a continuing constancy of everyday routine. There is no "everyday" anymore; every day is a different temporal unit of an overwhelming experiment of dehumanised and dehumanising nature of the new living. Citizens are turned into migrants, and their living space, from the city level to the apartment, is transformed into a camp. Living is turned into camping.

2 "BARRACKIZATION" OF THE LIVING SPACE

Camps are traditionally made up with barracks; their original function is to provide the army with shelter during military campaigns. Barrack proves to be the most progressive and at the same time sustainable type of dwelling of all times. The type of a barrack, where the rooms are arranged along the corridor, is still dominant in public spaces: from hospitals and hotels to prisons; from refugee camps to military dorms. Barrack is originally a temporal construction, which is assembled in the shortest time with the least expenses. The temporality of a barrack is its main feature, and it presumes its residents with the temporality of their presence in a barrack. In the hotels, hospitals, and in most cases in prisons residents intend, or at least, hope to stay for a period, not forever.

With the time barrack has become a constant construction. It is no more intended for the demolition or moves after its residents leave it, as it would happen before when the camps and field hospitals were taken down or transferred to another place together with the military or wounded soldiers. Now, in most cases, barracks are permanent constructions that accommodate different groups of people for a temporary stay. Those are hotels, hospitals, schools, office buildings, prisons, refugee camps, etc.

A barrack as a temporal construction has also become a strategic type of housing that is widely built by states in the times of crisis. Be it a war conflict, a natural catastrophe, or a revolution, the old neglected barracks that remain from previous disasters get renovated and the new ones are constructed to place as many people as possible and in the shortest time. The construction business of modern light module barracks and smaller individual units that can be easily assembled and transferred has been constantly growing since the rise of the international conflicts (e.g. Ikea's inventive types of flat box emergency housing that represent smaller individual units). The future of demand for temporal housing of all types is rather promising. Unlike traditional nomads that

Figure 2. Barrack for refugees in Stockholm after the fire. Photo by the author, 2016.

usually build even if small, but separate single-family units (tents, yurts, and wigwams), migrants, soldiers and refugees stay in barracks that assume a collective and communal way of living.

Barrack is a European invention of modernity and one of the symptoms of technological progress that stimulates migration of population. Present political and social climate attract closer attention to the study of Asian nomadic living space worldwide.

The Revolution of 1917 in Russia resulted not only in the huge emigration from the country but in the unseen migration within the new Soviet State as well. People were not only forced to leave their regions and move to other parts of the former Empire, but mass displacement happened within the towns, villages, and districts of the cities, and even further: within same houses and apartments through the process of Uplotneniye ("tightening").

The living space of all types that existed in Russia before Revolution was being massively turned into barracks. The old workers' barracks were re-inhabited to the most of their capacity; private apartments were turned into communal flats ("kommunalkas"), which possessed spatial organisation of barracks. The bigger apartments were re-planned inside in a way to create a "corridor system" with rooms that were multiplied by additional walls and that were placed along the narrow passages. The early-century condominiums along with the former public buildings such as theatres, hospitals, and different institutions from schools to administrative offices were turned into multi-storey barracks to accommodate the new population driven to the cities.

The total "barrackization" of space changed the lives of citizens. They were not running their households and they were not residing or living in their rooms anymore. In Moscow "indoors one only camps" (Benjamin, 1999:30). People were "camping", which meant that they were temporally settled in the occasional spaces, unrecognisable in their function and form from the glance of the previous age. The old spaces were being re-appropriated for the new space of a total barrack that the country was being transformed to. "Camping" and "remonte" – those were the terms that Benjamin used to describe the new living practices in post-revolutionary Russia and that defined them for the decades ahead. That transformation was strikingly obvious for Benjamin as for an outsider and a European man, who learnt to see bourgeois Berlin through seeing Soviet Moscow: "for this new life weighs on no one more heavily than on the outside observing from the distance" (Benjamin, 1999:30).

Barracks are the types of housing that provide with collective living. Bolshevism had not only abolished private ownership but as Benjamin remarked, it had "abolished private life" in the name of progressive living. Private spaces were exposed out to the streets, and public spaces were brought indoors: "through the hall one steps into a little town" (Ibid). The borders between inner and outer space were demolished: they were destroyed on ideological and physical levels. There were no more "inside" or "outside". Even those things that usually happened indoors in the most "indoor" furniture, such as beds, i.e. sickness or dying, were happening in the beds set outside in the "vast open-air hospital called Moscow" (1927:28).

The transformation of the space into the new milieu, into what was called the new "byt" (which loose translation from Russian would be a way of everyday being, a concept that was developed by such Avant-garde thinkers as Boris Arvatov), was greeted as the herald of the new life and as achievement of the progressive revolutionary thought.

The reformation of living was closely connected to the collective feeling of expectation that kept everybody in the state of not living his or her lives but camping on the front line of the upcoming future.

2.1 *Barracks as the ground for the new types of housing*

Constructivists aestheticised barracks, arguing for the return to an original hut, in which forms the function was purified from all unnecessary layers. The new type of housing that was re-invented by constructivists was dom-kommuna: a multi-storey barrack with small living cells that were placed along the corridor and that were meant only for sleeping. The collectiveness and communal nature of living that was inevitable in a barrack had been regulated and put in order through separation of each living practice into special zones: eating was to be done in the factory-kitchens, bathing in the collective banyas (bathhouses), children care was to be provided by the nurseries and kindergartens, etc.

Constructivists in the new living estates preserved both the collectiveness and spatial organisation of a barrack. Theoreticians avoided the very terms of "house" or "home", architects of architectural Avant-garde, as the new types of dwelling had little connection to what a traditional house, and household had looked like and from how they had functioned. The ideas of collectivism and demand for the communal nature of the new milieu were spread out in the air, and constructivists managed to catch the beat of the days, when, as Benjamin noticed, "there is no knowledge or faculty that is not somehow appropriated by collective life and made to serve it" (1999:29).

Figure 3. Dom-Kommuna of Narkomfin, arch. M. Ginzburg, 1928–30. Photo of the 1930-s by Robert Byron.

There is always a feeling of immediacy of presence during the stay in a barrack or in a camp. The permanency of living is substituted with the temporality of a stay. The country was transiting from the capitalistic forms of "byt" and ideology of being to the new "progressive" socialist state. The feeling of the proximity of the future as well as the very state of waiting and the transition was theorised by Avant-garde into the new aesthetics and materialised into the new living space.

The strict zoning of the living space did not allow tenants to choose the way they wanted to live. People, who resided within the new estates of "zhilmassivs" – the constructivist housing estates, were the workers of the large plants, but they could not make their living without the support from outside and without the infrastructure that provided the whole new system of living with sustainable functioning.

The existence of inhabitants of the new living space was supported by the services that were out of their control: factory kitchens were to be filled with food, palaces of culture with entertainment, schools with the staff and educational programs, hospitals with doctors, and bathhouses with hot water. Thus technological progress was to ensure and secure people's living practices as well as to establish control over their lives in the country that was undergoing a total "barrackization" of its space.

The living in a barrack cannot be self-maintainable; it requires a certain regulating organ be it a military rule for army camps, a state immigration policy and support for refugees, or any other sanctioning organ that controls people's stay in barracks. In other words, somebody has to allow for a barrack to be built and maintained. Barracks stand on the lands that are either appointed to a certain institution (plant, hospital), or they spot and trace the movement of the army or the route of the fleeing refugees. The public or state institutions usually run barracks. In either way, they keep records of the appropriation or the loss of land and experience.

When a private man comes to a new land that recognises the right of private ownership or which is not subjected to any state control, he builds a house; he does not build a barrack. He starts his household, and the combination of many households forms a village or a town. Through their very organisation, barracks reveal the cause of their construction – to provide a certain group of people with temporal shelter while they do a certain task, perform a certain function. The form of the barrack is the most adjustable and the most universal for any human activities from war making to imprisoning. In this sense, it is the most progressive type of housing.

Constructivists did not mean to reproduce barracks on industrial scale purposely, but it was inevitable. The state authorities demanded the new means of control over the population. At a certain moment, Avant-garde's searches for the new progressive forms of living and state housing policy merged into the production of communal barrack-like spaces. The state was a grateful spectator of constructivists' experiments with dwelling and continued "barrackization" of the country intentionally. Thus, the path taken by constructivists with the goal to create new liberating living space in the circumstances of the 1920s in Russia could lead only to a barrack, where people stayed waiting for the promised future. The expectation took lives of several generations but the promised future had never come.

3 THE DEATH OF AURA OF AVANT-GARDE IN THE MILLSTONE OF THE PROGRESS

The impoverishment of experience was achieved in the process of the constant movement, in the loss of the traditional living space, connections and practices; in the necessity to adopt and begin anew "and with few resources" (Benjamin, 1999:735) as well as in the will to adjust to reality where everything was different – from a sleeping place to a job.

There is no gain of experience in the constant moving, but rather its loss. A peasant, moved from the hut in his village to the third floor of the plant dormitory realised that his whole life experience of farming was useless in his new existence.

Benjamin provisionally saw economic reasons for the failure of communism already during the time of New Economic Policy (NEP), when those were still years before the system had acquired the forms of totalitarianism, and which was the only way to postpone its collapse. Benjamin had not expressed disappointment with what he had seen, but only its understanding, as he "grasped the concrete" of what was to happen soon with modernity and its faith into progress in both Russia and Europe.

Functionalism deals with the promises of the future. The regular living space in modern Europe and Russia is the blurred realisation of the promises of the 1920s.

The contemporary mass housing is what remains from those promises. The islands of the realised functionalism in Russia and Europe are the cradles for the contemporary living space. With time, the traces of the human presence penetrated the Avant-garde living space that was originally produced for the immediate experience and living. It became a monument to the faith in Progress as the reckless drive forward that was sweeping anything that stood on its way. That progression towards the future, accelerated by the theoreticians and masters of Avant-garde, reached its own end and turned inside out resulting in the perverted forms of Bolshevik Revolution.

The impoverishment of experience and the divorce with the past that was propagated by the drive of the Progress blows aura away. The sober glass and still surfaces of the modernist architecture do not keep traces of history, – another enemy of Progress that pulls it back and through which aura reveals its enigmatic and subjective nature. The abolishment of possession and private ownership prevents the establishment of connection with one's past through material things, as well as it does not allow a thing that has no possessor to acquire aura.

The generation of the 1920-s and 1930-s was an experimental material on the laboratory table of the new state construction. It had divorced itself not only with its past but also with its future. The unprecedented mobility was the only way to survive. The one, who stayed still, was swept off right away by the wind of the history that was blowing into her face.

BIBLIOGRAPHICAL REFERENCES

Benjamin, Walter. (1999) *Selected Writings*. Vol. 2. 1927–1940. Harvard University Press.

Benjamin, Walter. (2003) *The Arcades Project*. Harvard University Press.

Lefebvre, Henri. (2012) *The Production of Space*. Blackwell Publishing.

On progress: Remarks on the theoretical work of Adolf Loos

Jorge Nunes
University of Lisbon, Faculty of Architecture, CIAUD, Lisbon, Portugal

ABSTRACT: The idea of progress plays a key role in the thinking of the Austrian architect Adolf Loos. It is based on progress, translated into the idea of cultural evolution, that Loos justifies his refusal/condemnation of ornament. An idea of evolution influenced by the cultural context of Vienna 1900 and which in the results in a complex thinking that reflects the contradictions of the end of the century, simultaneously pessimistic and optimistic about progress, and which over the years has been the subject of contradictory interpretations. Nowadays, in a period once again marked by uncertainties and doubts about the future, we propose revisiting the theoretical thinking of this author.

Keywords: Adolf Loos, Ornament and Crime, Vienna 1900, Cultural progress.

The architect Adolf Loos began his career as a critic of the historical eclecticism in both architecture and interior design. In 1897, he began to develop, in conferences and in a series of articles published in the *Neue Freie Presse*, Vienna's leading newspaper, a set of ideas that would constitute the theoretical program of his own architecture. Loos's architectural theories were developed within the context of Vienna's aesthetic and architectural debates that occurred at the end of the 19th century.

In cultural and political terms, the Vienna fin-de-siècle or Vienna 1900, as it was also known, is the time that Hermann Broch characterised as the "joyful vegetation" of a "museum" city. A "Kitsch metropolis" dominated by the simulacrum culture in order to deceive the latent conflicts of the Austrian society. A city where the deeply ornate façades of the Ringstraße palaces were often just scenery that hid the true social condition of many of its inhabitants dominated by the cult of historical styles in the eagerness to emulate the ancient Catholic aristocracy. As António Sousa Ribeiro points out, behind this aestheticism hid the inevitable and ever closer reality of the last days of Habsburg power (Ribeiro, 1988: 143).

It is against this reality that a group of writers and artists reacted by seeking, through very diverse means, a more authentic artistic and literary expression that would allow escaping of the *status quo* of the bourgeois society.

This group nurtured an ideal of honesty and simplicity that was reflected in the architecture of Adolf Loos, the painting of Oskar Kokoschka and Egon Schiele, the poetry of Peter Altenberg, the philosophy of Ludwig Wittgenstein and the satirical texts of Karl Kraus. This author who exerted a decisive influence on all those who acted in this complex cultural context, becoming the leader of a moral and aesthetic crusade against the decorative "excesses" of the artists of the first generation of the Viennese modern movement that he associated with the "moral and artistic decadence", which had plagued Viennese society since the late nineteenth century.

Revealing at times the influence of Kraus's satire, Loos's writing aimed at reforming the habits and taste of Austrian society, which, in his view, were lagging behind the lifestyle of the Anglo-Saxon countries. In the Anglo-Saxon world ornament was a genuine quality but in Austria, said Loos, it is no longer related organically to their culture, it is no longer an expression of their culture (1993a: 345–346).

In articles such as "Die Potemkinsche Stadt" (The Potemkin Town, 1898), in which he criticised the palatial architecture of the Ringstraße, he began to question everyday objects according to their usefulness and social suitability. Gradually he ended up arguing that utilitarian objects should be freed from all artistic obligations and representative functions, because, Loos understood, the artistic pretension of these objects was "historically" obsolete.

It is this defence of the withdrawal of the utilitarian from the sphere of art that will form the basis of his campaign directed against useless ornament. In the article "Kulturentartung" [Cultural Degeneration] (1993b: 342–345) and the celebrated rehearsals "Ornament und Verbrechen" [Ornament and Crime] (1993a: 346–355) and Architektur [Architecture] (1993d: 23–35), Loos explains the rejection of the ornament from an anthropological context, arguing that the removal of the utilitarian from the sphere of art is a logical consequence of evolution and social progress. In his view, ornament had simply ceased to function in the cultural evolution of humankind.

In early times, in primitive cultures, people decorated not only the objects they wore but also filled their bodies with tattoos. "Nowadays" only criminals and degenerates resort to such ornamentation, so, concludes Loos, the ornament is somewhat regressive. It is a crime.

In the essay Architecture, Loos justified why the notion of applied art implies a contradiction in its own terms; the decoration had become an end in itself; More than an embellishment was a distorted view of the function of art:

> The path of culture is a path that goes from ornament to its renunciation. The evolution of culture is equivalent to the distance of the object of use from the ornament. The Papua covers everything within its reach with ornaments, from his face and his body to his bow and his rowboat. However, today the tattoo is a sign of degeneration that is only in use among delinquents and degenerate aristocrats. The educated person, unlike Papua, thinks a tattooed face more beautiful than a tattooed one, even if Michelangelo or Kolo Moser had made the tattoo. The nineteenth-century man wants to keep out of the reach of the new Papuans, not only his face but also his suitcase, his clothing, his habitation, his house! The gothic? We outgrow the Goth people! Renaissance? We overcame it. We have become more cunning and nobler. Today we lack the strong nerves that were necessary to be able to drink of a great marble cup where a battle of the Amazons was engraved. Have we lost old and old techniques? Thank God! We exchanged them for the spherical sounds of Beethoven. Our temples are no longer painted, like the Parthenon, in blue, red, green, and white. No! We learn to feel the beauty of a bare rock. (1993d: 23)[1]

In asserting that the evolution of culture is tantamount to moving away from the ornament of utilitarian objects, Loos is defending a progressive view of artistic facts that lead us to think that the abandonment of applied art is but one of the stages of the evolution of Art as it was theorised by Hegel. Like the German philosopher, Loos supposes the existence of a genuine historical continuity and even a form of progress.

Hegel thought that the energies of history had coincided at one time with the energies of art, but "now" that history and art had followed different directions, the existence of art had no historical significance. Art as such, or at least its highest manifestation, had come to an end as a historical step because it had reached, as Arthur Danto states (1995: 32–33), the final stage of absolute cognitive representation, that is, the end of art as a progressive discipline committed to a mimetic representation of reality (in particular painting) moving towards the gradual conquest of appearances.

Loos' perspective agrees, to some extent, with Hegel's view in that it admits an inner connection between nature and history and understands art as a necessary step to the advent of a certain type of knowledge. However, the distinction between art and crafts means that the end is only applied in the case of applied art; Art, on the contrary, maintains and even reinforces the role it has in the course of history, because, says Loos (1993d), art is great when it is timeless and revolutionary.

The artisan produces objects to be used at a specific time, the artist for men of any era. That is why Loos opposes the form of utilitarian objects – a collective product that results from their insertion in social life – as art, which on the contrary seeks to build the mind of man by diverting his attention from the monotony and harshness of everyday life to the realm of fantasy and spiritual values. In this sense, art is always revolutionary; aspires to transform man's view of the world and his attitudes towards his fellow man (Janik et al., 1973: 100).

The great spiritual value that Loos attributed to art placed an insurmountable division in what had until then constituted its field of application. The artefacts belonged only to one side of the barrier; One thing was art or it was not at all; Or was entirely spiritual or wholly material; Could not be both simultaneously because anyone who realizes that the function of art is to guide man upward, making him more and more like God, feels that the fusion of art with functional needs is a desecration of the highest degree (Loos, 1993d: 23–35).

Outside the sacred enclosure of art lie the utilitarian artefacts. Of these, only a few will be consecrated architectural works: the tomb, the monument and the urn. Monuments, like other works of art, are not completely without purpose, but, instead have a purpose, albeit one of a higher type, a type that cannot be fully articulated in advance and that was somehow connected with what Loos took to be the purpose of culture, as Roger Paden notes (2007: 51–52). For Loos, argues this author:

> Culture consisted in a set of activities that produce that balance between our physical, mental, and spiritual being which alone can guarantee sensible thought and action and suggested that the ultimate purpose of cultural objects is to help promote a type of critical thought and action that can lead to intellectual and artistic progress and, more importantly, to individual growth and development. Thus, works of art, on Loos's view have a transformative, even transcendental, purpose. This, according to Loos, is why they make people feel "uncomfortable": because such works are "revolutionary" and are "concerned with (a qualitatively different and still undefined) future" they threaten to disrupt our lives and our narrowly focused attempts to satisfy our existing needs through the creation of new ways of life. (Paden, 2007: 51–52)

Loos held that the house has to please everyone, unlike the work of art, which does not have to please anyone. The work of art comes into the world without the need

[1] Free translation provided by the author.

for it. The house fulfils a need. The work of art must not be accountable to anyone; the house to all. The work of art wants to rip people out of their comfort; the house has to serve the convenience. The work of art is revolutionary; the house is conservative. The work of art teaches humanity new paths and thinks about the future. The house thinks of the present. Man loves everything that satisfies his comfort; Hates everything that tries to pull him out of his comfortable position. Therefore, he loves the house and hates art.

So, the house has nothing to do with art because architecture should not be placed on the arts? Therefore, it is. There is only a small part of architecture that belongs to art: the funerary monument (the tomb) and the memorial monument. Everything else, which serves an end, must be excluded from the midst of art. It is only when the great misunderstanding that art is something that can be adopted for an end is overcome, only when the deceptive motto "applied art" has disappeared from the vocabulary of peoples, only then will we have the architecture of our time. The artist should only serve himself, the architect to the community. However, the mixture between art and manual labour brought endless damage to both and humankind. Humankind commits every hour an enormous sin, that cannot be forgiven; the sin against the Holy Spirit. Crime and robbery, everything can be forgiven. However, the various symphonies that humanity has made impossible by the persecution of artists will not be forgiven. The unlocking of God's plans will not be forgiven. Humanity no longer knows what art is. 'Art at the service of the merchant' was the title of an exhibition recently in Munich, and no hand was found to punish such a slight subject (Loos, 1993d: 32).

The exclusion of architecture from the realm of art became inevitable for those who, like Loos, advocated such a radical separation between the spirituality of art and the materiality of utilitarian objects produced by industry. Architecture should be confined to its role of 'good and honest construction'. In his opinion little by little, humanity is drawing a line of demarcation between spirit and matter. The struggle between the two tendencies ended in the nineteenth century. Until then, the artist and the artisan were one. The work of art was to be used. For today's man, this is barbarism. Over and over, all industries have moved away from the realm of art. The architecture was an art, today is no more than the tattoo, once the architectural work is made to be used and to deteriorate (Loos, 1993c: 161).

According to this distinction between artisans and artists, while artisans seek to produce satisfying, practical, and dependently beautiful objects, artists seek to create freely beautiful, transformational works of art. As stated by Paden:

> To do this, Loos believed, artists need to understand the living artistic tradition within which they are working and, he claimed (without providing much in the way of evidence or argument) that in the Western architectural tradition, ornamentation has come to play an increasingly insignificant role, while, correspondingly, purely spatial factors – such as geometric proportion, axial symmetry, and purity of form—have become increasingly important. (Paden, 2007: 52–53)

Implicit in this claim, says Paden (2007: 52–53), is the idea that cultures are necessarily dynamic: a culture does not consist of a collection of objects—a dead past to be endlessly repeated—but, instead, it is driven by an internal dynamic to more and more sophisticated and effective expressions. Good artists intuitively grasp this dynamic and by doing so they are able to create new, inspired, works which, in turn, transform the culture and, through it, both society and its members. True artists, unlike artisans, therefore, are not interested in creating need-satisfying functional objects, but are, instead, interested in advancing culture and, thereby, transforming life.

When working as "artists", architects should avoid ornamenting their buildings and focus, instead, on purely spatial matters. They should try to develop and purify classical forms, express simple geometric shapes, and display their harmonic proportions in a manner consistent with the living tradition of Western architecture. Only in this way, can they advance the culture and, thereby, help people, through the experiences provided by their buildings, to attain that balance that makes a truly human life possible.

If on the contrary, they apply arbitrary ornament to either practical or monumental buildings, they fail to understand the purpose of architecture and its role in Western culture. That is why, as Loos put it, in these cases, there is no longer any organic connection between architecture and our culture, architecture is no longer an expression of our culture. The architecture being created now bears no relationship to us, nor to any human being.

As a result of this process, says Paden (2007: 54), Loos thought that we are now living at a time "non-culture." Today, with the development of the new "non-culture" people have lost the "spiritual balance" that a well-functioning culture can provide and have become, therefore, incapable of "sensible thought and action"

For Loos, the death of culture had resulted in a materialism and an enslavement to existing desires that, in the end, produced a type of spiritual alienation characteristic of the modern age. The contemporary resort to architectural ornament, therefore, has helped kill our living culture, thereby putting an end to our spiritual and artistic development, fixing fast our current needs, and imprisoning us in our existing situation: this is ornament's true crime; make a truly human life impossible (Paden, 2007: 54).

According to Paden (2007: 56), Loos's theories were also a response to a more narrow debate that, in part, arose from the Ringstraße project, and that opposed Camille Sitte and Otto Wagner in a more widespread controversy within the disciplines of architecture and urban planning. Loos was able to adopt positions that could draw support from each while remaining at a critical distance from both.

Loos not only rejected Wagner's progressive search for a modern style, with the purpose of helping people adapt to social and technological change, but he also rejected Wagner's conservative view based on the Germanic traditionalist idea that the function of culture is simply to strengthen a people's national identity and their ability to work together within their community. For Loos, art transcends such narrow social needs. Against both planners, states Paden:

> Loos held that the (universal) purpose of culture is to help create a balance between our mental, physical, and spiritual being; it is to help advance the process of personal and cultural transformation. Thus, in Loos's view, for different reasons and in different ways, both Wagner and Sitte failed to understand the purpose of culture and were blind to the logic of its development. They both failed to appreciate the independence of culture and, as a result, failed to understand its spiritual nature. (2007: 56)

Nevertheless, Paden says that Loos could deploy a theory that, to a degree, could accommodate both Wagner's utilitarian and cosmopolitan concerns while still, to a degree, accommodating Sitte's concern for a culture that was free from technological determination:

> "In Loos's theory, Wagner's utilitarian concerns would be satisfied by the practical buildings of a city, while Sitte's cultural concerns (suitably modified) would be addressed by its monuments. This was possible, however, only because Loos simultaneously rejected both Wagner's and Sitte's theories of culture while substituted for them a cosmopolitan theory of the spiritual function of culture. (Paden, 2007: 61)

Like Paden, almost all commentators on Loos's work sought to clarify his complex and apparently ambiguous position vis-a-vis the dichotomy between tradition and modernity. Kenneth Frampton (1996: 15), for instance, claims that Loos was caught between the culture of the *Architekt* and the culture of the *Baumeister*, thus highlighting the contradictory situation in which the architect's condition lay at the end of the nineteenth century. However, although Loos's position can be interpreted from the transformations undergone by architecture as a discipline, there are inherent factors to the Viennese culture that underlie its apparent ambiguity.

At the turn of the century, the last days of the Habsburg power were lived. The Austro-Hungarian Empire was in internal disintegration due to the tremendous economic and social transformations that occurred during the last decades of the 19th century. The ethnic and religious divisions that plagued the vast imperial territory (which contained eleven nationalities and fifteen different languages) were exacerbated by Austrian nationalism and by the failure of the bourgeois revolution which, owing to the counterrevolution of the neo-absolutist monarchy, in 1848, lost its Hope of leading the various social and ethnic groups to the modern age under the leadership of a liberal elite.

These tensions generated in Austrian society an identity crisis that was interpreted as a breakdown of social order and traditional values. Hermann Broch dubbed this period of "happy apocalypse" and Karl Kraus in July 1914 described it as an "experimental laboratory for the end of the world".

The pessimism with which these events were faced was mingled with the suspicion that the "timeless" world of the Habsburgs would not survive these centrifugal forces. The permanence and stability evoked by the "golden age" of the empire were, after all, an illusion generating a pessimism that opposed the technical and scientific rationality with which modernity viewed the idea of progress.

Authors such as Allan Janik and Stephen Toulmin, Manfredo Tafuri and Massimo Cacciari are able to go further when they relate Loos' work to the fatalistic scepticism of Vienna at the turn of the century, characterized by a curious alliance between rupture and continuity, expressed in what António Sousa Ribeiro calls the anti vanguards of Viennese modernity: the place where the pathos of the new always appears in continuity with the assumption of the weight of tradition (Ribeiro, 1988: 141).

Frampton nevertheless reiterates this reading when he states that:

> We can quickly feel that after Theodor Adorno, in Schönberg as in Loos, cultural tradition and radical innovation are inextricably intertwined, so that none of them can be classified in the futurist sense, as vanguard. (Frampton, 1996)

Loos's relationship with the Modern Movement must be analysed in terms of its position in relation to the historical development of modernity itself. Loos wanted the idealistic and optimistic view of history (social utopias and the myth of technological and scientific progress, i.e. the completely ideological base of the modern movement) to be "reconciled" with the perspective of the historical continuity of cultural and civilizational processes. As Kurt Lustemberger asserts, his objections to faith in progress and his scepticism about technical developments were not shaped by the conservative critique of civilisation (popular at that time), but were essentially rooted in a humanist conviction that welcomed progress when it served people and vehemently rejected it when it explored them:

> For Loos it was incomprehensible that technological objectivity should become the basis of meaning and expression in architecture, and industrial building materials and their manufacturing processes should become the foundation of modern iconography. (Lustemberger, 1994: 25)

During the postmodernist debate of the 1970s and 1980s, Vienna fin-de-siècle was an alternative to the modernist belief in the utopian progress of a politicised vanguard. If this concept can be easily applied to cities like Paris, Moscow, or Berlin in Vienna, on the contrary, the crisis brought about by the failure of liberalism and the rise of nationalism produced an idea of exclusion that, paradoxically, allowed the flourishing of a cultural scene one of a kind. The subjects that occupied most of the artists who moved away from modernist dogmas in the last years – the limits of language, memory, body, self – had their origins in Vienna in the early years of the twentieth century.

The emergence of the term "Vienna 1900" is precisely a sign of what, as António Sousa Ribeiro points out, is no more than a desire to find in that Vienna an early projection of the post-historical and post-philosophical consciousness on our post-Modern Era (1988: 143). According to Ribeiro (1988: 140–141), there is no shortage of reasons, of course, for Vienna's notoriety at the turn of the century, with its amazing and multifaceted contribution to contemporary culture. The attraction it exerts, and the increasing market value it has achieved does seem, however, to me to be an often-blurred vision, a blur that can tell us something about the post-modernism. Trying to focus on this image, on the other hand, is also a good opportunity for the necessary rethinking of modernism.

Kurt Lustemberger reiterates this idea. In analysing the history of the critical appraisal of Loos's work, this author states that the interest that his work began to arouse from the fifties would play a key role in defining the postmodernist position during the 1960s and 1970s.

Realising that Loos's work was strongly related to the end of the hegemony of Modernism, which in Vienna was not restricted to architecture, allowed the modern movement to remain under postmodernist scrutiny. Loos and Kraus, like almost all the protagonists of the second generation of Viennese modernity, thus appear as the authors of an *avant la lettre* postmodernism that anticipates the ideological matrix underlying the critical scrutiny to which modern vanguards were subjected during the 70s and 80s of the twentieth century.

At the beginning of this century, Richard Calvocoressi and Keith Hartley, in a text included in the catalogue of the exhibition *Century City – Art and Culture in the Modern Metropolis* organized by the Tate Gallery, in London, point out that the "pessimism" underlying the thinking of Viennese artists is a portrait of the twentieth century as a whole – not only Vienna but the rest of the world (Calvocoressi et al., 2001: 231).

Today, 16 years later, we can say that this Viennese pessimism can also be used to portray the first two decades of the 21st century.

BIBLIOGRAPHICAL REFERENCES

Adorno, Theodor W. (2003) Progress *Can One Live After Auschwitz?* New York: Stanford University Press.

Aglietta, Michel. (1979) *A theory of capitalist regulation.* London: New Left Books.

Alexander, Ernest. (2016) There is no planning — only planning practices: Notes for spatial planning theories. *Planning Theory* 15 (1): 91–103.

Alford, Robert & Friedland, Roger. (1985) *Powers of Theory – capitalism, the state and democracy.* Cambridge: Cambridge University Press.

Allmendinger, Philip. (2002) *Planning Theory.* London: Palgrave.

Atkinson, Adrian. (2014) Urbanisation: A brief episode in history. *City* 18 (6): 609–632.

Barr, Alfred H. (1934) Foreword. *Museum of Modern Art. Machine Art.* New York: Museum of Modern Art. https://www.moma.org/documents/moma_catalogue_1784_300061872.pdf. (Accessed 8 June 2017).

Benevolo, Leonardo. (2002) *Historia de la arquitectura moderna.* Gustavo Gili.

Beunen, R, Van Assche, K & Duineveld, M. (2013) The importance of reflexivity in planning and design education. *Wageningen University Working Papers in Evolutionary Governance Theory* 14: 23.

Boyer, Robert H. (1988) Technical Change and the Theory of Regulation. In: Dosi, G (ed) *Technical change and Economic Theory.* London: Macmillan.

Bronowski, J & Mazlisch, Bruce. (1960) A tradição intelectual do Ocidente. Lisboa: Edições 70.

Bury, John Bagnell. (1920) *The Idea of Progress. An inquiry into its origin and growth.* London: Macmillan & Co.

Cabral, João. (1998) Spaces of regulation, the changing role of the state and the case for regional planning in Portugal. *XII Congress of the Association of European Schools of Planning.* University of Aveiro.

Calvocoressi, Richard & Hartley, Keith. (2001) Vienna 1908–1918. In: Blazwick, Iwona (ed) *Century City: Art and Culture in the Modern Metropolis.* London: Tate Gallery Publishing Ltd.

Campbell, Scott & Fainstein, Susan (eds). (1996) *Readings in Planning Theory.* Malden: Blackwell Publishers.

Castells, Manuel. (1977) *The Urban Question: A Marxist Approach.* London: Edward Arnold.

Colquhoun, Alan. (2009) Symbolic and Literal Aspects of Technology. *Collected Essays in Architectural Criticism.* London: Black Dog, 23–26.

Danto, Arthur. (1995) El Final del Arte. *El paseante* (23–25): 32–33.

Davoudi, Simi. (2012) The Legacy of Positivism and the Emergence of Interpretive Tradition in Spatial Planning. *Regional Studies* 46 (4): 429–441.

Després, Carole, Vachon, Geneviève & Fortin, Andrée. (2011) Implementing Transdisciplinarity: Architecture and Urban Planning at Work. In: Doucet, I and Janssens, N (eds). *Transdisciplinary Knowledge Production in Architecture and Urbanism, Urban and Landscape Perspectives.* 11. Dordrecht: Springer, 33–49.

Fainstein, Susan & DeFilippis, James. (2016) *Readings in Planning Theory.* 4th ed. New Jersey: Wiley & Blackwell Publishers

Frampton, Kenneth. (1996) Adolf Loos: The Architect as Master Builder. In: Schezen, Roberto (ed) *Adolf Loos: Architecture 1903–1932.* New York: The Monacelli Press, 14–21.

Gramsci, Antonio. (1971) *Selections from prison notebooks.* Hoare, Q and Smith, G. N. London: Lawrence and Wishart.

Griffiths, Paul. (1996) *Modern Music: A concise History.* London: Thames and Hudson.

Hall, A. Rupert. (2014) *The Revolution in Science*. Abingdon: Routledge.
Hall, Peter. (1992) *Urban and regional planning*. 3rd ed. London: Routledge.
Healey, Patsy. (2003) Collaborative Planning in Perspective. *Planning Theory* 2 (2): 101–123.
—. (1997) *Collaborative Planning, shaping places in fragmented societies*. Vancouver: UBC Press.
Janik, Allan & Toulmin, Stephen. (1973) *Wittgenstein's Vienna*. Chicago: Ivan R. Dee, Publisher.
Johnson, Philip. (1994) Preface to Sixtieth-Anniversary Edition. *Museum of Modern Art. Machine Art*. New York: Museum of Modern Art.
Kultermann, Udo. (1993) *Architecture in the 20th Century*. New York: Van Nostrand Reinhold.
Loos, Adolf. (1993a) Ornamento y delito. In: Opel, Adolf and Quetglas, Josep (eds). *Adolf Loos: Escritos I – 1897/1909*. Madrid: El Croquis Editorial, 346–355.
—. (1993b) Degeneración cultural. In: Opel, Adolf and Quetglas, Josep (eds). *Adolf Loos: Escritos I – 1897/1909*. Madrid: El Croquis Editorial, 332–334.
—. (1993c) Arte y arquitectura. In: Opel, Adolf and Quetglas, Josep (eds). *Adolf Loos: Escritos II – 1910/1932*. Madrid: El Croquis Editorial, 159–161.
—. (1993d) Arquitectura. In: Opel, Adolf and Quetglas, Josep (eds). *Adolf Loos: Escritos II – 1910/1932*. Madrid: El Croquis Editorial, 23–35.
Lustemberger, Kurt. (1994) *Adolf Loos*. Zurique: Artemis Verlag.
Masiero, Roberto. (2003) *La estética de la arquitectura*. Madrid: Machado Libros.
Merrifield, Andy. (2014) *The New Urban Question*. London: Pluto Press.
Mouffe, Chantal. (2013) *Agonistics – thinking the world politically*. London: Verso.
Nesbit, Robert A. (2009) *History of the Idea of Progress*. London: Transaction.
Paden, Roger. (2007) *Mysticism and Architecture*. Lanham: Lexington Books.
Pérez-Goméz, Albert. (1996) *Architecture and the Crisis of Modern Science*. Cambridge: MIT Press.
Ribeiro, António Sousa. (1988) Para uma Arqueologia do Pós-Modernismo: A "Viena 1900". *Revista de Comunicação e Linguagens* (6–7): 143.
Sager, Tore. (2016) Activist planning: a response to the woes of neo-liberalism? *European Planning Studies* 24 (7): 1262–1280.
—. (2011) Neo-liberal urban planning policies: A literature survey 1990–2010. *Progress in Planning* 76 (4): 147–199.
Sant'Elia, Antonio & Marinetti, Filippo Tommaso. (1971) Futurist architecture. In: Conrads. In: Conrads, Ulrich (ed) *Programs and Manifestos on 20th Century Architecture*. Cambridge: MIT Press, 34–38.
Schön, Donald A. (1983) *The reflective practitioner: how professionals think in action*. New York: Basic Books.
Struik, Dirk J. (1992) *História concisa das matemáticas*. Lisboa: Gradiva.
Tait, Malcolm. (2016) Planning and the public interest: Still a relevant concept for planners. *Planning Theory* 15 (4): 335–343.
Tatarkiewicz, Wladyslaw. (2013) *A History of Six Ideas: An Essay in Aesthetics*. Berlin: Springer.
Tiekso, Tanja. (2016) Luigi Russolo's Art of Noises and the Aesthetics of Musical Experimentalism. In: Berghaus, Günter (ed) *International Yearbook of Futurism*. Berlin: De Gruyter, 71–93.
Vairley, Francis. (2013) *Reflections on the Fall of Modern Civilisation*. London: New Generation Publishing.
Weber, Max. (1976) *The Protestant Ethic and the Spirit of Capitalism*. London: George Allen and Unwin.
Wittkower, Rudolf. (1998) *Architectural principles in the age of humanism*. 5th ed. London: Academy Editions.

The "anonymous" practice of construction and architectural theory: The tender age of the Form in the American industrial prototypes between 19th and 20th-century

Vito Quadrato
DICAR, Polytechnic of Bari, Bari, Italy
(Translation from the original Italian: Serena LEONE)

ABSTRACT: The intersections between the anonymous practice of construction and the development of Modernist architectural theories have provided some of the most interesting pieces of architectural history, determining a progress of built environment. In this regard, the relationship between the development of American industrial prototypes at the end of the 19th-century and the Modernist architectural theory in the first decades of the 20th-century seems to be exemplary.

Even though the early reinforced concrete construction systems were made in Europe, in the second half of the 19th-century, the American experimental practice led to the fulfilment of some industrial prototypes that fostered the aesthetic reflection and debate in the figurative research thanks to some leading figures of the Modern Movement.

In fact, the early concrete grain elevators and great flat-roofed frame structures arose in North America. Walter Gropius embraced the idea that "these forms are useful and powerful not only for the single but also for a big community because of their technical reproducibility" (Gropius, 2007: 211–214).

The proposed paper would analyse the relationship between this practice of construction and the European theory on the phenomenon at the beginning of the 20th-century. These analyses lead to questioning some of the contemporary paradigms about the roles of theory and practice in order to obtain progress.

The story of these structures shows how the anonymity of the practical experimentations on utilitarian buildings, produced in response to usefulness and economic standards, has different effects on an architectural theory.

Keywords: Modernism, anonymous process, industrial structures, form

1 THE RELATIONSHIP BETWEEN THE ANONYMOUS PRACTICE OF CONSTRUCTION AND ARCHITECTURAL THEORY IN THE 20TH-CENTURY

The history of architecture seems to promote a privileged status of buildings that arose from a representative need. There is no point in denying that architecture is one of the disciplines that best represents a social value system, with an emphasis on symbolical and communicative aspects. As Dejan Sudic (2005: 2) observed, "Architecture is the means to tell a story about those who build it".

Yet, there are some examples of architectures escaping that self-representative urge, which are rather committed to more practical and instinctive needs. This kind of architecture depends on construction practices which, although is not lead to a formal research, instils an archaic expressive power.

This already shows up on *Architettura rurale italiana* written by Giuseppe Pagano and Guarniero Daniel (1936: 11), which gathers examples of nameless houses, distinguished by a "wonderful primitivism". Pagano underlined the fact that buildings represent a "big source of human construction vitality"; they come out from a clear logic of construction and they are the result of an immediate executive purpose.

The comprehension of the anonymous architecture issue was cleared up by the important contribution of Bernard Rudofsky, who presented the exhibition called *Architecture without architects. A Short Introduction to Non-Pedigreed Architecture*, at MoMA of New York in 1964. The exhibition showed photographs sample cases of historical utilitarian structures and houses, coming from all over the world and forgotten by the history of architecture.

As Rudofsky said, they express the natural inclination of the human spirit to create beauty through the simple act of constructing. A comparison between the contributions of these two authors seems to be interesting because of the difference in the theoretical approach toward the same anonymous practice of construction. Pagano tried to recognize some of the

Modern features (like the flat roof or the purity of forms) in the Italian rural architecture as proof of the presence of a "perpetual Modernism" (Bragheri, 2012). On the other hand, Rudofsky proposed anonymous architecture as an idyllic alternative to contemporary architectural practice.

More recent works are expanding our understanding on the issue. With respect to new studies, "Evoked. Architectural diptychs" (2016) is a solid example of visual research about the informal Albanian buildings, the aim of which is to find some figurative inspirations in the spontaneous frameworks along the Albanian coast. These skeletons derive from the structural and expressive frame of the Maison Dom-ino by Le Corbusier and turn into the expression of unlimited variation on the theme of contemporary living in Albania, through the architectural drawing. By inducing a progress in the project's figurative instrument, therefore, the anonymous practice of construction can feed dynamically the architectural design theory in different ways, even without having any aesthetic purpose and these contributions make proof of that. The fitful search for "the tender age of the form" through the investigation of the "spontaneous" buildings seems to be fundamental in modern and contemporary architects' creative process. That holds true even if the outcomes often alter the very sense that brought to life those buildings.

2 THE "TENDER AGE" OF THE FORM: "THE MIRACULOUS AND THE UTILITARIAN" ON AMERICAN INDUSTRIAL PROTOTYPES BETWEEN THE END OF 19TH AND THE BEGINNING OF 20TH-CENTURY

In the essay *Mechanisation takes command*, Sigfried Giedion (1970: 30–40) illustrated the origins of the process of mechanisation with a focus on the consequences over human beings. He reports how, even since the 19th-century, the tendency to "the miraculous and utilitarian coexisted" in Europe, shaping the development of industrial automatons.

The author recalled this aspect in order to explain the American mechanisation process in the 19th-century. During this period, "a collective fervour for invention" seemed to be replaced by the idea of an "inventive urge possessed exclusively by a limited group of scholars". Thus, the masses were involved and contributed to the progress of the building practice.

According to Giedion, the great number of filed patents in the United States, during the '60 and '70, confirms the existence of a new "anonymous inventor". In the world of industrial production, those entrepreneurs who aimed to optimise time of realisation and storing of products embodied this kind of inventor. A strong development of the building sector, which we could define as spontaneous, caused a remarkable technical progress of some prototypes created in the American industry.

In order to understand the real effect of the phenomenon on the progress of the building sector we cannot dismiss to consider both the structure of American industry and the state of knowledge on a reinforced concrete system, which was as Siegfried Gideon pointed out, two different phenomena taking place in the USA. On one hand, the late development of American industry during the 19th-century forced the country to import European systems/models for production. On the other hand, however, the country had been shaping its own industrial appearance already since 1850, as "different problems asked for new solutions". (Gideon, 1982: 344–347). As a result, a large gap was arising between the new structure of American industry and the European old one, concerning the construction systems for industry. The inevitable outcome of this fact concerns the construction systems employed in the industrial structures. Unlike in the USA, where, until the 1880s, there had been a prevalence of steel or balloon frame and brick structures, in the Old Continent, the employment of reinforced concrete constructions was allegedly witnessed for the first time in France during 1860s, thanks to the contribution of Monnier who used a steel frame to build concrete bins.

This new concrete based system offered several qualitative advantages to the industrial structures, because it fitted for the industrial buildings market demand "to provide maximum strength and stability, at a reasonable cost, and with an acceptable degree of fire resistance" (Bradley, 1999: 133).

In the USA, the importation of the concrete building system presented an opportunity to carry on experimentations on the form of this new material. The construction of new industrial buildings constituted indeed the perfect testing ground for the reinforced concrete systems, especially the manufacturing factory. The concrete framework finally substituted the brick system.

The American engineer Ernest L. Ransome is believed to have a leading role in this technical revolution, according to the number of Patents on reinforced concrete system; among others, particularly relevant is the *Unit construction system* of 1906 that records the transition from a monolithic employment of concrete to an archaic shape of the framework.

As Reyner Banham said, Ransome laid the foundations for a new industrial prototype in manufacturing building, called *daylight factory*. (Banham, 1986)

The Pacific Coast Borax Refinery in Bayonne, New Jersey, built in 1904 came into sight as a full-grown version of the prototype[1]. This daylight factory consists of a 5-floor concrete framework and offers to the workers a copious lateral lighting, spreading through the great windows. The building turns out to be very compact and flexible to a lengthwise extension and the

[1] We have to consider also the work of Albert and Julius Kahn in the development of new Patents for "daylight factories". In particular, see the Packard Motor car company building n°10, in Detroit. (1905).

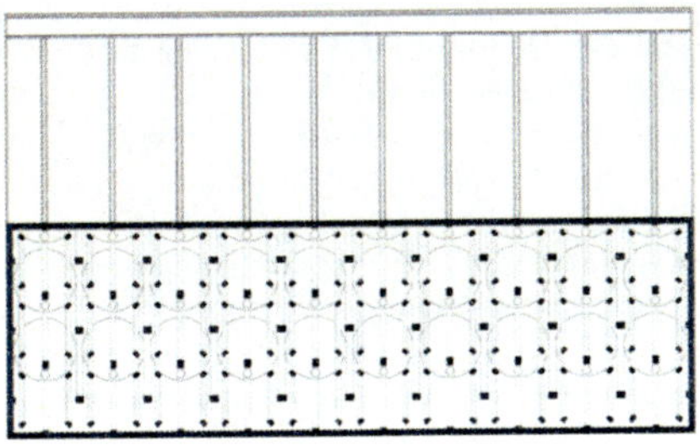

Figure 1. Axonometric view (bottom-up) of the Great Northern Elevator in Buffalo, 1898.

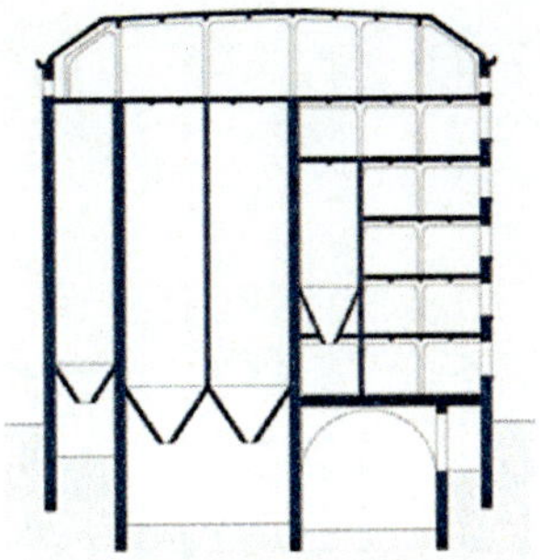

Figure 2. Cross-section of the "project par de moulin" François Hennebique, (application of the Hennebique's system 1893–1896).

open-space interior simplifies the production process, guaranteeing a better arrangement of the driveshafts and machines.

A second prototype developed in the USA is the grain elevator. Banham studied this model by looking at typological differences relative to the daylight factory because of its very different features. The dichotomy between these two kinds of buildings seems to mark out the building of industrial structures until the first decades of the 20th-century.

The grain elevator originated with the need to arrange the maximum quantity of grain in the minimum building area. Namely, the elevators arose in harbour areas or trading points, facilitating the transportations of stored wares. This aspect explains both the complexity of the challenge American engineers were dealing with, as well as what essentially boosted the technical advancement acceleration of building systems meant for this industrial prototype.

Until 1890, the constructive systems mainly employed were the balloon-frame also known as square bin. It consisted of a rectangular ground floor and a headhouse on the top hosting the conveyor belts, whose function was that of carrying the grain inside rectangular bins. In spite of a low production cost, the inadequacy of these systems is linked to the lack of fire resistance. This fact encouraged engineers to develop a steel system that gave guarantees to the skyscrapers buildings. The process of elevators formal evolution marks a real turning point thanks to the new steel system; the new cylindrical form of the bins permits to better resisting to the horizontal pressures from the grain.

The testing process on the new system had born witness to Buffalo, a harbour town near New York. Banham cited some examples, such as the Great Northern Elevator designed by Max Toltz in 1898. As we can see in the axonometric view (fig. 1), that building is provided with a curtain wall of elongated shape and consists of forty-eight cylindrical steel bins. The containers do not place on the ground thanks to the support of a great steel frame, which permits the grain dumping through tubular slipways. Seen from the outside, the building shows a tripartite elevation, hiding the typical head house of the previous architectures. From the first years of the 20th-century, the cases of steel frame system applied to grain elevators decreased in favour of ceramic tiles system (1895). What explains that transition to the new system was the fact that the steel frame system was not fireproof and it had a great coefficient of thermal expansion. The discarding of the curtain wall in favour of autonomous bins with their own coverage marked a significant turning point in the process of elevators formal evolution. Although it only lasted a few years, this phase led to Le Corbusier's idea of pure forms "assembled under light".

In 1892, François Hennebique filed a patent for the first frame structure applied to grain elevators (fig. 2). The American engineers probably studied this patent while Haglin was applying the Hennebique's system to the Peavey's Folley in Minneapolis, although making some modifications. Indeed, Haglin reduced implementation times by replacing scaffoldings with "a type of form worked called slip form that consists of two rings held apart by sturdy yokes" (Kowsky, 2007: 33–39).

Haglin's system is first enshrined in Buffalo, through the extension of the Washburn-Crosby in 1906. The iteration of cylindrical bins is so dense that it looks like the containers intersect each other but, actually, two triangular elements connect singular arcs of the circle. This expedient work out the problem of wasting space deriving from the juxtaposition of two cylindrical forms (fig. 3 C and E).

Even though the architecture historical approach needs to attribute a leading role to individuals (such as Ransome and Albert Kahn) who seem to direct a process of innovation, it still should be clear the importance of entrepreneurial dynamism and workers patient job in defining an anonymous practice of construction. Actually, the attribution of inventions depends on the speed of validating patents, (one of the few instruments of historical dating), but this figure does not to do justice to the actual scenery of innovation that North America was experiencing during those years. Reyner Banham (1982) grasped this issue when he talked about Ransome:

> His energy, creativity and especially the shrewdness and speed with whom he registered patents, ensured him a pass to the annals of concrete building.

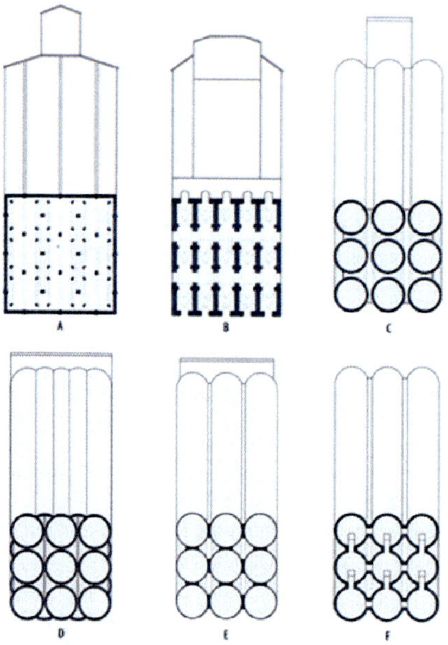

Figure 3. Taxonomic sample cases of bins. (Axonometric view). Constructive evolution of form. A. Steel frame structure (Great Northern Elevator, 1898) B. Curtain wall structure (Silo pre-squat Amsterdam, 1896) C. Haglin's system structure (1899) D. Concrete central elevator's system (Buffalo, 1915) E. Washburn-Crosby elevator's system (Buffalo, 1906) F. Chichibou cement company's silos (Yoshiro Taniguchi, 1956).

Consequently, the author seems to suppose that the evolution process of the daylight factory cannot be labelled as the exclusive product of few great creators, especially in an era in which the means of production were experiencing the first massification.

The American experience makes plain that innovation is the result of a constantly progressing collective process fostered by the practice of construction rather than the consequence of single participators' work.

3 THE EXPRESSIVE COMPONENT OF AMERICAN UTILITARIAN STRUCTURES. MODERNIST THEORIES UNTIL THE SECOND DECADE OF 20TH-CENTURY

The Werkbund association, which was created in 1907 with the specific goal of examining the relationship between artistic practice and industry, welcomes the launch of a debate on the meaning of American industrial structures in Modern architecture.

The association wondered whether it was possible to find an aesthetic dimension in a kind of building created without any expressive purpose. Even if all members agreed on the existence of an expressive component in the grain elevators, there still are different conflicting positions on the subject.

In Peter Behrens' speech (2007: 205–206) titled the "Aesthetic industrial building", the author rejected the idea that the beauty of these forms depends on "the absolute compliance to the useful" of the utilitarian structures. As Behrens sustained, a purely utilitarian building cannot exist because the built environment as a whole is the result of an aesthetic intention. Hence, the author justified the primacy of the *Kunstwollein* in the process of construction. This refuse of the "utilitarian form" seemed to comply with his projects of industrial structures: the steel frame system of the *Turbinhalle* in Berlin is emphasised in order to obtain expressiveness in the structure, a "sensation of aesthetic solidity" that, as Behrens says, "Is very different from the sensation of constructive solidity".

Muthesius and Walter Gropius tried to identify the features of American industrial buildings that might be compatible with their idea of the Modern project. Muthesius (2007: 215–219) underlined that "the transition from individual to typical" in the American experience is one of the crucial aspects explaining the tendency to an "objective form" concerning the Modern project.

Gropius (2007: 211–214) noted that

> The most brilliant ideas, thanks to the industrial reproducibility, are useful and efficient as for a single person as for a community.

The author seems to be in diametrical opposition to Behrens by affirming that:

> Germany is ahead of the rest of Europe as the ability to resolve in an artistic way, the problem of industrial buildings. However, in the USA, the homeland of industry, factories of such an impressiveness arose that even overcame out best works in this sector The American and Canadian grain elevators, the coal deposits of the great railway lines, the Modern factories of the American trust, bear a comparison with the monuments of the Ancient Egypt. Their architectural features have so much firmness that the meaning of the building shell reveals clearly to the viewer. The naturalness of these buildings is not based on the material superiority derived by their large-size –that is not the source of their monumental feature–but it is based on the workers' ability to preserve the natural essence of the compact form in an autonomous and authentic way.

Thus, Gropius conferred to the American industrial prototypes a sort of autonomous beauty that does not need any domestication by the architect, contrarily to what Behrens sustained. Furthermore, he conferred to the American builders, that is to say, to the practice of construction, the autonomous capability of preserving the architectural beauty in their works[2].

[2] Gropius seemed to agree with Le Corbusier (2007) who recognized, in American industrial buildings, a tendency to connection between "the universally abstracted and the rudely useful".

Werkbund's writings provide documentary evidence to the American industrial prototypes change of meaning. On one hand, it turns into a theoretical process that leads us to consider the building no more as mere utilitarian structures but as a real architectural work. On the other hand, this change of meaning gives us the opportunity to imagine the American buildings as icons of the "Modern Spirit" in a symbolic way, although not as design references[3].

Some exponents of the European Avant-garde seem also to recognise the importance of the American industrial prototypes even if in a different way than Werkbund did.

Even if Antonio Sant'Elia took as reference the new metropolitan infrastructures for his city planning without mentioning the industrial buildings, he still produced, in 1913, a series of drawings on the subject of great industrial structures. He represented electric power plants composed by monolithic and pure volumes, with the industry typical design criterion including the iteration, assembly and juxtaposition of basic shapes. The structural elements come also from the American buildings: the framework, the counterfort and the concrete monolithic mass. Therefore, Sant'Elia absorbed from the industrial structures the elements for the creation of a Futurist formal vocabulary. In such a way, the expressive component of the utilitarian form turns into a planning and operative action.

Erich Mendelsohn was so much impressed by grain elevators that he planned an American Tour in 1924. Contrary to Gropius, Behrens and Le Corbusier, Mendelsohn was not satisfied with the experience of buildings in image, but he searched a direct experience of grain elevators in two different ways: taking photographs and drawing sketches with the purpose of understanding these structures' essence. Mendelsohn (1993: 44) also wrote some letters to his wife revealing a sort of emotional zeal: "Childhood forms, clumsy, full of primaeval power, dedicated to purely practical needs". Upon his return from the USA, Mendelsohn built the red banner factory in Petersburg in 1926, through which he put into practice the American lesson.

4 CONCLUSIONS

To sum up, we can question the relationship between architectural theory and practice, as described by Carlos Martì Aris. He considered (2007) the theory as metaphorically similar to the rib that sustains (and thus determines) the arch, that is to say, the practice of construction. Consequently, the theory enjoys a certain supremacy over the practice.

This study would contribute to demonstrate that the progress is not directly linked to a unidirectional relationship between architectural theory and practice, but it welcomes the idea that an anonymous practice of construction can enrich the architectural theory.

The European architectural theory proposed two main theoretical positions on the issue of American industrial prototypes between 1860 and 1915. The first one considers the American buildings as the icon of an architectural movement. An approach influenced by the theorists' tendency to analyse the buildings as images, and considering the architectural object beyond its "corporeal" sense (Moreno, 2016: 25–35).

The second position attempts to understand the right way to transform the American phenomenon into design reference. In that approach, it is essential having a direct experience of buildings, in order to enlarge documentary reference with drawings and photos that can be essential elements in the architectural creative process.

What is it, therefore, the role of architectural theory in this perspective? Probably it is that of focusing and reflecting on the phenomenon of the built environment, to increase constantly new points of view about the architecture practice.

The progress takes place only when the theory returns to the practice of construction after an enriching process, with the precise aim of broadening practice horizons in favour of the creation of beauty.

[3] Banham (1982) wrote that the Fagus factory has no affinities with the "daylight factories" prototypes.

BIBLIOGRAPHICAL REFERENCES

Adorno, Theodor W. (2003) Progress *Can One Live After Auschwitz?* New York: Stanford University Press.

Aglietta, Michel. (1979) *A theory of capitalist regulation.* London: New Left Books.

Alexander, Ernest. (2016) There is no planning — only planning practices: Notes for spatial planning theories. *Planning Theory* 15 (1): 91–103.

Alford, Robert & Friedland, Roger. (1985) *Powers of Theory – capitalism, the state and democracy.* Cambridge: Cambridge University Press.

Allmendinger, Philip. (2002) *Planning Theory.* London: Palgrave.

Atkinson, Adrian. (2014) Urbanisation: A brief episode in history. *City* 18 (6): 609–632.

Banham, Reyner. (1982) *L'Atlantide di cemento*, Bari: Giovanni Laterza editore.

Behrens, Peter. (2007) Aestethik in industriebau In: M. De Benedetti, A. Pracchi ed., *Antologia dell'architettura moderna. Testi, manifesti, utopie.* (trad. it.). Bologna: Zanichelli editore.

Barr, Alfred H. (1934) Foreword. *Museum of Modern Art. Machine Art.* New York: Museum of Modern Art. https://www.moma.org/documents/moma_catalogue_1784_300061872.pdf. (Accessed 8 June 2017).

Benevolo, Leonardo. (2002) *Historia de la arquitectura moderna.* Gustavo Gili.

Beunen, R, Van Assche, K & Duineveld, M. (2013) The importance of reflexivity in planning and design education. *Wageningen University Working Papers in Evolutionary Governance Theory* 14: 23.

Blackburn, Edward Lushington. (1834) *An Architectural and Historical Account of Crosby Place.* London: John Williams.

Boyer, Robert H. (1988) Technical Change and the Theory of Regulation. In: Dosi, G (ed) *Technical change and Economic Theory.* London: Macmillan.

Bradley, Betsy H. (1999) *The works. The industrial buildings of United States*. Oxford University Press.

Braghieri, Nicola. (2012) Mito e sortilegio dell'architettura senza architetti. In: M. Bruzzone e L. Serpagli ed. *Le radici anonime dell'abitare contemporaneo*. Franco Angeli editore.

Branford, Victor. (1921) *Whitherward? Hell or Eutopia.* London: Williams and Norgate.

—. (1914) *Interpretations and Forecasts.* New York: Mitchell Kennerley.

—. (1893) The Edinburgh Summer Meeting. *PR* (12): 215–220.

Branford, Victor & Geddes, Patrick. (1919) *Our Social Inheritance.* London: Williams & Norgate.

—. (1917) *The Coming Polity.* London: Williams and Norgate.

Branford, Victor, Geddes, Patrick & White, J.Martin. (1908) New Crosby Hall. *Times.*

Branford, Victor V. (1904) A Sociological Approach Towards Unity. In: Hand, James Edward (ed) *Ideals of Science and Faith.* London: G. Allen, 104–156.

Bridges, J.H. (1904) Civics. *PR*: 198–203.

Bronowski, J & Mazlisch, Bruce. (1960) A tradição intelectual do Ocidente. Lisboa: Edições 70.

Bruce, Susan (ed). (1996) *Three Early Modern Utopias.* New York: Oxford University Press.

Bury, John Bagnell. (1920) *The Idea of Progress. An inquiry into its origin and growth.* London: Macmillan & Co.

Cabral, João. (1998) Spaces of regulation, the changing role of the state and the case for regional planning in Portugal. *XII Congress of the Association of European Schools of Planning.* University of Aveiro.

Calvocoressi, Richard & Hartley, Keith. (2001) Vienna 1908–1918. In: Blazwick, Iwona (ed) *Century City: Art and Culture in the Modern Metropolis.* London: Tate Gallery Publishing Ltd.

Campbell, Scott & Fainstein, Susan (eds). (1996) *Readings in Planning Theory.* Malden: Blackwell Publishers.

Carey, John (ed). (1999) *The Faber Book of Utopias.* London: Faber and Faber Limited.

Castells, Manuel. (1977) *The Urban Question: A Marxist Approach.* London: Edward Arnold.

Colquhoun, Alan. (2009) Symbolic and Literal Aspects of Technology. *Collected Essays in Architectural Criticism.* London: Black Dog, 23–26.

Corbin, Alain. (1992) *Le territoire du vide. L'Occident et le désir du rivage, 1750–1840.* Paris Editions Aubier.

Danto, Arthur. (1995) El Final del Arte. *El paseante* (23–25): 32–33.

David, JC. (1984) *Utopia and the Ideal Society: a Study of English Utopian Writing, 1516–1700.* Cambridge: Cambridge University Press.

Davis, J.C. (2010) Thomas More's Utopia: sources, legacy and interpretation. *The Cambridge Companion to Utopian Literature.* London: Cambridge University Press, 28–50.

Davoudi, Simi. (2012) The Legacy of Positivism and the Emergence of Interpretive Tradition in Spatial Planning. *Regional Studies* 46 (4): 429–441.

Després, Carole, Vachon, Geneviève & Fortin, Andrée. (2011) Implementing Transdisciplinarity: Architecture and Urban Planning at Work. In: Doucet, I and Janssens, N (eds). *Transdisciplinary Knowledge Production in Architecture and Urbanism, Urban and Landscape Perspectives.* 11. Dordrecht: Springer, 33–49.

Fainstein, Susan & DeFilippis, James. (2016) *Readings in Planning Theory.* 4th ed. New Jersey: Wiley & Blackwell Publishers

Fera, S. (1991) Aldo Rossi, rielaborazioni: viaggio e palinsesto. *Lotus* (68).

Figueira, Jorge. (2012) Fernando Távora, alma mater, viagem na América, 1960. In: Banderinha, José António (ed) *Fernando Távora Modernidade Permanente.* Guimarães: Família Fernando Távora, Fundação Instituto Arquitecto Marques da Silva, Fundação Cidade de Guimarães.

Frampton, Kenneth. (1996) Adolf Loos: The Architect as Master Builder. In: Schezen, Roberto (ed) *Adolf Loos: Architecture 1903–1932.* New York: The Monacelli Press, 14–21.

Geddes, Patrick. (1915) *Cities in Evolution.* Williams & Norgate.

—. (1911) *The Civic Survey of Edinburgh.* Edinburgh: Outlook Tower.

—. (1908?) *Chelsea.* London: County Council.

—. (1908a) *Town Planning and City Design – in Sociology and in Citizenship.* London: Sociological Society.

—. (1908b) Chelsea, Past and Possible. In: Hollins, Dorothea (ed) *Utopian Papers.* London: Masters and Co., 6–17.

—. (1908c) New Crosby Hall. *Saturday Review of Politics.*

—. (1904) *City Development.* Edinburgh: Geddes and Company.

Giedion, Sigfried. (1970) *Mechanization takes command. A contribution to anonymous history*. Oxford: Oxford University Press.

Giedion, Sigfried. (1982) *Space, time and architecture. The growth of a new tradition.* Cambridge, Massachusetts: Harvard University Press.

Glendinning, James. (1909) Utopian Papers (I). *Saint George* 12 (47): 176–185.

Godfrey, Walter H. (1913) The site of Beaufort House. *Survey of London: Volume 4, Chelsea, Pt II.* London: London County Council, 18–27.

Goss, Charles WF. (1908) *Crosby Hall: a Chapter in the History of London.* London: Crowther & Goodman.

Gramsci, Antonio. (1971) *Selections from prison notebooks.* Hoare, Q and Smith, G. N. London: Lawrence and Wishart.

Griffiths, Paul. (1996) *Modern Music: A concise History.* London: Thames and Hudson.

Gropius, Walter. (2007) Der stilbildende wert industrieller bauformen In: M. De Benedetti, A. Pracchi ed., *Antologia dell'architettura moderna. Testi, manifesti, utopie*. (trad. it.) Bologna: Zanichelli editore. pp. 211–214.

Hall, A. Rupert. (2014) *The Revolution in Science.* Abingdon: Routledge.

Hall, Peter. (1992) *Urban and regional planning.* 3rd ed. London: Routledge.

Hand, James Edward. (1904) *Ideals of Science.* Hand, James Edward. London: G. Allen.

Healey, Patsy. (2003) Collaborative Planning in Perspective. *Planning Theory* 2 (2): 101–123.

—. (1997) *Collaborative Planning, shaping places in fragmented societies.* Vancouver: UBC Press.

Hibbert, Christopher, Weinreb, Ben, Keay, John, et al. (2011) *The London Encyclopaedia.* London: Pan Macmillan.

Hollins, Dorothea (ed). (1908) *Utopian Papers.* London: Masters and Co.

Janik, Allan & Toulmin, Stephen. (1973) *Wittgenstein's Vienna.* Chicago: Ivan R. Dee, Publisher.

Johnson, Philip. (1994) Preface to Sixtieth-Anniversary Edition. *Museum of Modern Art. Machine Art.* New York: Museum of Modern Art.

Kitchen, Paddy. (1975) *A Most Unsettling Person*. London: Gollancz.

Kowsky, Francis R. (2007) Monuments of vanished prosperity. In: Lynda H. Schneekloth ed. *Reconsidering Concrete Atlantis: Buffalo Grain Elevators*. New York: Urban design project, pp. 33–39.

Kultermann, Udo. (1993) *Architecture in the 20th Century*. New York: Van Nostrand Reinhold.

Le Corbusier. (2007) *Towards an architecture*. Los Angeles: Getty Publications.

Loos, Adolf. (1993a) Ornamento y delito. In: Opel, Adolf and Quetglas, Josep (eds). *Adolf Loos: Escritos I – 1897/1909*. Madrid: El Croquis Editorial, 346–355.

—. (1993b) Degeneración cultural. In: Opel, Adolf and Quetglas, Josep (eds). *Adolf Loos: Escritos I – 1897/1909*. Madrid: El Croquis Editorial, 332–334.

—. (1993c) Arte y arquitectura. In: Opel, Adolf and Quetglas, Josep (eds). *Adolf Loos: Escritos II – 1910/1932*. Madrid: El Croquis Editorial, 159–161.

—. (1993d) Arquitectura. In: Opel, Adolf and Quetglas, Josep (eds). *Adolf Loos: Escritos II – 1910/1932*. Madrid: El Croquis Editorial, 23–35.

Lustemberger, Kurt. (1994) *Adolf Loos*. Zurique: Artemis Verlag.

Magris, Carlo. (2005) *L'infinito viaggiare*. Milan: Mondadori.

Martì Aris, Carlos. (2007) *La centina e l'arco. Pensiero, teoria, progetto in architettura*. Milan: Marinotti editore.

Masiero, Roberto. (2003) *La estética de la arquitectura*. Madrid: Machado Libros.

Meller, Helen. (1990) *Patrick Geddes*. Routledge.

Mendelsohn, Erich. (1993) *Erich Mendelsohn's "Amerika": 82 photographs*. New York: Dover publications.

Mendes, Manuel (ed). (2013) *Fernando Távora, Minha Casa – Uma porta pode ser um romance*. Porto: Fundação Instituto Arquitecto José Marques da Silva, Faculdade de Arquitectura UP, Reitoria UP.

Merrifield, Andy. (2014) *The New Urban Question*. London: Pluto Press.

Mesquita, Ana. (2007) O melhor de dois mundos a viagem do arquitecto Távora aos EUA e Japão – diário 1960. *Departamento de Arquitectura*. Coimbra: Universidade de Coimbra.

More, Thomas. (2012) *Utopia*. London: Penguin.

Moreno, Catalina Mejìa. (2016) Walter Gropius's silos and Reyner Banham grain elevators as art-objects In K. L. Thomas, T. Amhoff, N. Beech ed. *Industries of architecture*. London: Routledge, Taylor & Francis group.

Muthesius, Hermann. (2007) Der Werkbund gedanke in den Germanischen landern In: M. De Benedetti, A. Pracchi ed. *Antologia dell'architettura moderna. Testi, manifesti, utopie*. (trad. it.) Bologna: Zanichelli editore pp. 215–219.

Mouffe, Chantal. (2013) *Agonistics – thinking the world politically*. London: Verso.

Nesbit, Robert A. (2009) *History of the Idea of Progress*. London: Transaction.

Norman, Philip & Caroe, W D (1908) The architecture of Crosby Hall. *Survey of London Monograph 9, Crosby Place* London: Guild & School of Handicraft, 57–63.

Novalis. (1842) *Heinrich von Ofterdingen*. Cambridge: John Owen.

Paden, Roger. (2007) *Mysticism and Architecture*. Lanham: Lexington Books.

Pagano, Giuseppe and Daniel. Guarnero (1936) *Architettura rurale italiana*. Milano: Hoepli editore.

Pastore, Domenico et al. (2016) *Evoked. Architectural diptychs*. Bari: Giovanni Laterza editore.

Pérez-Goméz, Albert. (1996) *Architecture and the Crisis of Modern Science*. Cambridge: MIT Press.

Pevsner, Nikolaus & Cherry, Bridget. (1991) *London 3: North West*. New Haven: Yale University Press.

Preservation of Crosby Hall. (1832) London: London Crosby Hall.

RBKC. (2014) *CRH1203/DOC003 Heritage Statement, New Chapel at Crosby Hall, Cheyne Walk, London. SW3 5AZ*. London: RBKC.

Ribeiro, António Sousa. (1988) Para uma Arqueologia do Pós-Modernismo: A "Viena 1900". *Revista de Comunicação e Linguagens* (6–7): 143.

Rudofsky, Bernard. (2016) *Architecture without Architects. A Short Introduction to Non-Pedigreed Architecture*. New York: Doubleday & company.

Sager, Tore. (2016) Activist planning: a response to the woes of neo-liberalism? *European Planning Studies* 24 (7): 1262–1280.

—. (2011) Neo-liberal urban planning policies: A literature survey 1990–2010. *Progress in Planning* 76 (4): 147–199.

Saint, Andrew. (1991) Ashbee, Geddes, Lethaby and the Rebuilding of Crosby Hall. *Architectural History* 34: 206–223.

Sant'Elia, Antonio & Marinetti, Filippo Tommaso. (1971) Futurist architecture. In: Conrads. In: Conrads, Ulrich (ed) *Programs and Manifestos on 20th Century Architecture*. Cambridge: MIT Press, 34–38.

Schön, Donald A. (1983) *The reflective practitioner: how professionals think in action*. New York: Basic Books.

SKK. (1909) A Gospel from Cheyne Walk. *The New Age*.

SS. (1905) *Sociological Papers*. I. London: Macmillan.

SSCC. (1917–9) *PftP, 8*. London: Headley Bros.

Struik, Dirk J. (1992) *História concisa das matemáticas*. Lisboa: Gradiva.

Steiner, Hadas A. (2013) The grain of the image In Miriam Paeslack ed. *Ineffably Urban: Imaging Buffalo*. Aldershot: Ashgate Publishing Company. pp. 117–120.

Sweet, Rosemary. (2015) The Preservation of Crosby Hall, c. 1830–1850. *The Historical Journal*.

Swinny, S.H. (1914) Mr. Branford's Interpretations and Forecasts. *PR* (256): 91–94.

Tait, Malcolm. (2016) Planning and the public interest: Still a relevant concept for planners. *Planning Theory* 15 (4): 335–343.

Tatarkiewicz, Wladyslaw. (2013) *A History of Six Ideas: An Essay in Aesthetics*. Berlin: Springer.

Távora, Fernando. (1960) *Diário de "bordo"*. vol. 1. Porto: Associação Casa da Arquitectura.

Tiekso, Tanja. (2016) Luigi Russolo's Art of Noises and the Aesthetics of Musical Experimentalism. In: Berghaus, Günter (ed) *International Yearbook of Futurism*. Berlin: De Gruyter, 71–93.

The Unrestricted law of Supply and Demand. (1907) *NA*.

Vairley, Francis. (2013) *Reflections on the Fall of Modern Civilisation*. London: New Generation Publishing.

Weber, Max. (1976) *The Protestant Ethic and the Spirit of Capitalism*. London: George Allen and Unwin.

Wittkower, Rudolf. (1998) *Architectural principles in the age of humanism*. 5th ed. London: Academy Editions.

More's College: A case study in the eutopian praxis of deep heritage conservation

Matthew Wilson
College of Architecture and Planning, Ball State University, Muncie, IN, USA

ABSTRACT: This paper explores how Patrick Geddes and Victor Branford used Positivist sociology to recover the typological essence and historical significance of a medieval building called 'Crosby Hall' in Bishopsgate, London. We will see how they used such interpretations to save the structure from demolition and to substantiate a proposal to make it a future-focused centre for eutopian 'City Design' called More's College.

This essay thus begins with a discussion of Geddes and Branford's sociological theory of 'social formations'. Here the edifice will be presented as an index of alternating historical 'spiritual' and 'temporal' powers, from the point of its construction to the announcement of its imminent demolition. We will see that after coordinating its systematic removal to a new site in Chelsea, London, Geddes and Branford proposed to use their sociological survey method as the pedagogical basis of a eutopian school called 'More's College'. I will argue that Geddes and Branford aimed to link the building's former association to the utopian Thomas More to a eutopian future affiliated to the father of Positivist sociology, Auguste Comte. Effectively this paper examines an instance in which sociology was used in the early twentieth century to justify acts of deep heritage conservation, in which a structure's embedded symbolic meanings were inverted – to effectively serve as emblems of social empowerment.

Keywords: Heritage Conservation; Utopianism; Positivist sociology; Patrick Geddes; Victor Branford

1 INTRODUCTION

A beautifully-crafted piece of medieval architecture called Crosby Hall once stood in Bishopsgate, London. At this very place, the English civil servant Thomas More (1477–1535) supposedly penned[1] the final portions of his famous book, *Utopia* (1516). Built in 1466 and serving the moteliest of purposes down to the twentieth century the Hall was scheduled for demolition in 1907. Although it is rarely acknowledged today, the Positivist sociologists Patrick Geddes and Victor Branford vigorously and victoriously rallied to save the historic structure. The Hall was systematically dismantled and reassembled in the London borough of Chelsea. Before turning to an argument centring on Geddes and Branford's proposal for its new use, we will employ their method of the sociological analysis of 'social formations' to appreciate in what ways the edifice was of historical import. This kind of analysis informed their radical efforts in heritage conservation, to rekindle the typological essence of the Hall as a space of social empowerment. We will see that Geddes and Branford's calls to make the Hall the centrepiece of a new institution called 'More's College' were linked to their long-lost praxis of 'City Design' which they considered an 'art of polity-making' (BRANFORD, 1921: 97–98; 1914: 14; Geddes, 1908a). Along these lines I will contend that in the tradition of More they wished to cultivate the utopian imagination of the body politic; yet like the father of Positivism Auguste Comte they sought to employ 'applied sociology' to reconcile the ideal with the real.

I will present Crosby Hall as being central to Geddes and Branford's praxis of participatory City Design for making Chelsea, and perhaps Britain, into a eutopia or an idyllic real place. In terms of our congress theme of "PROGRESS(ES) – THEORIES AND PRACTICES", I will argue that this Positivist form of activist-led community engagement serves as a model for a future-oriented praxis of heritage conservation.

1.1 *Crosby hall bishopsgate*

Through the late nineteenth century, Geddes and Branford wrote extensively on sociological theory and practice. In the footsteps of Comte, their work concentrated on 'social formations' or the correlations between social power, citizens and architecture. Here they created a typology in which the social power of four groups of civic actors linked to four different types of urban spaces (Branford et al., 1919: 35–37). These civic groups comprised two spiritual and two temporal powers. The spiritual powers, branded as 'Intellectuals

[1] There are conflicting accounts as to whether More was a tenant in 1516, when writing *Utopia*, or if it was only between 1523–4 that he owned the Hall, and perhaps lived there.

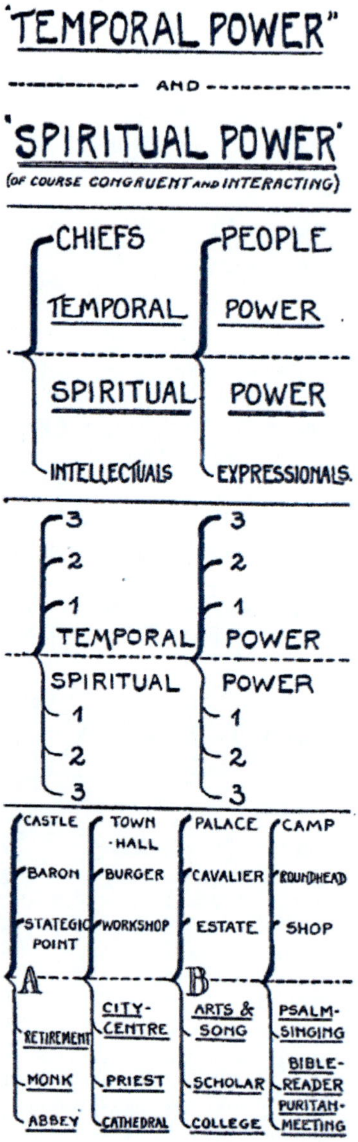

Figure 1. Scheme for the Study of Social Formations. Branford and Geddes, *Our Social Inheritance* (1919), p. 36.

and Emotionals', were philosophical or creative animals respectively. The temporal powers, known as 'Chiefs and People', had a political or economic nature. Geddes and Branford held that any 'fully orbed' or complete functioning community consists of these 'two co-relative hemispheres' of spiritual and temporal powers of Intellectuals, Emotionals, Chiefs and People. They had discovered these structural groupings in Plato's *Republic,* Aristotle's *The Politics,* More's *Utopia* and all the way down to the nineteenth-century works of Arnold Bennett and H.G. Wells (Branford, 1914: 378–379; SSCC, 1917–9).[2] Following Comte's *System of Positive Polity,* Geddes and Branford also used this sociological 'outlook of the world' to analyse historical and contemporary social relations. They traced the implications of how from the medieval to the late Renaissance period the intellectual authority of the abbey Monk had transferred to the college scholar, and the cultural authority of the cathedral priest had shifted to the bible reader of the Puritan meeting hall [Fig 1.]. As political chief, the cavalier of the Renaissance palace superseded the baron of the medieval castle; and the economic authority of the Burger and medieval town hall gave way to the Renaissance roundhead camp (Branford et al., 1919). How these spiritual and temporal power structures shifted over the course of time and how they shaped and reflected the evolution of the city-region were at the centre of Geddes and Branford's sociological research and practice. In chorus with Comte, they believed that sociology was a spiritualizing 'master-science' for coordinating civic groups to achieve structured social change. Additionally science and industry, they thought, represented the spiritual and temporal powers of the future Positive Era of society (Branford et al., 1919: 35–37; 1917). As such, 'Chiefs and People' would function as a temporal power of directives and operatives in banks, factories, fields and workshops. 'Intellectuals and Emotionals', on the other hand, would function as spiritualizing thinkers, teachers, artists and healers from within universities, schools, hospitals and homes.

Along these lines, Geddes and Branford portrayed Crosby Hall as an ideal 'spiritual' centre because of its utopian links to More. If we apply this Geddesian analysis of social formations to the Hall, however, we will see it empowered and represented different social structures throughout its existence. Crosby Hall has an extensive history of performance, action and event. The wealthy grocer and wool merchant Sir John Crosby built the Hall in 1466 as an adjunct of his courtyard mansion, Crosby Palace. In the Geddesian outlook, the Palace conveyed Crosby's status as a temporal economic power; Crosby was a patron of human-centred innovation and the artisan life of craft. Lost to Antiquity it is a matter of speculation as to whether the great London Fire of 1666 destroyed the magnificent Crosby Palace, or if it met its demise in a separate incident in 1672. Surviving against all odds as the 'last remaining relic' of medieval domestic architecture in London, Crosby Hall offered invaluable 'evidence of how sumptuously c15 merchants' built (Pevsner et al., 1991).

Crosby Hall at Bishopsgate was a thing of 'great richness and beauty'. The unbuttressed Reigate stone volume with gable roof sat elevated above ground level on a masonry under-croft (Norman et al., 1908; Pevsner et al., 1991). A row of double cinquefoil lancet

---

[2] For Geddes and Branford, More is the 'Chief' narrator of the story, Hythloday is represents the 'People', Holbein is the 'Expressional' and Erasmus is the 'Intellectual'.

windows complemented its most prominent external feature: a protruding, octagonal stone-vaulted oriel. Meanwhile eight arched trusses of heavy oak timber, from which hung ornamented pendants, structured the interior. Elaborately decorated brackets in-filled with frieze boards 'depicting grotesque heads, white roses, antique shields' and other signs of social power supported the trusses (Goss, 1908: 27–29). Linking ornament to social status the historian Edward Lushington Blackburn wrote that this composition made the Hall the apogee of architectural expression – the 'chief and most ornamental' space of Crosby Palace (Blackburn, 1834: 30; RBKC, 2014). It served as a precedent to 'modern' edifices such as the library at Arundel Castle and the hall at King's College, Cambridge (*Preservation of Crosby Hall*, 1832).

As indicated, for Geddes and Branford, Crosby Hall was a secular-spiritual power of London because More lived here while finishing his famous book, *Utopia*. Hence, they emphasised this connection to save the structure. Nevertheless, when More purchased the Hall in 1532 he considered it a vessel fit for the honourable representatives of the temporal power of England. Along these lines distinguished guests, including Elizabeth I, the Duke of Gloucester (later Richard III), and various Lord Mayors, enjoyed its pomp and ceremonialism; the edifice later appeared in the works of Shakespeare (*Richard III,* c. 1592). The Hall thereafter served as the headquarters of a chief temporal economic power, the East India Company, from 1621 to 1638. However, for over a century afterwards, until 1769, it was used for spiritual purposes – as a Presbyterian Meeting Hall. After being ruthlessly converted into a two storey commercial warehouse in 1770, which reflected Bishopsgate's burgeoning exchange economy, Hall's proprietors subsequently subdivided the space into apartments (Hibbert et al., 2011; RBKC, 2014). Then in 1831, the Hall became available for general letting, which for some meant its outright abuse and imminent demolition (Blackburn, 1834: 30; Sweet, 2015). After an enlightened, some say 'botched', restoration the Hall came to represent, once again in the Geddesian view, a premier spiritual space reflecting modern times. It served as a Literary and Scientific Institution offering a library and rooms for public lectures, evening classes and concerts (Goss, 1908; Saint, 1991).

When the Chartered Bank of India purchased the Hall, *The New Age* (The Unrestricted law of Supply and Demand, 1907) newspaper announced that it would inevitably be 'pulled down, and some disgusting building will reign in its stead'. The Bank had indeed scheduled this downtrodden structure for demolition. The rallying calls to save Crosby Hall appeared around mid-1907 after neither the City Corporation nor the London County Council censured the Bank's intentions to redevelop the site carte blanche. Geddes began lobbying the Council and lecturing the public about Hall's significance (Hibbert et al., 2011; Saint, 1991). He and Branford sought to set a precedent where, in their words, national memorials would be understood

Figure 2. Crosby Hall. *The Graphic* (1884).

as 'things apart from the merchandise of silver and gold' (Branford, 1914: 160).

There were, in fact, many appeals to save the building. Lord Leighton proposed to use the Hall as an art school in his garden, while Gordon Selfridge desired to use it as a reception space on the roof of his new London store (Kitchen, 1975; Meller, 1990). The press announced in 1907, however, that Crosby Hall would be re-erected in Chelsea with a 500-year lease as the 'first collegiate hall of residence officially recognised by the University of London', with Geddes as its warden (The Unrestricted law of Supply and Demand, 1907). The architectural historian Andrew Saint has claimed that Selfridge's daring feat of prospective engineering and other imaginative proposals caused Geddes and Branford to broaden their scope (Saint, 1991). The press, however, suggested its new purpose was merely to accommodate 100 graduates and students. Geddes and Branford (The Unrestricted law of Supply and Demand, 1907) maintained, however, that the re-erection of Crosby Hall on their watch was not an act of 'archaeological piety' or the desiccation of a 'dubious antique'. It was the civic 'utilisation of [an educationally useful] treasure', they wrote.

1.2 *More's college chelsea*

From 1909 to 1910, the architect Walter Godfrey oversaw the re-erection of Crosby Hall at Beaufort Street and Cheyne Walk in Chelsea [Fig. 2].[3] This site held tangible links to More and Geddes. After forgoing his freehold of Crosby Hall in 1524 More lived at this very site in Beaufort House for fourteen years until his attainder in 1535. Although demolished in 1740 the house was surrounded by a sublime landscape, later named More's Gardens, that overlooked the Thames

[3] Some aspects of the Hall could not be saved, but the original elements include the fireplace, roof structure, oriel window, pointed arch windows, postern door and a portion of the facing masonry. Later in 1925–6 Godfrey added to the north of the Hall a four-storey hall of residence for the British Federation of University Women.

.()(Godfrey, 1913). In the years of 1902 to 1904 Geddes and the Edinburgh architects Dunn and Watson had designed and built, on the western end of the gardens, a block of flats aptly called 'More's Hall' (Saint, 1991).

So, what was Geddes' reasoning for building More's Hall before Crosby Hall had presented itself as an opportunity? During this time, Chelsea was a borough of mixed character: patches of well-to-do middle class, the comfortable and the very poor, which the Positivist Sociologist Charles Booth would describe as those suffering from chronic want. Geddes saw this site in Chelsea as fertile ground to extend his practice of experimental urbanism. Since the 1880s and 1890s, he and his followers had incrementally transformed the squalid housing of the Edinburgh Old Town into refurbished self-governing university residences. This scheme was affiliated to Geddes' sociological institute and 'positivist school' called the Outlook Tower, which conducted sociological surveys before planning urban improvement schemes (Branford, 1893). The writer and reformer Sidney Webb remarked on seeing this work that Geddes was 'making an old Balliol!' (NLS, MS.19282, ff. 26). Geddes and Branford aimed to link this model of urbanism to the University of London reforms that were underway thanks to the efforts of Webb, RB Haldane and others (Saint, 1991).

In the Geddesian outlook, a modern institution such as the University of London represented a premier spiritual power, which should assume a socially engaged role like the ancient church. As an extension of this Geddes and Branford wished to establish an institute called 'More's College', which would be comprised of the freshly re-erected Crosby Hall and the university residence called More's Hall. The mission of More's College, it seems, was to systematically envision, plan and realise urban social reorganisation. While some scholars do not accept the complete contents of More's book as a committed illustration of a better world, this text, it seems, was in part Geddes and Branford's inspiration (Bruce, 1996; Carey, 1999: 38–39; More, 2012: xvii).[4] Scholars sometimes present utopia as a static place of exile for dreamers, a non-existent Eden where nothing ever happens, a place for and about the contemplative life (Davis, 2010). More's work may nonetheless be thought of as a humanistic lens for intellectualising the environment, for enabling one to critically evaluate what is liked and disliked about space and society – and to think, feel and act on its improvement. Geddes and Branford's activities suited this outlook.

During their effort to save Crosby Hall, Geddes and Branford joined a literary group called The Utopians, which served as the 'nucleus' of the Chelsea Association. The purpose of saving this space, they claimed, was to 'extend Utopia to a civic and public sphere'. Here The Utopians would conjure up 'idyllic spaces of Brotherhood, of Justice, and of Love'. Their intention was 'realizing something of Utopia in Chelsea here and now'. Yet critics asserted that there was not 'any real unity of outlook', 'manifesto or confession of faith' or even practical method to sustain their activities (Geddes, 1911: 537; Glendinning, 1909; Hollins, 1908: vii). Arguably, Geddes and Branford's sociological theory of social formations was quietly at work here. The Chelsea Association claimed that 'Utopia is to be visualised clearly and ultimately' by way of 'continuous co-operation of faith, reason and works'. This was 'the only way to reach any heaven, however mundane and humble' (Glendinning, 1909). In essence, Crosby Hall was being imagined as the place where idealists of all spiritual and temporal vocations would work to redeem London from its current state, which Geddes described as a 'strangely deformed, debased, mammonised, and conventionalised' locality (Hand, 1904; SS, 1905).

Envisioning a policy of culture Geddes discussed how aesthetically More's College would reflect its utopian past, and empower its users to work towards a eutopian future. The residence hall, library, common room, study areas, lecture and exhibition spaces would be adorned with murals depicting various utopias and utopians (Geddes, 1908?). Along with these utopian visuals, Crosby Hall would host music recitals and morality plays transmitting the spirit for improving life on Earth (SSCC, 1917–9: 20–21). This was not solely a place for social dreaming. In addition, Geddes and Branford stressed this point by linking More's College to the sociological principles of Comte. Of the many parallels between Comte's and More's schemes[5] the most important was that their small sovereign states idealised neither man nor nature, but the organisation (Bruce, 1996: xiii; David, 1984: 9). Following Comte, Geddes and Branford's theory of social formations idealised organisational urban systems in evolution. Here the interrelations between work, place and folk would be fine-tuned and calibrated over time via applied Positivist sociology. As such Geddes and Branford thought of Crosby Hall as a laboratory of civics and thus an offshoot of the Sociological Society in London (SS, 1905: 142). Branford, who was the Society's secretary, expressed his unfailing belief in this applied science. At the Society's first meetings he announced that the 'art of constructing idealistic Utopias' was

[4] In the story *Utopia* Hythloday attacks the 'injustices of European society,' describing a 'system in which greedy landlords … evict peasants who in turn must starve, steal and be hanged'. 'The social arrangements on the island of Utopia … can be read as a blueprint for righting these very real wrongs', as in Utopia 'there is no poverty, no exploitation, no luxury, no idle rich'. The humanist ambition was, after all, to 'link study and reflection to practical action'. Like Positivism, 'Utopian education is designed in the first place to produce good citizens: it is overseen by priests'.

[5] Both schemes were rooted in small self-governing places. Like More's Utopia, Comte's Occidental Republic focused on the 'spiritual' rather than the 'material' aspects of life. Also like More's vision, Comte's scheme included no lawyers or journalists. Along these lines, he called for the abolition of the monarchy, clergy and bourgeoisie. Certainly, however, More went to great lengths to show that utopia was not of the past or the future, but of the present.

> 'based no longer on poetic dreams and personal aspirations, but on a systematic study of immediate possibilities disclosed by scientific and historical investigation'. (Branford, 1904: 104–156)

Seeking to coordinate 'endeavours of citizenship' by organising a sociological survey of Chelsea, Geddes encouraged the 'University Militants' of More's College to take a 'more active and definite interest' in the borough (Branford et al., 1908; Geddes, 1908c). At 'the outset', Geddes wrote,

> 'we hardly . . . know our facts, and . . . we cannot even dream our Utopia . . ., much less define any single portion of it until we have come to know and understand something [of] our neighbourhood, our town'. (Geddes, 1908b)

Geddes, Branford and J.W. Slaughter proposed that More's College would accordingly offer seminars in sociology, civics and psychology (Geddes, 1908?). This course would serve as an intellectual introduction to the survey of social formations (Bridges, 1904; Geddes, 1915: 223; 1904; SSCC, 1917–9: 99; Swinny, 1914). This survey work intended to unite and coordinate the civic groups of Intellectuals, Emotionals, Chiefs and People in the creation of town planning exhibitions, seminars and debates, thus leading to participatory City Design (Geddes, 1911: 537). As such More's College was thought of as a place to synthesise sociology, social dreaming and practical policy in the 'useful action and re-action between student and citizen, college and borough' (Geddes, 1908?; 1908b).

In this way, Geddes and Branford envisioned More's College growing into a university and Chelsea into a City proper (Geddes, 1908b). They considered this approach – heritage conservation as urban intervention – a model for transforming towns across the world into sovereign city-states, not unlike Comte's *System* (SKK, 1909). Effectively they were offering a framework to create a 'doctrine of a realisable Eutopia everywhere instead', they maintained, 'of the illusionary Utopia of nowhere' (Branford, 1914: 378–379). In this way, Crosby Hall was both an exemplar of making future-oriented idyllic places and heritage conservation.

2 CONCLUSION

As we have seen Crosby Hall held various functions throughout its existence, empowering and being empowered by a succession of alternating, perhaps competing, spiritual and temporal powers. Geddes and Branford's act of heritage conservation, however, was driven by their social idealism. They wished to kindle the historical and sociological imagination, to initiate a local renaissance. Crosby Hall was thought of as one of the potentially myriad places where idealist activists could meet and collaborate on creating a more just and convivial world.

Despite their good intentions, Geddes and Branford lacked financial backing to see through much more than initial surveys and fleeting and symbolic acts that point to the creation of a true More's College. The Hall remained in public hands until the 1980s, and it is now part of a private residence. Ironically enough it is, as such, a 'no place' to the public at large. Notwithstanding Geddes and Branford's Positivist form of community engagement serves as a model for a future-oriented praxis of heritage conservation.

BIBLIOGRAPHICAL REFERENCES

Blackburn, Edward Lushington. (1834) *An Architectural and Historical Account of Crosby Place.* London: John Williams.

BRANFORD, Victor. (1921) *Whitherward? Hell or Eutopia.* London: Williams and Norgate.

—. (1914) *Interpretations and Forecasts.* New York: Mitchell Kennerley.

—. (1893) The Edinburgh Summer Meeting. *PR* (12): 215–220.

Branford, Victor & Geddes, Patrick. (1919) *Our Social Inheritance.* London: Williams & Norgate.

—. (1917) *The Coming Polity.* London: Williams and Norgate.

Branford, Victor, Geddes, Patrick & White, J.Martin. (1908) New Crosby Hall. *Times.*

Branford, Victor V. (1904) A Sociological Approach Towards Unity. In: Hand, James Edward (ed) *Ideals of Science and Faith.* London: G. Allen, 104–156.

Bridges, J.H. (1904) Civics. *PR*: 198–203.

Bruce, Susan (ed). (1996) *Three Early Modern Utopias.* New York: Oxford University Press.

Carey, John (ed). (1999) *The Faber Book of Utopias.* London: Faber and Faber Limited.

David, JC. (1984) *Utopia and the Ideal Society: a Study of English Utopian Writing, 1516-1700.* Cambridge: Cambridge University Press.

Davis, J.C. (2010) Thomas More's Utopia: sources, legacy and interpretation. *The Cambridge Companion to Utopian Literature.* London: Cambridge University Press, 28–50.

Geddes, Patrick. (1915) *Cities in Evolution.* Williams & Norgate.

—. (1911) *The Civic Survey of Edinburgh.* Edinburgh: Outlook Tower.

—. (1908?) *Chelsea.* London: County Council.

—. (1908a) *Town Planning and City Design - in Sociology and in Citizenship.* London: Sociological Society.

—. (1908b) Chelsea, Past and Possible. In: Hollins, Dorothea (ed) *Utopian Papers.* London: Masters and Co., 6–17.

—. (1908c) New Crosby Hall. *Saturday Review of Politics.*

—. (1904) *City Development.* Edinburgh: Geddes and Company.

Glendinning, James. (1909) Utopian Papers (I). *Saint George* 12 (47): 176–185.

Godfrey, Walter H. (1913) The site of Beaufort House. *Survey of London: Volume 4, Chelsea, Pt II.* London: London County Council, 18–27.

Goss, Charles WF. (1908) *Crosby Hall: a Chapter in the History of London.* London: Crowther & Goodman.

Hand, James Edward. (1904) *Ideals of Science.* Hand, James Edward. London: G. Allen.

Hibbert, Christopher, Weinreb, Ben, Keay, John, et al. (2011) *The London Encyclopaedia.* London: Pan Macmillan.

Hollins, Dorothea (ed). (1908) *Utopian Papers*. London: Masters and Co.
Kitchen, Paddy. (1975) *A Most Unsettling Person*. London: Gollancz.
Meller, Helen. (1990) *Patrick Geddes*. Routledge.
More, Thomas. (2012) *Utopia*. London: Penguin.
Norman, Philip & Caroe, W D (1908) The architecture of Crosby Hall. *Survey of London Monograph 9, Crosby Place* London: Guild & School of Handicraft, 57–63.
Pevsner, Nikolaus & Cherry, Bridget. (1991) *London 3: North West*. New Haven: Yale University Press.
Preservation of Crosby Hall. (1832) London: London Crosby Hall.
RBKC. (2014) *CRH1203/DOC003 Heritage Statement, New Chapel at Crosby Hall, Cheyne Walk, London. SW3 5AZ.* London: RBKC.
Saint, Andrew. (1991) Ashbee, Geddes, Lethaby and the Rebuilding of Crosby Hall. *Architectural History* 34: 206–223.
SKK. (1909) A Gospel from Cheyne Walk. *The New Age*.
SS. (1905) *Sociological Papers*. I. London: Macmillan.
SSCC. (1917–9) *PftP, 8*. London: Headley Bros.
Sweet, Rosemary. (2015) The Preservation of Crosby Hall, c. 1830-1850. *The Historical Journal*.
Swinny, S.H. (1914) Mr. Branford's Interpretations and Forecasts. *PR* (256): 91–94.
The Unrestricted law of Supply and Demand. (1907) *NA*.

Fernando Távora: The deontology of the journey as a form of cultural and personal progress

Raffaella Maddaluno
Department of Arts, Humanity and Social Sciences; Faculty of Architecture, University of Lisbon, Lisbon, Portugal

ABSTRACT: The importance of the journey has accompanied the architect Fernando Távora from the beginning of his architectural learning and continued as an element of the development of a method for evolving and the progress of the cultural formation itself and the update of his on research until the end of his life. The journeys themselves were numerous: several trips to Spain, Italy, and a trip to Europe in 1949, travelling on the CIAM (International Congresses of Modern Architecture). The first journey that will make him wander the ends of his Portuguese territory will be one in 1947, in which for the first time he will start feeling the tight limits of the superstructures themselves, both mental and cultural. The confrontation with different realities, in a crucial historical period such as the post-war, will lead him to an inexorable path towards a clearer conscience of what is really important in life and what place architecture and art hold in it. The journeys are accompanied by writings, in this case, letters addressed to his fiancé in Portugal. Few drawings, many thoughts.

His years of training culminated in a trip around the world in 1960, thanks to a scholarship from the Gulbenkian Foundation in Lisbon, which allowed him a four-month trip to the USA, Mexico, Japan, Thailand, Pakistan, Lebanon, Egypt and, finally, Greece. With constant discipline, he writes a voluminous diary of thoughts and drawings. In a slow yet constant progress, in his journeys, Távora evolves and his ideal of architecture evolves, always farther from the rigidity of predefined culture and always closer to the mutable and unstable human nature.

Keywords: Fernando Távora, Journey, Diary, Memory, Gulbenkian

1 JOURNEYING TO RETURN

Various ideas of travel exist. There is an idea of travel that goes beyond space and time, actually travelling with the intention of locating something that seems to have been lost or unjustly removed along the way. The territory to be explored along such a pathway is that of individual consciousness, if we consider the artist as an individual, collective if we consider a wider audience.

The work of art is the same as travels, a stranger to the place being visited. It is the romantic traveller, who is not seeking a "cultural pilgrimage", who does not proceed by analysing the landscape or comparing it according to a timely verification with the ancient text. No pleasure is found in the recognition or non-recognition of the distance from it: the aim of the journey is the perpetration of an individual dream where the imagination takes the place of reality (Corbin, 1992).

A fundamental condition for this type of travel is solitude. As Fera describes it, it is like a new country, a natural landscape, unknown customs, are a spectacle for four men, but for one man alone, an achievement. A solitary triumph over everyday things, of places far away in space and time, over the multiplication of meanings, over myths and stories: an achievement over time. (Fera, 1991: 73)

In *Heinrich von Ofterdingen*, by Novalis, the pilgrim asks a question and receives an apparently awkward answer:

> "Whither are we going?"
> "Ever homewards. (1842: 203)

The notion of travelling implies the idea of eternal return, which brings with it a constant sense of tragedy. The journey is thus experienced as the tragedy of returning, like Aeneas and Dido, or in contemporary culture with its inexorable loss of the centre. It is an endless journey that inevitably leads to the fear of return, which results in delay, for as long as possible, the moment in which you must return home and mature an awareness that everything will have changed. The great dilemma accompanies Ulysses throughout his journey. At the base of these two concepts the ancient question is expressed as to whether a life journey can be understood in the classical sense of remaining true to yourself, or whether can it be considered as a continuing sign of the impossibility of stemming the continuous mutation of the self, the loss of a centre, of identity, the danger and the desire to become another.

This is circular travel as opposed to linear travel. The traveller flows ever onward, he does not carry himself, his entire self in his progress, but each time annihilates his entire previous identity and throws it away (Magris,

2005). During the journey, a foreigner among foreigners, we learn in the great school of humility, which teaches the limits of our mental schemes and forces the destruction of the tools with which we have built the paraphernalia for continuous daily recitation. You learn the great sense of being no one.

2 THE DISTANT MEMORY

When individuals narrate the story of their journeys, they also depict something of the living world to which they belong. However, during the years of Távora's maturity and an inner peace that followed his trip around the world, Távora continued to use the trip overseas as a praxis in an essential experience for learning, and as a complement to the more general process of cultural development, which is constructed through study, teaching and professional activities. The journey for him became a consistent approach to continuous cultural and emotional renewal, towards the complete formation of both man and architect.

There were two ways of relating to travel, or of relating to Távora's travels. "Listen", to his travels, listen to him talk of travel or physically share with him the experience of the journey. Távora taught Theory of Architecture for many years at the Faculty of Porto, and one of the unchanging techniques of his lectures was recounting to his students what he had seen during his travels, adopting in his stories not a chronological sequence but a mixture of memories ordered according to the teaching needs of the moment. Excavations in Athens were thus approached by referring to his visit to ruins in Italy, or the experience of European landscapes complemented with a description of the spoken nature of Wright's Taliesin. His stories often continued beyond the classroom, recounting tales in his atelier to his staff, or during trips with friends. I discovered from interviews that his colleagues had many memories of his tellings, and the something happened with his former students.

In these cases, where experience is transmitted through a "story", those who "listen" use their imagination as a learning tool, while the teller uses memory. The concept of memory comes from anamnesis meaning reminiscence. It is an active function that begins with a multitude of sensations and develops towards a unity, understood through thought. Bringing images to mind comes from a tradition with a psychological connotation that defines memory as imagination with the connotation of time. Anamnesis literally means to bring images to mind.

The memory (Mneme) is, on the contrary, a passive function, the storing of the sensation experienced in the present, physically recording the sensation. Anamnesis is, therefore, to revive this sensation following the interval of a period. It refers, therefore, to the Platonic idea that learning is remembering, thus the process of knowing does not come only from the experience, but from remembering the experience.

Naturally, all these concepts connected to the way of experiencing reality and the way this experience becomes a part of our conscience are connected to the theme of time. A "measured" time, "fixed" on an event, and a time "lost" or "forgotten" in the memory of that event. In addition, Távora creates two categories of time, a *before* and an *after*. There is a time held a prisoner in the pages of his journeys' writings that are measured, made of numbers, names, and dates. It seems that he intends to communicate by "measured", "solid" time, transmitted by architectonic discipline and by life. Then when the lived experience starts to become a memory in his telling, in his lessons, in his didactical drawings, for him time sets itself free of the imprisonment of measurements and loses itself, diluted in an unchangeable impossibility of returning to the coded rule of chronology.

All of this makes us ponder on the theme of transmitting the architectural discipline, consider which are the most suited forms to teach or learn. Is there a more suited time form to "transmit" an "architecture class"? What does it mean to be a *teacher* of architecture?

We can assume two recognised/accepted ways to define its meaning. One can be a teacher by "imitating" a model: in this case, the teacher is simply a "channel" of transmission that brings a wisdom that somehow does not belong to him. Alternatively, one becomes a teacher by becoming the "example" ourselves: in this case, the transmission comes through the "Showing", showing by doing, or showing by telling. Távora's journeys can be a part of this way of transmission.

In Távora's own words:

> Nada pior do que ser inteligente, nada pior do que saber, nada pior do que andar com os sábios e com os santos, de os conhecer de perto – o ideal é não ser, é ser qualquer coisa como uma pedra, como um livro, como uma gota de agua, qualquer coisa onde a tristeza não entra, onde chover seja um paradoxo e para quem criar seja impossível. Porto, 27. VIII. 1947 (Mendes, 2013: 3)[1]

3 FROM PORTUGAL TO THE WORLD, PASSING BY EUROPE

> Espero que sintas a minha falta, mas creio que será até proveitos começares a habituar-te à ideia de que em breve teremos dois meses do mais completo afastamento. Ao certo, ao certo, não se sabe ainda a data da partida, mas suponho que lá para 6 ou 7 faremos as nossas despedidas. Ao fazê-las, resta-nos esperar que

[1] Translation of the original text: "There is nothing worse than being intelligent, nothing worse than knowing, nothing worse than being accompanied by wise or holy men, than knowing them closely – the ideal is not being, or being something similar to a rock, or a book, or a drop of water, something where sadness cannot enter, where raining would be a paradox and to whom creating is an impossibility".

os comprimentos de chagada sejam tão alegres como elas serão tristes. (Mendes, 2013: 3)[2]

In 1947, after the end of World War II, Távora begins a journey through Europe. A moment that coincides with a condition of discomfort and scepticism regarding his activity as an architect. It is a journey taken aboard in a Citroen, accompanied by his brother Bernardo and a friend. Porto, Guarda, Avila, Madrid, Lyons, Barcelona, Marseilles, Cannes, Zurich, Eindhoven, Delft, Rotterdam, Brussels, Anvers, Paris, Bilbao, are some of the European cities and Ventimiglia, Genova, Grosseto, Rome, Naples, Florence, Bologna, Venice, Verona, Rovereto, Milan, the Italian cities.

It is a journey that we take along with him, through his written words in the letters he sends to his fiancé, in Portugal. An attention to record reality pulls him to consider the testimony of his feelings as a powerful instrument that allows him to relive over again, reorganising and commenting with the care of a chronicler. He decides the how and when to read that experience and instructs his interlocutrice asking her not to consider true that what he writes, leaving an illusion that the absolutism of his experience does not exist:

> Há uma coisa que eu te peço: torna-te o mais alheia possível às minhas palavras para que elas não possam provocar em ti situações tão desagradáveis como aquelas que as produziram. Deixa que eu de vez em quando me abra um pouco mais, mas não tomes a sério aquilo que te digo; lême como lerias um romance muito mal escrito e muito maçador e a cujas emoções fosses absolutamente alheia. Bilbao, 13. XI. 1947 (Mendes, 2013: 63)[3]

The experience is lived in a self-centred way, and the telling and writing should provoke different emotions from those that have generated the written thoughts. It is like multiplying and chasing states of mind, in which the writer has a completely different experience than the reader. Behind a seemingly self-centered attitude, an awareness lies in Távora's writing that living is not telling, but the telling is re-living.

The fear of an excessive reworking of events accompanies him while traveling through Europe as well as during the Gulbenkian funded travel, but in this last one Távora writes in his journal every day, noting with journalistic rigor every single event, number, name, with a self-discipline that does not seem to allow time for reworking.[4] Távora fears this "distant memory", wanting a more organised sequence of data to let the goal of transmission to be more objective and simple. He even asks his fiancé to give him back the letters in order to review them thoroughly.

> Eu gostaria, quando chegar a Portugal, de fazer uns apontamentos com as impressões que fui colhendo nesta passeata, mas falta-me muita coisa porque a minha memória não aguenta tudo e eu não tomei notas de nada. A única coisa que me pode ajudar um bocadinho são as cartas que dia a dia te vou escrevendo e onde bem ou mal, muito ou pouco, as primeiras impressões ficam registadas. Importar-te-ás de emprestar-me todas essas cartas, certa de que não penso em ficar com elas? Creio que não te custará fazer-me esse favor sobretudo sabendo que a consulta do que te tenho escrito se torna indispensável para aqueles apontamentos. Peço-te por isso o favor de ordenares todas as minhas cartas (estão todas datadas) de modo a que possas entregar-mas logo que nos encontremos. Angouleme, 10-11. XI. 1947 (Mendes, 2013: 61)[5]

During the Gulbenkian journey, the goal of the diary was to write a final report requested by the Gulbenkian Foundation as mandatory for the scholarship. The same sense of loss is hidden allowing that the memory at a distance can reorganise the sense of reality. In the number of details and events is so large that he will never be able to write a definite summarised report, and kept postponing and presenting excuses with reasonable details, stating that he would send it to the Gulbenkian Foundation[6].

[2] Translation of the original text: "I hope you miss me, but it may prove profitable that you start getting used to the idea that soon we will live two months of the most complete remoteness. For sure, the date of departing is still unknown, but I think that we will say goodbye in the 6th of 7th. When doing it we only have to hope that the welcome upon arrival be as jolly as these will be sad".

[3] Translation of the original text: "There is something I ask you: distance yourself as far as possible from my words so they won't provoke in you unpleasant situations such as those that have generated them. Allow me to open myself a bit more, but do not take what I write seriously; read me as you would read a novel, badly written and very boring, to whose emotions you would be completely foreign to".

[4] In the diary written during Távora's journey, there are two pads of sketches of the places he visits. The drawings are rich in description and indications and become a parallel testimony equally as important as the narrations written in the diary.

[5] Translation of the original text: "Upon returning to Portugal, I would like to write a few notes on the reflections that I have gathered during this time but I lack many facts because my memory does not keep everything and I have not written a single note. The only thing that could help a little are the letters that I have written to you every day, where for better or worse, too much or too little, the first impressions have been registered. Would you mind lending me these letters, remaining assured that I will not want to keep them forever? [...] I ask you to put my letters in order (they are dated) in a way that you can give them to me as soon as we see each other" Angouleme, 10-11. XI. 1947.

[6] In the document, currently in the archives of the FCG dated September 17th, 1959. It is written: "Again we communicate that the Fine Arts Department is very interested in a detailed report of your journey and possibly in the production of one, or more, conferences regarding this theme". On June 7th, 1961 Távora receives a letter from the Foundation asking for the final report, yet to be presented, where we can read: "Dearest Távora, excuse us but due to bureaucratic matters we inform you that it would be extremely convenient that we "settle" the question regarding your scholarship that is linked to [...] your report." (Mesquita, 2007: note 366, 203). There

In fact, he experiences this specific travel with a strong feeling of persuasion, an urging possession, the will to live the moment, every single moment, not only those privileged or exceptional. Every encounter, every means of transport, every address is welcomed in his experience with the same will to register the presence, perhaps by fear of losing it in the memory's successive re-elaboration.

4 ARRIVING AT HIMSELF

Távora departs for the European travel full with uncertainties of not being at odds with his choice, that of architecture. These will accompany him during many moments of his own personal journey. Uncertainty and a lack of assurance of his skills, partly due to his young age, will make him determined, rigid, inflexible, and ready to renounce and abandon all distractions that could have mined the chosen path. This first big travel will redefine, polish, and allow himself a lightness that only maturity sometimes consolidates. One can understand that in the thoughts he confided to his interlocutor:

> Como sabes eu tenho, como todos os rapazes da minha idade, um futuro mais ou menos pensado, uma profissão mais ou menos escolhida e determinada; essa profissão só poderei exercê-la bem e só poderá garantir-me esse futuro se eu me entregar a ela tão inteira e completamente quanto possível, isto é, pondo de parte o que na vida eu possa considerar acessório, inútil ou prejudicial. Até hoje considerei sempre que qualquer namoro seria um obstáculo para a realização do meu programa de vida e por isso evitei sempre levar até esse ponto as minhas frágeis paixonetas. Sisteron, 24. IX. 1947. (Mendes, 2013: 15)[7]

However, the journey changes and makes feeble the barriers that one builds around one's personality, fearful of admitting and recognising that there are moments in life in which one needs to be exposed as much as possible to turbulence and the unexpected that will arrive from the outside. In addition, Távora became more intimate, more human, more consenting to admit such uncontrollable or well-defined feelings, like nostalgia:

> Fazes-me falta. Ainda agora depois do jantar fui deitar ao correio a carta que te escrevi em Marselha e me lembrei de que se aqui estivesses iríamos dar uma volta, embora o tempo não esteja muito convidativo para sair à noite porque o frio começa a sentir-se. Assim, como não estavas, deitei a carta e voltei para o Hotel, muito triste, muito pesaroso, muito saudoso, etc., etc. Cavalaire–sur-Mère, 26. IX. 1947. (Mendes, 2013: 20)[8]

Távora's state of mind in the 1960 journey is a lot different. Although he shows an open state of mind, the experience mostly American will always work as an antithesis and a parting point from his ideas, becoming more and more Portuguese.[9]

---

is a written sheet of paper, addressed to someone whose connections to the Gulbenkian Foundation are not clear, in which Távora excuses himself for the delay and tries to explain his reasons: "Mr. Dantas (?). Our common friend Master Carlos Ramos has asked me to present, as soon as possible, my travel report and has manifested his concerns for not having done it yet. First and foremost, I would like to give you my apologies and then to communicate that after several doubts, I believe to have identified a great outline for my report and that in this very moment I am developing the various points of that outline. Other than the fact that I am not an "easy" writer (or scribe) one can add the difficulty of condensing so many things seen in these four months and, as well as the circumstances surrounding the time of my absence that has complicated so much my professional life, consequences that I am suffering to this day. Therefore, I do not have enough time to dedicate with the intensity required by this report. I can assure you, however, of two things: the first is that I will deliver the report; the second is that this report will not shame me before the Foundation that so generously has granted me a scholarship"

[7] Translation of the original text: "…as you well know I have, like all the young men of my age, a sort of thought out future, a sort of defined and chosen profession; I can execute it well and it will guarantee me a future if I dedicate myself wholly and completely, setting aside everything that in life can be considered accessory, useless, harmful, prejudicial. Up until today I have always thought that infatuations were an obstacle to the accomplishment of my life program and because of this I have always avoided to take my fragile little passions seriously." Sisteron, 24. IX. 1947.

[8] Translation of the original text: "I miss you. Right now, after dinner, I went to leave a letter at the post office, that I have written to you in Marseille and I remembered that if you were here we would have taken a stroll, even though the weather is not that welcoming to strolling in the evening, because the cold starts to chill. Therefore, since you were not here, I have left the letter and went back to the hotel, very sad, very thoughtful, very nostalgic etc. etc." Cavalaire–sur-Mère, 26. IX. 1947.

[9] "Country in which the internship is to be carried out: United States of America. Course to be followed or scientific area to be studied: Architecture departments of various Universities and Institutes. Who will serve as a Reference for your studies? The professors of the respective departments. Has this already been agreed with the organisations indicated? No. (...) What is the program of study or work plan to be followed? Studies of teaching methods of Architecture and Urbanism in the following universities and institutes: Columbia University, Howard University, Harvard University, Massachusetts Institute ofTechnology, Institute of Design. What period is the Scholarship intended to cover? Four months. (...) What is the desired start date of the Scholarship? To be agreed following successful application. (...) What work projects are foreseen following the conclusion of the Scholarship? Teaching and freelance practitioner. References (who may assist in formulating a judgment as to the scientific and moral suitability of the candidate): School of Fine Arts of Porto, General Directorate of Urban Services, the Porto, Matosinhos and Gaia Council, Port Administration of Douro and Leixões".

> The more we advance, the more we become near the point of departure and the more we travel the closer we get to ourselves (Figueira, 2012: 39)

During the journey through Europe, everything was a reason for Távora's discomfort and a sense of loss, even the simple act of feeling, in the journey around the world, the bigger the disappointment the stronger becomes his sense of identity. Indeed, the "other", the "different" does not incite a crisis, instead, it confirms what one is, and the further and different the other is the more one becomes him/herself. In Távora's case, it is clearly a condition of acquired maturity, confirmed by a career as reputed architect. In Chicago, he writes:

> A América é um laboratório magnífico. É indispensável conhecer, principalmente saber exactamente o que se deve evitar a todo o custo. (Távora, 1960: fl. 266v)[10]

How does his journey in Europe contribute to his cultural and personal progress? It is 1947, World War II still has a massive and painful presence in many of the cities that he visits. It is a foreign condition to the Portuguese Távora, who has not lived it but has heard about it.

> Escuso dizer–te que ao lado de todos os luxos dos Monte-Carlos, etc., etc., existem as ruinas e as devastações que a Guerra deixou em toda a parte. […] Pode dizer-se que não há uma única casa que não esteja crivada de centenas de balas dando a certeza de que a luta aqui foi horrível. […] E o pior é que depois de tanto terem sofrido vê-se que os donos das casas nem dinheiro tiveram ainda para tapar os buracos produzidos pelas balas e pelas granadas. Ventimiglia, 27. IX. 1947 (Mendes, 2013: 21)[11]

The days pass and the emotions become more and more confused, the ideas less and less caged and a constant uncertainty towards a future more and present. This travel makes lose his way, driving him further, doubting the existence of a finishing line. Everything seems to be not enough; everything seems to be too small or too tight. Even one who comes from an extremely social, intellectual, economic privileged condition, feels rusted by the scepticism towards himself and towards the world. It is in Naples that Távora regains conscience of this terrible and human instability of the soul. That city, more than all others, has triggered in him negative impressions, on the other hand, changed his mind more than any other Italian and European cities. He leaves Naples recognising that the visible chaos on the outside was exactly the chaos he held inside himself. He said:

> Eu atravesso uma grande crise de cepticismo, cepticismo relativo aos outros e, o que é pior, relativo a mim; se duvido de tudo o que me rodeia, pessoas, coisas ou ideias, duvido também de mim próprio, a ponto de me manter numa permanente autodesconfiança. Esta situação, que eu não desejo a ninguém, vai-me corroendo pouco a pouco e eu não sei o que será de mim se não conseguir opor um dique a esta inundação de cepticismo. Vejo as certezas desaparecerem, as verdades a sumirem-se e em tudo e para tudo, uma dúvida permanente e pronta para substituir e aquilo que outrora foi atitude ou ideia segura e firme. Nápoles, 6. X. 1947 (Mendes, 2013: 32)[12]

Nevertheless, his irony and lucid sarcasm, always present throughout his life, were able to emerge even in this condition, such as when he granted war the merit of having solved some urban knots unresolved by man and in certain cases made worse. In Genova, he recognises the same problems that existed in Portugal when he saw that next to the highest tower of the cathedral the fascists had built skyscrapers in the centre of town, destroying and substituting some of the historical quarters.

> Enfim lá como cá, ou cá como aí, as mesmas desgraças, as mesmas questões, os mesmos problemas e com tal intensidade que chegamos por vezes a agradecer a certas bombas a "amabilidade" que tiveram em destruir certas casas e certos bairros. Génova, 30. X. 1947 (Mendes, 2013: 23)[13]

Architecture takes the second stage, it is not the protagonist of his written pages, and it makes part of the scenery that holds certain ways of life different to those, which he was accustomed to. He does not describe monuments; he cites them as if to define the surroundings of the tale that is centred on ways of life, about people and their thoughts and feelings:

---

[10] Translation of the original text: "America is a magnificent laboratory; it's indispensable to know, especially to know exactly what to avoid at all costs".

[11] Translation of the original text: "It is useless to say that right next to the luxury of Monte Carlo, there is ruins and destruction that the war has left everywhere. (…) one can say there is not a single house that has not been drilled with bullets, assuring that the fight was hard. (…) But the worst thing is that after having suffered so much you can see that the owners still do not have money to cover the holes made by bullets and grenades".

[12] Translation of the original text: "I am living a great crisis of scepticism, scepticism towards others and, which is worse, towards myself; if I doubt everything that surrounds me, people, things or ideas, I also doubt myself, up to the point that I find myself in permanent self-suspicion. This situation, that I do not wish onto anyone, is corroding me little by little and I do not know what will become of me if I cannot end this flood of scepticism. I see certainties vanished, truths submitted, and everywhere and anywhere a permanent doubt ready to substitute what had been an attitude or a safe and stable idea".

[13] Translation of the original text: "After all, here like there, the same misfortune, the same questions, the same problems and with such an intensity that sometimes we even thank the "kindness" that some bombs had when they destroyed some houses and some blocks".

> É por isso que eu me tenho prendido muito mais agora com o aspecto das cidades e com a maneira como vivem os seus habitantes, com as suas dificuldades e os seus prazeres, as suas tristezas e as suas alegrias, os seus problemas e as suas soluções. Estou mais perto de tudo isto, mais preso a tudo isto do que a qualquer velho quadro ou qualquer grande escadaria que me podem oferecer apenas problemas de forma ou de sensibilidade, mas raras vezes me colocam perante as duras realidades do momento que são aquelas que, pela sua natureza, mais nos devem interessar. Cada coisa no seu lugar e o passado não se pode sentar-se ao lado do presente. Génova, 30. IX. 1947 (Mendes, 2013: 24)[14]

His connection to art and with creativity changes radically after this journey through Italy. He sees a lot, visits buildings and works of art, studies people and habits, reflects and writes. Távora sees the best of European works of art, but they tire him, a hangover of forms, colours, lines, and such useless elements, according to him, to solve his personal problems. It is in Switzerland, that he understands what "creation" really is:

> Depois da Itália cheia de obras de arte, a Suíça cheia de obras de Deus. (Mendes, 2013: 43)[15]

He recognises in the works of Nature an eternity that art objects will never possess, terrorising and instilling a sense of loss:

> A Natureza é vida, é presente, é multiplicação, é juventude e frescura. Roma, o Coliseu, o Juízo Final, o Duomo de Florença, belezas sem duvidas, mas belezas passadas que o tempo corrói, uma árvore, um relvado, um rio, belezas eternas, sempre iguais, mas sempre vivas com os seus ciclos de crescimento e decadência, com o seu viço. Há mais possibilidades de vida num jardim do que num museu. Uma flor é mais fresca que a Vénus de Milo e não há pintor que conheça todos os verdes da Natureza. (...) Há em tudo que é obra do Homem um artificialismo, uma serie organizada de convenções que tornam fatigantes as obras humanas para aqueles que têm consciência das mesmas convenções. Meiringe, 19. X. 1947. (Mendes, 2013: 43)[16]

According to this analysis and reading the pages of his numerous journeys, not only those referred to in this text, one recognises a common element: for Távora it was fundamental not shutting our eyes towards the reality of the world, not to isolate, not to cut the ties with the "outside".

These were approaches that he always tried to keep strong, his awareness of how difficult it is to give time to time and not to barricade oneself inside the walls of self-reference. To allow the influences that come from the outside to change the culture one was fed on.

It was due to this perspective that Távora's journeys had an extraordinary value. One can end by saying that the outstanding benefit that the journeys brought Távora was born from the adopted perspective. Travelling is an irreplaceable way to build a living relationship, not only mentally but also physically, direct, intelligent but without the subjection to interpretation, with knowledge, a search for a non-specialist knowledge of places and people, considered essential for the discipline of planning according to one of the basic principles of Távora's philosophy: *"first live, and then draw"*.

BIBLIOGRAPHICAL REFERENCES

Corbin, Alain. (1992) Le territoire du vide. L'Occident et le désir du rivage, 1750–1840. Paris Editions Aubier.

Fera, S. (1991) *Aldo Rossi, rielaborazioni: viaggio e palinsesto*. Lotus (68).

Figueira, Jorge. (2012) Fernando Távora, alma mater, viagem na América, 1960. In: Banderinha, José António (ed) *Fernando Távora Modernidade Permanente*. Guimarães: Família Fernando Távora, Fundação Instituto Arquitecto Marques da Silva, Fundação Cidade de Guimarães.

Magris, Carlo. (2005) *L'infinito viaggiare*. Milan: Mondadori.

Mendes, Manuel (ed). (2013) *Fernando Távora, Minha Casa – Uma porta pode ser um romance*. Porto: Fundação Instituto Arquitecto José Marques da Silva, Faculdade de Arquitectura UP, Reitoria UP.

Mesquita, Ana. (2007) *O melhor de dois mundos a viagem do arquitecto Távora aos EUA e Japão – diário 1960*. Departamento de Arquitectura. Coimbra: Universidade de Coimbra.

Novalis. (1842) *Heinrich von Ofterdingen*. Cambridge: John Owen.

Távora, Fernando. (1960) *Diário de "bordo"*. vol. 1. Porto: Associação Casa da Arquitectura.

[14] Translation of the original text: "That is why I am ever more passionate by the aspect of cities and the way their inhabitants live in, with their struggles and pleasures, with their sadness and their joy, their problems and their solutions. I am always closer to all this, more attached to this than any old painting or big stairways that might at most give some problem of form or sensibility but rarely make me face the harsh reality of the moment, those are the ones, that by their nature, should interest me the most. Everything in its place and the past cannot sit alongside the present".

[15] Translation of the original text: "After Italy, filled with works of art, Switzerland filled with works of God".

[16] Translation of the original text: "Nature is life, is present, it is multiplication, it is youth and freshness. Rome, the Colosseum, The Final Judgement, Florence Domus, beauty beyond any doubt, but beauty corroded by time, a tree, a field, a river, are eternal beauty, always the same but always alive with their cycles of growth and decay, with its vigour. There are more possibilities of life in a garden than in a museum. A flower is much fresher than the Venus of Milo and there is no painter that knows all the greens of Nature. [...] There is, in all of men's work, an artificiality, an organized series of conventions that make human works tiresomely aware of these conventions".

Maquette-concept as project genesis in the teaching of architecture

Margarida Louro
University of Lisbon, Faculty of Architecture, CIAUD, Lisbon, Portugal

ABSTRACT: This article presents the potential of using the maquette-concept[1] in the creative process of space exploration and in the design genesis of architecture teaching.

Thus, from the reconfiguration of the concepts of space representation as methods of perception, understanding and design awareness, we try to present a learning process where these idea objects are configured as embryonic rationales. It is a kind of maquette-concept, where the argument or spatial poetics is succinctly expressed, and the basis that will lead the whole design process is formed, conforming the later contextual, functional and constructive constraints. Basically, the maquette-concept is chosen to represent an initial idea, where contextual, functional and constructive constraints are dismissed and the material is chosen as the guiding principle in the complex process that is the teaching and the learning of architecture.

In this sense, the approach is systematised on the following points: on the learning process in architecture; on three-dimensional visualisation of space; on the relevance of the maquette-concept; on some experiences and formalisations; and ultimately about final considerations of idea, object and space.

Keywords: Maquette-concept, sketch, architectural teaching, argument, project

1 ABOUT THE LEARNING PROCESS IN ARCHITECTURE

Defining the genesis of learning architecture as the understanding of space, the maquette-concept approach takes over as the primary element in the creative process of space exploration, proving itself the foundation of architecture teaching from a phenomenological view that is based on the following points:

Understand space (architectural and urban), as the main theme of the students' initial training, in their first learning phase. Promoting, from the outset, an approach to the visualisation and understanding of different scales that, centralised in the anthropometric reference, will allow the discussion of a range of themes, such as space, architectonic object, urban section and territory.

Space perception will also be a privileged approach, both due to the analysis perspective and work on intervention at various scales, as well as for the imposed methodologies and for the strong emphasis on the three-dimensional production of objects, ranging from the traditional level (maquettes) to the conceptual model (concept objects).

The conception of idea and space will be the primary theme in this context of general training. To conceive the space and the idea that underlies it, the design argument, encouraging in the design process, the interaction of thought and representation as a primary form of research.

Understanding, perception and design come together as the three strategies that underlie the learning process in architecture from the following critical parameters: the conceptual framework of architecture – critical judgment and spatial evaluation; the introduction of composition and context elements; and finally, the introduction to the design process from its design, spatial expression and representation.

The conceptual framework of architecture is thus assumed and privileged through an emphasis on critical judgment and spatial evaluation. In this sense, the re-framing of evaluative universes (aesthetic, ethical and ontological) are the main elements. Evidencing these evaluative systems, enumerating their universes, categories, and subcategories, fit into the privileged aspects of this initial level in architecture teaching, thus promoting a sustained and critical capability. Thus, the main message will be to understand "value" as the product of a judgment, a judgment of choices that are being built with several advancements and setbacks in the formalisation of the conceptualization process.

In this critical and creative process regarding evaluative systems and their practical applications, three-dimensional visualisations will obviously be featured processes. The various axiological orders and evaluative universes of both the aesthetics marked between the beautiful and the ugly, as well as the ethics between good and evil, and the ontology between true and false, are called upon to step in progressively, raising the learning process and the student to a critical judgment where one essentially seeks to think the "feeling".

Basically, the consolidation of an aesthetic judgment that is promoted by a kind of intellectualization of the sensitive. Going back to Raymond Bayer (1956), and his approaches to open experience, sensitive intuition, aesthetic organisation of space and time and intellectual intuition, where sensation to rationalisation, abstract to concrete, sensitive to intelligible are gradually exposed.

In addition, in this context, confrontation with various aesthetic subcategories (sublime, tragic, dramatic, pathetic, comic, poetic, picturesque, graceful and beautiful) emerges, marked by an operative consciousness that progressively constitutes a creative thought.

In this process, the composition and context elements are introduced, where function, construction and form of architectural composition appear as elements of primary focus. That is, from a consolidation of judgment and the critical incorporation of the different ways of feeling, students are confronted with the essential composition through basic transformations that explore morphological and spatial typologies, as well as modelling and principles of composition.

That is, exploring the essential basis of architecture: beauty (aesthetic demand), functionality (useful demand) and resistance (constructive demand) and through the enunciation of the Vitruvian principles: *venustas*, *utilitas* and *firmitas* transversal throughout the history of architecture and somehow omnipresent in the eternal condition of being and doing architecture, opportunities arise for basic transformations of formal composition that find systematic strategies such as division, inclusion, subtraction, addition, deformation, decomposition and intersection (Rob Krier, 1988).

This practical systematisation allows, from the stated evaluative consolidation, to establish parameters of artistic demand. Where, through formalisation strategies, students are progressively introduced to the design process, in its multiple levels of conception, expression and representation.

In fact, to the choice of dwelling is considered the fundamental quality of being. The dwelling is thus seen as the synthesis of desire for the world and its implicit ordering through the construction and the thought of this construction: emphasising the relationship between project and architecture. In this sense, students are confronted with the theme or question: what is architecture? Where, Build, Dwell, Think (Martin Heidegger, 1996) emerge as an epistemological rationale. Highlighting the relationship of construction, housing and beauty, which converges to the Vitruvian legacy. To dwell and the relation of being in space extols the fundamental anthropometric relation in the metric appropriation and the awareness of measurement. Finally, the typological exploration, through the consolidation of a discourse of space, poetry or an argument that will later mark the whole creative process and the spatial research.

From this embryonic approach on architecture teaching, the epistemological synthesis of design stands out through: the evaluative consciousness, an essential process in the formalisation of a judgment and in the assertion of choices (refusals and integrations); the compositional strategy, using different methodologies of composition; and finally, the construction of an argument of space and a conceptual logic that will dictate the whole subsequent process of spatial feasibility (functional, constructive and aesthetic).

2 THE THREE-DIMENSIONAL VISUALISATION OF SPACE

Following the conceptual, evaluative, critical and creative understanding, the visualisation of space is a fundamental approach, choosing three-dimensional design and representation as privileged strategies.

The introduction to the representation of the "idea" is established. An architectonic idea, as the project principle, the essence, the argument, the supposition of demand. Moreover, if in this strategy drawing is undoubtedly the method of choice, more accessible and operative, although the model and the three-dimensional visualisation are defined as the most effective strategies. Due to the artisanship involved, but above all, due to the overall spatial view of the whole, the model emerges as the great enabling artifice of a global visualisation. A visualisation that, directly, interacts with space, scale, light, aiming to "see" and consolidating this experience of space, empowering spatiality in the background.

Thus, from the various compositional methods addressed, and from the associated conceptual methodologies, it is possible to construct enriching forms of expression and representation that progressively "form" ideas. Representation thus assumes a demand status, an essential research strategy in the evaluative incursion, in the composition and in the conceptual strategy.

The drawing is the first abstraction of space. The domain of the design used both as an expressive end and a search method undertakes a very particular dimension in the learning process of architecture. If, in the early stages, free-hand drawing[1] is defined as the most straightforward skill, still poorly trained in general and somewhat constrained, architectural design,[2] which is representative of form, function or construction method, also acquires status in representation. Therefore, between the expressive freedom of

[1] Free-hand drawing: "Drawing that is executed without the use of instruments. Only the drawing instrument acts, by the hand's direct action, onto the medium." (Madeira Rodrigues, 1990: 105).

[2] Architectural design: "Representative design of the form, functions or construction method of a building. It is based on projection systems and is subject to norms and rules expressed in national and international standards, NP, ISO and DIN." (Madeira Rodrigues, 1990: 105).

the sketch and the rigorous systematisation of representation, students develop the ability to search, to draw and to embrace the space that is increasingly valuable from this perspective.

Moreover, as such, the maquette/model emerges as a fundamental field of exploration. Maquette defined as:

> Model, reduced in size and on a convenient scale, of a form or object that one intends to construct or make", in contrast to model, of the Latin "modus" – a measurement and that is expressed in three ways: "1, Matrix of anything manufactured or idealised. 2. Synthesis of the external purposes that lead to the appropriation of the type. 3. Instrumental mode in intentionality research. (Madeira Rodrigues, 1990: 178, 182)

Thus, model and maquette correspond respectively to the representation by free-hand drawing and by the architectural design. The first is less constrained and provides more freedom while the other is more standardised and systematic. Yet it is true that in both, the three-dimensionalisation appears as a fundamental approach on which the concept of measurement is projected.

In this process of metric awareness, students establish the relationship between measurement and architecture. A relationship which is defined as the primary relation of space quantification, perceiving through this abstraction, complex elements of understanding such as scale, proportion and rhythm. From these three points, a fundamental interaction between the concept and project is defined.

The scale assumes a privileged importance where the anthropometric relationship is undoubtedly the key element in spatial reading. Basically, the scale is a basic concept in spatial reading and design. On the subject of scale, several approaches are put forward in the course of the students' learning. Initially, the measure of one's own body, the span (vital in the relationship with the object), height (fundamental in relation to space), and step (fundamental in relation to the surrounding). Thus, the internalisation of measure as the first stage of understanding space in its various levels of intervention. As such, reference is made to Le Corbusier's Modulor (1977), as the synthesis of mathematical measurements and to the human scale. Where the abstraction of measurements is associated with a part of the body. A part of the whole. A whole that is articulated in proportional and coordinated parts.

Secondly, understanding proportion as the basic element of the composition. Architectural orders are the most insightful example of this relationship. On the one hand, for establishing the comparative relation between the various parts that constitute them and define them, and on the other hand, for the harmonious relationship they presuppose. Therefore, the concept of proportion emerges, simultaneously as inherent and as spatial reading, establishing the relationship with the third stage of understanding metrics: rhythm.

Spatial rhythms forebode succession, cadence, accentuations, peaks and voids. They are guiding predispositions that lead to conceptual matrices, to regulatory outlines. Mostly, steering the idea towards the spatial argument in the formalisation of space.

In this process of awareness, basic representational exercises are objectively valuable experimental exercises. This can be said of surveys, carried out using measurement, proportion and repetition strategies. Scale, proportion and rhythm are called to interact with systemization processes that, in an operational way, lead the student to represent the space. From this objectification, it becomes possible to act in more abstract fields of conceptualization related to the creative dimension of the project. An exercise of measurement and transposition for drawing as an equal form of understanding, reading and consequent visualisation.

The three-dimensional visualisation of space, therefore, presupposes a projection onto space itself. Through measurement, related to the various elements of scale, proportion and rhythm, we foresee our relationship with the world and objectively create the capacities to project in it. Moreover, as such, the three-dimensionalisation of the object takes on a new sensory dimension beyond sight: "touch". The touch that is so emphasised by Juhani Pallasmaa (2005), as the primary approach regarding the perceptual experience of space, as a counterpoint to the dominant sense of sight. Defending the idea that all senses are extensions of touch, where all experiences are, effectively, ways of touching the world.

"All senses, including sight, can be considered as extensions of the sense of touch, like specialisations of the skin." (Juhani Pallasmaa, 2006: 43)

3 ON THE RELEVANCE OF THE MAQUETTE-CONCEPT

Integrated with this privileged status of three-dimensional spatial visualisation, the subject of the maquette-concept has special relevance. Mainly, by the synthesis of two fundamental representational forms: the drawing and the maquette. The maquette concept is chosen on one side, by the emphasis on the aforementioned three-dimensional strategy and on the other, by the abstraction of the sketch or free drawing.

The sketch that associated with sketching and the first drawings of an idea has always been the lexicon of the architecture project, as a methodology of demand and conceptual research. In this primary process, the destitution of programmatic demands, constraints of place and contextual complexities make room for free formalisations that usually give rise to spatial development, being effectively in the sketch, and often in the re-reading of the original sketch- Examples of this magnificent uniqueness of line are the powerfully expressive sketching approaches of Oscar Niemeyer or Siza Vieira, that in a poetic, synthetic and effective manner "tell" their respective works.

> One day, Carlos Drummond de Andrade wrote in one of his verses: "Oscar draws his building in the sand." And our friend was right. Architecture is born from an initial line, and this can happen even in the sand. (Oscar Niemeyer, 1997: 12)

In this sense, it is from this efficacy of the sketch that, from very early stages it is imprinted on the students' path that the effectiveness of the maquette concepts used to express effectively the postulates of space. The maquette concept, in some way, establishes poetry for space and lays the basis for the foundation of a project methodology that, from the maquette concept, will be shaped by later developments and be increasingly more complex and elaborate.

Being aware of this process is a consistently hard task in architecture teaching. A kind of discovery, in a process of novelty and development of conscientious evaluation, where the real references: of the place or the programme, are replaced by relevant absences, emphasised by the critical search for aesthetic value.

The maquette-concept is the assumed as the construction of the idea...

4 ON SOME EXPERIENCES AND FORMALISATIONS

Framing this genesis of architectural learning and from the pedagogical experience undertaken as a teacher of design disciplines in the architecture course,[3] following are some reflections on experiences and formalisations performed within that scope.[4] These approaches fit into a year of basic education, where contact with the maquette concept is assumed as a fundamental pedagogical strategy. Therefore, based on diverse reasons and defined by the discipline's general syllabus,[5] the practical approaches explore multiple artistic formalisations that fundamentally predict the search for aesthetic value and consequent spatial definition. That is, from diverse causes, with distinct artistic expressiveness: a film, a painter, an exhibition or a poem, students are motivated by these stimuli and encouraged to produce a critical judgment of spatial value that will be materialised in the construction of a maquette concept. Through the election of value, they establish guidelines for the research of the idea, which will be formalised into an object and consequently define space.

The first studies presented (figure 1) stemmed from a very specific programme theme: a film.[6] Thus, in

Figure 1. Three-dimensional paper experiment and three-dimensional iron and ice experiment. Works done by Ana Catarina Costa and Nicolaas Leach related to Architecture Laboratory III subject, in 2010–2011 academic year. Photographs were taken by the author (2010).

both studies, movement and time are compositional strategies present and in some ways, triggered by a design strategy. If in the first study the articulation with the drawing is a compositional artifice, the second one has less conventional and much more experimental embodiments. The first one, presenting a three-dimensional paper experiment, shows an intrinsic relationship with the architectural design, the projection design, taking on the theme of volumetric composition in this abstraction. The second one represents a three-dimensional iron and ice experiment. The contrasted use of the nature of these materials explores the temporality, the fragility and time, values pursued and somehow effectively explored in this maquette concept, wherein transfiguration and temporality are explored beyond form and materialism.

The following experiments (figure 2) used as a pretext for reflection on a series of works by a painter.[7] Therefore, spatial exploration takes advantage of chromatic composition, and above all, of the modularity and repetition as elements of choice. The approach presented in the first image selects the module strategy and the compositional variations as the primary element. The second image, from the same compositional rationale: the module explores

[3] Integrated Master in Architecture at the College of Architecture of the University of Lisbon.

[4] The works presented were carried out by students in the 2nd year of the MSc in Architecture, in the discipline of Architecture Laboratory III - 1st semester, between the academic years 2010/2011 and 2015/2016.

[5] The syllabus content discussed and defined by the different teachers of the respective faculty of each academic year, was coordinated by Prof. Madalena Cunha Matos.

[6] *Aniki Bobó* by Manuel de Oliveira, 1942.

[7] Mario Rita: http://mariorita-home.blogspot.pt.

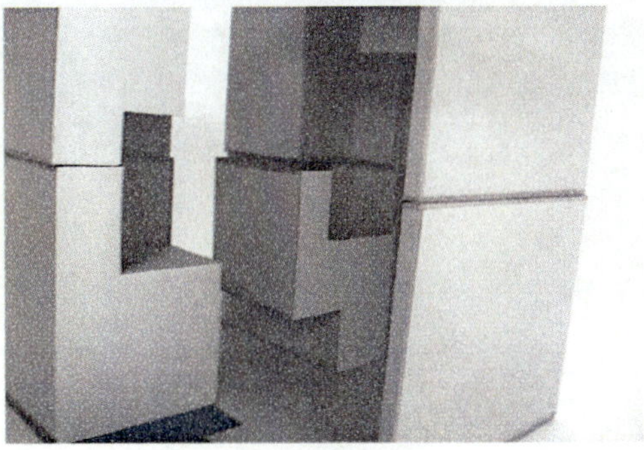

Figure 2. Repetitions and modularities and spatial sequences. Works done by Carolina Rocha and Alexandra Santos, related to Architecture Laboratory III subject, in 2011–2012 academic year. Photographs were taken by the author (2011).

Figure 3. Section and space and Extraction and spatial composition. Works done by Inês Almeida e Paiva and Rita Guerreiro, related to Architecture Laboratory III subject, in 2013–2014 academic year. The photograph was taken by the author (2013).

spatial sequences, deconstructing repetition and systematising composition strategies such as the plane, orthogonal, composition, linear, etc... Conceptual experiments are thus configured – maquettes concept, where space, scale, proportion and rhythm are used, formalising the "idea".

Other approaches (figure 3) came from a very different theme: a monographic exhibition of architecture.[8] As such, in addition to the conceptual reading of the various works presented and the creative discourse of the author, students were led to recreate a critical reading of these same experiments through the three-dimensional models. Because of the aforementioned critical and evaluative reading, diverse and complementary compositional strategies arise. In the first image, section and space are the main themes, added to a very rigid and orthogonal metric, where the section cut, as the primary spatial compositional element, redefines an array of full/empty, presence/absence, in/out relations... In the second image, using a different material (wood), the module and its extraction are explored, defining an interesting three-dimensional arrangement, fundamentally added to a deep anthropometric interpretation, where the human figure and expressiveness are introduced: scale and space.

Another series of experiments followed the same principle as the previous approach, the experience of visiting an architectural exhibition on a renowned architect.[9] In the two cases presented (figure 4), the matrix composition rationale is the main highlight. If in the first image, this composition matrix takes a convergence trajectory on which different dimensional planes are organised, the second image denounces volumetric plasticities organised by more orthogonal rationale that also explores different materialities. Matrix and model are the compositional pretexts and three-dimensional experiments, emphasising well-defined aesthetic and spatial values.

The last experiments presented (figure 5), originated from a poem.[10] The text abstract, and its use as an inspirational theme, thus embodies itself in a three-dimensional experience, which selects the interior/exterior relationship as elements of primary

[8] I am Fujimoto - Futurospective Architecture, Centro Cultural de Belém, Lisbon, 2013.

[9] Rafael Moneo - A theoretical reflection from the profession: archival materials (1961-2013), Centro Cultural de Belém, Lisbon, 2014.

[10] Poem by Herberto Hélder: About a poem.

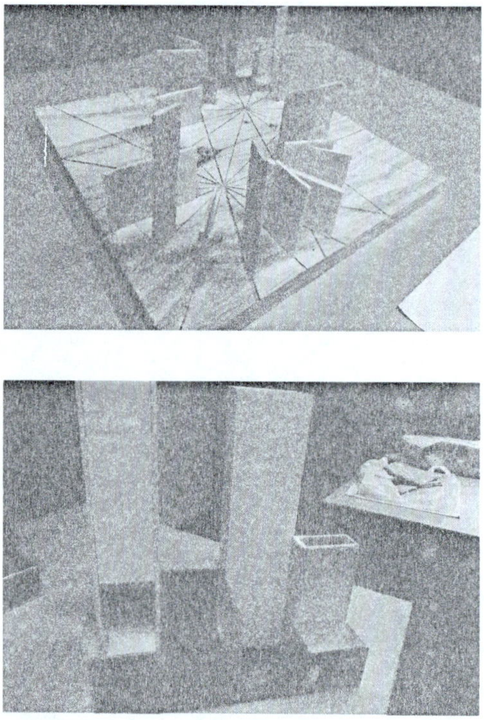

Figure 4. Matrix compositions and volumetric plasticities. Works did by Filipa Machado and Camila Oliveira, related to Architecture Laboratory III subject, in 2014–2015 academic year. The photograph was taken by the author (2014).

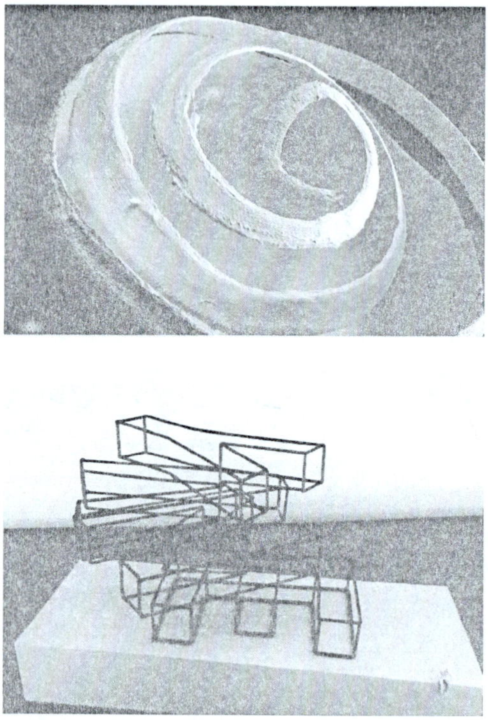

Figure 5. Interior/exterior and limits/surfaces. Works done by Catarina Carvalho and Joaquim Pereira, related to Architecture Laboratory III subject, in 2015–2016 academic year. The photograph was taken by the author (2015).

composition. The first image takes over in the plasticity of the form, the dematerialization of the surface creating a deeply organic space, where a deeply tense relationship between the interior and exterior emerges from surface disassembly (peeling). The last example highlights the question of boundaries and surfaces. In the same way that the relationship between the "outside" and "inside" is shown in the previous example, they are the main pretexts of the spatial definition. As such, the wire "letters" re-create the volume composition and the space it creates, articulating shape, materiality and structure and composing an interesting three-dimensional experience that the maquette-concept offers as a fundamental element of research.

The experiences and formalisations performed, summarise, from various developmental themes, approaches that mirror the potential of this research process. By elucidating highly expressive visual research, they show a coherence between the rationale, the representation and formalisation in object/space. Critical considerations that could obviously be made feasible in later projects, at this stage forming a basis of concepts, arguments or poetry.

5 FINAL CONSIDERATIONS: IDEA, OBJECT, SPACE

The given explanation and the approaches presented here, converge on the objectification of the maquette-concept as a significant strategy in architecture training and teaching. An embryonic strategy that sediments the creative act, the spatial criticism and the foundation of value.

From the foundation of value, confrontation and familiarity with several compositional reasoning, students, from several plastic experiments will configure objects representing "ideas" of space.

Therefore, the "idea" is born, understood as an object of critical thinking. That is the idea as a product of a valuation, a process that in a conscious and informed way establishes choices within ontological, ethical, and particularly aesthetic valuation levels. Like this, the idea as the intellectual creation's foundation is assumed as a starter for discursive development that in the case of architectural creation is then materialised in the project or rather, in the project's representation reflected in the "object", which is the case of the maquette-concept.

An "object" that is in itself assumed as a plastic research. It is this condition between the representative status of the idea and space and an object in itself that effectively adds to the maquette-concept a very significant exploratory field emphasising the representation process, promoting it as a research methodology and intrinsically of project search.

Therefore, "space", or rather the conception of space emerges from compositional logics, metrics, light and materiality. We then progressively achieve space, argued (by value) and objectified (by form). Idea and object articulate and support spatial reasoning. Moreover, the whole complex design process arises from this point onwards. Therefore, this point will be (the embryonic stage), some sort of inaugural moment, from which a path will be foreseen, and will be dismantled, sought, and finally confirmed.

Like this, "idea", "object" and "space", form the existential logic of the "maquette-concept", as a project genesis in architecture teaching. A "space" logic is driven by value considerations significantly expressed in the "idea", by compositional strategies formalised in the "object" and in their conformation, material and metrics. The creative process is consolidated, and space perception, apprehension and design are organised in a process limited by contrasting and complementary ends: rational versus irrational, tangible versus intangible, intuitive versus deductive, abstraction versus reality, etc.

Moreover, the idea of space, represented and inherent to the maquette-concept emerges!

BIBLIOGRAPHICAL REFERENCES

Bayer, Raymond. (1956) *Traité d'Esthétique*. Paris: Librairie Armand Colin.

Heidegger, Martin. (1996) "Bâtir, Habiter, Penser" in *Essais et Conférences*. Paris: Éditions Gallimard (Trad. de *Vorträge und Aufsätze, Pfullingen*, 1954).

Krier, Rob. (1988) *Architectural Composition*. London: A.D. Academy Editions.

Le Corbusier. (1977) *The Modulor – A Harmonious Measure to the Human Scale Universally Applicable to Architecture and Mechanics*. Trad. Peter de Francia and Ana Bostock. 2 ed. London: Boston: Faber and Faber.

Madeira Rodrigues, Maria João et al. (1990) *Vocabulário Técnico e Crítico de Arquitectura*. Lisboa: Quimera.

Niemeyer, Oscar. (1997) *Conversa de Arquitecto*. Porto: Campo das Letras - Editores S.A.

Pallasmaa, Juhani. (2005) *Los Ojos de la Piel - La arquitectura e los sentidos*, Barcelona: Editorial Gustavo Gili, SL, 2006. (Trad. T*he Eyes of the Skin, Architecture and the senses*. Chichester, West Sussex: Wiley-Academy).

Buildings and regenerative thinking: A forward perspective for sustainability

Emilia Conte
DICATECh, Politecnico di Bari, Bari, Italy

ABSTRACT: In this paper, the author reflects on the contribution of sustainable buildings to the built environment and people in it.

The first section introduces the emergence of sustainable buildings in the general scenario of sustainability, highlighting that greening buildings has been the main operative way to practice sustainability in the built environment.

From such way of operating, the design activity has become a key to improve the quality of buildings in terms of environmental friendliness and human well-being. However, in the second section, the paper emphasises that the greening process of buildings has not changed conventional practices of building design, failing to improve the person-environment relationship and give adequate answers to sustainability demands beyond the short-term.

Regenerative thinking is then suggested in order to overcome these limitations. Drawing on the scientific literature on the subject, the third section describes rationales of regenerative thinking as well as its essential characteristic of working in a continuous engagement of natural and human systems for the sustainable development of the built environment in a long-term perspective.

Conclusions highlight that a new generation of sustainable buildings based on regenerative thinking would be able to nourish complex interactions in the built environment for a mutually beneficial development of all components in it and the progress of society.

Keywords: Sustainable Buildings; Built Environment; Regenerative Thinking; Natural/Human Systems.

1 THE EMERGENCE OF SUSTAINABLE BUILDINGS

The introduction of sustainability in development discourse in the late Eighties (Zaccai, 2012) started to inspire also the construction field, sensitising architects and engineers and the other actors in the building design and construction process to consider the many consequences of the built environment on the natural one. The first shared outcome of this awareness dates back to the mid-Nineties when Kibert (1994) gave a seminal definition of the objective of 'sustainable construction'

> The creation and responsible management of a healthy built environment based on resource efficient and ecological principles.

By this definition, an attention emerges for processes, not only for products, and for the healthiness of the built environment, due to a new concern for environmental issues.

However, the implementation in practice of the objective of sustainable construction was neither simple nor obvious. In the search for meaning and visions of sustainable construction, multiple case studies were suggested as 'better practices' (CIB, 1998). These experiences showed two aspects affecting further developments in the sustainable built environment: (i) sustainable cities and buildings as the outcome of the sustainable building process and (ii) lesser or higher quality of life respectively related to the degraded or improved environment.

In the late Nineties, interpretation of sustainability in the construction sector was still an on-going process in which the built environment assumed a prominent position. As Kibert (1999: 3) argues:

> At its very roots, sustainability demands attention to the environmental and social contexts of the built environment. The interconnectedness of the built environment with the community and the health of its citizens can no longer be neglected by architects, engineers, building owners, and developers.

The sustainable development of built environment has been a means for improving the three dimensions of sustainability –environment, economy, society- in interconnection. For this purpose, many actors of the building and construction process are involved and asked to renovate their knowledge, competence and skills in order to deal with the challenge of complexity posed by sustainability (Conte et al., 2009).

However, taking this challenge and implementing practical developments in the built environment is neither easy nor clear. Langston and Ding (2001: XIV) acknowledge that:

> The concept of sustainable development has stimulated the search for construction solutions that do not result in a clash between rising living standards and environmental protection. The essential ingredients for sustainability seem clear, although how they are brought together in practice is undeniably complex.

We have then entered the new millennium with the consciousness of sustainability demand to the built environment and perception of what we should have to do but without a structured complex of practical suggestions and tools on how to operate.

1.1 *Greening as the operative answer*

This situation, on one side, originated by sustainability principles asserted in relation to development of the built environment but on the other side made also by vagueness, fuzziness, and uncertainty of translating theory into practice, largely explains why the 'greening' of buildings became the usual operating practice to meet sustainability requirements for the built environment development.

Greening was the means to reduce or even reverse the negative impacts of the built on the natural systems without compromising health and living conditions of citizens while, at the same time, stimulating (green) economy. It is possible to argue that the habit of focusing on building functions and performances in addition to the attention to health and comfort of inhabitants, naturally slipped architects and engineers and other actors of the construction process towards green buildings.

Buildings have proven to be the privileged scope of the sustainable building process, from which also cities benefit in terms of sustainability; quality of life of inhabitants and citizens results enhanced by greening that aims at improving the environment. Moreover, technological advances and design directions have supported this process by making available more and more choices and solutions for green buildings. Several authors have increasingly offered handbooks and guidelines for green buildings, both professionals and researchers (for example, see Attmann, 2010).

2 DESIGNING THE QUALITY OF BUILT ENVIRONMENT

What emerges from this way of operating is the building design activity as a fundamental key to improving the quality of buildings in terms of environmental friendliness and human well-being, i.e. a way to relate natural and human systems. The focus on design invites to consider its quality and the role it can play as a value generator, particularly in a period in which increasing the efficiency of the construction industry and enhancing the design of the built environment receives much attention in governmental policies given their contribution to sustainable development (Macmillan, 2004); good design produces a valued architecture that impacts on the urban level contributing to its quality. What is a 'good' design is questionable, and sustainability raises new challenges in that direction, but there is no doubt that it is the aspiration of every architectural theory.

2.1 *Postmodernism and sustainability*

Dealing with architectural theories is not the purpose of this paper, but drawing a basic connection between sustainability and Postmodernism can be useful.

Koslowsky observes (1991: 148):

> The critique of Functionalism in architecture converges today with a critique of functionalism in the social and human sciences.

Further, he explains (149):

> The postmodern axiom, "The idea determines form," opposes the modernist axiom, "Function determines form." A person or thing that corresponds to its idea always fulfils more than a function and cannot be reduced to its relational determinations. It represents itself and its idea and contains in its form metaphors that refer to the entirety of the cosmos.

On this basis, Postmodernist thought became a useful cultural substrate for the interpretation of sustainability in the built environment through 'exemplary' buildings, best practices, creativity and innovation of design solutions, that is expressions of design ideas managed to find a practical application, although unfortunately they have often been singular not replicable examples because overly symbolic, experimental or technological.

The postmodernist axiom 'the idea determines form' recalls the importance of the 'form' in architectural discourse and design; however, the architects' concern is mainly related to its physical structure and beauty than to its functioning in relation to people. Lang and Moleski (2010: 13) underline that:

> [a] major criticism was and still is that architectural ideologies are based on an unrealistic model of the impact of built forms on human activities especially social behavior, self-esteem, and, more generally, quality of life. The layout and aesthetic qualities of the world matter but the way the built world functions was misunderstood. It was believed to be deterministic in shaping human behavior. It was assumed that the users of a building would use it sensibly as the architect intended.

Such misunderstanding in the person-environment relationship mainly depends on assumptions used by architects when designing with regard to "how people use and derive meaning from the world around them" (Lang and Moleski, 2010: 12).

Even sustainability with its stimuli and challenges has not been able to provoke a substantial turn in architectural practices improving the person-environment relationship by using more complex models of people and most of all considering people's aims and aspirations while designing buildings and developing the built environment. The greening process of designing, producing and operating buildings has based on the assumption that the sustainable quality of buildings supports a sustainable quality of life of their inhabitants while respecting the environment.

2.2 *Building performances*

Improvements in technology have supported this process by making available more and more efficient solutions to raise increasingly the building performances. Such focus on the performance of buildings has sparked the need to identify minimum threshold values and indicators, thereby directing design activity and choices of architects.

Consequently, high performances of buildings, heavily related to save energy, lessen the use of non-renewable resources and draw upon renewables, have largely equated the sustainable quality of buildings; design a new or an existing building to gain such performances has become the specific goal of the design activity almost disregarding other aspects of sustainability concerns.

At the very end, the greening process of buildings has not changed conventional practices of building design as well as consumption patterns, though in search for higher and higher performances it has increasingly promoted eco-efficiency, shifting from 'pure' green as minimisation of impacts on the natural environment -use of non-renewable resources, raw materials and water, pollution, waste–, to achieve neutral impacts maximising ecological efficiency. In the construction field, this mainly corresponds to the nearly zero- energy buildings, which now represents the new target to reach for building design and operation at least on the short-term (in Europe, see for example, EU Commission Recommendation 2016/1318): the 'new' sustainable practice for the quality of buildings and built environment.

2.3 *The need to overcome the short-term approach*

Nevertheless, even the eco-efficiency process does not show the power to challenge conventional practices of building design, so to give adequate answers to sustainability demands beyond the short-term (Jenkin and Pedersen Zari, 2009). A major question still remains open: are practices deriving from an interpretation of sustainable development as environmental conservation "enough, sufficiently widespread and in time to counteract the continuing path of unsustainable growth?" (Blowers et al, 2012: 4).

Indeed, sometimes practised solutions are falsely sustainable, as Rees (2009: 304) maintains:

> Most mainstream approaches to sustainability today – hybrid cars, green buildings, smart growth, the new urbanism, green consumerism – do not, in fact, address the fundamental problem. Instead, they attempt to reproduce the status quo by other means . . . through technological innovation and greater material and economic efficiency.

Even recognising that greening buildings and cities reflects widespread initiatives that are commendable, it is also indispensable to recognise with Blowers et al. (2012: 4) that sustainable development interpreted as a process of ecological modernisation where

> Environmental conservation [is] achieved through green technologies promoted through market based mechanisms, partnership between business and state and the incorporation of civil society through participation in policy making

presents the substantial flaw of using a short-term approach to the future. Something more is then necessary to overcome current limits and change theory and practice for achieving a true sustainability. As Leman Stefanovic (2012: 18) speculates:

> Important as these initiatives [greening] are, however, there is the risk that something essential is missing. . . . Moving forward will require more than assembling a compendium of such discrete initiatives. More important will be to consider a repositioning of fundamental values, paradigms, and world views that sustain these efforts in the long term.

3 REGENERATIVE THINKING

Our inability to build a sustainable future shows the need for a quantum leap in the approach to sustainability in order to change radically the assumptions used so far (Hes and du Plessis, 2015). This will deeply influence the concept of quality of the built environment and related activities to achieve it, i.e. decision-making, knowledge construction, citizen involvement, building design, technological implementation, and more.

Along with this direction, regenerative development and design offer a promising scenario to reflect theoretically and act at the operational level (Cole, 2012). Introducing the necessary shift from sustainability to regeneration, Reed (2007: 675) asserts:

> It is time to change our mental model to one that better reflects the new understanding of how the universe actually works, and also enables us to design, build and heal with the whole system in mind – a deeply integrated worldview.

Using an integrated worldview, natural and built environments are an interconnected entirety (2007: 675):

> A world activated by complex interrelationships – natural systems, human social

systems, and the conscious forces behind their actions.

The greening process has missed such view by focusing on technical and economic systems to develop the built environment for society, i.e. it has not exploited the potential included in the sustainability approach to build a new integrated vision of quality in the built environment. On the contrary, it is necessary to focus on the primary systems that produce all the others, i.e. the natural and the human systems, the earth and the people (Reed, 2007), considering that "all things are alive and in a process of 'becoming'" and realising that "the purpose of sustainability is sustaining life-enhancing conditions" so to understand "how we as a culture and planet might come to be in conscious, participatory relationship" (2007: 676).

3.1 *Rationales*

Some rationales of regenerative thinking are important in this discourse.

First, there is the shift in worldview that becomes a prerequisite to a (sustainability) paradigm shift. Regenerative thinking bases on the whole systems worldview and develops theories of value and action which allows overcoming limitations of the two more conventional views used so far: the mechanistic view that characterises the prevailing sustainability paradigm and the holistic ecological view that enables to deal with complex living systems. At the same time, regenerative thinking uses tools and insights of both those views, considering them not in opposition but rather evolutionary (du Plessis and Cole, 2011).

Changing values and consequently, actions depend on the cultural background of a place; hence, people and the specific context have high relevance.

In fact, a second significant element in regenerative thinking is related to consider humans strictly interconnected with the natural system in a whole living and evolving system, whose development aims at a mutually beneficial relationship (Jenkin and Pedersen Zari, 2009). More explicitly Pedersen Zari and Jenkin (2010: 3) explain:

> Regenerative development acknowledges humans, as well as their developments, social structures and cultural concerns, as an inherent and indivisible part of ecosystems. Regenerative development investigates how humans can participate in ecosystems through development, to create optimum health for both human communities (physically, psychologically, socially, culturally, and economically) and other living organisms and systems.

A third critical element concerns the 'role' of humans; it is different from any other approach to development in that humans contribute to "repair the capacity of ecosystems to function at optimum levels without ongoing human intervention" (Jenkin and Pedersen Zari, 2009: 8) whereas other approaches require active human management to improve ecosystem health (Jenkin and Pedersen Zari, 2009). As Mang and Reed (2012: 26) argue, greening or eco-efficiency processes result insufficient since they miss

> The real potential that arises out of the human presence on this planet ...It is not enough to aspire to mitigate the effects of human activity – people need to take their place again as a part of nature.

From this, a fourth compelling element of regenerative thinking emerges strictly related to practice: the co-evolutive character of regenerative development and design 2012: 26) which

> Means the reconnection of human aspirations and activities with the evolution of natural systems – essentially co-evolution. It means shifting human communities and economic activities back into alignment with life processes.... This is not preservation of an ecosystem, nor is it restoration. Instead, it is the continual evolution of culture in relationship to the evolution of life.

Such meaning recalls the importance of both the consciousness of development processes and the participatory relationship cultivated in the specific context where development is implemented, which is "the only way to sustain sustainability" (Reed, 2007: 677).

The fifth element of relevance in regenerative thinking is, in fact, the significance of 'place'. In a whole systems perspective, whether a place is more natural or more constructed, its character or essence derives from its development over time influenced by the relationships between all components constituting it, including not only physical parts but also human communities and cultural experiences (Pedersen Zari and Jenkin, 2010).

Through the 'story of place,' it is possible to interpret and understand the complex relationships between natural and human systems in a specific context and how it has evolved over time and might evolve for mutual beneficial life-enhancing conditions. The story of the place is the context of a comparison of nature and human stakeholders' aspirations, a way to learn, understand and find common values on which to base the search for win-win solutions (du Plessis and Cole, 2011). The story of the place as the context of development works as a metaphor for communicating ideas and is, therefore, a powerful tool to engage quickly lay people who can understand their role in an ecosystem (Reed, 2007).

Build the story of place means structuring knowledge of the place and its history, collecting the many stories and expectations of old and new inhabitants of that place, nurturing the ability of 'inventing' design solutions for a built environment co-evolving with nature in a specific context; these are all factors necessary for making decisions and implementing actions able to translate local potentialities into real capacities for a regenerative development. By regenerative development and design, in fact, we build capacity, not things (Reed, 2007).

3.2 *A forward perspective*

Regenerative thinking offers a suitable perspective to overcome the reduction of building and urban complexity to eco-efficiency or effectiveness. Following Mang and Reed (2012: 26), it is important to underline that:

> Regeneration depends on a developmental process that improves the value of the whole, works to take systems to the next level, evokes a set of higher-order aims and develops the capacities to pursue them.

Then, it is a never-ending process aimed at engaging rather than providing a set of outcomes; for these characteristics it shows the potential to enable a "fundamental rethinking of architectural and urban design" (Pedersen Zari and Jenkin, 2010: 4) in order to shift from

> Human-only oriented design approaches which have a focus on improving efficiencies, to a more inclusive system approach to designing the built environment, which focuses on maximizing mutually beneficial interactions between human and non-human elements.

Using a whole systems worldview, considering the complex network of interactions in the built environment and being a developmental process, regeneration goes beyond a temporal perspective of short-term to "likely ensure a continuous suitable environment" (Pedersen Zari and Jenkin, 2010: 8) for all.

On the basis of these premises, the quality of built environment becomes the integrated quality of life of the whole natural-human system whose peculiar features in the regenerative perspective are: adaptability and flexibility, self-organisation, non-predictable statuses, non-linear dynamism; the system is constantly in transition and its features are determined by properties of its components and their interconnections as a whole, questioning the effectiveness of optimising the behaviour of a single component (du Plessis and Cole, 2011: 438).

Effective regeneration concerns the whole system and all its components are engaged, be they (Reed, 2007: 678)

> Our home community, a corporate campus, a small lot, or a building. When we start from a whole systems understanding, any of these entities is an entry point into the whole system. Each is an integral part of a living system and a key role can be found for anyone and any system.

3.3 *The building as a complex open system*

Therefore, when we design buildings, we cannot consider them as individual objects, rather we must design them (Jenkin and Pedersen Zari, 2009: 17)

> As parts of larger systems allowing complex and mutually beneficial interactions between the built environment, the living world and human.

A shift is necessary from the building perceived as 'complicated' system, the artefact that we produce mainly focusing on performances, to a 'complex' system. As du Plessis and Cole (2011: 438) suggest:

> Complexity is introduced at the building level when the building is no longer perceived as merely a physical artefact, but rather as a process that involves interactions between natural laws, biophysical systems and the actions of their human occupants.

Consequently, any time we design a new building or the renovation of an existing one, we are also designing the regeneration of the built environment in which that building is.

The quality of a building is no longer related to its ecological, health and well-being performances as an individual object, a complicated 'closed' system, but rather depends on the mutually beneficial interactions that it is able to trigger and participate in the built environment, i.e. the building becomes a complex open system, an agent among others in the built environment (Conte and Monno, 2016).

These interactions are not only physical but also immaterial, as ethical and behavioural, and determine the quality of space and life in it; they are culturally dependent and then specific of a place though evolving over time. Hence, when we design building patterns in an urban environment and decide distances between buildings, for example, we are not only defining physical aspects as lighting, shadowing, thermal gaining, etc., but also interpreting systems of values and behaviours more or less close to past, present and future perceptions and attitudes.

Considering each agent as an integral part of the whole living system that is the built environment, regenerative thinking allows recognising the multi-dimensional configuration of a place, usually ignored by both conventional and green design practices (Benne and Mang, 2015), thus stimulating the reconstruction of knowledge related to the built environment and enhancing the design activity. In order to recognise and develop the regenerative potential of a place, narratives of it become essential; so the regenerative approach requires a continuous involvement of inhabitants and all the actors of building and development processes even beyond the implementations: in this way the regenerative potential of a place can become a real capacity expressed even without human intervention.

4 CONCLUSIONS

How can then buildings contribute to the sustainable quality of the built environment in the regenerative perspective? The most likely answer is: through relationships and interactions between other agents – buildings, infrastructures, natural elements and forms of life, inhabitants and even passers-by, systems of norms and values, aspirations and expectations.

Buildings shall participate to imagine and realise adaptability and flexibility of the built environment in which they are, to help self-organisation of the whole system, to accompany the constant transition from a status to another non-predictable status enabling diverse non-linear dynamics of the built environment over time. In this way, buildings as an 'entry point' among others in the whole living system, will play their role in nourishing complex interactions in the built environment for a mutually beneficial development of all components in it and the progress of society.

BIBLIOGRAPHICAL REFERENCES

Attmann, O. (2010). *Green Architecture: Advanced Technologies and Materials*. New York: MacGraw-Hill Professional.

Benne, B. and Mang, P. (2015). Working regeneratively across scales – insights from nature applied to the built environment. *Journal of Cleaner Production*. 109, pp. 42–52.

Blowers, A., Boersema, J. and Martin, A. (2012). Is sustainable development sustainable? *Journal of Integrative Environmental Sciences*. 9(1), pp. 1–8.

CIB report 225 (1998) *Sustainable Development and the Future of Construction*. CIB Working Commission 82 "Future Studies in Construction". [Accessed 8 March 2017]. Available from http://www. sustainable-design.ie/sustain/cib225.htm

Cole, R.J. (2012). Regenerative design and development: current theory and practice. *Building Research and Information*. 40(1), pp. 1–6.

Conte, E., Kersten, G.E. and Vahidov R. (2009) Architects decision station and its integration with project-driven supply chains. In Zarli, A. and Scherer, R. eds. *e-Work and e-Business in Architecture, Engineering and Construction*. London: CRC Press, Balkema, Taylor & Francis, pp. 317–325.

Conte, E. and Monno, V. (2016). The regenerative approach to model an integrated urban-building evaluation method. *International Journal of Sustainable Built Environment*. 5(1), pp. 12–22.

du Plessis, C. and Cole, R.J. (2011). Motivating change: shifting the paradigm. *Building Research and Information*. 39(5), pp. 436–449.

EU (2016) Commission Recommendation 1318 on guidelines for the promotion of nearly zero-energy buildings and best practices to ensure that, by 2020, all new buildings are nearly zero-energy buildings. *Official Journal of the European Union*, L 208/46, 2/8/2016.

Hes, D. and du Plessis, C. (2015). *Designing for hope: pathways to regenerative sustainability*. Abingdon and New York: Routledge. Chapt. 1 Why do we need regenerative sustainability? pp. 11–22.

Jenkin, S. and Pedersen Zari, M. (2009). *Rethinking our built environments: Towards a sustainable future*. Wellington, New Zealand: Ministry for the Environment, New Zealand Government, Research document, pp. 55. [Accessed 8 March 2017]. Available from https://www.researchgate.net/publication/261476724_Rethinking_our_built_environments_Towards_a_sustainable_future

Kibert, C.J. (1994). Final Session of First International Conference of CIB TG 16 on Sustainable Construction, Tampa, Florida, 6–9 November.

Kibert, C.J. (1999). *Reshaping the Built Environment. Ecology, Ethics, and Economics*. Washington DC: Island Press, Introduction, pp. 1–6.

Koslowsky, P. (1991). The (De)Construction Sites of Postmodern. In Hoesterey, I. ed. Seitgeist in *Babel: The Postmodernist Controversy*. Bloomington and Indianapolis: Indiana University Press, pp. 142–154.

Lang, J. and Moleski, W. (2010). *Functionalism revisited. Architectural Theory and Practice and the Behavioral Sciences*. Farnham, Surrey, England: Ashgate Publishing, Ltd., The Inheritance: Architectural Practice and Architectural Theory Today, pp. 3–26.

Langston, C.A. and Ding G.K.C. (2001). *Sustainable practices in the built environment*. Oxford: Butterworth-Heinemann, 2nd edition, Introduction, pp. xiii–xvi.

Leman Stefanovic, I. (2012). In Search of the Natural City. In Leman Stefanovic, I. and Scharper, S.B. eds. *The Natural City. Re-Envisioning the Built Environment*. Toronto, Canada: University of Toronto Press, pp. 11–35.

Macmillan, S. (2004). *Quality and Value in the Built Environment*. London and New York: Taylor&Francis, Chapt. 1 Design as a value generator, pp. 1–12.

Mang, P. and Reed, B. (2012). Designing from place: a regenerative framework and methodology. *Building Research and Information*. 40(1), pp. 23–38.

Pedersen Zari, M. and Jenkin, S. (2010). Redefining cutting edge sustainable design: From eco-efficiency to regenerative development. Paper presented at Sustainable Building Conference. Wellington, New Zealand, 26–28 May, pp. 10. [Accessed 8 March 2017]. Available from http://www.branz.co.nz/cms_show_download.php?id=2ab15295f01d3078189c96855a6fa055226d58b4

Reed, B. (2007). Shifting from 'sustainability' to regeneration. *Building Research and Information*. 35(6), pp. 674–680.

Rees, W.E. (2009). The ecological crisis and self-delusion: implications for the building sector. *Building Research and Information*. 37(3), pp. 300–311.

Zaccai, E. (2012). Over two decades in pursuit of sustainable development: Influence, transformations, limits. *Environmental Development*, 39(4), pp. 79–90.

Progress, energy and architecture – The building as a power cell

Francisco Oliveira
University of Lisbon, Faculty of Architecture, CIAUD, Lisbon, Portugal

ABSTRACT: Nicola Tesla's research induces an intrinsic capacity to work on the materiality of things to produce clean and free energy. Since the early days of the energy revolution, we have witnessed the suppression of ideas, to maintain inefficient models of production, consumption, and distribution of energy.

The scientific community has not been able to conclude about the real causes of the climate changes. We must understand that all these phenomena have substantially altered the context in which we live and at what cost.

It will be up to architecture to allow the promotion of free access to energy, just as we obtain the air we breathe, both ubiquitous and of free fruition.

Since the beginning of the twentieth century, the electromagnetic potential of the atmosphere has been known and has been intensely worked on by multiple researchers, which resulted in several discoveries and inventions that enable the current lifestyle.

The non-return point, where we may find ourselves, is forcing the emergence of new technologies that will change the socio-economic model of the planet.

The liberation of energy contingency would be protected, assuming the architecture itself as a self-sufficient energy cell, where the components of energy capture and consumption, are ubiquitously integrated into the materials and construction systems.

Keywords: Free energy; Nicola Tesla; Climate Changes; architectural engineering.

1 INTRODUCTION

Over the ages of civilizational evolution, Architecture, as the art of modelling space, retained an intrinsic capacity of modelling societies and collective life.

The real correlations between design, engineering, and biology, about the functioning of the biological component of the living beings, introduces an operational capability, opening doors, to a quasi-divine understanding of man's actions on territory allowing both parts to coexist in harmony. The evolution of knowledge, in these areas, opens opportunities for new creative chimaeras, in which architecture cannot be unheard.

Over the ages of civilizational evolution, Architecture, as the art of modelling space, retained an intrinsic capacity of modelling societies and collective life.

The real correlations between design, engineering, and biology, about the functioning of the biological component of the living beings, introduces an operational capability, opening doors, to a quasi-divine understanding of man's actions on territory allowing both parts to coexist in harmony. The evolution of knowledge, in these areas, opens opportunities for new creative chimaeras, in which architecture cannot be unheard.

The growth of energy consumption stems from the growing need to support information systems, that ubiquity circulates through space and things. This ambition has led to an extreme phenomenon recognise as the "climate change", which forces us to develop adaptive processes, to be able to endure, on an environment always stepping away from what we are "programmed" to live in.

Biology is no longer a static science, aimed at observing and recording what we consider alive and natural. Today, we deal with the fundamental blocks of life, reinforcing the idea, that the biological design will be a widespread reality, and that the incorporation of communication systems between beings, and a world based on complex communication networks, storage, and processed information, will be the basis of a new post-human era. The chemical computing, based on the processing and storage of information system, allied with the power of quantum computing, may ultimately change the technological basis in which we have previously operated (Gohd, 2017).

This reality led us to the design of context, and in the limit, it will guide us to the amazing idea that architecture can also be reinvented, in the light of new materials, structural models and hybrid shapes. Where the physical and the virtual, fuse themselves in the process of complementary inheritance, filling the spaces with additional dimensions, where the physical, is only an interface to another form of dwelling.

Enhancing longevity and releasing artefacts from complex operating systems that tend to confuse the interaction of man with the actual leading technologies. The promise of ubiquity will be to ensure a high degree of technological reliability, where the current complexity of computers will be replaced by the simplicity of distributed operations.

In this context, it is estimated that energy consumption needs will continue to increase. Not only because of environmental comfort needs, but by the increasing incorporation of dynamic empowerment of built structures, and the management and transportation of information, required for its performance and functioning.

Today, we have hybrid materials with memory, with the capacity for self-regeneration, self-cooling[1], among many others that we observe in the structures that we recognise as natural (Wang et al., 2017) These materials will progressively incorporate other mimetic capacities, taken from nature and that inevitably point us the possibility of an integrated production of energy, through the conversion of the energetic potential existing in the planet's atmosphere. Thus, making possible one of the greatest paradigm shift in the recent history of humankind.

The pathway is long and has been crossed by many researchers and inventors around the world, proving that it is really the possibility to collect and use the electromagnetic energy present in the atmosphere. This idea will give shape to the pioneering works of one of the most prominent and, eventually, one of the most forgotten men of all time – Nicola Tesla (Trinkaus, 1988). Along with names such as Leonardo DaVinci, Isaac Newton, and Albert Einstein, Nicola Tesla risked exploring the unknown, inventing a future that is now our present in transformation.

2 TESLA AND QUEST OF PROGRESS

The work that Nicola Tesla developed during his life (1856–1943), in the obsessive search for an understanding of nature, in the most fundamental of its essence, led to the invention of alternating current, polyphase motors, generators, reversible magnetic field, radio transmission, television broadcast, remote control, and robotics. An inventor, on whose patents were mostly based the twentieth century, and still allows structural progress in science and technology today. He worked alone, in search of explanations for the cosmic processes for decades, pursuing the desire to combine materially and theoretically all his practical discoveries.

To transpose, for the architectural thought, this method of approaching reality is one of the fascinating challenges of a future that we desire to be more intelligent and harmonious. Activating multiple areas of knowledge for this art of fusion that once again may be assumed as a civilizational synthesis. The growing trend towards the incorporation of integrated management systems of the creation process, with extreme virtualization and simulation capabilities, allows looking at the phenomenon of construction and architectural establishment in an entirely new way.

The trail drawn to the evolution of communication, materials, and technologies, allows us to imagine that, in a short period, it will be possible for any human being, through embedded communication systems, to interact with another being or artefact without physical, technological, or cultural obstruction. There is, of course, different ways of looking at the production of design and architecture in ways never considered before, where the traditional means of presentation of products becomes something where the design of forms becomes the design of ideas.

This way, it transfers its formalisation to something dependent only on the user's desires, in a search for an interactive identity between the consumer and the product. The dematerialization of the physical appearance of objects is something that potentiates profound changes in the way we relate to the physical world and consumption in general.

Conceiving the future is the major task for architects. Those who are allowed to go beyond what is established are entrusted with the mission of making a better, more just and adapted world. The concerns about climate security today, in the sense of the uncertainty generated by the possible inadequacy of humankind to an accelerating environmental change, are exacerbated by the exponential increase of the world population occurred in the last decades. The instability arising from the scarcity of resources and its unproportioned territorial distribution contributes to the need to consider new values in architectural and urban thinking.

In this field, the original works of Nicola Tesla (Tesla, 1901)induce an intrinsic capacity to work on the materiality of buildings for founding new systems, which can promote the production of clean and free energy (Trinkaus, 1988). These have not only not yet been sufficiently studied, and, even less, potentiated as generators of architectural and city transformations.[2]

Energy self-sufficiency, using clean and renewable sources, does justice to a secular desire for freedom. The current state of knowledge allows us to believe that it is possible to balance, locally, the consumption and energy production needs. The problem lies in the irrational interference, from political and economic factors associated with the industry and trade, which encourages "obsolete" technology, which reinforces world economic model and underlies the so-called "globalisation".

---

[1] Reference to the research of the Tangible Media Group at MIT Media Lab, where the exploration of new frontiers led to the creation of a new biological tissue "BioLogic" that incorporates genetically modified bacteria that mixed with inert fibres allow to produce a fabric programmed to respond to complementary needs to the Biological performance of those who wear them.

[2] *Planned obsolescence or built-in obsolescence* in industrial design and economics is a policy of planning or designing a product with an artificially limited useful life, so it will become obsolete, that is, unfashionable or no longer functional after a certain period.

The planned obsolescence[3] of technologies is, as we know, one of the most subtle systems to support the global consumption model. Since the early days of the energy revolution, brought by the advent of electricity distribution, we have witnessed the suppression of ideas and filing of patents, to maintain inefficient models of production, consumption, and distribution of energy, but very efficient in generating the revenues that feed into powerful industrial and economic conglomerates.

Elon Musk[4], the media-conscious entrepreneur, founder of Tesla Motors, a company whose name pays tribute to the genius of Nicola Tesla[5], runs one of the most innovative businesses in the field of clean energy. His activities associated with the construction of vehicles and systems that produce and electricity storage of solar origin points us a driveway for shifting from the era of fossil fuels to an era of sustainable energies.

At present, human activity adds 33 Gigatons per year of CO2 to the natural carbon cycle of the planet (The State of Greenhouse Gases in the Atmosphere Based on Global Observations through 2015, 2016) This fact justifies the inevitability of changes in the context, which manifests daily in many places on the globe, in the form of extreme events, always associated with phenomena of climatic origin.

Musk realised that the potential of photovoltaic solar energy is tremendous. From the available technology, and considering only the production of electricity of solar origin, it would only be necessary 496,805 Km^2 of solar panels, to feed without any emission of CO2, the total needs of energy consumption of the planet in 2030[6]. This area corresponds to only 9.74% of the globe's surface. In addition, if we consider the possibility of diluting this area into the existing built surfaces, it is easy to understand that it would not be necessary to build a single square meter of dedicated infrastructure, simply by merging the corresponding systems into the roofs and façades of existing buildings. Moreover, if we want to promote areas of high solar potential, as are some of the desert areas of the planet, it becomes clear that the possibilities in this field are remarkable.

The idea that energy is infinite but difficult to obtain has generated in public opinion, a predisposition to accept the inevitability of having to pay too much for its availability. In the case of electricity, the production profile, given the clear increase in demand, continues to be based on very inefficient and incredibly impactful in the environment. The transition to a model, purged of intermediaries between energy sources and consumers, would generate a change in the way society perceives it and could release extensive financial resources.

To speak about progress in architecture implies a large range of concepts that are difficult to recognise because architecture is omnipresent in almost all actions of man. The human species is increasing in a way never observed before. Moreover, it is expected, according to projections of the United Nations (UN), that in the coming decades we will be over 8.5 billion inhabitants and it is projected to exceed 11 Billion at the end of this century[7], multiplying by seven the inhabitants of the planet since the beginning of the 20th century. The urban population worldwide has grown even more rapidly, from 746 million in 1950 to 3.9 billion in 2014 and is expected to exceed 6 billion by 2045[8]. Today, it is already clear that this phenomenon will generate movements of rebalancing with unpredictable consequences in the current models of governance.

The survival of societies in the future, even more so than today, will depend on access to energy in all its forms of presentation. Beyond the availability of food resources that guarantee energy for the biological survival of populations, we will have to find ways to support food production systems with other energy models. Because the current ones will end up destroying the atmosphere and the environment, possible making, in a short period, the planet unsuitable for the existence of human life.

This (un)consciousness assaults the media today in the form of big buzzwords, acronyms, terms and phrases that are part of the global marketing of consciousness numbness. Climate change, oceans rising and polar melting are presented as the visible face of a potentially lethal phenomenon for humankind, creating a sense of conditioning, sympathetic to the control of the masses. The scientific community has not been able to conclude about the real causes of these phenomena that have been proven to occur on the planet in other eras, supposedly without human interference, but which have proven to lead to the extinction of the dominant species in due course.

More important than converging on the causes, we must understand that all these phenomena together have substantially altered the context in which we live

[3] *Planned obsolescence or built-in obsolescence* in industrial design and economics is a policy of planning or designing a product with an artificially limited useful life, so it will become obsolete, that is, unfashionable or no longer functional after a certain period.

[4] Elon Musk is the CEO and product architect of Tesla Motors and de CEO/CTO of Space Exploration Technologies (SpaceX).

[5] It is not widely known that Nicola Tesla built and tested an electric automobile in the early 1930s. Although the exact technical details are now obscured by time, it was known that a special electrical unit containing a vacuum tube array, which appeared to be a radiant energy converter, powered a large electric motor.

[6] LAGI Infographics. Areas are calculated based on an assumption of 20% operating efficiency of collection devices and a 2000-hour per year natural solar input of 1000 watts per square meter striking the surface. It is based on the US Department of Energy statistics of worldwide Btu consumption with data from the *International Energy Agency.*

[7] Dados da UNO - http://www.unric.org

[8] In 1900, at the beginning of the 20th century, the world population was 1.65 billion people.

and the question today is whether we can persist at the same and what cost.

The answer has to be global because the problem is global. From an abstract point of view, the conviction emerges that the final balance of the so-called "laws of nature" will tend to find harmony on stability, in an endless quest for balance. In this sense, we are convinced that this will be the case and that this balance will be reached. The question is whether we participate in it, or whether we will be eliminated as a factor of entropic destabilisation.

The supreme disease of humankind is the breaking of the consciousness of belonging to a larger system. Political-religious dogmas contribute to the strengthening of barriers incompatible with the current state of widespread access to information and will undoubtedly lead to an insurrection of the masses, in a way never known before. The wars of the past, commanded at the service of the ambition of few, are no longer even an option; such is the state of its inconsequence, due to the change of the geopolitical contexts. Today it no longer makes sense to dominate an energy source as a model of authority statement, and it is already present that we are moving towards an era where energy will be free, clean and inexhaustible. Sustaining the ultimate leap of progress, and eventually, grating the future of humankind.

It will also be up to architecture to build this future, consolidating models of the design of buildings, cities, transport, and logistics of resources that allow the promotion of free access to energy, just as we obtain the air we breathe, both ubiquitous and of free fruition.

The world economic model will have to change. Gradually, we are witnessing the obsolescence of hydrocarbons and the growth of the solar, hydroelectric and wind resources, duly accompanied by a revolution in the transportation models as a support for a new industrial era. The construction industry and architecture already incorporate identical values in their ambitions, and they are creating procedures for new ways to approach materials and constructive systems. The liberation of the energy burden, releasing the weight on the ecological footprint, will reduce it to the point where the balance between existing resources and their widespread use can be achieved.

On the other hand, governments will have to find different ways of financing public services, heavily taxing all those who resist moving to clean systems without the use of non-renewable energy sources, and charging taxes only on the temporary use of reusable raw materials. In this scenario, States would rent the right to use reusable natural resources, controlling their life cycle and their environmental impact. Thus, the sense of the state would be fulfilled, and the management of the collective patrimony of nations guaranteed.

3 INDEPENDENT AND CLEAN ENERGY DEMAND

From the beginnings of humankind, the control of energy has been building the power of civilisations. For centuries, access to energy sources has conditioned progress, and their diversity has altered their impact on the environment. The most obvious consequence of this phenomenon is the rate of carbon dioxide particles present in the atmosphere and oceans, and the evolution of these aspects now shows the rise of unique environmental factors.

Apparently, due to human activity, the presence of CO2 in the atmosphere was in 2015, already 60 times higher than in 1975. According to the latest studies of the International Energy Agency (IEA) were removed from the ground and have been released into the atmosphere since then, 935 billion tonnes of CO2 equivalent to 476,850,000,000,000 m^3, 4.4%[9] of the total volume of our planet. (*Energy efficiency indicators*, 2016)

This sudden change in the atmospheric conditions will certainly have consequences throughout the living world, with which we will have to deal for many years. The possible acceleration of natural processes, or the recognised cycles of increasing of this chemical element in the air is a useless discussion when it is verified that there is no record of such a significant increase of these particles, nor in such a short period. This fact means that we are facing something, which the consequences of it may still surprise us, and its long-term impact is unknown.

It becomes tough to accept that in a context in which science has reached a level of control and understanding of matter beyond the atomic scale, with surprising discoveries, validation the existence of subatomic particles essential to understanding the functioning of the universe. That the problem of widespread access to a clean and renewable energy type, and without the need for storage, has not yet been solved. Moreover, there is the awareness, for example, that the amount of potential energy present in the atmosphere due to the earth's electromagnetic field is, for this purpose, inexhaustible.

The intermittent manifestations of the tremendous force of this energy, made possible by the electric discharges during the storms, are a slight and traditional evidence of this energy potential. Since the beginning of the twentieth century, the electromagnetic potential of the atmosphere has been known and has been intensely worked on by Nicola Tesla, from which resulted in most of the discoveries and inventions that enable the current lifestyle.

However, the most revolutionary of all them, which refers to the prototype of a system for the atmospheric capture of clean energy and the medium to transmit it, without the use of wires, was abruptly threatened, through actions of sabotage and disposal of equipment. (Trinkaus, 1988) Resulting in the destruction of Tesla's main laboratory in an incomprehensible fire. Historically, the interference by one of the major barons of

[9] The molecular weight of the carbon dioxide is 44 grams per mole. In one ton of CO2 there are 22,727 moles and one mole equals 22.4 litres. Thus, one ton of CO2 will occupy about 510 m^3. Which would correspond to a volume of a cube of 8 meters of side.

the emerging steel and oil industry, JP Morgan, prevented it from being entirely replaced by a new model of absolute sustainability, just at the beginning of the current energy era. At that time, it was apparently considered premature and inopportune from the economic and financial point of view[10].

In fact, further research developed in this area continues, in many cases, to be suppressed and discredited, ensuring the maintenance of the dominant system. The environmental non-return point, where we can find ourselves, requires a change in this process, forcing the emergence of technologies that make it unnecessary to use polluting energy sources and with this, completely change the socio-economic model of the planet.

Most systems of the so-called free electromagnetic energy[11], in fact, do not create energy. (King, 1994) Rather take advantage of natural sources of electromagnetic energy, present in the atmosphere and which, through various forms of induction, manifest themselves in the form of electricity, continuing to maintain intact the coherence of the fundamental laws of thermodynamics[12]. Unlike solar or wind power systems, these systems require little or no energy storage capacity, since they are based on a model where the energy is "harvested" according to the needs of immediate consumption and only when necessary.

Solar energy, from direct radiation, has the obvious disadvantage of being dependent on the position of the sun. Which clouds, trees, buildings, or the Earth itself often block, or has its intensity reduced by mist or smog, or even by the atmosphere, thicker at lower altitudes or higher latitudes. Similarly, wind speed is variable and often non-existent. Neither solar energy nor wind power is suitable for direct feeding of vehicles or buildings, due to the need for implicit storage (batteries) and the uncertainty of their availability. Systems and devices based on the principles of this new energy, properly designed, developed and integrated, would not have such limitations, instantaneously providing energy, without territorial limits and according to the necessity of the instant and without recourse to the required storage and distribution infrastructures (Trinkaus, 1988).

Today we are witnessing the growing demand for alternatives to the current status quo. It is curious that, now, and with the astonishing increase in energy demand, the price of oil is at such low levels in the face of more than proven scarcity of that raw material. This fact is due, among other things, to the desperate attempt by large producers to keep this type of energy competitive. If the price rises, demand would dramatically fall, as there is the awareness that there are currently energy sources that can replace it[13]. The changes in the taxation models and the penalization of CO2 production through the burning of fuels will have to accompany this transformation because the hidden subsidisation of the high polluters has helped the persistence of this environmentally unsustainable energy model[14].

The long-term effects of this paradigm shift would be extensive and profound. The amount of energy required from conventional energy producers (coal mining companies, oil companies, and producing nations) would drop to zero (King, 1994). Energy resources as coal, uranium, oil, and gas, would also remain on the ground. Prices for conventional energy supplies would fall to near zero, while the price of direct energy capture systems and equipment would decrease in price as demand increases and their production will go global. The production costs of products that require large amounts of energy to be produced would fall along with final prices for consumers.

That is, energy could be progressively removed from the formulation of the production costs of any good or service. The fossil fuel exporting nations, whose economy is based on this economic model, would be profoundly affected by the abrupt drop in demand, with unpredictable consequences for global geopolitical stability, which allow us to understand the reasons for the permanent postponement of the of paradigm shift momentum.

The gradual transition to a halfway energy model, using a multitude of alternative sources of production, is underway and is part of a subtle evolution. This fact gives the society time to adapt to new players. The speed of the changes will be imposed by nature's response to modifications already introduced in the system. The possible climatic disruption may inflict more damage than all the wars and conflicts that have been fought together, and this awareness is now part of an alternative discourse accepted by many of the world's actors and decision-makers.

In this context, the role of architecture will always be to shape original models for the use of resources, to create a habitat beneficial to human happiness. Thus, it will be imperative to transpose this to the multiple

[10] It is significant to note that Nicola Tesla began to fall into disfavour with the establishment of the time, when he proposed the development of the direct transmission of electricity based on his Colorado Springs experiments of 1899.

[11] The concept of electromagnetic "free energy" should not be considered to be the same as natural free energy sources such as solar, wind, hydro or geothermal energies, because these new EM machines usually require an input energy in order to obtain an increased energy output, which the natural sources do not require.

[12] Reference to the works of Nicola Tesla, Bruce DePalma, Don Smith, D.A. Kely among many others who since 1890 have sought to break free energy.

[13] In this regard, natural gas is emerging as an immediate alternative to petroleum, especially since its reserves are so large that they can meet the needs of the current fuel civilization for at least another 250 years.

[14] By way of analogy, Elon Musk, at a conference in 2016, at the Sorbonne, refers to this as equivalent to non-payment of charges for the collection and processing of domestic waste in our cities. It is estimated that this hidden subsidy of CO2 reaches an annual value of 5.3 trillion US $ to the oil industry, according to IMF data.

scales and contexts of intervention, potentiating different approaches, considering the local cultures and the available material resources. Thus, the liberation of energy contingency would be enshrined, assuming the architecture itself as a self-sufficient energy cell, where the components of energy capture and consumption, are ubiquitously integrated into the materials and construction systems, from the simplest house to the most complex industrial building.

This fact would be the ultimate progress, balancing opportunities and eliminating the artificial barriers to development, imposed by the current global energy model.

4 ARE WE READY?

At one time, Nicola Tesla was banished from the best technical school in Yugoslavia because he believed he could build a generator that would extract electricity directly from a stream of water in a continuous and perpetual way (Trinkaus, 1988). His teacher then called him and said that the "Perpetual Motion" was a scientific impossibility! Years later, in 1895, Tesla under the sponsorship of George Westinghouse built the first hydroelectric plant in Niagara Falls in the US and began electrifying the planet. Probably the largest civilizational transformation, since the invention of the wheel (Trinkaus, 1988).

This fact and many other claims of scepticism were, in their turn, uttered by the greatest authorities of each time, fortunately without ever impeding the ambition of those who wish to see beyond the unknown. It seems evident that science is in permanent maturation and that it is the dream and creativity of some that allow us to break the barriers of their understanding. Nature is fixed, and it is our understanding of this that unfolds the horizons. Sometimes this knowledge has been painful and often has provided great disruption in the instituted system. The fear of change is not an adjusted excuse for not pursuing actively the development of our capabilities as human beings, as individuals or as groups.

Architecture and the art in general, have always embodied the spirit of the invention and the breakdown of established barriers. Today, more than ever, we are experiencing times of fusion and sharing of resources and strategies. The future of humankind depends on objective responses to inescapable phenomena of a scale never observed before. The challenge of creativity emerges once again, as hope to overcome obstacles and unexplained prejudices.

So, let it be...

BIBLIOGRAPHICAL REFERENCES

Eurostat, 2016. *The EU in the world – 2016 edition*. Luxembourg: Publications Office of the European Union.

God, C., 2017. Futurism.com. [Online] Available at: https://futurism.com/researchers-have-just-devised-a-way-to-store-information-in-chemicals/[Accessed 21 may 2017].

King, M., 1994. Vacuum Energy Vortices. *New Energy News*, July, Volume 2, p. 20.

—. 2001. *Quest for Zero-Point Energy – Engineering principles for "free energy" inventions*. 1st ed. Kempton (Illinois): Adventures Unlimited Press.

Tesla, N., 1901. *Apparatus for the Utilization of Radiant Energy*. USA/NY, Patent No. 685957.

Trinkaus, G., 1988. *Tesla: The Lost Inventions*. Portland (Oregon): High Voltage Press.

Valone, T. F., 2007. *Zero point Energy: The Fuel of the Future*. First Edition ed. Beltsville (MD): Integrity Research Institute.

Wang, W. et al., 2017. Harnessing the hygroscopic and biofluorescent behaviors of genetically tractable microbial cells to design biohybrid wearables. *Science Advances*, 19 may, Issue BIOENGINEERING, p. 8.

World Meteorological Organization – Atmospheric Environment Research Division, Research Department, Geneva, 2016. The State of Greenhouse Gases in the Atmosphere Based on Global Observations through 2015. WMO GREENHOUSE GAS BULLETIN, 24 October, p. 8.

Executing progress through the design-build platform

Kate O'Connor
Architecture, Marywood University, Scranton, PA, USA

ABSTRACT: By exploring moments of inspiration between designers and ecologists, we can investigate how both practices intersect and remain intertwined pragmatically and philosophically within the still emerging field of sustainable design. The goal of sustainable architecture is to unify holistically art and science in order to find practical solutions to environmental problems while addressing the social and economic aspects of sustainability.

Understanding how sustainable design benefits a community introduces the notion that architecture is not only an academic discipline but also one that serves social consequences. Students from the Marywood University School of Architecture explored the design/build platform for a not-for-profit animal sanctuary in Mehoopany, Pennsylvania by providing a structure for two growing pigs rescued from slaughter as piglets (Fig. 1). The student work examined the capacity of design to the betterment of the farm, and ultimately, community. The process provided the framework for an engaged and comprehensive learning opportunity, developing industry skills and accumulating material knowledge while emphasising citizenship as part of the curricular standards.

Emphasising design that values human and local research, students approach learning through creative problem solving, while the development of skills results in a full understanding of how ideas become real. The holistic approach to sustainable design must be addressed as part of curricular standards, especially where the survival of the individual is dependent on the community, and the community profits from the individual.

Keywords: Socially Responsible Design, Design/Build, Citizenship

1 STUDYING PRECEDENT IN RESPONSIBLE DESIGN

Post World War II Bauhaus designers were concerned about functionalism in terms of the full human being, including rational, emotional and environmental needs. By setting the human condition at the centre of their projects, they came to guard the humanist tradition. Their adaptation of the ecological sciences illustrates that the humanist legacy, derogatorily labelled "anthropocentrism" by some, has offered more to the history of environmental debate than the chauvinism toward nature that critics point out. Instead of such false duality, the London Bauhaus designers illustrate the importance of humanitarianism to environmentalism as it pursued in their work.

Public Architecture, a non-profit design firm started by John Peterson in 2002, and his colleagues might be making some of the boldest statements in design today. Their declaration could not be simpler: Good design can drive social justice. It is a message that all of the groups share, with varying results, and borrows elements from progressive design movements of the past such as the London Bauhaus. They are challenging their entire profession to take the high design standards usually reserved for elite clients and systematically deliver them to society's most vulnerable. In other words, they convince ordinary people and those on the margins that architects do not just make giant, radical shapes; they can make giant, radical change.

Figure 1. Our Clients- Duncan and Nugget

Similarly, the mission statement of Project H uses the power of the design process to catalyse communities and public education from within. A team of designers, builders, and teachers engage in improving quality of life. The design process results in simple and

Figure 2. (Left) The Gorilla House, Tecton Group (Right) PigHAUS under construction. Both utilise light and fresh air as design stimuli, and concrete and rammed earth as structure.

effective design solutions that empower communities and build collective creative capital.

> There is no design without (critical) action; We design WITH, not FOR; We document, share and measure; We start locally and scale globally; We design systems, not stuff; We build.

The "London Bauhaus", established after fleeing Dessau from Nazi-occupied Germany, proffered a belief that Bauhaus design could solve social as well as environmental problems. Former Bauhaus faculty, including Walter Gropius and Laszlo Maholy-Nagy, were able to meet as a group in London and set forth to re-establish the Bauhaus school in London. Furthering their concept, they formed the Modern Architectural Research Group or MARS. The MARS group became advocates of environmental sensitivity: "There must be no antagonism between architecture and its natural setting", they pointed out in their exhibition manifesto of 1938. For them, modern architecture, with its focus on light and fresh air, represented an advancement in public health as well as a remedy that could halt environmental destructions of past developments.

Less than a year after his arrival in London, and incorporating the ideology of MARS, Berthold Lubetkin joined forces with a group of architects to form a group named Tecton, an abbreviation of Architecton, the Greek word for architecture. The zoo staff used Bauhaus design to promote a socialist-inspired view of the connection of the between animal and human nature. They saw a close connection between animals and humans and consequently an evolutionary development from the animal house to the Bauhaus, which offered health, welfare and peaceful relationships between humans and the natural world. The cumulative design philosophies from the London Bauhaus, MARS and ultimately Tecton embodied the evolution of the design process for the pigs.

Lubetkin was able to focus on the principles of natural light and fresh air and incorporated movement into buildings by designing sliding and revolving screens at the Gorilla House at London Zoo in Regent's Park (Fig. 2). This, in turn, created a controlled environment to protect the gorillas from humans and humans from gorillas as necessary. He built the house in his favourite form, a circle, as a Corbusian *machine à habiter* for animals. He also incorporated reinforced concrete in his design, a new material he had experimented with while living in Paris prior to moving to London.

1.1 *Designing with inspired intent*

The pigHAUS offered the opportunity for the students to experience design sensibility driven by context determined research allowing learning through creative problem-solving. The process provided the framework for more engaged and holistic learning opportunity and emphasised citizenship as part of the curricular standards.

The circular shape of a corrugated metal grain silo, a nod to the Corbusian machine à habiter for animals from Lubetkin's motivation originally inspired the design. The circle was to be halved and then slid off-centre to create two "C's" that faced each other. The original name of the project was "Pigs in a Blanket", emphasising the enclosure of the walls around the pigs, tucking them in. Unfortunately, the design had to be modified, as the silo was not obtainable.

The parallels of how people interact with animals and how animals live in a contained space are comparative and are explored when examining the design intentions and achievements of the Tecton Group and the London Zoo, with the students from Marywood and the Indraloka Animal Sanctuary.

The account of the ecological sciences and design is an interdisciplinary history. Here, the word ecology refers to BOTH science and architecture. On one hand, ecology is widely hailed among environmental thinkers as a view of the world in which humans are not at the centre; on the other hand, its origin is the most anthropocentric object on earth, namely the human house, oikos. This can be further explored within the rural community, where the centre of that universe is the (a Greek tongue in cheek reference to the pig house) oink-os. This allows a study of the connection between animals and humans and consequently an evolutionary development from the animal house to the Bauhaus, which offered health, welfare and peaceful relationships between humans and the natural world.

1.2 *Indraloka Animal Sanctuary, inspiring citizenship*

At the Indraloka Animal Sanctuary, animals in desperate need are provided a lifetime of the safe sanctuary. The organisation generously takes in abused, neglected, or abandoned animals of every species, both large and small. Animals with severe medical and behavioural issues that would otherwise render them un-adoptable receive veterinary care, retraining and nurturing, ultimately bringing each animal back to full health.

The Sanctuary also offers a multitude of programs to foster citizenship with the animals as a resource. They provide a safe, educational and fun destination for field trips with the purpose of educating both children and adults regarding animal welfare, caring for the earth and the benefits of volunteer service. The Sanctuary survives with the assistance of donations, and a brigade of dedicated volunteers to help care for the farm and

Figure 3. The Penguin Pool, Tecton Group. (Left) Formwork (Right) Ramp

the animals. The community sustains Indraloka, and Indraloka in turn services for the community.

1.3 *Designing for a sustained environment*

The Tecton commissioned Penguin Pool at the London Zoo opened in 1934 as the culmination of a series of new buildings designed according to the Bauhaus style. It was immediately recognised as a masterful piece of the avant-garde, and its double helix ramps have been a pilgrimage for admirers of modern architecture. Lubetkin saw this building as another opportunity to explore creatively the possibility of a new building material available in 1934 – reinforced concrete. Having studied the habits of penguins he created a penguin enclosure and pool that provided an interesting environment for the penguins, a multiplicity of viewing angles for the spectator and a Modernist building of true clarity and style. The overall elliptical shape is referenced in the spiralling intertwined ramps that connect the different levels; these sloping cambered ramps give a dynamic feel to the building.

Emphasising design sensibility that values human and local research, students approach learning through creative problem solving, while the development of skills results in a full understanding of how ideas become real. A majority of influential educational experiences happen through tactile and demonstrative processes. The students diligently researched the region and compiled information about the pigs. They saw the design as a vehicle to promote an inspired view of the connection of the between animal and human nature.

The "Butterfly" roof enclosure of the pigHAUS serves numerous functions. The design allows for water collection along the inverted ridge beam, pooling the run-off into a natural wallow for the pigs to relax and frolic in. The students found that the mud from the wallow is a necessity for the sensitive skin of the pigs as it prevents sunburn, and deters insect bites. Additionally, the overhang of the roof along the south exposure reduces the sun exposure during the summer months when the sun travels high, and maximises the exposure in the winter months to heat the mass when the sun is lower in the sky. The clean profile and sensitive design of the building allow the animals to experience fresh air, natural light and a temperate atmosphere. Most importantly, the architecture was the vehicle to create a comfortable atmosphere for the pigs. (Figure 4)

Knowing from their research that large farm animals need to be cool in the summer months and warm

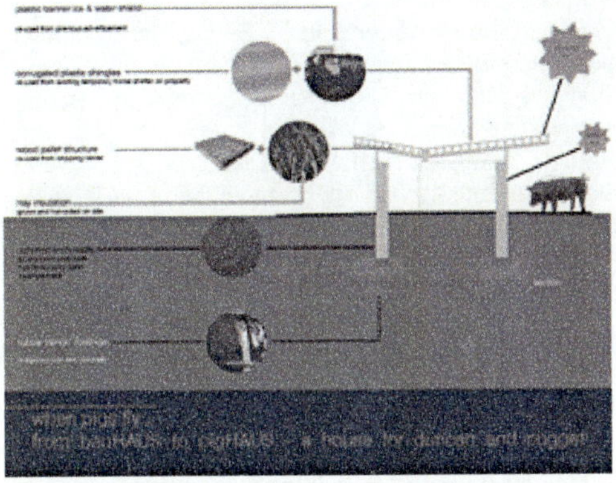

Figure 4. Building Section, Material Research

Figure 5. Completed Design

in the winter, a rammed earth wall system was chosen for its insular qualities. For example, on a 92-degree day, a 12-inch wide wall exposed to the sun will keep the interior of the structure 17 degrees cooler. The walls were constructed with the assembly of Symons Forms, a sustainable, reusable formwork system typically used to form concrete foundations. It was also chosen due to its structural integrity. Not knowing how large the animals would grow, it would provide a strong enclosure (Figure 5).

1.4 *Sustaining – society*

Zoologist and Secretary of the Zoological Society of London, Peter Chalmers Mitchell promoted Lubetkin and the Bauhaus design, as he believed that "evolutionary biology gave example to a cooperative model of social behaviour and that of peaceful coexistence was the best strategy for evolutionary survival among animals as well as humans".

The same would hold for workers who were in desperate need of being liberated from their "natural" condition of slums. He arranged for unemployed miners to perform a large part of the heavy work in constructing the zoo buildings since this allowed them to be in "good, open-air conditions until they had recovered sufficiently mentally and physically to pass out into the ordinary labour world". Taking both penguins and miners out of their usual environments and placing them in a fresh and peaceful setting proved to him that humans and animals alike could prosper. Peoples' and animals' health and welfare were of equal concern and the new architecture was to promote them. The zoo, he believed, should be a place where humans could have a "close up view" and reflect in the instincts they shared with other animals".

Applying their acquired design skills to a building project, students start to see the implications not only of the built structure but those inhabiting the structure. Many of the students had not been to a sanctuary, let alone a farm and were educated on farm life, as well as the philosophy of Indraloka. They gained perspective on a day in the life of a pig; eating, sleeping, and socialising.

A healthful environment at the London Zoo was to be a model for healthful living in the future. The London Times noted that the penguin pool was to have "architectural unity and pleasing effect, and thoroughly hygienic, give the birds what they require and afford ample space for visitors". A 1936 popular history of the London Zoo described it as a "very good indicator as regards the nation's material, and indeed spiritual health" and said that its new architecture should serve as a model for the "ultra-modern human dwelling-house," complete with "access to the outer air" and opportunities for a "sunbath".

The most essential data collected for the pigHAUS was how THEY (the students) responded to their interaction with animals. I concur when Chalmers-Mitchell compares the act of taking the students out of their classroom "environment" and "placing them in a fresh and peaceful setting" to promote their "health and welfare". This enabled them to design thoughtfully the most sensitive structure for the daily habits of the pigs while enjoying the surroundings of the farm.

1.5 *Socially responsible outside of the classroom*

As an solution to rigid, antiquated top-down instruction that does little to activate the type of critical creative thinking that today's design student needs to succeed, the interrelated design/build process is introduced as an alternative to problem solving armed with fresh and inspirational conceptual design thinking that the student and the rural community needs to thrive.

My firm conviction is that such learning in support of broader social transformation is best served by direct common engagement. As educators, we must make citizenship part of the beginning design students' curricular standards, especially where the survival of the individual is dependent on the community. Innovative community development projects enable current day students to develop the creative capital, critical thinking, and citizenship necessary for their own success and for the future of their communities. By examining design problems with the needs of the clients AND the community, educators and students start to address the broader definition of sustainability. It is logical to start the macro ideology of public and community responsibility in the micro setting of the classroom. Ultimately, the goal of sustainable architecture is to unify art and science in order to find practical solutions to environmental problems within contemporary education.

BIBLIOGRAPHICAL REFERENCES

Anker, Peter. (2010) *From Bauhaus to Ecohouse, A History of Ecological Design.* (Baton Rouge, LA: Louisiana State University Press.

Brightwell, Leonard Robert. (1936) *The Zoo You Knew?* Oxford: Basil Blackwell.

Crook, D. P. (1989) Peter Chalmers Mitchell and anti-war evolutionism in Britain during the Great War. *Journal of the History of Biology* 22 (2): 325–356. http://dx.doi.org/10.1007/BF00139517.

Grove, Richard H. (1995) *Green Imperialism.* Cambridge: Cambridge University Press.

Hughes, Robert. (1988) *The Shock of the New.* New York: Alfred A. Knopf. http://www.galinsky.com/buildings/penguin/.

Latour, Francie. (2008) A Blueprint for Good, A New Movement Aims to Change the World through Free Architecture. *The Boston Globe.*

Project H Homepage. (2011) *Mission Statement.* Available at: http://projecthdesign.org/index.htm.

The void concept in building design

Jorge de Novais Bastos
Department of Technologies in Architecture, Urbanism and Design, School of Architecture, University of Lisbon, Lisbon, Portugal

ABSTRACT: Adequate building design using sustainability requirements recommends that building "thickness" may not exceed 15.0 m to 20.0 m in-plan.

Natural comfort conditions (lighting, ventilation, thermal control, sound limitations) will be attained through the contour openings in exterior façade's walls. When the building in-plan size becomes larger, the designer has to create "openings" in the plan layout by inserting empty space areas. These "voids" can be either open air or to be roof partially covered with a roof canopy. This study demonstrates throughout different ages that the strong "void" (cloister, courtyard, patio) concept still influences contemporary architecture, particularly, in Louis I. Kahn building solutions.

Keywords: architecture, void, cloister, courtyard, patio

1 INTRODUCTION

Current large-scale building design rules, both public and private, require that they must incorporate all the available data regarding comfort conditions (ex. light, temperature, ventilation, noise control and visual contact), from the very early design stages until the final approved proposal. These practical requirements are dependent on the building location and its solar exposure throughout the year. The winter and summer solstices provide the extreme boundaries angles for sunlight incidence on the building's façades. The light rays that "pierce" through the façade's openings (ex. windows, doors, gates).

Supposing that the during the winter solstice, shallow light rays penetrate through an assumed 3,00 m tall window into the building interior space, they will hit the pavement at a maximum depth of about 7.50 to 10.00 m, if an incidence angle of approximately 30° is considered. This half-length building plan dimension indicates that the maximum building in-plan "thickness" between adjacent parallel exterior walls façades planes lay within the 15.00 m to 20.00 m range, see Fig. 1–a, b.

Considering that building comfort design needs impose a limit on its maximum thickness of e(t) = 15.0 m, for different types and proportions of quadrangular plans (ex.: 1:3, 1:1.5, 1:1) a "void" is required when the façades exceed the 15.00 m boundary, see Fig. 2.

In Fig. 2–a, the gross plan area increases from A(top) = 45.0 m × 15.0 m = 675 m^2, to A(interm.) = 45.0 m × 30.0 m = 1350 m^2 up to a maximum value of A(bot.) = 45.0 m × 45.0 m = 2025 m^2.

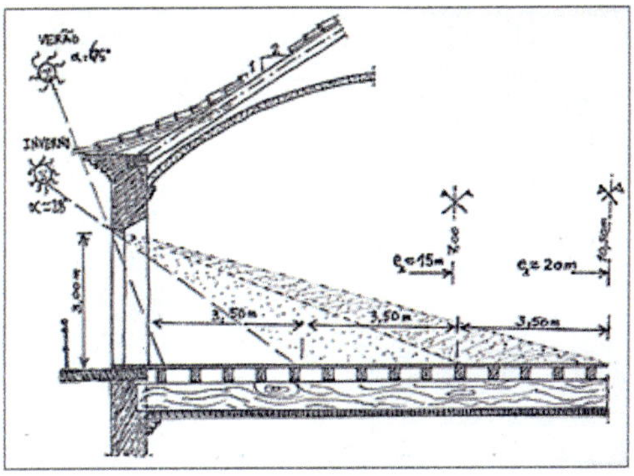

a. Façade cross-section – solar lighting

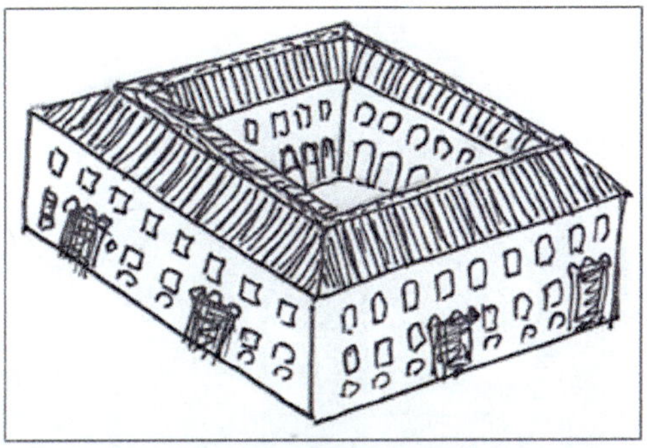

b. Building interior space

Figure 1. The void – building "thickness" vs. solar insolation.

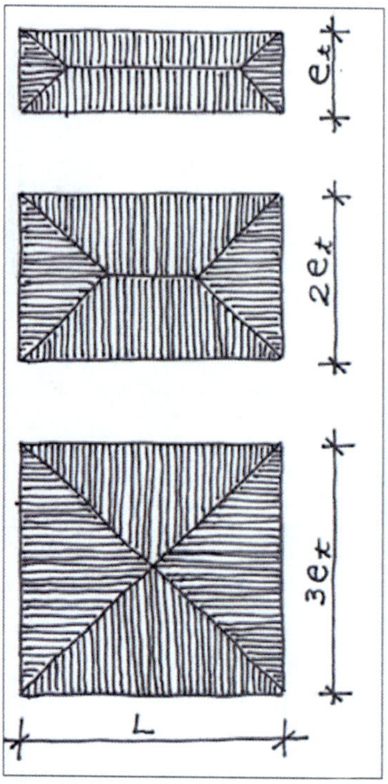

a. Building plan proportions - 1:3; 1:1.5; 1:1.

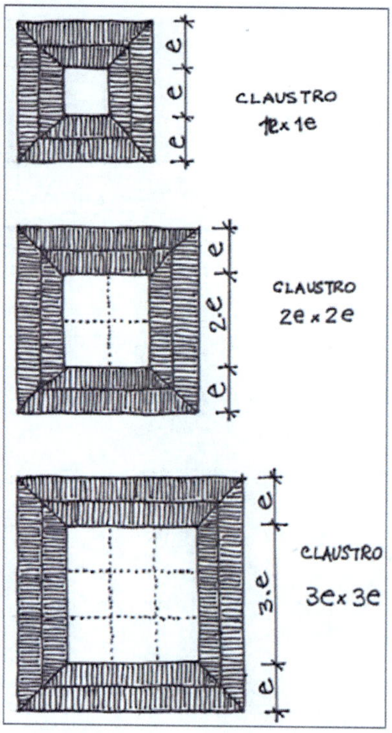

b. Cloister plans and sizes (e = 15.0 m).

Figure 2. The quadrangular building and the void.

In the Western cultural tradition, the extensive use of cloisters (ex. religious buildings), courtyards (ex. civil estates), and patios (ex. housing units), may have its origins from the Roman *impluvium* with both its functional (water collect) and comfort requirements, Fig. 2–b.

In the Iberian Peninsula, through the Arab civilisation, the interior patio concept was transmitted to the Christian world. This "void" creates an intimate, family area within the family dwellings.

In the second millennium, both concepts were included into large size building design plans. In religious buildings, the cloister becomes a functional space to be used for meditation and to link the all the adjoining areas.

In Fig. 2–b, the total plan area A(tot.) occupied the construction site grows successively from 2,025 m^2 (top), to 3,600 m^2 (interm.) to 5,625 m^2 (bott.). However, the actually built plan area varies (from top to bottom) with 1,800 m^2, 2,700 m^2, and 3,600 m^2, whereas the interior voided space increases from 225 m^2, 900 m^2, and 2,025 m^2.

The cloister space was conceived to unite all the adjacent spaces in the monastery – church, refectory, library, dormitories. Current design approaches applied to rehabilitate long abandoned religious buildings are eager to absorb these available cloister areas and to use them as an enclosed space. Some design solutions with ingenious roof systems are well achieved, although large energy costs have to be paid in order to attain adequate comfort levels – temperature, lighting, humidity, ventilation, and noise control. In this study, some important aspects of "void" concept/design solutions are presented: (1) singular cloister shapes; (2) actual rehabilitation designs where the "void" is enclosed by a canopy; and, (3) the "void" concept being used in two American buildings by a unique architect – Louis I. Kahn.

2 THE CLOISTER

2.1 *Changes in the architectural space*

During the building planning phase, the designer's team tries to provide an adequate answer to all possible different construction challenges they can face. The final design solution must include several unique aspects: (1) an iconic concept – a singular and original solution must be reached; (2) tectonics – local materials and construction techniques should be used; (3) site integration – the respect for Nature and local landscape deserves to be valued; (4) natural comfort conditions – major design aspects have to be observed; (5) structural stability – the human life safety has to be guaranteed; and, (6) economy – existing limited resources and funding must be used wisely in accordance with real social conditions.

These multiple requirements have their roots in the Roman builder Vitruvius and his classical words – *venustas*, *vetustas*, *et firmitas*. Sometimes, they are overlooked, partially ignored, leading to poorly designed buildings without character or, worse, without a soul.

2.2 *A unique cloister shape*

In 1537, the Portuguese King D. João III following strict Counter-Reformation recommendations ordered the royal builders João de Ruão and Diogo de Castilho to design a new monastery to be built in the Serra do Pilar hill, just overlooking the river Douro and the Porto north bank. The foundation works were still going on in 1542, and the church (first phase) was completed in 1567. The circular cloister was only completed in 1576, to be disassembled a few years later (1598), to build a new choir building behind the altar. The final building ensemble construction is only completed around 1669 – 72, more than one hundred years later, after it began.

In 1526, Emperor Carlos V marries Isabel de Portugal in Seville. The royal couple moved to the Alhambra, Granada, where they lived for several months. The Emperor orders to Pedro de Machuca, to design a new palace. The magnificent square building with a 63.00 m long façade encloses a unique circular courtyard that was, probably, the source of the Porto inspiration. The Serra do Pilar Monastery when completed (1580), had a 30.0 m diameter church with an adjacent similar size cloister, with a 22.0 m diameter void. Both circular shapes created the "Perfect Infinity" ensemble, see Fig. 3–a, b.

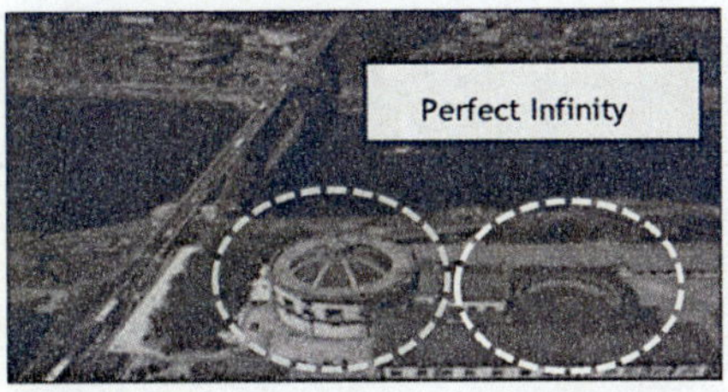

a. Serra do Pilar Monastery - air view.

b. The Mannerist-style cloister.

c. S. Bento da Vitória Monastery - enclosed cloister.

Figure 3. Porto – cloister spaces and uses.

2.3 *The enclosed cloister shape*

The requirement to meet the architectural brief needs and the existence of an available space ("void"), leads the design team to propose its enclosure with a permanent roof canopy and, sometimes, to enclose the surrounding ground floor spans with curtain glass walls, see Figs. 3–c and 4.

In Porto, the S. Bento da Vitória Monastery that started to be built over the Old Synagogue and the Olival Jewish Quarters, under the design of Diogo M. Lucas, was only completed in 1690.

In the 1980s, the building undergoes extensive consolidation works. In 2000, the 24.0 m × 24.0 m in-plan cloister with is enclosed by using a structural steel glass roof supported on four interior slender steel columns. This space was created as a concert hall for the Porto Metropolitan Orchestra. It required extensive comfort needs (ventilation, humidity, thermal control). Now, it has little use as the orchestra moved into a new building.

In Spain, two singular enclosed cloister designs are presented. In Madrid, the cloister of the San Jerónimo El Real Monastery underwent extensive damages during the Spanish Civil War (1936–39). In the process of expanding the Museo del Prado installations (the 1990s), Arch. Rafael Moneo proposed the use of the ruined cloister with 24.0 m × 24.0 m and the underground space. The lengthy and careful rehabilitation process required the cleaning of each stone block, the pre-assemblage at the stone works factory and the following installation in the construction site. The final design was used to install a sculpture area, with an excellent active climate control and lighting. The singular structural steel roof canopy with shading answers to the Museum demanding conditions, see Fig. 4–a, b.

In the University of Alcalá de Henares, the School of Architecture is installed in the old Convent of Carmen Calzado (1563). In 1657, is created a University College in this building located on the Calle Santa Úrsula. In the 1980s, the building was extensively rehabilitated and the 18.5 m × 18.5 m cloister space was enclosed with a striking timber glulam grid with twenty-five skylights, see Fig. 4–c, d. This University space is currently used by architecture students for drawing classes, due to the excellent zenith light that floods this ambient. In the stone arcades that surround the courtyard were inserted glass walls in order to control noise and to allow visual contact between the passing users and the students. Under an extensive period of continuous use, rainwater leakage problems had to be solved.

2.4 *The courtyards of Louis Kahn*

Louis I. Kahn (1901–74), born in Parnu, Estonia, and naturalised North American was one of the foremost contemporary architects where the teachings

a. Museo del Prado - new wing.

b. New wing - enclosed cloister.

c. Alcalá de Henares - cloister skylights.

d. Cloister interior with a timber glulam grid.

Figure 4. Spain – two enclosed cloister spaces.

from ancient buildings were clearly inserted into his designs.

Kahn's original building designs pursue a constant search for quality and for meeting the Owner's needs. More than forty years later, the constructed are actual and they never cease to wonder the observer. The careful treatment given to the exhibition spaces, the use of classical concepts, e.g., the courtyard is essential to understand the “light and silence” existing in this spaces (Johnson et al., 1978). The search for a geometrical cycloid – shaped roof shell, with a continuous longitudinal slit that allowed light coming from the roof to flood through the walls, was a major design solution to allow a better view of the exposed objects, see Fig. 5.

The quality in the construction details leads Louis Kahn to express his thoughts metaphorically:

> When one person composes music he is not attracted by what he sees, but by what he listens [. . .]. His compositions represent the sound, and how it is organised. The architect must be able to read the life concept that exists inside his projects through his drawings. (Johnson et al., 1978)

The use of a “void” in large-size buildings can be observed in two museums: (a) the KAM - Kimbell Art Museum built between 1967 – 72, in Fort Worth, Texas; and, (b) the YCBA – Yale Center for British Art built from 1969–74, in New Haven, Conn..

In the KAM, three design solutions were needed for the architect to understand better the owner's needs regarding this special project. The careful choice of materials (ex. travertine) and detailing, the high construction standards are well visible in this excellent building. The concrete post-tensioned cycloid shell was the best solution for the interior natural light access. The interior voided spaces were important, see Fig. 5–a, b.

The cycloid concrete shell has generous dimensions in-plan – 31.7 m × 7.0 m, and it was used as the space generator in both plan directions, by repeating this modulus. The total building plan has 55.0 m × 100.0 m.

When natural light is introduced into the interior of a museum space, major architectural design concerns arise (Siza Vieira, 2009: 319):

> Nos museus, a luz faz-se doce, cuidadosa, de preferência, impassível e imutável. É preciso não ferir, é preciso não ferir os cuidados de Vermeer, não se deve competir com a violenta luz de Goya, ou a penumbra, não se pode desfazer a quente atmosfera de Ticiano, prestes a extinguir-se, ou a luz universal de Velasquez ou a dissecada de Picasso, tudo isso escapa ao tempo e ao lugar no voo da Vitória de Samotrácia.[1]

At the YCBA, one of Kahn's last masterworks, the extensive use of a modular shape combined with precast concrete elements allowed major construction

[1] “In museums, the introduction of light must be soft, well designed, preferably, discrete and timeless. It must not hurt, not hurt the brush strokes of a Vermeer, it must not compete with Goya's violent light, or its shade, it cannot disassemble Titian's warm atmosphere, almost vanishing, or the universal light of Velazquez or the light dissected by Picasso, all those facts escape to the time and the place, under the Victory of Samothrace flight”.

a. KAM - the cycloid concrete shell.

b. KAM - an interior courtyard.

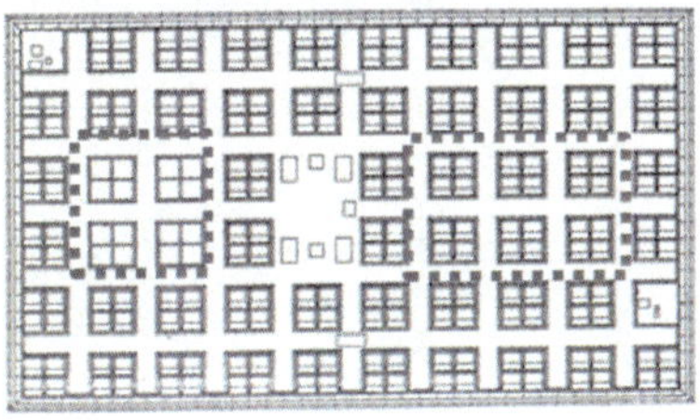

c. YCBA - roof skylights and courtyards location

d. YCBA - multi-story interior courtyard

Figure 5. Louis Kahn – courtyard spaces.

savings and time reduction in order to build this art museum. The building has an approximate plan size of 60.0 m × 20.0 m, with four 3.60 m high floor levels above ground (Hubert, 1992). The main entrance is located at a corner zone in the shadow, where the visitor's eyes adjust to the transition of to an interior, darker space. After passing by the reception, the observer is caught by surprise, in a sunlit interior courtyard covered with skylights. The square plan, with a side of 12.0 m and a height of 14.40 m, creates a unique moment in the construction of this architectonic space. The location of this "finding" may be considered the announcement of a masterwork (Heidegger, 2010; Robinson et al., 1997).

The visitor is led to an architectural promenade, where not only the fine arts collection on display can be admired, as well as, the rational use of the building materials (concrete, glass, wood). A second covered rectangular courtyard, 12.0 m × 24.0 m in-plan, rises along three storeys to a height of 10.80 m, being covered by the skylights that flood with light this space, see Fig. 5–c, d.

3 FINAL OBSERVATIONS

The importance of Light and its relationship with the material tectonics being used to make it real the thoughts underlying the design action is essential in Kahn's mind (Johnson et al., 1978: 11):

> Silence to Light, Light to Silence
> The threshold of their crossing
> Is the Singularity, is Inspiration
> (Where the desire to express meets the possible)
> Is the Sanctuary of Art,
> Is the Treasury of the Shadows.
> (Material casts shadows, shadows belong to light).

The different connections between Light, Shade, and Silence are constant in major artistic endeavours – architecture, sculpture, painting, and design. The light–shade contrast allows the observer to understand better the object spatial dimensions. In large built masses, the void – patio, courtyard, cloister, is a major powerful element in the architectural composition. Its action has a polarising effect on the observer thoughts and the understanding of a building rationale. Different examples are presented in this study to provide a better understanding of the design–building process. The existing ancient building heritage is a powerful learning tool for currently designed solutions.

BIBLIOGRAPHICAL REFERENCES

Heidegger, M. (2010) *A Origem da obra de arte.* Lisboa: Edições 70.

Hubert, B. J. (1992) *Louis I. Kahn. Le Yale Center for British Art.* Marseille: Éditions Parenthèses.

Johnson, N. E. and Kahn, L. I. (1978) *Light is the Theme – Louis I. Kahn and the Kimbell Art Museum.* Fort Worth, TX: Kimbell Art Mus. Ed.

Robinson, D. and Finn, D. (1997) *The Yale Center for British Art: a tribute to the genius of Louis I. Khan.* New Haven: Yale University Press.

Siza Vieira, Á. (2009) *01 TEXTOS.* Porto: Livraria Civilização.

Graphic dialogues: The progress of knowledge in design in the architecture studio

Artur Renato Ortega & Silvana Weihermann
Architecture and Urban Planning, Universidade Federal do Paraná, Curitiba, Brazil

ABSTRACT: This paper aims to address one of the most common activities in the teaching of architecture: the counselling that professors provide to students in the architectural design studio using drawings. For us, these actions are known as graphic dialogues. The purpose is to remind, reinforce and, above all, promote the idea that drawing is an essential instrument for the graphic dialogue learning method and for the dialogue between students and professors, who are faced with the task of constructing apprentices, advancing their knowledge, and ensuring the attitudes required to learn design. It considers drawing as learning the reflective practice and structuring architectural thinking, and as an instrument for the development and representation of ideas shared autonomy in the articulation of theory and practice, and the construction of knowledge, skills and competencies in the act of designing. Finally, it encompasses the progress of syncretic thinking to synthetic thinking in the design process. It states that the teaching of architectural design is based on the continuous designing in which the encounter between professors and students seeks to enable an exchange of knowledge that favours the performance of this activity and, evidently, the progress of students and their promotion in the course.

Keywords: graphic dialogue, drawing, progress section

1 THE ARCHITECTURE STUDIO: DRAWING, DIALOGUE AND DESIGN

In the scope of the studio of architecture, drawing is the main language[1] of dialogue between professors and students. In this domain, students describe and appreciate the consequences of their actions, and the professors, while drawing, learning, and guiding, cooperate toward the progress of mutual knowledge. This dimension of the drawing practice goes beyond the simple representation of ideas; it covers the development of intelligence from the perception and analysis of one's own drawing, a medium that is the reflection in action (Schön, 2000). It means the freedom to learn by doing in a deep and conscious action, and refers to two distinct, albeit interrelated mental processes, considering that:

> (. . .) reflecting during the action is to ask yourself what is happening or what we should do, what is the best tactic, which detours and precautions should we take, what risks are we taking, etc. Reflecting on the action is something quite different. In this case, we perceive our own action as the object of reflection, whether to compare it with a prescriptive model, what we could or should have done, or to explain it or criticise it. [. . .] Therefore, reflection is not limited to an evocation, but it is subjected to criticism, analysis, rules, theories, and other actions, imagined or performed in an analogous situation. (Perrenoud, 2002: 30).

As the students prepare their drawings, qualities and relationships that were not previously imagined reveal themselves; their errors and successes help them develop skills and assessment capabilities (Fig. 1).

To achieve this purpose, students must learn to design by drawing, and thus view and explain the world of their ideas through drawing. This is the case, explains Schön (2000: 70), in which the practice

> resembles research, your investigation is a transaction with the situation, in which knowing and doing are inseparable. Therefore, practice is not merely a complementation of the teaching-learning process.

Practice as an active process of reflection in action and reality as the object of knowledge enables the configuration of the learning process as an interactive action between professors and students, and this "learning how to learn" (Martins, 2006: 98). It also prompts the cognitive and affective abilities of students to understand, control, and decide their learning,

[1] The term language is understood as "(. . .) any system of signs that serves as a means of communication between individuals and can be perceived by various sense organs, which allows us to distinguish, for example, an oral language (speaking, singing), a visual language (mime, dance), a tactile language (Braille, engraving), an auditory language (music, nature sounds), or a graphic-visual language (drawing, writing, painting)" (Gomes, 1996: 21).

Figure 1. Student drawing development (dialogue with oneself). Source: author's photograph.

Figure 2. Graphic dialogue between professors and students in the drawing activity. Source: author's photograph.

thus placing them as the historic subject. This does not imply minimising the responsibility of the professor's duties, but rather to invite them to assume their roles as educators in a permanent confrontation of conflicts, contradictions and unusual situations that the social determinants generate in a confrontation with the possibilities of human achievement based on education. The learning environment, that is, the design studio, is characterised as:

> (...) a collective living space, a space of unique and original relationships, similar to an ecosystem, for the intensification of learning, in which the bonds of pupils and professors with knowledge are accentuated (Romanowski, 2006: 105).

The production of a drawing is therefore always an action and reaction that provides professors and students with the necessary path in the complex process of bringing forward an image of the imagined building and, consequently, drawing after drawing, reaching the desired solution or possible solution for the moment. Therefore, drawing, in the design activity of the architect, is understood as the structurer and structuring of architectural thought, and as the cognitive "artefact" (Fig. 2).

This cognitive path, which also contains a strong emotional component, so to speak, of the students through the drawings in the discipline of design, depends on a lot on the nature of the dialogue with the professors during counselling.

2 THE GRAPHIC DIALOGUES AND THE CONSTRUCTION OF DESIGN KNOWLEDGE

We use the term graphic dialogue to define the dialogues mediated by drawings produced in and for the design studio, that is, the graphic dialogue from: (a) drawings made in a private dialogue with the students with themselves, as they conceive their ideas and, thus, connect with the professor(s); (b) drawings made in moments of counselling, by the professors or by the students; and (c) drawings made on the board for a particular explanation, also by the professors or by the students (Fig. 3 and 4).

Figure 3. Graphic dialogues: drawings made on the board, by the professor or by the students. Source: author's photograph.

In fact, it was Schön (1983) who suggested the metaphor of "graphic conversation" to identify the relationship of drawing with the thought in the design studio. However, the word "conversation" encompasses several actions of the relationships between subjects, such as talk, lecture, seminar, panel, dialogue, debate, and discussion.

Figure 4. Detail of drawings made on the board. Source: author's photograph.

The word "dialogue", however, suggests a broader approach regarding the meeting between students and professors in the design studio since this relationship occurs, in accordance with the thought of philosopher Martin Buber (1982), between You and I, which he defines as interplay because it involves dialogue, meeting, and responsibility between two subjects and/or the relationship between the subject and the object:

> [(. . .) I have no teachings to transmit. I merely point something, I indicate something, in reality, something not seen or barely detected. I take whoever listens by the hand and take him or her to the window. I open it wide and I point outwards. I do not have any teaching, but I have a dialogue (Buber, 1979: s/p).]

The term dialogue comes from the Greek words *dia* ("through") and *logos* ("reason"). The dialogue is a way of circulating senses and meanings, that is, the words circulate among people without necessarily being agreed upon, disagreed upon, reviewed or judged because they present the immediate experience of the participants. However, it is a cooperative activity of reflection and observation of ideas and experiences. Thus, it becomes a pedagogical practice that allows the subjects to fecundate the thoughts of one person to the other, exchange data that arise from this relationship without searching for it, at least at first, and analyse or judge these data since they aim to produce new ideas with shared meanings.

To dialogue is, above all, to learn to listen to others without interruption, whether to agree or to disagree. Silence is part of that process and, in silence, the subject observes and self-observes. Knowing how to listen allows the subject to pay attention to the phenomena that present themselves, and places the subject as a participant in the process rather than a mere observer. As stated by Buber (1982: 40):

> The dialogue is not limited to the traffic of men among themselves; it is – and this is how it proved to be to all of us – a behaviour of men with each other, which is only represented in his traffic.
>
> Therefore, even if one can do without speech or communication, there is an element that seems to belong inextricably to the minimum constitution of dialogue, according to its own meaning: the reciprocity of inner action. Two men who are connected through dialogue must evidently be facing one another; therefore, they should be turned towards one another, regardless of the extent or awareness of the activity.

The dialogues that take place in the design studio favour and determine this educational practice of the professors. The very structure of the disciplines in the studio points to this fact since the adopted work method is counselling. In this context, the dialogues between professors and students are also affected by the mutual representations, i.e. by the ideas that one has to the other.

Thus, the dialogue between professors and students in the design studio are understood as, according to Freire (1967: 66) "[. . .] a horizontal relationship. It is nourished by love, humility, hope, faith, and trust". It is in this relationship of trust and sharing that the dialogue occurs, in the counselling of the design studio, and where the professor is not entitled to dictate right and wrong as an absolute truth since he or she cannot impose guiding maxims. However, the professor should not evade the exercise of his or her authority and take responsibility for "what is right and what is wrong in a given situation that is his task" (Buber, 1969: 53).]

In the case of the design studio, this is crucial, since, in order to learn, students "dive" into the activity of design, and try, from the very start, to do what they do not yet know how to do, while essentially, like a compass, counting on the dialogues with their professors.

We understand that students can or should find a sense in the tasks, thought out and implemented in the drawings, that they must complete solving an architectural problem based on suggestions and criticisms, within the routines of graphic dialogues, in the counselling with professors.

From the reflections of graphic dialogue, students present results – whether from the return to the drawing to make changes, whether from the implementation of new ideas, whether from the almost imitative use of suggestions of professors – to understand the solution of the problem established as a design activity.

Therefore, we can see that: (a) when students express themselves through drawings, they build [and rebuild] a line of thought of their ideas; (b) consequently, the dialogue with professors improves; (c) the dialogues are more thorough; and (d) these subjects gradually turn into [progress as] designers.

It is in this context that the graphic dialogue process enables shared autonomy and becomes a form of theory and practice, in which students and professors build knowledge and skills in the context in which the action occurs: the action of designing by drawing; and, consequently, the development of this knowledge.

This proposal agrees with the proposal of Anastasiou (1998: 194), that:

> (. . .) radically changes this relationship, indicating a contractual relationship of deliberate partnership, focusing on a process of adding efforts, acquiring the knowledge [. . .] Thus, we propose the construction of a new relationship: {(professor + students) × knowledge} [. . .] The opposition (symbolized by the × sign, or versus), explains the challenge placed in the conquest and/or field of knowledge, a challenge that must be confronted by the professors in the partnership processes, departmentally with their peers, and in the classroom with their students.
>
> For this process to become effective [. . .] the words of the professor in the classroom do not suffice: it is necessary to intentionally enable a new methodological process to consider the approach of knowledge as a result of reality, therefore, taking it as a point of departure and arrival, passing through the construction process of thought with that of the student.

The processes of dialogue between professors and students, in the studio, as in the case of knowledge, are also a form of progress. Students need to develop the condition to expose their ideas and defend them in the same proportion that they accept the arguments and suggestions of the professors, as a pact for the joint construction of knowledge.

3 FINAL CONSIDERATIONS

In the process of learning to design in the studio, from the production of drawings, the students see themselves confronted with constant challenges, both for the resolution of their own design problems and for the critical work of the assessors. The theme (s) proposed by the professors is always a scenario among several possible scenarios to promote with and for the students' interactive processes of knowledge that they can use consistently and flexible in design, and, therefore, enables students to dive into the process of designing architecture.

The construction of design knowledge in architecture implies the recognition of the language of drawing, in all its traditions and limits of the revelation of significant structures, since drawing is the structure and structurer of architectural thought.

The knowledge built in the graphic dialogue occurs by both the action of drawing and the mental action, which is essential for the representation of the object in space. The recognition of students of the importance of drawing and the progress of this knowledge as a special form of learning is fundamental to their education and their capacity to develop the graphic dialogue.

The production of drawings made by the designer students to unleash their ideas is, in fact, a facilitator for the development of the design action in architecture. It facilitates or, even better, allows both the individual process of designing their ideas and the graphic dialogue with professors.

Thus, the graphic dialogue sustains the reflections, orders the thoughts, examines, analyses, and reviews the decisions of thinking and doing of the subjects who are the accomplices of the design practice, that is, in turn, dynamic and transforming.

The graphic dialogue process enables shared autonomy, in which, differences aside, there is no winner or dispute. The success of the students is the same as that of the professors. The action is interactive, communicative, qualitative, and collaborative.

Therefore, the graphic dialogue is understood as a sort of exchange that enables reflection among professors and students. For these reasons, we believe that the educational process in the design studio is essentially dialogic and occurs through the meeting of professors and students. However, the drawings are the main thruster of the dialogue between these agents and, consequently, the process of learning how to design architecture.

From this perspective, the key characteristic of graphic dialogue is to foster the student who thinks about what he or she is doing, that is, about the process of designing architecture. In other words, teaching, in the design studio, that focuses on teaching and prompting students to design through drawings, also benefits from a focus on formal and operative capacity-building, as opposed to the transmission of content, and, therefore stimulates the subject designer toward permanent and progressive learning, that is, to learn how to learn, learn how to think in the action of doing, of designing by drawing.

BIBLIOGRAPHICAL REFERENCES

Anastasiou, Léa das Graças C. (1998) Metodologia do ensino superior. Curitiba: IBPEX.

Martins, Pura L. O. (2006) As formas e práticas de interação entre professores e alunos. In: VEIGA, Ilma Passos Alencastro (Org.). *Lições de didática*. Campinas: Papirus.

Buber, Martin. (1982) *Do diálogo e do dialógico*. São Paulo: Perspectiva.

—. (1979) *Eu e tu*. São Paulo: Cortez e Moraes.

—. (1969) *Reden über Erziehung*. Heidelber: Lambert Schneider.

Freire, Paulo. (1967) *Educação como prática da liberdade*. Rio de Janeiro: Paz e Terra.

Gomes, Luiz V. N. (1996) *Desenhismo*. Santa Maria: Ed. da Universidade Federal de Santa Maria.

Perrenoud, Philippe. (2002) *A prática reflexiva no ofício de professor: profissionalização e razão pedagógica*. Tradução de Cláudia Schilling. Porto Alegre: Artmed.

Romanowski, Joana. (2006) Aprender: uma ação interativa. In: VEIGA, Ilma Passos Alencastro (Org.). *Lições de didática*. Campinas: Papirus.

Schön, Donald. A. (2000) *Educando o profissional reflexivo: um novo design para o ensino e a aprendizagem*. Porto Alegre: Artes Médicas Sul.

Housing as a reflection of the human unconscious; the permanent flexibility in housing

Jorge Jerónimo
University of Lisbon, Faculty of Architecture, Lisbon, Portugal

ABSTRACT: Nowadays there is a time of exaltation of the identity of the place as a concept repository of values, feelings and collective memories, which transmit to the individual a feeling of belonging to a certain place. In this search, the potentialities of the house, as a compositional and indispensable cell of the identity of the place, seem to have been forgotten by the great majority of architects, notwithstanding the concern about the subject taken by the social sciences, namely at the level of psychology, anthropology and sociology. Its role in the social and psychological development of man has been the subject of studies that conclude that the intervention of man in the space of the house, as a cell of society, is fundamental for its expression in the world (Marcus, 1995).
In spaces where identity significance may be greater such as historical neighbourhoods of spontaneous and organic development, the interior of the house continues to remain immobile under rigid rules of architectural design.
The decor has been, systematically and with few exceptions, the only way for Man to intervene in his home and this limits his ability to understand himself.
This article intends to demonstrate that the more consolidated the identity of a place is, the greater the solutions for the intervention and personal identification of the interior space of the house. The first gives individual a sense of belonging in the world and the second allows him to understand himself justifying his own existence.

Keywords: Place identity, individual and collective unconscious, house space, flexible house

1 INTRODUCTION

From the Cave to the contemporary dwelling, Man uses the space of dwelling for the most basic functions of everyday life. The space he chooses to fulfil these inherent needs of his condition is assuming a fundamental importance in his own identity because the way he interprets and fills it is an unconscious way of expressing his feelings and the way he sees the world around him. Each man has a conscience and individual identity associated with his origins and his collective family memories which, by manifesting and interpreting individually, are reflected in a personal and non-transferable way in the dwelling. This appears as the intimate and reclusive space where man releases part of the social masks that he assumes in the interpersonal relationship with the outside.

Symbolically, the house represents our unconscious and in this context the house – here understood as home – as well as the mind, expresses the cognitive and emotional content that constitutes us as individuals distinct from the group, making it a repository of our physical, affective and intellectuals experiences (Cardua, 2011).

Our memory and our life history find a favourable place of expression in the house. Because of this, the way we organise it can tell us a lot about how we feel, how we think, and how we act in the world. It is symbolically an image of ourselves, the outer representation of our experiences and one of the most profound and unconscious expressions of our individuality. In other words, by being so much a part of what sets us apart from others, the house, seen in the light of a privileged space of communion with our unconscious, can offer a place of recognition of our identity, creating a sense of well-being and fullness.

Associated with the identity of the place, vastly reinterpreted by some philosophical currents of the second half of the twentieth century, the house is a collection of our deepest experiences. Essentially because it is in the private and intimate space that best formulates an interpretation of who we are and what we want to be (Cardua, 2011).

At the bottom and at the top of the social pyramid man is allowed to intervene in outer space and in the design of his own home. So it happens in the house of illegal genesis and in the detached house customised by the architect to the taste of the owner. For most, the space of the dwelling is acquired with the formatting to the median taste and it seems like an immutable place, leaving to its inhabitant the decoration to personalise it and in it to mirror a series of values.

This article seeks to deepen the relationship of the inhabitant to the dwelling, in its most intimate and personal dimension, which we call home, and in the way that this process can be achieved through architecture, so that Man expresses his unconscious in the interpretation of space.

2 THE HOUSE AS A SPACE OF TENDENCIES

The information age has made the house a space of trends, a fashion space that perishes with time. In fashion, we leave aside part of who we are, implied by signs that we mistakenly recognise as ours. The image we want to convey of what we socially aspire to is often of medium or bad taste, just because we are inundated every day by the media with images that virtually transport us to associations with a higher social status.

Therefore, turning the house into a space of tendencies will be one of the problems that prevent people from obtaining genuine satisfaction in their dwellings. Nowadays many people deal with the house as if it were a fashion showcase where the must-have the way of belonging to a particular group invariably associated with a given social status. With this, it is common to forget to prioritise the personal taste, associated with their value system, and the comfort necessary for the house to function for the purpose it has unconsciously always meant: our space of seclusion and protection from a threatening outer world.

At the forefront of combating this trend, we have witnessed in the field of Interior Design a movement back to simplicity, the search for a more human and less object space.

Simplicity can, in a certain sense, contribute to the solution of a common problem in contemporary domestic spaces, which prevents proper use of the advantages of being at home: excessive information.

Houses full of furniture, textures, colours, objects and with plenty of sensory stimuli end up suffocating our perception of the environment. This suffocation inhibits our ability to appreciate the details, to absorb appropriately the visual, tactile, olfactory, and other information that constitutes the environment, and which are necessary for the way of perceiving and enjoying space.

The sensuous enjoyment allows us to integrate into the environment, and not to be submerged in it. Simplicity, understood as the environment thought from what is essential and pleasing to the senses of those who inhabit it, creates conditions so that the focus of our attention turns to ourselves. When this happens, we can properly interpret information from the environment, processed by the brain, without excess, and it becomes easier to pay attention to our feelings, thoughts, sensations and emotions.

The interaction of the inhabitant in his domestic space is circumscribed, with some exceptions, to the decoration, either in the colour we choose for the walls, in the paintings or photographs that we hang on the walls or in the furniture with which we furnish the interior. Although very few have the possibility of building a home during life and thus perceive how they relate to the surrounding environment, we all have the possibility to create some space that we can call "ours" and "ours" consciously or unconsciously, we can Shape as an expression of our identity.

According to Clare Cooper Marcus (1995) for some people the possibility of intervening in the house, whether in the decoration or the constant realisation of works of alteration, is not a mere whim or just a hobby, but rather a fundamental part of its existence. The house assumes itself as an evolutionary extension of itself and not as a mere object.

Contrary to the view of contemporaneity in which the distance from the problem to its own resolution is required, in this relationship with the house the higher the level of individual/environment interaction the smaller the distance between the observer and the object. Therefore, the reduction of this distance, which makes the understanding of the messages of our unconscious easier and immediate, can be achieved the greater the freedom of intervention of the inhabitant in its surrounding space.

The decoration of the house is now widely recognised in psychology as one of these processes, just as the clothes, the haircut or the car we drive are intrinsic expressions of our values and the image that we intend to design. This kind of interpretation of space, consciously or unconsciously, is undeniably a vehicle of communication with ourselves and with those who enter our private space.

However, will it be the decoration, enough intervention so that our intimate space can fully reveal our unconscious? Part yes, but not enough.

To go further in understanding this process, it is important to conceive at the outset three fundamental premises:

1. The economic value of a home, and in turn the value of its decoration, is not related to the well-being that it can psychologically afford;
2. The house is the space that contains the deepest reflection of ourselves as a sensory way of realising where we are and where we want to go;
3. Reorganising space to a form that satisfies us – and this stage is changeable throughout life – is a universal necessity (Marcus, 1995).

The understanding of these three premises allows us to conclude that the mere decoration of a house is only part of the possible representation of the unconscious of its inhabitant and that it is possible to go ahead in the manipulation of the space in search of the diminution of the distance between the observer and the object.

3 THE PLACE IDENTITY: HOUSING AS THE EXTENSION OF HUMAN INDIVIDUALITY

Some strands of philosophical thought of the second half of the last century, like structuralism, have based their theory on a model in which social reality is deduced from a considered elementary (or formal) set of relations. For sociology, anthropology and linguistics, structuralism is the methodology by which elements of human culture can be understood in the face of its relation to a larger, more comprehensive system or structure. Post-structuralism overcomes this view and understands that reality is a social and, mainly, subjective construction.

The concept of place identity as a substructure of the identity of the self, built in the relationship with the socio-physical environment in which the individuality of each person is forged, begins to be apprehended and gradually reaches other areas of the social sciences (Proshansky, 1983). Man ceases to be the number of the Modern Movement and the importance of the relation of Man to his surroundings, which inevitably influence, begins to acquire expression.

From the perspective of Carl Jung (2011), the feeling of place identity is constituted by matter that has been genetically inherited and is where the functional traits, such as virtual or unconscious images, that would be common to all human beings reside. As a superior collection of experience, this collective unconscious is not derived from individual experiences, and by this, Jung designated all the functional traits of the unconscious as archetypes that were not observable among themselves but only through the images they provide.

The place identity can be a division of a dwelling, a dwelling of a neighbourhood, a neighbourhood of a city or a city of a country. However, regardless of its size, an identity place is a home in its broad sense. It is a place of security in an insecure world, a space of certainties in a world of uncertainties, a familiar place in a strange world, a sacred place in a profane world. It is convincingly a place of our autonomy and power in a heterogeneous world where others make the rules (Dovey, 1985). The perception of increasing autonomy and power will indirectly be a way of reaffirming our sense of control in the world.

Housing reflects personal identity as the gathering space of intimate moments in which the human being is freed from social masks and assumes himself in his true essence, in his true Self. As Marc Fried (1963) states, "being in our home does not It's just being in an apartment or a house, but in a territory where some of the most significant experiences of life are lived." For Bachelard (2008) the importance of housing is understood in the light of the concept of "maternity of the house", in which it states that, it is necessary to go beyond the descriptive aspects in order to achieve the first function of dwelling: the shell. In this relation are included ideas, values, feelings, attitudes, meanings and conceptions of behaviour related to a complexity of places that define the daily existence of each human being. In this perspective, the identity of place must be understood as a personal construction in which direct experiences with the physical environment modify it. It has a meaning for the subject that it incorporates into the identity of the Self. For each role that the individual plays in society, there are dimensions and

Figure 1. "The house of the homeless" from Elaine Pedreira Rabinovich. Source: http://www.scielo.br/scielo.php?script=sci_arttext&pid=S1414-98931992000300004 (Accessed in February 2016).

characteristics of the physical environment that help to establish that identity. The house gives the address that, in turn, identifies the subject in the space of the neighbourhood and the city. In this sense, place identity (its meaning) is a specific component of the subject's own self-grounded in the interaction with its surroundings.

The meaning of dwelling goes back to the very history of our evolution as a species. For primitive man, the first shelter, the cave, was the space in which he felt protected from external threats, in which he felt comfortable to rest, to procreate and to care for one another without fear of the imminent attack of a predator or the fury of Mother Nature. The primitive feeling that the place to which we return when we leave the impersonal outer world is our shelter and our defence against unwanted threats persist unconsciously in our brain. The sense of comfort we feel upon returning home is triggered by this emotional inheritance, inherited from our ancestors and preserved by our limbic system (Cardua, 2011). So we can say that if the house offers us this sense of comfort, protection, relaxation and recognition, it will necessarily contribute to physical and emotional well-being; Feeling good in a space that we recognize as our own, in our image, will contribute to combat the daily wear and tear that is required in public spaces, where we need to adapt to the impersonal.

The sense of identity appropriation of housing is something inalienable of the inhabitant himself at the ends of the social pyramid that is to say in the economically more and less favoured classes. Only these have the condition to reflect, within their economic condition, their experience through the plasticity of a volumetry or the functional organisation of the interior, whether in clandestine precarious construction or in a personalised single family dwelling. By contrast, in the real estate market, the quality of the house as an identity space is diluted, leaving the inhabitant to live according to typological models that are very experienced and, in the majority of cases, rigid, which refer to their decision-making power only the decoration of the space, as Way of expressing individual identity. Colour is discussed to paint a wall, the location of a picture on a wall, but usually, the wall itself is not discussed. It exists and the individual takes it as immovable.

In this architectural paradigm of mediating proportions, compartmentalization and, why not, aesthetics, to increase the range of potential interested, flexible housing acquires special importance.

4 THE INTRODUCTION OF PERMANENT FLEXIBILITY AS A PROMOTION OF THE PERCEPTION OF THE UNCONSCIOUS

Nowadays, there are theoretical alternatives, such as that of structuralist and post-structuralist architects, who, without denying contemporary forms and thinking, seek to learn from environments, traditional urban fabrics, given that architecture is a process of participation that is based both on culture as in the history of the place. As Edward T. Hall best explains in "The Hidden Dimension" (1986), architecture is as rich and diverse as the society itself, which serves as the seat of exchange for human activity, and which is far from being untouchable if Bridge between the city and citizenship. Architecture will always be a cultural act, first structured in a spatial and historical context.

As Edward T. Hall best explains in "The Hidden Dimension" (1986) architecture is as rich and diverse as society itself. It serves as the seat of exchanges for human activity, and that, far from being untouchable, it places itself as a bridge between the city and citizenship. Architecture will always be a cultural act, first structured in a spatial and historical context.

The answers to the elaboration of a conceptual methodology of a project that aims at the full satisfaction of the inhabitants seem to be natural since the points of contact between the architectural imaginaries and the living space are evident. In the first place, it must be recognised that the urban environment is eminently territorial, consisting of a hierarchy of thresholds, interstitial spaces and other elements of spatial delimitation that are essential parts of our environment. Historically this territorial hierarchy has provided a sense of security and identity by inspiring a large part of the architectural language of a place.

Secondly, the importance of the value system must be recognised. The traditional urban forms that we admire were generated because builders and inhabitants shared an understanding of typologies, patterns, and styles. The traditional city that we learn to admire is endowed with an ancestral repository of collective memories, feelings and values that dominate a part of our unconscious and give us a sense of belonging that is no more than the identity of the place, which the city Modern does not contain.

Paradoxically, it is in the traditional neighbourhood where the identity of place is stronger, that the house is notoriously less flexible. In the place where the social man best interacts with the community and the surrounding neighbourhood, where you experience the street in a way that modern/contemporary settlement does not allow, is where you are mostly limited to the possibility of intervening within your intimate space.

If we have already noted the importance of human intervention in its primordial space, it increases when added to the feeling of place identity. The conjugation of this identity, culminating in the identification and sense of belonging to a group, associated with a house full of effectively identitarian symbols that recognise each other and recognise us as a whole, is, therefore, a conjugation that must be enhanced by widening the possibilities of the inhabitant to intervene in space.

Human beings make the society and these are different from each other. Therefore, Hall's (1986) expression in which he says that Architecture is as rich as society itself seems to confirm that there is space in social diversity for the multiplicity of architectural forms. Or rather, that for each Man there may be a different architectural solution. Thus, it is up to the architect to understand that the space that makes available to the inhabitant must take into account solutions for the expression of the individuality of each one. The more forms or possibilities that exist in the interior configuration of the house the greater will be the fullness of the inhabitant and therefore the better the society.

That being said, the partition wall cannot be unchanged in the urban regeneration of traditional hulls. Technically, the restoration of a building in this type of settlement today responds to a different program from that which the building responded organically throughout its useful life, mainly because the dimensions of habitability are different from the evolution of the anthropometric and because the ventilation and the lighting of the house are essential aspects in the salubrity of the living space. If the architecture introduces technical parameters by virtue of the technological evolution and the increase of the quality of lives of the inhabitants, it can also introduce solutions that respond to the concerns of some social areas that are concerned with the personal development of the inhabitant.

The permanent flexibility in the interior of the house, as a multiple and constant possibility of man adapting his private space to his individual needs, as his life develops and goes on experiencing other stages, is the same as saying that the house follows the development of its inhabitant and in this symbiosis the symbol of the Self is undeniably closer to the Self.

5 CONCLUSION

In a city that regenerates as a living organism and identifies place identities associated with values, experiences and collective memories, Man finds the ideal space to develop himself as a social man and to satisfy fully his needs for interaction with other members of the group. His vision recognises the places, his smell identifies the smells, his touch knows every route, his hearing perceives the sounds that surround him and everything is familiar. This proximity creates in Man a sense of well-being that is unparalleled in the endogenous potential of his dwelling. In this, strict rules of compartmentalization limit their creativity and, therefore, curtail their personal development in their most intimate space. If it is proven that the way we interact with the interior of the house tells us a lot about our personality (Marcus, Fried, Bachelard, Jung) and it turns out that this is practically limited to the decoration of the house, except for the fringes of the social pyramid, then increasing that interaction is a way of bringing man closer to a gradual process of self-understanding and acceptance. The permanent flexibility, in which the house adapts to the Man, in the different phases of its life, to the detriment of the abandonment of the place where it feels socially inserted, seems to respond fully to this demand. Rigid compartmentalization can be neglected in urban regeneration by introducing new architectural techniques that adapt the building to contemporary habitability parameters with the adoption of various architectural solutions where the creativity of the architect is called and recognised. Thus, the architect does not stray from his mission of designing for Man, as an artist that effectively is, nor does it dilute the very diversity of architecture, but understands and accepts that beyond the basic functions of the house there is an invisible space of fomenting the sensorial perception of the inhabitant on the way to the understanding of his unconscious.

BIBLIOGRAPHICAL REFERENCES

Bachelard, Gaston. (2008) *A Poética do espaço [Poetic of Space]*. São Paulo: Martins Fontes Publishers, 2.ª Edition.

Cardua, Angelitas. (2011) *A casa e o seu significado para o bem estar físico e emocional [The house and its meaning for physical and emotional well-being]*. Available in: https://angelitascardua.wordpress.com/2011/08/02/a-casa-e-seu-significado-para-o-bem-estar-fisico-e-emocional/. (Acessed in April 2016).

Dovey, Kimberly. (1985) *Home and homelessness in Human Environments; Human behaviour and environment: Advances in theory and research.* Vol. 8. Plenum Press, New York.

Fried, Marc. (1963) *Grieving for a lost home.* Leonard Duhl, ed. The Urban Condition. New York: Basic Books.

Hall, Edward T. (1986) *A dimensão oculta [The Hidden dimension]*. Lisbon: Relógio d' Água Publishers, Lda. Antropos Colection.

Jung, Carl. (2011) *Os Arquétipos e o Inconsciente Coletivo [The archetypes and the collective unsconscious]*. Rio de Janeiro: Publisher Vozes.

Marcus, Clare Cooper. (1995) *The house as a mirror of self. Exploring the deep meaning of home.* Berkeley: Conary Press.

Proshansky, H. M., Fabian, A. K. and Kaminoff, R. (1983) *Place-identity: Physical world socialization of the self. Journal of Environmental Psychology.*

Living the contemporaneity: The idea of progress from the origins of the house

Carla Chiarantoni
DICATECh, Polytechnic University of Bari, Bari, Italy

ABSTRACT: The problem of the home has always been a social issue of utmost importance. Home is one of the priority needs for men, to be sanctioned as a right in the Universal Declaration of Human Rights (1948). Housing and the right related to it are the assumptions for many aspects of life, work, family, health, physical security and well-being. The possibility of exploiting this right conflicts with various issues related to legislation and programming of development's processes and anthropization of the territory. These circumstances have always influenced the idea of home that reaches to this day as a certain and consolidated axiom over time. However, the society is currently undergoing profound changes. So, how does the paradigm of the past combine with the needs of contemporaneity? This paper intends to describe, from what we could call the prodomi, the evolution of the dwelling model until today. The intention is not to start from origins of the house but to go through some fundamental steps that have led to the formation of the great building heritage, gave shape to the Italian cities, to arrive, if possible, to the analysis of the needs of contemporary life and its housing model reference.

Keywords: council housing, space and form of housing, living contemporaneity

1 COUNCIL HOUSING: ORIGINS

The housing problem or better the workers' housing started to emerge, in the whole Europe, starting from the 19th century. This type of "house" was supposed to be thought and designed for the "masses". The intellectuals of the time were entrusted with this task; they had to respond to some needs, adjusted to the whole working class, respecting the fundamental individuality of the family and the maximum economy of the construction.

The necessity of the "home for workers" was born from a series of historical events that, after the industrial revolution and scientific progress, led to the large migration of the population from the country to the city with the subsequent uncontrolled growth of the urban population. The cities, assaulted, were affected by uneven growth and in some areas, the home within it became a pernicious problem. It was the start of the first debates, which involved the city and its form, on the one hand, the housing and their healthiness on the other.

The "workers' house" became, substantially, one of the most important themes and large field of experimentation in the whole Europe. If the theoretical basis was, one can say, substantially common on a European scale, the different result derived from a different attention to the administrative and economic policies from the various countries. In fact, by moving differently, these obtained different results. The situation that occurred in Italy was particular. The scarce availability of capital and the lack of fossil fuels caused a deceleration of the industrial revolution, which only started towards the end of the 19th century. The housing problem emerged, therefore, with a delay in relation to the European context and manifested itself in a double aspect: the absence of housings and the renovation of the existing ones. The existing housing estates in the cities, in fact, were not suitable since they lacked, in the majority of cases, hygienic facilities. The response to this state of distress regarded only a few Italian cities particularly afflicted by epidemies or those in which the government took interest in the improvement of the living conditions [law n. 2248 of 30/03/1865 and law n. 5849 of 22/12/1888].

The renovation actions could not be random. The attempt to uniform the Italian conditions occurred, in early 1900, with the introduction of Luttazzi Law [Law n. 251 of the 31 May 1903]. Among the many initiatives, the most important were, undoubtedly, the introduction of credit and tributary subsidies in favour of specific building subjects of Council housings. This condition led to the institution of organisms meant exclusively for the construction of houses destined to "the weaker bands of the population". The process started before the two World Wars, saw a moment of stall between the two conflicts. In the aftermath of the end of WW2, the situation was undoubtedly more problematic due to the scenery of great destruction and the "birth" of other afflicted categories within the middle class. Among the recipients of the interventions were, therefore, also included the middle-low social classes which were, equally, in a deranged state of housing. Substantially, two typologies of houses were included within possible fundings: the Council housing ("workers' house") and the "economic house". Differently, from the Council one, the "economic house" was realised for social groups belonging to certain institutions not necessarily poor but certainly in poor economic conditions, i.e. the middle class; vice-versa, the "Council housing" was aimed at those who could not afford a house, i.e. the low poor class. In both cases, all the hygienic and health conditions had to be met as required by the Council regulations and have certain building features: every house required its own direct access, be equipped with a latrine, water supply. The two categories were supposedly distinguishable in terms of the economics of the construction, the size of the single quarters, the number of rooms, the quality of the materials employed, and the finish.

This is the period of creation of institutions for the development of housing as the National Insurance Institute (INA-Casa) and the strengthening of the existing ones, as the Autonomous Council Housing Institute (IACP), destined to make the history of the "economic and Council housing". The idea of a house having to host an increasingly growing population started to be conceived as a collective house with a medium or high number of floors, involving the preference of the typological model of the terraced house. The idea of the city, instead, was accompanied by disembowelling policies of the town centres and the creation of densely inhabited outskirts lacking services and connections.

2 THE INA-CASA PLAN. THE FIRST STEP TOWARDS MODERNITY

The reconstruction panorama on national scale finds its origin with the INA-Casa Plan. This will mark the beginning of an important phase of the post-War economic policy, which will leave the most important testimonies in the field of social housing in Italy with the construction of more than 350.000 houses. The INA-Casa Plan was aimed at bringing the construction of low cost houses for employees (a right then extended to all the citizens regardless of their job, as long as they were in need of a house and met certain requirements) and the incentivisation of occupation through the employment of non-specialised workforce for the construction of new buildings.

The entire programme was articulated in two seven-year plans, operative between 1949 and 1963 and this period represents, undoubtedly, one of the most important stages of urban building and social development in Italy.

The institution was organised in two fundamental branches: the Executive Committee and the Management INA-Casa. The first branch reported directly to the Ministry of Labour, the second had operational functions. The latter, indeed, was responsible for the coordination of technical and planning aspects. The Architecture office internal to the Management branch defined the unitary references for the urban architectonical and technological aspects, was responsible for the centralised approval of all the projects and the monitoring of the construction stages. The guidelines were collected in a series of files and had the purpose of stimulating and granting high-quality levels in the planning and construction stages. The files were divided mainly into two macro areas: "Suggestions, norms and schemes for the elaboration and presentation of the projects – calls for tender", aimed at the definition and the typological, functional and distributional planning"; "Suggestions, examples and norms for urban planning – typical projects", aimed at tracing guidelines for the formation and urban organisation of the quarters.

Surely these files were fairly innovative for the time: on the one hand, reintroducing concepts originated from northern European countries (England and Scandinavia, in particular), tried to propose models for urban development based on the construction of autonomous and isolated quarters; on the other, they introduced the constructive categories to be built and the related constructing techniques to employ. The vade mecum provided punctual directions on the kind of building to favour considering its height, its surface in relation to the habitable rooms, the functional-distributional organisation, the exposure, illustrating everything with drawings and particulars. The rooms had to be sound and modern, positioned in newly constructed urban areas, and offer the possibility of thousands of families to improve their living conditions.

The operation surely tended to try providing a middle-high habitable standard, in both constructive and urban terms, but the urge to act and the scarcity of the resources, along with a possible misinterpretation of the models of reference, led to the acquisition of average peripheral areas, which lacked collective services and equipment. Another important issue regarded the non-differentiation between the settlements to be constructed in agricultural towns and industrial ones, generating, consequently, urban outskirts, similar to one another and without an identity.

3 THE ARCHITECTONIC FOOTPRINT: THE SPACES, THE FORMS, THE MATERIALS AND THE CONSTRUCTIVE TECHNIQUES

The experiment INA-Casa was based on a global idea of planning that ranged from urban scale, up to the details of the building. The founding idea was to create quarters of public buildings through the construction of a primary social residential unit in which the importance was given not only to the housing spaces but also to the spaces external to the building (streets, squares, etc...). These outdoor spaces, through the creation of hierarchical places and connections, should have caused an increment of the housing space.

The general idea was, therefore, to create an entire public quarter, in a proper urban context with new urban development procedures. Some of the experiments in this direction involved, for instance, the rupture of the relationship building facade-street facade, with the creation of ground floor apartments, sometimes correlated by a small space in the front (first seven- years period); the introduction of the balcony, inspired in northern European countries; the formulation of public spaces as systems building-place of relevance, in a sequence of public and private spaces, opened and closed.

Quite different was the matter of the construction techniques and materials. In the rest of Europe the first technological experiments had started on the dry layered construction, or the steel structure, in Italy it was the "Piano Fanfari" itself to lead the alternative reconstruction.

As Di Biagi (2001: 114) explains: The plan, in fact, identifies in the construction of houses for the workers, long before the goal to satisfy the housing need, the means to increment the employment of the workforce with a low-mechanised manner of construction and high employment of labour, the innovation standing in the typification of the constructive elements. In addition, Di Biagi continues referring that it was a singular strategy, one based on the block of construction innovation, different from those adopted for the post-War reconstruction in other European countries.

The technical and technological specifications were delineated, still in the INA-Casa files, and distinguishable in a more general part and in a more detailed one inside the special sections.

Still following Di Biagi study it was a system composed of a balanced combination of masonry elements and armoured concrete elements, all made mostly along the way. In the two or three stories buildings, the supporting function was usually ascribed to brick walls or stone blocks with an innovative effect due to the possibility of the attic masonry to perform the function of bracing, with the consequent prospect of lightening the non-supporting walls. Considering the taller houses or towers, Di Biagi refers the stability was ascribed to an independent armoured concrete skeleton structure even if the typical conformation of the wall construction was maintained.

Anyway, the INA-Casa plan became a great opportunity for all the Italian construction specialists, from building surveyors to engineers and architects on the territory.

The study and research based on the proposed housing model were interesting: the idea was very simple, providing housing to a family (be it working class or middle class) whose demands were thoroughly analysed, but substantially common. As already mentioned, the difference consisted almost exclusively of the number of locals, in the materials employed and in the finishing.

In the first INA-Casa experimentation (first seven-year period), the size of the houses was fixed in relation to the number of beds: a double bedroom involved a 30-sqm housing. This surface could be incremented by 15 sqm for every added bed up to a maximum of six beds, to which corresponded 90 sqm.

In the second INA-Casa seven-year period, the idea of calculating the surface according to the number of beds was abandoned and the concept of the number of locals prevailed: the minimum housing became two rooms, to which corresponded 50 sqm, to the three rooms corresponded 70 sqm, four rooms 90 sqm, five rooms 110 sqm.

The typological model that dominated was the terraced house, 3 or 4-storey on average, with a staircase, which distributed two or three flats per floor, able to house a high number of people. The distributional functional organisation was founded on the necessity to separate the living from the sleeping area and with the idea of making the living room independent. In the attempt to minimise the passageways, it had often been decided to distribute the various locals of the flat. The day area was meant to be organised guaranteeing the health principles of the locals and so, guaranteeing a transversal ventilation. The sleeping area, located in succession, was meant to be organised with bedrooms and hygienic services.

A conception of housing was defined which privileged the clear separation between the areas dedicated to living together (living area) and those dedicated to resting (sleeping area). Ultimately, the optimisation of the surfaces caused the disappearance of the "representative" locals, which, for example, consisted of the "fancy living room". All the locals were thought, studied, and optimised to be lived.

Inevitably, the realisation of the model to follow, provided by the vade mecum of the "INA-Casa notebooks", presented the problem of repetitive and subsequent depersonalisation of the housing solutions. Among the proposed resolutions the one that emerged was the rotation of the standard module (of the housing with respect to the staircase or the entire building with respect to the blind side) sided, wherever possible, to the altimetric variation of the floor plan of the building.

Undoubtedly, the experiments of this period, which were achieved in the realisation of entire neighbourhoods of economic and popular buildings, have constituted the precursor of what has been the conception of housing for the whole modern period and a great part of the contemporary times.

The research of the standardisation of the needs for the average family, regardless of the class it belonged to, was undoubtedly important in this period. The housing estate heritage, up to present days, has been so conceived and realised for apparently "unchanged and constant" demands.

4 LIVING THE CONTEMPORANEITY

Since the 1980s, the cultural landscape described, historically dominated by the housing theme, has seen the rise of other themes and interests on the constructions in which the single artefacts and their uniqueness prevailed. To this situation, the energy crisis was associated (in the early seventies), definitely switching the housing problem to a matter of mere energetic performance.

The research, therefore, switched mainly to the cladding of the building, in order to obtain a renewal and optimisation of the energetic performances. In addition, to achieve a normative regulation of the technological, implant and performance characteristics of the artefact, it has continued to conceive the house with logics of aggregation and organisation derived from the past. In this whole system, therefore, the housing quality has been considered almost exclusively connected to an energetic performance, rather than to a global view.

However, society has changed in the meantime. The demographic and social-cultural changes of the last decades have deeply affected Italian society and are not easily reversible. The modifications of familiar structures, the relationship between generations, the relationships between genders and the mobility of human capital due to various causes have deeply changed the demand for "housing".

Therefore, with respect to the years in which the majority of the existing building heritage was constructed, the society and in particular the family model has strongly changed, fragmenting itself, and giving birth to multiple distinct models. One of the possible interpretations of this phenomenon (Rosina et al., 2014) lies in the emancipation of women, in the structuring of the family in a reduced number of components, in the increase of people living on their own.

Other factors are to be found in the later emancipation of youth, connected to an economic model in crisis, in the greater self-sufficiency of the elderly, as well as the changes in the interpersonal relationships and the domestic and working demands. All this and more have contributed to the creation of an innovative paradigm based on the characters of flexibility and lack of hierarchisation of the housing.

Therefore, in order to understand the demands of this "renewed population" that now enters the housing market, it is necessary to understand the possible grouping. A categorization, which can be derived, regards mainly two macro areas: the family in its different aggregations and the new social groups.

Within the macro area of the family can be found (ISTAT, 2011):

- Single person nuclei within which all the economically autonomous people aggregate with a desire to maintain their independence and, therefore, youths and self-sufficient elderly. This category includes an interesting phenomenon whose acronym is LAT (Living Apart Together) which regards people involved in stable relationships, but continue to preserve their own independence (Arcaleni et al., 2007);
- Nuclei of recent origin chiefly related, but not only, to the economically independent young couples without children;
- Reduced nuclei: constituted by a maximum number of components of four, of which only two are economically independent. This is the category, which includes the "new formation of the family". This category is due to a late emancipation of the parental component or the economic impossibility to sustain a wider nucleus;
- Extended nuclei constituted by a maximum number of components of five, of which only two are economically independent. Similarly, to the previous case, also this category can be defined as the evolution of the traditional family of the 60s–70s;
- Single-parent nuclei, composed by a single parent figure, generated for different reasons, among which separation, divorce, widowhood or the arbitrary choice to become a single parent;
- Complex nuclei, also known as reconstituted nuclei (Martorell, 2006), composed of previously fragmented family nuclei;
- Units without a nucleus: constituted by individuals among whom no blood and/or sentimental relationship exist and for whom the economic power on the housing is equally held.

The demands expressed by these variously articulated nuclei are numerous and are shown not only in the demand for a "house" different from the building heritage received from the past but also in a different conception of the relationship with the city, the housing times and ways.

If in the past the house and housing, in its general conception, were seen in a static dimension, permanent in the lives of the individuals, currently the concept of housing has inexorably launched itself towards a transitory and dynamic dimension. Hence, the development of some housing and working experiments: co-housing, co-living, social hotel, co-working, which originated in northern European countries and are spreading in Italy. These models try to conjugate the diffused research of individual privacy to the opposed tendency of sharing spaces and services typical of the "traditional house". In these models, it is not only the flat that must be rethought but also the whole building. The codification of design strategies becomes fundamental to be able to repair the fracture between the public and private parts and respond to the demands of the new housing ways.

5 CONCLUSIONS

The strong variability of the research on the theme of housing is mainly concentrated on the technical and technological character of the building artefact, on its qualities in terms of performance and energy efficiency, and little on the dimensional and functional quality and the nature of the spaces. The important cultural and social modifications, vice-versa, are imposing a thorough reflection on the housing ways and times. The ambition of the users has substantially changed with respect to the idea of the house. If in the two INA-Casa seven-year periods the typical, the idea of standardisation of the housing, the idea of a vade mecum was successful to overcome the building impasse, it has however been a rigid and standardised route.

Currently, we can no longer rely exclusively on standardisation thought to guarantee a dimensional minimum. This idea with the intent to guarantee a requirement has actually blocked and stiffened a process that has repeated itself even though the society asked for something else. For the "INA-Casa route", it is important to save and try to repeat, currently, the same synthetic view of the cultural and social reality, trying to guarantee a very high design flexibility. It is not important to exclusively provide a number, a piece of qualitative data (similar to the energetic matters), but a process, a project path which takes into account the changed ambitions of the users.

The idea of the regenerative design (Cole et al., 2013) ultimately, in its holistic view, places great attention on the multi-directional component of the project and its ability to provide a better quality of life to the individuals, not exclusively visible in a higher technicity of the forms. If in the first place, in the second post-war, there was an exclusive concentration on the research of the form, to move subsequently to an energetic view of reality and sustainability, the current tendency is a search for satisfaction, which sees the cultural and social needs united to the technic and technological ones. The technologic aspect could paradoxically be defined as "slow" since it affects the energetic management of the building, the housing, but not the quality, the size and the nature of the inhabited spaces.

BIBLIOGRAPHICAL REFERENCES

Arcaleni, Emilia & Baldazzi, B. (2007) *Vivere non insieme: approcci conoscitivi al living apart together.* Roma: Istat.

Cole, R.J., Robinson, J.B. & Oliver, A. (2013) Regenerative design, socio-ecological system and co-evolution. *Building Research and Information.*

Di Biagi, Paola. (2001) *La grande ricostruzione. Il piano INA-Casa e l'Italia degli anni Cinquanta.* Roma: Donzelli.

ISTAT. (2011) 15° Censimento generale della popolazione e delle abitazioni. Roma: Instituto Nazionale di Statistica.

Martorell, Josep María Montaner. (2006) Vulnerabilidades urbanas: separar, olvidar, deshabitar. In: Font, Joan Nogué I and Romero, Joan (eds). *Las otras geografías.* Valencia: Tirant lo Blanch, 353–368.

Rosina, Alessandro & Migliavacca, Mauro. (2014) Investimento in formazione e qualità del lavoro. In: Toniolo, Istituto Giuseppe (ed) *La condizione giovanile in Italia. Rapporto Giovani 2014.* Bologna: Il Mulino, 25–49.

Progress and the happiness ideal: Materialisation of a utopia with the fortified enclave: The case of Casa Forte

Tatiana Fonseca
Faculty of Architecture, University of Lisbon, Lisbon, Portugal

Alexandre Fonseca
Maxwell School, Syracuse University, Syracuse, USA

Fernando Moreira da Silva
Faculty of Architecture, University of Lisbon, Lisbon, Portugal

ABSTRACT: The advance of construction techniques and the increase in urban violence led to a process of transformation in Casa Forte, a traditional neighbourhood in the Brazilian city of Recife. Single-family residences are being replaced by large vertical condominiums thanks to their leisurely appeal and security. They became the ideal of happiness for part of the residents of that vicinity, which greatly affected the urban life of that neighbourhood.

It turns out that living in one of these vertical condominiums, with characteristics of fortified enclaves, brings the resident happiness by living in an apartment that not only provides social prestige but it is also considered modern and safe and offers leisurely activities with peers from the same socio-economic point of view. These skyscrapers have also brought a civilizational setback to the neighbourhood. This contributes to the security problem, increasing the traffic and infrastructure problems within the vicinity and generating socio-spatial segregation, which leads to the denial of the city as an area of freedom. Public policies are needed in order to prevent these harmful effects from undermining the collective happiness of the neighbourhood.

Keywords: progress, happiness ideal, fortified enclave, urban violence, Casa Forte.

1 INTRODUCTION

The neighbourhood of Casa Forte is one of the most traditional vicinities of the Brazilian city of Recife, which assumed its name from a sugar cane mill – Engenho Casa Forte. A green area, with relatively narrow streets and avenues, the neighbourhood carries with it a symbology of tradition and status, which is reflected by the high cost of real estate.

Figure 1. Typical house in the neighbourhood of Casa Forte. Credit: Author, 2017.

Originally formed by single-family horizontal residences (figure 1), the neighbourhood witnessed, since the mid-90s, the growth of urban violence and a concomitant process of replacing houses by vertical condominiums of more than 30 floors (figure 2). The vertical condominiums' leisurely appeal and security have become symbols of social status and the object of desire for many of its residents.

The growth of these bubbles of safety raises questions not only about social inclusion and territorial justice but also about the impact on the urban life of the neighbourhood and its quality of living.

2 THE URBAN VIOLENCE GROWTH, PROGRESS AND THE FORTIFIED ENCLAVES

The process of producing gated residential spaces is not a recent one or a privilege of Brazil, Recife or of

Figure 2. Maria Clementina Viana Building – A skyscraper in the Casa Forte Square. Credit: Author, 2017.

Figure 3. A view of the Alameda Casa Forte Building recreational area. Credit: Author, 2017.

the Casa Forte neighbourhood. Sposito (2013) points out that these residential spaces are already found in the United States in the 19th century and that in the 1920s these condominiums appear in Venezuela and Argentina.

Gated communities have been defined in different ways but they normally include residential areas with restricted public access, with gates, walls and fences. They often include security systems and services such as leisure facilities. (Atkinson and Blandy, 2005)

Although housing enclaves are seen as residential spaces characterised by the security and isolation of a specific group within society, there are different international contexts and drivers. Atkinson and Blandy (2005) found the desire for status, privacy and the investment potential of gated dwellings as others motivations to live in gated communities. Eshuis et. al. (2011) studied the case of the Netherlands and found that Dutch housing enclaves are predominantly without gates and little segregation and that the main reason for its inhabitants to demand this type of a housing enclave is the search for a pleasant living environment – and not for security.

In the Brazilian context, analysing the growth of horizontal condominiums in the interior of the State of São Paulo in the 1990s, Sposito (2013, p. 65) found great empirical evidence relating the emergence of these condominiums due to the "significant increase in insecurity".

Supporting Sposito's research, Caldeira (2000, p. 260) finds that while the 1970s buildings used to be open to the streets, the new condominiums, in the 1990s, both vertical or horizontal, tend to ignore the urban network, with large areas, many types of equipment of collective use and great security devices, which led the author to denominate them as fortified enclaves.

In the case of Recife and the neighbourhood of Casa Forte, the evidence is that, just as in São Paulo, violence is also the driving force behind these fortified enclaves. In an interview in a local newspaper, José Simón, the vice president of the Union of Civil Construction Industry in the State of Pernambuco, explained that the builders build what the population wants or asks for and, for the same reason that the walls have risen in size and buildings became gigantic, with the leisure areas being built inside the buildings. "It's the lack of safety in the streets", the entrepreneur concludes. (Alves, 2105, n.p.)

On the other hand, in the case of the neighbourhood of Casa Forte, in the absence of large areas available for horizontal condominiums and given the advances in architecture and construction technologies, the option was to demolish houses, rematch lots and replace them with large high rises far from the street, which include the characteristics of fortified enclaves reported by Caldeira.

On Figure 3, one can see the swimming pool, part of the playground, the games building and the children's recreational area of the 40-floor fortified enclave Alameda Casa Forte, which also includes a party room, fitness centre, jogging path and tennis court.

It should be noted that in Casa Forte, a very traditional neighbourhood, apartments were initially rejected and seen as tenements, a point of view that was changed when small buildings with one apartment per floor were developed and then when high skyscrapers with large apartments came about, which have become object of desire and a symbol of social status.

Although in a completely different historical context, the towers of Casa Forte, some with 40 floors, like the one in figure 4, bring us back in symbolic terms to the medieval towers that existed in several

Figure 4. Freguesia de Casa Forte Building – Skyscraper of 40 floors located at Praça de Casa Forte Square. Source: Author, 2017.

Italian cities, such as Bologna and San Gimignano. As well as in Casa Forte, in the medieval Italian context, the towers were not only restricted and well-defended buildings but also symbols of power, so that the family that owned the highest tower was the richest and most powerful in the city.

At this point, it is important to note that the process of fortification observed in Casa Forte and elsewhere in Recife and Brazil is sometimes compared to a process of feudalization. Even though the walls try to isolate the inhabitants from the threats of barbarians or thieves in both the medieval and current contexts, there is a crucial difference. While in the medieval city, there was a socially heterogeneous coexistence, the new walls of the condominiums seek to isolate the community from the socially unequal.

At the end of the 1990s, with the intensification of the process of verticalization, the neighbourhood was experiencing a worsening in traffic conditions, sewage and water supply problems. Because of that, the City of Recife proposed and approved the Law 16.179/2001, which limited the height of buildings in some neighbourhoods to a maximum of 6 to 20 floors and reduced the construction parameters in these areas.

Despite the law, and due to high demand for this type of housing, the market continues to offer, although to a lesser extent, these types of apartments. In Casa Forte, the appearance of these fortified enclaves, including in the Casa Forte Square, a plaza that was designed by the renowned Brazilian landscaper Roberto Burle Marx and a symbol of the neighbourhood, has not yet caused the disappearance of single-family houses, which still predominate in the neighbourhood, but has generated new social relations and impacts that we analyse below.

3 A NEW PARADIGM OF HAPPINESS AND THE DENIAL OF THE CITY

Starting with the micro point of view, for the individual who chooses to live in a fortified enclave, living in this type of housing brings him happiness as it bestows a social prestige, is considered modern and safe, and offers leisurely activities to enjoy with peers from the same socio-economic point of view.

When we analyse these fortified enclaves from a macro point of view, considering the neighbourhood and the city as a whole, as these fortified enclaves spread there is a civilizational setback due to the problems generated in the neighbourhood.

Although the high walls of these condominiums protect their residents, they end up increasing the insecurity in the neighbourhood as these buildings are unrelated to the street, which have fewer people watching the movement of the public area. The attempts of the City of Recife to harmonise these enclaves with the public space requiring condominiums to use metal fences or walls that allow part of the interior of the building to be seen by the outside and exterior gardens between the wall and the sidewalk were insufficient. Note that this process generates a circular cumulative causation movement where insecurity generates enclaves that result in increased insecurity, which increases the demand for fortified enclaves and so on.

In an unplanned neighbourhood, such as Casa Forte, where the infrastructure was designed and built for single-family housing, the replacement of these homes with large buildings is aggravating the traffic conditions and overtaxing the water supply and sewage collection infrastructure. In addition to these infrastructure issues, there is research (Gregolato and Reis, 2012) that correlate high skyscrapers to changes in the local microclimate and to negative changes in the landscape of the city.

In figure 5, the change in the landscape from the construction of a fortified enclave can be noticed. Note the traffic in the area and the contrast between the building and the house.

These fortified enclaves moved away from the unsafe street and began to offer alternatives to the public space. Thus, within the walls, the playground and the swimming pool, the street by the soccer field of the condominium, the gym of the neighbourhood by the fitness centre of the building, replaced the square. All of these alternatives to public facilities increase the socio-spatial segregation, generating a new paradigm

Figure 5. Fortified enclave close to a typical house of the Casa Forte Neighbourhood. Credit: Author, 2017.

and the denial of the city itself, understood as an area of freedom.

These new vertical condominiums are also generating a standardisation of the architecture, leading to a loss of identity and personality within the neighbourhood making its appearance more similar to the other districts of the city.

By generating costs and loss of quality of life for other city dwellers, these mega-projects generate negative externalities to the city, which justifies, from an economic point of view, the intervention of the government through the adoption of public policies to mitigate or compensate for these impacts. (Weimer and Vining, 1998)

The growth in the demand for housing in fortified enclaves, where each resident in choosing this type of housing seeks the best for himself but brings the worst for the community, can be seen as a non-cooperative game, which reinforces the need for government intervention. In this case, the free market is generating social costs.

Given the increase in violence and insecurity that Brazil is witnessing in recent years, which is also happening in Casa Forte, there is little hope that the demand for this type of enterprise will diminish, reinforcing the need to revise the policy of land-use rights in the city.

4 CONCLUSIONS AND RECOMMENDATIONS

The insecurity of the streets and the technical progress of the architecture and construction techniques led to largely fortified enclaves in the neighbourhood of Casa Forte, which have become the ideal of happiness for several residents of the neighbourhood who seek leisurely activities, security, modernity and social status that these new building offer.

It turns out that the progress represented by these high-fortified enclaves was accompanied by a civilizational setback in the neighbourhood represented by socio-spatial segregation, increased street insecurity, increased traffic and problems of water supply and sanitary sewage.

In other words, this new ideal of individual happiness provided by security, social status and leisurely activities with people of the same social-economic status within the walls of a self-contained condominium complex is leading to the denial of the city as a fruition place, an area of freedom and demanding public policies that prevent its harmful effects from generating collective unhappiness throughout the neighbourhood.

BIBLIOGRAPHICAL REFERENCES

Alves, Cleide. (2015) *A arquitetura contemporânea do Recife e seus espigões*. available from: http://jconline.ne10.uol.com.br/, accessed 15/02/17

Atkinson, Rowland and Blandy, Sarah. (2005) 'Introduction: International perspectives on the new enclavism and the rise of gated communities. *Housing Studies*. 20(2), pp. 177–186.

Caldeira, Tereza. (2011) *Cidade de muros: crime, segregação e cidadania em São Paulo*. 3rd Ed. – São Paulo: Editora 34.

Eshuis, Jasper, Klijn, Erik, and Twist, Mark. (2011) Housing enclaves: Security in gated communities or mainly a pleasant living environment? *Journal of Town & City Management*. 2(1), pp. 9–19.

Gregolato, Debora and Reis, Antonio. (2012) Os edifícios altos na percepção do usuário do espaço urbano. *Cadernos do PROARQ* n. 19, dezembro. Rio de Janeiro: Universidade Federal do Rio de Janeiro.

Sposito, Maria and Goés, Eda Maria. (2013) *Espaços fechados e cidades: insegurança urbana e fragmentação socioespacial*. 1st. ed., São Paulo: Editora Unesp.

Weimer, David and Vining, Aidan. (1998) *Policy analysis: concepts and practice*. 3rd ed. New Jersey: Prentice-Hall.

Towers in the contemporary city

Eneida Kuchpil
Department of Architecture, Course of Architecture and Urbanism, Federal University of Paraná, Curitiba, Brasil

Andrezza Pimentel dos Santos
Department of Architecture, Course of Architecture and Urbanism, Positivo University, Curitiba, Brasil

ABSTRACT: The vertical building, the typology of urban buildings that were products of industrialisation and were defined architectonically by modernism, evolved to become a key component of the complex mechanism known as the contemporary city. The architecture of these buildings is an ideal field for demonstrating the cultural and technological changes of the recent decades in the design process. The current international scenario points to the importance of the vertical building, as an image of modernity in cities and its role as an agent of urban qualification, either in the formation of the landscape and construction of public spaces or in the formation of identities. This article discusses the enduring contributions of the modernist project as a paradigm to the individualities of contemporary conceptions.

Keywords: Tower, vertical building, contemporary city.

1 INTRODUCTION

The vertical office building, besides its role as an icon, is also the result of a particular interaction of economic forces and land appreciation that has an indubitable influence in the verticalization of cities with the introduction of the multiple floor building. The construction of the vertical building in cities has been the image of modernity since the process of industrial revolution in the 19th Century, demonstrating the importance of the city as origin (which demands spaces), objective (to which buildings turn to transform) and also place (in which architecture happens) to the elaboration of buildings design.

The introduction of new technologies at the end of the 19th Century, with the Chicago School, that allowed a new vertical building language, going through the years after World War II to the end of the 20th Century, when the technological, typological and urban evolution of construction in height of the building was consolidated in the current mixed-use structures and the search of new paradigms for the contemporary towers.

Even though the debate on tall buildings during the past century has been characterised by extreme positions, for or against this typology, the current international scenario points to a change in this matter. The economic and urban pertinence of these buildings is addressed, discussing insertion policies of vertical buildings, weighing patrimony and preservation against economic growth, image and competitiveness.

The comprehension of the whole period of evolution of the vertical building construction, going through the transition between modernism and the contemporary condition is fundamental to define a link between the founding model and the contemporary scene, because, in some way, this diverse and plural experience of architecture in the 20th Century allows, through many approximations, to undo the complexity intrinsic to the modern experience itself and glimpse the ways of contemporary architecture. According to Solà-Morales (1998), a diffuse heterogeneity comes to the world of architectonical objects. Each work comes up from a cross between partial and fragmentary discourses.

To Abalos (2003) the study of the evolution of office spaces in terms of their topological organisation evidences the distance between modernist and contemporary conceptions of the vertical building.

2 FRAGMENTATION AND INDIVIDUALITIES IN THE CONTEMPORARY CITY

However, in the era of the digital revolution, who will need skyscrapers?

To the French philosopher Baudrillard, the current culture tends to present itself in a completely decontextualized way; "as atoms whose meaning is lost in space" (Cerver, 1998), but from the city, of urbanistic thought and architecture that marked the 19th and

20th Centuries, maybe we can construct a vision of the future of cities and of architecture in the 21st Century.

According to De Masi (1999), the industrial society is a brief passage of human history. Changes in the pattern of consumption; in the ways in which society comprehends the social times; in the ways of production with the change of the mechanical technology of motors to the intellectual technology of information, of codified knowledge and microprocessing and the capacity of interacting with the communicative world mark the passage to the so-called post-industrial society. The communication infrastructures are more important than the transportation and energy ones; the more technical and professional sectors begin to be more fundamental.

Contemporary society has become immensely complex in all its meanings, and the current context and the great social fragmentation, which demonstrate a renewed force of different identities and cultures. The world is constituted by symbolic ways of communication. The ways of communication assume an important role in social transformation because they massify not only information but also promote a fragmentation of cultural identities.

This way, according to Vattimo (1989), technical civilisation propitiates an intensification of the communicative phenomena, provokes a change in human relations with nature. The world then is image and spectacle, becoming as a whole simultaneous and transparent.

In this context, the reflection of Habermas (1999) comes up, considering that contemporary society is determined by the multiplicity of voices that impose a plurality of languages, pointing to a metaphysical unit in the conception of media.

3 CONTEMPORARY CITY: THE PLACE OF ARCHITECTURE

In this social context, the sophisticated techniques used in the construction of contemporaneity, at the same time high-tech and traditional mark the physiognomy of the city. To a certain extent, knowing and recognising the diverse contemporary ways of production and consumption of the physical exterior image of the centre of traditional cities is the indispensable condition to the comprehension of some relevant transformations of the urban space.

The production of the urban space in the contemporary city according to Abascal (2007) demonstrates an evident individualism and fragmentation, opposites that seem to live together as assumptions of the thinking of today's city. In this space of opposition, contemporary architecture comes up.

From the classical studies of Saskia Sassen (1991), it can be affirmed that the known references of a historical point of view for the analysis of urban spaces have transformed themselves. We watch a new imaginary to define urban contemporary space, in which the city, historical source of settlement and permanence, constitutes today more than ever the place of mobility.

Mobility and flows configure paradigmatic images of digital networks and acceleration, which concretize with the emphasis and the investment in infrastructures and mass interconnections. Contemporary life becomes strongly urban and expresses the coexistence of diversity and of inequalities occurring in a fragmented territory, which is the negation of territoriality.

There is a problem that is then defined: the one of knowing which the role is and which are the characteristics, limits and possibilities that contemporary architecture executes in the new places that are at once absence and territory.

Transformations that occurred in the short interval of time of the 1980s and 1990s, lead the main urban centres to an intense expansion of their metropolitan character, in a process of diffuse growth, evidencing a fluidity or dissolution of the urban limits.

Accompanying the productive changes, a digital technological revolution allowed a communication system that changed equally the conception and the production of architecture, which has been absorbing the impact of these digital innovations.

Significant alterations of the work relations and the industrial work stability derive from this set of changes. The tertiary and advanced service sectors were promoted, concentrated in areas endowed with infrastructures and architecture capable of housing companies and qualified professionals.

From increasingly more diverse work notions, the tendency for flexibility is found. Some recent, relatively established, phenomena indicate how this evolution in the work patterns is been expressed: workers working from their homes (Suburban), or telework (work from distance connected through computers); the tendency to transfer corporate offices from urban centres to the periphery, where quick accesses and the construction of great complexes of multifunctional buildings, that replace the monofunctional modernist office building, are facilitated. The monofunctional building implicit in the modernist conception of the city tends to disappear. (Abalos, 2003)

The flexibility of these spaces was accompanied by another transformation that does not refer to the building itself, but to its position in the city. The typological subjectivity of the office was followed by a more ample subjectivity: because the necessity for contiguity has disappeared, it was again possible to understand the relationship between work and the city in terms of mobility and location. Now, as work has become mobile, the notion of flexibility associated with the workstation was extended to the kind of building and its location in the city.

The vertiginous development of Communications: satellite television, the internet, mobile phones and the tendencies of current development in cities, which extend and connect, will maybe lead to the disappearance of cities, as we now know them. Nevertheless, most urban centres continue to be where companies

Figure 1. Tokyo, London and New York Skyline www.ise.ac.uk.

are headquartered and where the possibility of personal contact remains a decisive factor. The traditional central area has the capacity to offer a rich variety of support elements – stores, cultural activities, restaurants, hotels – creating a complex urban environment, with multiple opportunities, that not even the most sophisticated video conference systems can replace.

Solà-Morales (1998) emphasises the fact that, in this turn of the millennium, architectures that allow interconnections and exchanges between different networks come up. The current urban experience is much more than a simple agglutination of the architectures present in the city, architecture transcends the solitary condition of autonomy and answers the urban and social needs.

The model of ordered space, taking nature as the framework of buildings "disposed in the green", has been reconsidered. Today, not even the landscape constitutes a background in which inserting and integrating the architectonic object can be thought. Powerful processes, which the philosopher Gilles Deleuze calls deterritorialization, place architectonic objects in non-places, in non-landscapes. The contemporary architectures come up unexpectedly, surprisingly. Their presence is not connected necessarily to a "place". (Solà-Morales, 1998)

In this scenario, in cities as diverse as New York, Kuala Lumpur, London and São Paulo, the rhetoric of capital and power still appeals to height to express itself (fig 1). The interest in the form of construction in height is increasing. When authorities and counsels evaluate their options to meet the need of future expansions of their communities, they continue to be interested in the typology of tall buildings as an inevitable solution to urban growth.

In cities that are already dominated by vertical skylines, more mature views of the tall building, as "one" of a group is also evident, creating inter-relationships that can also mitigate the concept of a tall building as an isolated tower.

4 THE TOWER

The office building and vertical building, typologies of urban buildings that were products of industrialisation and were defined architectonically by modernism, evolve to become key components of the complex mechanism known as the contemporary city.

Tower designs demonstrate that, after the end of the 1980s, a new generation of tall buildings is being developed around the world. In these last years, the field of architecture was characterised by a series of experiments opened to multiple directions and that in many occasions knew how to break dichotomies established during the last century.

If the formulations of the 1960s are monothematic and coherent – Rossi, Venturi, Archigram, etc. – on the contrary, the most representative architects of the 1980s and 1990s present a more hybrid vision, apparently contradictory, in which many concepts are conciliated. Gehry, Koolhaas, Hollein, Siza, Ito, are not explainable through restricted and single ideas, but as an evolutionary synthesis of many tendencies: minimalism, post-modern hedonism, vernacular references, technology, ecology, all that assumed at the same time. (Montaner, 1978)

While the great cities were extending themselves ever more, occupying entire regions with their network of infrastructure, making neighbouring populations disappear, some constructions that were many times more city than the city that means, that were that in a more intense manner, as if they were concentrated cities, came up with their own interior.

This is a phenomenon that would seduce Borges' imagination: a superposition of scales and times. Simultaneously, the city plays the leading role in its end and its beginning. In a certain way, it constitutes a logical situation, because if the city is, in essence, concentration and density, in opposition to the countryside, it seems normal that, when an excessive extension of urban fabric is produced, when territory and metropolis identify themselves, then the city tends to redefine itself as what it is: a concentrated enclosure inside a disperse fabric (Cerver, 1998).

Because of that, more and more frequently we are witnesses to the construction of buildings of great dimensions that include many activities simultaneously: commercial areas, internal communication systems, leisure spaces, offices, restaurants, transport stations, internal public squares, gymnasiums, store streets, hotels, garages, etc.

If we consider that a city is fundamentally a concentration (of services, activities, information, etc.) then these mega centres are, in fact, current cities. However, in substituting the city by a building, its surrounding is converted to the periphery and in front of the concentration of services, the rest begins to be a subsidiary and dependent territory.

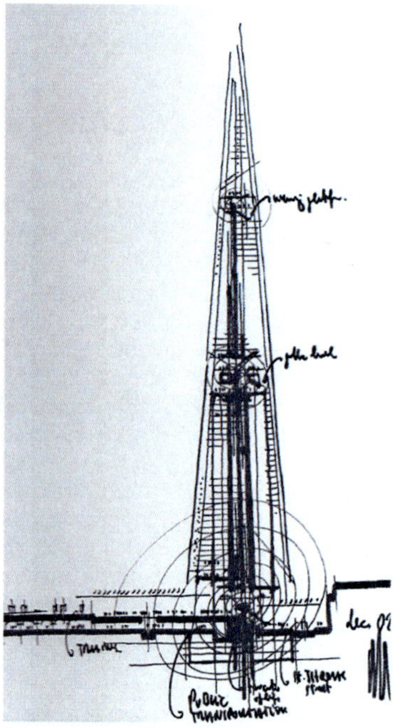

Figure 2. London Bridge Tower, London. Riley, T., Nordenson, G. (2003) Tall Buildings. New York: The Museum of Art Modern.

Figure 3. Swiss Reinsurance Headquarters, London. Riley, T., Nordenson, G. (2003) Tall Buildings. New York: The Museum of Art Modern.

Many contemporary buildings are conceived to expand and increase the public domain, and that way is more permeable and transparent. Much attention has been given to the urban life at the street level, with its pedestrian circulation, public spaces, or commercial activities, beyond simply establishing towers in squares. Formal and prohibited lobbies designated the line between public activities in the ground level to the private space on the floors above; now many recent buildings also seek to extend street activities inside the structure, not only for placing structures over existent transport centres, connecting the building with local and regional transportation. (Fig. 2)

Despite the security problems, today's architects recognise that tall buildings should not be separated from the permeable fabric of the city and the public activities from which it is born.

Mixed used structures create a model radically opposed to the modernists. Not only its functional configuration contradicts the principles of orthodox zoning (since the multifunctional structures contain the functions that orthodox modernism dispersed throughout the whole metropolis), but, as high-density detailed events, they acquire a high degree of urban autonomy, serving as eloquent symbols of the metropolis and its dynamism.

These structures can be away from business centres and the urban centre and promote the appearance of multiple centres capable of changing the hierarchical topology of the modern city. The hybrid building is that way the paradigmatic expression of the contemporary conception of the workspace in relation to the high densities, the logical result of technical and functional transformations that altered the modern vertical office building.

Both, the workspace and the vertical office building, started to abandon the urban model of functional zoning that was crystallised in the Chicago Loop and determined the theories of modern movement and ideas about the city. The centre of the city stopped acting as just a magnet for businesses, and consequently the same happened to the building. The characteristics of isotropy of the building repeated floors, and the universal replication (the typology associated with modernism) not for very long resonated sufficiently to construct an ideal. What developed, however, was a building linked to an urban topography, its functions and sections discontinuous, with a unique presence.

One of the main factors of innovation in buildings in height continues to be technology. In terms of structure, some of the most significant recent advances have been made possible using sophisticated computer modelling, making the analysis and the design of complex structural systems (Fig. 3). Emphasising the technology architecture relation is nothing new. There is a true architecture tradition that manifests itself by the technique in which it is grounded. Since the middle of the 19th Century, a part of architecture has developed based on the formal possibilities of utilisation of new materials and the support of new technologies, generating symbols such as the Crystal Palace in London (1851), and the Eiffel Tower in Paris (1889). In 1923, Le Corbusier publishes the book *Vers une Architecture*, which is structured in a complex thinking in which the problem of new technologies occupies a central place in the definition of a new architecture. Abandoning the style discourse, modern time's architecture seems like it should be characterised by its capacity to utilise the innovations that science and the

Figure 4. Lloyd's Bank, London. Nakamura, T. (1988). Richard-Rogers: 1978–1988. Tokyo: A+U.

technique of present time offer as a specific conquest of this same modernity.

This architecture based on high technology rejects any historicist return or any formality or decoration. It is a reductive architecture, which tries to solve the biggest number of questions with the smallest number of forms, relying on the synthesis capacity that technology possesses.

In this panorama, we find different postures: one high technology architecture that is expressed in a careful drawing, free of the modern brutalism, as in the work of Norman Foster; and other works in which the aggressiveness itself and the presence of structural and installation elements predominates, such as what happens in the headquarters of Lloyd Company of London (1979–1986) by Richard Rogers (Fig. 4); tendencies in which this technological language abandons the strict rationality conditioner to explore formal fields based in the evocation or in the proximity to other arts, as Jean Nouvel accomplishes; or in cases such as the one of Santiago Calatrava in which the confidence in the scientific world is unfolded in sculptural and expressive forms and still, in other cases, such as Jacques Herzog and Pierre de Meuron's, the aspiration to monumentality, which was almost always linked to the use of high technology, is abandoned, and a more modest and renovating way in which the material of each work is essential is sought (Montaner, 2001).

Innovation is not only constructive, as the application of new mechanical possibilities, but especially of communication and management. The more highlighted emphasis is put on the efficiency with which new artefacts explain their functions, exhibit their objectives, and show their tectonic logic. It is the triumph of communication through images of an architecture of growing transparency and immateriality.

To Solà Morales (1998), high-tech architecture is not something that closes upon itself but the announcement of a way through which the social objectives of a highly developed culture are achieved by the intelligent application of technical rationality. The result is always a rhetorical exaltation of great corporations in which the high-cost production of these architectonical artefacts finds its strongest allies.

Technological architecture started to respect much more the pre-existences, adapting to the urban lots where it is inserted, many times rejecting the designation of "high technology" and assuming that it is an "ecological" architecture.

Montaner (1978) considers high-tech architecture a rhetorical posture, is evident that great part of high technology has the same scenographic and visual role of post-modern architectures. What the great buildings of multinational companies intended, with their historicist languages, – define a commercial image, a business logo –, is similar to what technological architecture proposes: being a symbol and an expression of power.

Union, tension, lightness, the juxtaposition of intervention scales, a total environment, continuity, and transparency are the predominating criteria that seem to present themselves like a metamorphosis of the Vitruvian principles of *Utilitas, Firmitas, and Venustas*. Definitely, the buildings of these last decades show us the persistent search for surprise elements, for traces of contemporaneity in the use of new technologies.

That way, another aspect of technological innovation in recent vertical buildings is the design of structures that consume less energy than the conventional ones, complying with the European laws relative to energy efficiency and environment preoccupations. Engineers and architects are using, once more, sophisticated computer analysis to reduce dependency on mechanical systems in tall buildings. The innovating use of double glasses to reduce solar gain, strategies of passive air circulation, computational modelling to better air circulation around buildings and a great quantity of other small, although effective, measures mean that recent tall buildings are more efficient than last generation ones.

Contemporary architecture is made to propose abstract objects, this unusual heterogeneity of languages and its unfoldings characterise what the Catalan critic Josep Maria Montaner (1998) denominates architectonic posture, which means the development of distinct and personalised design methodologies by the architects that act in the global circuit.

The unusual possibilities open for architecture concerning language and expression enable unexplored volumetric manipulations that allow another degree of abstraction and an enrichment of urban landscape. An architecture opposes the *genius loci* because the character of the place is the main source of inspiration for the project. In this sense, it is distanced from contextualism and the memory of the place.

Through the notion of character and by means of images, coming especially from plastic arts experiences, current works present an autonomy of the interior regarding the surrounding. It frees the modern orthodox ties between programmatic content and form and, besides, establishes an unequivocal relation between geometrical form and image, with that enriching the significant presence of the building and putting into question the autonomy of architecture in opposition to meaning.

Through an empirical method, it succeeds in that each work is singular and unrepeatable. In a way apparently paradoxical and with high technology material, a work of art is made, a sculpture in a world that is an incessant consumer of images.

5 FINAL CONSIDERATIONS

If the vanguards of the interwar period emerged in a context of tensions between the bourgeoisie, technology and capitalism, coming up in various central European metropolises, the current context is that of post-industrial societies, the world of image and of globalisation. Thanks to the systems of information transmission, the verticalization experiences can be promoted both in very different metropolises and in places considered peripheral. The use of high technology is a common element in these works, however, from consistency, force and efficiency, it was turned to lightness, transparency, intelligence and density of information. On the other hand, changes in architectonical practice tended to formal plurality and dynamism, reflecting the interrelation between heterogeneous references.

Inside their diversity, all those architectures stand out by a common *avant-garde* characteristic: the laborious effort to construct its own formal universe, its own language, in part with references to the circle of art and of avant-garde architecture, but also with the ambition to create something completely new and unprecedented, used always in an open and ambiguous way in the context of urbanism of density, of congestion, of speed and of chaos.

We are facing another "history" of architecture, maybe also submitted to cycles of variation, increasingly shorter and perhaps in front of a close diversity of mercantile "offers" of forms and architectonical concepts.

How to propose buildings in a city that seems like it has lost all meaning? How to elaborate architectonic thought when the fruitful shelter of poetical images, that precedes thought itself, seems to have ceased already to exist? In many of the contemporary buildings, what establishes its meaning is not context, or tectonics, the meaning of place, the typological or figurative references to other architectures from the past. Its architecture is, in a way, immediate, direct, and perceptible through the kinaesthetic experimentation of who contemplates it.

The vertical buildings explore architectonical forms to the limit of the possible in a constant challenge to the new technologies and the laws of equilibrium. The tower learns the sociological and cultural mutations, redefining the relations between public and private space and conducting new forms of urbanity, leading us to think about the fascination that it brings like a utopia of the contemporary city.

BIBLIOGRAPHICAL REFERENCES

Äbalos, I, Herreros, J. (2003) *Tower and Office: from modernist theory to contemporary practice*. Cambridge: The MIT Press.

Abascal, E.H.S. (acesso 2007) *Cidade earquitetura contemporânea: uma relação necessária*. www.vitruvius.com.br/arquitextos/066.

Cerver, F.A. (1998) *Imágenes de arquitectura contemporanea*. Barcelona: Arco Editorial.

Curtis, W. (2008) *Arquitetura moderna desde 1900*. Porto Alegre: Bookman.

—. (1987) *Le Corbusier: Ideas y Formas*. Madrid: Hermann Blume.

De Masi, D. (1999) *A sociedade pós-industrial*. São Paulo: Editora SENAC.

Habermas, J. (1999) *Pensamento pós-metafísico: estudos filosóficos*. Rio de Janeiro: Tempo Brasileiro.

Hall, P. (1995) *Cidades do Amanhã*. São Paulo: Perspectiva.

Koolhaas, R. (2001) *Delirio de Nueva York*. Barcelona: Gustavo Gili.

Meyer, R. (2007) *Os desafios contemporaneos da metropole: O caso de Sao Paulo*. Sao Paulo: Fauusp. Montaner, J. M. (1978). *Depois do movimento moderno. Arquitetura da Segunda metade do século XX*. Barcelona: Gustavo Gili.

Sassen, S. (1991) *The global city*. New York: Princeton University Press.

Solà-Morales, I. (1998) *Diferencias. Topografia de la arquitectura contemporânea*. Barcelona: Gustavo Gili.

Vattimo, G. (1989) *A sociedade transparente. Biblioteca de filosofia contemporânea*. Lisboa: Edições 70.

The classical progress: French neoclassicism as "modern architecture", a symbol of progress in the city of Recife, Brazil, in the 19th century

Gisele Melo de Carvalho
Faculty of Architecture, Lisbon University, Lisbon, Portugal

ABSTRACT: This article analyses the concept of "progress", in architecture, in the context of research into new rational solutions that would support the construction of the great urban centres of the European modernity at the end of the XVIII century and beginning of the XIX. It marks the moment when "progress" in architecture arrived in Brazil, including the city of Recife, as "modern architecture", adopting classicism not only in buildings but in all social domains. French neoclassicism is one of the facets of classicism that is taken as a reference for the production of a variety of designs of buildings, as well as interiors of houses. The illustrations demonstrate how the patterns of classical aesthetics were incorporated, suggesting that it is still the reference adopted in the current architectural composition.

Keywords: Progress, classical aesthetics, interiors, neoclassical, housing

1 INTRODUCTION

Just as progress is relative, so is the concept of modern. Modernism, having progress as its driving motor, comes from it. The modern term, as Habermas writes, was first used in the fifth century to differentiate the new Christian times from the Roman pagan past (Arantes, 1992: 100), and always appears throughout history to designate something new, but relating to the past. Here, "modern times" is considered for our reflection to begin in the eighteenth century, "the century of lights," characterised by a new posture of man in the face of nature and its purposes in the world. It is at this moment, through empiricism, and in the search for better times through reason, that the proposals of a "modern architecture" begin to materialise.

We identify the beginnings of the materialisation of "illustration" in architecture and the arts in the preceding century when the concept of Mimesis is altered. When the perfection of the naturalistic copy, for artists, yields to the "Intellectual passion of the craftsman (Diaz, 2000: 283)[1], bequeathing thought-provoking works, and opening the way for a creation free of ordinary formal results but without abandoning the perennial classical concepts of composition. After all, the Vitruvius' *Architecture Treat* had always been the landmark of the architectural compositions until then.

We will see that, for certain artists and architects of the XVIII century, "What motivated them was not simply to copy the ancients, but to obey the principles on which their work was based" (Frampton, 2003: 4)[2]. Thus, in a revolutionary France enlightened by the contributions of the excavations of Herculaneum and Pompeii, the precursory theories of Perrault, Le Roy, and the Abbot of Cordemoye, which challenges the "Vitruvian orthodoxy", contribute to the formation of the neoclassical movement's field of investigation of different forms that could respond to the new social demands in a more appropriate way.

One of the strands of French Neoclassicism that will continue with this position, emphasising the expressiveness of form, will be the one initiated by Jacques-François Blondel, the professor of the so-called 'visionary generation' of architects"[3], which included Étienne-Louis Boulée, Jacques Gondoin, Pierre Patte, Marie-Joseph Peyre, Jean-Batiste Rondelet and... Claude-Nicolas Ledoux. (Frampton, 2003: 6). Blondel will exhibit, in his "Cours d'architecture" of 1750, architectural concepts for architectural constructions, relative to the composition, the type and character, supposedly fundamental for this creative practice.

The historiography of modern architecture demonstrates, through illustrated graphic schemes, how Jean-Nicolas-Louis Durand, professor of the "École Polytechnique" of Paris, and disciple of Boulée, would also focus his work to obey the classicist principles of composition, within a conceptual perspective. In his work "Précis des Leçons Données à l'École Polytechnique", he would develop, in a perspective of economy and functionality, a series of "model combinations" in

[1] Original text: "pasión intelectual del artífice".

[2] Original text: "o que os motivava não era simplesmente copiar os antigos, mas obedecer aos princípios em que a obra destes se baseara".

[3] Original text: "mestre da chamada geração "visionária" de arquitetos".

the plan, which also included classicist compositional parameters such as order, symmetry, and module, but worked to fit the new programs.

For Frampton, it will be from the "Structural classicism", initiated by Cordemoy, Laugier, and Soufflot, freelance architects of the historicist's formal patterns, who emphasized in architecture its structural part, where there would be projections of works such as prisons, hospitals and railway stations (Frampton, 2003: 10), new programs for industrial cities with significant population growth at that time. Another characteristic of the above, which emphasized the formal aspect, was called by him, "Romantic classicism", and was more directed to projects of representative institutions such as museums, libraries and monuments, making use of historicist elements such as colonnades, pediments, arcades, capitals and cornices, now much more faithful to the models of the ancients, since exhaustively studied in the "Century of Lights."

2 FRENCH NEOCLASSICISM IN BRAZIL AND RECIFE

However, what was the link between the work of French neo-classicists and modern architecture in Brazil, and particularly in Recife? In our view, there was a significant connection. Both groups served as teachings for the architectural community in general at the time, and represented a new form of design, which was being developed in Europe, with the "École Polytechnique" and the "Académie de Beaux Arts" as major centres of studies. These new teachings were within the context of researching new rational solutions to meet the challenges of the eighteenth-century man. They meant "progress" in architecture, and would support the new constructions of large urban centres around the world, including Brazil and the city of Recife, which would not be left out of this "wave of modernity", despite the great social and economic differences between the Brazilian cities and the centre emanating from these new ideas, the city of Paris.

Thus, in nineteenth-century Brazil, "modern architecture" will be the architecture of European industrial cities. From these countries came "progress" for Brazilians, who abandoned their Lusitanian references of a colonial past, and absorbed the "modern" in all domains, including the social. Gilberto Freyre comments on the predominant view of Portuguese colonial architecture, with an appearance called "Ugly and strong . . . disdainfully by the Frenchmen"[4] (Freyre, G. apud Carvalho, 2002: 98). On the same question, the author also points out the saying of Fernando Azevedo:

> It is true that, in the twentieth century, the pressure of other similar influences on the old colonial culture, which, having moved from the orbit of the Jesuits, fell in the next century under French language and literature which, by becoming the most humane and universal in Europe, operated intellectual penetration in the world and whose ideas and tendencies infiltrated our spirits more than those of other nations. (Azevedo, F. apud Carvalho, 2002: 56).[5]

And Mauricio Rocha Carvalho:

> . . . the ideals of development and progress, which characterised the nascent industrial societies, contaminated the environment of the young nation, internally instituting the desire to participate in the "modern world" whose urban image was associated with historicist aesthetic criteria." (Azevedo, F. apud Carvalho, 2002: 84)[6].

First, two ways are identified, as to how the new ideas of "progress" and "modern" would reach the city of Recife, in Northeastern Brazil. One of them was direct; brought by the numerous Europeans settled in Recife from the nineteenth century onwards, but also by the Recifenses who went to Europe to study and then returned with the knowledge acquired in schools and academies. The second, indirect way, occurred through the influence of the Portuguese Royal Court in Rio de Janeiro, upon its arrival in 1808, which brought the entire necessary infrastructure for D. João VI to continue his reign over Portuguese lands, from South America. Major urban and social reforms were undertaken by the Portuguese court installed in the city of Rio de Janeiro, and particularly in architecture and the arts, "Escola Real de Ciências, Artes e Ofícios" will be set up at the beginning of the century in 1816, the future "Academia Imperial de Belas Artes". In that same year, D. João VI will contract the French Artistic Mission to direct the teaching of architecture and the arts in accordance with the French schools, and governed by classicism. In architecture, it is important to emphasise the decisive influence of the work done by French architect Grandjean de Montigny, who is largely responsible for the "neoclassical phase of Brazilian architecture" (Rios Filhos, apud Carvalho 2002)[7].

In Recife, throughout the nineteenth century, the absorption of the "progress of the civilized world" in architecture, dwelling habits, and interiors will be

[4] Original text: ". . .feia e forte. . .desdenhosamente pelos afrancesados . . .".

[5] Original text: ". . .é certo que, para isto, devia concorrer, no século XX, a pressão de outras influências semelhantes sobre a velha cultura colonial que, tendo-se deslocado da órbita dos jesuítas, caiu, no século seguinte, sob o domínio da língua e literatura francesa que, tornando-se as mais humanas e universais da Europa, operavam a penetração intelectual no mundo e cujas ideias e tendências se infiltravam em nossos espíritos mais do que as de outras nações".

[6] Original text: ". . .los ideales de desarrollo y progresso, que caracterizaron lãs nacientes sociedades industrialles, contaminaran el ambiente de la joven nación, instituyendo internamente el deseo de participar en el "mundo moderno" cuja imagem urbana estaba associada a los critérios estéticos historicistas".

[7] Original text: "fase neoclássica da arquitetura brasileira".

noticeable, and will have the Count of Boa Vista, governor of the province of Pernambuco from 1837 to 1844, become known as the great promoter of these changes. He employed the French engineer Louis Léger Vauthier (1815–1901), from 1840 to 1846, to be the Engineer of the Public Works of the Province, and to be responsible for the modernization "à la française" of the city of Recife, with Reforms, measures and projects that were built.

From that period, we learn of a series of examples of classicist parameters in the building projects in Recife in the nineteenth century, beginning with him, and then by his successors, where one can perceive a clear influence of "classicist currents" named by Kenneth Frampton. These buildings will become references to the new "modern" aesthetic for the city, which will make it "the most classicist of the great Brazilian cities"[8], in the words of Alberto Souza (Souza, A. apud Moreira, 2000).

It is to him that we also owe the following description, of the Theatre Santa Isabel, a project of Louis Léger Vauthier, inaugurated in 1850, and remaining as a prominent location for spectacles in the city:

> On the front of the block of the foyer were the French influences that predominated materialised in wide openings with semi-circular yards and a decoration in lioz, in the form of a segment of a Doric peristyle superimposed on the facade on the first floor, along with the extension of the front portico. (Souza, A., apud Carvalho, 2002).[9]

We can verify that the Teatro Santa Isabel, with the adoption of classical elements in its facade, such as modulation, symmetry, arcades, triangular pediment and columns, emphasizing the formal aspects of the classic, is inserted in what Frampton names as "Romantic Classicism", where the set of buildings of representative institutions for government and society were included. In Fig. 1 it is also observed how the module was applied in the composition of the main façade of the theatre, proving the link to the classical compositional parameters.

A second significant project for the introduction of classical compositional parameters in Recife was the project for the House of Detention, by José Mamede Alves Ferreira (1820–1865), who also studied at the "École de Ponts et Chaussées", and took the post of Director of the Bureau Of Public Works in 1850, succeeding Vauthier in coordinating the main projects currently underway in the city. In the House of Detention, which can be included in the list of examples of "Structural Classicism" we can see a symmetrical composition of the wings, the rhythmic fenestrations, and the absence of decoration, tending towards a more neutral plastic, and a functional relation with the "Panopticism" developed by Jeremy Bentham, circa 1791, as shown in Figs. 2 and 3:

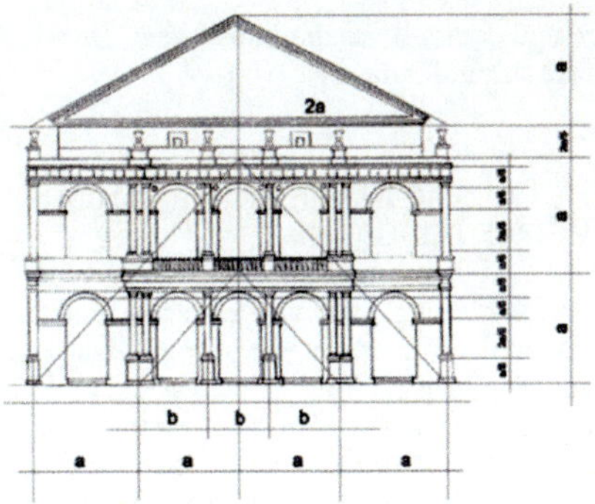

Figure 1. Study of the facade of the Teatro Santa Isabel proposed by Vauthier. Source: UCIO, 1990, p. 14.

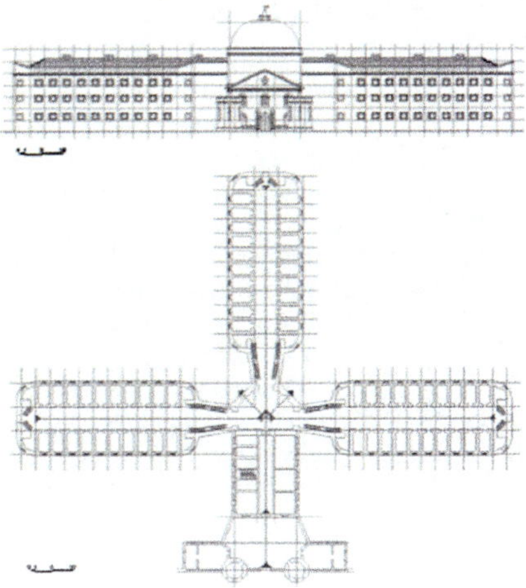

Figure 2. Old plant-house of Detention, current House of Culture Source: Sobral Filha, 2009: 50.

Figure 3. External view of the former House of Detention, current House of Culture. Source: https://en.wikipedia.org/wiki/Casa_of_Cltura_(Recife).

In the context of architecture, these two examples clearly illustrate the two currents described by Frampton as representative of the Neoclassical Movement in

8 Original text: "a mais classicista das grandes urbes brasileiras".

9 Original text: "Na face anterior do bloco do foyer eram as influências francesas que predominavam, materializadas em amplas aberturas com vergas semicirculares e numa decoração, em lioz, na forma de um segmento de um peristilo dórico sobreposto à fachada no primeiro andar, ao longo da extensão do pórtico frontal".

France and demonstrate the impact they have had on a diverse cultural milieu.

3 THE CLASSIC IN THE HOUSES AND IN THEIR INTERIORS

Concerning housing, there would also be adaptations, reforms, and constructions, following the new "modern" and classicist patterns, as well as the interiors of the houses, and the way of composing them, or, more precisely, the need to compose them within the new social concerns that were part of the daily practices of the wealthy bourgeoisie. We see in the following text an example of how the concept of "progress" was instilled in the Municipal Postures that imposed these reforms, and that by law, defined the classicizing contours in the buildings:

> The empanadas that the establishments of that city place in the respective doors do not bring any utility to its owners and perhaps can only serve for fraud and deceit; Are inconvenient and harmful to the public transit, undoing the city and oppose to progress and civilization, being that for all these considerations, such practice should cease and for it to disappear we are seated to make an article of the posture revoking the no. 8 of the 8th position of the positions of June 30, 1849, which we present to you. And we ask that you kindly approve it provisionally so that it can be observed. (Câmara Municipal. Códice 44 – F1 N° 10 APEJE apud Rocha, 2003).[10]

Like the empanadas, the "mucharabis" that composed the façades were also forbidden, stripping them of the Lusitanian references, which would be replaced by the iron balconies, glazing and classic plaques. These constructive reforms and directives began to be governed by the aesthetic postures established by the Recife City Council, which dates back to 1839, and comprised the "Addictions of Architecture, Regularity and Reforesting of the City"[11], proving its classical framework and rigour in the discourse of J. Lamas:

> The Renaissance principles are applied to facades as pictorial works, in search of balance, the façades of the houses of Recife are drawn from the aesthetic postures of 1839, with symmetry, proportion and rhythm as their principle. Architecture is thus integrated with urbanism, as the architectural treaties in force from the Renaissance to the XIX century. (Lamas, J. apud Sousa, 2002).[12]

It is seen that, involuntarily or voluntarily, the houses in Recife were to be modernised, from the rich mansions to the simpler houses, and the classicist patterns will be the tonic in the constructions. Alberto Souza, a student of what he calls "Imperial Classicism" (Moreira, R. 2000), points out as a significant reference in Recife the residence of Barão Rodrigues Mendes, a project by the French engineer Pierre Victor Boulitreau, who in 1860 proved his classicism through rigor of composition, portico in pronounced arcade at the entrance, and triangular pediment with central oculus.

> "The great growth and prosperity of Recife at the time made the taste for the classic contagious the public, who from the simplest to the most affluent, began to build or reform the appearance of their residences (Souza, A. apud Carvalho, 2002: 104).[13]

Within the dwellings, a multitude of rooms begins to emerge, echoes of a more complex social structure centred on the new "enlightened" and "academic" ideals.The living room, a social ambient par excellence, is now the stage where a whole structure of representation of power, erudition and the possessions of its owners will be processed, through furniture, objects and ceremonials. These new habits of living, which have their original influences from the eighteenth century in Europe, had arrived through the theoretical contribution of Jacques-François Blondel, who proposed a kind of internal division of housing, in articulated spaces of apparatus, society, and comfort, necessary to an increasingly numerous and influential bourgeoisie, where intimacy should be protected from the eyes of strangers. (Franco, 2014: 135).

Thus, the interiors of the Brazilian colonial period, devoid of greater compositional preoccupations had accentuated a utilitarian character, with furniture and utensils freely arranged since they were not yet structured as "visiting" rooms. These interiors, from the nineteenth century, will enter into the rigors of classicist composition; first as a rebuff of the classicist buildings, but also as a synonym for "good taste" and the modern, where symmetry, balance and hierarchy

[10] Original text: "As empanadas que os estabelecimentos dessa cidade colocam nas respectivas portas não trazem utilidade alguma aos seus donos e talvez só possão servir para fraude e engano; são incovinientes e prejudiciais ao transito publico, desfeião a cidade e se oppoem ao progresso e civilização, sendo que por todas essas considerações, deva cessar semelhante pratica e para que ella desapareça assentamos de fazer um artigo da postura revogando o de no. 8 do fit 8° das posturas de 30 de junho de 1849, o qual apresentamos a V. Exa. e pedimos se digne a aprovar provisoriamente a fim de que possa ser observado".

[11] Original text: Posturas Addicionaes de Architectura, Regularidade e Aformoseamento da Cidade".

[12] Original text: "Os princípios renascentistas são aplicados às fachadas como obras pictóricas, na busca do equilíbrio, as fachadas do casario do Recife são desenhadas, a partir das posturas estéticas de 1839, tendo por princípio a simetria, a proporção e o ritmo. A arquitetura integra-se, assim, ao urbanismo, como dispõem os tratados de arquitetura em vigor da Renascença ao séc. XIX".

[13] Original text: "O grande crescimento e prosperidade do Recife na época fez com que o gosto pelo clássico contagiasse o público, que desde o mais simples ao mais abastado, passou a construir ou reformar a feição das suas residências".

Figure 4. Frontal facade of the Barão de Beberibe site (already demolished), following the new aesthetic recommendations for the city. Source: Carvalho, 2002: 208.

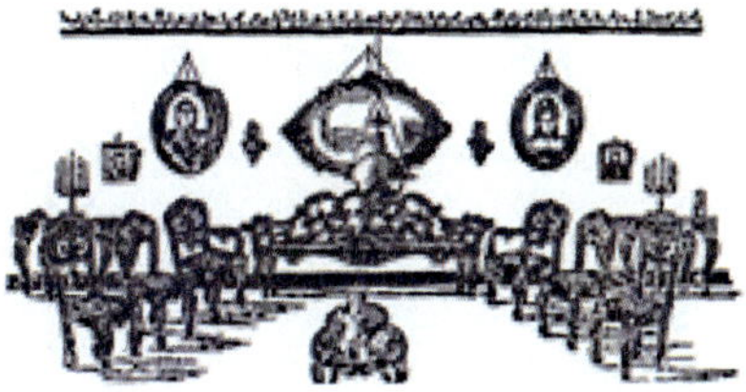

Figure 5. Illustration of 19th-century living room in Recife. Source: Carvalho, 2002: 208.

will be present, besides a whole universe of furniture, adornment objects and new coatings, representative of the new times. Figure 5 illustrates a scheme of a living room of this period, where one can perceive the rigidity of the composition. A clear hierarchy in which the focal point is the canapé, on the axis of the room, from where women would sit on the right, and men on the left, the furniture aligned and symmetrical.

4 CONCLUSIONS AND RECOMMENDATIONS

In spite of the "illustration" of modern man, which would allow him to respond to the challenges that the practice of architecture imposed from the eighteenth century, the classicist guidelines remain, in both neoclassical currents, either "Structural", and "Romantic". Linearity, symmetry, equilibrium, modulation, and regulatory trajectories, would guide the architectural practice of modern man, and give continuity to the precepts initially theorized by Vitruvius, from Classical Antiquity, and applied in the design of the city, its architecture and its interiors, according to what was demonstrated in the Brazilian architecture, and finally for the architecture of the city of Recife. This continuity will extend as a conceptual basis to the "Modern Movement" in architecture and the arts of the early twentieth century, to the present day, as Peter Eisenman attests in his reflection on everyday practices, "The End of The Beginning, The End Of The End, "(Eisenman, 1984), which suggests ways for other reflections in the search for what would ultimately progress in modern architecture, and whether it would have already occurred.

BIBLIOGRAPHICAL REFERENCES

Arantes, Otília Beatriz Fiori and Arantes, Paulo Eduardo. (1992) *Um Ponto Cego no Projeto Moderno de Jurgen Habermas. Arquitetura e Dimensão Estética Depois das Vanguardas e Duas Conferências de Jurgen Habermas*. São Paulo: Brasiliense.

Carvalho, Gisele Melo de. (2002) *Interiores Residenciais Recifenses – A Cultura Francesa na Casa Burguesa do Recife no Século XIX*. Dissertação de Mestrado em História. Recife: CFCH, UFPE.

Díaz, Marta Llorente. (2000) *El saber de la arquitectura y de las artes*. Barcelona: Edicions de la Universitat Politècnica de Catalunya.

Eisenman, P. (1984) *The End Of Classical: The End of the Beginning, The End of The End. Available* from: http://isites.harvard.edu/fs/docs/icb.topic136722.files/Week_4_Repetition/Eisenman_End_of_Classical.pdf in 03/04/2017.

Frampton, Kenneth. (2003) *História Crítica da Arquitetura Moderna*. São Paulo: Martins Fontes.

Franco, Carlo. (2014) *Casas das Elite de Lisboa – Objetos, interiores e vivências. Tese de Doutoramento em Estudo do Património*. Lisboa: Universidade Católica Portuguesa.

Lucio, Silvana Tercila Maria Pettinati. (1990) Teatro Santa Isabel. Análise de projeto. Article in *V Seminário de História da Cidade e do Urbanismo "Cidades: temporalidades em confront*. Campinas: Pontifícia Universidade Católica de Campinas.

Moreira, Rafael. (2000) Correspondência – "O Classicismo Arquitetônico no Recife Imperial" – Article in AU – 06/30/2000.

Rocha, Artur Gilberto Garcéa de Lacerda. (2003) Discursos de uma modernidade: as transformações urbanas na Freguesia de São José (1860–1880). Dissertação de Mestrado em História. Recife: CFCH, UFPE.

Sobral Filha, Doralice Duque. (2009) Lazer, Saúde e Ordem – Principais programas desenvolvidos na arquitetura do século XIX no Rio de Janeiro e no Recife. Dissertação de Mestrado. Rio de Janeiro: Programa de Pós-Graduação em Arquitetura – PROARQ – FAU – UFRJ.

Souza, Maria Ângela de Almeida. (2002) Posturas do Recife Imperial. Tese de doutoramento em História. Recife: CFCH, UFPE.

Will stage and exhibition design save contemporary theatre?

Santi Centineo
DICAR – Dipartimento Scienze dell'Ingegneria Civile e Architettura Politecnico di Bari, Bari, Italy

ABSTRACT: In the last fifty years, two main factors have pushed theatres and music venues to change.

Foremost is an inherent trend to new forms of theatre and music production. Traditional forms of musical production are currently being overtaken by a new aesthetic sense and different forms of fruition, in which musicians and players (likewise listeners and spectators) demand new technical potential and layout.

The second factor is a combination of new visual and acoustic contemporary scenarios, in which we are surrounded by images (television, cinema, advertising, industrial design, graphic design, fashion) and sounds (jingles, ringtones, signature tunes), an uninterrupted visual and incessant wallpaper music background.

This new tendency entails that the traditional task of theatres (a *medium* role in aesthetic terms) has been replaced and that if theatres want to survive they must necessarily change.

These changes are slowly taking place in two different ways. On the one hand, the theatre has taken refuge in intellectual elitism, becoming increasingly cryptic, closed, narrow and uncommunicative. On the other hand, the theatre is giving way to the pressure of a new social context and turning towards more catchy and showy performance genres.

In architectural terms, there has always been a correlation between the preordained destination of a building and staple designs required for that function. Today, on the other hand, attempts to change have concentrated largely on massive technological innovation, which in turn supports a hyper-functionalist trend culminating in specialist buildings that do not correspond to any particular dramaturgical form.

Keywords: Theatre, exhibition design, function, Type

1 INTRODUCTION: NEW SCENARIOS

The development of music halls has always corresponded to a specific function, expressing the relationship between music and society. However, today the effects of globalisation and new forms of visual and music production call for music halls to change rapidly.

Building music halls according to pre-established forms and features no longer suffice in today's context (full of all kinds of sounds and images), which is divesting theatre of its traditional historical role. If they wish to survive into the future, theatres must necessarily adapt to the multiple purpose functionality that began with some experimentations carried out in electronic music in the 1950s. As Luciano Berio wrote in *La musica elettronica*, "electronic music today is not a novelty because it is everywhere" (Berio, 2006: VII).

While historic performance venues continue to be used in their original function, it is also true that new genres often embracing more than one medium require new buildings that have yet to be conceived.

One possible answer could come from stage and exhibition design. It remains to be seen whether the necessity to reconfigure concert halls for different purposes will present new architectural opportunities. The scheduling of mass events such as rock concerts in stadiums, or television shows in warehouses, appear to show that this is a viable trend.

The conditions for listening to contemporary 'serious' music today are changing. The term for this genre is becoming increasingly unsuitable because the contemporary musical production panorama is a varied landscape in which different genres coexist freely.

2 CORRESPONDENCE BETWEEN FUNCTION AND TYPE

In past centuries, each period yielded a theatre type in compliance with the main performance features and dramatic forms of the time. The historical analysis allows us to understand how the development of theatre types has always been related to a specific function expressing the relationship between music and society (Panizza, 1996: 79). Today, however, theatres are experiencing a design crisis because they are torn between past forms and contemporary needs.

Up until today, buildings designed for musical performance abide by architectural established forms. In the last fifty years, however, a number of factors have put pressure on theatres and concert halls to adapt. This pressure is both internal and external in origin.

The pressure from within is decreed by the cornerstones of theatre production: new forms of music production, new ways of listening to it, new spatiality, new visual quality, new physical gestures that condition the dramaturgy.

External pressure comes from a combination of extrinsic factors. First is the unprecedented visual and sound quality of contemporary life where we find ourselves continuously surrounded by images (television, cinema, advertising, industrial design, graphic design, fashion) and sounds (jingles, ringtones, signature tunes), an uninterrupted visual and incessant wallpaper music background. Within this force fall other elements such as the role of the artist, the role of media, and the role of the technicians. In short, we are dealing with the effects of globalisation with regard to aesthetic fruition. It would seem that our contemporary, chaotic environment is impeding the natural recognition of a single type of theatre or building for music.

Furthermore, some of the inputs from our chaotic environment demand an increasing interaction from users. With their captivating nature, these environmental inputs are divesting theatre of its traditional historical role, with which it shared a visual monopoly with the major arts.

From this new perspective, it follows that the theatre traditional role of the medium has been replaced.

A blatant example can be found in the quasi-shamanic cult power exercised by rock and pop stars in their concerts, who give physical embodiment to the role once played by the architecture of concert venues of the past.

3 NON-CORRESPONDENCE BETWEEN FUNCTION AND TYPE

In architectural terms, these two different approaches have, followed largely well established typological directions or guidelines. The fact that classical and traditional architectural repertoire survives in ancient buildings, whose distinctive characteristics are so familiar to us, prompts contemporary architects to base their designs for new theatres according to well-known features and schemes.

On the other hand, today's popular entertainment for the masses such as rock and pop concerts takes place in sports grounds and halls or stadiums, while the more intellectual and highbrow art forms find it hard to identify suitable venues, often resorting to experimental spaces such as abandoned industrial sites. Now considered of archaeological value, the full potential offered by these sites in terms of interior design has yet to be explored in their adaptation or reconversion. These settings offer scope for staging all kinds of new forms of performance, thereby testing novel set-designs, and seating for the public.

Figure 1. Renzo Piano, Auditorium Parco della Musica, Rome 1992–2003.

Contemporary performance is following a worldwide tendency towards a new configuration between individuals in terms of aesthetic and musical fruition, be they the emitter, the medium or the receiver.

Studied by many leading philosophers, this traditional scheme has recently evolved into a new theory that defines contemporary aesthetic fruition as a "continuous circular flow" of information (Barthes, 1977 & 1992; Lotman, 1985).

4 TECHNOLOGY AND FLEXIBILITY AS A SUBSTITUTE FOR GENRE

As McLuhan wrote in 1967 and according to the theory of globalisation, "the medium is content" itself. This means that the actual contents of a given venue (drama, opera, theatre, music, etc.) are progressively becoming subordinate to different forms of staging. It is equally true to say that, so far, the only original attempts at innovation have been in terms of significant technological updating, which has driven a hyper-functionalist trend culminating in specialist buildings that are unsuitable for any single dramatic genre.

In this on-going specialisation process, many theatres rely heavily on technology for their day-to-day running, justifying this in terms of scientific progress. A number of examples confirm a trend in which reliance on technology, in fact, hides a question of typological choice. Technological support would have been more effective had it been employed to explore and produce new staging alternatives. Although some experimentation was attempted in the 70s and 80s, new concert halls and theatres still present traditional staging and seating arrangements.

Renzo Piano's concert halls are a case in point. As well as the Lingotto Auditorium (Turin, 1983–1995), the IRCAM (with Richard Rogers, Paris, 1971–1977) and the "Parco della Musica" Auditorium (Rome,

Figure 2. Aldo Rossi, Teatro La Fenice, Venice, 2001–2003 (section).

1992–2002) stand as hyper-specialist buildings in which high technological standards, in fact, conceal an interpretative vacuum. New ways of improving listening quality have taken precedence over conceiving the building's architectural role. Most theatres and auditoriums worldwide, built in the past few decades, have followed this trend.

It would be logical to conclude that while in the past it was meaningful to consider buildings destined for musical performance exclusively from a typological point of view, today it makes more sense to detect and identify new dramaturgical and musical forms specific to our time.

In recent decades, it has become more commonplace to refer to a multitude of crossover genres rather than to a single new one.

Consequently, there is now an urgent need to consider methodically music and buildings for musical performance within new parameters and their flexibility (Centineo 2006).

Through this new analytical approach, four different levels of flexibility can be identified.

4.1 *First level: specific buildings*

The tendency with buildings closely in keeping with a confined, restricted and specific function is particularly evident in the major restoration works carried out in two Italian opera houses, restored to their original scheme and purpose: La Fenice (Aldo Rossi, Venice, 2001–2003) and La Scala (Mario Botta, Milan, 2002–2004). This trend is also visible in buildings that have a precise focus in terms of function, such as Gregotti's Teatro degli Arcimboldi (Milan, 1998–2001) or in Renzo Piano's IRCAM (Paris, 1987–89).

4.2 *Second level: multi-purpose buildings*

Such buildings are generally made up of two or more halls, depending on the various uses for which they will be employed. The halls, therefore, differ in seating capacity and arrangement. Besides Piano's Rome Auditorium, where there are three halls each with a different seating capacity, other significant examples are Rafael Moneo's Kursaal (San Sebastian, 1990–1999), Julio Cano Lasso's Teatro de la Galicia (Santiago de Compostela, 1988–1989), Herman Hertzberger's Teatro Chassé (Breda, 1992–1995) and Jakob MacFarlane's Fanal Theatre (Saint-Nazaire, 2007).

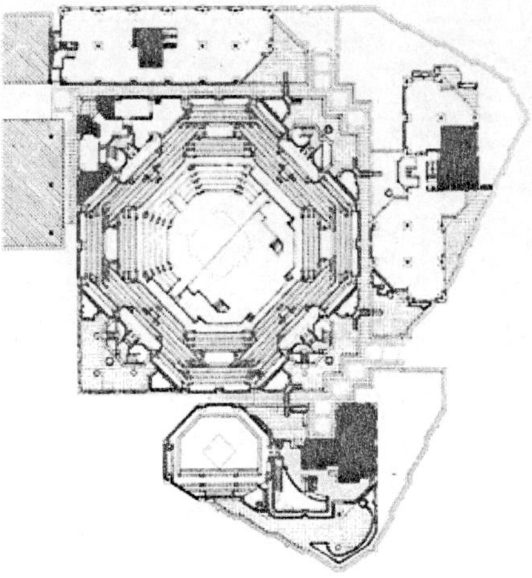

Figure 3. Herman Hertzberger, Muziekcentrum Vredenburg, Utrecht, 1977 (plan).

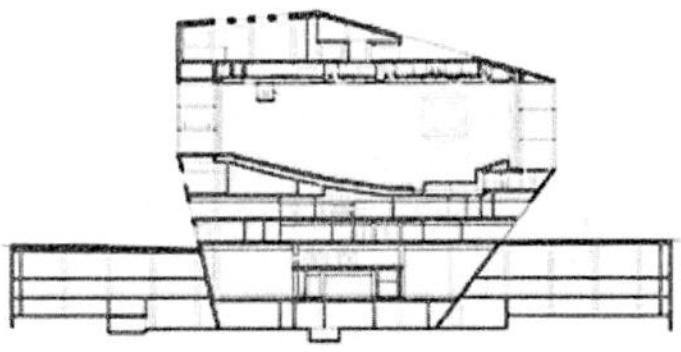

Figure 4. OMA – Rem Koolhaas, Casa da Musica, Porto, 2005 (section).

The culmination of this trend is the so-called "music towns", building complexes built over a vast area, developed for various functions and serving different purposes. Such complexes are being erected chiefly in large cities, such as with Christian de Portzamparc's Cité de la Musique (Paris, 1984–1995), Rafael Viñoly's Tokyo International Forum (1990–1996), Peter Eisenman's Ciudad de la Cultura (Santiago de Compostela, 1999–2007), Paul Andreu's Oriental Art Centre (Shanghai, 2004) and the National Theatre (Beijing, 2005–2008), Foster & Partners West Kowloon Cultural District (Hong Kong, 2008).

4.3 *Third level: flexible buildings*

These buildings, designed as flexible halls that are easily and quickly adaptable to different purposes, are generally small in scale and frequently towable from a larger, main building. Examples include Carlos Ott's plan for the Modular Hall at the Opéra Bastille (Paris, 1981), Peter Moro's Wilde Theatre (Bracknell, 1984), Arata Isozaki's Nara Centennial Hall (1992–1998), Shizuoka Convention & Arts Centre (1993–1998), or OMA's Casa da Musica (Porto, 2005). One example worthy of particular note is the Forum (Ghent, 2004) designed by Andrea Branzi and Toyo Ito, where the flexible element is in the architectural concept itself – a sponge shape that allows sound to spread throughout the entire experimental space (Branzi 2006, p. 174).

Figure 5. Marina Gariboldi & Partners, Fonderie Limone, Moncalieri, 2003–2005.

4.4 *Fourth level: "Theatrum extra theatrum"*

There is a need today for theatres and concert halls to move to another level. In 1922 Gordon Craig, one of the most important theatre renovators of the first half of 20th century, wrote:

> I feel a need: theatres must be empty spaces enclosed by a floor, a ceiling, and four walls. Inside it is necessary to build a kind of scenery or temporary house, right for the purpose: by this way, we will discover new theatre forms. (Balò, 2007: 118).

This idea can be regarded as a starting point for defining a new horizon. The so-called "fourth level" could be a pre-arranged space in which, thanks to the role of exhibition forms, pre-existing scheme-types could be recognised and adapted to new theatrical necessities, happenings and events.

Resorting to making use of empty, re-organized spaces has already begun with the conversion of industrial archaeology sites into cultural and musical spaces. One of the most outstanding examples of this is Gariboldi & Partners' restoration of Moncalieri Fonderie "Limone" (2003–2005), where the vast spaces of an abandoned foundry have been turned into two interesting multi-purpose stages complete with a system of modular seats.

The ultimate definition of this question has yet to enshrine a new role for interior design, one in which the difference between stage design, audience space and exhibition and pre-existing space no longer exists.

Good examples of this include Maurizio Sacripanti's plan for the Cagliari opera house (1965), where he conceived a carpet and ceiling made up of modular structures (two series of ram pistons) that can re-draw the overall plan of the building by sliding up and down.

However, exhibition resources in planning new fruition schemes are also evident in other experiments. Frank Owen Gehry for example, with his plan for the Jay Pritzker Pavillon (Chicago, 1999–2004), came up with an answer to the contemporary practice of holding concerts in stadiums by planning an open, multi-purpose "concert box" in his unmistakable "Guggenheim-style". Renzo Piano, with his historical set for Nono's *Prometeo* (1983–1984), planned a skeletal structure in which the positions of audience and performers are switched, with the audience inside and the players all around.

5 CONCLUSIONS: THE NEW TASK OF EXHIBITION AND STAGE DESIGN

It would seem we have now reached a crossroads, with period buildings unsuitable for contemporary music and contemporary buildings unsuited for use as both classical and modern music venues. While feeling the urge to stage traditional repertoire, we are unable to identify a new dramatic or musical form capable of representing our time. We build new theatres with the same specifications and features of the past while countless period theatres similar to the new ones that are being built lie unused, unrestored, neglected or even demolished.

These historic theatres are part of our architectural heritage, could be ear-marked for staging traditional repertoire while disused industrial areas could be converted with a view to providing planning opportunities and reappraising their future use.

With his Total Theatre, Walter Gropius got to the significant part by stating that concert halls were exclusively an interior design problem. Craig, on the other hand, contends that architects must explore another way, one that has already in part been undertaken, that of the empty space – a black box that each time is prepared differently and which embodies the fourth level mentioned above. This path could represent a true reform in theatre building, elevating stage and exhibition design (and consequently interior architecture) to new heights in terms of task and role. Innovative seating plans, stage layouts and orchestra dispositions could be arranged anew with each performance in such a space, which thanks to its flexibility represents both an "un-typological" path and a means of exploring, as indicated by Craig, new forms of theatre. This option is seldom explored by contemporary architects, possibly because of speculative factors but most likely because of the programming demands made by today's affluent and consumer-oriented society.

There are many examples of architectural forms in which buildings are part of the exhibition itinerary, becoming showy exhibition pieces in their own right. Frank Owen Gehry's Guggenheim Museum in Bilbao or Daniel Libeskind's Jüdisches Museum in Berlin are typical examples of this trend. Similarly, opera houses and concert halls are rising to a new aesthetic and fruitive scheme.

Overall, we can regard this process as forthcoming – even incoming. Rock and pop concerts, for example, clearly highlight that in neutral (and most frequently unsuitable) spaces such as sports grounds and stadiums, exhibition and stage design play a crucial role in creating a catalysis between performers and audience. In our everyday lives, a prime example is television, which constantly airs shows with temporary, short-lived staging that is nonetheless impressive and effective. The stage designs conceived for television studios are generally arranged within neutral spaces similar to empty boxes, which are continuously rearranged with the aid of powerful technology.

The contribution of the technology is also evident in many cases of recovery of unused industrial buildings turned into centers for the dance, favoring so the multipurpose needs of dance staging, for instance, Maurizio Zamboni's Fonderia (Reggio Emilia, 2004), where an empty box has been transformed, thanks to the use of high-technological equipment, in the ideal building theorized by Gordon Craig.

BIBLIOGRAPHICAL REFERENCES

Balò, Maurizio. (2007) Progetto per un teatro all'italiana. In: *Architettura & Teatro*, Il Saggiatore, Milan.

Barthes, Roland. (1977) "Ascolto". *Enciclopedia Einaudi*, vol. 1, Turin.

Barthes, Roland. (1992) *Elementi di semiologia*. Einaudi, Turin.

Berio, Luciano. (2006) *Un ricordo al futuro (lezioni americane)*. Einaudi, Turin.

Branzi, Andrea. (2006) *Modernità debole e diffusa*. Skira, Milan.

Centineo, Santi. (2006) *Representamen. Re-examining interior design's task among the connections in architecture, music and society*. Politecnico di Milano.

Lotman, Juri. (1985) *La semiosfera. Marsilio*, Venice.

McLuhan, Denis. (1967) *Il medium è il messaggio*. Einaudi, Turin.

Panizza, Mario. (1996) *Edifici per lo psettacolo*. Laterza, Bari.

Piano, Renzo Building Workshop. (2002) *Architettura & Musica*. Milan, s.e.

The Evolutive Design; the interpretation of the structure of the physical space in the architectural design

Calogero Montalbano
Polytechnic University of Bari, Department of Architecture and Civil Engineering Science Department, Bari, Italy

ABSTRACT: This essay introduces the theoretical and application principles of a model of analysis and spatial architectonic design called Evolutive Design.

It is aimed, in the first place, at defining the "elemental components of space", attributable to the historical-architectonic and environmental heritage that belong to a circumscribed "landscape" and are prevalently identifiable as a "homogenous spatial region". Subsequently, the model is useful to detect the degrees of the relationship existing between the various "validated elemental components" that, therefore, can be identified as "resources".

This way it is possible to come to comprehend the more or less complex structure that characterises every single territory. At the same time, it is possible to comprehend the multiplicity of the (physical and immaterial) behaviours, which are to be controlled for a correct design of the anthropic space, it is related to the building, the urban space, or the extended territory.

From here, the study model offers the potential to assess the degrees of bond that the project can assume with the surrounding resources as well as the effects that the project might generate on the surrounding spatial dynamics.

Regardless of the choice to respect, enhance or alter some orders or degrees of connection between the "resources" of a specific "spatial region", every project (of either restoration or new intervention), ends up altering the internal "status" of such system. Hence, the validity of a model capable of providing the designers and the public administration with an evolutive tool of knowledge and control of the territory.

The paper is conceptual and foundational, synthesising fragments from several different disciplines, including philosophy, software engineering, design theory. The paper is intended to raise awareness of the far-reaching implications of the architecture compared with the cultural context of belonging.

Keywords: Evolutive (Architectural) Design, cultural landscape, topology, spatial regions, boundary.

1 APPLICATION SPHERE

The Evolutive (Architectural) Design (EAD) is meant to be a "method of design support aimed at the simulation of the behaviours of the cultural landscapes"[1] and, more specifically, of the existing relationships between the "spatial elemental components", of anthropic or natural kind, which are recognised as "resources" of a specific "homogeneous landscape"[2], based on their physical (historical, archaeological, architectonic, landscape) and cultural evidence.

[1] To favour the overcoming of a monumental vision of the landscape the World Heritage Committee nominated a work group. The result was the "definition of the concept of cultural landscape" adopted in 1992 and inserted in the operational guidelines in 1994 for the application of the European Landscape Convention. Here the "cultural landscapes" were defined as the "combined action of nature and man", as well as specified by the article I of the European Landscape Convention. The cultural landscapes are, therefore, "... representative of the evolution of the society and human settlements over time under the influence of the physical constraints and/or the opportunities presented by the natural environment and subsequent social, economic and cultural pressure both internal and external." (par. 47 and par. 6 Annex 3, of the Guidelines for the application of the European Landscape Convention).

[2] The "homogeneous landscape" is here intended in the dual dimension of "homogeneous spatial region" and "place". The first refers to a delimited physical space, characterised by the unity of the morphic-structural and identitary (historical-procedural) behaviour of that specific "spatial region", thus called homogeneous. The second refers to a "homogeneous spatial region" to which are added a specific "organisation, connection and fruition" of the resources associated; a characteristic "perception" (intended as individual perception and social sharing) of the former; and their peculiar "historical-narrative memory" (intended as collective and individual memory of such space through time). All this highlights the identitary value (physical and psycho-perceptional) of that space, so called "identitary homogeneous spatial region" or "place" (Norberg Schulz, 1996; Augè, 1993: 43–70).

Through it, the relevance of historical-cultural features of the landscape can be highlighted and measured. Overcoming the traditional dualism between nature and culture (Pannel, 2006), the model keeps together the complex spectre of resources that range from the purely anthropic features to the natural ones. Connecting the history, topology and consistency of the existing "resources" in a given "place"[3] the EAD aims to highlight the various levels of interconnection existing between the above mentioned resources and detects the new bonds that a correct design can activate with these and further "spatial elemental components" of potential relevance (potential resources).

The EAD is, therefore, a useful tool for the monitoring and enhancement of the identity of the places (and their social, economic, cultural, and environmental factors) which operates through the synchronisation and valorisation of the cultural and environmental heritage of specific territories[4]. In this sense, given a determined project spheres, it can allow the identification of constraints and boundary conditions that are useful to the development of projects for the enhancement of specific resources and strengthening of the connections towards further available or potential resources.

The EAD acts on differentiated scales that range from an extended territorial context to one of the small urban compartments, reaching, in specific cases, the building scale, and its relations with the context. At the highest scales, it enables to identify the value of a specific "landscape enclave" (complex landscape systems and homogeneous areas) with respect to the adjacent ones as well as its ability to grow as a social and cultural attractor of the territory. At the lowest scales, it enables to define priorities and typologies of project intervention until the identification of specific functional and formal prerogatives that must be assured by the project.

Regardless of the scale in which it is called to operate and the complexity of the context of reference, the architectural design becomes, in this sense, a form delegated to solve specific orders of the bond of the complex space in which it operates. Therefore, the project here conceived cannot, be autonomous nor self-referential and needs to meet precise functional, historical, perceptive, and organisational requirements of the area of reference. At the same time, it has to be able to satisfy optimally the renovated demands of the community it belongs to, in the belief that these constantly change in time, and require a continuous adjustment of the orders of bond existing between the various components of the space.

The employment of EAD assumes, finally, particular relevance in presence of settlement systems distinguished by a particular state of alteration (e.g. abandonment, decay or depletion) or however characterised by a conspicuous presence of cultural and environmental resources, even if not particularly relevant or sufficiently structured (in this case denominated "minor resources"). Starting from this definition, the concept of "weak areas" can be introduced. These are intended as areas characterised by the presence of widespread "sparse cultural and environmental resources or resources of reduced entity or force, to which a stable relationship between local communities and natural/constructed surrounding environment is however associated in time"[5]. This typology of areas constitutes, on the other hand, the vast majority of "historicized territories"[6] of the European and Mediterranean context (Carta, 1999: 107–127; Montalbano, 2011). Hence, the not simple acknowledgement of some families of resources, due to a different cultural sensitivity or the negligence of institutions or single individuals, determines an elevated risk of loss and uncontrolled consumption of such heritage in particularly short temporal cycles. The elevated degree of risk connected to the loss of these "minor resources" and their connected landscapes leads, thus, to identify such territories or parts of them as "fragile areas" (Montalbano, 2015). In all these cases, the EAD assumes significant importance as a useful process to monitor the territory and as a design tool able to highlight specific project areas for the enhancement of the cultural and environmental heritage of the places as well as their management and synchronisation (Montalbano, 2016).

2 THE THEORETICAL PARADIGM

The EAD adopts the principles of the anthropogeographic and Muratorian view of the urban space and the historicized anthropic space, perceived in the historical dimension of becoming (Muratori, 1967; Caniggia 1976), and connects to the more contemporary reflections of the "Regenerative Thinking" (Cole 2012), the "Urban Space Design" (Broadbent, 2016; Banerjee and Loukaitou-Sideris, 2014; Sternberg, 2000) and the theory of "identity of places" (Hauge, 2007) in the attempt of translating the historical-cultural connotations of the places in structural elements of design.

Leaning on the theoretical and technical models hitherto illustrated and translating these experiences in the field of architectural design, the EAD aims at the realisation of spatial simulation models of

---

[3] See note 2.

[4] The text "Transforming our world: Agenda 2030 for sustainable development" establishes 17 Sustainable Development Goals (SDGs, #GlobalGoals) for the period 2015–2030. Among the the varuous SDGs section "11. Cities and sustainable communities" specifically refers to the heritage and, in particular, section 11.4. to the "Initiatives aimed to strengthen the protection and safeguard of the cultural and natural heritage" to make cities and urban settlements inclusive, safe, resilient and sustainable.

[5] They are, for instance, easily associable to the more properly rural areas.

[6] Structured territories whose overall pattern is the product of a prolonged and sometimes millenary interaction between nature and man.

the behaviour of specific regions, through which it enables to:

- "delimit areas of project intervention attributable to well-defined homogeneity criteria"; where the adopted criteria will become a fundamental element of the successive project choices;
- "isolate specific spatial behaviours to define punctual project actions" referred to a given place or resource, which falls within the prefigured intervention;
- "design the places prefiguring their possible future permutations", and trying to take decisions that can be reconsidered in time;
- "simulate and visualise the systemic effects generated by a hypothesis of design" over a homogeneous context;
- "communicate to various interlocutors", both in the "bottom-up" participative processes and in the downward stages of project transcriptions (Top-down), the effect, and the consequences of determined design choices.

On a different plane, the EAD tries, moreover, to face the problem of interconnections between the various design scales and the modifications made to the territory over time, in order to introduce a principle of continuous update of the simulation model.

In modern times, in fact, it is not possible to think about settlement spaces (either metropolitan areas, towns or small villages) if not in relation to a wider (urban and landscape) geographical context than in the past, where the flows and ramifications created by the urban places with the extended territory can be read. At the same time, the settlement spaces are not static over time and space but subject to dynamic forces that act in the economic, social and cultural sphere and affect the modification of the physical space. For these reasons, the contemporary architectural design cannot be entrusted a priori to the sole definition of data correlated to limited spatial regions and temporally static data.

In terms of modelling, the EAD is therefore addressed in the processes of "continuous integration" and "continuous design" which, in the field of software development process (Fowler's, 2004), enables to create and modify the design of a system during its development, rather than demanding to completely specify the system before starting to develop it, as it happens in the waterfall or sequencing models.

In this sense, not only does the EAD deal with the definition of sets of data correlated with a specific field and a specific resource but also with the possibility of "regroup"[7] these data in function of the progressive modifications the same resources undergo over time. Such condition induces, therefore, in presence of every new project or physical modification of the analysed space;

- "the regrouping of the data related to the behaviour of the boundary and the extension of the homogeneous region of reference" in order to validate its consistency and modification over time (contraction, extension, re-merging with regions, fragmentation), varying the (historical-cultural, social, economic) conditions of the orders of bonds between the resources taken into consideration;
- "the regrouping of the data related to the resources belonging to the analysed homogeneous region" in order to achieve a constant readjustment of the project processes and a continuous improvement/adjustment of the various projects to the (historical-cultural, social, economic) demands of the context.

All this poses the attention on the extreme importance of the construction of reliable data by specialised technicians, but also on the necessity of an "open access to the classification data of the territorial resources" through their archiving in "territorial institutional repositories" to guarantee the maximum interfunctionality of the data between public and private actors who operate in the field of physical transformations of the landscape. Through all this, the birth of management and virtuous transformation of the territory would be favoured and it would be possible to overcome the traditional tools of the static design of the territories and the urban spaces.

3 THE THEORETICAL MODEL AND ITS DEFINITIONS

An indispensable condition for the functionality of the EAD model is the detection, at the input, of the "interrogation scale of the model" (territorial, urban, neighbourhood, internal place, scale) on which it is decided to operate. Such choice allows defining, suitably deriving them from a first contextual analysis, the main "homogeneity factors" that can be referred to the analysed context. Based on these "factors" it is, therefore, possible to proceed to an initial detection:

- of those "elemental components" of the analysed context (monuments, specific buildings, archeologic areas, roads, squares, environmental contexts, ecologic corridors, ...) that best answer, due to real conditions or direct/indirect capability of the detected space, to the adopted homogeneity

[7] The refactoring is the procedure, which ought to allow the modification and the update of the code of the single classified resources, so that the architectonic simulation model can provide better design responses. Where by "better responses" it is intended "more effective", or which leave room for further design choices of transformation or space management that enable to "consume less", "improve the social or economic processes" or simply provide "more original and modern formal responses" able to guarantee the durability of the places.

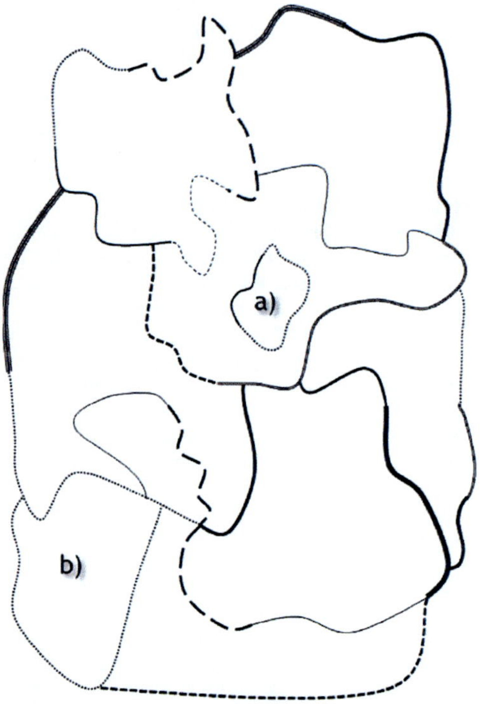

Figure 1. Characterization of the spatial regions and the territorial continuum. Every spatial region can show strongly diversified behaviours. Among the extreme conditions, the behaviours of a specific type can be observed: a) "spatial regions", completely contained in other regions (called "intercluded regions), or type b) "spatial regions", characterised by a boundary with homogeneous behaviour along its entire outline.

criteria and that are therefore classified as "real" or "potential resources";[8]

- of the "system of reference" (or "homogeneous spatial region"), suitably derived from the geomorphologic and topographic perimeter of the site as well as the perimeter defined by the various "families of resources" previously identified.

Given the "system of reference" on the other hand, other external environments interact in various ways with it. Moreover, it is always possible to detect the existence of macro-fields that contain the detected "system of reference" as well as micro-fields into which it can be further split out.

Every "system of reference" is, therefore, a spatial region embedded in spatial regions of higher order ("over system or spatial organisms") and it is composed internally of further spatial regions ("subsets or spatial structures") that can be further decomposed into "microsystems" or "spatial elements" (e.g. roads or squares) (Montalbano & Neglia, 2014).

Based on an organic view of the territory it is, therefore, possible to imagine the terrestrial space as a "continuum" in which there are no empty interstices (Purini, 2000: 118), but in which every spatial region, working as a cell, is distinct from the adjacent ones and contributes to compose wider and more articulated spatial organisms.

Therefore, every properly defined "spatial region" can be intended as:

- "closed", because contained within a continuous perimeter;
- "non-autonomous" or "correlated", because it is necessarily connected to other spatial regions in order to be able to contribute to the construction of complex "spatial organisms" (i.e. urban or territorial organisms).

Every "spatial region" results, moreover, constituted by:

- a "boundary" or "margin". This, with respect to other adjacent spatial regions, can assume the role of passage (or connection) or, on the contrary, of threshold (or net separation) between the adjacent spatial regions. In any case, the boundary of a spatial region belongs contextually, in the common portions, to the two adjacent spatial regions that delimit it. The boundary, also space itself, can be intended as a place of a continuous and almost always inhomogeneous succession of (natural, e.g. ridges, cliffs, valleys, or anthropic, e.g. fences, walls, buildings) "resources";
- an "internal region", characterised by the presence of "real" or "potential resources";
- a "limit region". A subset of the internal region previously defined, which directly interacts or is subject to the effect of the "boundary" or part of it.
- an "internal region" a), uniquely defined by the homogeneous behaviour of space and its "family of resources";
- a "boundary" b), articulated in multiple traits with diversified behaviour;
- some "limit regions" c), belonging to the "internal region" which, although characterised by the homogeneity of the region they belong to, are subject to (and induce) an alteration determined by the interference with the specific "boundary portion" which they come into contact with;
- some "limit regions" d), external to the identified spatial region, belonging to the "adjacent regions" which share specific "boundary traits" with the specific "portion of boundary" which they come into contact with;
- some "spatial crossroads" e), transition points between a boundary "portion" and the following one.

Assuming the physical condition of a spatial fence as "continuous and closed boundary", it is necessary, on the other hand, to grasp its extreme mutability in

[8] Having to take into consideration the particular nature and variety of such "resources", as well as their different ability of interfere with the spatial region of reference, they are moreover reclassified, based on their topological features, in "punctual", "linear" and "spatial" resources.

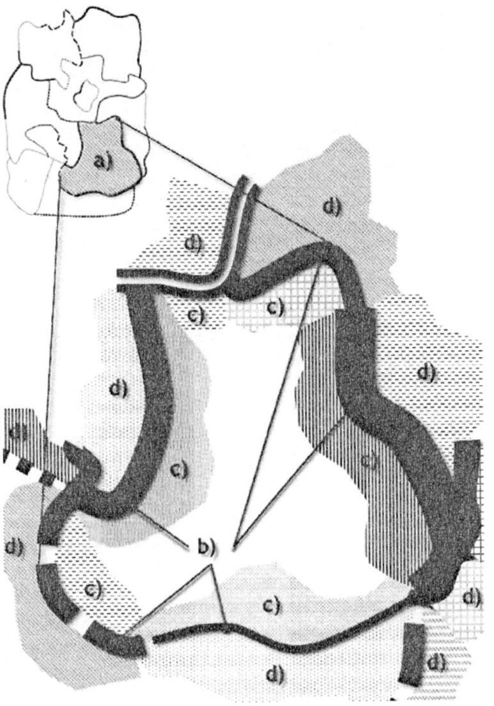

Figure 2. An example of characterisation of a "homogeneous spatial region" a) from which it is possible to notice the presence of:

terms of formality, consistency (density and texture) and especially thickness. In this sense, the fence or boundary of a certain spatial region is characterised as a further continuous spatial structure whose physical connotations and whose thickness are extremely variable along its own development. The cause of such variations is connected to the single perturbations undergone by the "internal regions" that contain it. In particular, the single variations undergone by a "boundary" are connected:

- to the relevance and entity of the interaction of the "resources", belonging to the two "spatial regions" adjacent to a specific "boundary" portion, and concentrated within the "limit region" of that specific "boundary" portion.

To sum up, a boundary is always, simultaneously, part of two and only two spatial regions. It inevitably undergoes the perturbations of two and only two "limit regions" which interfere with it from two opposite sides of the boundary. With respect to it, the specific point of transition between a "portion" of boundary and the following one assumes, even more, relevance: the single physical point of space in which three or four distinct "spatial regions" can cross and that can be interpreted as a "crossroad".

Such condition makes the "boundaries" of very particular interest from an architectonic point of view since they are points of extreme mobility and spatial tension. Places in which the design can operate until determining the opening of the boundaries and the aggregation of distinct "spatial regions"; or, on the contrary, places from which to start resorting to the fragmentation and re-articulation of the semantics of predefined "spatial regions".

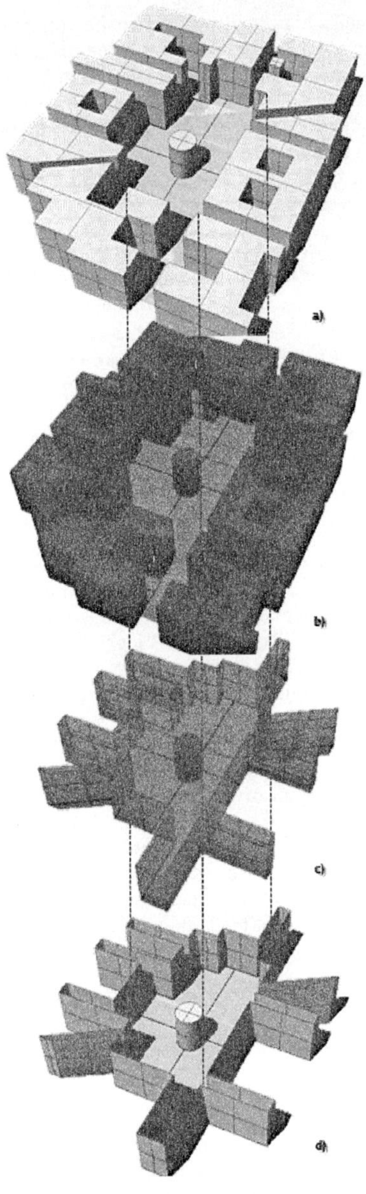

Figure 3. Example of "spatial region" declined on the spatial model of a square.

4 APPLICATIVE PRINCIPLES

Although the exposed concepts can be applied to scales and multiple views of the territory, an easy summary of the model hitherto illustrated can be exemplified in an urban square and in the role that the single building façades that surround it assume in relationship with the definition of "boundary" of the "internal region of the square". In this sense, both the square and the building fronts constitute a single "spatial region". At the same time, but in an opposed sense, every single façade, even constituting a "portion" of the "square boundary", appears as a "portion" of the "building boundary" it belongs to. That is how the "limit region" of the square ("limit region") which controls the façade interferes on its texture and spatial structure (giving it proportion, measuring the entrance, regulating the outlooks, the protrusions, ...) and undergoes, in turn, a re-modulation of space (in the positioning of the seatings, stairs, staircases, elevated paths, ...). Nonetheless, the "limit region" of the building related to the façade on the square, undergoes an alteration in the regulation of the relationship with the latter: as, for example, in the interference between entrance and structure/sequence of the rooms and the internal spaces that interfere with the mentioned façade; in the alignment of the courtyards; in the positioning of halls; in the hierarchical arrangement and organisation of the rooms, ... In this sense, the "scope" of the square related to that façade and the "scopes" of the interior spaces of that building that overlook the square, interact, through the above-mentioned façade/boundary, undergoing its specific vibration and interferences, as if the fence generated a magnetic field. For the same reason, the corner of the building takes into account the tensions of four specific "scopes" belonging to the three "spatial regions" which border it (the building, the square and the side street) introducing, in relation to the nature of these tensions, the problem of the "corner solution" of the building.

5 CONCLUSION

The principles hitherto exposed lead to highlight the important repercussions that the EAD spatial model can cause in terms of the thorough investigation of the territorial and urban spaces. Furthermore, it offers the possibility of transposing such analysis in concrete mathematical models that enable to highlight complexities and levels of resolution of specific aspects.

In this sense, therefore, it is possible to find, in typical mathematical concepts such as "metric space" and "topological space"[9], a perfect alignment with the model and spatial aspects of architecture introduced in the approach to the "cultural landscape", through the articulation of the concept of "spatial region".

[9] More in detail, within the mathematical reflection on "metric space" and "Euclidean space", such study assumes a deep correlation with some fundamental mathematical concepts, specific of the "topological space", such as "closed set" (according to the property that "a set is closed if and only if its boundary is contained in the set and open if and only if it is disjoint from its boundary") and the "boundary, margin or border", "limit", "compactness", "accumulation point", "isolated point", "density", "separability", "contraction or application of contraction".

BIBLIOGRAPHICAL REFERENCES

Augé, M. and Rolland, D. (1993). *Non luoghi*. Milano: Elèuthera.

Banerjee, T. and Loukaitou-Sideris, A. (2014). *Companion to urban design*. London: Routledge.

Caniggia, G. (1976). *Strutture dello spazio antropico: studi e note*. Firenze: Uniedit.

Carta, M. (1999). *L'armatura culturale del territorio: il patrimonio culturale come matrice di identità e strumento di sviluppo*. Milano: FrancoAngeli.

Cole, R. (2012). Regenerative design and development: current theory and practice. *Building Research & Information*, 40(1), pp. 1–6.

Fowler, M. (2004). *Is Design Dead?*. [online] martinfowler.com. Available at: https://www.martinfowler.com/articles/designDead.html [Accessed 5 Apr. 2017].

Broadbent, G. (2016) *Emerging Concepts in Urban Space Design*. London & New York: Taylor & Francis.

Hauge, Å. L. (2007). Identity and place: a critical comparison of three identity theories. *Architectural science review*, 50(1), p. 44–51.

Montalbano, C. & Neglia, G. A. (2014). Sviluppo sostenibile del paesaggio del Mediterraneo. Una ipotesi progettuale. In: Castorani, A., D'Amato Guerrieri, C., Andria, G., Di Sciascio, E., Camarda, P., Monno, G., Nuzzo S. eds. *1st Workshop on the State of the Art and Challenges of Research Efforts at POLIBA. Groups research on KETS and SCS, 3–5 December 2014, Bari*. Roma: Gangemi, pp. 667–671.

Montalbano, C. (2011). Un Metodo Progettuale per il Paesaggio Urbano: Il caso di Grottaglie (Taranto). In: D'Amato, C. ed. *1st RETE VITRUVIO International Congress Architectural Design between Teaching And Research, 2–6 May 2011, Bari*. Bari: Poliba Press, 6, pp. 2441—2450.

—. (2015). The Identity of Places as a Tool for Economic Empowerment of the Territories, Project Wealth Local Sustainable Economic Development Research Group [online] http://www.lsed-wealth.org. Available at: http://www.lsed-wealth.org/cgi-webaxy/item?406 [Accessed 5 Apr. 2017].

—. (2016). THE SENSE OF PLACE. Ildiri: the bond between the urban fabric and its history. In: Gambardella, C. ed. *WORLD HERITAGE AND DEGRADATION, Smart design, Planning and Technologies. XIV Forum LE VIE DEI MERCANTI, 16–18 June 2016, Napoli*. Napoli: La scuola di Pitagora editrice, pp. 819–828.

Muratori S. (1967). *Civiltà e territorio*. Roma: Centro Studi di Storia Urbanistica.

Norberg Schulz, C. (1996). *Architettura: presenza, linguaggio e luogo*. Milano: Skira.

Pannell, S. (2006). *Reconciling nature and culture in a global context?*. Cairns, Qld.: Rainforest CRC.

Purini, F. (2000). *Comporre l'architettura*. Roma-Bari: Laterza, pp.118–121

Sternberg, E. (2000). An Integrative Theory of Urban Design. *Journal of the American Planning Association*, 66(3), pp. 265–278.

Urban restoration for territorial development

Giacomo Martines
Department of Science of Civil Engineering and Architecture, Polytechnic of Bari, Bari, Italy

ABSTRACT: The proposed text summarises the research conducted by the Department of Science of Civil Engineering and Architecture on the theme of enhancement of tissues and urban areas through the recovery of the history, of the identity and at least of the architecture. The study is carried out by analysing several case studies in the area of the Apulian Murgia (inner area of the province of Bari) with a particular focus on the site of Altamura, key cities in this territory. The research work has been articulated since the architectural, urban and landscape restoration for the cultural enhancement of the site.

The first operation was the recognition of the heritage that is spread throughout the country and largely out of the urban area. It is characterised by values of different nature: natural heritage, landscape, environmental, heritage related to the world of traditional products of the territory and eventually linked to the artistic activity.

It is often very difficult to put in relation these goods, but, without a connection net, the goods are not able to produce sufficient attraction for autonomous self-maintenance.

Keywords: Urbanism, history, cultural heritage, Bari

1 NET VALUES AND WEAK QUALITIES

Convinced that congeries of values distributed throughout an area, each one too weak to be individually a "point of excellence", the idea was to insert them in a network could set up a value widespread in relation to devoted paths of fruition.

The runtime system must be constituted with different depth, different length, so have to be intermodal and articulated in different fields of interest. As many fields as many specificities are presents in a territory rich of historical, cultural, landscape and production values, related to the tradition and the values of health.

The proposed work develops an analysis sample downstream of a large enumeration of heritage elements.

It defines a potential relationship program that can simultaneously ensure the protection and exploitation of the heritage.

In synthesis, it is intended to facilitate the attraction capacity of each site through an integrated communication system, a more simple accessibility with soft mobility systems and a more efficient link capacity of goods belonging to different categories of interest, hoping that they can support each other.

2 RESEARCH METHOD

The Bari Polytechnic has been working, in the last few years, on the recovery of the landscape values of the Murgian area.

This work started in 2014 with an agreement between the Department of Civil Engineering and Architectural Sciences and some municipalities: Acquaviva delle Fonti, Gravina, Altamura. These three areas belong to the same landscape system although they present different typologies, especially in terms of agricultural use and anthropisation.

Our research aims to highlight the intrinsic values of the territory that detect the identity relationship between the populations and the areas. Are there legible traces of social history whose memory is intended to recover the identity engine of a territory?

3 THE MURGIAN AREA

The anthropic territory of Alta Murgia is an immense, little-known, dry, and arid heritage, characterised by mild karstic slopes, obviously inhospitable, except for sporadic vegetation, specifically selected for centuries of decisions. It is almost Inhospitable to human life.

About Alta Murgia, Tommaso Fiore wrote, everywhere stonewalls, not ten, not twenty, but more, much more, aligned on the hips of each relief, horizontally, even a few yards away, to hold the ground, to pick and hold it a bit between so much limestone. You will wonder how so many people have done to dig and align so much stone. I think that would frighten a people of giants. This is the Murgia most harsh and stingy; . . . did not want less than the labour of a people of ants.

Rarely in Italy is it possible to observe such a landscape made by the work done by nature and man in a continuous synergy that last for centuries: Here the populations have worked continuously, generation after generation, to make liveable and productive a hostile landscape.

For these reasons, the Apulia Region, identify the Murgian landscape as a Homogeneous field in the Landscaping Territorial Plan. Moreover, in 2014, it was instituted the national park of the "Alta Murgia" to protect specific environmental values.

4 THE CITY OF THE MURGIAN AREA

Altamura, Gravina in Puglia, Santeramo in Colle, Acquaviva delle Fonti and other urban centres of the High Murgia constitute a local identity unit linked to the territory that, crossed by the Appia Way, a pilgrimage path to the Holy Land, (lacking historical accuracy) inserted in the itineraries of the "Francigene streets" and the streets of the tratturi. It is a meeting point of very heterogeneous cultures.

The Altamura history (the most populous of the High Murgia centres, and the one that provides the largest area of the National Park), claim the rebuilding of the city linked to Frederick II of Swabia. In reality, it is certain that there was a fortified settlement to control the transit over the south alley of the city starting from the time of Peucets, in the same place in which, in the Roman time, was built the Appia way. A trail axis of particular importance because it allowed the best connection between the Ionian and Adriatic gulfs.

The Federici "Foundation" corresponds to a historical phase in which the configuration of the Murgian ridge, dotted with defensive sighting points, is substituted by a virtuous formula. The Federici Idea of cultural syncretism, Altamura offers soil to build houses on an urbanised system, rich in wells and public cisterns, walled (with the only cathedral founded by Frederick II) and farmlands. However, the land needs to be transformed to be cultivated, and only a unified community, overcoming the babel of languages and cultures from which it is composed, could make that.

This accompanies the genesis of Altamura's historic centre in its Jewish, Saracen, Arabic, and Latin districts centred in the Frederik cathedral.

The division in the four Altamura districts, seemingly planned in four unequally quarters around the cathedral, corresponds to a different conformation of the neighbourhood units with essentially two forms of aggregation. One around micro-community collective spaces each characterised by spaces for social life and by the presence of at least one well for public use. The "Greek" clause is characterised by "squares" spaces, while the "Latin" is shaped like "budello" or "line" and has smaller dimensions for a smaller number of dwellings, but consequently, generates a denser urban fabric.

Both signal different social uses of the common spaces.

In the first one, the main community activities are carried on in the open space of the square; in the second one, the social life is carried on inside the houses of the most important families of the clause.

5 THE RECOGNITION OF THE VALUES FOR ALTAMURA (BA)

The first operation was the definition of the consistency of a widespread and multi-segmented patrimony in the urban, peri-urban and rural areas characterised by a multiform presence of values that are often difficult to relate and which spontaneously fail to form a network capable of developing real productivity. The presence, though frequent, of "values" located in the territory, which individually cannot form "excellence points", without connection cannot create a process of valorisation. The same elements can be inserted, on the other hand, into an auto-incremental relationship cycle, a diffused value in relation to interscale and intermodal fruition paths articulated with different interest profiles.

Once the recognition and classification of this heritage have been carried out, the research has aimed at defining the criteria of alignment: the problem was that different elements of the heritage, landscape, cultural values (that sometimes have no precise location), have to be incorporated into the same database.

Our criteria have to be settled in relation to the establishment of relational networks between the identified assets and their accessibility, also in a revaluation key of the overall system of cultural linkage of the "items" present on the territory, in particular within the urban area, with a specific optic in relation to their cultural and social function in the system.

The research has been carried out on different levels: from the territorial analysis of the city's connectivity capabilities to the existing itineraries and public transport systems and in general the existing models for the use of sites of interest and their attractiveness. This aims to understand the vocation of the territory. At the same time, to fully understand in which cultural areas each well-cited "item" could best express its networking ability, assuming that almost all of the "items" spread across the territory adopt multiple cultural characteristics that can be attributed to different Areas of interest and all are important to be preserved.

The necessity we want to answer to is about the preservation of the smallest cultural values in the territory because they are representative and emblematic of an anthropic model in relation to the productive and agrarian landscape, which becomes, in its historical identity and mosaic stratification, the sign of a specific local culture.

The preservation of cultural roots in material traces becomes a fundamental element not only for the study but also for the memory of the settled population. However, the preservation capacity of these traces, beyond their cultural and identity needs, is predominantly linked to their ability to be re-established in economically active cultural circuits whose capacity for self-denial is linked to some of the driving themes of the territory and represented and disseminated in a much wider area of the territory analyzed at this stage of research.

After the definition of the mobility capability and cultural attractiveness of the spatial sector, there follows the cataloguing of resources belonging to the municipal territory. The analysis, divided by sectors of interest, aimed at enumerating its morphological characteristics and analysing its level of attractiveness, relationality with other attractors with the aim of identifying those "items" that could represent nodal values for more than one profile of interest.

They have been identified because they are considered functional in the structuring of a narration and visit system of some specific areas: Archaeological History, with specialised paths and the Naturalistic Area, with hiking trails.

6 THE CULTURAL ITINERARIES

Within the main theme, there have been identified some "itineraries", which can be subject to sustainable value-orientation and have a specific "story-telling" related to their main themes:

- Medieval itinerary: for the remains of the Norman-Swabian period
- Archaeological itinerary: centred on the stratification of the Peucet period;
- Palo-anthropological and palaeontological itineraries: related to the recapture (and the recent monumentalization) of the most complete and ancient Neanderthal man skeleton found and an important reservoir (also recent acquisition of the public patrimony) of dinosaur prints found in a cave;
- Itinerary of the rock civilisation: linked to the troglodytism and the knowledge of the "lamies" inhabited until the first half century;
- Naturalistic itinerary: oriented to the paths that wind up in the Natural Park of the High Murgia, also equipped with paths of considerable commitment or to be carried out with soft mobility (by bicycle or on horseback);
- Enogastronomic itinerary: linked to the typical local products deriving directly from the agricultural and pastoral area in which the paths are grafted.

The planned itineraries have (and must have) a very high level of overlapping.

They have a very close frequency of points of interest of different levels of importance and functionality and, starting from the centre of the Park, can connect to adjacent urban realities, exploiting both rapid mobility through the provincial and state roads and the "slow" mobility From paths that allow a more detailed and thorough visit of the territory.

To complete this kind of "itineraries" in the typing of "items" and defining the most compatible qualities from the vast assets of masserie, largely abandoned (almost entirely private property), with the consequent complexity of management, necessarily through a coordination of activities and investment opportunities to be developed through adequate incentive planning.

The last lens of the analysis, faced with the aim of facilitating the "city-to-country" correlation and the permeability of the territory, was the urban mobility model. Here the elements related to the cultural heritage identity are radically different. In the first instance, it becomes much more difficult the collectivization of the concept of identity, where elements belonging to the cultural patrimony that has been dismantled and degraded for too many years, in the collective mentality, are consolidating elements of urban degradation, difficult to re-associate with new potential for local development.

The particularity of the survey carried out, in relation to the methodologies and the themes of the urban restoration discipline, in the opinion of the writer, resides in the fact that the specific case of the municipal administration of Altamura concerned a networking of assets of the cultural heritage located outside of a consolidated context.

This is more evident in the case of the urban context because, in this region, the rural areas suffer very low aggressions and so the landscape has fewer changes, and it allows maintaining, if not exactly the same use, almost the same image.

On the other hand, the historic centre, although in its foreseeable contradictions, presents a strong unitarily and a remarkable state of conservation, both of the common parts and of private buildings with isolated few cases of abandonment.

For the purposes of our study, therefore, the "homogeneous" area of the old town, need not to be broken down into its individual components because of its absolute importance the old town became the major attraction, representative of the abundance of functions and areas of interest, and as such preferential destination of almost all planned itineraries.

Precisely because of this almost exclusive location of the historical centre and the lack of recognition of the cultural identity of the "items" placed in the external urban areas, grew the need to develop tourist information programs for knowledge, re-nationalization and reintegration into the circuit of the life of this city heritage.

It has also been decided to implement cultural circuits at the expense of the historic centre, to favour the green areas intertwined in the city that, besides acting as an environmental lung, in view of their formal characteristics, can constitute at least two landscaping points of urbanized excerpts of Murgian landscape, one with panoramic opening towards the National Park and the Pulo, the other towards Matera.

Their position can also serve as a point of access to the urban system by deflating the historic centre during periods of increased tourist flow. These "cultural hubs" of the "Altamura system" would summarise the nodes where the urban cultural itineraries would converge to connect to the territorial network.

7 THE TOOLS

The tools for the valorisation of each "item", in almost all the cases, do not use the instrument of design, which is limited to providing communication tools and deepening knowledge of the main elements identified.

There are two exceptionally unusual poles of a different kind, but also strategic for the functionality of the exposed program: The first is the area of the ex-aqueduct building.

An eclectic style building dating back to the first third of the twentieth century located in an area of great urban decay, even though it is part of a cardinal portion of the city's eighteenth-century planning and could constitute a first cultural centre in direct relation with one of the cultural hubs for the city access.

The other pole, in an extra-urban position, is the Cava Pontrelli, where the discovery of more than 25,000 dinosaur steps, complete the paleo-anthropological itinerary opened by the recognition of Neanderthal man by making Altamura one of the most important sites in Italy for scholars of this discipline.

BIBLIOGRAPHICAL REFERENCES

Angelastri Marcello. (1990), Altamura: Segni e percorsi di un'evoluzione urbana, Ed. Torre di Nebbia.

Carbonara Giovanni. (2012), Restauro architettonico: principi e metodo, Ed. M.e. architectural Book and review.

Centoducati Michele, Losurdo Vincenzo. (1999), Altamura: Lo sviluppo della città dal secolo XIII al XIX in relazione agli avvenimenti economici, politici e sociali, Ed. A.R.S.A.

Desio Ardito. (1989), Geologia applicata all'ingegneria, Ed. Hoepli.

Laureano Pietro. (1993), Giardini di pietra, Ed. Bollati Boringhieri.

Locapo Paolo. (1996), Altamura nei secoli, Ed. Cultural Association "Piazza".

Licinio Raffaele. (1998), Masserie medievali: masserie, massari e carestie da Federico II alla dogana delle pecore, Editor: Adda.

Mirizzi Ferdinando (1982), Note su insediamenti rurali nell'Alta Murgia, Ed. Leo S. Olschki.

—. (1990), Tra la fossa e le lame – Territorio, insediamenti, cultura materiale dell'Alta Murgia, Ed. Congedo.

Mongello Luigi. (1996), Le masserie in puglia, Editor: Adda.

The green way of the Apulian Aqueduct. A regional corridor for the enhancement of the environmental heritage and local culture

Calogero Montalbano
Dipartimento DICAR, Politecnico di Bari, Bari,Italy

Luigi Guastamacchia
Sezione Tutela e Valorizzazione del Paesaggio, Regione Puglia, Italy

ABSTRACT: The research operates as part of the investigation that the interdisciplinary Laboratory of the Polytechnic of Bari, "Cultural Landscapes Analysis and Design", has been conducting for several years in terms of the enhancement of the Cultural Landscapes. The method of Analysis & Design here applied, called Evolutive (Architectural) Design (EAD), and operates within the creation of specific models for the architectonic and spatial simulation of the territory. It is meant to show how, after the identification of some typologies of resources belonging to a specific territorial area, their value can be enhanced, interconnecting already existing resources on various levels or creating new ones.

The object of the present study is the network of the Apulian Aqueduct. A water duct among the most ancient ones in Europe, it has so far enabled the transport of drinking water to almost the entire Apulian territory, contextually becoming a "path stitching" of numerous typologies of "cultural landscapes". This, traversing the majority of the regional territory, has started to intersect, almost without realising, an extraordinary range of historic, environmental and landscape resources. The essay attempts to demonstrate how the "aqueduct route", attracting, sometimes in a disorganised way, a wide range of more or less complex settling phenomena (from parts of rural settlements to entire urban structures) and crossing sensitive points of the territory for specific technical reasons, has become, in fact, an anthropic corridor of extraordinary historic, cultural and environmental importance. Through the application of the EAD model, we intend to prove that the idea of a transformation of the "aqueduct route" into a cycling path, as proposed over the last years, is likely to be ineffective if the deep degree of connection with the various "cultural landscapes" which it traverses and contributes to define is not understood.

Keywords: Cultural landscapes, Apulian Aqueduct, ecosystemic services, green infrastructures, Evolutive (Architectural) Design

1 INTRODUCTION

This essay attempts to illustrate the important design implications, which might result from a correct action of "Analysis and Assessment of the Places" in which specific design choices are declined. Here the main route of the Apulian Aqueduct is analysed with the intention of proving how the choice of transforming it into an important cycling path of regional crossing could cause much greater scenarios and design repercussions than those associated with the realisation of a simple tourism-recreational itinerary. Such resource, of relevant interest for nearly the entire Apulian regional system, both for its technical value and for its evident historic-architectonic value, has assumed considerable interest over the last years. The nearly complete accessibility of many of its portions and its capacity to traverse a number of landscapes typologies of elevated environmental and cultural value, that are difficult to connect through other routes, has highlighted its extraordinary value and its capacity to be a slow crossing system of the entire regional landscape. Such condition, although significant in terms of tourism development and local economic enhancement, does not take into consideration, on the other hand, an important range of already existing connections and resources in the Territory. These might transform the Aqueduct network in a real supporting structure for the "slow access" to the Regional Territory and to its specific "homogeneous regions" (recognisable as distinct cultural landscapes). In this sense, the "Aqueduct Way" would become a "green road" able to catalyse local "ecosystemic services"[1] and become a gate of access to the homogeneous areas of the territory, defining itself not only as a model of tourist fruition of the territory and its resources, but also as a new model of "human habitat" and "settling resilience". We intend to demonstrate how the correct Analysis of a resource, which operates on a vast area, could systematically affect the start of important processes of local and territorial regeneration. At the same time, the enhancement process of specific points of this complex territorial structure (often identifiable as significant environmental and cultural resources) can determine, on a local scale, important effects on the enhancement of specific "homogeneous regions" and on the well-being of the local communities.

2 METHODOLOGIES

The Research & Design method here proposed is denominated "Evolutive (Architectural) Design" (EAD)[2]. It proposes itself as a method of design support aimed at simulating the behaviours of the cultural landscapes (UNESCO, 2005: 83; Montalbano, Neglia, 2014). Through its application, we intend to define a sequence of processes useful to highlight specific dynamics of the project sites. The first phase of investigation starts with the collection of some data related to the space to analyse and to the typology of intervention to apply. It entails:

a) The assumption of a specific "scale of reference" (identified as "interrogation scale of the model" it can assume territorial, urban, neighbourhood, local ... scale);
b) The definition of a "preliminary context of reference" (a non-definitive "spatial region" in its exact boundary, but symptomatic as a macro-area of design), with respect to which, although at a preliminary level, a specific degree and typology of homogeneity can be identified;
c) The identification of one or more "primary resources"[3] of the "context of reference". These have to be identifiable for cultural, historic, architectural, landscape, reasons, of priority interest for the analysed territory;
d) The identification, alternatively to the previous point, of a "site" or a "resource" (not necessarily primary) with respect to which it has been decided to take action for political, technical or cultural reasons.

This stage is followed by the identification of the planned typologies of intervention in relation to the assumed "context of reference":

- Restoration, conservation or reconstruction of the sites;
- Retrieval of the context;
- reprograming and readdress of areas with re-identification of spaces;
- Construction of new places and identities.

The definition of the typology of intervention enables to predetermine the "strength" with which to act on specific contexts and resources.

In order for the model to work properly, such direction should not be intended as totally restrictive, but employed to evaluate, at a final assessment stage, the real effectiveness of the envisaged design actions and, in case of need, to suggest more appropriate and specific typologies of intervention.

After this stage, "real" or "potential resources" can be identified as those belonging to the considered "context of reference" and that better respond to the necessary criteria of homogeneity and interoperability for the enhancement of the "spatial region" and the "resources" identified in it[4].

The identification and characterization of the "resources", as well as the progressive adaptation of the "boundary or margin"[5] of the "context of

---

[1] The ecosystemic services can be subdivided in the typologies of "provisioning services" (services/interventions aimed at the optimisation/enhancement of the food, hydric, wood, ... resources), "habitat services" (services/conducts aimed at the enhancement and the conservation of the habitats and bio-diversity) and "cultural services" (services/actions aimed at the enhancement of the cultural, educational recreational offer.

[2] The methodology illustrated in the following operates within the investigations that the interdisciplinary Laboratory of the Polytechnic of Bari "Cultural Landscapes Analysis and Design" has deepened, in reference to the identification of the processes of architectonic, urban and landscape design aimed at the enhancement of the Cultural Landscapes. In particular, in terms of the "Sustainable and Integrated Urban and Rural Development"; "Cultural Districts & Regional Tourism"; "Territorial Recovery".

[3] Main resource of a determined "spatial region", with respect to which strong hierarchisation are already active towards further "resources", called "secondary". Enhancement and ramification actions can be taken in relation to such primary resources, in order to widen its degrees of connection with further "sites" ("homogeneous regions") and surrounding "resources.

[4] They are, in particular, those "elemental components" (monuments, specialist and non specialist buildings, archaeologic areas, roads, squares, environmental contexts, ecological pathways,...), belonging to the "scale of reference" assumed, whose (own, positional or topological and relational) "value" can be finalised (due to own characteristics, for direct or indirect capacity to interact with other resources, or for potential condition of a "site"/"spatial element") to increment the value induced by the "homogeneous spatial region" or the "primary resource" prevalently identified.

[5] Every "spatial region" appears to be constituted by:

- a "boundary" or "margin" which can be intended as a site of continuous succession, and almost always inhomogeneous, of (natural, such as crests, ridges, compluvia, ... or anthropic, such as fences, walls, cortines, buildings, ...) "resources";

reference", assume fundamental importance for the following stage of design due to the significant role they play in the processes of identification of the sites; permanence of the memory; sense of belonging to the local communities.

In particular, in order for the "resources" to be recognisable in terms of value and order of bond, they need to be grouped in specific classes of belonging, to highlight possible levels of interoperability. Firstly, it is necessary to distinguish them based on their dimensional characteristics and their capacity to assume, within the "spatial region" they belong to, the role of connectors (spatial vectors) or, rather, of centres (hubs) of the new spatial system. The distinction between:

- Punctual resources. With prevalently "polarising" behaviour;
- Linear resources. With prevalently "connective" or "separation" behaviour;
- Aerial resources. With prevalently "autonomous" behaviour[6].

Allows, therefore, the redistribution of the resources according to their "capacity for interconnection".

At the same time, it is necessary further to distinguish them in relation to their specific "characterization". Such subdivision enables to read the "dominance" of a "spatial region" and, therefore, its "vocation". It contextually enables to compose possible frameworks of coherence between resources of the same kind or, in the framework of their systemic integration, to envisage actions aimed to interact different resources so to favour interferences and stimulate specific economic and social-cultural repercussions.

This way, starting from the topological characterization of specific "resources", as well as the arrangement and characterization of the "boundary" of a "spatial region", a wide spectrum of design actions, can be activated: ranging from the facilitation of processes and bonds existing in that region, to the alteration of specific bonds to favour the birth of new spatial hierarchies. In all these cases, the EAD enables to predict the direct and indirect consequences of the single design actions adopted.

The principles hitherto illustrated have been subsequently applied to the principal route of the Apulian Aqueduct.

- "internal region", characterised by the presence of "real" or "potential "resources;
- a "limit region", subset of the internal region, which directly interacts or is subject to the effect of the "boundary" or one of its parts.

[6] They are very particular resources whose aerial extension is such to characterise them as sub-sectors of the identified "spatial region". Often derived from the fragmentation of large scale linear resources (see the phenomena of rarefaction of environmental pathways which produce circumscribed environmental reserves), tend to appear autonomous or correlated to "elemental components of space" which are often characterised as detractors or restricting the degrees of connection with surrounding "resources".

Table 1. Possible characterization of the resources of a spatial region.

Possible characterization of the Natural System	Possible characterization of the Anthropic System	Possible characterization of the Productive System
Environmental (resources connected to flora-fauna aspects of the marine or land systems)	**Architectonic** (building, compound, settling system or part of it)	**Agricultural**
Landscape (areas connected to landscape-perceptive aspects)	**Historic-Monumental** (sites, buildings or compounds, ...)	**Crafting**
Ecological (areas of environmental services or eco-systemic equipped)	**Archaeological** (sites, buildings or compounds, ...)	**Industrial**

3 THE APULIAN AQUEDUCT: FROM SERVICE NETWORK TO HISTORIC-CULTURAL "RESOURCE"

Apulian economic, social and cultural asset, a symbol of the development of Southern Italy, the Apulian Aqueduct, started in 1906 and ended in 1915, represents one of the major works of hydraulic engineering existing in Europe. Taking 22,000 workers, 60 engineers and 400 technicians, active over 100 building sites, 99 galleries were realised (for a total of 110 km), 91 canal-bridges (in total 6.8 km) and a total route, of the main duct, of no less than 244 km.

The ridge of the main Canal[7] of the Apulian Aqueduct[8] marks the articulated territory, traversing it and, at times, heavily transforming it in order to bring drinking water to a territory before then organised with entirely traditional hydric management methods.

Starting from this moment, the aqueduct set out the rules of the anthropisation of the territory. To it can be attributed the transformation of many small rural settlements in real towns of tens of thousands inhabitants and the further enhancement of the great urban centres as well as the rural ones.

---

[7] The two historic ducts of the aqueduct compose the route: the Principal Canal, from Caposele (AV) to Villa Castelli (BR), which in just nine years (1906–1915) managed to take water to Bari; and the Great Lecce Siphon, which from the termination of the former reaches Santa Maria di Leuca.

[8] The aqueduct starts from the Basilicata hydrogeological basin from which originates the sources of Sele.

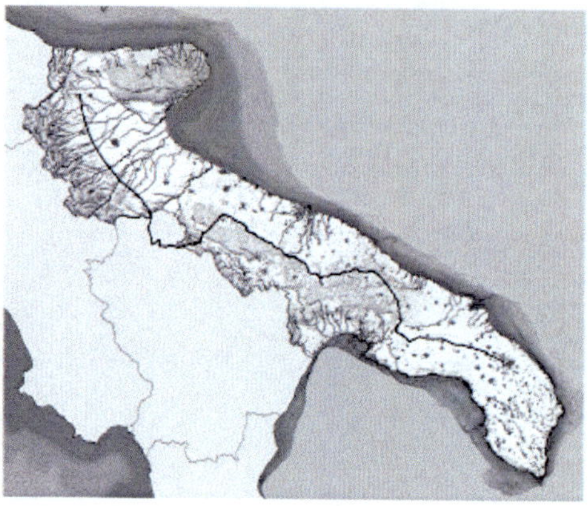

Figure 1. The Apulian Aqueduct and the crossing of the regional hydromorphological ridge.

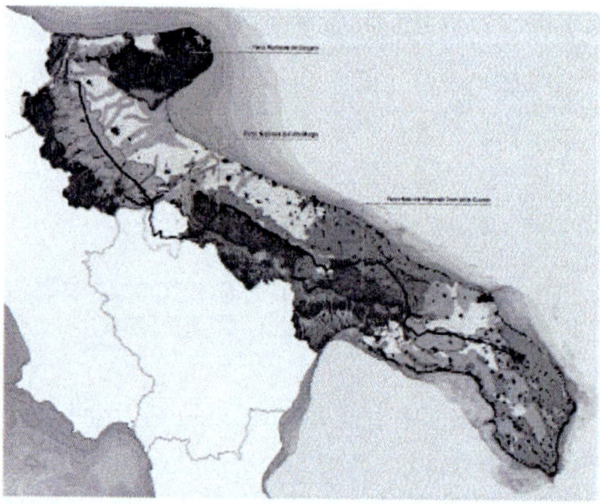

Figure 2. The Apulian Aqueduct and the crossing of the regional environmental and natural systems.

The necessity to flow along ridges and particularly elevated areas of the territory, indispensable to guarantee constant slopes and pressure along the hydric distribution network, as well as the necessity to guarantee an optimal distribution both along the Adriatic front and the Ionic one, have favoured its transformation into a proper territorial watershed.

For this reason, its particular location has very often led it to traverse the great natural areas (identified in the EAD model as aerial resources of natural and landscape type) and, more generally, the less anthropised spaces of the internal part of the region.

The changed living condition, favoured by the presence of the hydric network, have also led to a growth of the settlement cores (intended in the EAD model as punctual territorial resources) and a further compression and fragmentation of the natural resources, forcing them to concentrate prevalently along the central regional corridor innervated by the Aqueduct network.

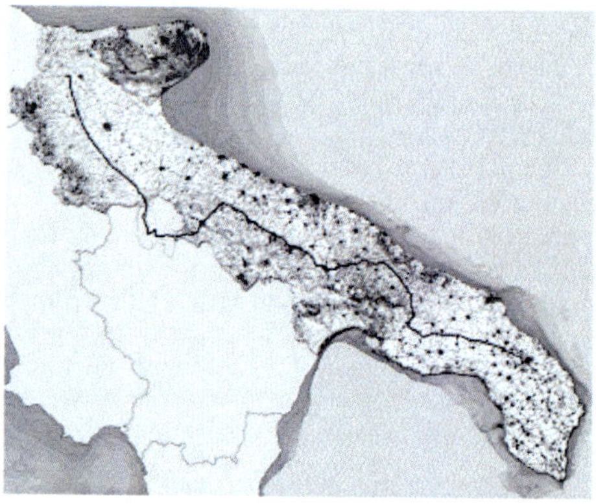

Figure 3. The Apulian Aqueduct and its interconnection with the regional linear and punctual resources.

This way, a very particular territorial situation is generated, in which the Apulian Aqueduct is featured as a territorial watershed, but also as a proper environmental and landscape corridor, whose width influences an increasingly higher number of settlements as it reaches the coast. Such settlements, variously distributed on the regional territory in terms of size and mutual distance, also as a result of specific historic-territorial dynamics, have led to the creation of particular territorial enclaves (homogeneous territorial regions), which result intertwined in various ways by the infrastructural corridor of the Apulian Aqueduct.

4 THE APULIAN AQUEDUCT: FROM HISTORIC-CULTURAL "RESOURCE" TO CYCLING PATH OF THE REGIONAL SYSTEM

Without perceiving the bonds hitherto observed, the idea to think about the paths of the Aqueduct's Main Duct for cycling purposes was conceived in 2008 through a protocol[9] of agreement between the Apulian Region and the Apulian Aqueduct. This has been followed by the realisation of a first 10-km segment of the cycling path in Itria Valley, between Figazzano and Ceglie Messapica, and with further 4 km along the last segment of the Main Duct until Villa Castelli. Today, the project intends to integrate the "Aqueduct cycling path" with Route n. 11 of Bicitalia's national cycling path, which originates in the National Park of Casentine Forest (between Tuscany and Emilia Romagna) and, running along the entire Appennino, ends up in Puglia.

[9] Such project originates from a feasibility study and an agreement protocol between the Region and its hydric company in 2008 regarding the cycling and cycling-pedestrian accessibility of the service ways of the Aqueduct (Project CYRONMED).

In the project of the Aqueduct Cycling path, a few characteristics that easily associate it with the consolidated model of the Greenways can be found (Padovani, 2013):

- "the safety", since they are physically separated routes from the ordinary road network exclusively dedicated to non-motorised users;
- "the accessibility", for all typologies of users with different characteristics and abilities;
- "the slow mobility", which allows the "slow fruition" of the routes offering a different point of view of the surrounding landscapes;
- "the multi-user usage", since the "greenways" are generally open routes to every typology of users (pedestrians, cyclists, horse riders, etc.);
- the retrieval of existing infrastructures and structures, such as paths, historic roads, trails, dismissed railways, minor rural roads, etc., for the realisation of the routes and the service structures (resting and refreshment sites, information points, etc.);
- the integration with the natural landscape, which enables the greenway to offer a respectful access to areas of particular natural importance and carry out an important educational function allowing the sustainable awareness and fruition of the territory.

The importance and the potential offered by the infrastructure of the Aqueduct to the Apulian territory, on the other hand, has been thoroughly investigated in the Strategic and Sector Planning on different scales:

- On a national scale. Through the Law of Stability 2016 which has inserted the Apulian Aqueduct Cycling path among the four projects with national priority together with the Verona-Florence Sun cycling path; the Venice Turin VenTo cycling path; the Roman Grab;
- On a regional scale. Through the Implementation Plan of the 2015–2019 regional Plan of transports and the Regional Territorial Landscape Plan; both aimed at seeing the aqueduct routes, also on the basis of already completed projects and proposals, as slow and cycling mobility or rather, as naturalistic routes (Green Way) included into projects of regional ecological networks (within the implementation of the policies and the norms in matter of biodiversity and, more generally, of natural conservation);
- On a local scale. Through the funding of projects aimed at the cycling-pedestrian mobility (within programmes connected to the Cultural Environmental Systems – SAC; to the planning of Vast Areas; of the Local Tourism Systems. . .);

Notwithstanding, the tools hitherto analysed and put into action by the various actors called to define proposals and projects for the Aqueduct Cycling Path seem to confine this infrastructure on a regional scale at a bi-dimensional condition. This is characterised, in fact, as a constant and homogeneous strip passing through the territory; it is not interconnected or placed in relation to the eco-systemic flows and at the various territorial environmental systems or to the articulated monumental and archaeological heritage of local contexts. Moreover, the design practices activated so far have been concentrated, on a local scale, on isolated and textbook project realisations, prevalently aimed at the fruition of specific paths without bothering to define particular connections or relations with surrounding sites of social, cultural or recreational interest. All this, although not questioning the value of the transformation of the Aqueduct routes into a "cycling greenway", in fact, prevents the favouring of a true ramification of such territorial network in the local systems and, moreover, obstructs a real and direct interaction of the local population with the network itself and the urban and eco-systemic facilities potentially interconnected to it.

5 THE TRANSCALAR APPROACH AS PERSPECTIVE OF DEVELOPMENT OF THE APULIAN AQUEDUCT GREENWAY

Using a different perspective, it can be proved that, starting from the integration of the different tracts of the Apulian Aqueduct, an imposing landscaped corridor of around 500 km can be composed, able to become place and space of a privileged territorial crossing and, especially, gate of access of extraordinary interest to the multiple enclaves of "cultural landscape" it traverses and partly contributes to define.

In fact, due to its connotations, this corridor would compose one of those infrastructural green networks (Green Way)[10] of the regional territory, which the London "Town And Country Planning Association" (Cullingworth et al, 2015) identifies among the multifunctional resources able to offer important benefits to the enhancement of the landscape quality; to favour a better ecological functioning, an improved quality of life and, in brief, the achievement of a more elevated social, urban and environmental sustainability. In particular, the design and management of these green infrastructures[11] can contribute to protect and improve the character and the identity of the places, safeguard the various kinds of existing landscapes and habitats, mitigate the fragmentation of the territory and make accessible and useful land resources, including through alternative road routes. The green infrastructure of the Apulian Aqueduct, rethought as a cycling

[10] The green infrastructure composed by several territorial networks: the hydrographic network, the ecological network, the cultural goods and sites of tourism interest network, the agricultural network and the light mobility network (pedestrian and cycling pedestrian) which guarantees accessibility and public fruition (Socco, 2008).

[11] The green infrastructures, according to the Communitarian definition, are "networks of natural and semi natural areas, strategically planned with other environmental elements, designed and managed in order to provide a wide spectrum of ecosystemic services".

path, can ensure wide benefits to the populations involved, among which:

- An important contribution to the development of the rural contexts traversed, bringing wealth and incentives to the creation of new entrepreneurial activities;
- The promotion and diffusion of outdoor activities, with beneficial effects on the citizens' health;
- The development of a new form of active, responsible and sustainable tourism;
- The awareness of the nature and respect for the environment;
- The awareness and the enhancement of the historic, artistic, architectonic, cultural, environmental and landscape heritage, as well as the traditions and the **typicality** of the traversed areas.

The interest in this extraordinary, potential, interaction is clearly visible in the increasingly popular models of "Slow Town" which associate with the fundamental concept of "slow mobility", the similarly significant one of "slow economy" (prevalently related to the circular economy) and "slow society" (related to the wellness and ecosystemic services). (Heitmann et al., 2011; Pink, 2008).

Starting from all this and attempting to define a systemic design vision of the "Greenway" of the Apulian Aqueduct, a transcalar territorial vision has been assumed, which has operated on three distinct levels of regional analysis, of the homogeneous territories (homogeneous region) and the minimal territorial units, highlighting the potentials and the different degrees of interconnection of the Aqueduct network with each one of these levels.

From here, the first level of investigation, analysed in the present essay, has led, through the consideration of homogeneous territorial features (morphologic, environmental and historic-cultural), to a mapping and weighing of the regional resources and their subsequent grouping in terms of punctual, linear or aerial typologies (Montalbano, 2012). This way, the Simulation Model has enabled to identify, within the investigated territory, specific homogeneous landscapes (as identified in Fig. 4 based on the homogeneity of the environmental, geo-morphological, historic, architectonic and cultural processes which have characterized its transformations), recognizing a role and a field of reuse for the rich real estate assets once at Aqueduct service (such as the road worker houses).

All this highlights how the system of the "Green Way of the Apulian Aqueduct", traversing great part of the regional territory, connects an extraordinary variety of historic, environmental and landscape resources of elevated cultural value. In this sense, the Apulian Aqueduct cycling path becomes a starting point of an articulated reflection on the capacity of this light infrastructure to "traverse" and "tie" a multitude of typologies of the landscape. At the same time, its extraordinary tendency to be constituted as a privileged corridor emerges, through which enter in heavily identitary landscapes. These last have the

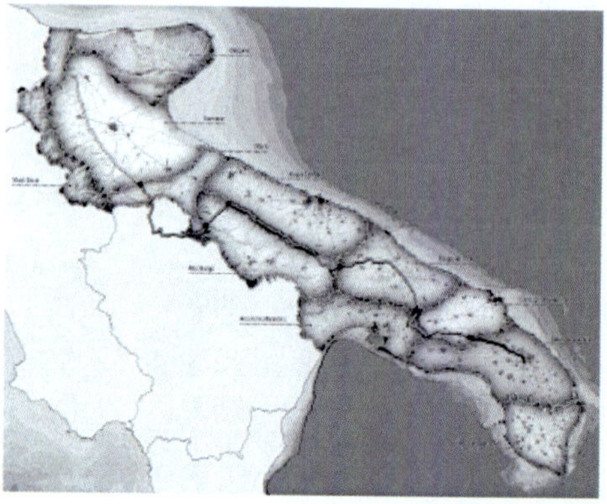

Figure 4. The Apulian Aqueduct and its interconnection with the homogeneous territorial spaces.

opportunity to find, at the points of contact with the green infrastructure, important eco-systemic facilities (such as ecomuseums, environmental parks, green equipped areas, archaeological parks and, more in general, leisure and sport areas), useful to the functioning of the settlements and the well-being of the local collectivity. (see Fig. 4).

In particular, this system of analysis was followed by the next level of investigation. This through a parameterization, classify and evaluate on the basis of a multi-criteria matrix, the potential binding orders and degrees of importance of specific resources of the environmental and cultural heritage, in order to determine their degree and interference capabilities in context and in relation to the Aqueduct heritage.

6 CONCLUSION

Recurring to the image of a territory composed of corridors (connections) and rooms (homogeneous spatial regions), it is difficult to think that the Apulian Aqueduct system could behave as an isolated element with respect to the territory. Rather, its real value can be highlighted in the capacity for catalysing flows of users and specialised services that increasingly identify it as a green infrastructure of service and entrance door to the various territorial spaces.

In particular, the present research attempts to demonstrate the deep influence that the route of the Apulian Aqueduct generates on the landscapes traversed and its favourable impact that can be demonstrated on different project scales, on the enhancement processes of the local territories, enabling also to:

- define integrated programs of design actions apt to create conditions and the attractiveness of the investments connected to the durable development of the territory;
- identify specific areas of intervention from which to begin for the definition of particular design interventions;

- identify specific categories of stakeholders and facilitate their aggregation around common projects.

ACKNOWLEDGEMENTS

The considerations in this work are the result of personal reflections and do not involve the Apulian Region. Part of the studies and research here summarised have been conducted through the development of a degree thesis, from the working group coordinated by prof. C. Montalbano and which, apart from Arch. L. Guastamacchia has seen the contribution of the professors of Polytechnic of Bari, prof. S. Bisciglia, prof. G. Martines, prof. V. Castagnolo, and the collaborators Doc. Andrea Capuano, Doc. S. Cascella, Doc. V. Cifarelli, Doc. S. De Pascale Doc. P. Mennoia, Doc. P. Palumbo.

BIBLIOGRAPHICAL REFERENCES

Cullingworth B, Nadin V, Hart T, Davoudi S, Pendlebury J, Vigar G, Webb D, Townshend T. (2014) Town and Country Planning in the UK. Routledge, London & New York.

Firenze, Alinea Editrice UNESCO (2005) Operational Guidelines for the Implementation of the World Heritage Convention. UNESCO World Heritage Centre. Paris. p 83.

Heitmann, S. Robinson, P. & Povey, G. (2011) Slow food, slow cities and slow tourism. Research themes for tourism, 114–127.

Montalbano C & Neglia G A. (2014) "Sviluppo sostenibile del paesaggio del Mediterraneo.Una ipotesiprogettuale", in A Castorani, C D'Amato Guerrieri, C G Andria, E Di Sciascio, P Camarda, G Monno, S Nuzzo (eds) 1st Workshop on the State of the Art and Challenges of Research Efforts at POLIBA. Groups research on KETS and SCS.Gangemi, Roma, pp. 667–671.

Montalbano, C. (2012) METODO e PROGETTO per la costruzione del paesaggio. Ugento: un atlante progettuale. Bari: Caratterimobili.

Padovani, Laura Maria. (2014) in Italian Ministry of Environment, Sea and Land Protection. Federico, T. et al. (2014). Processo partecipativo attivato dalla Conferenza di Roma sulle infrastrutture verdi. Le infrastrutture verdi, i servizi ecosistemici e la green economy. Conference "Nature of Italy" Rome, December 11 and 12, 2013.

Pink, S. (2008). Sense and sustainability: The case of the Slow City movement. Local Environment, 13(2), 95–106.

Socco C. (2008) Osservatorio del paesaggio dei parchi del Po e della collina torinese. L'infrastruttura verde del parco del Po torinese, Firenze, Alinea editrice.

"Rural villages" as engines of territory sustainable growth

Tiziana Basirico
Faculty of Engineering and Architecture, University of Enna "Kore", Enna, Italy

Mário Say Ming Kong
CIAUD (Centro de Investigação em Arquitetura, Urbanismo e Design) – Faculdade de Arquitetura – Universidade de Lisboa / CHAM, FCSH, Universidade NOVA de Lisboa, Universidade dos Açores

ABSTRACT: This paper presents results on a project that involves the realisation of a touristic-cultural route that virtually and physically connects the rural villages realised in the 1940s by the Italian totalitarian regime, following the policy of colonisation of the *latifundium* in southern Italy. This project aims at introducing innovative social and environmental solutions, presenting some specific cultural goods (material and immaterial) as drivers for the sustainable growth of underdeveloped rural areas.

These rural villages constitute not only the evidence of an architectonic and technological age but also what remains of the 20th-century peasant civilisation. The aim of the project is to go far beyond conservation, restoration, physical rehabilitation, or re-use of such villages, and to demonstrate how cultural heritage is a powerful economic, social and environmental catalyst for regeneration, economic growth and economic improvement of people and places.

This project is a model of sustainable development that can be replicated, with appropriate contextualization, in other similar territorial environments in other parts of Italy as well as in European countries such as Portugal and Spain that have adopted between the 1930s and the 1960s the same agricultural policies.

Keywords: Rural villages, cultural heritage, rural tourism, sustainable development

1 INTRODUCTION

The economic and social vitality of a rural area relates to agriculture and tourism, allowing spontaneous production and generation of rural capitals. From this starting point, the project integrates agriculture and tourism with another variable – cultural heritage – creating an eco-tourism route including the rural villages scattered in the Sicilian hinterland. These villages were built between the 1940s and 1950s.

These rural villages, built in the 1940s by the totalitarian regimes of the Mediterranean (Italy, Portugal, Spain) and in the colonies (Greece, Libya, Ethiopia, Eritrea, etc.), are an important evidence of the period, as well as architectural heritage to be protected, projected by well-known architects of the period.

Unfortunately, many of these villages are included and surrounded by territories with economic and social problems that have led to depopulation over the years and the loss of traditions.

It is, therefore, necessary to implement regeneration models going beyond the simple recovery and re-functionalization of architectural heritage in rural areas.

The aim of the project is to activate, through the enhancement of the building heritage, a system that can combine cultural, tourism and productive activities and reconnect population with the resources of the land (agricultural, naturalistic, archaeological, historical and cultural) for the sustainable development of rural areas.

2 ORIGIN, EVOLUTION AND ACTUAL STATE OF RURAL VILLAGES IN MEDITERRANEAN COUNTRIES

In the economic climate of the Great Depression, which began with the financial crisis in 1929, whose effects have been felt all over the world in the 1930s and 1940s, first in Italy and then in Spain and Portugal, totalitarian regimes promoted agrarian reclamation programs for land transformation and population settlement in the countryside.

This transformation concerned the improvement of the productive capacities of the land, first through the drainage of wetland areas and the implementation of a series of infrastructural interventions (roads, aqueducts, water dams) and then by the fragmentation of extended and badly cultivated land, the property of few landowners.

After the end of World War I, this policy responded to the demand for the occupation of the war veterans.

In order to implement this policy, it was necessary to develop peasants' settlement in the countryside for the construction of farmhouses and agglomerations. Several necessary services were thus built: civil assistance, health care, school assistance, religious assistance (church and rectory), logistics (inn, several shops: the blacksmith, the carpenter, the cartwright, the shoemaker, the saddler), connection (post office and telephonic point) and public order protection (Carabinieri station). The projects of these villages were entrusted to some renowned architects of the various countries (A. de la Sota, C. Arniches Moltó, J. Borobio, C. Petrucci, R. Silenzi, L. Piccinato, etc.).

The settlement system adopted in the various European countries, albeit following the same principle, was slightly different as the actual state of these rural villages.

In Spain and Portugal, unlike in Italy, the persistence of the totalitarian regimes and agricultural policy until the 1970s has led to consolidation and progress of communities in most rural villages.

2.1 *The Iberian Peninsula*

The agricultural policy adopted in Spain and Portugal aimed at ensuring food self-sufficiency to the countries.

In the colonization plans developed in Portugal by the "Junta de Colonizacao Interna" (JCI – est. 1936) and in Spain by the Instituto Nacional de Colonización (INC – est. 1939), the technicians referred to similar agrarian operations such as the Italian Land Reform by the "Opera Nazionale Combattenti" (ONC – est. 1917) or the experience of the Belgian Société Nationale de la Petite propriété terrienne.

Therefore, the Spanish and Portuguese rural villages show the same urban and architectural features of the Italian ones.

In both cases, the apportionment system was based on two types: concentrated farmhouses and scattered houses.

In Spain, the new settlements, about 300 during Franco government (1939–1975), were mainly built in the river basins, creating a regional structure around the main ones: Duero, Ebro, Tagus, Guadiana and Guadalquivir (Monclùs et al. 1988).

The more isolated villages were active for a few years and then they were progressively abandoned (remaining only residential agglomerations for no more than 50–100 people) as for the village of Guma, Águeda del Caudillo (Salamanca) and Foncastín (Valladolid), similar to some of the Italian ones; these villages need such a regeneration strategy.

The villages near the rivers, such as those of the Guadalquivir valley (crossing the provinces of Jaén, Cadiz, Seville and Córdoba) and those of the Béticas irrigation areas, have managed to intensify their production or develop strategies helping to consolidate people permanence (e.g. in the smaller El Viar and the bigger Isla Mayor) (Villa Díaz and Ojeda Rivera, 2005).

The same destiny shared the Portuguese villages. For Faias and Pegões, the apportionment was made along the existing rivers (farms between 15 and 20 hectares), while in other cases, as Figueiras, the apportionment was in concentrated form, consisting of an orthogonal mesh of roads for the division of properties.

Among them, Colonate of Cela and the colony of Santo Isidro de Pegões, which is the largest and most complex development carried out throughout the country, bringing together 207 farmhouses of the 500 nationwide.

The success of the "agricultural colonies" seems to have been greater in the concentrated settlements, compared to the dispersed ones. In the first ones, similar to the Italian rural municipalities, there were stable residences. It has led to the creation of social relations and a spirit of characteristic rural community – as in the case of Santo Isidro de Pegões (nowadays counting 1500 inhabitants), Boalhosa and Barroso. In the second ones, there was isolation and abandonment, such as Martim-Rei ou do Alvão (Paulo Lima, 2013).

Some of these new settlements have undergone the depletion of their natural resources due to the opening of the country to international markets, but others have consolidated their activities, diversifying their production strategies.

However, all rural villages are interesting to investigate representing the evidence of a lost epoch of Spanish and Portuguese history. A differentiation among the villages must be done: the villages, which do not always have the original features, due to adaptations and scaling of the original elements and the abandoned ones that retain the original architectural features beyond the size.

2.2 *Italy*

The "colonisation of the latifundium" was extensive in Italy, affecting several regions, including Lazio, Tuscany, Sardinia, Puglia, Sicily, Friuli Venezia Giulia and Istria.

In Lazio, the "Bonifica dell'Agro Pontino" and in Puglia, the "Piano per il Tavoliere" saw the same settlement policy-oriented by the ONC, responsible for the management of the latifundium colonisation.

The settlement system adopted in Puglia was very similar in Lazio. It was a scattered apportionment system with individual farmhouses set in a territory of about 10 to 20 hectares, near to a rural village with the most important public services and buildings.

These rural villages, in addition to being a service centre for 3000 inhabitants spread across the territory, gave the opportunity to accommodate about 500 permanent residents (Pennacchi, 2008). In addition to these rural villages, there were also real "rural municipalities" with stable residences for 3000 inhabitants and reference point for 9000 settlers.

The apportionment system was hierarchical (farmhouses, rural villages and rural municipalities). In some cases, new rural municipalities, surrounded by several rural villages, were built around existing cities. This is the case of the "Piano per il Tavoliere" in Puglia, providing for the construction of four rural municipalities (Segezia, Incoronata, Daunilia and Arpi) around

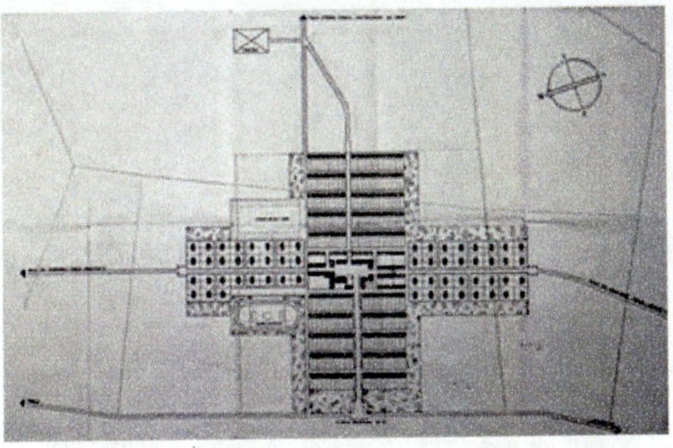

Figure 1. Plan of Segezia, 1939. The service buildings in the central square and the residences around (Cucciolla, 2006).

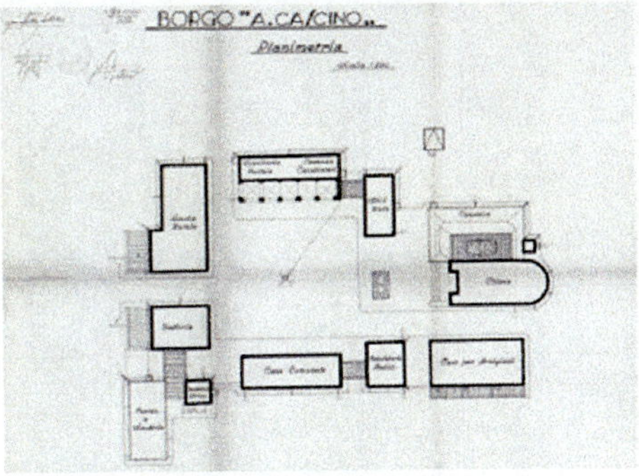

Figure 2. Project of Cascino village, 1940 consisting of the service building around the square (ESA archive).

the existing city of Foggia and six other rural villages (Cioffi, 2017).

All the villages and rural municipalities, both in the first phase of the colonisation in the 1930s and in the second one in the 1950s, have the same plan imposed by ONC, including the central core of service buildings around the square. In some cases, especially in the larger centres, there are two squares, one for religious power with the church and one for civil power with the house of the Fascist party. Many rural municipalities were already designed with a modular structure for future extension (fig. 1).

The various rural villages were active for the first years then some were developed further and some were abandoned.

Many villages, due to the war, were never completed (e.g. Daunilia and Arpi in Puglia), many others were occupied by Anglo-Americans and left at the end of the war in a state of desolation and degradation.

As in Portugal and Spain, the rural municipalities were developed into important territorial presences such as Aprilia and Pomezia in the Agro Pontino, with more than 60,000 inhabitants and Segezia and Incoronata in Puglia.

Most of the small service centres, near the farms, were often abandoned, especially those villages where the planned houses were not completed (such as Borgo Domizio and Borgo Appio in the Volturno plain).

While in Puglia and Sicily, with the fall of fascism, the settlers left the farms to return to the towns and cities, in Lazio this depopulation did not occur because settlers were not local people, but came from other regions of northern Italy, especially Veneto and Emilia Romagna.

In the 1950s with the "Agricultural Reform", both the construction of rural villages and the allocation of farms continued, but these ones were small land plots (2.5–4 hectares), which were not sufficient for the sustenance of the farm family. This brought again emigration to the industrialised areas of the country and to the abandonment of such land, which lately, was again grouped for a largely cereal-rich broad-growing type.

Recently, these villages are being rediscovered for historical and cultural reasons. Unfortunately, there are only a few examples of re-functionalization, especially by private initiative.

2.2.1 *The Sicilian rural villages*

Among the various Italian regions, the urban-architectural phenomenon that led to the realisation of rural villages was particularly widespread in Sicily. In the course of 30 years, 68 villages (with more than 25,000 farmhouses) were built. This explains how rural villages, of varying sizes and types, are a typical feature characterising Sicilian land. They are also a testimony of an era, both from the political, social, architectural and urban point of view. In fact, the villages built in the early 1940s, just 13 (Fazio, Bassi, Schirò, Borzellino, Rizza, Lupo, Bonsignore, Callea, Cascino, Petilia, Giuliano, Guttadauro e Ventimiglia), are examples of architectures of Italian rationalism with hints to the traditional rural architecture, made by the most renowned engineers and architects of the time, and therefore recognized of cultural interest to be safeguarded (Basiricò, 2009).

In Sicily, unlike the other Italian regions, the apportionment was not managed by the ONC, but by the "Ente di Colonizzazione del Latifondo Siciliano" (ECLS, which was specially established in 1940). No rural municipalities were built, but only rural villages as service centres for the peasants widespread in the territory. In addition, although these centres had three possible dimensions (small, medium and large), they did not have stable residences for settlers, but only residences for the owners of the various services (Fig. 2).

Although in the 1950s with the agricultural reform, managed by the new regional institution with legislative and operational autonomy, plans were made for the expansion of each village with buildings for the settlement of peasants but were not completed, due to the state of abandonment of some of these villages.

From the analysis of the villages in Sicily between the 1930s and the 1960s, it was found that most of them, nowadays, are abandoned, suffer from severe structural degradation.

The use, though improper, of some villages, was the only factor that allowed an acceptable conservation

Figure 3. Fazio village (author image, 2008).

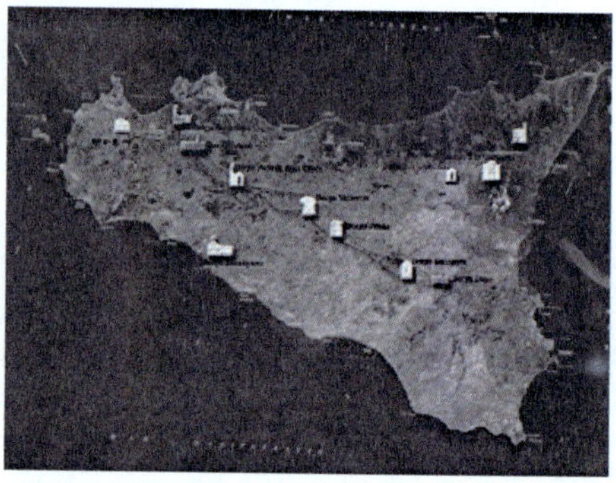

Figure 4. Project of tourist-cultural itinerary "Via dei Borghi" (ESA archive).

status despite the changes due to adaptation to new functions and the relative anthropogenic degradation.

Some villages located in proximity to road or rail infrastructure with their stations, small towns or city centres, has remained active as a residence for no more than 50 people, but without any service. Some of these examples, enclosed in western Sicily, are the Bonsignore, Petilia, Cascino, Buseto and Manganaro villages.

None of these villages has become, as in other cases in Italy and abroad, a source of building and economic development in the surrounding area.

In other cases, however, the village is completely abandoned, mostly far from roadways and remaining an isolated sign in the landscape, still remarkable only by its location most often set on low hills. In the long list of such villages, there are Fazio, Schirò and Borzellino completely abandoned and almost completely ruined (Fig. 3).

Unfortunately, the location of such villages in areas not easily accessible to the Sicilian hinterland and above all the abandonment, have led to the loss of memory of such a vast and valuable heritage.

Only recently, some building recovery initiatives have taken by the competent administrations (by means of European funds), but no major-scale intervention has led to a real upgrading and regeneration of its rural areas.

A strategic project is under development at the request of the "Ente Sviluppo Agricolo" (ESA) in collaboration with the University of Enna Kore, with industry and other stakeholders in the phase of submission to regional planning calls (POR), European Funds 2014–2020, as follows (Basiricò, 2016).

3 THE DEVELOPMENT MODEL OF THE "VIA DEI BORGHI"

More than 70 years have passed since the first rural villages had been built, and social, cultural, agricultural and industrial development has been remarkable. It is necessary to think of alternatives that meet new economic policies in line with the 21st-century logic, recognising the values of diversity, mobility, social integration, globalisation, etc.

It is necessary to face the recovery of these villages from two points of view, both for the Sicilian villages and for the villages with the same characteristics and conditions in other parts of Italy and other countries affected by the last century phenomenon of the "Colonization of the latifundium".

The first point of view is cultural. Indeed, through the recovery of these villages, tangible and intangible assets, essential to understanding and preserving the history of the second half of the 20th century in Italy, can be re-evaluated.

The second one is economic. Recognising in general terms, the importance of these villages is necessary to carry out a detailed analysis of how each of them could be re-evaluated. Their peculiar characteristics, exploiting the recognised value from the standpoint of location, architecture and landscape can become an input for the social and economic re-evaluation of the area.

The project developed for the Sicilian rural villages is a strategic (inter-municipal and multi-sector) development project that combines tourism, culture and agriculture synergistically. The project aims at revitalising some underdeveloped territorial contexts through the recovery of some rural abandoned or underused villages and their networking for tourism and cultural purposes.

The planned new tourist-cultural itinerary, known as the "Via dei Borghi", which would cross Sicily from west to east, would actively promote the Sicilian hinterland tourism currently unaddressed and fragmented away from traditional tourist circuits, aimed more at the monumental heritage of the great cities, coastal centres and smaller islands (fig. 4). Moreover, each village stands near monumental, archaeological, anthropological and naturalistic sites, and therefore can be a reference and connecting point with existing regional itineraries ("wine roads", "path of flavours", trails of historical trains, archaeological and naturalistic itineraries, literary park itineraries, UNESCO routes, etc.).

The strategic lines for the development of the project are:

- complete the historical knowledge of villages and organise a network of knowledge available to citizens, experts and administrators;
- promote the valorisation and protection of such architectural-urbanistic-landscape complexes by the competent authorities;
- valorising cultural and territorial realities of obvious marginalisation with respect to the main and well-established tourist itineraries;
- identify the potential for using this cultural heritage in the current political and economic landscape.

The project creates a path of both knowledge and enjoyment through the transformation of Sicilian rural villages into recreational and tourism facilities and their networking. The link between the villages would not only be achieved virtually, but physically through the link between the existing roads and any new greenways.

The new cultural tourism, once initiated, will be not only seasonal but thanks to the mild weather conditions in any season, could extend almost year around, linked to the specificities of the places in the various seasons.

There are many cultural, social, environmental and economic benefits of the project, such as set-ups, start-ups of new productive activities in various fields (cultural, tourism, construction, agriculture, computer science, etc.).

Cultural Benefits:

- The recovery of material assets through the restoration of rural villages, the architectural heritage of the Sicilian Region;
- The recovery of immaterial assets such as the memory of the Sicilian peasant civilisation of the first half of the 20th century through the testimonies of folk traditions, festivals and folk songs, but above all handicrafts, agricultural productions and peculiarities of food and wine.

Social Benefits:

- New users of the internal territories and small centres currently not familiar with the national and international circuits of emerging and Mediterranean countries;
- the enjoyment of high-quality of life places, away from urban centres, pollution, automobile traffic and stressful rhythms of city life, surrounded by nature, healthy, responsive, environmentally friendly rhythms.

Economic Benefits:

- Training of small farmers towards organic production and training of agricultural entrepreneurs on the productive, administrative and marketing activities;
- opening up new market opportunities through the networking of small existing companies, which can make their specificities and biodiversity strong points for increasing production and sales; opening to new national and international markets with the use of Information and Communication Technologies (ICT);
- Incentives for self-entrepreneurship and innovative start-ups through renewed interest in rural settlements and innovative agricultural productions.

Environmental Benefits:

- Territory control;
- Conservation of biodiversity.

The project will be supported by ICT, providing for the creation of an internet website that will enable the implementation of current national and international trends in so-called on-line or do-it-yourself participative and shared tourism.

The project the "Via dei Borghi" is the repetition of a "model of development" consisting of a small network of rural villages of different sizes, used to attract and enjoy a new tourist path, intertwining with local tourist trails, local production activities, unique scenic and naturalistic features.

The model consists of one or two complete full cultural tourism service centres, located within entire villages to be completely restored, linked to their history and image and other smaller centres using only a few buildings of the villages destined for the tourist services essential for the "Via dei Borghi".

The choice of complete service centres was based on the architectural characteristics of the villages, their location, conservation status and the presence of naturalistic, archaeological and monumental presences.

A territorial study was carried out, allowing the identification of future destinations linked to cultural and food and wine tourism compatible with the specifics of the territory and the strategies of revitalization and marketing of the territory itself.

In the complete tourist cultural service centre, some facilities are provided for reception and short stays (bar, restaurant, forestry), buildings at the service of the greenways (hiring bicycles, horses, parking and assistance), exhibition and sale spaces for typical local products, information point on all the initiatives of the area and all other villages on the net (info point, etc.). There will be also recreational centre for video demonstration installations regarding with the production of farms near the village, museum of peasant civilization of the 19th century with a thematic section on the history of the colonization and realization of the Sicilian rural villages (with photos, drawings, films, etc.) with the enhancement of the ESA's photographic, iconographic and bibliographic material.

In the smaller tourist service centres, the recovered buildings will be used as tourist information and reception, promotion of typical agricultural products in the surrounding area, through the valorisation of farm products far from the village about 30 km (at least one room for the sale of the products and for demonstration activities of the respective productions). The other buildings of the village will be assigned in free use with the commitment of restoration to public or private entities for activities compatible with those of the village.

Figure 5. Borzellino village (author image, 2008).

As an example of "development model" we refer to that of the western Sicily area. It consists of two full-service centres (Fazio and Borzellino villages) and two other smaller centres (Bruca and Portella della Croce village) located in the territories of two provinces (Trapani and Palermo). To bridge the excessive distances, in the case of alternative means of transport (horse, bicycle, etc.), the "model" also includes the use of some farmhouses (converted by private citizens, now owners) to bed and breakfast model.

Borgo Borzellino, in the territory of Monreale, is among the fascist villages, the one to be recognised historical and cultural value by the Superintendence of the cultural heritage and therefore itself constitutes an emotional tourist attraction due to its history and its architecture (fig. 5). In addition, the village, in full state of abandonment, allows the recovery and use of all buildings. Its strategic position on the Palermo-Sciacca highway, only 8 kilometres from San Giuseppe Jato, San Cipirrello and Camporeale, and therefore in a highly valued area for its proximity to the archaeological park of Monte Jato (with the important Ruins of Ietas) and the archaeological sites of Val di Bella, makes it an excellent centre for the promotion and enjoyment of these areas.

The well-equipped large central square could become a place for outdoor events, fairs, exhibitions and festivals to promote the area.

Due to the considerable presence in the proximity of wineries, Borzellino village could experience considerable tourist flows of the "wine routes" and in particular that of "Monreale Doc". In fact between the municipalities of Roccamena, Santa Cristina Gela, Corleone, San Cipirello, San Giuseppe Jato and Camporeale, and of course in Monreale, vines are being cultivated for the realisation of the "Monreale Doc".

Similar scenarios can be considered for Bruca village, located near the archaeological site of Segesta (temple and Greek theatre). The village could also intercept other existing regional routes such as the "Strada dell'Alcamo doc" or the "castles route" (the castles of Calatubo, Inici, Norman-Arabian castle of Castellammare del Golfo and Eufemio).

4 CONCLUSIONS

The "development model" for the valorisation of Sicilian rural villages has to be really tested to evaluate its effectiveness in terms of tourism, cultural and entrepreneurial impacts. It may be exportable and repeatable in other villages with similar settlements according to their specificity. This would allow creating a widespread national and international tourist network, due to the presence of similarities in other Mediterranean countries, such as Spain and Portugal.

The success of the project requires the involvement of local and regional authorities and local communities in a partnership with well-defined roles and responsibilities.

With this project, the conservation of architectural heritage would be seen not only as a cost but also as a driving force as it uses its potential for the economic growth of underdeveloped rural areas with high unemployment.

BIBLIOGRAPHICAL REFERENCES

Basirico', Tiziana. (2016) Sicilia è partito il recupero della via dei borghi, Ananke, n. 76, Firenze: Altralinea.

—. (2009) Architettura e tecnica nei borghi della Sicilia occidentale, Palermo: Fotograf.

Cioffi, Gianluca. (2017) Architettura e paesaggio agrario, Roma: Aracne editrice.

Cucciolla Arturo. (2006) Vecchie città/nuove città. Concezio Petrucci 1926–1946, Bari: Edizioni Dedalo.

Lima, Paulo. (2013) A Colónia Agrícola de Santo Isidro de Pegões, Montijo: Câmara Municipal do Montijo.

Monclus, F.J., Oyon, J.L. (1988) Historia y evolución de la Colonización agraria en España, Madrid: Ministerio de Agricultura, Pesca y Alimentación.

Pennacchi, Antonio. (2008) Fascio e martello. Viaggio per le città del Duce, Bari: Editori Laterza.

Villa Díaz, Águeda A., Ojeda Rivera, Juan F. (2005) Paisajes coloniales en el Bajo Guadalquivir. Origen, evolución y carácter patrimonial, PH Boletín del Instituto Andaluz del Patrimonio Histórico, n° 52.

Studying and living in the city

Carla Chiarantoni
DICATECh, Polytechnic University of Bari, Bari, Italy

Maria Immacolata Marzulli & Manuela Persia
DISAAT, University of Bari Aldo Moro, Bari, Italy

ABSTRACT: The paper tries to discuss the theme of students' temporary housing in light of the structural changes that are affecting the contemporary society. It is widely believed that the future of the communities goes through the development of the human and social capital; moreover, the capability to attract this capital lies within the construction of what we can define as "learning communities" (Longworth, 2006). These are societies that can invest in culture and stimulate in the citizens a strong desire to improve the future of the place where they live, learn and create. The role of the university and its supporting structures – such as student residences – is very important. Historically, these buildings have been the first symbol of the existence of a "studiorum" within the cities; nowadays they constitute an important element of experimentation of new dwelling models. One of the main characteristics of the contemporary society is a widespread tendency to the mobility of people; students constitute a significant part of this moving population. The features of the latter are particular in terms of permanence and expressed demands. Therefore, it cannot simply be labelled as "temporary population". Starting from a brief historical overview, we come to illustrate the new setting and distributional-functional trends that underlie two different ways of conceiving the life of a student. The substantial switch is the formulation of living paradigms that see the student living away from home as a sedentary city dweller, part of a population in which to invest.

Keywords: Student housing, student living, student well-being

1 STUDYING AND LIVING IN THE CITY THAT CHANGES: TYPOLOGICAL-FUNCTIONAL EVOLUTION OF UNIVERSITY RESIDENCE

The student residence starts, in its origins, as a mediator of the architectonic form of the convent, being the instruction reliant upon the ecclesiastical orders. Therefore, the residence was organised through a set of single rooms (denominated cells) and communal areas (study rooms, refectories and external bathrooms). In some cases, the structure could include the rooms destined to the lessons. The succession of political changes and the extension of knowledge to other institutions, such as the political ones, did not bring about substantial changes in the development of the building, which continued to have essentially the same functional distributive features. The change started with the establishment of a new approach: the "university residence" and the classrooms were separated and found their location in distinct structures, although internal to the city. The first substantial diversity that followed was a different urban setting. The prevalently Anglo-Saxon towns opted for the realisation of proper citadels, outside the urban centres, which incorporated the residential and the more specifically University related functions. This model was conceived with the possibility of expansion according to the demands and with the aim to create a proper University community. On the contrary, in Italy, the prevalently urban character of the university and residences was established. There were multiple causes, probably dictated by a mostly historical tradition, its origins, but also by the ease to find already available structures to be destined to these functions. Therefore, it was already affirmed a naturally "rental" model, which, unable to expand in continuity with itself, started to be inorganic and dislocated in various parts of the city. It should be underlined that until the end of the '800 the instruction was imparted by "masters" who moved from one town to another according to the patron of the moment, consequently determining the movement of the whole student population. This phenomenon — peregrinatio accademica – has undoubtedly generated a tendency to give scant attention to the "student" and, in general, towards the "student population", considering it as a temporary population ever since. The resources and structures had to be located within the city or in suitable buildings. The result of this situation was the tendency to think that the only demand of the student was to have a room for the night.

A real overturn occurred, with a noticeable time leap, during the fascist period when giving a new impulse to the scholastic and university education, the University hubs started to be erected (comprehensive of university venues and student houses) in the expansion areas of the cities. The residence was structured as a sequence of single rooms with external bathrooms; the common areas were located between the basement and the mezzanine floor. In this period, the colleges start to be configured as support structures to the student life. In them, the student was supposed to find a response to his needs, which would not only involve sleeping. This phenomenon represents in Italy what also in the rest of Europe has led to the interest of the Modern Movement in the formulation of a new way of housing (the traditional residence) and in the Student Residences, making them a real field of experimentation. The experimentation was concretised in a renewal and update of the housing space and its relationship with other service areas of the residence. The ultimate objective was to find a universally recognised code, able to satisfy all the possible demands of final users and become a housing model. The result was undoubtedly a break from the way of thinking about these places, which saw the arrangement of the building no longer as a cloister or corridor, but as an organised ensemble of places that stimulate the intellectual and social liveliness. In these experiments, as for example Le Corbusier's realisations, the break of the separation between the private (the rooms) and the public space can be clearly seen; the room continues to be in a row and repetitive but works in relation to the public space obtained from the enlargement of the distributors. Other experiments, such as Alto's, try to provide the students with their own living space, recognisable in terms of both shape and enjoyable panorama. To this heavy experimentation period during the second half of the '900 (starting from the post-war period, the restoration period) is opposed a progressive inattention to keep the interest in these structures alive. The buildings, especially in Italy, are realised with prefigured housing models and mediated by other typologies, such as the hotel, which allowed the allocation of a large number of users. The result was the application of rigid quantitative standards and already known distributive-functional logics, which were consolidated and economically verified. The imposing model was the double room with external toilets and common areas (prevalently kitchens) obtained from the resulting spaces.

In the 90s, the situation in Italy was extremely messy. The existing structures on the national territory were inadequate from a normative point of view and did not respond to the lifestyle changes that the students required. The first fundings were intended to correct these situations. A further incentive occurred with the inclusion of a specific normative that identified its fundings and blocked spaces and functions, introducing the concept of the satisfaction of student's demands.

2 UNIVERSITY STUDENTS: A NEW COMMUNITY IN THE COMMUNITY

The student residences are counted among the new forms of living because they express a new way of life at a young age outside the home. The university residence, then, is no longer understood as a simple lodging place, but as the place to live the transition period that leads from the family of origin to the adulthood. The residence assumes, therefore, the features of a house in which to reside for a period not expressly defined and definable. In this perspective, the realisation of this type of building has to lose the characteristics that have defined it in the last periods, or rather as an emergency solution: find a home for students. This condition is often materialised, especially in Italy, in solutions not really conform to the needs of the student (for the functional organisation of the buildings and the contextualization of the building from universities or services) and a poor quality of buildings. Let us dwell particularly on the practical question. The utility of such structures is mutable in terms of cultural and geographical and temporal; this leads to a social mix that is difficult to obtain in other living situations. This fact was often overlooked as a winning combination using the uniformity of the "room" solution with the repetition of the same. The result was undoubtedly a monotony and a lack of enhancement of social and cultural differences, which is also linked to a low propensity to search for technical and technological solutions that supported the choices.

The university students' population is a society that by nature changes dynamically and suddenly. Social interactions, methods and learning opportunities have profoundly changed compared to ten years ago, and the change cannot simply consist in the attribution of a private bathroom or just inserting a broadband connection. If there arises the question of what students want and what are their expectations in order to create a community, the actions and the sites on which to operate would be substantially schematized in three categories: the space of individualism "room"; the space of the play sharing; the learning spaces. Hence the attempt to identify two ways in the creation of this community that has to be thought as such: a small city within a city.

3 NEW WAYS OF "STUDENT LIVING"

Considering Petruccioli's statement (...): "The birth of a type is conditioned by the fact that a series of buildings share an obvious functional and formal analogy among themselves. In the process of comparing or selectively superimposing individual forms for the determination of the type, the identifying characteristics of specific buildings are eliminated and only the communal elements remain which then appear in the whole series. Type is depicted as a scheme deduced through a process of distillation from a group of

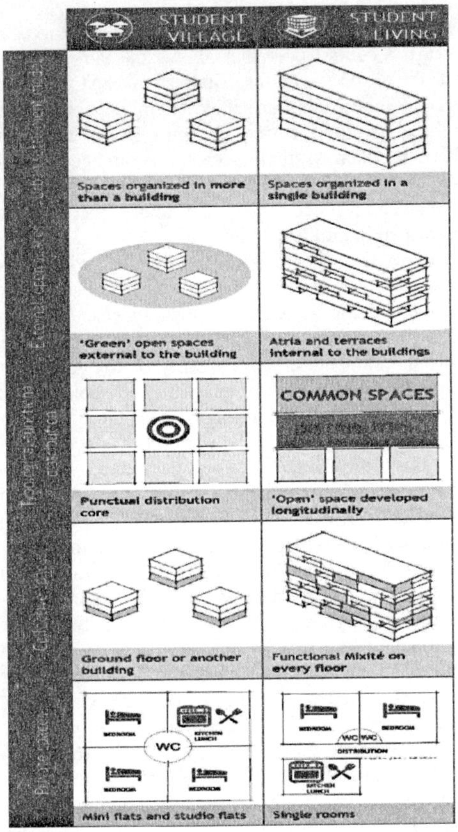

Figure 1. Matching scheme between the models. Image elaborated by the author.

formal variants to a basic form or communal scheme", regarding the buildings housing university residences, the typical morphological classification is not immediate. The difficulty arises from the duty to combine the shape of the building to the typological distribution, with the way of life that comes with it.

Hence, the necessity to identify two macro categories in which to group the structures dedicated to "student housing". These two groups have been analysed and associated through the identification of five characteristic elements that determine substantial design differences. The philosophy of life resulting from the two models is substantially different, as will emerge in the following review. The two models have been denominated Student Village and Student Living. The analysis has been conducted, as mentioned, through the identification of five characterising elements, which are based on both models: urban settlement model, external/open spaces, typological–functional distribution, collective spaces, and private spaces.

Regarding the "urban settlement model", in the Student Village, the university residence spaces are organised in an ensemble of buildings, mainly dedicated to specific functions (residential and semi-private), and/or in buildings which house collective spaces; all of these buildings are located on the university campus or nearby. It is restored the conception of the American campus model. The idea of Student Village allows, therefore, the development of a small "university village". On the contrary, in the Student Living, the residential functions and services are organised within a single independent structure.

For what regards the external/open spaces, the residences identified in the Student Village model are immersed in wide green spaces. This characteristic allows the students to have continuous contact with nature, the existing vegetation; the open spaces are located completely outside the buildings. Consequently, the buildings' internal surfaces remain completely dedicated to prevalently residential function, guaranteeing vast surfaces for the outer open spaces. As highlighted in the study by Hipp et al. (2016) the availability of green spaces within the university campus improves the quality of life, increasing the perception of green.

Mensah et al. (2016) underline that the green spaces provide a place for leisure time and recreational activities, the latter associated with a higher quality of life. Moreover, the green spaces positively contribute to the physical and mental health of the individuals, as well as favouring the interaction, integration and social cohesion. Conversely, within the Student Living model, not able to guarantee vast open spaces outside the building, these are obtained with terraces and courtyards, trying to satisfy the demands of the resident students in the best way possible.

From the distributive-typological-functional point of view, in the Student Village, the distribution functions rely on punctual spaces that determine the total removal of long corridors. Numerous social-psychological studies support this choice. The observations made by David Halpern in his book "Building of Happiness" expose the negative influence in the classic "double-loaded corridor" on the psychological and emotional state of the users. In the Student Village, the distribution of the spaces is therefore entrusted to a central hub that connects the various floors of the structure, leaving to other areas the task of favouring the socialisation and integration between the users. This way, it is possible to have some spaces in which dwellings alternate with common spaces on every floor, generating a "more familiar" environment. The Student Living model seeks the right mixture between distributive, private and collective spaces. The distribution of the spaces is ascribed to an "open" space characterised by atria, glazings and common terraces, which are particularly suited places for the socialisation and integration of its users. This represents an evolution of the traditional model that involves the distribution of the spaces along corridors, which, as already highlighted, may have negative repercussions on the social-emotional condition of the residents.

The functional organisation within the residence, or rather the relationship between common spaces and the residential area, is of paramount importance. The necessity of a student to develop simultaneously a private and social life within the residence determines a classification of the spaces; it is necessary to guarantee the privacy of the private spaces and the socialisation in the public spaces. The private spaces refer to activities performed by the single individual in the room (primary functions and studying) or in its immediate vicinity; the public spaces refer to collective activities carried out by a restricted group of friends (secondary functions, e.g. collective study).

The Student Village model creates flats (which house about six students) equipped with common spaces, such as dining and living room, and private spaces, such as (single or double) bedroom with en-suite. –The common spaces for the whole residence are instead located on the ground floor of the single buildings or within another building. This way there is a clear separation between the private space (bedroom), semi-private space (living room and kitchen) and the prevalently public areas of the residence. In the Student Living model, functions connected to living are integrated to the collective ones through the functional repartition of the internal spaces in private and semi-private areas. The collective spaces (common living rooms and areas for preparation and consumption of meals, to be shared with more numerous groups of users than those involved in the "Student Village") are located on every floor. This way, the collective space is developed in the entire structure, acting as a place of exchange, place to study and to share interests. For what regards the private spaces, the "student village" model is characterised by completely independent housing units which form actual mini-flats constituted by single rooms which refer to a common space (kitchen/living room), or studio flats that include a kitchen. In the "student living" the functional arrangement of the housings in the building clearly defines the possibility of having exclusively single rooms (with or without kitchen) that, in groups, refer to a single space for the preparation and consumption of meals and a common living room, shared by a higher number of students.

4 CASE STUDIES

In support of the study of the models described in the previous paragraphs, four European student residences have been analysed, two for each model. The study has been conducted analysing each residence in function of the characterising elements of the abovementioned models, highlighting the positive and/or negative aspects.

4.1 *Student village*

The student residences studied in reference to the Student Village model are the Student Village in Hatfield (England, 2012) and the Student Housing DUWO in Delft (Netherlands, 2009). Hatfield Student Village presents an urban settlement model characterised by eight square pavilions whose height does not exceed four floors. The buildings are paired, joined together through a central core that allows the vertical connections of the structures. Similarly, DUWO Student Housing shows an urban settlement model constituted by three six-storey cubes, realised next to each other with a modular system. Both the residences, therefore, create an actual "university village" in which the residential functions and the support facilities are aggregated in building hubs.

The two residences studied are equipped with external spaces that constitute the places for leisure time and recreational activities in open spaces. As highlighted in the previous paragraph, these external spaces allow the students a better quality of life favouring the contact with nature, interaction and socialisation (Hipp et al. 2016; Mensah et al. 2016). Furthermore, internal spaces remain entirely for residential functions. Regarding the typological-functional distribution, Hatfield Student Village's consists in the organisation of each floor in flats (five on average) with a vertical connection located outside the buildings. On the contrary, in DUWO Student Housing a central internal body that distributes the various existing housing typologies to each floor constitutes the typological functional distribution. This way, the total removal of the corridor, and of the negative aspects associated with it, create a more familiar environment, closer to the needs of a fresher student.

The collective spaces in support of the residence, in the case of Hatfield Student Village, are located far from the pavilion cluster; this way, the entire community of the university campus can also use them. In the case of DUWO Student Housing, instead, the collective spaces are located on the ground floor of each block, where the area for the preparation and consumption of meals is located. This single attribution of the ground floor to the facilities limits the relationship between public and private spaces because it determines a clear separation between the area destined to the sharing and the purely individual one. This, however, allows the students a higher level of privacy, limiting the possibility to socialise and/or relate to students outside their own housing hub.

The private spaces in the analysed residences are arranged in different ways. Hatfield Student Village entails flats equipped with a bedroom with en-suite and common kitchen/living-dining room; in DUWO Student Housing, instead, the articulation of the private spaces allows the user to choose between studio apartments and housings constituted by two or three rooms, all equipped with a private kitchen. It is evident how Hatfield Student Village aims to create proper flats that simulate the concept of familiar life, while in DUWO Student Housing there are numerous studio apartments, which limit the creation of interpersonal relationships but at the same time, guarantee the privacy of the users who require it.

4.2 *Student living*

The residences studied in reference to the Student Living model, both located in Copenhagen (Denmark), are

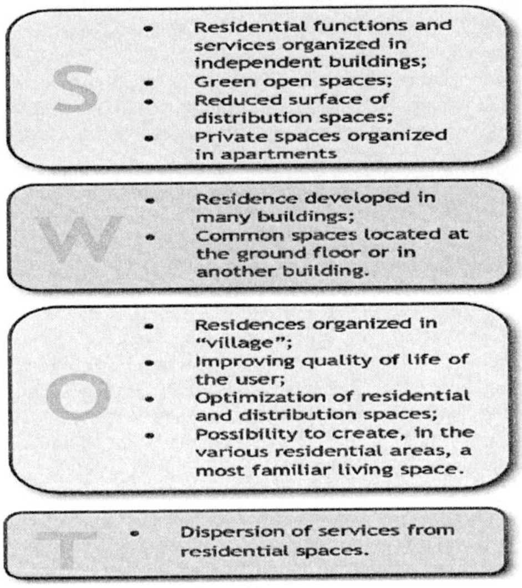

Figure 2. Comparative analysis of Student Village. Image elaborated by the author.

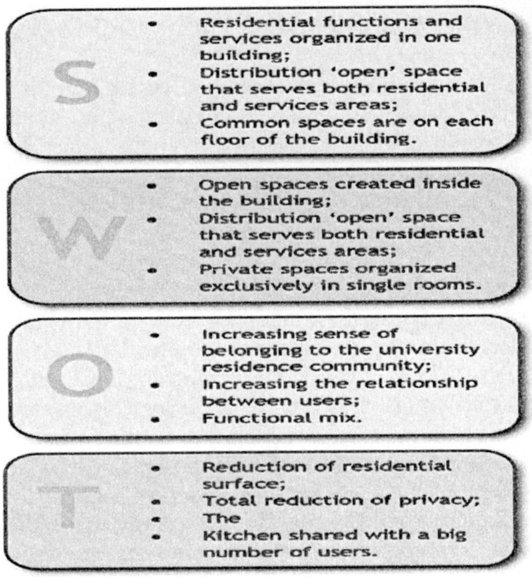

Figure 3. Comparative analysis of Student Living. Image elaborated by the author.

the Tietgen Residence Hall (2005) and the Bikuben Student Residence (2006). Tietgen Residence Hall appears as a single cylindrical seven-storey volume, while a single cubic six-storey volume constitutes Bikuben Student Residence. The university residences, therefore, in agreement with the urban model of Student Living, are entirely developed within a single structure, whose housing functions appear to be integrated with the facilities connected to them. Both the residences are included in an already consolidated urban context. Consequently, the "open spaces" for socialisation are obtained within the buildings with the creation of terraces and central courts, core of the structures. Although this way the surface destined to housing is reduced, these spaces constitute the only places for outdoor activities.

The typological functional distribution of the spaces, in both case studies, relies on a route that re-proposes the shaped plan of the building, dividing the private spaces, which overlook the city (outwards), from the public and semi-public spaces, which overlook the internal court. The resulting distribution is almost completely "open" thanks to the employment of glass systems and opaque closures (walls or doors). This way the distributional organisation loses its "classic" corridor setting and organises the space in a stimulating environment that gives the dwellers the possibility to relate themselves. As the same designers of Bikuben Student Residence (AART Architects) state, in the idea of the design, they tried to avoid loneliness and the lack of interpersonal relationships. This issue is highlighted by many students in their own current housing situation and very often is seen in the classic design solutions. The collective spaces, as prescribed by the Student Living model, exist on every floor. In Tietgen Residence Hall's case, jutting volumes, which mainly house kitchens and living rooms, constitute these spaces. These kitchens and living rooms, shared with small housing hubs (each should serve a group of twelve rooms), allow the creation of those spaces of sharing, external to the rooms but semi-private, in the view of sharing within a restricted group of residents. In Bikuben Student Residence's case, each floor is substantially rotated and, consequently, the spaces are never piled on top of one another on the various floors; there is no distributive functional monotony. In both case studies occurs an undoubted advantage. The floors are arranged as if they were a vast dwelling: the room (privacy) overlooks a space which is already a social space and which, owing to its structure, is organised in various ways to stimulate social relations. Regarding the private spaces, the two residences arrange the rooms in different ways. In Tietgen Residence Hall there are single rooms with en-suite grouped in small hubs, formally divided from the vertical distribution hub. In Bikuben Student Residence, there are single rooms equipped with toilet and kitchen. The difference in the spatial organisation of the two private environments (the room) underlies two different conceptions and possibilities of life. In Tietgen Residence Hall the absence of a kitchen in the rooms forces the residences to share at least the space and time for the preparation and consumption of the meals, and therefore "forces" socialisation; in Bikuben Student Residence the presence of a cooking corner offers a greater feeling of privacy because it leaves to the student the possibility to choose whether to develop social relationships, but contextually might complicate socialisation, in this case attributing it to other spaces.

5 CONCLUSIONS

The ideas outlined in this paper allow pointing out how it is possible to group and identify trends in the construction of these buildings, strongly influenced by the culture and philosophy to approach to specific cultural areas. In contrast, it could be underlined that some countries have focused their attention on the definition of minimum size standards (including Italy), while others on architectural forms that may assist in mediating the relationship between individual and collective space. Some studies tend to emphasise that, like an unstoppable process of globalisation of training models, residences are increasingly entering the individual and community is training process. Therefore, University residences, designed as places of cultural education, act directly on the individual's development and the quality of the entire university system. The perspective that opens is an interdisciplinary research in which the reflection on the type, morphology and technology of these buildings must necessarily be put in relation with sociological issues. However, the experimental aspects should not be limited only to the definition of functional and distribution models, but should also concern technological innovations. These buildings can become places of research applied, in which to experiment with innovative construction solutions from a technical and technological point of view.

BIBLIOGRAPHICAL REFERENCES

Bellini, O. E. 2016. Studenti quegli animali! La questione degli "utenti" nella progettazione delle residenze per studenti. In: Del Nord, R., Baratta, A., Piferi, C. Residenze e servizi per studenti universitari. Firenze: Centro interuniversitario TESIS, pp. 285–296.

Brandon, A., Hirt, J.B. and Cameron, T. 2008. Where You Live Influences Who You Know: Differences in Student Interaction Based on Residence Hall Design. The Journal of College and University Student Housing. [Online]. 35(2), pp. 62–79. [Accessed 02 February 2017]. Available from: http://eds.b.ebscohost.com/

Chiarantoni, C. 2008. La residenza temporanea per studenti: Atlante italiano. Firenze: Alinea Editrice.

—. 2012. Residenze universitarie. Bologna: Marconsu Editore, pp. 10–19; 136–147; 166–173.

—. 2016. Le residenze universitarie. Contributi alla ricerca sull'abitare. In: Del Nord, R., Baratta, A., Piferi, C. Residenze e servizi per studenti universitari. Firenze: Centro interuniversitario TESIS, pp. 327–338.

Hipp, J.A., Gulwadi, G.B., Alves, S., Sequeira, S. 2016. The Relationship Between Perceived Greenness and Perceived Restorativeness of University Campuses and Student-Reported Quality of Life. Environment and Behavior. [Online]. 48(10), pp. 1292–1308. [Accessed 15 February 2017]. Available from: http://journals.sagepub.com/doi/abs/10.1177/0013916515598200

Jaglarz, A. 2016. New Concepts in Arrangement of Living Spaces for Students. Advances in Human Factors and Sustainable Infrastructure. [Online]. 493, pp. 63–73. [Accessed 27 January 2017]. Available from: https://link.springer.com/ chapter/10.1007/978-3-319-41941-1_6

Kjlicaslan, H. 2013. Design of Living Spaces in Dormitories. Procedia- Social and Behavioral sciences. [Online]. 92, pp. 445–451. [Accessed 27 January 2017]. Available from: http://www.sciencedirect.com/ science/article/pii/S1877042813028309

Longworth, N. 2006. Learning Cities, Learning Regions, Learning Communities: Lifelong learning and local government. Londra: Routledge.

Mensah, C.A., Andres, L., Perera, U., Roji, A. 2016. Enhancing quality of life through the lens of green spaces: A systematic review approach. International Journal ofWell-being. [Online]. 6(1), pp. 142–163. [Accessed 01 February 2017]. Available from: http://dx.doi.org/10.5502/ijw.v6i1.445

Petruccioli, A. 1998. Exoteric – Polytheistic – Fundamentalist Typology: Gleanings in the Form of an Introduction. In: Petruccioli A. Typological Process and Design Theory. Cambridge, Massachusetts: Aga Khan Program for Islamic Architecture.

Rodger, S.C. and Johnson, A.M. 2005. The Impact of Residence Design on Freshman Outcomes: Dormitories Versus Suite-Style Residences. The Canadian Journal of Higher Education. La revue Canadienne d'enseignement supérieur. [Online]. 35(3), pp. 83–99. [Accessed 25 January 2017]. Available from: http://journals.sfu.ca/

Sanni-Anibire, M.O. and Hassanain, M.A. 2016. Quality assessment of student housing facilities through post-occupancy evaluation. Architectural, Engineering and Design Management. [Online]. 12(5), pp. 367–380. [Accessed 30 January 2017]. Available from: http://dx.doi.org/10.1080/17452007.2016.1176553

Research and culturalist practice as a matrix for urban and architectural rehabilitation in Lisbon

António Santos Leite & Ana Marta Feliciano
University of Lisbon, Faculty of Architecture, CIAUD, Lisbon, Portugal

ABSTRACT: The work we are presenting reveals the importance, both in the Portuguese case as generally in the European context, of interventions developed in recent years in architectural matrices of the historical city. It also sustains the present research and concrete intervention on buildings and urban groups of recognised patrimonial value, entities that carry an effective cultural meaning for the valorisation of the collective memory. At this point, it seems, therefore, pertinent to reflect on the reason for the meaning and value of these interventions on building, a process that during the time of the research project and the variability of its situations, allows a comprehensive reading of significant urban fragments in the city of Lisbon.

One of the basic premises was the awareness of the multi-secular constitution of our territory, confirmed by successive areas of urban growth that characterised it as a puzzle of different temporal environments, but also of the importance of successive layers, which over the centuries have built a palimpsest of experiences characterised by successive centuries of construction. The realisation of this reality, coupled with a new awareness of our contemporaneity, has been providing the architectural discipline with the possibility of a critical moment of theoretical and practical reflection on the issues and priorities that are fundamental to our territory.

Keywords: Rehabilitation, heritage, memory, project, Lisbon

Our recent research and practice on the theme of project intervention in culturally qualified architectural matrices, which has been centred on Lisbon, has allowed, through the monitoring of different works, the consolidation of a theoretical support and a set of practical projects, sustained either by a successive revision of fundamental theoretical works associated with this subject – reference bibliography, heritage letters and instruments of legislation – or the practical project strategies developed for problems.

Figure 1. Aerial view of the historic centre of Lisbon – 'Castelo, Baixa Pombalina, and Tejo River'.

The work we are presenting reveals the importance, both in the Portuguese case and generally in the European context, of the interventions developed in recent years in architectural matrices of historical cities, and it can sustain both the present research and concrete intervention on buildings and urban groups of recognized patrimonial value, entities that carry an effective cultural meaning for the valuation of the collective memory.

At this point, it seems, therefore, pertinent to reflect on the reason for the meaning and value of these interventions on buildings, a process that during the research project time span and the variability of it situations, allows a comprehensive reading of significant urban fragments in the city of Lisbon, or by analogy, in similar situations, and can contribute to a more sustained reflection on the theory of architecture at the present time.

First, a basic premise was the awareness of the multi-secular constitution of our territory, confirmed by successive areas of urban growth that characterised it as a puzzle of different temporal environments, but also of the importance of successive layers, which over the centuries have built a palimpsest of experiences characterised by successive centuries of construction. The realisation of this reality, coupled with a new awareness of our contemporaneity, has been providing the architectural discipline with the possibility of

Figure 2. Referential elements of Lisbon ('rambling sketches') publishing in *"The Illustrated London News"*, 5 Jan 1984.

a critical moment of theoretical and practical reflection on the issues and priorities that are fundamental to our territory. This search for a more conscious intervention in our territory probably could lead to a more qualified urban and architectural practice about our relevant cities and buildings.

In addition to the obvious incomings arising from economy and sustainability, using a whole set of infrastructures already built, the actual possibility of intervention on the existent buildings, be it understood in a more or less urban scale, seems now to be increasingly associated with the rediscovery of the cultural value and memory present in our consolidated urban matrices.

This research is particularly fruitful in the general context of the numerous contributions presented in recent years on the value of specific cultural fields. However, more objectively on the intentional appropriation of the ubiquitous cultural values in architectures of the consolidated city, realities that tacitly reinforce, or can reinforce a sense of belonging and the consequent creation of an effective cultural identity.

Due both to the difficulties it presents, and because of the challenges it poses, the present moment seems to challenge us to produce a critical reflection on contemporaneity and an understanding of the determining factors for the characterization of our society, its needs, values, and aspirations.

The comprehension of the previous reality, necessarily multiple, complex and even contradictory, and the development of a new awareness of our role as active elements in this process of rereading, renewing and reframing was, therefore, responsible for the investigation that we now present in all projects submitted. That is, how can we effectively intervene in these pre-existing matrices, assuming their cultural value, but also introduce the questions inherent to our contemporaneity?

Reinforcing this idea, according to Francisco de Gracia:

> . . . almost everything that this work contains reminds us, in the last instance, of this city which, since the end of the eighteenth century, profoundly altered by the industrial revolution, superimposes the effects of a Historical double mission. On the one hand, it will remain the supreme manifestation of Western culture; on the other hand, it will assume its condition of the centre for a new system of production and distribution of material goods. (1992, p. 17)

Taking into account the different situations considered during the theoretical and project reflections addressed throughout the research project, it tended to consolidate the value inherent in our cities, which, in the case of Lisbon, surcharged by centuries of cultural accumulation, can be considered a paradigmatic example of Western urban culture. It is a situation, which necessarily implies fundamental identity values for future generations.

On the other hand, the investigations carried out have allowed us to observe that, in keeping with the importance of these urban centres in the contemporary world, one needs to place in the process of evolution the question of how they can assume their status as identity centres, reflecting the current system of production and distribution of material goods and services, thus requiring effective awareness of how the preservation and modification missions may be made compatible.

Specifically, the question inherent in the previous dual mission of the western city, necessarily a process with doubts and uncertainties, has been approached successively over the last decades, using the advantage of the particular path created by the theory and practice that the architectural discipline developed after the World War II.

In fact, the many attempts in recent years to organize architectural theory according to fragmentary currents are, in fact, the outcome of a broad process of awareness and criticism unleashed in the context of a feeling of loss emanating from modern, abstract and massified post-War reconstructions, as well as the profound socio-cultural transformations that mark our contemporaneity.

Therefore, due to the exceptional nature of the situation, the period after World War II has seen the concentration of a set of principles of Modern Culture. It also witnesses the beginning of a new critical awareness that was to be found in the background of a cultural change we would call post-modernity, and which characterises the contemporary matrix, despite

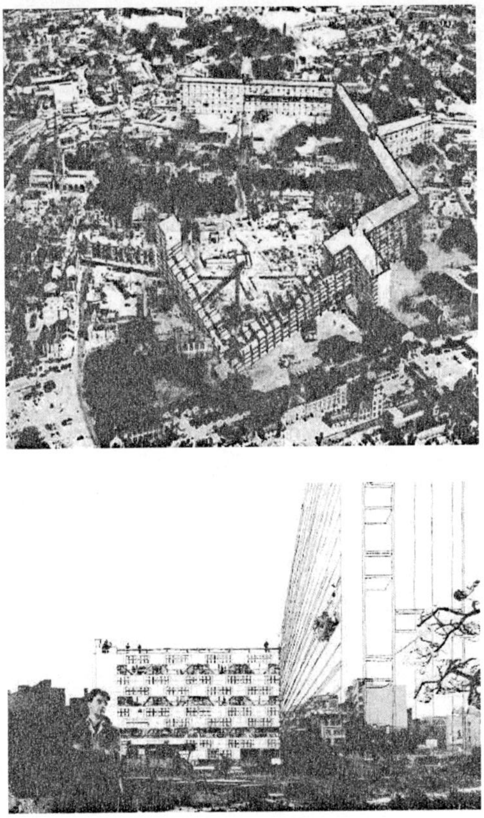

Figure 3. Alison and Peter Smithson, perspective drawings for Golden Lane Housing, 1952.

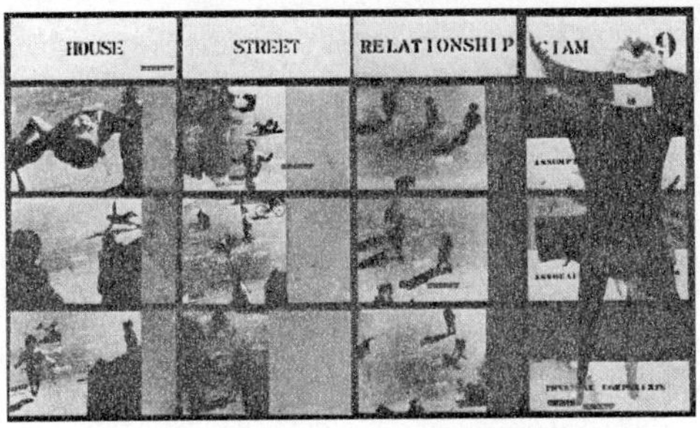

Figure 4. Reproduction of the 'Urban Re-identification grid' presented at the ninth CIAM in Aix-en-Provence, 1953.

the digital revolution and the amplitude of globalisation. In fact, due to the urgent need for the reconstruction of core European cities and the breadth of intervention, the doctrinal principles of a Progressive Urban Model, deeply supported by the principles of the Modern Movement, did unexpectedly find a field of Implementation in different scales and intervention programs.

In the twentieth century, since the end of the 40s and during the following decade, there was a systematic willingness to put into practice a new way and scale of construction. This largely reflected the principles that came to be systematised throughout several CIAM meetings, after World War I, by some of the most influential architects of the new architecture of the Modern Movement.

A whole set of principles found, at the time, the opportunity to be put into practice in a wide range of interventions. These were associated with an effervescent cultural scene predominantly associated with contemporary advances of science and medicine.

Noteworthy is the new role reserved to body culture and heliotherapy the new urbanism of the Modern Movement sought to put into practice through a de-densification of the 'old' urban structure, traditionally compact, in order to isolate under the sun, and in new and ample green spaces, the buildings that were no longer connected with each other, becoming autonomous units. Similarly, driven by a 'positivist view', the Architecture of the Modern Movement also based itself on an understanding of Humanity as an ideal, pure, perfect, generically abstract individual, abolishing cultural and identity differences. Therefore, successive doctrinal documents proposed a generalisation of human complexity to the restriction to four primordial functions, and the different spaces of everyday life sought in this way to respond to the same formatted range of needs and aspirations, forgetting in the process human complexity. The complex human characterization, its cultural background, its collective memory, and its particular identity.

Thus, as urban and architectural facts were materialised in the post-war stage, expanding the pre-existing city and intervening 'without restrictions' in its historical centre, a whole set of problems of segregation and loss of identity began to emerge in its inhabitants. So begins an important critical awareness, which, following the broad movement of the restructuring of thought and art was to consolidate a new, more eclectic and permissive postmodern culture, requiring a different reflection on emerging urban and architectural phenomena.

In this way, fundamental for the construction of the present moment and for the recovery of a broader set of values that had been subjugated by the Modernist doctrine and by an implicit progressive idealisation of the future, in the Fifties it became discernible a whole set of transformations in society. Actually, during this period, a new awareness of contemporaneity and its realities began, and a new approach to architecture was increasingly affirmed by a new idea of a differentiated individual in a deeply hedonistic society. This led to the progressive desire to approach architecture and the cultural context, the common individual, taking note of his tastes and preferences. Alongside went the emergence and consolidation of phenomena such as the increase worth of cultural diversity, context, environmental pre-existence, tradition, communicative language, as well as the awareness of a tacit idea of heritage that emerged as fundamental to our contemporaneity.

From this moment on, there began a process that, supported by a new Post-Modern Culture, with broad reflexes in varied fields of thought, from Philosophy to the Arts, would start to propose a new, more eclectic, permissive and inclusive attitude, in architectural theory. In this way, recovering the legacy of the early Culturalist Model[1], the question of the relationship between Architecture and the concrete individual arose, defending the need to recover the urban and architectural qualities of contrast, variety, and cultural diversity, which had been devalued in innumerable and dogmatic post-war reconstructions. In fact, one can say that:

> . . . in architecture, the influence of the *avant-gardes* plunged many into a position of discomfort and placed others in a dogmatic attitude. Nothing consolidated by history was worth be it the *integritas*, the *euritmia*, the *ordinatio*, the *conjuctio*, or the *proportio*, much less the *similitudo*. However, the idea of modernity had existed clearly since at least the Renaissance, and since then it had been a cultural constant throughout the course of history without ever denying its own continuity. Moreover, the past was used as a revealing instrument for the future, especially during the second half of the eighteenth century. The aesthetic vanguards of the first half of the twentieth century, clad in anti-historicism, risked perhaps too much in their prophetic and regenerating messianism. (Montaner, 1997, p. 30)

Thus, in the second half of the twentieth century, after the dogmatism of a first period, a progressive return and evaluation of urban and architectural matrices present in the historical city was witnessed, seeking solutions, reinterpretations and, more specifically, looking for design in a more organic way that would allow cities to preserve their differences and identities. Thus, emerges a completely new intellectual posture due to new theories about the city, housing, and the different ways of living it, the theoretical reflections and the projects developed from the 60s by the members of the Team X. In fact, this group began a set of reflections to understanding of contemporary citizens, society, and the possible intervention of the architect in them. Now, far from the revolutionary will to change and cut with the past, architects suggested to work according to a new sensibility "equating a utopia of the possible, accepting the tastes and needs of the populations" (Gracia, 1992, p. 30). Consolidating the desire for change and a criticism of the basic assumptions of the CIAM, Team X sought to introduce in the architectural debate concepts that would allow the discipline to reflect more accurately the emerging diversity of the new cultural post-war society. They brought up concepts like 'identity', 'model of association', 'neighbourhood,' and 'living organism', thus globally recognising the new conception of complexity in urban models.

Figure 5. Aldo Rossi, Drawings that show the layers of 'temporal strata' of the city.

From this context, and already distanced from it in time, we can affirm that the post-war period has, indeed, assumed an important catalytic moment – a moment of change – of a broad debate, advocating the triggering of a new critical awareness on the interpretation and intervention on urban and architectural matrices of the consolidated city.

In spite of the continuous and inevitable critical revisions of the matrices and languages of the Modern Movement, it now seems to make sense to accept the presence of a new identity – time. We necessarily recognise the heritage of the Modern Movement but also the heritage and the consciousness that values of the historic city tend to emerge as inevitable worth and ordinary.

As Francisco de Gracia writes:

> . . . these fascinating contaminations translated into architectural facts allowed us to corroborate the following: continuing history means doing the new with absolute contemporaneity, but accepting that the historical material of our memory will intersect in the process. It is not possible to undo the journey or to isolate architecture from its own biography, much less dispense with the context as a place, in the Heideggerian sense of the term, as a situation unveiled through its historical construction. (1992, pp. 107–108)

[1] According to Françoise Choay, the starting point for a definition of the 'Culturalist Model' "…is no longer the situation of the individual, but the situation of the human group, the city. Within this, the individual is not an interchangeable unit as in the progressive model By its particularities and by its own originality, each member of the community is constituted as an irreplaceable element. The historical scandal from which the partisans of the culturalist model react is the disappearance of the old 'organic' unity of the city under the disintegrating pressure of industrialization. (…) Thus, to realize the beautiful cultural totality, conceived as an organism in which each has its original role, the city of the culturalist model must also present a number of spatial determinants and material characters. (…) Morris proposes the irregularity and "asymmetry' which are the mark of an organic order, or inspired by the creative power of life, in which the highest expression is given by human intelligence." (1965, p. 21)

In this context, we can assume that we are today in a moment that simultaneously understands the utopian and culturally decontextualized inheritance of a heroic Modern Movement and indelibly affirms an understanding of Post-Modernity, thus requiring a complementary process of operating with both matrices present in our contemporaneity; that is:

> Nowadays, after the debates on the notion of Post-Modernity, the possibility is presented to redefine what is or can be modern in architecture. It seems to be equivalent to an expression that is in tune with time and place so that local cultures can effectively mediate the search for a cultural synchrony with the history of architecture. (Gracia, 1992, p. 20)

In fact, the previous finding, tended to prove that the significant arrival at the end of the century and the period of transition to the twenty-first century, in many significant projects on consolidated historical zones tended to manifest its own cultural time, with a new reading and appreciation of the notion of place, basic condition of any cultural identity. The actual moment seems to validate a clear withdrawal from the formal axioms of the Modern Movement, and a quest for the reinvention of architecture, based not only on the perception of its time but on the evident meanings of its cultural place, reinterpreting them according to the functional and identity needs of contemporaneity.

Thus, the question initially posed – how can we intervene today on the constructed city, on pre-existing matrices of recognised cultural value?

It seems, therefore, urgent to offer an answer not only conscious of our own time, with its scientific and technical progress, economic and sociological challenges, but also to offer an answer marked by a broad awareness of the cultural values inherent to the place and the specificity of its culture and architecture, seeking, in this process, to reconcile coherent solutions linked with its context.

Based on a new 'culturalist time', that emerges with the rise of Postmodernity, one can try to answer consistently to the question initially posed. This we have sought to materialise in the projects that issued from the works of this research. Namely, how can we intervene today on the constructed, on the pre-existing matrices, assuming their cultural values, but also solving the questions inherent in our contemporaneity?

In spite of the apparent simplicity and clarity of the question, we have always confronted ourselves with the tacit difficulty of objectively realising it. However, if we do not close it in a dogmatic context – a condition that would not be proper to a time of postmodernity – the situation begins to sketch a clarifying scope of the question, and the answer will tend to emerge through a broad understanding of the cultural and identity issues that have been structuring the main theories and practices of architectural rehabilitation. This reality seems to require, consciously, an open and sensitive attitude to forms of dialogue between the recognised need to preserve the past and the urgency of values and uses required by our contemporaneity.

Figure 6. Alvaro Siza, aerial view of the connection of courtyard B to 'Largo do Carmo', Lisbon; Joana Diniz, perspective view of the rehabilitation of the 'Convento do Desterro', Lisbon (Santos Leite and Feliciano, 2016).

BIBLIOGRAPHICAL REFERENCES

Choay, F. (1965) *L'Urbanisme: Utopies et Réalités, Une analogie.* Paris: Éditions du Seuil.

Gracia, F.d. (1992) *Construir en lo construido; La arquitectura como modificación.* Madrid: Ed. Nerea, S.A.

Montaner, J.M. (1997) *Después del Movimiento Moderno – Arquitectura de la Segunda Mitad del Siglo XX* Barcelona: Editorial Gustavo Gili, S.A.

Santos Leite, A. and Feliciano, A. M. (eds.) (2016) *Memória arquitectura e projecto: reflexão e propostas para uma reabilitação sustentada do património urbano e arquitectónico.* Lisboa: By the Book.

Architecture – a product for retail sale?

Marta Germano Marques
Mestrado Integrado em Arquitectura, Faculdade de Arquitectura, Universidade de Lisboa, Lisboa, Portugal

ABSTRACT: There's an urge to keep up with the global rhythm, to accelerate at all cost and to transform, manipulate, characterise, "carnivalize" our countries, our cities, our lives even, in order to respond to the consuming reality we currently live in.

Society is manipulated by unknown entities that the common citizen can no longer recognise and therefore criticise. And as an answer to the hallucinated management, people are only able to give delirious results back: being forced to react to something they don't have enough data about at the same time that common knowledge is more and more centralized in virtual and global investments and relations; as all relations become more and more similar to business management.

The following focuses on the general effects that globalisation and the wider market has had on the idea of identity, establish by singular places and their people and the way architecture contributes or is manipulated into having a major role in the transformations that contemporary cities are suffering.

How has all this changed the formal and informal character of these cities?

How can small-scale relations subside in between the gaps of such a global plan?

Keywords: City, consumption, identity, representation, visitors

1 THE CONTINUOUS WONDERS OF GLOBAL CONTROL MANAGEMENT

Do global interactions imply the disappearance of the local scale? Does the architectonic model cloned at a worldwide scale better reinforce an international unit or does it destroy single, national and local identities? Is a certain political, social and economic behaviour always related to a way of thinking about the city? Alternatively, do the material actions applied to the city (even if manipulated in accordance with its surroundings), always contribute to a positive transformation?

After de rationalist model generated in the first half of the 20th century, the idea of the "global village" has been disseminated as a definition of the current situation. This is an interpretation of the result generated by a century of material dramatic alterations, regarding territorial conditions that have redefined our way of life.

Zaida Muxí refers to globalisation as a process characterised by three essential aspects:

- Market and major companies' control of strategic points that generate privileged places for the domain of global activities.
- The belief that decisions generated for the consumption market are also sufficient to deal with predominant social issues and the class struggle.
- Prioritisation of tertiary services production as the main resource according to the urban life which is characterised as a continuous party (pt. um evento continuamente "festivalizado").

According to the previous premises and the idea that all these assumptions are deeply connected (especially in the virtual world), we know that these are bonds with the possibility to establish much stronger interdependence relations than some physical connections.

What happens then in all the territories that are not as much or not at all exposed to these connections, the actual lands and places where no points of control exist, where none of these interactions take place? (Even if some make tremendous efforts to feel part of the game or to have a share in this global system).

Are these places completely extraneous to the market management?

Currently, they seem to depend entirely on the same support possibilities which they don't have the ability to control. On the other hand, these are the same places, many times countries that see their natural and working resources exploited by foreign (maybe unknown) identities, taking them the possibility to take any advantage of those same resources themselves.

> "In economic, social and politically more fragile societies, the negative and perverse effects from the present situation are more exaggerated and therefore, more clear." (free translation)

> ("En sociedades económica, social e políticamente frágiles, los efectos negativos y perversos de la situación actual son más exagerados y, por tanto, más nítidos." – Muxí, Zaida, 2004)

The nations that occupy the final place in the economic chain are also used as experiment stage for different products, seeking to relate to the "first world" in terms of appearance.

Well, if the same general production system that forces the tertiary sector to depend directly upon the secondary sector and the secondary to the first, then it would be expected that the investment of human resources and economic means would be proportional to the reported interdependence. However, with records of the exponential growth of the industry and service offer allied to need less and less for specialised human resources, especially in the primary sector, the equation has been inverted.

At a different scale...: if the economic relations among sectors and activities used to be managed in the limits of a group of articulated actions and the application of different efforts in a city, country or a group of articulated nations (that were usually related by actual territorial proximity or practical needs); when the interaction becomes global, the equation is necessarily modified for there is no need to maintain certain systems of production in places where it will never have the same range of profit than in another. Certain businesses or productions are localised exclusively where they can contribute with the biggest advantage (either it is for natural conditions, human, technological, etc.). The investment is concentrated in different and specific conditions in order to achieve the best rate of return.

That is obviously why the primary sector is largely localised in developing countries, the secondary sector most widely in Asia and some developed countries and the tertiary sector, in its greatest range, in the developed countries. These are statements that simply translate and statistically resume an established circumstance but which is often referred to as a simple statement and not a premise to critical analysis or even hypothesis of reformulation.

The capitalist model as propelled the transformation of the local scale economy to the worldwide scale, without it having an intermediate moment of reflexion. The consequences have been measured in an economically critical and socially acritical plane of action.

A rupture between reality and abstraction is verified, opening a growingly bigger gap between the bigger layer of the population and the possibility to understand whom or what constitutes the control of the global economy. This is an entity with no particular location associated, it is just impossible to recognise: it is a complicated and intricate system that has integrated itself in all actions regarding human life and for which continuity we are inevitably contributing.

Some places or non-places might be called as a director starting point, where some decisions are made regarding highly important options. Options that would be impossible to take in consideration if local effects counted to the matter.

2 THE DISPOSABLE CITY

If the city can be seen as the material mirror or representation of political, social and economic conditions of a civilisation then... Looking at a group of disorganised and disconnected examples, the representation that we have access to be one of an extremely complex virtual set that is forced to adapt to the physical world. Therefore, the physical world that we know becomes a monstrous caricature of what is happening in a different dimension that most of us can't have access to. The "global village" is a scenario or a set of scenarios that remind and compete directly against each other, generating questionable relations with different places and cultures.

So, if there is an hypothetic urge to maintain local identities even with a sense of global consent that is always present (the need to answer to touristic demands or to generate attraction points), there are big chances that this will become such a paradoxical matter that the result doesn't consist in a "differential singularity" ("singularized differential") but a carnival that gives privilege to the mercantile relations and virtual interactions.

> "For millions of people in the United States, the rights to buy and possess have been converted in much more meaningful expressions of freedom than the possibility of going to the ballot boxes to vote." (free translation)
>
> ("Para millones de estadounidenses, los derechos de comprar y poseer se han convertido en expresiones de la libertad individual mucho más significativos que acudir a las urnas a ejercer su derecho al voto." – Muxí, Zaida, 2004)

In the 1970s, the architectonic culture has revealed a critical attitude regarding the functions of the city and proposing the possibility to recuperate the idea of those territories as a local of encounter and gathering.

This recuperation or attempt to re-convert has been almost exclusively directed into the tertiary sector and in areas of commercial interest. Some, which became economically privileged places, are also mostly artificial, composed by buildings that disconnect their drawn reality from any possibility of connection with the environment, working as ephemeral novelties: volumes that isolate themselves from the exterior reality in order to generate an exclusive internal speech.

Buildings or areas where these characteristics contribute for the city to work as a museum ("museificação da cidade"). Architecture becomes one more "disposable object", one more product for regular consumption.

Figure 1. Tourists in the centre of Lisbon, Portugal. http://expresso.sapo.pt/economia/2016-07-03-Muitos-turistas-nao-iriam-a-Lisboa-sem-a-Airbnb.

Figure 2. Tourists in the centre of Lisbon, Portugal. http://expresso.sapo.pt/economia/2016-07-03-Muitos-turistas-nao-iriam-a-Lisboa-sem-a-Airbnb.

3 THE WAY WE REMAIN CITIZENS

In a context where market tendencies and values have a big range of oscillation, it is frequent that an amount of population with fewer resources will never have the chance to keep integrated.

Very often, it is imposed a reorganisation of the city and the ones that have no possibilities to keep up with the rhythm are dislocated to the surroundings, either it is the suburbs or other areas at a bigger distance from the business centre and most services.

The city becomes a place of occasional visit for its citizens and a festive meal for external visitors.

If visiting becomes the only purpose of the city, either it is from its country's nationals or the internationals, then what's left for the original streets and the neighbourhoods and local commerce and small cafes, to subside in between the dominant reality of globalisation and international tourism and local "traditional products" merchandise?

> "... the cultural exterior "happenings" which suffocate, exclude and take space from the people and the expression of local rites: walls are coloured in "joyful" and cheerful tones; the vulgar kitsch explodes, the "fast food's", the atrocious "typical" product shops, actually the beginning of the extreme vulgarization of everything that used to be unique and essential."
>
> ("... os "happenings" culturais exteriores que abafam, excluem e retiram espaço à gente e à expressão de ritos locais: pintam-se as paredes de tons "alegres" e garridos; explode a vulgata do kitsch, dos "fast-food's", das lojas do atroz produto "típico"; enfim, inicia-se a extrema banalização de tudo o que era único e essencial." – Aguiar, José, 2004)

Figure 3. Massamá norte, Sintra, Portugal http://cidadaniaqueluz.blogspot.pt/2009_11_01_archive.html.

In fact, there are two major issues that come hand by hand and that keep feeding each other at a non-stopping pace: while a massive layer of the population is pushed from the centre for the city's commercial benefit; then historical centres and tradition is sold for the mere reason of a financial purpose. When national finances are above all applied into maintaining and polishing what's there to be photographed and easily disposed of; less and fewer efforts are applied to any kind of long term inhabiting relations – for these become place that exists further and further from the foreigner's sight.

Having such a massive group of the population being forced to move, some territories that are usually characterised with fewer possibilities of regeneration are exactly the ones where, by what is imposed by circumstances, social relations actually take place. In addition, it is in those places that have always lacked in history and timeline logics, where the majority of the population keeps reinventing itself.

The suburbs, the outskirts of a major European city have always been characterised by the pendular daily movement that takes many people to their works into the city, where most needs can be satisfied and where the majority of services are reunited. These massive gatherings of housing buildings have subsided, until now, in direct dynamic with a business district that they relate to. However, at the same time, that the centre frees itself from its original dynamic to make room for touristic entrepreneurship and all the accommodation services; many activities that used to support the daily basic needs lose their sense of existence.

With growingly more difficult access to the city, the outskirts have to subside by themselves, finding ways to provide interiorly.

Figure 4. Avenida dos Missionários, Cacém, Portugal. http://mapio.net/s/30859881/.

In addition, as it has always been it is in places that lack in the definition that some possibilities remain available, that some daily private life is still allowed. It is when not everything is under control that people have the chance to imagine different sets and make decisions. The city is always as much what is built as the spaces in between.

The more people have access to the city, the more we have the chance to see it being softened by its use, by its necessary manipulation and constant reconstitution. The more it is transformed, the more it exists. The more it is recognised, the more it is cared for, parts are added and taken, stone, wood and metal exist more through the conversations among them than in the mechanic relation they establish.

That way, what happens to the layer of culture and history that is this way kept under the touristic glass panelled sites, the viewpoints, the packed public gardens and the obstructed sidewalks in the main city?

How can new generations be able to recognise themselves (inventing themselves culturally), if their city smells and feels so similar to all others?

If all that used to be essential and dominant means now one more business opportunity? If weekend-tourists get to ride the tramway around the best places in Lisbon and for you, it becomes unbearable?

How is it possible still to recognise a city that you hear stories about but that is under constant airport dynamic?

4 STRANGERS

The city makes a promise as a primary offer. The city allows us to believe that the possibility of being anonymous is possible, that it is possible to live a private life at the same time as being part of a crowd, a massive crowd that accepts personal defects among all its ridiculous features, a crowd that one can follow for no reason and end in a reasonable place, a crowd to which you can repeatedly ask something having no answer back, the same that might also give all the information you need, the one where you can lose yourself with no constraints.

One is given the promise of anonymity, the promise to be looked as a stranger, repeatedly, and the promise of looking at others as strangers as well.

BIBLIOGRAPHICAL REFERENCES

Aguiar, José – Projecto de Conservação: da importância do método à oportunidade das surpresas, 2004, Article.

Jacobs, Jane – Muerte y Vida de las Grandes Ciudades. 3ª edição, Madrid, Capitán Swing Libros, S. L., 2013. ISBN: 978-84-938985-0-2.

Lynch, Kevin – A boa forma da cidade. Lisboa, Edições 70, Lda, 2015. ISBN 13: 978-972-44-1330-3.

Muxí, Zaida – La arquitectura de la ciudad global, Barcelona. Editorial Gustavo Gili, SA, 2004 ISBN: 84-252-1560-9.

Pacheco, Carlos; Richards, Colin; Zaqtan, Ghassan; Barrento, João; Frow, John; Simpson, Moira; Gilroy, Paul; Sloterdijk, Peter; Rennó, Rosângela; Kovadloff, Santiago; MunshI, Surendra; HuI, Wang. – O estado do Mundo, Lisboa. Edições Tinta-da-china, Lda., 2007 ISBN: 978-972-8955-27-4.

Sloterdijk, Peter – Regras para o Parque Humano, Coimbra: Angelus Novus, Editora, 2007. ISBN: 978-972-8827-39-7.

Steiner, George – A Ideia de Europa. 1ª edição, Lisboa: Gradiva, 2005. ISBN: 989-616-022-8.

Tanizaki, Junichiro – Elogio da Sombra – Edição, Lisboa: Relógio D'Água Editores, Colecção arquitectura, 2008. ISBN: 978-972-708-968-0.

The Portuguese city, 1930–1960: The progress of thinking as displayed in the architecture magazines

José Cabral Dias
CEAU – Centre for Studies in Architecture and Urbanism, Faculty of Architecture, University of Porto, Porto, Portugal

ABSTRACT: This article will be focused on the Portuguese city, between the 1930s and the 1960s. Its approach will rely on the most important Portuguese architecture magazines of that period. It will be also taken into account the fact that professionals and academics were then unaware of any theoretical question. In spite of such lack of theoretical concerns, the spread of knowledge regarding international trends was only possible through the pages of that kind of publications, considering its large impact among architects and urban planners. It will be also taken into consideration that that kind of publications reflected, almost in real time, the relevance given to the published themes and subjects.

This article aims, therefore, to go deep into the debate about the Portuguese culture regarding urban design. The article goal is to find out if themes published by those magazines gave a contribution to the disciplinary debate. In other words, could they have contributed for overcoming the theoretical absence among architects, which was undeniable within the considered period? In short, how did urban design ideas evolved between the 1930s and the 1960s in Portugal, under the ideas spread through the architecture magazines?

Keywords: Portugal, Portuguese culture on urban design, Portuguese city, Portuguese architecture magazines, 20th-century urban concepts

1 INTRODUCTION

Without surprise, one can say that in the first two decades of the last century the thinking and practice in Architecture in Portugal relied on the 19th-century ideas. In fact, the Portuguese architects were eclectic in the beginning of the 20th century (Tostões, 2006:118). In other words, they stood between the nostalgia for national authenticity and the Beaux-Arts eclecticism, or between romanticism and cosmopolitanism (Grande, 2006: 62).

Differently, the 1960s broke with the previous time on how to face architecture issues. Addressing new disciplinary approaches, which were characterised by concerns about contextual features, history, tradition, and spaces for the relationship between people, rather than by abstract principles, the new decade, is of plurality: the pop style, the Nordic neo-empiricism, as well as Italian and Brutalist influences were different possible trends in Portuguese architecture. Significantly, political and social conflicts broadened the limits of debate and the ones of creativity in that period.

On the contrary, between 1930 and 1960, architecture principles showed signs of continuity. That unity was expressed under the stability of modernist ideas, even though the linguistic codes assumed different expressions over the time. Even if such modernist principles had been replaced by the monumentality on nationalist architecture during the 1940s, they resumed on a deeper basis by the work of the second generation of Portuguese modernist architects, at the end of that decade.

In any case, even if it is possible to trace the evolution of Portuguese architecture during almost thirty years, it is instead difficult to understand the way in which the ideas on urban design changed during the same time. Despite the efforts made by different authors so far, there are still not many systematised studies to do so. However, some of them gave us some clues, which show that the issue is of complexity. For Nuno Portas (1973: 724), the lack of theoretical concerns among the architects was the main reason for turning architecture into monumental expressions during the 1940s. According to him, this kind of absence was already responsible for the hesitations on all the different trends on urban design in the 1930s. Besides, he also underlined (2006) that, after 1934, Planos Gerais de Urbanização (General Urbanization Plans) took the urban spaces of Lisbon as a role model as they equally remained faithful to tradition almost all over the country. In sum, it is true that the incipient idea of urbanism, which was real in the first period of the Portuguese dictatorship (1928–1974), relied on the continuity of the former settlements (IDEM: 59). On the contrary, the 1950s came as a time of changes upon modernist principles. However, such disciplinary framework evolved quickly. Actually, new ideas focused on the canonical city space principles arose in the 1960s. Such ideas were based on rediscovering the street, the square and the block.

In spite of the above-mentioned lack of theorization, which was as strong among architects as within the academia and among the critics, this paper will try to find out if any sign of a cultural debate on urban design and urbanism took place in Portugal between 1930 and 1960. An answer to this question can be found upon the impact of architecture magazines on Portuguese architectural culture. Would it have been possible to overcome the lack of theorization with the spread of knowledge on topical international trends through the pages of the most important and popular magazines such as those ones? Taking into account the Portuguese architectural context between 1930 and 1960, the aim of this article is to find out if those magazines had an active role in promoting theoretical progress over the years and, therefore, if they could have filled the theoretical gaps during the mentioned time frame.

Regardless of who had been the producer of knowledge in the articles that will be considered (either the individual whose work is published in a specific article or the author of such article), this approach will take into account an undeniable truth: the exchange of knowledge between the individual and a wide audience (which is always the target of the knowledge spread through the media) can also be a factor of progress within the society, in many different fields. On the other hand, the notion of progress will be considered as a concept of both gradual developments of civilisation and improvement or enhancement of something. In other words, this paper will try to understand whether the mentioned magazines have led to an improvement in the way of thinking the city space because of that kind of exchange between the public sphere and the individual.

2 CONTEXT

As mentioned above, Nuno Portas (1973: 741) established a link between the lack of theoretical support for understanding the inner energy of the city and the unsuccessful attempt for renewing the Portuguese architecture like it was tried by the first generation of Portuguese modernist architects between the late 1920s and the 1940s. The lack of theoretical basis was responsible for the disciplinary superficiality and, therefore, for the later stylistic inflexions. This weakness affected the urbanism issues with the same violence that struck the architecture. As said above, not even the academia was well aware of the issues related to urbanism according to the intellectuals' attitude, who underestimated such kind of themes (Idem: 728). In addition, Pedro Vieira de Almeida (1986) denied that the artists were inherently resistant. For him, the assumption that the artists did not want to get involved with the political power does not fit the facts. Many artists and architects not only reacted positively to the conservative ideas of the new political regime but also tried to win its confidence (Idem: 131). Still, in accordance with Pedro Vieira de Almeida, the political authorities did not create an architectural expression as an official "directory" in the 1940s

> One of the major inner conflicts for the architects of this generation, if not the biggest one, was the one that led them to their own doubt about which of the trends, traditionalist or modernist, would be strong enough to impose itself as an official architectural style, since traditionalism and modernism were not concepts clearly understood by them. (IDEM: 122)

Stressing the inconsistency on criticism even later, Rafael Botelho (1959) addressed the urbanistic theme in *Binário* magazine by the end of the 1950s: "One rarely talks about subjects related to urbanism" (Idem: 1).

One might say that if theoretical concerns had led both architecture and urban design beyond superficiality, those disciplines would not have been merely understood as aesthetic issues. This assumption creates room for the aim of this paper has, like it was said before, its goal is to give a contribution to the understanding of the role played by architectures magazines in the cultural framework and among architects, between 1930 and 1960, knowing, as also mentioned above, that such specialized periodical publications reflect almost in real time the relevance given to the covered subjects.

Accordingly, the followed methodology will rely on the main and most significant magazines of the period: *Arquitectura Portuguesa: revista mensal de arte arquitectural antiga e moderna; A Arquitectura Portuguesa e Cerâmica e Edificação Reunidas, revista mensal técnica, industrial e artística; Arquitectura, revista de arte e construção; Binário, arquitectura, construção e equipamento; e Urbanização, revista do Centro de Estudos de Urbanismo e Habitação Engenheiro Duarte Pacheco*[1].

The articles, which are deemed paradigmatic over the years, will provide a broader perspective for understanding the subject. Thus, every significant article will be considered, and, furthermore, the most significant ones will be commented on these pages.

[1] *A Arquitectura Portuguesa: revista mensal de arte arquitectural antiga e moderna* published by Mário Collares in Lisbon, between 1908 and 1934; *A Arquitectura Portuguesa e Cerâmica e Edificação Reunidas* published by Sociedade Editora Frace in Lisbon, between 1937 e 1951.

Arquitectura, revista de arte e construção, Lisboa, published between 1927 and 1984 (by Francisco Pereira da Costa until 1947). The number of March of 1947 is already published under the ownership of Iniciativas Culturais Arte e Técnica. In 1969, its title was changed to *Arquitectura, planeamento, design, artes plásticas*.

Binário, arquitectura, construção e equipamento published in Lisbon between 1958 e 1976.

Urbanização, revista do Centro de Estudos de Urbanismo e Habitação Engenheiro Duarte Pacheco published in Lisbon from 1966 onwards.

In other words, the next pages focus in finding out how the set of articles published between the above-mentioned years contributed to the progress of thinking and criticism in a period as rich as complex[2].

3 MAIN IDEAS

The Urban design did not get the attention of Portuguese architecture magazines until 1929. Before, the articles on city development relied on irrelevant references. The approach to the problems of car traffic in Paris, in this same year, in the pages of *Arquitectura Portuguesa* magazine (1929), was still irrelevant. This article is merely about solutions to face the problems posed by crossings regarding the increasing number of cars: tunnels and viaducts. The article presented them all as unrealistic due to their cost, and, as said, it did not surpass superficiality.

In addition, the first articles about specifically designed proposals in *Arquitectura* magazine relied on academicism. The article entitled *"Prolongamento da Avenida da Liberdade atravez do Parque Eduardo VII"* ("Extension of Avenida da Liberdade through Parque Eduardo VII") in 1931 (Arquitectura, 1931a), which was designed by the architect Luiz Cristino da Silva, is a strong example of such trend as, above all, it explores the aesthetic effects of the proposal. One can say, therefore, that the city was then solely understood as an exercise in composition.

Six years later, in 1937, the framework of thinking was almost similar. An article defended the picturesque against what was considered futurism in urban design – "unjustifiable futurism" (Colaço, 1937: 7). That kind of arguments reflects the lack of basic training for urbanism in Portugal even in the end of the 1930s, or at least, a strong conservative point of view towards this issue. Even though the author's words attest by themselves the lack of theorization on urbanism by the time the article was written, the superficiality on the arguments strengthened by the little graphical quality of the drawings included in the text[3].

Significantly, the editorial in the issue No. 46, written in January 1939, is violent to either Modern Architecture or the one influenced by modernist principles (Arquitectura PORTUGUESA, 1939c): *"Façam-se Casas Portuguesas em Portugal"*, that is to say, the houses in Portugal should be built in a Portuguese style. Despite such objection, different projects in a modernist style were published in the exact same issue. Even more paradigmatic of such a contradiction, the modernist Filipa de Lencastre High School building (1932–1938; by Jorge Segurado, 1898–1990) was highlighted for its relevance in the magazine issue No. 49 (Idem, 1939d). Such antinomy is exemplary on how the magazine was swinging between modernity, upon modernist projects, and more conservative architecture and ideas. Besides, the editorials tended to be even more conservative, showing how deep the lack of theorization was. In fact, this undeniable antinomy shows how contradictory the editorial line was. Considering the different magazines, it can be said that the matters on urban design and town planning were not sequentially published nor importance until 1949. The change came with the Athens Charter. It was published in eleven *Arquitectura* magazine issues, between February 1948 and September 1949, opening a new era in both doctrinal and methodological terms. The defence of the Athens Charter principles did really intend, also in Portugal, to break with the way of organising the urban space in canonical basis. Supporting such a new way of thinking, the article entitled "Conjunto Residencial para a Av. dos E.U.A" (Housing in USA Avenue) is a role model of the new phase. Through a project for the core of the Lisbon urban fabric, it shows the impact of the Modern City in Portuguese culture, too (Idem, 1953b).

However, *Arquitectura* magazine faced significant changes in the editorial line because of a new direction board[4]. Carlos S. Duarte (1958) triggered an inflexion in *"Introdução a Howard"* (Introduction to Howard). Concepts like the uniqueness, the singularity, the unrepeatable, the not systematised and relying on contextual circumstances filled the magazine pages from then on, expressing a different time in accordance with new concerns. The enthusiasm about Howard (1850–1928) and the Garden City (Mumford, 1958) reflected an entirely different mindset which was opposed to *the four functions of the city* in constructing and reconstructing the cities (Lima, 1959). Significantly, the Portuguese delegation on CIAM 10 – Viana de Lima, Octávio Lixa Filgueiras, and Fernando Távora – proposed a new urban planning method in opposition to the CIAM grid. Accordingly, this team testified the closeness of Portugal to the international and broader concerns with respect to urbanistic matters (Idem).

> It is clear that whatever relates to organising human life escapes through the mesh of the four functions – and it is in fact beyond any analytical thinking" (FILGUEIRAS, 1959: 3)

It should be noticed that the Smithsons had already stated, in 1953, in the context of the CIAM meetings as well, that the Athens Charter was no longer a useful or relevant tool for organizing the city space: *the four*

[2] In the late 1920s as in the 1930s, the Portuguese dictatorship (1928–1974) adopted the Modern Architecture as a sign of a new time in politics. Later, in the 1940s, the architecture expression turned very often into a nationalist expression under the neoclassicist codes giving a physical expression to the conservative regime. After 1948, because of the first Portuguese architects Congress, the professionals both in architecture and in the city would confidently adopt the modernist codes.

[3] The magazine began the debate on technical matters in this issue. This article, which was written by an engineer, was once more dedicated to crossings. (Colaço, 1937: 7–11).

[4] Frederico Santana succeeded Alberto José Pessoa in No.59, in 1957. Later, Rui Mendes Paula took office as the magazine's director in No.68, in July 1968.

functions of the city – dwelling, recreation, work, and transportation – were not proper features to express urban complexity, nor to find the solutions to deal with it. Therefore, it is to be noted that the editorial was in accordance with those ideas, even with a delay. Accordingly, the article *"Considerações Sobre Urbanismo"* (Considerations on urbanism) (Campos, 1960) set the idea that the urbanism and its practice must reflect a precise context and a specific time. Reinforcing the uncertainty within the design process of the city, the article also states that it is useless trying to foresee the evolution of an urban settlement without considering the peri-urban areas. Accordingly, in *"A Incógnita da Cidade Moderna"* (The uncertainty on the Modern City) (Boróbio, 1960), *Binário* magazine did not support Howard's or Le Corbusier's ideas. On the one hand, the landscape dimension in the Garden City was insignificant, and the means of transport were ineffective. On the other hand, Le Corbusier set the nature free, instead of the man (Idem, ibidem). In 1961, the same approach against Modern City prevailed (Idem; ibidem). The urban form based on a grid was then repudiated considering that it imprisons the car and kills urban life. Thus, Le Corbusier's ideas were rejected (Idem, 1961). In addition, the urban heritage arose as a major issue in *"O Problema dos Bairros Antigos"* (The Problem about the Old Quarters). Regarding heritage values and concerns about preserving city authenticity, the article reproaches demolitions for freeing up space within the surrounding area of monuments. It rejects also building reconstruction on architectural styles of the past, as well as the loss of scale in the former urban spaces under street and square enlargements. Such approach took into account not only isolated buildings but also the relevance of certain sets of buildings, (Auzelle, 1961).

Such awareness became stronger over the time. In fact, the need for historical studies on urban design was emphasised some years later (Zerebska, 1968). This kind of concerns also arose in a paper presented in the *Colóquio sobre Salvaguarda da Paisagem e dos Lugares Históricos* (Symposium on the Protection of the Landscape and Historical Places), and it seemed to be a sign of conceptual advances facing such issues: besides pointing them out as highly topical issues in an introductory note, the article provided methodological criteria to proceed to the strategy proposed by it.

In this process of approaching a new thinking framework regarding urban space, the disciplinary limits became wider. By including other fields of knowledge, the debate on the city turned more complex also in Portugal. By giving voice to Roland Barthes (1968), in *"Semiologia e Urbanística"* (Semiology and Urban Planning) the magazine set out that the significance of space was no more either understandable or foreseen solely through disciplines such as Architecture and Urban Design (Branco, 1967)[5]. Biology, Ethnology, Semiology, History, Geography, Demography, Economy and Politics should be gathered together, theoretically and in the decision-making process (Auzelle, 1968). The urban plan should be flexible and adaptable to the "unforeseeable circumstances of life" (Idem: 214). The diversity of urban contexts – completely different from Detroit to Venice – should prevent urban planners from designing proposals on standardised principles, as Robert Auzelle said before (1960). According to other authors, there was not a specific time in which urbanism can perform a precise control of territorial space. Accordingly, the urban plan should not be designed as a finished object since it would be useless: instead, it should be adaptable to local and regional changing circumstances.

Breaking with former times, the chances for modernist speculation had finished. Walter Gropius (1971), one of the most significant architects during the heroic period of the Modern Movement, was the one who provided such a strong argument. In "El Futuro de Nuestras Ciudades" (The Future of Our Cities), he states, "today there are no more utopias" (IDEM: 100). In his opinion, it was the time for a flexible urbanism, which should be adapted to the social, technical and economic context in each moment. Reinforcing such idea, "The city is no more a condition, but a process". Hans-Jochen Vogel said it in the same issue of *Urbanização* (1971: 103). According to him, a lot of information inputs in urbanism could only lead to the understanding of the city as a process, which must be in constant adaptation to reality. Multi-disciplinary is, therefore, the key for reading this article. Besides, the author defends clearer criteria for organising the city, not abstractly but for the man, that is to say, different from being designed as a result of mere technical and quantitative concerns (Idem: 105–106).

In short, the change was irreversible at the end of the 1960s. Besides a more realistic approach, different from the dogmatic urbanistic ideas developed after the World War II, a stronger awareness with regard to the complexity of urban design had been increased over the time. Confirming such realism, the plans would only become effective if the municipalities had financial means to put in motion a land-use policy, claimed an article (Araújo, 1969). In other words, such realistic approach was already far from the academicism based solely on aesthetics, which was, oppositely, the main feature over the 1920s and the 1930s in the pages of Portuguese architecture magazines.

"A Cidade Deixou de Existir?" (The City No Longer Exists?) (Delfante, 1969) is the epilogue of this debate. In such assumption, the city was shaped because of an author's work and, therefore, by expressing his authorship, was set in opposition to the contemporary city, which was considered to be dispersed, decentralised and not conceptualised as a whole. According to the article, Charles Delfante believed that he was not facing an alternative type of city

[5] Because of a more complex physical distribution of economic, social and administrative factors, other activities became important in territory, challenging its balance. (BRANCO, 1967: 211–213).

still defined under the traditional urban concepts, but, instead, another reality with a very different meaning. Therefore, the trend that the pages of Portuguese magazines had been defining through the years was firmly confirmed in the late 1960s.

4 CONCLUSION

Considering urban design subjects, the first modernist era did not have the attention of the Portuguese magazines. Instead, the magazine's pages were filled with academicism still in the end of the 1940s. In fact, the magazines were solely focused on aesthetic effects. As a metonymy of that trend, the new concerns to adapt the urban spaces to the increasing car movement (which was taking place outside of Portugal since the 1920s), were then completely forgotten. In its long path over the years, the knowledge that was spread through the magazine pages in Portugal relied on incipient ideas during the first years of the *Arquitectura Portuguesa* magazine. On the other hand, it expressed the lack of technical training on urbanism, which the magazines spoke about in 1937. Indeed, the ideas related to urban design were only acknowledged as a major issue later on. It is not surprising, therefore, that the modernist principles were solely fully understood between 1948 and 1958.

Nevertheless, a third way arose by the end of the 1950s, similarly to what happened with architecture. Accordingly, the 1960s definitely broke with the previous time. The new trend was of plurality among the architects, city planners, and the society.

Despite all that, it should be emphasised which is likely to have been the most significant constraints within this cultural framework. Due to inadequate training, remoteness to the misunderstood contexts, respect to tradition, common sense, and realism – what Alexandre Alves Costa refers to as "(...) the effectiveness that evolves with the changing circumstances and on the basis of case-by-case (...)" (Costa, 2007: 74) – the Portuguese architects tried to stand between tradition and progress, instead of being avant-garde. By attempts to make the synthesis between the national context and the ideas that came from abroad, the Portuguese architects stood also between either the monumental or traditionalist architecture and the modern style. Accordingly, the knowledge that was spread through the magazine pages, in Portugal, stood between progress and atavism. It also stood between the ones who defended the rural landscape and the attempts of some political minorities to industrialise Portugal. Similarly, it stood between a country of villages and a most progressive one defended by Duarte Pacheco years before[6]. In accordance with the context displayed in these pages, Portugal remained between tradition and progress on the urban design ideas, which were spread through the magazines. Even within the scope of a discourse set free from real needs, and not defined by everyday pragmatism or political compromises – like it was the one expressed under those ideas in the architecture magazines – the articles on urban design stood close to the architecture practice as it was testified in this paper. It can be said, therefore, that the above-mentioned magazines did not express a progress because of international trends. Alternatively, instead, it might be possible to make a very different assumption. It may be said that, between the 1930s and 1960s, those magazines expressed a Portuguese way of thinking on spatial issues relying on Portugal's own themes and pace, giving voice to its own context and tradition. Thus, this kind of distance to avant-garde themes can really be seen as a different way of progress.

[6] In his role as a minister of Public Works and Communications (from 1932 to 1936, and after from 1938 to 1943), Duarte Pacheco (1900–1943), intended to modernize Portugal. (Costa, 2012: 98).

ACKNOWLEDGMENT

Assignment co-financed by the European Regional Development Fund (ERDF) through the COMPETE 2020 – Operational Program Competitiveness and Internationalization (POCI) and national funds by the FCT under the POCI-01-0145-FEDER-007744 project.

BIBLIOGRAPHICAL REFERENCES

Almeida, Pedro Vieira de (1986) Arquitectura Moderna. In História da Arte em Portugal. Lisboa: Alfa, Vol. 14, 2012.

Araújo, Gonçalo Nuno Araújo. (1969) "Meios de Acção Necessários e Realização de Empreendimentos Urbanísticos". In Arquitectura, revista de arte e construção. Lisbon: Iniciativas Culturais Arte e Técnica. No.108, March-April, year XXXVIII, 2nd series, p. 75–78.

A Arquitectura Portuguesa. (1929) "O problema da circulação em Paris". In A Arquitectura Portuguesa, revista mensal da arte arquitectural antiga e moderna. Lisbon: Mário Collares. No.1, June, year XXII, p. 8.

—.. (1939c) "Façam-se Casas Portuguesas em Portugal", in A Arquitectura Portuguesa e Cerâmica e Edificação Reunidas, revista mensal técnica, industrial e artística. Lisbon: Sociedade Editora Frace Lda. No.46, January, year XXXI, 3rd series, p. 9.

—.. (1939d) "Liceu Filipa de Lencastre", in A Arquitectura Portuguesa e Cerâmica e Edificação Reunidas, revista mensal técnica, industrial e artística. Lisbon: Sociedade Editora Frace Lda. No.49, April, year XXXII, 3rd series, p. 18–19.

Arquitectura. (1931a) "Prolongamento da Avenida da Liberdade atravez do Parque Eduardo VII". In Arquitectura, Revista de arte e construção. Lisbon: Grupo editorial de Arquitectura. No.20, August/September, year IV, p. 72.

—. (1948a) "Carta de Atenas", in Arquitectura, Revista de arte e construção. Lisbon: Pereira da Costa. No.20, February, year XX, *2nd* series, p. 23–24.

—. (1948b) "Carta de Atenas". In Arquitectura, revista de arte e construção. Lisbon: Iniciativas Culturais Arte e Técnica. No.21, February, year XX, 2nd series, p. 23.

—. (1948c) "Carta de Atenas". In Arquitectura, revista de arte e construção. Lisbon: Iniciativas Culturais Arte e Técnica. No.22, April, year XX, 2nd series, p. 23–24.

—. (1948d) "Carta de Atenas". In Arquitectura, Revista de arte e construção. Lisbon: Iniciativas Culturais Arte e Técnica. No.23–24, May-June, year XX, 2nd série, p. 39–40.

—. (1948e) "Carta de Atenas". In Arquitectura, Revista de arte e construção. Lisbon: Iniciativas Culturais Arte e Técnica. No.25, July, year XX, 2nd series: p. 19–21.

—. (1948f) "Carta de Atenas". In Arquitectura, Revista de arte e construção. Lisbon: Iniciativas Culturais Arte e Técnica. No.27, October-December, year XX, 2nd series, p. 17–18.

—. (1949a) "Carta de Atenas". In Arquitectura, Revista de arte e construção. Lisbon: Iniciativas Culturais Arte e Técnica. No.28, January, year XXI, 2nd series, p. 15–16.

—. (1949b) "Carta de Atenas". In Arquitectura, Revista de arte e construção. Lisbon: Iniciativas Culturais Arte e Técnica. No.29, February-Março, year XXI, 2nd series, pp. 15–16.

—. (1949c) "Carta de Atenas". In Arquitectura, Revista de arte e construção. Lisbon: Iniciativas Culturais Arte e Técnica. No.30, April-May, year XXI, 2nd series, p. 15–16.

—. (1949d) "Carta de Atenas". In Arquitectura, Revista de arte e construção. Lisbon: Iniciativas Culturais Arte e Técnica. No.31, June-July, year XXI, 2nd series, p. 17–18.

—. (1949e) "Carta de Atenas". In Revista de arte e construção. Lisbon: Iniciativas Culturais Arte e Técnica. No.32, Agosto e Setembro, year XXI, 2nd series, pp. 15–16.

—. (1953a) "Centro Urbano Presidente Juárez", in Arquitectura, revista de arte e construção. Lisbon: Iniciativas Culturais Arte e Técnica. No.50–51, November-December, year XXV, 2nd series, pp. 8–18.

Auzelle, Robert. (1960) "A cidade e o automóvel". In Binário, arquitectura, construção e equipamento. Lisbon: A. Palmares. No.18, January, p. 75–78;

—. (1961) "O Problema dos Bairros Antigos", In Arquitectura, Revista de arte e construção. Lisboa: Iniciativas Culturais Arte e Técnica. No.71, July, year XXXVII, 2nd series: 3–10;

—. (1968) "O Ordenamento dos Aglomerados". In Arquitectura, Revista de arte e construção. Lisbon: Iniciativas Culturais Arte e Técnica. No.105–106, September-December, Ano XXXVII, 2nd series, p. 211–14.

Barthes, Roland. (1968). "Semiologia e Urbanística". In Arquitectura, Revista de arte e construção. Lisboa: Iniciativas Culturais Arte e Técnica. No.105–106, Septembre-December, year XXXVII, 2nd série, pp. 180–82.

Boróbio, Luiz. (1960). "A Incógnita da Cidade Moderna". In Binário, arquitectura, construção e equipamento. Lisbon: A. Palmares. No.22, June, p. 205–206;

—. (1961) "A Incógnita da Cidade Moderna". In Binário: arquitectura, construção e equipamento. Lisbon: A. Palmares. No.28, January, p. 33–36.

Botelho, Rafael. (1959) "As novas cidades Inglesas". In Binário, arquitectura, construção e equipamento. Lisbon: A. Palmares. No.10, January, p. 1–6;

Branco, Duarte Castelo. (1967) "Implicações Urbanísticas do Lazer". In Arquitectura, revista de arte e construção. Lisbon: Iniciativas Culturais Arte e Técnica. No.99, September-October, Ano XXXVI, 2nd series, p. 211–213;

Colaço, Thomaz Ribeiro. (1937) "Do Cruzamento de Ruas, nas Cidades e Vilas". In A Arquitectura Portuguesa e Cerâmica e Edificação Reunidas, revista mensal técnica, industrial e artística, A Arquitectura Portuguesa. Lisbon: Sociedade Editora Frace Lda. No.22, Janeiro, year XXIX, 3rd series, p. 7–11;

Correia. M. (1967) "Descontinuidades Espaciais e Temporais dos Plano de Urbanização". In Binário, Arquitectura, Construção e Equipamento. Lisbon: A. Palmares. No.110, November, p. 225–227.

Costa, Alexandre Alves. (2007) "Portugal, Cidade e Arquitectura". In Textos Datados. Coimbra: e|d|arq, p. 59–74.

Costa, Sandra Vaz. (2012) O País a Régua e Esquadro: urbanismo, arquitectura e memória na obra pública de Duarte Pacheco. Lisbon: IST Press.

Delfante, Charles. (1969). "A Cidade Deixou de Existir?". In Binário, arquitectura, construção e equipamento. Lisbon: A. Palmares. No.124, January, p. 368–373.

Duarte, Carlos S. (1958) "Introdução a Howard". In Arquitectura, revista de arte e construção. Lisbon: Iniciativas Culturais Arte e Técnica. No.63, December, year XXVI, 2nd series, p. 5–18.

Filgueiras, Octávio Lixa. (1959) "Aditamento à Grille CIAM d'Urbanisme". In Arquitectura, revista de arte e construção. Lisbon: Iniciativas Culturais Arte e Técnica. No.66, December, year XXVII, 2nd series, p. 3–12.

Grande, Nuno (2006). Campo Magnético. Polaridades e Tensões na Arquitectura Portuguesa do Século XX. In Inquérito à Arquitectura do Século XX em Portugal. Lisboa: Ordem dos Arquitectos, p. 61–64.

Gropius, Walter. (1971) "El Futuro de Nuestras Ciudades". In Urbanização, Revista do Centro de Estudos de Urbanismo e Habitação Engenheiro Duarte Pacheco. Lisbon: C.E.U.H.E.D.P., No.2, Setembro, Vol. VI, p. 100–101;

Lima, A. Viana de et al. (1959) "X Congresso CIAM". In Arquitectura: revista de arte e construção. Lisbon: Iniciativas Culturais Arte e Técnica. No.64, January-February, year XXVII, 2nd series, 1959, p. 21–28.

Mumford, Lewis. (1958) "A Ideia de Cidade Jardim e o urbanismo Moderno", In Arquitectura: revista de arte e construção. Lisbon: Iniciativas Culturais Arte e Técnica. No.63, December, year XXVI, 2nd series, p. 19–22.

Portas, Nuno. (1973) A Evolução da Arquitectura Moderna em Portugal: uma interpretação. In História da Arquitectura Moderna. Lisboa: Arcádia, p. 687–746.

—. (2006) IAPXX – A Oportunidade do IAP XX e uma Interpretação dos Anos 40. In Inquérito à Arquitectura do Século XX em Portugal. Lisboa: Ordem dos Arquitectos, p. 51–60.

Tostões, Ana. (2006) IAPXX – Sob o Signo do Inquérito. In Inquérito à Arquitectura do Século XX em Portugal. Lisboa: Ordem dos Arquitectos, p.17–38.

Vogel, Hans-Jochen. (1971) "La Ciudad de Manãna". In Urbanização, revista do Centro de Estudos de Urbanismo e Habitação Engenheiro Duarte Pacheco. Lisbon: C.E.U.H.E.D.P., No.2, September, Volume VI, p. 101–108;

Zerebska, Teresa. (1968) "Estudos Históricos para a elaboração do Plano de Urbanização". In Arquitectura, revista de arte e construção. Lisbon: Iniciativas Culturais Arte e Técnica. No.105–106, September-December, year XXXVII, 2nd series, p. 224–26.

Drawing progress within the design process

Ana Moreira da Silva
University of Lisbon, Faculty of Architecture, CIAUD, Lisbon, Portugal
Lusíada University, Faculty of Architecture and Arts, CITAD, Lisbon, Portugal

ABSTRACT: This research focus aims to analyse nowadays progress on the importance of Drawing within the Design process. As the designer has to conceive and develop solutions for precise problems which may be of very different nature, Drawing presents itself as an operative support for problem-solving and critical analyses in the Design conceptual process. The fact of nowadays-wide use of new technologies does not invalidate the important role played by hand drawing as a stimulating instrument when sketching the first ideas and as a critical verification of the several hypotheses. Through the study of several statements from various authors, we intend to investigate the permanence of this important role since mid-last century to present. From this theoretical approach, we also intend to verify the relevance of drawing teaching in the formation of the future designers, despite the paradigm changes that emerge from times changes. Based on a survey methodology with two different moments, being the first an inquiry based on a questionnaire applied to undergraduate students, and the second a group of semi-structured interviews applied to designers. We intend to verify the permanence and the importance of hand drawing in their daily design work, although the new technologies constant progress.

Keywords: Design, drawing, progress, design process

1 INTRODUCTION

This paper stems from a current post- doc research project motivated by the need of producing more knowledge on design thinking after previous researches developed for the Master and PhD degrees obtained by the author, about the relevance of drawing in the conceptual process in Architecture and in Design.

Our main research question is: Which is the importance of hand drawing as a critical instrument and as an operative support within the conceptual process in Design?

The main objectives of this current research are: to verify the relevance of the drawing teaching in the formation of the future designers, and to verify the permanence and the importance of hand drawing in designers' actual daily work, although the new technologies constant progress.

We aim to achieve contribution on the conceptual design process, enhancing hand-drawing important value for Design students and professionals.

2 DRAWING AND DESIGN

In this study, we consider the close relationship between drawings and design the relevance of drawing in the act of designing as stimulating instrument when sketching the first ideas and as critical verification of the several hypotheses. "Designers are inextricably associated with drawings" (Lawson, 2004:31).

We find reference to drawing in the very origin of the word design. "Designare", is a Latin verbal form, which covers both the meaning of "to designate something? and of "to draw something".

In its origins, the term has already a double sense: a more abstract aspect of "to attribute, to conceive, to project and another one of more concrete nature of "to register, to configure, to form". (Denis, 2000:16). Design adds intention to drawing: to imagine, to conceive, and to assist in the achievement of solutions for practical problems, which can be of very different nature.

Manzini refers the issue between the "thinkable" and the "possible", which is always underlying the design process, considering it neither simple nor linear (1993:5). Lawson also characterises this process as "highly complex and sophisticated" (2005:67).

During all the complex process which goes from one imagined object to its implementation, the designer has one essential media to help him on developing the idea – drawing.

The complexity, characteristic of the transition from the idea to its materialization, is related to the actual extent of design where a number of factors come into play, such as the ability to understand the context and to imagine the solutions, the ability to know and take advantage of the processes and materials as means or vehicles through which the solution is materialized,

Figure 1. Sketches for a chair. Daciano da Costa (1994). Source: Costa, 1998:154.

the ability to transform ideas into appropriate forms, having in mind the limits and material possibilities. Drawing holds, in itself, the making visible quality of the whole mental process underlying the conceiving stage, from the first sketches of an object vague idea to the creation of its final form.

Drawing's transcendental importance lies in the capacity that it gives the designer to materialise abstract conceptualizations and to create the ideational basis for design forms and objects.

Within the complex process which takes place from how to imagine an object till its materialisation, the designer can use hand-drawing as an indispensable mean to help him to develop the idea, as an operative support in the conceptual process. This attitude comes across the concept of drawing as a mental activity, which uses the hand as an extension of the brain.

Campo Baeza writes in the introduction of one of his books, that its title – Thinking with the Hands – is intended to make clear that "the creative work requires both hands and head.

This "thinking with the hands" is inspired by the bright image of Saramago's novel, *The Cave*, when he states that creators have a sort of little brains in their fingertips" (2011:11).

The search for solutions, even for simple problems, implies that drawing studies in detail each phase of the process for obtaining the result we seek for. The complexity of the required drawings emerges as the designer approaches the solutions, which he considers ideal or that is possible.

During the conception phase, we should not disregard the importance of a trace, of a scribble, of a sketch. They can all contribute to the idea development.

For a designer, the sketchbook is not just a place to draw but also a place to order thoughts, to gather

Figure 2. Sketching allows the exploration of a wide variety of ideas all at once. Source:[http://alistapart.com/article/sketching-the-visual-thinking-power-tool] accessed 2016.12.2.

Figure 3. The idea generation and development stages through drawing. Source: [https://www.google.pt/search?q=idea+generation+through+sketches&sa] accessed 2017.01.10.

graphically and visually information and to develop a design response.

It emerges as a way of thinking through drawn lines as a creative process, revealing the central role that drawing can play within the development of any design product.

> The real goal of sketching is about generating ideas, solving problems, and communicating ideas more effectively with others. (Rohde, 2011)

The act of drawing allows that the reasoning and thoughts we have developed can be gradually translated and decoded throughout the drawn lines. Somehow, we debate ourselves with our own ideas in the paper space. We scratch, we draw, we overwrite features, we configure, we represent, we visualise, giving physical form to our thinking.

There is a direct link between the thinking and the hand that performs the drawing: the hand as an extension of the brain, of the reasoning.

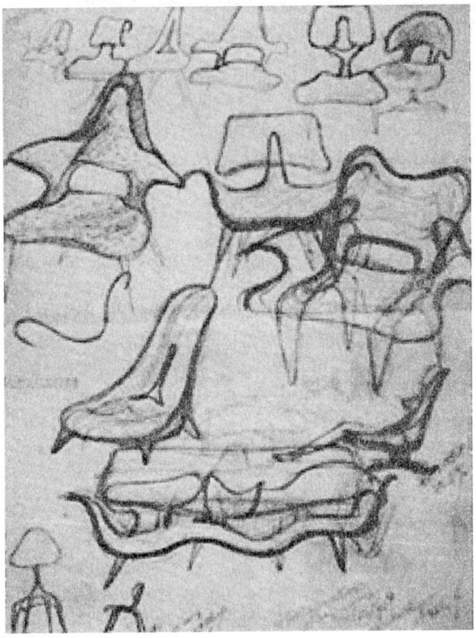

Figure 4. Charles Eames chairs sketches (1943–46). Source: [https://www.google.pt/search?q=Charles+eames+sketches&tbm] accessed 2017.01.10.

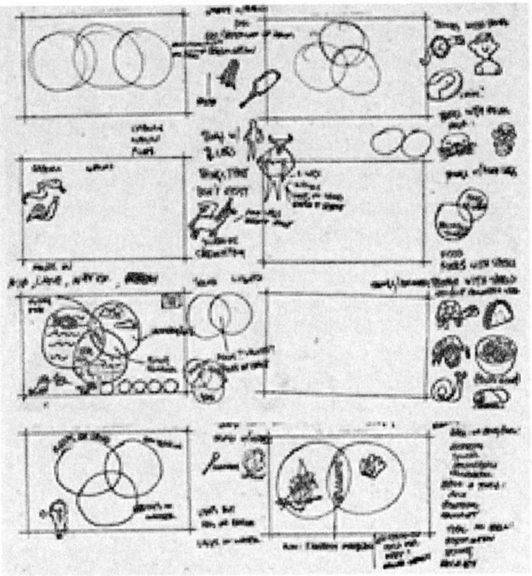

Figure 5. Mike Dutton's sketches for Google doodle (2014). Source:[http://www.google.com/doodles/john-venns-180th-birthday] accessed 2016.12.21.

If drawing is central for conceiving, it is also central for defining how that conception is managed as it moves from its initial stages through its actual development and realisation as a material form.

Drawings illustrate various aspects of the design process.

Drawings are used by designers in their own thinking and design processes as a kind of internal conversations and as a way to record, test, and reflect on a design. (Robbins, 1994:5)

Drawings have been used for generations of designers. The use of sketches not only helps a great deal in starting the ideas generating process, it also enables high-quality concepts, to begin with as initial ideas. These ideas can be accessed, combined, selected, and developed.

Drawing can be the key instrument of creation and control, also playing a critical role in the design process.

Nigel Cross states that:

> drawings are a key feature of the design process. At the early stages of the process (...), they are communications with oneself, a kind of thinking aloud.

and concludes:

> The conceptual thinking processes of the designer are based on the development of the ideas through their external expression in sketches. (2005:9–20)

There are a number of ways of looking at design drawings. We can see them as representations or a language, or even see what ideas each drawing embodies. To look at drawings in these ways is certainly critical to any understanding of what role it plays in the creative and communicative processes that design entails.

Tom Porter and Sue Goodman (1999) claim that in the wake of rapidly computer-graphics technology, drawing by hand remains undisturbed as the central activity in the process of design.

The impact of new materials and production processes, along with the changing nature of society and the alliance between technology and design, present designers with considerable challenges (Sparke, 2004:3).

Lawson states:

> So important has drawing become in the design process that every contemporary design curriculum places considerable emphasis on the acquisition of skills in drawing. This is usually thought so important and basic that it invariably starts right in the very beginning of the course. (2004:32)

According to Donald Norman, the recent profound changes in design nature, due to the contemporary historical context, justify an enlarged discussion on the important role played by drawing in design courses, adapted to the new paradigm (Norman, 2010).

The fact of using new technologies does not invalidate the important role played by manual drawing, both at the initial stage of recording the first ideas and during their subsequent development and in the critical analysis of the different hypotheses.

Mike Rohde states:

> Adding sketching to the design process is a great way to amplify software and hardware tools" and concludes that the use of sketching

> in the designer's creative process, reveals the central role drawing plays within the development of a design and the importance of drawing by hand, particularly in an age dominated by digital media. For Rohde, sketches can work like a visual thinking tool as a primary language for capturing thoughts, exploring ideas, and then sharing those ideas. "Sketching provides a unique space that can help you think differently, generate a variety of ideas quickly, explore alternatives with less risk, and encourage constructive discussions with colleagues and clients. (Rohde, 2011)

Given the difficulties of the project process as an ability to imagine what doesn't exist, the drawing appears to be the project tool that enables the idea to become visible. Drawing mediates between thought and action, expresses the thought by images that allow us to visualise the future conception.

To study drawing understood as a bearer and an intermediate of the idea and of the project means to deepen its values in order to explore the multiple meanings contained in graphic language, understanding, directly, through the "reading" of the drawing, the generating idea of the project.

The design is inseparable from project methodology, and drawing assumes itself as an indispensable tool in the design process. The issue focuses on the relationship between drawing and project, and the importance that drawing acquires in the conceptual act of the design project, as a booster instrument when registering the first ideas, and as verification of the several hypotheses.

Being aware of the project complexity as an intellectual ability for imagining what does not exist yet, the drawing appears as the project essential tool that enables the possibility of making the idea visible. As the designer has to conceive and develop solutions for precise problems which may be of very different nature, drawing presents itself as an operative support for problems solving in the design conceptual process. Due to its operative characteristics, the importance of drawing assumes itself as a wide sense, providing the act of drawing with the ability to be a means with multiple resources and a basis for the conceptual process underlying the practice of design.

In a previous research (Moreira da Silva, 2010:101) approaching the subject, we concluded that drawing accomplishes the most different objectives, such as:

- a way to communicate;
- a means of discovery;
- a process of interiorization;
- a graphic method of study;
- an observation and registering process;
- a research tool;
- a privileged mean for the communication of ideas;
- a link in the mental and creative process.

Therefore, the importance of drawing assumes a broad sense, conferring to the act of drawing the ability to become a means of multiple resources to practice the discipline of design.

> Drawing today serves as a primary medium for generating, testing and recording an individual designer's own creative and conceptual musings about a design. (Robbins, 1994:27)

In The Inexorable Rise of Drawing, an article that analyses contemporary thinking about drawing, Annabel Tilley, try to identify the reasons why, nowadays, the drawing subject became so relevant. According to Tilley, this happens because "(. . .) we are still trying to define what it is, and, imagine what it might become." Drawing interdisciplinary nature

> has been acknowledging, while the tangible has given way to drawing's recognised facility to allude to or describe the intangible way of articulating the inarticulate. (Tilley, 2008)

Several authors have written about the importance of drawing in design methodology. Their general conclusion points drawing as an essential tool for design research, as it allows investigation of several alternative solutions in the design process (Lawson, 2004; Cross 2005; Rohde, 2011).

The constant progress of the new information technologies offers to the designer new ways, allowing him/her to save time and facilitating many of the daily practice tasks, even becoming indispensable in many work phases. However, they do not replace hand drawing, which continues to assume a "chameleon" shape during the various historical periods, in a constant time adaptation, and especially by incorporating a critical dimension in design's conception process.

3 CONCLUSION

Despite the paradigm shift required by the changing times, we believe that hand-drawing remains inseparable from the designers training and professional practice, assuming an essential operating support in the project activity, which continues being the design basis.

Various authors have researched and written about the importance of hand drawing in the design process and several designers refer the permanence and the importance of sketching in their daily work.

Drawing presents itself as an operative support for problem-solving and critical analyses in the design conceptual process.

Although technological development has created several tools and techniques of graphic representation, we can conclude about the importance of the hand drawing for the design creative process. Freehand drawing is used to facilitate the development of projects and show them more quickly and efficiently, and it is an essential tool for any designer, regardless if he works in informational, graphic, product or fashion design areas.

In the future, design education should pass through a systematic approach to hand drawing in order to highlight and analyse the flexibility with which it adapts itself to various purposes, fulfilling effectively a wide range of intentions, in a constant and vital adaptation to the continuous changes in the teaching and in the practice of the student and of the designer, given the new techniques and technologies.

In order to validate these assumptions, the current post-doc research can constitute a contribution to the understanding of the importance of hand drawing's permanence as the basis of the design conceptual process, although the constant progress of the new information technologies.

ACKNOWLEDGMENTS

The author would like to acknowledge the support given by CIAUD – Research Centre for Architecture, Urbanism and Design, Faculty of Architecture, University of Lisbon; CITAD – Research Centre in Territory, Architecture and Design, Lisbon Lusíada University, and FCT – Foundation for the Science and Technology, Portugal.

BIBLIOGRAPHICAL REFERENCES

Campo Baeza, A. (2011). *Pensar com as Mãos*. Lisboa: Caleidoscópio.

Costa, D. (1998). Design e Mal-Estar. Lisboa: CPD.

Cross, N. (2005). *Engineering Design Methods, Strategies for Product Design*. Chichester, England: John Wiley & Sons Ltd.

Denis, R. C. (2000). *Uma Introdução à História do Design*. São Paulo: Blucher.

Lawson, B. (2005). *How Designers Think The Designers Process Demystified*. Oxford: Elsevier.

Lawson, B. (2004). *What Designers Know*. Oxford: Elsevier.

Manzini, E. (1993). *A Matéria da Invenção*. Lisboa: CPD.

Moreira da Silva, A. (2010). *De Sansedoni a Vasari: um contributo para o estudo do Desenho como fundamento do processo conceptual em Arquitectura*. Lisboa: Universidade Lusíada Editora.

Norman, D. (2010). *Why Design Education Must Change*. Retrieved from [http://www.core77.com] accessed 18/12/16.

Porter, T. & Goodman, S. (1999). *Designer Primer for Architects, Graphics Designers and Artists*. London: Butterworth Architecture.

Robbins, E. (1994). *Why Architects Draw*. Cambridge Massachusetts: The MIT Press.

Rohde, M. (2011). *Sketching: the Visual Thinking Power Tool*. Available from [http://alistapart.com] accessed 11/12/16.

Sparke, P. (2004). *An Introduction to Design and Culture*. London: Routledge.

Tilley, A. (2008). *The Inexorable Rise of Drawing*. Retrieved from [http://www.transitiongallery.co.uk] accessed 10/12/16.

Sustainable design and technological innovation; New perspective for the traditional sector of the pottery

Annalisa Di Roma
DICAR Department, School Industrial Design, Polytechnic University of Bari, Bari, Italy

ABSTRACT: The present paper will describe a critical approach for the product design field, related to the local identity of a territory and with its specific material culture. This approach concern both with the human dimension of the innovation processes and with the competitiveness of manufacturing systems based on the culturalization of the Economy, particularly in the traditional sector.

The current studies on sustainability for product design are mainly related to carbon emission reduction; this paper will highlight a new critical and practical approach based on how: the cultural identity of a place and the specific material culture could address traditional manufacturing sectors toward a competitive strategy for global market; the progress in science and technology could be addressed to the enhancement of the communities' identity; the technological progress in the field of the enabling technologies could upgrade the traditional manufacturing sectors.

The paper will describe first the state of the art and then, with some practical experimentation, the results of technological advancement in the traditional artisans' sector of the ceramic pottery.

In conclusion, the present paper aims to define possible approaches related to the cultural heritage, in order to enhance the competitiveness of the creative industries, balancing innovation through tradition, from the "hand made" to the "enabling technology made" manufacturing.

Keywords: Sustainable design, digital design and manufacturing, digital ceramics, cultural economy

1 INTRODUCTION

Till a few years ago the answers given by the design disciplines to the issues of sustainability applied to the built environment has been the definition of practical methods and guide lines mainly addressed to the long cycle assessment of the product, the use of low-impacts and bio-based materials, and so on. These approaches have been developed in order to reduce carbon emissions and the digital revolution has had a positive impact on the manufacturing systems.

Nowadays the tendency coming from the economic theory, known as the cultural economy (Throsby, 2001), is fostering a new approach to the design disciplines the economic theory defines a cultural responsibility in the mediation of the progress achievement between economy – social needs and cultural identity.

Given that premise this paper will describe a methodological approach for the product design, trying to bring the gap between traditional methods of the production, directly connected with the local identity of a territory, and the advanced manufacturing systems, able to approach the innovation and the economic competitiveness: from the traditional craft to the digital craft.

The structure of the paper will follow these topics: could the cultural identity of a place and the specific material culture address traditional manufacturing sectors toward a competitive strategy for the global market, balancing tradition with innovation? Could the progress in science and technology be addressed to the enhancement of the communities' identity, rising social inclusion? Could the technological progress in the field of the enabling technologies give a new face (2.0) to the traditional manufacturing sectors developing new production?

In order to give an answer to this questions, the paper will describe first a critical scenario, and then, with some practical experimentation, a methodological approach for product design in the field of digital design and fabrication.

2 BACKGROUND AND STATE OF THE ART

The process of human settlement of a place tells the materials operations developed to conquer and settle the land, identifying the characteristics of the territory and the authentic language of the place. A sort of an anthropology of the techniques and materials, developed in order to produce, accompanies the cultural landscape and manufacturing that kind of "thinks" useful for everyday life needs.

In this historical scenario, the "object of use" is connected to the needs and to the user, together with the implications of the dynamics of use, the collective experience, and the impact of production on the environment.

At this connection, Tomas Maldonado has written:

> Non ci sembra superfluo ricordare che in ogni società esiste un punto nevralgico in cui ha luogo il processo di produzione e riproduzione materiale, cioè un punto in cui, secondo le esigenze dei rapporti di produzione, vengono man mano sancite le corrispondenze tra "stato di bisogno" e "oggetto di bisogno", tra bisogno e fabbisogno. Il Disegno Industriale, in quanto fenomeno che si situa precisamente in tale punto nevralgico, emerge come un "fenomeno sociale totale. (Maldonado, 1987: 15)

In this light, the scientific progress applied to material culture, intended in its technological transfer to the material production, also defines a strength relationship between manufacturing processes and cultural value of the making.

At this connection, the art historian G. Kubler alluding to the "history of the things" in it is essay *The shape of time*, identifies the meaning of the shape of time as a synthesis between material culture and scientific culture, in a continuous and perpetual renewal of forms and matter.

> Science and art dealing with certain human needs of both the mind and hands satisfy producing things. Tools and instruments, symbols and expressions are equally as needs and all must be first planned and then carried out [...] Today, it is again clear the artist is a craftsman and he belongs to a human group distinguished as homo *faber*, whose job is to evoke a perpetual renewal of forms of matter. (Kubler, 1962).

Nowadays the themes of the technology transfer meet the instances to share technical and technological world learned of artistic excellence with the world's advanced technology: mix artisanship with industrial skills; the ability of technologists and managers with those, extraordinary, technicians and artisans. See for example the on going revolution that gave rise to the movement of the "makers" of those who do things for themselves, which is gaining support all around the world.

This aspect of "material culture", as detected by Renato De Fusco, is topical in the current debate on the role of craft practices in the stream of creative industries, giving a positive judgment to small series productions, traditional and local referred, replacing also technological progress in a new cultural dimension. Cultural Industries, in fact, are industries that use culture as an input and have a cultural dimension, although their outputs are mainly functional. They include architecture and design, which integrate

Figure 1. A 3-D printing process for digital manufacturing: FDM system.

creative elements into wider processes, and subsectors such as graphic design, fashion design or media.

Particularly in 2016, the sector with the best performance in Italy is the industrial design (with +10.8% in value added and +13.8 in employment). Tourism is fine: 37.5% of the national tourist expense is in fact "activated" by culture[1]. The most promising alliance is the one that sees culture joining the world of food and wine, for a cuisine that is increasingly linked to local traditions.

The impulses coming from the so-called movement of "digital craftsmanship", contributed to foster the interest of the design methods against the renewal of the traditional knowledge of crafts, raising a new sustainability for product design and for the cultural economy. It means, "stir the craft skills with the industrial skills, the ability of technologists and managers with those, extraordinary technicians and craftsmen" (Micelli, 2011).

Small series production, which is typical for the craft, is often a differentiator key also for the most developed manufacturing sectors, able to meet the growing demand for customised production, an alternative to the mass production. This aspect can be integrated into the 3-D processes of digital design, with generative parametric software. The word craft in the digital systems, therefore, assumes the value of a selection of materials, production methods, assembly logics corresponding to the practical methods for of sewing, weaving, turning, and so on.

3 DIGITAL DESIGN: NEW FOCUS ON MAKING

The digital revolution in its first expression has had an immediate effect on the distancing between the designer and the material objects. Therefore, it has produced a new dimension for the user experience (from the virtual prototyping to the photorealistic

[1] At this connection see the 2016 "Io sono cultura" report by Symbola.

representation, from the virtual movies to the augmented reality) and new forms of social communities remotely organized (on line and off line) and focused on common interests, on the open design processes, on the sharing of the results. This process has been characterised by the *open sources* philosophy (Boldrini, Levine, 2012; 22–24) that has opened to a new form of social cooperation in the design processes. The fourth industrial revolution has been in effect, properly, when productive system has started to be connected with the design processes with a new possibility for designer, expert or not, and user, expert or not, to interact; as a result nowadays a new era of the material culture defines a role for the expert designer as a "mediator between technological demands and social and cultural issues (Manzini, 2015, p. 49), giving rise to a new era of creativity (2.0) that has a new form of gratification in *making* for itself and for the others (Gauntllet, 2013: 19). In this light, a new page of the material culture, that contains the history of techniques, materials and making processes, (De Fusco, 2012, p. 42) could be written. In fact, as Hauser intends:

> the history of the Art could be defined as the continuous renew, enlarging and improvement of the tool for its expression, [...] as a harmonious equilibrium between the power and will, between the instruments and the artistic intention. (Hauser, 1956: 114).

This equilibrium between technical and intellectual evolution could be upset by the social organisation of the work and not just from the equipment in itself. The material and technical aspects of the creative processes, in fact, could feed creativity and could weave a dense network of connections with the physic environment, intended as the productive places, placed in the context of the cities, re-establishing the new connection between social life and work places. Making implies a connection, with materials, other people and the world (Gauntllet, 2013: 56). Nowadays, besides the "virtual plaza", the digital fabrication places (Fab Labs) are playing a new role in connecting people and in giving a new expression of the *joy of labour*, where the value of making is in the creative process and a new democracy could take place. Branzi says:

> New robotic technologies have reduced the distance between consumer and producer and allow producing customized small series of craft objects, but also arms – such as drones – genetic engineering, nano – technologies, robotic organisms; post Fordist factory seems to entrust to new technologies and new materials, the role of opening new dimensions to democracy. (Branzi, 2015: 14–15)

In fact, as traditional manufacturing district (crafts districts, pottery districts, textile districts, etc.) were placed in the centre of the ancient city, where cultural and leisure activity also took places, the new fab lab are today organized, from the smaller one to the bigger one, in the context of the city.

The industrial revolution, since it is beginning, defines an ideological and physical separation and specialisation of the city areas: manufacturing areas, civics areas, cultural area, etc.

The fab labs, for example, are now a day playing a new role as "workshops" places in which creativity and manufacturing are re – connected as new spaces for socialisation and share culture.

Now a day, *when everyone design* and every one makes, the new intersection between creativity and production allowed new models to be explored by design disciplines, not just from the skills and tools point of view, but also from the social dynamic point of view, also based on co-design processes. As Ezio Manzini intends:

> Obviously, the notion of co-design intended here is not standard one found in formalised processes where a team of stakeholders discusses round a table, trying to create a common language and share visions and strategies. What I mean by co-design, is more like a vast, multifaceted conversation among individuals and groups who set design initiatives rolling at the nodes of the networks they are part of social conversion in which different actors interact in different ways (from collaborating to conflicting) and at a different time (in real time or off-line). (Manzini, 2015: 48–49).

4 DIGITAL MANUFACTURING: TECHNICAL RENEW

This section presents the results of some experiments aimed at the implementation of parametric computer systems for the development of a ceramic container system. In particular, the course has focused on the understanding and innovation of artistic crafts processes, mediating the scope of unique artefacts (or limited series) with the industry of large-scale industrial production.

Specifically, with reference to the material and fabrication processes of the product, the focus was on traditional production, reaching the integration of technologies enabling 2.0 craft productivity.

The very meaning of the term "manufacture" is historically tied hand in glove with that of "artisan"; shops and large manufacturing industries often have a common past of creativity, inspiration and dedication manual labour.

Usually the distinguishing feature between industry and handicrafts is expressed in the size and production capacity on the one hand and the other in the uniqueness and specialisation.

Digital manufacturing brings us back to the change induced by the manufacturing technologies. This change, when applied to reality in which the creation, customization, and work on the individual product are particularly relevant, stimulates and opens the way to digital artisanship.

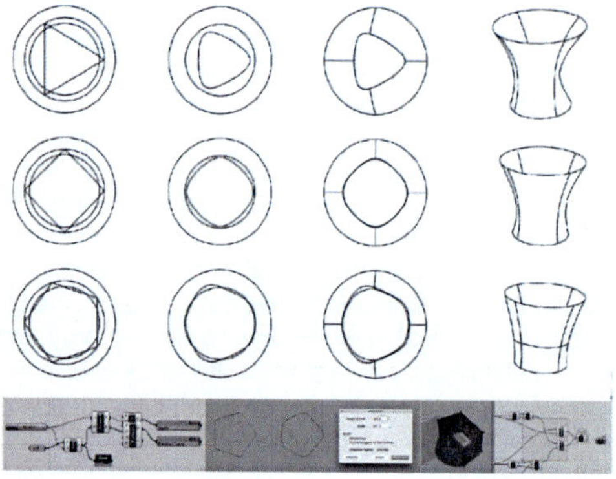

Figure 2. The figure showns the informatics parametric process for the three-dimensional modelling of the digital ceramics pottery set.

Figure 3. Details of the digital pottery set, traditionally hand made decorated.

This essential process of evolution starts from the adoption of 3-D modelling software for design, digitising, plotting and cutting components. The considerable development had also adopted new materials, among which the polymer was taken over since the 60s as innovative materials for all the components for accessories, and a whole range of technical fabrics and synthetic specially developed. Until you get to the present, where the introduction of 3D printing is making a real "new industrial revolution", including in the Fashion Sector.

Among the many techniques used today in the industry of fashion design, laser-cutting techniques are having a great development and application. Laser processing and 3D printing are contributing, in fact, not only to redefine the aesthetic canons of the formal production in that area but also to bring innovation to a new level of complexity.

As for the traditional ceramic pottery, for example, the conformation rule followed the setting of a revolutionary geometry around the lathe axis, and the formal repertoire has been differentiated and articulated based on the sequence of curved lines bound in the sequence for the condition of internal and/or external tangency. The assumption of a three-dimensional print production allows the project to be realised by means of a new geometric sequence that binds to the axis of the parameterized cross section. In the experiment shown in the figure (Fig. 02) the parametric process has set the three-dimensional conformation process through the stem of three cross-sections. The median section has been parameterized by changing the number of vertices of the polygon (3, 4, 5) the degree of curvature (2, 3, 4), the number of control points (from $n + (1 + m)$.

Prototyping was performed using a fixed extruder with an FDM type clay machine, producing a superficial characterization comparable to that produced by Colombian technique.

This approach allows us to hypothesise new results for the ceramic manufacturing industry based on the definition of the three-dimensional form through the specific transition of adaptable geometric geometries. Generative modelling, thanks to the contribution of coding, has given great development to the theme of the new concept of design and the same digital artisanship. This new modelling process has the character of computer programming, but with an easy-to-use approach through visual components.

The system allows normal operation of substantive generation, transformation, and evolution of models through reversible processes, preserving the memory of the starting geometries.

Combining tradition the hand made painting and finishing of the pottery have been made using ancient techniques and colours.

The Persian blue, the multiple fire painting and the glass painting have been used connecting the most recognisable ceramics identity of the place.

The advantage of this technique is:

1) Introducing new geometrical morphology in the panorama of the traditional hand made pottery;
2) Possibility to substitute the casting technique with a direct moulding system;
3) The combination of traditional methods and innovative one
4) The development of a new customization process.

In fact, serial, standard, mass production is the first assumption expressed by the "after war" industrial design.

Nowadays the technological aspects originated by the fourth industrial revolution completely sets down the "mass production" meaning, standard according to the 4.0 processes.

For this reason the customized production, paradox for the mass production logics and for the deterministic role of the design approaches devoted to the unique solution and to the simple interaction between a generic expression of the user experience (globally defined), is today a principles that gives new meaning to the product design discipline, going toward such "informal" processes typical for the handicraft in which the user could customize each component of the product, from the dimension to the style.

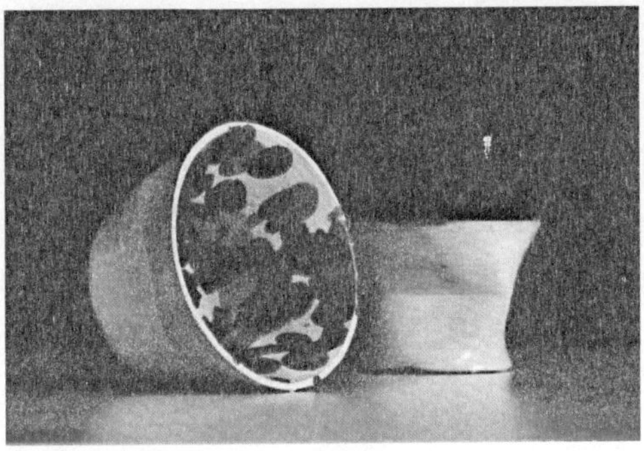

Figure 4. Details of the digital pottery set, traditionally hand made decorated.

5 CONCLUSION

The 2.0 craft re-connects the tradition of material culture that stratifies the technical knowledge related to materiality and materials, techniques and technologies, artefacts and works of art, and contribute to strengthening the relationship between productive places and socialisation. In fact, craft and design are based on the close connection that creativity is able to establish with the production places, with its political and social context. Focusing on the cultural value of creative processes a new form of social innovation could be enhanced.

These aspects are integrated into an innovation process that regards on the one hand technology and production systems, and on the other with the sense of the artefacts, responsive to social needs more than to market needs.

So, relocating the hand made know how and the traditional industrial methods in the context of the advanced production methods and processes, is aimed to the enhancing of the cultural matrix of a territory, able to define a new cultural economy approach, enabling new competitiveness keys for footwear market. The contemporary scene shows that artisan work is now one of the most important figures of culture and the Italian economy; bet on it, contaminating it with "new knowledge" technological and opening it to the international market would be reflected in his hands a powerful tool for growth and innovation. As evidenced by some Italian companies (Geox, Tod's, and so on) "know how" remains an indispensable ingredient for the entire Italian manufacturing. That, in the end, is one of the few vital sectors of our economy (Micelli, 2013).

The parametric design combined with the production of models by electronically controlled print output and/or processing is a phase of evolution of the contemporary art industry related to the digital outcomes.

Vector and bitmap software applications through generic logic offer new operating tools and the ability to develop innovative formal languages, consistent with mathematical models that support them through algorithms. In accordance with the Kunstwollen Riegliana, the instrumental technical equipment supports the will of art and sets new operating modes in which the designer, the computer technician and the machine operator share the artistic process of transforming matter.

BIBLIOGRAPHICAL REFERENCES

AAVV, (1990). *History of Industrial Design: 1851–1918 The Great Emporium of the World*, Milano: Electa.

Boldrini, M. Levine, D. K., *Abolire la proprietà intellettuale*. Bari: Laterza, pp. 22–24.

Branzi, A. (2008). Introduzione. In *Introduzione al Design Italiano, una modernità incompleta* Milan: Baldini e Castoldi, ed., pp. 10–11.

Branzi, A (2015). "Design after design", in *Interni* n 10, 2015, pagg. 14–15.

De Fusco, R. (2012). *Filosofia del Design*. Milan: Einaudi.

Di Roma, A. (2008). L'ornato digitale. In: Di Roma A. *La produzione in serie dell'ornato architettonico: dall'industria artistica ellenistica alla prototipazione con processi*. Rome: Aracne.

—. (2013). Ornato digitale: strumenti e tecniche del design digitale. In Di Roma A., *L'ornato di serie. Dalle arti tecniche al design digitale*. Firenze: Alinea ed. pp. 50–60.

Di Roma, A. Scarcelli A., (2016). *Manifattura digitale e produzione su misura a distanza. Nuovi orizzonti dell'esperienza d'uso e dei processi progettuali,* [Digital fabrication and on remote on measure production. New horizons of the user experience and of the design methods], in A.A.V.V., *Material Design Journal: Synapse Design and Connectivity,* Ferrara: Media MD, vol 2., pp. 58–71.

Focillon, H. (2002). *Elogio della mano postfazione a La vita delle forme* (*Vies de Formes suivi de Eloge de la main*, 1943, Press Universitaire de la France, Paris), Milan: Einaudi, p.113.

Friel, M. (2012). "Artigianato, industrie creative ed economia della creatività" in *Quaderni di ricerca sull'artigianato*, n° 59, pp. 35–45, Il Bologna: Il Mulino.

Gauntellet, D. (2013). *La società dei makers. La creatività dal fai da te al web 2.0.* Bologna: Marsilio.

Hauser, A. (1956). *Storia sociale dell'arte*. Torino: Einaudi.

Kubler, G. (1962). *The shape of time.* New Haven: Yale university press.

Maldonado, T. (1987). *Disegno industriale: un riesame.* Milano: Feltrinelli.

Manzini, E. (2015). Design when everybody design, Cambridge: MIT Press.

Menichelli, M. (2012). "Artigiani e designer: per una comunità dei makers", in *Domus Web*, 14 Marzo.

Micelli, S. (2011). Futuro Artigiano, l'innovazione nelle mani degli italiani, Marsilio ed.

Sennet, R. (2008). *L'uomo artigiano*, Milan: Feltrinelli.

Thorosby, D. (2001). *Economics and Culture*. Cambridge University Press, Cambridge.

Vitta, M. (2000). "Un secolo e mezzo di design", in: *Il progetto della bellezza*. Milano: Einaudi, pp. 5.

CACO: Promoting the progress of joinery in Brazil

Juliana Cardoso Braga
Federal University of Uberlândia, Faculty of Architecture, Urbanism, and Design, Uberlândia, Brazil, University of Lisbon, Faculty of Architecture CIAUD Lisbon, Portugal

Fernando Moreira da Silva
University of Lisbon, Faculty of Architectures, CIAUD, Lisbon, Portugal

Luis Carlos Paschoarelli
São Paulo State University, Faculty of Architecture, Arts and Communication, Bauru, Brazil, University of Lisbon, Faculty of Architectures, CIAUD, Lisbon, Portugal

Leonor Ferrão
University of Lisbon, Faculty of Architectures, CIAUD, Lisbon, Portugal

ABSTRACT: Studies in the area have presented alternatives in sustainable strategic design into local contexts. This paper presents the partial results of the CACO research project, from the diagnosis phase performed in five custom furniture companies in the city of Uberlândia, Minas Gerais State, Brazil. It is believed that the insertion of a set of ecodesign strategies in MSEs can generate greater competitiveness for companies generating new business opportunities, creating more favourable conditions of competition among companies of the sector in the region, improving their income while simultaneously promoting commitment and the sharing of responsibility (social, environmental, cultural) by all stakeholders. In this case, it is noted that the progress of furniture MSEs depends fundamentally on minimising the waste generated and the wastage of materials.

Keywords: Ecodesign, sustainability, environmental labelling, social design, furniture manufacturer MSEs

1 INTRODUCTION

In Brazil, the wooden furnishings industry is comprised of approximately 16,000 companies of various sizes (Movergs, 2014). It is estimated that most furniture companies (84%), consist of MSEs – Micro and Small Enterprises[1] (Fiemg, 2002; Garcia, 2005) many of them are cabinetmakers performing custom work (Galinari *et al.*, 2013). Currently, these companies have used MDF – Medium-Density Fibreboard as their main raw material. Brazil is the world's seventh largest producer of reconstituted wood panels in the world; in 2015, the production was 7.98 million cubic meters. A large part of this production, 95%, goes to the domestic market (IBÁ, 2015).

An expressive number of custom furniture Micro and Small Enterprise – MSEs, around eight hundred[2], is located in the city of Uberlândia (669,672 inhab.), in Minas Gerais State, Brazil (IBGE, 2015).

Figure 1. Outskirts of the city of Uberlândia (Minas Gerais State, Brazil), sites used for irregular waste disposal. Source: Cardoso Braga, 2015.

The production process of the joinery is not sustainable and results in a large volume of discarded waste

[1] In the industrial sector, micro businesses are those with up to 19 employees, whereas small businesses have 20–99 employees (Sebrae, 2013).

[2] Information provided by Ruy Mendes Gouveia (President of SINDIMOB – The Union of Furniture Industries from *Triângulo Mineiro* and *Alto Paranaíba*) in an interview done on August 11, 2015.

on the outskirts of the city (Nunes, 2013). In July 2015, there were ninety-nine points of irregular disposal of waste of various types (verbal information)[3]. This practice occurs, mainly, due to the low social and environmental awareness of cabinetmakers, and due to the lack of proper management of industrial waste by the municipality.

Within the empirical part of the CACO project, five case studies were selected and a study case methodology was implemented. Six phases were defined: diagnosis, analysis, development of strategies, action planning, implementation of strategies, and verification, being presented in this article the results of the first phase– the diagnosis phase.

2 CACO PROJECT

The CACO[4] project, which is a PhD research project in Design, involves the development of a sustainable strategic design model aiming to minimise the waste produced by furniture manufacturers in the city of Uberlândia.

Strategic design proposals should generate direct and indirect benefits for companies. With the reduction of waste, it is expected that there will be an increase in the firm's profitability. Therefore, environmental and economic progress, if achieved, can also generate social and cultural benefits through the maintenance of jobs, the preservation of this millenarian craft (expertise), and the permanence of some cabinet makers in the market.

The model will be comprised of a set of eco design strategies that will be implemented systemically in all phases of the product's life cycle (pre-production, distribution, use and disposal). The strategies are being defined from the conclusion of the diagnosis and will be based on the eco design strategies proposed by the authors Behrendt *et al.* (1997); Brezet and Hemel (1998); Fiksel (2009); Gertsakis *et al.* (2001); Manzini and Vezzoli (2008); and Tischner *et al.* (2002).

The model intends to supply, firstly, the needs and collective problems – already diagnosed as being common issues among the companies in the sector. However, according to Freitas e Merino (2011), it must also be considered that each company has individual needs and, therefore, the implementation process must meet the geographic, economic, social and cultural characteristics of each one of them.

[3] Information provided by Silas Guimarães (Advisor of the Municipal Urban Services Secretariat) in an interview on August 11, 2015.

[4] The word "CACO" (shard) was proposed due to its meaning: a piece, small fraction, part or remainder of something (Michaelis, 2009). The use of this term is meant to represent, symbolically, the need to connect the different areas necessary for the development of the proposed project and integrate "fragmented" knowledge in favour of more sustainable actions for the furniture sector.

3 THE DIAGNOSIS PHASE

The diagnosis was made in five MSEs (with a minimum number of 06 and a maximum of 20 employees) who produce customised furniture (furniture for living room, bedroom, kitchen, shops, among others) and use MDF as the main raw material. The data was collected by the researcher through (i) semi-structured interviews; (ii) direct observation with mechanical support; (iii) documentary research; and (iv) data collection (toolkit) on the volume of waste generated during the production process.

The production processes for the manufacture of a closet and a kitchen in each of the companies was analysed, as these are types of furniture commonly produced by these joineries. For this process, the investigator stayed approximately 15 days in each of the companies.

According to Rampersad and Hubert (2007), waste is usually hidden by the workers because it is considered as a normal symptom in the workplace. Therefore, the diagnosis was made to detect the main causes of waste generation in the joinery by analysing all stages of the life cycle of the customised furniture.

In order to evaluate the generation of residues during the production of furniture, a toolkit tool was created with the objective of detecting the amount of used material (material input). In this sense, the employees were guided and monitored during the use of the toolkit, which aimed to register: (i) the form used for the planning of the cut; (ii) and the consumption of material (new and reused MDF sheets). In addition, during the furniture production process, the generated waste was separated, identified, weighed and measured in order to identify the percentage of material loss during the production process waste (waste exit).

In an interview, the owners of the joinery declared that they use an average of 110 MDF sheets monthly (white and coloured), which are purchased from resellers. According to the carpenters, the MDF is responsible for approximately 42% of the furniture manufacturing costs. This data demonstrates the importance of MDF in economic spending and reinforce the need for optimisation and better use of MDF sheets even for the financial health of the companies. The diagnosis resulted in the identification of several common causes for MDF waste. In this paper, we present some of these main causes, namely:

1. Pre-production:

All the companies work with approximately five different MDF brands (*Arauco, Duratex, Eucatex, Guararapes* and *Masisa*) from different regions of the country. The wide colours and brands variety has increased losses, reduced the possibilities of reuse, and contributed to the generation of a large volume of waste.

Figure 2. Waste generated due to production errors identified during the diagnostic phase. Source: Cardoso Braga, 2017.

Figure 3. Pieces of MDF cut from entire plates, demonstrating that there is no reuse practice for leftovers. Source: Cardoso Braga, 2017.

Figure 4. MDF remains arranged in a joinery in the city of Uberlândia – Brazil. Source: Cardoso Braga, 2017.

2. Production:

2.1 Several production errors have caused rework in the companies studied. The following were mainly observed: lack of concentration of employees due to the use of the cell phone while working; lack of employees' commitment; absence of verification protocols (of material, tools and procedures) for assembling the product.

It was verified that all the owners of the studied companies accumulate a significant amount of activities (elaboration of the budget, negotiation with clients, surveying of measures, the realisation of production order, production of furniture and, when necessary, administrative activities). The accumulation of functions also contributes in part to reworking caused by measurement error and communication problems between management and production. Some direct consequences of rework are highlighted: wear and tear on the relationship between the company and the users, deadlines, economic losses due to loss of time, loss of labour, material, and increasing of fuel consumption.

2.2 It was observed that 80% of the furniture companies have a production process per project, where each joiner is responsible for the complete execution of the furniture (from the cut of the plate to the assembly). It became evident that the practice of production by the project is causing a lot of waste due to the lack of planning, especially during the cutting stage of the sheets. In 100% of cases, MSEs cut the material without performing at least surface calculations or analyse the best way to make good use of the sheet. Only one of the companies uses software for sheet optimisation.

2.3 Although employees are systematically advised to reuse MDF scraps to produce smaller items (drawers, shoe racks, slats, etc.), MDF leftovers accumulate very quickly. Due to overlapping MDF scraps, disorganisation, dust accumulation, and lack of inventory control, woodworkers prefer and often make use of new sheets, even to make smaller objects.

2.4 In 100% of the companies there are no spaces or shelves organised to receive leftovers. The most common practice consists of stacking pieces that are pressed against the walls of carpentry (taller pieces are placed on the bottom and lower ones on the front). However, the arrangement of pieces without any classification by size, thickness, or colour makes it difficult and doesn't encourage reuse, since it prevents the quick identification of the desired part. Added to this is the fact that leftovers accumulate a large amount of dust, which aggravates the difficulty of identifying colours and textures.

3. Distribution:

The transportation of goods, in 80% of cases, is carried out by own truck. The finished products are packed with blankets, cardboard and/or bubble wrap. With the exception of the blankets, the other packaging materials are usually used once or twice and then are sent to the municipal landfill. These packages do not always protect furniture sufficiently during transport, causing havoc and rework. In addition, recyclable packaging is usually mixed with other non-recyclable ones (such as MDF), making it impossible to dispose correctly of the final product.

4. Use:

At the stage of use, it was verified that the companies do not offer guidance to the users on the correct ways of maintenance and cleaning of the furniture. This is done verbally, and only when asked by users. In addition, the studied companies do not offer maintenance or repair services for furniture. This type of service is only provided, in special cases, to established clients. It is concluded that the lack of information on the correct maintenance of furniture and the lack of repair services can lead to a reduced shelf life.

5. Disposal and/or recycling:

5.1 Among the consulted companies, four joineries carry out the final destination of the waste in the municipal landfill. One of the companies disposes of its waste irregularly – one part is dumped on land on the outskirts of the city and the other part of the waste is usually destined for incineration in bakery ovens. These practices cause public health risks arising from the toxicity of urea-formaldehyde or phenol-formaldehyde additives used in the production of reconstituted wooden boards, these substances are on the list of carcinogens issued by the International Agency for Research on Cancer (IARC).

5.2 The companies do not carry out the separation of residues (MDF, plastic, cardboard, screws, etc.). There is also no suitable place to receive recyclable materials in the municipal landfill for recycling. Therefore, in general, companies have been using this fact to justify the lack of separation of furniture residues.

4 PRELIMINARY RESULTS

According to PNRS – National Policy on Solid Waste, management should be prioritised by the following order of priority: non-generation, reduction, reuse, recycling, solid waste treatment and the environmentally safe final disposal of waste (Brazil, 2012). So, to encourage the non-generation, reduction and reuse of waste, some preliminary strategies have already been proposed:

- To develop, in partnership with designers from the region, annual catalogues of colour trends and MDF standards, with the objective of significantly reduce the number of colours recommended by the woodworking shops to users and, consequently, increasing the possibilities of reuse of raw materials. Prioritising the local economy through raw materials;
- To improve the layout of the company and the organisation of workbenches in order to reduce production time, increase worker concentration, and reduce material losses caused by errors;
- To create a checklist for measures surveying in order to minimise errors that occur during the measurement conference;
- To designate a single employee to perform the sheet cutting, facilitating the flow optimisation and control of new and/or reusable raw materials;
- To provide training on cutting planning to optimise the use of MDF panels; or even using software to reduce losses and waste: to encourage the use of software which enables wood panel planning and optimisation of cut-outs, reducing waste and the cost of purchase of equipment and material;
- To carry out the planning of cutting of the furniture from a modular approach (when possible) aiming a greater use of the sheet;
- To create a suitable shelf for facilitating reuses with stock organisation and classification of the MDF pieces according to the dimensions and colour of the pieces of panels;
- To replace disposable packaging, used during transport, with reusable packaging;
- To create a checklist of delivery and assembly of the furniture in order to avoid errors and rework, reducing unnecessary travel, fuel consumption, CO2 emission and delays in delivery;
- To create an instruction manual with the purpose of informing users about the correct maintenance of the furniture (optimising life span);
- To indicate the collection system (environmentally correct final disposal), reuse and/or recycling (donation and/or renovation) of the furniture after use;
- To guide companies on disposal methods for environmentally sound waste;
- To provide repair services;
- To promote the recycling of materials such as plastic, cardboard, aluminium and others.
- To create a communication platform for buying, selling and exchanging material between manufacturers. The platform will be developed and made available to MSEs in order to sell debris material with potential for reuse, for example, medium or large indentations reconstituted wood plates. This communication channel intends to confirm the reduction of waste, encouraging the separation and proper storage of materials, while demonstrating the value of products that are commonly disposed of by such factors as reduced demand for very specific colour panels, and improper storage.

5 CONCLUSION

The diagnosis made by this investigation verified that MSEs of custom furniture in the city of Uberlândia have discarded annually a large volume of residues composed mainly of MDF. The average amount of waste per company is 14 tonnes per year, so if we consider the eight hundred companies in the city, the volume of discarded waste is approximately 11,000 tonnes per year.

It has also been found that this waste of materials has generated, not only serious environmental problems but has also hampered the development and progress of these joineries. The diagnosis pointed out that the highest cost of the companies (around 42%) is related to the acquisition of raw material (MDF).

According to the Moveleiros (2015), the serious political and economic crisis faced by Brazil since 2015, affected the companies in the sector and caused a significant slowdown in growth. In the Uberlândia cabinetmakers, the investigation found that the crisis has considerably reduced the production of furniture by order, has caused layoffs and is putting at risk the very subsistence of these micro and small companies.

After the diagnosis, it is intended to carry out a set of measures of sustainable strategic design in the MSEs of furniture of the region, encouraging the change of culture with the objective of reducing the consumption of resources and to minimise the waste. As Fiksel (2009), Jeswiet and Hauschild (2005) state, design-driven behaviour change can be economically advantageous because there is a strong connection between environmental and financial progress. According to this, the entrepreneurs are motivated to implement the strategies proposed by the CACO Project in the joinery.

ACKNOWLEDGMENTS

This work was carried out with the support of the National Council for Scientific and Technological Development (CNPq), Brazil; the Research Centre for Architecture, Urbanism and Design, Faculty of Architecture, University of Lisbon and FCT, Foundation for the Science and Technology, Portugal.

BIBLIOGRAPHICAL REFERENCES

Behrendt, S., Jasch, C., Peneda, M. C. and Weenen, H. van (1997). *Life cycle design: a manual for small and medium sized companies*. New York: Springer.

Brazil (2012). Institui a Política Nacional de Resíduos Sólidos, Pub. L. No. Lei nº 12.305 de 02 de agosto de 2010. Available from: http://www.planalto.gov.br/ccivil_03/_Ato2007-2010/2010/Lei/L12305.htm. [Accessed 23 January 2017].

Brezet, H. and Hemel, C. van (1998). *Ecodesign: a promising approach to sustainable production and consumption*. Paris: UNEP.

FIEMG, F. D. I. D. E. D. M. G. (2002). Diagnóstico do pólo moveleiro de Ubá e região. Belo Horizonte: IEL/MG.

Fiksel, J. R. (2009). *Design for environment a guide to sustainable product development*. New York: McGraw-Hill.

Freitas, R. F. de and Merino, E. A. D. (2011). *A gestão de design como estratégia organizacional*. Rio de Janeiro: Rio Books.

Galinari, R., Júnior, T., Rodrigues, J. and Morgado, R. R. (2013). A competitividade da indústria de móveis do Brasil: situação atual e perspectivas, *BNDES Setorial*, 37 (mar.), pp. 227–272.

Garcia, A. de A. (2005). *Estudo de mercado e pesquisa de benchmarking para o setor moveleiro DF*. Brasília- DF: SEBRAE.

Gertsakis, J., Lewis, H., Sweatman, A., Grant, T. and Morelli, N. (2001). *Design + Environment: A Global Guide to Designing Greener Goods*. Sheffield: Greenleaf.

Ibá (ed.) (2015, Ago). Cenários Ibá: Estatísticas da indústria Brasileira de Árvores. Renata Coltro.

IBGE (2015). Cidades. Available from: http://cidades.ibge.gov.br/xtras/perfil.php?lang=&codmun=317020 [Accessed 22 September 2015].

Manzini, E. and Vezzoli, C. (2008). *Desenvolvimento de produtos sustentáveis: os requisitos ambientais dos produtos industriais*. São Paulo: EDUSP.

Michaelis (2009). Michaelis: moderno dicionário da língua portuguesa, *online*. São Paulo: Companhia Melhoramentos.

Movergs (2014). Relatório setorial 2014 polo moveleiro do Rio Grande do Sul. IMEI.

Nunes, V. G. A. (2013) Design pilot project as a boundary object: a strategy to foster sustainable design policies for Brazilian MSEs. (PhD Thesis). Polytechnic of Milan, Milan.

Rampersad, H. K. and El-Homsi, A. (eds.) (2007). Tps-Lean Six Sigma: Linking Human Capital to Lean Six Sigma – *A New Blueprint for Creating High Performance Companies*. Charlotte, N.C: Information Age Publishing.

Sebrae (ed.) (2013). *Anuário do Trabalho na micro e pequena empresa*. Brasília- DF: DIEESE.

Tischner, U., Schmincke, E., Rubik, F., Prösler, M., Dietz, B., Maßelter, S. and Hirschl, B. (2002). *How to do ecodesign? a guide for environmentally and economically sound design*, Agency, G. F. E. (ed.). 1 edition. Basel: Birkhäuser.

Non-object; designing a conceptual model for the design process

José Silveira Dias
Research Centre for Architecture, Urbanism and Design, Faculty of Architecture, University of Lisbon, Lisbon, Portugal

ABSTRACT: Since the 1990s we are seeing a global change in society. From the acceleration of this change and the erosion that its perception provokes, it is noticeable to search for new ethical and moral values. The artistic communities reflect and create strategies about their idiosyncrasies and interactions, developing a responsible experience where the human being responds to his own actions through processes for the resignification of contents.

The transformation of viewers into producers makes them aware of their strength as agents of change, coupled with the appropriation of the strategies of the media, the global market, and the entertainment industry. Understanding the creative experience can significantly change the design process and design behaviour.

The Participatory Design creates new processes to stimulate creativity through participatory practice in the creation of objects and the humanization builds knowledge among individuals responding to their needs. The intention (design) precedes the creation of the object not by lack but by the desire to respond to needs through the process.

The space between the subject and the object is where the experiences are lived and provide emotional and cognitive relevance making its retroactive effect felt for a new ethical meaning of knowledge through a conceptual model in a research Design context: the Non-Object.

Keywords: Non-object; Design Process; Participatory Design; Aesthetic; Self

1 INTRODUCTION

In the last two decades, there has been a global change in society, economy, politics, and culture. The perception of change is a constant but aggravated by increasing acceleration. Currently, the telematics network[1] connects all human beings in a global and real time. The material and immaterial production is transformed faster and the culture is produced and consumed by anyone. Every human being can be both a producer and a consumer.

It is imperative to develop an experience where the human being can acquire, produce and disseminate information that responds to his actions or others in a knowledge society that demands processes that give significance, meaning and value to the contents produced through the mediation of Electronic devices and social media, in order to hold the author or authors involved accountable.

In this paper, we approach the ambition of design for the integrity of the human being in detriment of the functional technicity that was proposed in the genesis of the industrial era, developing a critical thinking for the conceptualization of a model for the process in design.

2 TECHNOLOGICAL REVOLUTIONS [IDENTIFICATION]

In the first Technological Revolution, the hand is replaced by the tool. A new form of human existence takes form. The human being, when surrounded by tools, ceases to be the centre of the world.

In prehistory, the human being used only the hand and produced anywhere. The human being is the constant, irreplaceable, while the tool is the variable, because fungible. In the first Technological Revolution, the human being is expelled from nature, becoming protected and subjugated by the culture.

In the second Technological Revolution, which occurred just over two hundred years ago, the tool is replaced by the machine. The machines are more effective, faster and more expensive than the tools. In this revolution, the machine is the constant and the human being becomes the variable element. As long as the

[1] Telematics Network is the set of information and communication technologies resulting from the combination of telecommunications and information technology that enables the processing, compression, storage and communication of large amounts of data (text, image and sound) within a short period of time and between users anywhere in the world.

machine remains, man is replaced. The human being is expelled from his own culture.

In the third Technological Revolution that began in the mid-1940s, machines designed and produced according to scientific theories, especially of physics and chemistry, give way to electronic devices that also encompass applications, theories and hypotheses of neurophysiology and biology (Flusser, 2015: 37–38).

In fact, the hand is emulated empirically by the tools, then mechanically by the machines, and then neurophysiologically by the devices. The inheritance of simulated and increasingly perfected genetic information also becomes an object.

The Palaeolithic man seized nature with his hands in order to appropriate things to transform them. It did not need a particular place to manufacture and produce anywhere. When the human being begins to use the tools, he needs to delimit spaces for production, being in the centre of production (factory), surrounded by his tools, which, in turn, are surrounded by nature. With the invention of the machine, and attentive to its longevity, it acquires greater value than the human being itself, becoming the centre of production. Machines absorb nature and the human being. The machines are grouped in industrial complexes linked together, forming links of a circulation network and creating the industrial parks that extend throughout the world, developing the structure of the industrial architecture of the XIX and XX centuries. With electronic devices, this structure changes completely, not only because the devices are more adaptable to use, smaller and cheaper than the machines, but also because they are inseparable as extensions of the human being.

While in the Palaeolithic man grasped nature with his hands and could fabricate at any time and place without any mediation, now equipped with tiny and even invisible electronic devices, man is always ready to fabricate something, anytime, anywhere. The machines become obsolete and the factories superfluous. Devices connect everything and everyone to an invisible network, anytime, anywhere, in real time.

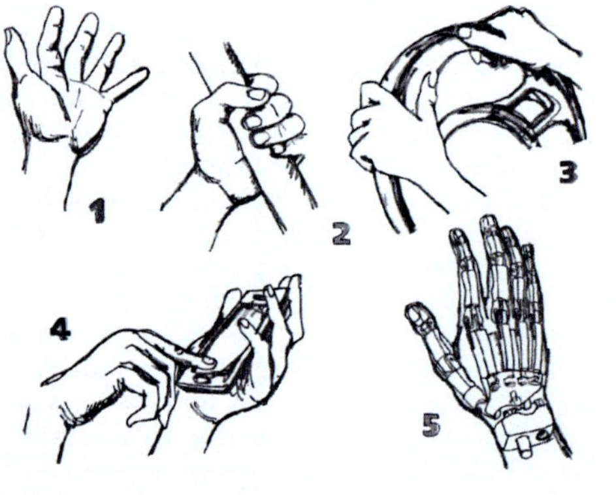

Figure 1. Hand's revolution: hand (1); hand-tool (2); hand-machine (3); hand-device (4); ex-hand (5), 2017, author's drawing.

The human being appropriates nature by transforming it into the culture through devices, which in turn promote the transformation of nature by replacing human beings themselves (Flusser, 2015: 39–42).

The human being submits voluntarily to his domain. The electronic device only does what the human being wants, but as the human being wants ever more, he creates increasingly autonomous and intelligent systems to produce other devices even more intelligent than the human being himself does. In this way, the domain of the electronic device is expanding. Even without wires, man is increasingly attached to devices, and wherever he goes or wherever he is, he takes them or is carried away by them (Fig. 1).

This generalised submission of the human being leads us to feel the need to counteract this "obedience", promoting the experimentation as a laboratory of learning and community teaching.

3 CONSTRUCTED SITUATION [EXPERIMENTATION]

The author's role is undergoing a major process of transformation.

The only creator who produced his work in isolation lost credibility.

Nowadays, anyone has the possibility to produce their own interpretations of everyday life and the need to share them immediately, provoking synergies on countless fronts.

New forms of cultural production of collective and participatory production appear. The internet globalises and with its interactive and participative potential feeds the democratisation of production and cultural consumption.

The notion of art production by communities is ancestral. However, today's collective art concept refers to the avant-garde movements of the last century such as Dadaism, Futurism, Constructivism, Bauhaus and Situationism. From the 1960s on, the rise of the critique of representation, from postmodernism through relational aesthetics to the present post-digital generation (Spampinato, 2015: 07).

There is a common desire to turn the spectators into producers by making them aware of their strength as agents of change. There are innumerable motivations, resources and goals that lead artists to share their experience, involving anonymous participants in these collective artistic practices. Approaching the strategies of the media, the global market and the entertainment industry, these communities are able to denounce and even collapse structures of social, political and cultural power. These collectives related to the visual arts and design often come from pre-existing communities of theatre, music, fashion and the media. They develop around the concept of participation in the relationship with visual art, with design and architecture. They practice a variety of multidisciplinary activities ranging from performance through advertising to urban art, to organising community projects and interventions, to curating temporary spaces, real or virtual, and experiencing new forms of cultural production.

The concept of community as a social unit, whose members share common values based on religion, tradition, location, or lifestyle, is a conducive interface for discussion and debate among its members, interconnected by the same cultural and historical heritage. In these artistic communities, they reflect on their idiosyncrasies and interactions, which guarantees them visibility and growth in relation to the outside world.

The collective art uses the repository of images as a tool of direct communication, of manifesto or protest. It draws on the recognised symbols and signs of pop culture and the media in order to highlight mechanisms of representation that reside in their origins.

The "constructed situation" advocated by Debord in his report on the construction of 1958 situations is at the origin of most of these groups. The situation is experienced by its constructors as a unit of temporal behaviour constituted by actions and events created and circumscribed in a given environment. These actions result from the environment and from them and produce other forms of the environment and other actions. The spectator is led to intervene in the situation deliberately constructed by the community in order to attribute new values to the street, the city and the world, through the construction of emotionally intense environments and temporary events.

Some of these communities move according to a political agenda, developing a close relationship with the economy, simulating their language, transposing symbolism and identity. The relationship between art and economics defines an "imaginary economy" by appropriating the techniques and strategies of Branding[2] in its cultural actions (Spampinato, 2015: 10). Some communities use this strategy to express themselves about the current dissimulations that sustain the policies and means of social communications in force. The continuous growth of collective art and design is a reflection of the instability of the political, economic and social structures of the last decades, and is the result of the democratisation of art by the mass diffusion of technology that allows anyone to become a producer. This phenomenon is revealing not only of a redefinition of the author's role in society but also of the paradigm shift in understanding the generalised and expanded application of art. Thus, free from the search for contemplation in the visual arts and functionality in design, any image, event or object that allows us to reflect on the role that images, events and objects play in life begins to be defined by art and forces us to inquire about Who produces them and for what purpose, also involving us in the process. Whether it is a publication, a performance, or an installation, we need these collective productions as critical simulations of realities that seem close to us. Anonymous ones with transdisciplinary or even self-taught formation, who experience democratic and participatory dynamics (Spampinato, 2015: 13), often implement these simulations. These dynamics can be applied in other areas and appropriated by other communities willing to learn this interaction - it is the behaviour of each individual as a stimulus to the other as a phenomenon that allows the constitution of communities by this reciprocal influence.

The human being must relate to the other to reveal his true essence through the constructed situation and thus create experimentation interfaces to develop responsible content that adds value and meaning.

4 SPACE BETWEEN THE SUBJECT AND THE OBJECT [REFLECTION]

Assuming that a subject relates to an object, whatever its nature, the distinction between subject and object is real and apparent. In the realm of knowledge, the real distinction expresses the split of the human condition (Arendt, 2001), something that arises inevitably through the forms of life that man imposes on himself to survive. This separation cannot be hypostasized nor invariant because it is apparent (Adorno, 2005). We cannot do without analysing these concepts separately, but this separation exists to the extent that both are reciprocally mediated. This opposition between the subject and the object is the domain of the theory of knowledge. Radically separated from the object, the subject reduces it to itself. The subject absorbs the object by rejecting how much he himself is an object.

The ideal conception of the subject is, at the outset, unsuccessful as something objective as the being-in-itself, which precisely the subject is not. According to the dimension of being, the subject is condemned to be nothing. According to Adorno (2005), the subject is all the more the less he is, and the less he believes himself to be, the more he deludes himself into being objective for himself. As a moment, it is inexhaustible. By eliminating the subjective moment, the object diffuses diffusely and likewise the subjective life.

The object also does not exist without the subject. If the subject does not appear as the moment of the object, objectivity loses meaning. When the subject thinks about the object, he/she must print objectivity, so that it, as its extension, responds to its needs.

Since the object can be an extension of countless subjects, it is the renter of subjectivity. Through subjectivity, freedom and autonomy can be achieved in the subject/object relationship. In this way, as the object is responsible for the expansion (mental and material) of the subject who experiences it, the mediated rational construction returns to the subject to consolidate and

[2] Branding is a marketing approach to marketing products and services under a specific name – a brand. As a strategic management process between suppliers and consumers in their construction, a successful brand is one that strikes the balance between the "profit" of the organization and the "profit" of the consumer. A brand can be a product, a service or even a person. Anything that can be bought and sold as an idea or product can be subject to a Branding process.

increase the mind. Spirit expands because of experience, rationalises, transporting objectivity back to the object. It adds new meaning and in a circular movement (Hegel, 1992), this interaction is enriched the greater freedom there is in the experimentation granted to the subject for the concretization of the object.

Currently, developing the object by the subject is a static and non-evolutionary process. One does not give time to reflect on the process because the acceleration of the succession of things does not allow the delay for its mental processing. In this way, objectivity based on the object and subjectivity on the subject is considered. In the inversion of the taking of these positions, this pretension is not a crystallising process, because there is an effect of increment, value and meaning if the creative subject is permeable to the experience in the production of the object that is developed through experimentation.

As such, one feels the need for the creation of the non-object to define the interaction between the subject and the object, as a free form approach, without formal or other prejudices, reflecting on the plurality of their functions and in the forms, that they can give way to new meanings because of guiding strategies of the process.

The non-object is not intended to denote the negation of the object as an object with features contrary to the recognised objects, much less is it to counterpose the material object to the immaterial object.

The non-object will be another process where it is intended to realise the synthesis of material and mental experiences, as a body inductor for the phenomenological knowledge, integrally perceptible. It incorporates the meaning of the work or lived experience. The non-object is revealed in its construction, transformation or modelling, it departs from the conventional and adopts an aesthetic autonomy whose origin is linked to its authors.

The non-object denounces the representation of other objects of the world that lose meaning at first contact. The problem discusses the meaning imposed by the form and function that are inherent to it. The form imposed by its representation to the perception gives resignification to its function and its integration with space where it is. It will then be necessary to move from the representation of the object to the presentation and experimentation in order to expand the sensorial possibilities, investing against the conventional aesthetic of contemplation and taking into account the involvement of all the senses of human perception, and not only that of vision.

The realisation of the object in space can also lend to this space, through the appearance of the object, new meaning and transcendence, because by performing outside the limits of the conventional it recovers this need of absence of impediment as the fundamental intention of its appearance. The common object is the material thing as we know it, connected with everyday designations and uses. In this condition, the object exhausts itself in the reference of use and meaning, whereas the non-object does not happen because it does not fit into the useful condition and common designation. This approach stems from Gullar's (2008) Non-Object theory, which argues that there are products that reside beyond the object itself in the emotional connection with the user. In essence, the purpose of the process is not only the object, but also the space between the object and the subject, where the experiences are lived and can provide an emotional and cognitive relevance to the user, that in this way makes the cognitive return in the consolidation of the ethics of knowledge through experience with the non-object.

The non-object opens new paths, allows going beyond the obvious, creating the involvement for later redetermination. By revealing the displaced object of its function, it is possible to establish new relations between this and other objects. The non-object resulting from this process passes through the transformation of the object so that it merges less in its formal qualities than in its meaning, and in its relations of daily use and habit. The non-object takes place outside the formal and functional conventions, reaffirming the creative attitude, triggering the mechanisms of experimentation and discovery.

It will be in the relationship established with the fantasy of the subject that the object is defined. The same object can express different meanings for different subjects or for the same subject at different times. The affective charge of the subject lends meaning to it, losing the condition of the simple object. Imbued with affection it becomes part of the subject, it is subject to the object.

The creative subject enters into a relationship with the non-object so that it participates in its creation, and this, transcending itself, reports the subject to the fullness of being (Clark, 1980: 21). The subject finds his precariousness, discovers the character of the ephemeral as opposed to the perpetuity through this relation. Space belongs to time continually metamorphosed by action and thus subject-object is identified in the act. According to Clark (1980: 27), we exist before the after, but the latter anticipates the act. What results from the act is embodied in the perception of absolute time. The instant of the act is not renewable. It exists by itself. The act has no perception of the past. It is just another time. The moment it unfolds, it becomes the thing itself. The awareness of the moment is already being in the past. The perception of the act is the future of doing. The past and the future are inevitably implied in the present, the now of the act.

This reverses the proposition of the human being whose work is increasingly automated, losing the expressiveness he once had when he was an artisan and dialogued with his work. In addition to the manipulation and explorative participation of the discovery, it is sought that the work is completed in itself and impel the liberation of the author. It will be at the instant that the subject practices the act that he will simultaneously perceive the meaning of his own action. The process identifies the possibilities of action, allows the choice, the unpredictable, the transformation of the abstract into a concrete enterprise.

Although the current insistence on individual reference or individualization of references, it is necessary to pay attention to the factor of singularity (of objects, groups, belongings, places), counteracting with operations that establish relations of acceleration and relocation, such as homogenization or the globalization of culture (Augé, 2012: 39).

If the loss of individuality is imposed on man by merging in the masses, the non-object proposes in this way the liberation of the subject as an author, extolling the singularity through the process and granting it new expression and signification.

5 EXPROPRIATION OF THE SELF [RESIGNIFICATION]

As a process of manipulation through data, Big Data[3] transforms the action into a transaction that takes place without autonomy or dramaturgy on the part of the subject. Actions become automatic through data and information when operations are only commercial, submitted to calculation and control (Han, 2016: 20). Information lacks the interiority that characterises knowledge. While knowledge exists in a permanent tension between the past and the future, information persists without time, without duration, the result of undifferentiated moments of the present, without events and without direction.

Sharing, connecting, and "real-time instant access" destroy the fantasy of appropriation, ownership, and power. We live increasingly surrounded by digital devices that allow the autonomy of the objects that host them, withdrawing the possibility of acting and becoming mere interfaces in the global communication network (Han, 2016: 26).

The digital media is artificial, without authenticity. The images are faster than the feelings, they just affect and do not fondle. The affectation does not recognise the delay of the contemplation nor the necessary susceptibility to the revelation. It only provokes anxiety, excitement and stimulation to the moment (Han, 2016: 51).

It eradicates the aesthetics of the event. As stated by Han (2016: 55), all events are beautiful because they expropriate the self.

In the aesthetics of the event, the subject loses possession of himself, he removes his interiority, loses his subjectivity, and happiness seizes him.

Beauty is an aesthetic form. It is not a stimulus. In the present aesthetic regime, numerous stimuli are produced. Between stimuli and excitations to which we are constantly exposed, the beauty disappears because no contemplative delay in relation to the object is possible and relegated to consumption. (Han, 2016: 59)

In fact, it is understood that the non-object is beautiful because it is something with which the subject establishes a free relation, in which there is no form of dependence and coercion. The beauty remains free of any extrinsic determination of use. The subject renounces his interest in it as a concept or purpose, suppresses its ends in relation to the object and works it as if it were autonomous in itself. The subject before the beauty takes disinterestedly, without having to possess or use it. In the presence of the beauty, the separation between subject and object also disappears. The subject plunges contemplatively into the object, joins the object. The self-liberates itself because it reconciles itself with the object.

The beauty does not advertise for itself. It cannot be appropriate, nor enjoy, only contemplated, inviting the delay and duration.

In the aesthetics of Hegel (2001: 128), the beauty promises freedom and reconciliation, does not sympathise with consumption, although art is currently subject to the speculation of capitalism (Han, 2016: 70). The acceleration and density of the present communication also do not allow the possibility of contemplation. The beauty has its own time and space, where it stands and arouses stimuli and tension to the senses.

Faced with an object of consumption, the subject is compelled to assume a central, individualistic position. Consumption does not allow alterity, it destroys the heterogeneity of the different. Transcendence, meaning, and value are lost beyond the aesthetic (Han, 2016: 77). The beauty detaches itself from the ethical-moral judgment. In the experience of the beauty, the subject expropriates himself. The subject withdraws from the centre and gives the place to the other. This aesthetic experience benefits the other as an ethical act – the resignification is established.

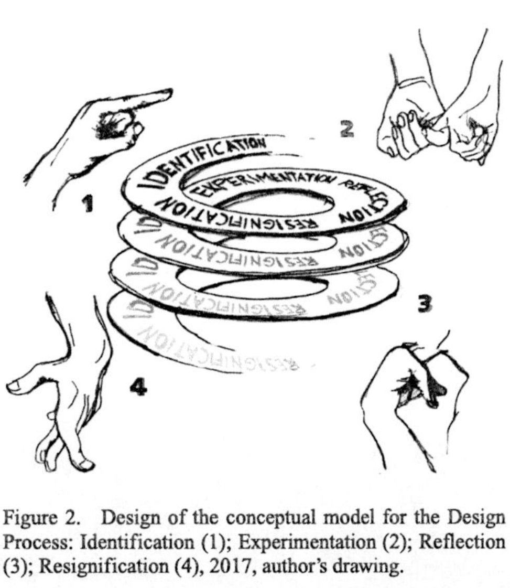

Figure 2. Design of the conceptual model for the Design Process: Identification (1); Experimentation (2); Reflection (3); Resignification (4), 2017, author's drawing.

[3] Big Data describes data sets of great size or complexity that traditional data-processing applications cannot work with. Challenges in this area include analysis, capture, data protection, search, sharing, storage, transfer, viewing and data privacy information. Big Data refers to the application of predictive analysis and methods for more rigorous data in order to be able to make resolutions with greater confidence.

6 CONCLUSION

Through the development of a responsible experience, the human being will be able to work the information responding by his or her actions or of another in a knowledge society whose processes confers meaning, sense and value to the contents produced through the mediation of electronic devices and the media.

Collective productions, as critical simulations of realities that seem close to us, are exploratory interfaces for these democratic and participatory dynamics, which, when recognised, can be applied in other areas and appropriated by other communities willing to learn this interaction for responsible content development interfaces.

By taking advantage of personal experiences in interacting with each other, the design process and design behaviour change significantly. Design ceases to be only for and for consumption, applying itself through the learning of these experiences. Participatory design as a collaboration of all through design draws new methods to stimulate the creative process and experience through participatory practice in the creation of objects as acquired knowledge.

It is believed that design, by continually learning to apply its own creativity in order to stimulate creativity, designs the production process with a view to its application in other areas such as education, health, economics and politics.

The non-objects conceptualization comes to draw the interaction between the subject and the object, as a free approach, without formal or other prejudices, reflecting on the plurality of its functions and in the forms that they can have, always giving place to other or new meanings.

If the loss of individuality is imposed on the human being by mass production, the non-object proposes the liberation of the subject as an author, extolling the singularity through the process and granting it new expression and signification.

In the aesthetics of the event, the subject loses possession of himself, is torn from his interiority, loses his subjectivity, and happiness seizes him.

As in the experience of the beauty, the subject expropriates himself. The subject withdraws from the centre and gives the place to the other. This aesthetic experience benefits the other as an ethical act. This act has an aesthetic dimension, in that it gives duration (the time of the delay to the feeling) to the transformation of the prescriptive matter in cognitive form of the knowledge, in an upward circular movement that designs a model in 4 times: identification, experimentation, reflection and resignification of the subject (Fig. 2). In a context of design research, Non-Object is defended as a conceptual model for the design process.

BIBLIOGRAPHICAL REFERENCES

Adorno, T. 2005. Critical Models: Interventions and Catchwords. New York: Columbia University Press. ISBN: 978-0231135054.

Arendt, H. 2001. A Condição Humana. Lisboa: AntropoAS. ISBN; 972-708-637-3.

Augé, M. 2012. Não-Lugares. Introdução a uma Antropologia da Sobremodernidade. Lisboa: Letra Livre. ISBN: 978-989-8268-14-3.

Clark, L. et al. 1980. Lygia Clark. Rio de Janeiro: Funarte.

Debord, G. 1958. Questões preliminares à construção duma situação. Internationale Situationniste n.° I. http://guy-debord.blogspot.pt/2009/06/questoes-preliminares-construcao-duma.html – accessed in 23 July 2016.

Flusser, V. 2015. O Mundo Codificado. 5ª edição. S. Paulo: Cosac Naify. ISBN: 978-85-7503-593-1.

—. 2010. Uma Filosofia do Design. A Forma das Coisas. Lisboa: Relógio d'Água. ISBN: 978-989-641-036-0.

Gullar, F. 2008 Teoria do Não Objeto. Deposit. https://poars1982.wordpress.com/2008/02/28/teoria-do-nao-objeto-ferreira-gullar/accessed in 23 July 2016.

Han, B. 2016. A Salvação do Belo. Lisboa: Relógio D'Agua. ISBN: 978-989-641-617-1.

Hegel, G. 2001. Cursos de Estética. Vol.1. 2ª edição. S. Paulo: Edusp. ISBN: 85-314-0467-3.

—. 1992. Fenomenologia do Espírito. 2ª Edição. Petrópolis: Vozes. ISBN: 85.326.0687-3.

Spampinato, F. 2015. Come Together. The Rise of Cooperative Art and Design. New York: Princeton Architectural Press. ISBN: 978-1-61689-268-5.

Knit and technology: A long lasting friendship

Gianni Montagna
University of Lisbon, Faculty of Architecture, CIAUD, Lisbon, Portugal

Luís Santos
University of Lisbon, Faculty of Architecture, Lisbon, Portugal

ABSTRACT: Knitted garments are nowadays an unavoidable product for users of any age, covering any type of use. The use of knitwear is increasingly present in all types of clothing, be it classic or sporty, and gains more and more space in the user's wardrobes thanks to its multi-level comfort features.

The knitted material is a complex textile surface that brings together in its making, the technical capacity of the one who produces it, the characteristics of the raw material used and the capacity to perceive the best result that may better serve its intended needs.

Knitwear has been part of many world cultures due to its ability to create constantly new surface structures and its great adaptability to the users' body. Even today, knitwear is being massively used and reinterpreted in its intrinsic relationship with casual dress, active wear and sportswear. Its close relationship with technology makes knit an ever-innovating fabric that can answer the present day productive and usage challenges.

The aim of this paper is to propose an integrated reading of the different elements that compose this textile material, and the reasons that drive its use, through the opposition of different authors who, as a whole, can offer a more careful reading of this material that always had the capacity to reinvent itself over time.

Keywords: knitwear – knit design – knit technology – Material culture

1 INTRODUCTION

Knitwear is part of the everyday wardrobe of all people of all ages and status. To talk about knitting is to touch different concepts and variables that, working together and mixing themselves constantly, offer multiple products and structures that are constantly being reinvented.

Among the different characteristics that made knit one of the most used elements for the production of clothing, its ability to constantly create new surface structures and its great adaptability to the users' body are probably the ones that contributed most to its mass use.

Made by hand or using a knitting machine, with craft techniques or industrial ones, knit retains its unique characteristics of adaptability, which are very different from other types of textiles used for clothing. Reinventing the knit and its structures are a constant over time and seasons, in a persistent search for tactile and visual sensations that have always accompanied this type of material and clothing.

There is a natural interplay between craft, design and new technology, refers Juliana Sissons (2010). Knit gives the designer the possibility to work on the textile structure and at the same time to work on the garment form and function. Knit also generates a great array of experimental possibilities to the designer, allowing him to discover and work with different variables, which are able to create new possibilities and aesthetic combinations. In this way, the designer can choose, together with the previously variables, a great combination of stitches for a specific project and a defined colour palette in order to personalise his work.

Manual production is a slow process that cannot meet all user needs in terms of lead times, new product creation, and production costs, presenting the product uncompetitive in most markets. This is why the role of technology in the development of knit structures, and new knit products in general, has always been central to its growth.

2 STATE OF THE ART

Looping techniques date back to the 7th millennium BC and remained present in the human history until today. They started to be used because of their lightweight and flexible construction, characteristics

Figure 1. A child's wool sock from Oxyrhynchus, Egypt, made in the 2nd century AD. In Children and Youth in History, Item #208, http://chnm.gmu.edu/cyh/primary-sources/208 [Accessed 23 April 2017].

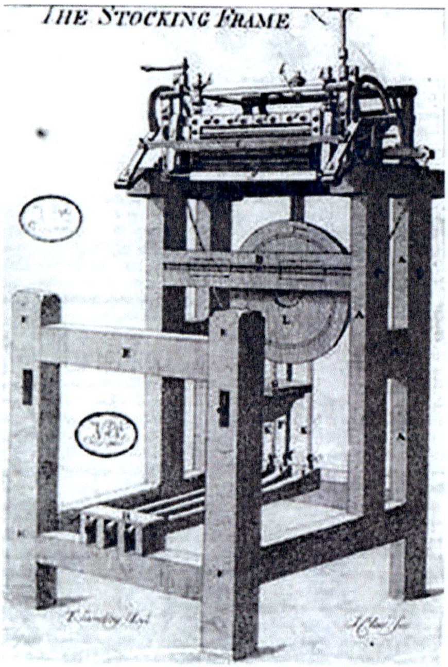

Figure 2. The stocking frame knitting machine — the first knitting machine —, invented by William Lee in 1589, for the production of stockings. In http://www. sussexvt.k12.de.us/science/The%20History%20of%20the%20 World%201500-1899/William%20Lee.htm [Accessed 23 April 2017].

that were maintained over the centuries, reinforcing the importance and use of these textile techniques (Schoeser, 2003).

As an ancient craft with its origins dating back to 1000BC, knitting originated from hand knotting or twisting of yarns using fingers. The word 'knit' traces back to the old English term 'cnyttan', and the German 'knütten', which developed to knot. Hand knitting rapidly advanced from fingers, to hand tools – known as pins –, into a skilled and complex craft where, by the fifth century AD, knitted objects such as socks are recorded, combining fashioning, seaming, and circular knitting together with patterning techniques simultaneously (Steed, 2016).

Cross loop knitting (or cross-knit looping) – a single-needle technique – dates back to the 2nd century AD, being found in civilisations like the Egyptians, Romans and Scandinavians (Fig. 1). Pattern knitting, known in Arabia and India from the 9th century, is introduced in the 13th century in Iberia and southern Italy (Schoeser, 2003).

As the most of these handmade techniques were expensive and time-consuming, William Lee creates the first knitting machine in 1589 (Fig. 2), as a way to produce in England, a bigger quantity of silk stockings that, at that time, were imported from abroad due to an increase of consumption of this product. In that epoch, the production of socks was handmade and the fashion of shaped socks – dictated by Spain – in England, was an important part of the economy in all Europe. Knitting was very important at that time as a social economy, helping thousands of families to gain an extra income – using a simple and inexpensive equipment –, spinning yarns by hand and at home, to make wool or other fibre. In fact, during the 17th-century, England exported handmade wool stocking through Germany, France, Italy, and Holland (Steele 2005; Schoeser, 2003)

As said by Schoeser (2003) and Steele (2005), the development of knitting has been responsible for the most striking and fundamental changes in textiles and clothing since 1600.

This deep connection with the consumer is maintained as one of the strong abilities of the knitted textile over the years, and the use of knitted textiles has been increasing year by year since the seventies, due to the introduction of specific garments for sports activities.

Home-knitting machines that had already been introduced to the international market by the Japanese in the 1950s provide the knit market with benefit by the development of electronics and the introduction of computerised machines. In the 1980s, home-knitting machines lessened the time required to produce tailored patterns. These machines, which used computer-controlled patterning, counteracted standardisation (Schoeser, 2003), permitting a high standard quality in home making knits.

With the new millennium, the production of state-of-the-art industrial machines allowed the production of sophisticated knit textile substrates, allowing the creation of very complex structures, different gauges and the introduction of multiple different yarns on the same substrate. In addition to this, machines have become capable of producing whole pieces with minimal waste. This rapid prototyping approach meets the needs of consumers in terms of customization, immediacy, sustainability and respect for the environment.

According to Sinclair (2015), knitwear will have a massive growth in the fashion industry, related to its intrinsic relationship with casual dress, active wear and sportswear, crescent trends in our society. Recent statistics of the European Apparel and Textile Confederation (Euratex) show that in 2015 more than 1/3 of the fashion market was made of knit garments.

3 KNITTING AND CULTURE

As a highly human-related textile surface, knitting evolves with culture.

According to Quinn (2010):

> A textile's surface is more than just a façade — it's a curious layer of aesthetics and identities and a contentious site of exploration and resistance. (...) Surfaces are also manifestations of social norms and moral codes, (...) [and] have perceptual qualities, their mutability enabling the body to take on chameleon-like characteristics. Because surfaces are interpreted by sight and touch, physical contact with the surface initiates a complex multisensory, emotional and cognitive experience that provides a uniquely individual interpretation of the world.

In the following topics, it is sought to present the connection points of the actual cultural environment and the ancestral process of knitting.

3.1 *Connection with the true self*

The crescent virtualization of our world and the growth of impersonal cities where our life is anchored in seamless daily routines take the individual to feel displaced and disconnected from the reality.

Understanding his lost humanity, the individual seeks to reconnect with the world, with his community and with his history. According to Steed (2016), knitting has this connecting ability, restoring the "sense of self and locality by better connecting people to places and history".

Because of this, knitting, an accessible, portable, adaptable and simple craft, is experiencing a revival interest. It is thus, embraced as a personal self-discovery experience where the making process is an integral part of it.

Knitting is rising as a collaborative experience. Groups of people get together (physically or digitally) to share experiences and knowledge. These groups are not motivated by commercial practice but in personal satisfaction, community values, and intrinsic gratification experienced in the act of 'making'. (Steed, 2016)

One example of these knitting communities is 'Ravelry' that "connects knitters and crocheters with each other" and where they can communicate their experiences with materials and patterns (Tricksy, 2015).

3.2 *Sustainability*

In a world where the main resources are decreasing on a daily basis, hand-in-hand with a crescent intoxication of the environment, Udale (2008) suggests that:

> Developments in the creation of textiles seem to be following two paths – ethically driven by the environment and future-technologies driven by scientific advances.

Knitting has the ability to respond ethically to apparel production, talking back to a more demanding, more informed consumer. Brands like Benetton, known for its iconic knit designs, are rethinking their production strategy in order to accommodate more sustainable solutions. Their 'Wholegarment' technology, capable of constructing a complete, seamless garment, eliminates yarn wastage and has a reduced environmental impact (Knitting Industry, 2016).

This holistic approach to design, found in knitting, allows the designer to create a product directly from raw materials. Therefore, knitting can be more than a method of production, acting as "a process of investigation", exploring the design problem as a whole. (Steed, 2016)

Another example is Kniterate (2017), an affordable and compact digital knitting machine that allows turning digital designs into knitted garments ready to be worn, where there is almost no waste involved, because "the product is knitted to shape and you can reuse the yarn". Re-knitting strategy – as seen in Amy Twigger Holroyd's work – is another way to enhance the potential of knitting as a strategy for sustainability (Steed, 2016).

3.3 *Social awareness*

As a craft firmly rooted within society, knitting has always been a method for "expressing oral history, facilitating community engagement and expressing deep personal attachments". In a society where the individual seeks to be heard, knitting emerges as a powerful social and political tool (Steed, 2016).

The 'Knit Graffiti' movement in 2005 is an example how knitting can act as a mean of communication from the self to the community. In this initiative, "'guerrilla' knitters anonymously 'tag' street lamps, public and private property" (Steed, 2016).

Another example of social connectivity enhanced by knitting is the News Knitter (2007), presented by Seymour (2008) as:

> A data visualisation project, which focuses on knitted garments as an alternative medium for displaying large-scale data. The system converts information, gathered from live RSS news feeds, and incorporates it into a knitted sweater".

3.4 *Responsivity*

We live surrounded by textiles from birth to death. They embody our second skin, which we use for

Figure 3. The Oricalco shape-memory shirt uses a Nitinol (Nikel and Titanium) memory metal that can be manually distorted, then regain its shape when heated. In http://advancedtextilessource.com/2016/01/22/fashion-grounded-space-inspired/ [Accessed 21 April 2017].

protection, ornament and self-signification. The close relationship that we have with clothes raises our expectations on what they can do for us, regarding our daily life's necessities.

Sportswear, with its relaxed and easy to wear garments, is gradually influencing our everyday clothing, which we require to be practical, due to our overactive life. Knitting, because of its adaptive abilities, has always been used in active wear and sportswear. In the future, we will see a growing migration of high performance knit textile surfaces – used on both of these areas – to everyday life. Daily life will embrace clothes that use new varieties of micro fibres and nanotechnologies, able to generate ultra-lightweight, breathable, quick drying and supportive garments that are able to offer UV protection and excellent fit to body contours. (Sinclair, 2015)

Future knitting will surely evolve to connect even deeper with the user, providing high levels of responsivity to his demands.

We can acknowledge this from projects like Space-Bra, a wearable vital-functions monitoring system that can detect changes in body form which may indicate a potential health problem (medical), monitors heart rate (sports), and can change shape to provide varying levels of support (fashion). Another interesting project is the Oricaloco shirt, whose fabric used for the sleeves can be programmed to shorten immediately as the room temperature becomes a few degrees hotter (Fig. 3) (Seymour, 2008).

3.5 *Customization*

Consumers are tired of the standardisation imposed by the fast fashion phenomenon and will be looking for customised alternatives. According to Sinclair (2015), individually designed garments that are tailored to specific consumer needs will be at the forefront of fashion in a near future.

'Rapid Prototyping' is used to describe a range of technologies created to make quick prototypes of products in the early stages of development. These methods evolved and now allow the direct manufacture of consumer products. The generalisation of these technologies is making them widely available for designers and users, promoting a change in the consumer market.

Knitting has the ability to transform raw material into three-dimensional forms, suggesting close synergies with complex emergent technologies, such as 3D printing (Steed, 2016). This makes knitting a promising area of investigation, in the design and production of customised apparel.

In order to potentiate customization, knitting can be approached in the following three ways.

3.5.1 *Adaptation*

In order to suit the individual needs, the final product has to adapt to each consumer, during the use phase. To fully achieve this, designers need to create products that interact with the consumers, offering multiple permutations of use. This kind of approach is only possible when multiple factor combinations can be built into the product (Gilmore et al., 1997).

One example of "adaptation" in knit design is Issey Miyake's 'A-POC (A Piece of Cloth) collection', which utilises knitting technology to produce a tubular fabric with integrated garment shapes that can be modified by the wearer to create customised body pieces (Fig. 4). In addition to this, Miyake created a retail laboratory environment, engaging the wearer as a co-designer (Steed, 2016).

3.5.2 *Infrastructuring*

For Dantec and DiSalvo (2013), through infrastructure, users should be able to resolve current and future design situations on their own, with the capacities provided by the design team.

This approach seeks to empower the user with the tools to create their own designs, responding to inherent individual needs.

Rapid Prototyping technologies are becoming widely available, not only for designers but also for users, shortening the gap between design and production.

One infrastructure example is 'Kniterate', an affordable and compact digital knitting machine (Fig. 5), that easily allows the user to turn digital designs into knitted garments ready to be worn. Most of the process is made automatically, allowing the operator to focus on the creative work (Kniterate, 2017).

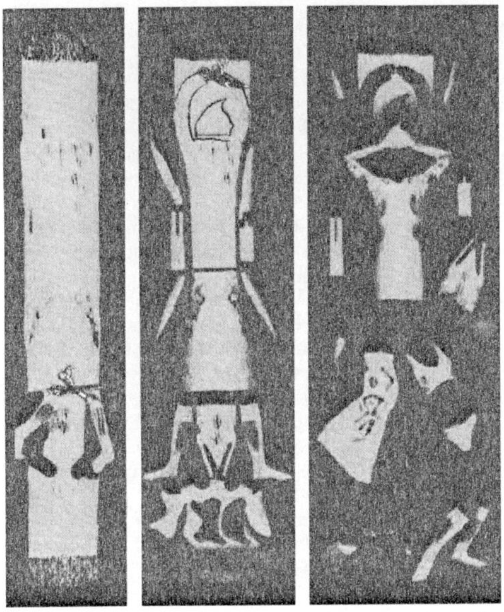

Figure 4. Issey Miyake's A-POC (A Piece of Cloth) collection. In http://www.wearehereatlast.com/2011/12/defile-mocap/ [Accessed 21 April 2017].

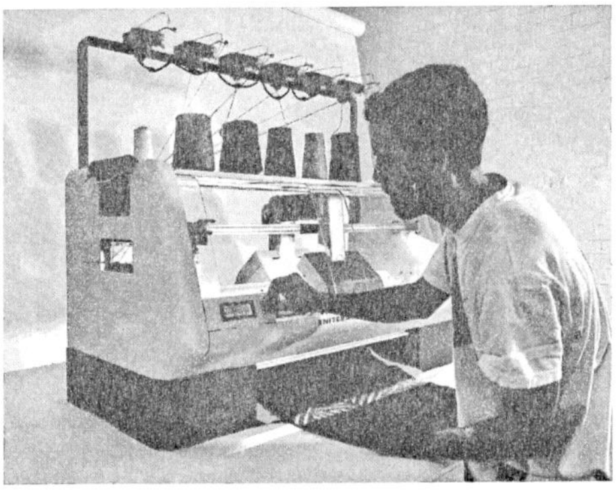

Figure 5. Kniterate, an affordable and compact digital knitting machine. In http://www.kniterate.com/index.php/about/ (Accessed 21 April 2017).

3.5.3 *Collaboration*

In response to the generalisation of mass production, there has been an increasing appreciation of 'slow clothes' and desirable, one-off garments that are more personal to the wearer. (Sissons, 2010) In order to attain more personal, customised clothes, brands are allowing the user to collaborate in the design process, by creating "in loco" production methods.

One way to achieve these goals is by merging innovative designs with old, traditional techniques. New ways of knitting production are being experimented with designers and brands, in the attempt to re-appropriate knitting technologies, "hacking" them (Steed, 2016).

We can see this new knitting experience in the project 'Trikoton' (2007), where the human voice is converted into a binary code for knitting patterns, produced by a knitting machine from the 70s that was hacked to become interactive (Seymour, 2008).

Another example is 'Unmade', a customised knitwear brand that uses standard industrial knitting machines with a special software, that makes them suitable for mass customisation (Hobson, 2015).

However "hacking" the old is not the only way of achieving user's collaboration. New flexible manufacturing techniques – for instance, 3D scanning and printing – allow a large amount of customization and are set to further revolutionise traditional methods of production, and have major implications for knitting in the future.

These technologies use a three-dimensional body scan to gather the user's precise body measurements. This data is then translated into a two-dimensional computer-aided design system, which is integrated with industrial knitting machines. The final product is a direct, three-dimensional, seam-free knitwear, with enhanced fit and customization for user requirements (Steed, 2016).

This 3D knitting technology allows shoppers to personalise items to their specifications, meeting the growing demand for customised products. Unlike cut-and-sew garments, which are assembled from flat pieces of fabric, a seamless 3D-knit garment is designed and produced in one piece and in three dimensions, matching the shape of the body. The knitting machine is able to create details such as articulated elbows and shoulders that allow a built-in range of movement. It also produces less waste than a traditional cut-and-sew garment, where leftover fabric scraps end up being discarded (Bain, 2017).

Bain (2017) identifies the use of these innovative technologies in brands like 'Adidas', who are already experimenting with "3D knitting on-demand" in stores. This allows customers to walk in, get a 3D body scan, and have a merino-wool sweater knitted to their size and colour specifications, washed, dried, and ready to go within four hours. Another example is 'Ministry of Supply', who has permanently installed a 3D-knitting machine at its flagship store in Boston. Aman Advani, a co-founder of the brand says that:

> This isn't a niche product, (...) this is step one of a longer route to a sustainable and strong production method that's here to stay. At this point we are pretty convinced that we can get one-quarter, maybe even one-third, of our production done in-store, or even to serve the online market within the next couple years.

4 CONCLUSIONS

Knitwear is an activity that has survived for centuries in the development of man, and continues to fill the one who develops it with a gratification, experienced in the act of 'making'. Knitting is a complex, manual and industrial activity, which fills different cognitive domains, interpersonal and social levels. For these reasons, knitting is used on many levels, from the commercial production of textile products to the act of therapeutic knitting.

In manual production, knitting transforms the produced object into a social object, which, as a form of individual expression, is shared publicly, admired and commented, raising that same object to the sphere of singularity and therefore to the field of artistic creation, becoming part of a cultural history and heritage.

The creative and productive activity of manual-crafting textile objects develops and helps the conservation of cognitive domains such as attention, memory and reasoning. In the production of this type of objects the development of perception and creativity, together with the handling of the material and the recognition of its plastic characteristics, are one of the most valuable expertise of designers and producers.

In social terms, knitting is nowadays also experimented in working groups that share their practice and, with the aim of helping each other, meet in cafes and public places to share this social experience.

With the development and democratisation offered by digital platforms, different ways of facilitating the creation and production of textile products in general, and knitwear in particular, are born. Even in industrial production, the process of creating knitted garments allows a 'holistic design methodology', by offering the possibility of interfering with the production of almost all the elements that belong to the object of study.

The connection between technology development on knitwear production and the creative possibilities – brought by designers –, has always been a close and highly productive relationship that continues to manifest its influences on the creation of the final product.

The intrinsic link between knitting technology and design is clearly seen in the knit textile products, which presents themselves as sophisticated objects, involving interdisciplinary working groups, which, from this point of view, create a product that requires continuous exchanges of knowledge and social relations, among the different participants of this same creative process.

BIBLIOGRAPHICAL REFERENCES

Bain, Marc (2017) Brands see the future of fashion in customized 3D-knitted garments produced while you wait [Online]. Available: https://qz.com/949026/brands-including-adidas-uniqlo-and-ministry-of-supply-see-the-future-of-fashion-in-on-demand-3d-knitting/ [Accessed 21 April 2017].

Dantec, C., DiSalvo, C. (2013) Infrastructuring and the Formation of Publics in Participatory Design. In: Social Studies of Science. vol. 43, no. 2, pp. 241–264

Euratex: The European Apparel and Textile Confederation "index" http://euratex.eu [Accessed 15 April 2017].

Gilmore, J.H., Pine II, B.J. (1997) The Four Faces of Mass Customization. In Harvard Business Review

Hobson, B. (2015) Mass customisation "can be the Future of Fashion", says Knyttan co-Founder [Online]. Available: www.dezeen.com/2015/02/18/movie-mass-customisation-future-fashion-knyttan-bespoke-knitwear-ben-alun-jones [Accessed 21 April 2017].

Kniterate (2017) Knitting machines made for everyone [Online]. Available: http://www.kniterate.com/index.php/about/ [Accessed 21 April 2017].

Knitting Industry (2016) Benetton goes back to its roots with seamfree knitwear project [Online]. Castrette, Treviso. Available: http://www.knittingindustry.com/benetton-goes-back-to-its-roots-with-seamfree-knitwear-project/ [Accessed 21 April 2017].

Quinn, B. (2010) Textile Futures: Fashion, Design and Technology. New York, Berg

Schoeser, Mary (2003) World Textiles: A Concise History. London, Thames and Hudson Ltd.

Seymour, Sabine (2008) Fashionable Technology: The Intersection of Design, Fashion, Science, and Technology. Vienna, Springer Wien New York

Sinclair, Rose (2015) Textiles and Fashion: Materials, Design and Technology. Cambridge, Woodhead Publishing Limited

Sissons, Juliana (2010) Knitwear. London, AVA Publishing Lda.

Steed, Josephine (2016) Hand Knitting in a Digital Era. In Kane, Faith; Nimkulrat, Nithikul; Walton, Kerry (eds.) Crafting Textiles in the Digital Age. London, Bloomsbury Academic.

Steele, Valerie. (2005) Encyclopedia of clothing and fashion. 1st Edition. Detroit; London: Thomson Gale.

Tricksy (2015) Ravelry Patterns [Online]. Available: http://www.tricksyknitter.com/ravelry-patterns [Accessed 21 April 2017].

Udale, Jenny (2008) Textiles and Fashion. London, AVA Publishing

Prisoners of progress? Women, body and fashion in the 19th century: A reflection on city, society and conspicuous consumption

Maria João Pereira Neto
University of Lisbon, Faculty of Architecture, CIAUD, Lisbon Portugal

ABSTRACT: Progress is the basis of the material and social transformations on a century of revolutions as the nineteenth century. It assumes its claims on the structures of social life and in the private sphere consequent to the industrial revolution, capitalism and the redefinition of codes of the modern life determinants of the nineteenth-century society.

The power of tradition will give place to new patterns of social and cultural relations with reflections on new practices promoted by the city and urban life. All these transformations are also reflected in the emergence of the fashion industry throughout the nineteenth century, with emphasis on women's body, rather than what happened during the eighteenth century in which men's clothes, prevailed as a form of representation and affirmation of power,

The bourgeois society of the nineteenth century affirms and sublimates status and social differentiation through the conspicuous consumption of objects and goods.

Fashion is a form of affirmation of power. The adorned and caged woman is a consequence of this kind of power and the city its main stage. This article aims to emphasise not only an approach to nineteenth-century society and emerging modernity but also in what will become one of the most flourishing industries of western societies: Fashion relating to women, main producers and consumers, its biggest victims.

Keywords: Progress, women, fashion, body, city

1 FRAMEWORK

Richard Sennett (1978) questioned the impact of industrial capitalism on public life and its changes on the nineteenth-century city' inhabitants personality, especially on the bourgeoisie and public goods. He emphasised the Marxist approach to commodity fetishism: an appearance of things expressing the buyer's personality, or how secularisation led to the introduction of personality in public as a form of belief in the immanent meaning of the world or even in the essence of private life: in the family.

According to Goffman (1983), the public man of the nineteenth century is an actor and also a spectator, and in this condition he assumes the pleasure, creating the discipline of silence – of order – and the absence of social interaction. Therefore, the characteristics of the collective personality that makes the individual personality triumph over social classes, stimulating the creation of what he calls the destructive communities or ghettos in which it is clear the logic of a legal personality in the purge, meaning its self-destruction.

According to this author and also Santos (1994), these destructive communities are expressed in the analysis of the Dreyfus affair and the Zola manifesto that shaken the nineteenth century in France. Narcissism will thus be the Protestant ethic of the modern world.

It is therefore assumed the gender distinction: the woman imprisoned/and "prisoner" of progress as an object of ostentation and submission, while the man is sober in wearing but dominant in society.

With this text, we intend to contribute to the debate about how attire and fashion can, not only inculcate values and social belief systems but also demonstrate aspirations and desires. In the mid-nineteenth century, men will fully assume their disdain for the excesses of the costumes of the pre-revolutionary era and dress mostly in black, in a properly sober way for the "serious" and austere trade and business world universe.

By contrast, their wives and daughters are adorned with flaming tissue and sometimes-bright colours and unique patterns, wrapped and literally caged in tight corsets and heavy and complex crinolines in uncomfortable costumes, lavishly adorned by impractical cufflinks, to determine and emphasise the social status of the families they represent.

It also arises the habit of shopping (Lysak, 2008) by the women who will dedicate their free time to keep the appearance of an emerging and powerful class or even to promote a new social order through conspicuous consumption of goods (Veblen, 1965). In the nineteenth century, Paris will become the Fashion and habits of the bourgeois dressing capital – a new way of life that soon will extend to other places in Europe and the US and to the cities of the great colonial empires.

According to Perrot (1994), this is what is termed "the levelling process" when the upper classes try to maintain their status and their rights to difference by promoting and maintaining what will be the emerging haute couture industry.

The nineteenth-century bourgeoisie in search of its affirmation revolutionised the world by violating the sumptuary laws, and thus affirmed the emergence of hedonism through the consumption of pleasures, goods, and fashion (Veblen, 1965, Bell, 1976).

As stated by Baudrillard (2005) the system of needs becomes less coherent than the system of objects that not only enforces consistency but also gains the power to manipulate a civilisation.

According to Breward (1995) technological innovations, the reorganisation of trade and industrialisation led to profound changes in the design and expression of fashion as a phenomenon and industry in the second half of the nineteenth century. The very meaning of modernity as a progress and future symbol will foster the spreading of taste for fashion and attire, assuming the change to social codes that are no longer based on the hierarchy of traditional powers but rather on the interest in consumption and expression of fashion.

Figure 1. Worth's look: 1875 cartoon by Bertall. Source: https://en.wikipedia.org/wiki/Charles_Frederick_Worth#/media/File:Bertall.

2 THE NINETEENTH-CENTURY WOMAN

We would like to emphasise the role of women, especially the bourgeois woman and her role as representative/showcase of the social status of the family. The literal woman imprisoned by a progress that confines her to the domestic universe and to a fashion that oppresses and shapes her body, but also the ordinary woman: the other women who contribute a lot to the progress as workers and manufacturers of a new industry that soon will move the world: Fashion.

From its relevance on the household and in management and organization activities of private life to the representation of social life and affirmation of a new power: the bourgeois society, with explicit translations in the various fields of art and architecture and the city, a society of excesses decades later so criticized by authors like Adolf Loos (2004), himself the fruit of an elitist and cosmopolitan society: Vienna.

The nineteenth-century cities and architecture are reflective of this. Paris, Vienna, Berlin, New York (see Olsen, 1986 and Charle, 2008) will undergo major changes that will make them assume as the privileged places for the social representations of the new emerging classes and a new social order that highlights ostentation. There is also a certain feminization of the aesthetic culture of the nineteenth century, which exemplifies not only some imagery but also the various generalist and fashion journals and other publications addressed to the female audience, as well as the importance of a conspicuous consumption addressed towards the feminine gender.

The phenomenon of entrepreneurship and strategic vision that was Charles Frederick Worth's (1825–1895) are also highlighted, as he was the man who promoted fashion to a global phenomenon, enhancing the beauty of women, while imprisoning and shaping their bodies like a demiurge, through artificial frames that shape female body, torturing them by determining a model of beauty that is contrary to nature itself.

The woman's entanglement with the corset and the crinoline assumes not only the taste for the modern, the frivolous but also an intricate form of power and control from the man-dominated society.

Until the 1860s, any woman adorned and walking unaccompanied on the streets was considered amoral. From then on, and especially after the revolution proposed by Worth, the use of fashionable attires and accessories is inseparable from the city and urban life, bringing women together, and also accentuating differences. Here we highlight the reference to coquetterie, the pleasure and the sexualisation of women through attire and fashion that leads to the game of seduction and claiming the conquest.

Authors such as Simmel (2014) theorized about the role of fashion and social convention of nineteenth-century women in their quest for a partner (struggle for survival?), highlighting the role of man, protector, seducer and seduced that through women, wife, lover,

Figure 2. Women and 19th-century fashion. Source: http://www.victorianlondon.org/clothing/crinoline.png.

daughter, intends to relieve his social status and assert his power.

Women and their body are thus exploited to affirmation on the public sphere, but also as vehicles for the status claim. In this field, it is of particular importance, as a social document, the great nineteenth-century literature and its strong and tragic heroines.

3 PROGRESS AND FASHION

> Every woman from the Empress on her Imperial Throne down to the slavey in the scullery, wears crinoline, the very three years old wear them…Crinoline has become a vast commercial interest. It is no longer a matter affecting merely a few work girls in the Lonson Shops. It extends itself to the forge, the factory, and the mine (….)

Henry Mayhew, The Shops and Companies of London and the trades and manufactories of Great Britain, 1865 (David, 2015: 160).

Fashion in these contexts promotes and is produced, through innumerable technological innovations such as the sewing machine and the intricate corsets and innovative crinolines that have shaped the S-shaped woman's body since the 1860s, translated into several patents registered in France, England and USA.

The colours and patterned textiles also promoted a revolution in the chemical industry: not only through new dyes but also in the revolution on the textile industry world, which also becomes part of the new intimate places like the interiors of the new housing.

New industries, new markets, new and large metal mining companies, dyes production companies, crinolines and corsets plants, as well as new larger or smaller factories for fashion accessories, and new Media mainly aimed at a new target audience: a woman who manages, controls and organises the interior spaces.

The new fashion consumer women in the last decades of the nineteenth century are different from the demure and resigned wife of Balzac who was not only embodied in his fiction but also in his *Traité de la vie élégante* (1830) about fashion and frivolity. It is a woman who is also trying to promote herself, though not always successfully, as suggested by Krista Lysack in an extraordinary work that addresses the reflections of the narratives on female consumption through the great literature of the nineteenth century.

Fashion consumption is decisive for the emergence of a new Women's role in the nineteenth-century society: fervent consumer, eager for news but at the same time yearns for freedom and independence. It is also for its producer as a worker, as small industrial (seamstress and milliner) and as a saleswoman. All these women wander through the cities scenes/spectacles of the nineteenth century that aggregate and define themselves according to their places of representation. As Parsons (2003) suggests, we can highlight here the role of grand avenues in the nineteenth century as well as the performance spaces and also the large gardens and urban parks as scenography elements of representations.

The bourgeois woman, "honest" wife or a mother, and the other women, more or less anonymous, run the same places of an expanding city. In the gardens, they walk with their children accompanied by their nannies and socialise with their counterparts with whom they cross. In the parks, sometimes they opt for riding a horse in a group or by carriage.

Sometimes they dispense means of transport, real social status symbols, preferring to have a walk, which provides social interaction while practising some exercise enjoying the outdoors benefits in accordance with the recommendations of the modern doctors and hygienists.

From mid-nineteen century, bourgeoisie women leave their homes towards modern shops and department stores that dominate the city centres of the old and the new world in search of the latest market news, walking and wandering in the great avenues and in the urban parks also a novelty of contemporary urbanism. Working class women attend these places in their leisure time, or free from work, usually on Sundays. They will also be directly or indirectly immortalised in spare time activities or as determinants of leisure or pleasure, as mentioned by Clayson (2003). The other women, less respectable, also circulated in these places, being immortalised by their painter's friends and lovers. See for example the beautiful portrait of Valtesse de la Bigne (1848–1910) – the most famous demimondaine of Paris, inspiration/ model for the Nana by Émile Zola – immortalised by Henri Gervex (1852–1929) in 18979: a portrait of a young woman dressed à la mode, between gardens and flowers.

Eager to be fashionable, Women of the high bourgeoisie and the others may be confused by pageantry: both can be decorated as cocottes and vice versa, what could lead us to a wider reflection about social roles, ownership and/or its subversion.

It is also noteworthy the adverse effects of progress and fashion in women bodies that are deeply altered and deformed, contributing to the high female mortality rates and to the many deaths by fire consequent not only by the use of highly flammable fabrics, but

Figure 3. Valtesse de la Bigne by Henri Gervex. Source: https://fr.wikipedia.org/wiki/Valtesse_de_La_Bigne#/media/File:La_Bigne,_Valtesse.jpg.

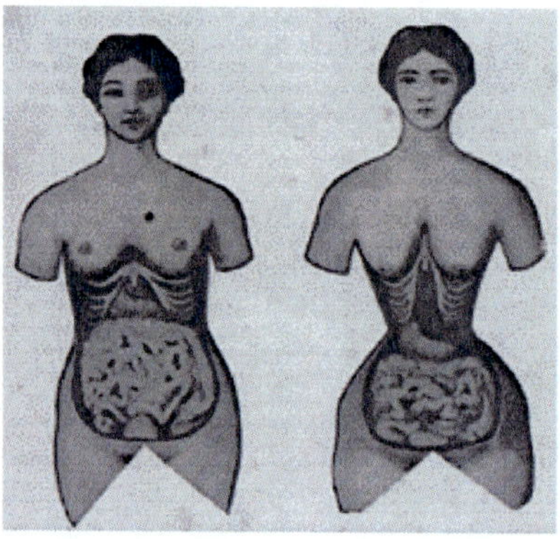

Figure 4. Deformed women's body. Source: https://s-media-cache-ak0.pinimg.com/736x/a2/ea/9c/a2ea9c5ccbba20be610a154eb3bcc1ee.jpg.

especially by excessively voluminous crinolines, in a society where gas lighting or candle lights and fireplaces were prominent; at home or in public places, such as theatres.

4 PRISONERS OF PROGRESS?

The social chroniclers of the time, authors, painters, caricaturists and photographers, do not miss the opportunity to "portray" women in their diverse contexts of daily life taking on the "apparent" domestic stillness but also their moments in the open air. Through their creations, they reveal us a woman present in activities apparently outside their private daily lives confined to their homes, assuming a wider domestic sphere

Figure 5. Dressing a crinoline. Source: https://en.wikipedia.org/wiki/Crinoline#/media/File:Crinoline_joke_photograph_sequence_04.jpg.

that includes the city and its extensions. Their works were an inexhaustible source for the cultural history of modernity.

This includes small bursts of bourgeois everyday life, and reports of small practices of sociability such as the social affirmation of new habits as gardening, reading[1] or various new ways of socializing with friends or family like the nightly exits to Theatre and the Opera where they exhibit their new and sophisticated costumes adorned and customized to fashion.

They will also attend the Normandy beaches such as the Empress Eugenie, one of the icons of fashion at the service of a new industry promoted by Charles Worth who, after the fall of the Empire, in 1871, will put his services in favour of the new social order in Paris by exporting to places, such as the US or the new colonial cities, while encouraging the French textile industry to the new laws of the market.

By the end of the 1880s, the structure of a modern *Maison of haute couture*, with two seasonal collections and the extension of the brand, had been consolidated, with a personalised label for the franchising of patterns and fashion designs.

Worth understood the magical value of his own creations and the power of woman as fashion consumer. He innovated the process of creating luxury clothes and defined the Haute Couture system for the following centuries.

Women will remain trapped in a system that guarantees quality rather than functionality by sublimating

[1] We highlight the role of nineteenth century women writers such as the Countess of Ségur, George Elliot, Louise May Alcott, the Brontë sisters, Elizabeth Gaskell that became true phenomes of sales and notoriety, moulding the behaviour and social role models that persist in time.

their symbolic value, their cultural and prestige capital, for being unique and exclusive, inaccessible in its essence to others, to the minorities that are the true engines of progress.

BIBLIOGRAPHICAL REFERENCES

Balzac, Honoré. (1922) *Traité de la vie elegante suivi de la théorie de la démand.* Paris: Editions Bossard.

Baudrillard, Jean. (2005) *The system of objects.* London: Verso Books.

Breward, Chistopher. (1995) *The culture of fashion.* Manchester: Manchester University Press.

Charle, Christophe. (2008) *Théâtres en capitales. Naissance de la société du spectacle à Paris.* Paris: Albin Michel.

Clayson, Hollis. (2003) *Painted love: prostitution in French art of the impressionists.* 2nd ed. Los Angeles: Getty Research Institute.

David, Allison Matthews. (2015) *Fashion Victims the dangers of dress past and present.* London: Bloomsbury.

Goffman, Erving. (1983) *A representação do eu na vida cotidiana.* Petrópolis: Editora Vozes, ltda.

Loos, Adolf. (2004) *Ornamento e crime.* Lisboa: Cotovia.

Lysak, Krista. (2008) *Come buy, come buy shopping and the culture of consumption in Victorian women's writing.* Athens: Ohio University Press.

Olsen, Donald J. (1988) *The city as a work of art.* New Haven: Yale University Press.

Parsons, Deborah L. (2003) *Streetwalking the metropolis women, the city and modernity.* New York: Oxford University Press.

Perrot, Philippe. (1994) *Fashioning the bourgeoisie- a history of clothing in the nineteenth century.* Translated by Richard Bienvenu. Princeton: Princeton University Press.

Santos, Andrea Paula. (1994) Sennett, Richard. O Declínio do Homem Público – as tiranias da intimidade. *R. História* (129–131).

Sennett, Richard. (1978) *The Fall of Public Man.* London: Penguin Books.

Simmel, Georg. (2014) *Filosofia da Moda, portuguesa.* 2nd ed. Lisboa: Texto & Grafia Lda.

Veblen, Thorstein. (1965) *A Teoria da Classe ociosa.* S. Paulo: Livraria Pioneira Editora.

Part III

Arts

The art and architecture of inner progress: Four significances of contemporary creations

Sarah Frances Dias
University of Lisbon, Faculty of Architecture, CIAUD, LabCor – Laboratório da Cor, Lisbon, Portugal

ABSTRACT: Progress, understood as a growth or movement towards a goal or higher state of existence, is deeply connected to the visual arts, including architecture. Art and architecture, responding to human needs, allows the progress of the world shape itself and then moves beyond it, promoting and accommodating new views, leading the way into new heights of the self and new depths of the spirit. The work analyses four inner progresses, embodied in the significance of contemporary creations. Firstly, a brief analysis of the progressive journey of the arts is presented, observing how its essence expanded and clarifying that contemporary art still strives to embody new ideas (beyond the depiction of the world) that relate to the self and its significant growths. Secondly, the significant inner progresses are presented, analysing both its aims, origins, and contemplating examples. These are: (1) the openness of existence and being, (2) the wholeness of the self, (3) connectivity and integration, and lastly (4) expansion and infinity. By connecting man to himself and to the world (cosmological and metaphysical dimensions), and by promoting an integrated open wholeness, art and architecture guide existence into new depths, defined not only by outward creations, but also and mostly, by the profound ideas that define our spirits.

Keywords: Progress, Art, Architecture, Significance, Principles

1 INTRODUCTION: PROGRESS, THE ARTS AND THE REFORM OF SOCIETY

Progress, which can generally be understood as a growth, movement or development towards a goal or higher state of existence, is deeply intertwined with all the visual arts, including architecture. That there is a great connection between progress and art is unquestionable, however, the question is not whether there is progress in the art (since some of the world's greatest art was produced centuries ago), but mostly which progress is incorporated into artistic creation, in contemporary art (and architecture) and how it is shaping and changing our world. Progress can be found in every dynamic growth: in the fields of knowledge and creation (philosophically, scientifically, and artistically), and internally (emotionally, psychologically and spiritually). So to understand progress itself, naturally, one has to ask: which comes first? Do philosophies and creations occur first, or is it the transformations of the self that guides new outward developments? It is impossible to know, for every change on the outside world, represents an inner change too. And so hand in hand, progress occurs in the world, reforming and transforming man and the world that defines him: simultaneously, interchangeably and co-dependently, expressing inward significances, inner growths, inward aspirations of progress. It is in this way, that although sprung from it is time, art rises above and beyond it, promoting ideas that allow society and man to propel into further stages of growth and expansion. The work will thus analyse four ideas of progress, embodied in the significance of contemporary creations. But first, a brief analysis of the progressive journey of the arts is of the utmost necessity, to clarify the deep connectivity between both.

Figure 1. Detail of a column in the Acropolis, Athens, Greece. Source: Picture taken by author; Date: June 2014.

2 THE PROGRESSIVE JOURNEY OF THE ARTS

Classical art embodied ideals of perfection derived from a long tradition rooted in reality, however this

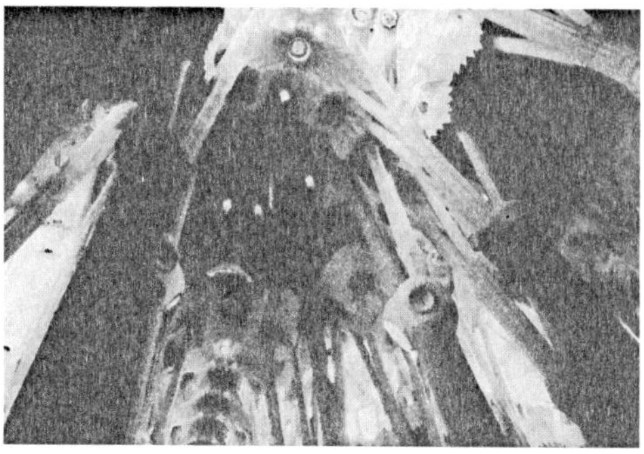

Figure 2. The Sagrada Familia, by Antoni Gaudi, Barcelona, Spain. Source: Picture taken by author; Date: February 2014.

changed drastically, and a reformation of the arts occurred led mainly by a series of technological progresses. This is explained by Arthur C. Danto (2014) that points to the invention of photography in the 1830s by William Henry Fox Talbot, and to the screening of photographs into movies in 1895, as the main aspects that made the traditional idea of art (a realistic depiction of the world) redundant. Since photography and film captured motion and 'real' life, art revolutionised itself and begun to capture other essences and significances. Main examples that led this transformation include, for Arthur C. Danto, Picasso's Les Demoiselles de Avignon (1907), the works of the Fauve movement and the paintings by Henri Matisse, which all tried to create something other than realistic depictions of the world. Indeed, it is clear, that with fauvism, expressionism, cubism, and constructivism, a new era in the visual arts begun, leading the way for contemporary art which today still strives to embody new ideas (beyond the depiction of the world) that relate to the self, the very nature of art and of existence, and aim to add further steps towards new significant growths. It is Umberto Eco (1962) whom unifies these 'progressive' ventures with an understanding of the world, explaining that the difference between artistic movements should be understood not only as different ways of creation, but as different ways of perceiving and understanding the world, "(...) which lies under these different aesthetic experiences. '' (Eco, 1989, p. 7). On this issue, author Robey (1989) further adds that since traditional art followed conventions (rules of creation that followed specific limits), meanings that emerged were also conventional, and thus conventional views of the world were brought forth. Whereas in modern art, unconventional means (no rules of expression), allow for unconventional meanings and new views of the world to emerge. Now that we know how these new progresses are materialised through (using unconventional means brought through by new technologies and material advances), it is important to briefly understand how the essence of art intrinsically relates to progress. Because classical theories no longer accommodate the broad expansion of the arts, it was necessary to define in a new way. In The Transfiguration of the Commonplace (1981), Danto identifies two conditions that define art's essence and that define art: its 'Aboutness' (what the artwork is about, what it refers to and what it means), and its 'Embodiment' of Significance which generates interest through interpretation. Cynthia Freeland in her book 'But is it Art?' (2002), echoing Danto's views, claims that art is that which expresses a significance, a meaning, a message within itself, that expands consciousness (proposed by Robert Irwin) and promotes growth, which is, in her view, the main driving aim of contemporary art. She summarises "Irwin's definition emphasises art's role in enhancing our awareness both of ourselves (expanding our perceptual functioning) and of the world." (Freeland, 2002, p. 207). It is thus clear, that the very essence of art has roots in a deep desire for progress. Thus, these are the aspects that will be analysed bellow, trying to understand what progressive understandings contemporary art is embodying: what transformations in society is it trying to define.

3 PROGRESS AND SIGNIFICANCE

3.1 *Openness of existence and being*

The first significance that relates to progress that will be analysed, is the idea of openness of existence and being, which was first pointed out by Umberto Eco (1932–2016), in the book 'The Open Work' (1962), but is still very actual and accurate in contemporary art. Eco first identified this quality 'openness', to better describe the new forms of art which defined his time. Openness of a work, that it is open ended, means that the creator leaves the work 'unfinished' letting the viewer, observer or listener create his own 'ending' and interpretations. It deeply opposes classical creations of art and architecture, which were unambiguous, directing one specific response and one form of interpretation. Open works, modern creations, are characterised by their ambiguity, connected to their formal innovation, allowing for a great variety of different meanings to co-exist. Possessing a high degree of ambiguity, structural vitality and creating a sense of the 'unlimited', allows for drama and a plural communication through half-perceived infinitudes that can never be fully resolved defining an endless cycle of interpretation, which allows for the viewer or user to experience a metamorphosis through contemplation. One first example of this can be found in Baroque art and architecture, which aimed to be dynamic conveying an idea of space opening up in a kind of kinetic movement, which deeply echoed a psychological and scientific shift: the world was no longer fixed, but in a fluid state of movement. Eco further announces works that portray a sense of movement as other good examples. Echoing Einstein's ideas of the universe, these works possess an inherent mobility, which constantly presents itself in renewed ways. The sculptures of Alex Calder (whose forms move with the air, continually creating new space and new connections), the

Impressionists (which aim to capture movement), and Jackson Pollock (that uses disorder and disintegration) are all examples of such movement and openness. More examples in art can be announced with the works of Pablo Picasso, specially his abstract portraits that use broad lines and gestures to capture an intangible essence of the person he is portraying. Another example can be the works of Henry Matisse, specially his line drawings, whose loose gestural strokes evoke movement and a fleeting open quality. These artists strived not only for a sense of progress in art, but also of the self, promoting a more broad, open and expanded understanding of existence. Another example of this property, in contemporary art and architecture, can be found in the works of Anselm Kiefer, whose creations are highly open ended and endowed with a high degree of indeterminacy, which are accentuated by his cosmological themes and aim to expand the viewer's awareness. In architecture, it is of reference, the spaces of Peter Zumthor, whose spatial atmospheres echo an intangible and subtle presence, always somewhat indeterminate, mysterious and highly intangible, leaving the user in a constant state of search and unresolved contemplation. It is through these types of open art forms (ambiguous, unpredictable and improbable structures), that man creates a better understanding of his true self and his relation to the world: searching for significances, formulating new understandings, and integrating within himself the idea that all existence is, indeed, highly open ended. Also of reference are the architectural organic forms of Frank Lloyd Wright, specially the Falling Water, where Wright uses dissociated volumes overlapped and connected to each other, to create an architectural space that appears to be floating in space. The house itself seems to be moving with the waterfall, as it shifts and changes with the colours of the landscape around it. Wright aimed not only to promote an interaction of architecture with nature by echoing its principles (open and always changing), but also, metaphorically, to echo a more open and connected sense of existence for its inhabitants.

Figure 3. Anselm Kiefer Exhibition, Pompidou, Paris; Source: Picture taken by author; Date: January 2016.

3.2 *Wholeness of the self: body, mind and soul*

One second significance embodied in contemporary art and architecture which relates to progress is the idea of wholeness of the self (mind, body and soul), which can be traced to Merleau-Ponty's (1908–1961) phenomenological ideas of perception being rooted in the body and derived from sensation. In this way, it is the body that 'creates' the experience in the construction of meaning and significance. His theories deeply influenced multiple contemporary artistic and architectural theories and architects that begun to focus on the senses as the primary fundamental aspect of experience. Architect and theorist Juhanni Pallasmaa (1996), for example, explains that the qualified artistic and architectural creation evokes a multi-sensory experience (derived from all the senses in combination) that creates such an intense experience with the 'imaginary' world it projects, allowing man to become more aware and whole. Steven Holl (1994), likewise, understands the phenomenological aspects of the artistic and architectural experience as a condition for significance to emerge. For him emotions and sensations are triggered by what he describes as 'exhilarating perceptions', and derived by what he identifies as 'phenomenal zones' (light, shadow, textures, sound, among others), which all add to generate a significant architectural encounter. In art this is ever so clear with the multiple examples of more tactile creations: the works of Anselm Kiefer, the works of Art Povera, of Lucio Fontana, of Gerhard Richter, and even of Jenny Saville, who all, either use diverse materials that promote an immersive experience, or try to echo a textural quality that directly appeals to the senses. In architecture, where these ideas were more deeply understood and explored (in a duality of theory and practice), it is of reference the space of Steven Holl which materialize his theories, combining multiple visual and sensorial elements into one single space in an intense way. Another example can be found in the spaces of Peter Zumthor, who deeply relies on the phenomenological aspects of materials, colour and light, combining the senses, emotions and perception, to create an atmospheric immersive environment that engages man's inner world. Mirroring both Merleau-Ponty and Heidegger's ideas of experience, existence and being in the world, his spaces allow man to find a sense of existence deeply rooted in life and sensation. In this case, it was philosophical progress which led the way for new artistic and architectural understandings, which define new approaches for the understanding of space and for man's existence altogether. It is a phenomenological discourse in art and architecture that integrates human life and creates the understanding that man is not a fragmented and distorted material being, but an integrated whole (physical, emotional, and spiritual self). It is, by promoting the integration of all the senses, through art and space, that man can find a wholeness that was once forgotten.

3.3 *Connectivity and Integration*

The third significance brought forth, is the idea of connectivity and integration. Contemporary art and architecture has not only expanded in its depth (relating to the essence of what art is), but also in the means by which it achieves its own essence. One main significant way it has been using is through the integration of different artistic fields, which together, aim to express new significances in new ways. The traditional artistic distinctions between artistic fields (architecture, painting and sculpture) is no longer possible, but a new realm of artistic creation, where the fields merge together. Painting is also sculptural (as in the case of Anselm Kiefer), sculpture becomes architectural (as in the case of Richard Serra who uses architectural space to emphasise his artistic pursuits), and architecture deeply takes into account other artistic forms promoting a union between all the arts (as is the case with Le Corbusier's La Ronchamp, which can be considered a cornerstone example, or Barragán's house and studio, who designed the interiors in consonance with the art in each room). These architects and artists also promote a connectivity and an integration of the self in yet another way: for them architecture and art is a means of connection of the self to itself and to the world around it. Kiefer emphasises through his large scaled works, a connection between man to a cosmological time (connection of man and the universe), Le Corbusier emphasises a connection between man and poetry (promoting a unity between man's poetical and emotional faculties). Another example of this 'self' integration through architecture can also be found in the works of Tadao Ando, who creates spaces that are minimal, through pure geometrical forms (scale, poetry and light), in order to enhance a sense of integration and connectivity in the user. Ando sustains an architecture that can enhance emotional and poetical encounters creating a sense of harmony in the self (progress towards a more spiritual existence).

Metaphorically and symbolically, this new integration represents an abandonment of the traditional fragmentation of existence into defined and confined means of understanding the world, and promotes a new sense of cohesion and connectivity. At the root of these ideas, it is possible to announce an emotional and spiritual progress. Separation, which is now avoided, gives way to unity; deeply expressing the idea that everything is connected and part of the same essence, art now sees the world as a continuum. Another example of these forms of integration that promote connectivity and integration between the self and the world at all levels of existence, can be found in land art or environmental art, which, instead of unifying art forms, transforms the natural world by adding elements that create visual, emotional, psychological and spiritual associations. In some cases, the elements added promote an architectural experience: producing a new sense of space, as in he installation 'Sun Tunnels' (1973), by Nancy Holt, comprised of four large cylinders of concrete (architectural forms with the dimensions of the human being), placed in a vast arid landscape that faces the movements of the sun, creating different experiences throughout the day. Another reference could be made to the work of Nicola Yeoman, specifically the work a 'T falls in the woods' (2010), which used ephemeral materials such as cloths and sticks to create architectural forms (arches) placed within spaces and nature. Creating spaces within the undetermined nature of landscape, these art forms aim to promote not only an integration between the different art forms, but also between man and nature and, more profoundly, man within himself.

Figure 4. Richard Serra Exhibition, Reina Sofia Museum, Madrid, Spain. Source: Picture taken by author; Date: January 2015.

3.4 *Expansion and Infinity*

Lastly, the fourth significance brought forth, is the idea of expansion and infinity. Probably a result of all the previous ones, man finds himself searching for his place within the universe, which intertwined with a sense of expansion, leads him into uncharted territories of the self. This idea can be found in the most varied contemporary artistic and architectural experiences, but specially in the new art forms that promote complex experiences, of all the senses, through architectural spaces.

A first example can be found in the artwork ofYayoi Kusama (whose artistic progression moved from painting to a three dimensionality) in her art installation entitled the 'Infinity Mirrored Room – The Souls of Millions of Light Years Away' (2013). The installation aims to echo the vastness and expansion of the cosmos and of the universe by creating a mirrored dark room filled with a countless number of small lights. The immersive experience allows man to be placed in the midst of what seems to be an infinite endless sky, allowing him to expand his own awareness of his place within the vast cosmos.

Multiple artworks have explored the concept of infinity itself, which in itself was already a step towards this idea of expansion and infinity of the self. A cornerstone example can be found in the works of William Turner, whose abstract landscapes aimed to evoke God's endless and infinite power and presence, thereby promoting a spiritual development and growth in the viewer. It is also important to emphasise the

Figure 5. Rodin Sculpture, at the Victoria and Albert Museum, London, UK. Source: Picture taken by author; Date: May 2015.

Figure 6. Detail of a temple, Bali, Indonesia. Source: Picture taken by author. Date: June 2016.

works of Escher, who used architectural forms and mathematical concepts to promote this same idea of infinity, trying to convey a sense of progress (moving beyond the illusions of limitations of the mind).

Another significant example can be found in the work of James Turrell whose perceptual light and colour immersive experiences create an existential transformative shift to the viewer, transporting him into a sort of spaceless time or timeless space. Either in his multiple Skyspaces (round small observatories which open to the heavens), or in his Perceptual Chambers (rooms flooded with intense colour and light), the impact of the experience aims to 'awaken' the viewer from his usual day-to-day condition, expanding his awareness by creating a deep impact in the being at other levels of existence (psychological, emotional and spiritual). Turrell aims not only to awaken the viewer but also to help him understand his own place within the vast universe, a goal that is clearer in his architectural and artistic project entitled Roden Crater. Placed next to a volcano, multiple successive rooms form a path, where each creates its own perceptual experience connected to the infinite cosmos, thus allowing the viewer to understand his own place in the vast universe. It is this broad understanding of a vast universe that allows man to understand his own cosmological and metaphysical dimensions.

And lastly, another example that is vital, are the works of Anselm Kiefer, who uses the concept of infinity throughout many of his works: the endless landscape portrays a sense of an Arcadia (a lost paradise in time), the endless sky scenes with countless stars, and even the use of the ancient pyramid, all contribute to this exploration and understanding of the infinite (of the world and of the self).

4 ENDING REMARKS

At this point, it is clear, that the previous analysed progress concepts not only support contemporary creations but also echo universal and timeless aspirations that might likewise be embodied in artistic creations of the past. (Were not the ancient pyramids of Egypt a means to perpetuate man's place in the universe?) So are they really progress? One may ask. To which the answer is always, yes. It is through the progress of technology (that allows for new methods and materials) and philosophy (that creates new understandings of the self) that these are revived and conveyed in new ways, reforming man towards a new future, even if that future includes the same longings of the past. Understanding openness, wholeness, connectivity and infinite expansion, is part of what defines our contemporary artistic and architectural phenomena. Through metaphors, techniques, symbols, spaces, emotions and ideas, artists and architects convey deep and profound ideas of an inward progress that shapes society and reforms our world. Connecting man to himself and to the world around him, promoting an integrated wholeness and an open type of existence, and helping him to understand his own cosmological and metaphysical dimensions, art and architecture guide man into new ways of life, defined not only by the creations that fill the world around us, but also and mostly, by the profound ideas that define our spirits. Although somewhat utopian, striving endlessly towards an ideal, they turn broad and ample scientific, philosophical and spiritual ideas into something concrete and tangible that changes man and finds deep roots within his soul.

BIBLIOGRAPHICAL REFERENCES

Battcock, G. (1995). *Minimal Art: A Critical Anthology.* California: University of California Press.

Berger, J. (1972). *Ways of seeing.* London: Penguin Books.

Butterfield, J. (1993). *The Art of Light + Space.* New York: Abbeville modern art movements.

Danto, A.C. (2014). *What Art Is.* New Haven and London: Yale University Press.

—. (1998). *The End of Art: A philosophical defense.* History and Theory: Art History, Historiography and After the End of Art, 37 (4), pp. 127–143. Available at: http://www.jstor.org/stable/2505400 [Accessed 12 Dec. 2015].

—. 1981). *The Transfiguration of the Commonplace: A Philosophy of Art.* New York: Harvard University Press.

Dias, S. F., DURÃO, M. J. (2015). *Architecture and Art: La Ronchamp's symbiosis as a 'total work of art,* Le Corbusier, 50 years later International Congress, Universitat Polite'cnica de Valência, Valência, Spain.

Eco, U. (1989). *The Open Work.* Cambridge, Mass.: Harvard University Press.

Elkins, J. (2000). *What Painting Is.* New York: Routledge Publications.

Freeland, C. (2002). *But Is It Art?: An Introduction to Art Theory.* Oxford and New York: Oxford University Press.

Goldstein, E. (2009). *Sensation and Perception.* New York: Cengage Learning.

Heidegger, M. (1989). *A origem da obra de arte.* Lisbon: Ediçẽs 70.

Holl, S., PALLASMAA, J., GOMEZ, A.P. (1994). *Questions of : phenomenology of architecture.* 2nd edition. New York: a + u Publishing Company and William Stout Publishers.

Merleau-Ponty, M.(2009). *O olho e o espi'rito.* 7th ed. Lisbon: Editora Vega.

Pallasmaa, J. (2011). *The embodied image: imagination and imagery in architecture.* Chichester: John Wiley & Sons.

Pallasmaa, J. (1996). *The Eyes of the Skin: The Architecture and the Senses.* London: Academy Editions.

Schielke, T. (2013). *Light Matters: Seeing the Light with James Turrell,* in: ArchDaily [online]. Available at: http://www.archdaily.com/380911/light-matters-seeing-the-light-with-james-turrell/ [Accessed 11 May. 2015]

Scruton, R. (2009). *Beauty.* Oxford, England.: Oxford University Press.

—. (1979). *The Aesthetics of Architecture.* Princeton, New Jersey: Princeton University Press.

Shirazi, M.R. (2011). On Phenomenological Discourse in Architecture. Available at: http://www.arch.ksu.edu/seamon/Shirazi_phenomenological_discourse.htm [Accessed 21 Oct. 2013].

Rendell, J. (2006). *Art and Architecture: A Place Between.* London: I.B.Tauris.

Turrell, J. (2014). *James Turrell* [online]. Available at:http://jamesturrell.com/about/introduction/ [Accessed 25 Oct. 2015]

Zumthor, P. (2006). *Thinking architecture.* Basel, Swizterland: Birkhäuser.

—. (2006b). *Atmospheres: Architectural Environments – Surrounding Objects.* Basel and Boston: Birkhäuser,

Zybaczynski, V.M. (2013). *The Colour of Architecture: Past and Presen, in: Urbanism, Arhitectura?, Construcţ ii,* 4 (3). Available at: http://uac.incd.ro/Art/v4n4a08.pdf [Accessed 24 Nov. 2014]

Note

The pictures embodied in the text aim to establish a parallel means of contemplation from the text: drawing further examples from both classical and contemporary creations.

Progress and return: Chaos, action and aesthetic contemplation, or knowledge never begins at the beginning

Ana Santos Guerreiro
University of Lisbon, Faculty of Architecture, CIAUD, Lisbon, Portugal

ABSTRACT: Knowledge and aesthetic contemplation are intimate as modes of access to the experience of reality. There are two great roads to produce knowledge and to know it: the one that is achieved through science and the one that is achievable by poetry making. Either one is ways of being in the world, of understanding chaos, and daily living is usually done in the mixture of these two ways of being in the world. Everyday life, by means of the experience we acquire in the world, also promotes a certain subjectivity, by the simultaneity of these perspectives and experiences that are being constituted and acquired. The aesthetic experience, however, is not exclusive to poetic knowledge or artistic contemplation: It is also intrinsic to the scientific experience and the composite mode with which reality is re-established.

Keywords: Dynamics of knowledge, science, poetic thinking, aesthetic experience, action

1 MYTHS AND UNKNOWNS

Gianni Vattimo refers in *The Transparent Society* (1992: 36) that contemporaneity uses the myth, or its term, in imprecise form. He refers more to its common, non-technical use, entering into that imprecision of the notion we call reference. Its presence is therefore related to a vague and somewhat indefinite meaning of understanding myth as contrary to analytical thinking, but akin to the narrative, linked to the fantastic, related to the emotions and with few pretensions of objectivity. Mentioning Cassirer, Vattimo says that myth:

> Tem a ver com a religião e a arte, com o rito e a magia, e a ciência nasce, pelo contrário, em oposição a ele como desmistificação, 'desencanto do mundo'. O saber racional sobre a realidade, "onde quer que procure constituir-se como consideração teorética e explicação do mundo, vê-se oposto não tanto à realidade fenoménica imediata, mas à transfiguração mítica desta realidade. Muito antes de o mundo se apresentar à consciência como um complexo de 'coisas' empíricas e de propriedades empíricas, apresentou-se como um conjunto de potências e de acções míticas. (Vattimo, 1992: 36)[1]
>
> [Has something to do with religion and art, with rite and magic, while science was born fighting it, as demystification, 'disenchantment of the world'. Rational knowledge of reality, 'wherever it tries to establish itself as theoretical reflection and explanation of the world is seen not as opposing not so much as to immediate phenomenological reality, but rather to the mythical transfiguration of that reality. Long before the world presented itself to conscience as a compound of empirical 'things' with empirical properties, it presented itself as an ensemble of mythical powers and actions.]

Vattimo also considers that, nowadays, the metaphysical conception of myth, linked to the historical evolution, is finished, lost. In turn, the words continue to relate to a non-current form of knowledge, which, to scientific knowledge, is characterized by less objectivity or less technological effectiveness. In turn, the metaphysical crisis of historical evolutionism, in which scientific rationality itself is also revised, increases the conception of myth as primitive, 'prior' to the scientific one. At the same time, cultural and philosophical approaches are hybrid regarding the *archaisms* of non-explicit return to myth; the *cultural relativism*, which does not assume a definite position; and, finally, the third position, which Vattimo considers to be that of *mitigated irrationalism* or of *limited rationality*, in which myth is understood by its narrative structure (which does not oppose itself in a balanced way to scientific demonstrability and objectivity). This approach can be found in several domains, such as in psychoanalysis, in the theory of historiography, which makes possible its irreducible plurality and also in the sociology of the mass media.

[1] For convenience to the English reader, all sources used in Portuguese versions will have a free translation attached below.

It seems pertinent to refer here Gorjão Jorge, regarding the question of different world representations between myth and objectivity:

> [*ldots*] toda a representação é um produto da 'ordem sincrónica' vigorando no momento em que esta surge, e como cada nova representação constitui uma reinterpretação da(s) realidade(s) que, por seu intermédio é (são) representada(s), não deverá surpreender-nos o facto de tal proliferação de representações se tornar um sintoma de aceleração do processo de 'mutação de significados', no respeitante ao conteúdo das 'mensagens espaciais'. E isto acontece, por um lado, em resultado da revisão de critérios de verosimilhança e das regras de inteligibilidade que, perante cada nova alteração da 'ordem sincrónica' (provocadas pelas alterações do código geral) se vêem renovados, e por outro lado, dos vários tipos de tensões que se geram no interior de uma sincronia e acabam por caracterizar a 'ordem sincrónica' pela qual essa sincronia se exprime. Finalmente isto significa que existe uma dupla alteração na significação do espaço representado – e, por inerência, do espaço a que este se refere –quando transitamos de uma 'ordem sincrónica' para outra. (Jorge, 1993: 189–190)
>
> [All representation is a product of the 'synchronic order' being active in the moment when this one appears, and as each new representation is necessarily a reinterpretation of realities that are so represented, one should not be surprised by the fact that the proliferation of representations becomes a symptom of the acceleration of the 'mutation' process of meanings in what concerns 'special messages'. And this happens, on one hand, as the outcome of the revision of the verisimilitude criteria and of the rules of intelligibility that, according to every new change of the 'synchronic order' (caused by transformation in the general code) are thus renewed. On the other hand, this happens also because of a different kind of tensions that are generated inside a synchronicity and end up by characterizing the 'synchronic order? By which that synchrony expresses itself. Finally, this means that there is a double change in space representation meaning and consequently of the space to which it reports whenever one moves from a synchronic other to other].

2 "DIAPHANOUS PERSPECTIVES" BETWEEN PERSONAL AND SCIENTIFIC

Walter Watson (2008) refers to "Diaphanous Perspectives" in the way culture and gnosiology are constituted through various disciplines and theses, which in themselves manifest the process of assimilation and plurality, even when the objectification is key. According to Watson one can avoid, in individuals the relativity of personal perspectives not only eliminate subjectivity in an objective perspective but also subordinate each individual subjectivity to a vaster and more inclusive perspective and finally to an absolute perspective where all other perspectives are included. And Watson continues stating that individual subjectivity can disappear not only if we transform ourselves into mere mirrors, but also if we become transparent to an absolute perspective, as to an omniscient God. However, since we are not gods, the author of a text transmits truths that transcend his own perspective, or he talks instead of the superior author. He becomes a conductor, a means and his perspective becomes oracular, inspired, enlightened. According to Watson, this can be considered a *revealed* perspective, but due to the religious connotations, Watson prefers to name it as *diaphanous*, meaning it shows through itself (Watson, 2008: 64).

Science made ourselves familiar with an operative, active way of being in the world and observing it, developing a movement of a horizontal sense, in which action integrates the attainment of stages by an inquiring person to the enigma of the object of science. About progress as a continuous renewal of the present, about things rapid obsolescence as a movement, Vattimo remarks:

> A modernização não acontece através do abandono da tradição, mas através de uma espécie de interpretação irónica desta, uma "distorção" (Heidegger fala, num sentido não muito distante, de *Verwindung*) [*ldots*], que a conserva, mas também em parte, a esvazia. [...] o progresso tem uma espécie de natureza nostálgica, como o classicismo e o romantismo dos séculos passados nos ensinaram. Mas o significado desta nostalgia só se manifesta [*ldots*] quando a desmistificação é revelada como mito, o mito recupera legitimidade, mas só no quadro de uma experiência "enfraquecida" da verdade. [Nesta] já não há evidência apodíctica, aquela em que os pensadores da época da metafísica procuravam um *fundamentum absolutum* et *inconsussum*. O sujeito [contemporâneo], se olha para si à procura de uma certeza primeira, não encontra a segurança do cogito cartesiano, mas as intermitências do coração proustiano, os relatos dos media. (Vattimo, 1992: 47–49)
>
> [Modernity does not happen through the rejection of tradition, rather through its ironical interpretation, a "distortion" (Heidegger speaks, in a similar meaning, of *Verwindung*) [. . .] that preserves it, though partly emptying it. [. . .] the process has a kind of nostalgic nature, as classicism and romantism of previous centuries taught us. But the meaning of this nostalgia only manifests itself [. . .] when the demystification reveals the myth that thus recovers its legitimacy but only as an "enfeebled" experience of truth. [In this] there is no longer an apodictic evidence, the one that

metaphysics thinkers searched for as a *fundamentum absolutum et inconsussum*. If [contemporary] individual looks at himself in search of a primary certainty, he no longer finds the security of Cartesian *cogito*, but rather the intermittency of the Proustian heart, the media reports.]

Science is active and observant. It is carried out as a body of knowledge which, formulated methodically and rationally, is systematized and acquired through a method that integrates observation, identification, research and explanation of certain categories of phenomena and facts. However, despite the intrinsic rationality of science, Gaston Bachelard mentions that scientific language is, in principle, a metalanguage:

> Se concentrássemos a nossa atenção nesta actividade de tradução, muitas vezes encoberta, aperceber-nos-íamos de que existe, na linguagem da ciência, um grande número de termos entre aspas. Poder-se-iam comparar estas aspas com os parenteses dos fenomenologistas. (Bachelard, 2006: 201)
>
> [If one concentrates one's attention in the translation activity, so many times veiled, one would notice that there is in scientific language a great number of words in inverted commas. These could be compared with the parenthesis used by the phenomenologists.]

Continually to surpass these steps reshapes the limits of human knowledge. Epistemology highlights this precise movement (in what resembles the design of a chronological movement), continuously evaluating the cognitive validity of the sciences described in their varied evolutionary trajectories, in the constant reassessment of their structural paradigms.

Scientific knowledge is one that seeks transparency. In this sense, the transparency metaphor acquires a value of *cultural indicator* that relates to the thirst of science and to the place in which, in certain historical moments, cultural metaphors act as interpretative keys.

Scientific knowledge makes the enigma invisible, and its project is one of disarticulation. The anxiety to reveal, to unveil, to know, characterizes, from the Renaissance onward, science as a great instrument of objectification that begins to guide, in a modern sense, individuals against nature and history.

Reason dominates and scientific enlightenment seeks to make clear and transparent what is obscure; Heinemann wrote:

> Excluem-se os milagres e as ligações orgânicas dão lugares a deveres livremente aceites. (...) alarga-se o horizonte no espaço e no tempo. É certo que há ainda carência de autêntico espírito histórico; no entanto, nascem os primeiros passos de história universal, independentes da teologia, com fundamentos pragmáticos. Tenta-se a exposição dos progressos da humanidade em todas as esferas da cultura para determinar o ponto em que o homem se poderia inserir nesse processo. (Heinemann, 1993: 208)
>
> [Miracles and organic connections are eliminated giving place to duties freely accepted. [...] Space and time horizons are expanded. For sure there still lacks a true historical spirit, nevertheless, the first steps are done towards a universal history, independent from theology and with pragmatic foundations. There is an attempt to expose humanity's progress in every cultural sphere to determine where the individual can insert himself.]

Transparent vision assumes itself as an achievement of the rationality of sight and as the achievement of the rational knowledge of the world. Action characterizes the attitude of science from Renaissance to Illustration. The need to investigate, to know the core, to dismantle the enigma, goes through many epistemological fields, of which medicine and anatomy have been exemplary as a way of listening to the world (and the world was, in the Renaissance, a place under human perspective) and its mystery. The scientific method, based on observation, invests in the ability to look at the delirium of knowing and engenders the prosthesis for the poles of the infinitely great and the infinitely small. A whole array of visual artifacts extends and amplify the power of individual's gaze of science. From the telescope to the microscope, it is important to understand that, at present, the inheritors of these prostheses, such as the electronic microscope or the scanner, translate the visual image into non-visual activity, accessing an abstract knowledge that departed from and out of sight, from visualization.

This situation is related to the triggering of the demand for transparency in Western culture. The context of the 'demoniac explicit', in the sense used by Sloterdijk (2006: 6) – "all very explicit becomes demonic" – in relation to the euphoria of understanding, the author states that anatomy has proved to be one of the most important branches of science, in the thirst for transparency. This is understood by the author as the greatest revolutions in the context of science, in the inner dive of man as a process of knowledge of the world through self-knowledge.

3 CONTEMPLATIVE AND DYNAMIC: HYBRID AND TRANSLUCENT

This movement is in a certain way antagonized by the contemplative attitude. There is in this attitude a vertical movement of ascension that also characterizes religious contemplation (the religious *acedia* – a kind of spiritual aboulia that is related antagonistically to the exercise of the virtues, namely with regard to worship and communication with God, which can be understood as a prostration, can be related to contemplation).

The contemplative attitude (reverberating between observation and the construction of an inner world, giving it meaning) is also a way of seeking knowledge, as a dimension between the individual and the world, intrinsic to poetic knowledge. This attitude is found in art, in poetry, in philosophy – it is *par excellence* an aesthetic attitude.

It seeks not transparency (for the solution of a problem, understood as an enigma to be solved, as in science), but a mode of knowledge that is not acquired by a rational, intuitive and sensitive way that does not disregard the mystery. The reverie being part of the poetic thought expressed in images of translucency:

> The diaphanous, the light and the sonorous determine a kind of conditioned reflex of the imagination. These conditioned reflexes, linking imaginary qualities, are the ones that specify the different poetic temperaments (Bachelard, 1990: 53).

The feeling produced by the reverie is a door to the poetic knowledge and also to his subjectivity. About this, Mario Perniola refers:

> Que a dimensão estética, habitualmente considerada a mais estranha e afastada da realidade, se tenha tornado a mais efectual, é algo que surpreende quem está habituado a pensar que saber é poder, que agir é poder: no entanto, hoje, mais do que nunca, sentir é poder. (...) Que o sentir, que Hegel desvalorizava enquanto subtraía o indivíduo à comunidade da razão e o fechava na sua particularidade isolada, se tenha tornado o lugar por excelência da socialidade e da efectualidade, é certamente uma prova da distância que separa a nossa época da dele. (...) O segredo da efectualidade, no fundo, é sempre o mesmo, aquele que foi identificado por Hegel no início do século XIX e por ele considerado o motor da cultura moderna. O movimento de alheamento. Só o que se aliena a si próprio adquire efectualidade: o que, pelo contrário, preserva a sua identidade, e a sua pureza é impotente e não efectual. (Perniola 1993: 18–20)
>
> [The aesthetical dimension, usually considered the strangest and the furthest from reality, becoming the most effective surprises one who is accustomed to thinking that knowledge is power, that acting is power. However, nowadays, more than ever, the feeling is power. [. . .] That feeling, which Hegel devaluated because it subtracted the individual from the community of reason and enclosed him/her in their respective isolated particularity, has become the place *par excellence* of sociability and effectivity and it is certainly a proof of the distance that separates our time from his. [. . .] The secret of effectiveness is, actually, the same that was identified by Hegel in the beginning of the 19th-century e also by him considered the engine of modern life - the movement towards alienation. Only those alienate themselves attain effectiveness. Who, on the contrary, preserves his/her identity and purity is powerless and noneffective.]

One may, following Mario Perniola idea, consider the connection that, through feeling, the poetic knowledge engenders with the translucent, regardless of the contemporary character that the author confers to it.

Poetic knowledge relates to the "variables" and "varieties" that can be established in diversity and in the abyss of chaos. Among the infinite diversity, the Deleuzian meaning articulates interestingly the two great strands of infinite *variabilities* that are offered to thinking as well as the need to retain them a little through the association of ideas, as a possibility of necessary order:

> What the philosopher brings back from the chaos are variations that are still infinite but that have become inseparable on the absolute surfaces or in the absolute volumes that lay out a secant [secant] plane of immanence [. . .]. The scientist brings back from the chaos variables that have become independent by slowing down, that is to say, by the elimination of whatever other variabilities are liable to interfere, so that the variables that are retained enter into determinable relations in a function [. . .]. The artist brings back from the chaos varieties that no longer constitute a reproduction of the sensory in the organ but set up a being of the sensory, a being of sensation, on an anorganic plane of composition that is able to restore the infinite. [. . .] The three disciplines advance by crises or shocks in different ways, and in each case it is their succession that makes it possible to speak of "progress" (Deleuze et al., 1994: 202–203).

These are, through the relationship with the abyss of chaos, similar to an ordering, which is achieved through the establishment of disciplinary areas, respectively scientific, philosophical and artistic, as ways of accessing chaos, the unknown, in the sense of capturing from it a bit of order to protect us from chaos. It is the way in which we relate to it that makes possible the various types of knowledge.

These subjective processes of ordering are the outcome of our mental dimensions – the cognitive, the volitional and the affective – and from their articulation with the constant establishment of a particular cosmos. It is also through these processes, as ways of ordering ideas that we have access to the world that presents itself in its complexity, giving shape to what we call Reality.

BIBLIOGRAPHICAL REFERENCES

Bachelard, Gaston. (2006) *A epistemologia.* Tradução de Fátima Lourenço Godinho e Mário Carmino Oliveira. Lisboa: Edições 70.

Bachelard, Gaston. (1990) O Ar e os Sonhos. tradução de António de Pádua Danesi. São Paulo: Martins Fontes.

Deleuze, Gilles & Guattari, Félix. (1994) *What is philosophy?* Translated by Hugh Tomlinson and Graham Burchell. New York: Columbia University Press.

Heinemann, Fritz. (1993) *A Filosofia no se'culo XX.* 4th ed. Trad. Alexandre F Morujão. Lisboa: Fundação Calouste Gulbenkian.

Jorge, José Duarte Centeno Gorjão. (1993) A noção de sincronismo na leitura e representação do espaço. Lisbon: Faculdade de Arquitectura, 209.

Perniola, Mário. (1993) *Do sentir.* Trad. António Guerreiro. Lisboa: Editorial Presença.

Sloterdijk, Peter. (2006) *Esferas III: Espumas: Esferologia plural.* Trad. Isidoro Reguera. Madrid: Ediciones Siruela.

Vattimo, Gianni. (1992) *A Sociedade Transparente.* Trad. Shooja De Hossein e Isabel Santos. Lisboa: Relógio d' Água.

Watson, Walter. (2008) La Arquitectura del Significado: Fundamentos del Nuevo Pluralismo. Trad. de Javier Garcia Rodriguez y Maria José Móran de Diego. Córdova: Ed. Berenice.

Am I always drawing the same drawing?

Ana Leonor M. Madeira Rodrigues
University of Lisbon, Faculty of Architecture, CIAUD, Lisbon, Portugal

ABSTRACT: Drawing establishes itself in Kairos, a propitious moment outside chronological time, where progress becomes a foreign concept.

Although the drafter may identify progress in his or her workmanship, it is actually a simultaneous evolution of the ability to control one's own perception and the ingenuity to register one's intention on the paper.

This text will observe the phenomenon of drawing, and analyse the relationship between contour and drawn lines, time, and space.

Keywords: Drawing, line, contour, Kairos, eternity

> A Klee painting named 'Angelus Novus' shows an angel looking as though he is about to move away from something he is fixedly contemplating. His eyes are staring, his mouth is open, and his wings are spread. This is how one pictures the angel of history. His face is turned toward the past. Where we perceive a chain of events, he sees one single catastrophe, which keeps piling wreckage upon wreckage and hurls it in front of his feet. The angel would like to stay, awaken the dead, and make whole what has been smashed. However, a storm is blowing from Paradise; it has been caught in his wings with such violence that the angel can no longer close them. This storm irresistibly propels him into the future to which his back is turned, while the pile of debris before him grows skyward. This storm is what we call progress. (Benjamin, 1940)

At a time when history and progress were in the process of annihilating him when the future ceased to exist for him, Walter Benjamin wrote his last texts: *On the Concept of History*, eighteen paragraphs and two addenda. These catapulted him utterly into the future and catapulted him into my moment of the present.

Progress — a concept that emerged during the 18th-century Enlightenment and became mainly related to technological and economic evolution — was at the core of the destruction he was enduring.

In the name of progress, the 20th-century witnessed two of the most terrible wars; in the name of progress, technology showed its real efficiency in destroying humans, body, and soul.

In the name of progress, the war was even considered a necessity, something that would solve problems, clean society, bring back a golden age paradise for the chosen ones, those that, like progress, were clean, obedient, and pathetically machine-like.

Figure 1. Angelus Novus, Paul Klee, 1920, drawing.

For Benjamin, time was ending, and his creativity was a transitory escape from the hell from which these intense and lucid texts sprung.

In this excerpt, Benjamin questions history annihilates *Chronos* time—chronology—to plunge completely in *Kairos* time: the moment, the right moment.

Kairos, the younger son of Zeus, represents nonlinear time, eternity, and the moment.

All creative activity is in some way a step outside *Chronos*, outside the progression of time, existing in the forever-moment of Kairos.

Looking at definitions of time we find ourselves before an immense body of information and knowledge from which we can conclude that time exists in the plural, "times", but also that it is a human structure, which explains the progress of existence.

Seen as a one-way, apparently irreversibly movement, beginning in the past and flowing through the present into the future, time contains in itself the concept of an endeavour to understand the presence of eternity. In eternity, we are always present, always returning and becoming in a chain of never-ending possibilities.

Zenon of Elea explains this concept of movement without progress, this standstill, with the parable of the tortoise and Achilles:

> Aquiles corre dez vezes mais depressa que a tartaruga e dá-lhe dez metros de vantagem-Aquiles corre esses dez metros, a tartaruga corre um decímetro; Aquiles corre esse decímetro, a tartaruga corre um centímetro; Aquiles corre esse centímetro, a tartaruga um milímetro; Aquiles o milímetro, a tartaruga um décimo de milímetro, e assim infinitamente, de modo que Aquiles pode correr para sempre sem a alcançar. É este o paradoxo imortal. (Borges, 1989: 252)[1]

The idea of an action existing within a "standstill", or trying to attain a goal that is never reached are concepts that I aim to retain when observing the phenomenon of drawing, particularly line drawing from life.

Imagine that you are looking at a cup, and you decide to draw this cup registering with a line what you consider its edges. The cup, being a round surface, does not have precise edges, but rather there is a moment when your eyes capture the limits, or edges of its material.

You begin to draw a line on your paper that curves in an equivalent manner to the limits you perceive in the material object. When you finish your line drawing of the cup, if your actions were successful you will recognise in the drawing the cup you have in front of you.

However, the line your eyes discovered when looking at the cup defines a border between something that exists and the empty space around it, a precise location where something ceases to be, an end. When you transpose this perception to paper, the line you draw is a completely new object, coming into existence from nothing (the emptiness of the paper) to something (your drawing); it establishes a beginning.

Therefore, the border you observe has a constant relationship to the line of your drawing, even if the first approaches zero as the end of the object, and the second withdraws from zero as the beginning of the drawing. Zero, or nothingness, is the common trait of both objects, cup, and drawing, and the movement of each to or away from zero will be forever equidistant in the same way that Achilles and the tortoise forever maintain the same ratio between them.

Figure 2. Still life with jugs, Giorgio Morandi, 1956, etching.

The act of drawing investigates both the instant of perception and the stimulus emitted from the perceived object. But drawing establishes other relations with time and self, and you may draw the same object over and over, like Morandi ceaselessly painting his cups and bottles, and end up, every time, with a different result.

Returning to our moment of drawing: we observe the drafter looking at the outline of the cup, closing one eye to ignore volume and erase the blurred contour. Micro-decisions begin to come into play: from choosing the relatively random or intentional place where the pencil touches the paper, activating its surface with an infinite number of possibilities of existence, to the continuing registration of lines, intentional or accidental.

The line that registers or describes the contour of the cup is also a personal record of the perception/observation of the drawn object by the drafter so that the drawing of the cup is a story of the senses, mainly the eyes, as they observe the cup.

However, the object observed is not a single shape in a non-place, without context.

It might exist in the morning light at your table, projecting a shadow, an umbra, some of its edges disappearing due to the intensity of the illumination, while at other points reflecting a shining light that blinds you. You drink the coffee that is inside and while the caffeine intensifies your perception allowing you to focus completely on the subject you draw, you will see that the cup is now stained with brown, diminishing the reflection.

Keeping this example of the cup as a model, as a way to limit the unending possibilities of observations, I ask you to imagine that the hand of someone else picks up the cup and drinks from it, causing your static

[1] "Achilles runs ten times faster than the tortoise and so gives it a ten-meter advantage. Achilles runs those ten meters, the tortoise runs one; Achilles runs that meter, the tortoise runs a decimetre; Achilles runs that decimetre, the tortoise runs a centimetre; Achilles runs that centimetre; the tortoise runs a millimetre; Achilles the millimetre, the tortoise a tenth of the millimetre, and so infinitely, so that Achilles can run forever without catching up. Hence the immortal paradox". (free translation).

and stable object to acquire a morphing character. In addition, you witness a sequence of different shapes becoming dissimilar from the original.

Now it is night and the cup is lit by different sources: a lamp from the ceiling and a lamp at your table, as well as the light from your computer. Now you will see not only the umbra, but also the half shadows of the penumbra, and you will not be able to resist including this entire feast of lines and shadows.

Your drawing is no longer a clean contour of the cup, nonetheless it registers other attributes of the cup and also of an instant of observation, of your mind perceiving what your senses tell it, your emotions being activated and moved by the play of light, and an excitement that propels you to draw, not only to register what you see but to create something that, maybe, maintains a dialogue with the beautiful moment you are experiencing.

I would like to go even further and change the material of the cup. Now it will be made of glass instead of the usual earthenware, and you will go on drawing it.

If the cup goes through all the stages described before, other changes will happen as glass plays a different game with light. Not only you have all the variations of shadows, but you will also observe transparencies, reflections, and refractions.

Glass, because it is transparent, seems impossible to draw.

How will you draw the glass panel of your window?

You do not draw the glass. You will draw its characteristics mainly through the fact that you can see beyond it even if the window is closed, that you see your reflection in the glass, that the house outside you window seems cut vertically when the window is half-opened, that it casts a slight shadow.

Your drawing is saying that there is a something that is transparent, that reflects, refracts, and that is substantial enough so as to casts some shadow.

Your glass cup could react to light in all these ways, and when you draw it, in that moment, they all become possibilities for your drawing.

The choice is inherently present in our bodies, including at the cellular level, in the unceasing decisions cells, in fact, make, so that we exist in a world where Schrödinger's cat may be alive and dead at the same time.

Chris King (2003) states that living systems consist of bifurcations of information coming from the past and the future, in a constant state of choice.

Like Schrödinger's cat, which you do not know if it is alive or dead until you open the box, your drawing will have all the possibilities of existing or not, to be there or not, only when you draw it.

The succession of actions resulting in the final drawing is not pre-established and cannot be predicted. For every detail or general image observed, there will be choices that may constantly diverge. Every aspect that is ignored, every gesture that emphasises a line or even the decision to integrate an accident into the general register, makes every drawing, any drawing, the proof of the unending possibilities of parallel existence.

Every intention that initiates a gesture provoking a mark on the paper contains an infinite possibility of other gestures that would result in other drawings. In the moment, a drafter decides to draw he/she is activating an infinite possibility of drawings-to-be.

Within the complexity of our conscious states, artistic creativity appears as a way to acknowledge reality or create worlds, taking parts or stimuli, from the world or from within the self in order to create new realities.

These new realities or artistic objects stimulate our senses, perception, and empathy—some would say our sensibility—to attain particular moments of clarity and universal consciousness. The artistic object is a time-traveller in the sense that it steps out of the progression of chronological time, attaining an ability to communicate highly complex information that may be experienced and perceived in very different situations.

The artist is the one that accepts this "standstill", but goes on forever outpacing both Achilles and the tortoise, as he or she brings from out of nowhere a new existence: the drawing.

Once again, you begin drawing and once again time becomes your moment, the forever-moment of your perception when it is irrelevant if you are making the same drawing or a different one.

However, your drawing may well have acquired that quality found in *Kairos*, which is a link, I would like to say a line that carries its observer back to a moment where shadows, or lines, or textures communicated endlessly and directly to your mind-body unity.

BIBLIOGRAPHICAL REFERENCES

Benjamin, Walter. (1940) *On the concept of History*. Available at: http://members.efn.org/~dredmond/ThesesonHistory.html

Borges, Jorge Luis. (1989) A perpétua corrida de Aquiles e da Tartaruga. *Obras Completas 1923–1949*. Lisboa: Teorema, 252–256.

King, Chris. (2003) Chaos, Quantum-transactions and Consciousness. *Neuroquantology* (1): 129–162.

Progress and regress: The current status of art in two post-communist countries

Gizela Horváth
Department of Art, Partium Christian University, Oradea, Romania

ABSTRACT: The idea of progress is strongly associated with the Enlightenment belief that humanity is capable of improving its condition through reason. "Progress" is the central concept of the 19th century, whose importance diminishes in the 20th century and is disappearing from contemporary social discourse. My paper examines the position of the idea of progress and the related system of ideas in the current political discourse of two post-communist countries, along with the way in which this is reflected in the results of the artistic activity. In Romania, progress represents a legitimate, but not central and dominant, the idea of politics, while the political leadership of Hungary views progress and its ideological context as harmful and favours a view centred upon the nation, which is oriented towards the past and builds exclusively upon tradition. In Romania, due to the fragmented political spectrum, there is no coherent political position relating to art or a developed and uniform support system; thus, artists have to organise themselves. The Hungarian government has recognised the potential propagandistic value of art and therefore centralised its organisation and financial support through the Hungarian Academy of Arts. Although one might think that a better organisation and financial support would strengthen art, we find that while the Romanian art scene is in full bloom, Hungarian art is having trouble finding its place in the international arena.

Keywords: Progress, anti-progress attitude, Hungarian Academy of Art, The Paintbrush Factory

1 THE REGRESS OF "PROGRESS"

The idea of progress gained its central importance at the beginning of the Modern Age. The roots of the secular form of progress reach back to the Enlightenment, and it can be interpreted as advancement achieved with the aid of reason, based on man's capacity to change his conditions. According to the idea of progress, we can become happier and freer with the aid of reason. The intellectuals have developed and popularised the idea of progress. Already at the beginning of the Enlightenment, the "philosophers" were working on transmitting the light of reason to all segments of society. Progress was related to every important field: social emancipation, the extension of human rights, the development of science and technology, and the improvement of living conditions.

In the 19th century, the discourse of progress was integrated into the wider perspective of evolution and became one of the central ideas of positivism. At the same time, intellectuals had ceased to be the main representatives of progressivism and gave way in this respect to the newly established nation-states. The idea of progress "was thus institutionalised in the policy and planning functions of the state" (Eyerman, 1990: 98). Based on the positivist approach, progress seemed natural and almost self-evident as the sole possibility for the future.

Among the first critics of the concept of progress were George Sorel, Durkheim, Freud, Weber and Wittgenstein in the interwar period (Alexander, 1990). Due especially to the disillusionment which followed the First World War, these thinkers refused the linear model of progress, yet continued to strengthen the world-modelling power of rationality. The two World Wars have weakened the faith in progress; the economic crisis and genocide have called into question the direction of the world's development.

The idea of progress, although weakened due to the two World Wars, reappeared stronger after the Second World War. The development of Western welfare states and the liberation of the colonies gave hope for the spread of modernization and foreshadowed its positive effects. Already in the seventies, this optimism disappeared, and hope was substituted by apocalyptic fears (Alexander and Sztompka, 1990).

The postmodern turn of the eighties has also established theoretically the dethronement of the idea of progress, since – borrowing Lyotard's famous phrase – "the grand narrative has lost its credibility" (Lyotard, 1984: 37).

Nowadays, it seems that the idea of progress has lost its society-shaping power, politicians seldom invoke it, and the social sciences have renounced it:

> In the last third of the 20th century, sociological, cultural anthropological, and other social-scientific researches have abandoned the idea of historical progress" (Sárkány and Somlai, 2003: 3).

The idea of progress was substituted by the idea of "sustainable development". At the end of the 20th century, along with the environmental aspects, many other phenomena had to be faced which has shaken the faith in progress: globalization, the fall of the socialist regimes, followed worldwide by ethnic and religious confrontations, wars, aggressive fundamentalism, and the ideas and aspirations known as "tribalism". Sárkány and Somlai (2003) call attention to an interesting paradox: notwithstanding the critiques directed at the concept of progress, the way of life has not been radically changed, and the different domains of social life (production, economy, science, technology, education, medicine etc.) continue to use the vocabulary of progress, and 'progress' is still a valid objective for the economies of developed countries.

2 "PROGRESS" IN THE POST-COMMUNIST WORLD

"Progress" was an important part of the official ideology in communist countries. The communist leadership legitimised its absolute power through the – sometimes real, but more often imaginary – progress achieved in various domains. "Progress" was used as an indispensable magic word in official party documents, while it became increasingly evident that the communist regimes drive these countries into an ever-deeper crisis. "Raising the fatherland to new heights of civilisation, well-being, and progress" was also a constant slogan of the discourses of the Romanian dictator Nicolae Ceausescu. The reference points of communist rhetoric were wealth, technology, literacy, high culture, hygiene, and class inclusion, while ever more people realised "that almost right from the beginning this revolutionary society has made of the hopes for perfection a bitter joke" (Alexander, 1990: 23).

The characteristic duality of communist regimes manifested itself also in Romania: the official discourse had little to do with either reality or the general opinion of the population. The population rightly regarded political discourses as empty phrases. Thus, although widely used at the official level, the idea of progress has become empty here just as it did in the Western world.

At the same time, viewed from the Eastern Block, the West continued to appear as an earthly paradise and the homeland of progress. Consequently, the collapse of the communist regimes in '89 has revived the faith in progress and historical optimism for a short period.

The difficulties of economic recovery, the falling behind of certain social segments, the controversial aspects of European integration, and the newly formed distrust of the political class have shaken the faith in progress within post-communist societies.

Although the countries of the Eastern Block were in a similar situation at the collapse of communism, very different answers emerged in the domain of cultural politics and, more specifically, in the post-communist art world, depending on whether the political class pursues a conservative and anti-progress strategy (such as in Hungary) or an actually directionless, laissez-faire attitude (as in the case of Romania).

3 CULTURAL POLITICS IN ROMANIA AND HUNGARY

Although our analysis is concerned with the cultural politics of these two countries, regarding primarily the arts, we find it necessary to present briefly the characteristics of the two political spheres.

Hungary and Romania are neighbouring countries struggling with the same communist heritage. However, after '89, fundamentally different political environments evolved in the two countries.

In Romania, a multi-party system was established, new political organisations continue to appear before every election, political groups come together, and parties change their names. Although there are parties that use terms such as "social democratic", "liberal", and "national" in their names, there are no real ideological differences between them; at most, one could say that they would solve certain specific problems differently. This is why it is so difficult to orient oneself within Romanian political life since parties with strong ideological profiles have never been formed. The political class aspired generally to European integration and there are no strong critical attitudes toward the EU. "Progress" is present in the political vocabulary and even in the name of a smaller political party ("the National Union for the Progress of Romania").

In contrast with this situation, the opposition between the liberal and the conservative side was soon established, and the number of political parties is quite small in Hungary. The most important political actor of the past two decades is the FIDESZ, which defined itself as a liberal party at the beginning, but a conservative turn took place within the party in 1992. The FIDESZ formed governments thrice thus far: 1998–2002, 2010–2014, and after 2014 to the present day. Sharp political differences resulted in clearer political relationships than in Romania; however, they also led to the strong polarisation of the population and to animosities between people of opposite ideological orientations. As attested by the results of the elections, the majority of the population holds conservative, religious, and national values. Due to the strong opposition, "liberal" has become a curse word for the majority of the public, and it is invoked as an accusation and a shameful option. "Liberalism" is the carrier of historical sins also within the rhetoric of the FIDESZ. The statements of Prime Minister Orbán Viktor, according to which Hungary is an illiberal state, have received wide media coverage. The FIDESZ frequently criticises the decision makers of the European Union and feeds hostility against the EU in its adherents. In the political rhetoric of the FIDESZ, "progress" is a negative concept. At the 2015 Congress, Orbán Viktor stated that Europe is in danger of annihilation due to spiritual causes, because "the European spirit and its representatives today believe in shallow and secondary things, such as human rights, progress, openness, new kinds of families, and tolerance", but do not believe in "Christianity, common sense, military virtues, and national pride." (Orbán, 2015: 28).

It is no surprise, then, that the different political climates of the two countries give rise to different kinds of cultural politics, also influencing the state of the arts in different ways.

In Romania, one cannot identify any well-defined and consequent policy concerning art in the past twenty-five years. In fact, the political class is indifferent towards the arts. After the collapse of communism, the art scene started to organise itself: galleries, artistic groups, and novel private initiatives appeared. The most successful of them is the gallery "Plan B", opened in 2005 by Mihai Pop and Adrian Ghenie in Cluj. As their website also states, the name of the gallery,

> Is a reference to the general situation of the insufficient public resources supporting art spaces in Romania and the constant need to find alternative solutions. The 'plan B' strategy is animated by individual involvement, a different understanding of the institutional, and a coherent managerial policy.

The Romanian state applied the policy of "laissez-faire" in the domain of art: it did not intervene in the realm of cultural production, and it did not consequently support it financially. Small existing grants may be obtained for projects, which are assessed by professional bodies. Thus, artists started to organise themselves, and Plan B has become a regular participant at international fairs and exhibitions. In 2012, the San Francisco MOMA has chosen six artistic centres deemed important for contemporary art, one of which was Cluj (SF MOMA 2012). In a volume published in 2013, presenting the 12 most dynamic artistic centres of today, one such centre is Cluj (Byrd and Shier, 2013).

The founders of Plan B have also established further spaces for artists. At the site of a disused paintbrush factory, also in Cluj, they have established a colourful space for artists, galleries and organisations. The Paintbrush Factory project started at the beginning of 2009, as an independent initiative to bring together ideas, events and projects of cultural organisations, galleries, producers, and independent artists in Cluj, and as a reaction to the local lack of production and exhibition spaces in the city.

In Romania, an association (Uniunea Artistilor Plastici) which defines itself as a non-governmental organisation and an apolitical association provide the official professional representation of artists.

In Hungary, the political sphere is much more preoccupied with the arts. The ideological difference, which divides the country, can also be felt within the artistic society. This fact can also be observed at the institutional level.

In 1992, two art academies have been established in parallel with each other: the Széchenyi Academy of Letters and Arts (SZIMA) and the Hungarian Academy of Arts (MMA). SZIMA has been established by the Hungarian Academy of Sciences as an independently associated institution. Its budget is part of the Hungarian Academy of Sciences' budget. The members of the SZIMA do not receive any remuneration, and they do not pay any membership fees. According to the statement, which can be read on its website, "the Széchenyi Academy of Letters and Arts embodies the idea that the spiritual freedom of the artist and the independence of creative artistic work are inviolable values." (SZIMA website). The number of simultaneous members below the age of seventy is limited to sixty.

The Hungarian Academy of Arts has been formed at the private initiative, with the participation of twenty-two artists who held conservative ideas and belonged to the circle of the world-famous organicist architect Makovecz Imre (since then deceased). They have chosen the name "academy" for themselves and are not in any relationship with the Hungarian Academy of Sciences.

Functioning as a private association at the beginning, the Hungarian Academy of Arts acquired an extraordinarily strong position during the second FIDESZ government. In autumn 2010, the Hungarian government decided to include the Hungarian Academy of Arts in the new Constitution as a public body concerned with the arts, having similar attributions and importance as the Hungarian Academy of Sciences. The founding members of the newly established public body had to be regular members of the original MMA, established as an association in 1992. Although its statute includes the provision that "regular members have to be artists who have demonstrated outstanding spiritual, creative, or performative achievements in Hungarian artistic life" (MMA website), the president of the MMA, György Fekete mentions three conditions:

> Membership has three criteria. The first is that the MMA is not an organisation of young talents… our members have fulfilled artists, recognised by society and professionals alike". The second is an "above-average public activity", and the third, "an openly stated national commitment". (Fekete, 2013)

As one can see, the interview does not mention "outstanding spiritual, creative, or performative achievements", but it mentions the requirement of public activity and also the orientation of this public activity: national commitment. György Fekete has defined the second criterion, i.e. the requirement of public activity in the following way in one of his interviews: "One has to be involved in the transformation of the structure of society, public education, and cultural PR." (Fekete, 2015). The third criterion is the most problematic; first, regarding its measurability: how can it be decided whether someone has the national commitment? In addition, does national commitment exclude criticism? Yet, these aspects do not bother György Fekete; for him, the picture is clear: "By a nationally thinking creator I mean someone who is concerned with the problems of the Hungarian nation." (Fekete, 2013) In answer to the question whether left wing or liberal artists can also be members of the MMA and if this is compatible with national commitment, György Fekete answers in the following way: "I don't know because we never had any request like this. And let me be the one to decide what I personally mean by commitment!" (Fekete, 2015).

The ideology-based exclusivist practice of the MMA's President is obvious. He establishes (ideological) criteria, which cannot be measured and, as such, permit arbitrary selection. It is also quite telling that, although its membership is today at 276, no left wing and liberal artists have so far requested admission to the organisation.

The Hungarian Constitution guarantees the authority of the MMA. Membership has several advantages: annuities (600–800 Euros monthly, depending on the character of the membership), grant applications accessible only to members, and exhibition opportunities. In addition, many representative buildings have been placed at the disposal of the MMA, which also reigns over one of the most important Hungarian art institutions, the Hall of Art (Mücsarnok). In 2016, the budget of the MMA was about 22 million Euros, and in 2017, it will be raised to almost 34 million Euros. From 2016 onward, the government nominates one-third of the Board of Trustees of the most important official tenderer, the National Cultural Fund (Nemzeti Kulturális Alap), one-third is appointed by the MMA, and one-third by other professional institutions. Thus, the absolute majority of the conservatives the money-granting institution is granted.

The MMA rejects the accusations levelled at it by posing as a representative art institution. However, knowing its history and the statements of its president, its ideological commitment is obvious.

As far as concerning the aesthetic preferences of the MMA, (progressive) modernist art is frontally attacked, and traditional, figurative, "object-making" artists are preferred. Even the famous 'new sensitivity' praised by Lorand Hegyi has been forgotten, it has not been included in the current canon. (Hegyi 1983). One of the favourite sculptors of the regime, Parkányi Raab Péter talks about an "art mafia", meaning modernist art, which allegedly marginalised talented artists for decades in Hungary (Parkányi Raab, 2014). He believes that there are two separate camps in the art world: "artists who create objects" and those who deny the *raison d'étre* of art objects. As a closing argument for his case, he demands the restoration of a situation, which is the continuation of the past: "Let those artists back into the exhibition spaces who believe in the art of the past two or three millennia". (Raab Párkányi, 2014)

The new director of the Hall of Art, György Szegö commented similarly: that which is considered contemporary by the Western world today is nothing but a soap bubble that will eventually burst, and the institution run by him would like to distance itself from art which is defined by "technical media", since these can only look back at a past of 20–25 years, while traditional arts, especially painting, has a tradition of 8–10 thousand years. Additionally, he also stated that art has to delight and not criticise. (Földes, 2015)

To conclude, Romanian politics does not intervene into the self-organization process of the arts: Romanian artists do not benefit from annuities, and they have to apply for tenders to realise their projects. Here, we have an ideologically neutral advocacy association for artists, and there are numerous private initiatives. Hungarian politics, however, uses art as an ideological weapon, and it has created the institutional and financial background to do this. With what results?

4 RESULTS

The artists working according to the principles of the MMA obviously prefer national and traditional themes, which they present in a nonconflictual form. Thus, chances are slim that anyone else besides their co-nationals would interest themselves in their work.

Another characteristic is the prioritisation of the "object-making" branches of art, such as figurative painting, sculpture, and interior design (!). In recent years, numerous controversial public statues have been erected. Some of them are contested for formal reasons (e.g. the Historical Memorial Park of folk artist Pál Czupp, established in a township with approximately 700 residents), others for their ideological background (e.g. the statue of Hóman Bálint, who was one of the proponents of Hungary's anti-Jewish legislation), or for both reasons (e.g. the Memorial to the victims of the German occupation).

The MMA has also organized conferences funded by the state with the following titles: "The current situation and future direction of Hungarian applied and design arts" (2013), "Metalwork as an auxiliary art of architecture" (2013), "Rural mythologies and realities in fifty years of Hungarian literature" (2014), and "Conference on pastoral art" (2015). In 2013, a memorial conference was held for Tormay Cécile, who in the twenties and thirties discussed the topic of worldwide Jewish conspiracies and openly considered herself a fascist.

In the Hall of Art, exhibitions prepared by curators have been replaced by the 19th-century system of the art salon, which displays the annual production of numerous artists.

The international visibility of Hungarian art is poor. Artists left to their own devices are incapable of entering the international scene. The Venice Biennale offers the possibility of becoming more visible for national artistic productions. Curators try to present here quality artworks, but the national pavilions of the later editions did not reap any great success. There were some blogs, which discussed them, but none of these pavilions received any great press coverage. The website of the 2015 Hungarian Pavilion only contains six references. In the central pavilion of the Biennale, for which the main curator can choose artists from any country, Hungary was not represented by anybody at the last three editions.

In contrast, several Romanian artists have acquired international recognition. Adrian Ghenie is the best-selling contemporary Romanian artist, and he is among the top ten best-selling young artists worldwide. He was hardly 39 when his painting entitled "Sunflowers in 1937" was sold for 4.5 million dollars in 2016. Other Romanian artists, such as Mircea Cantor and Dan Perjovschi, regularly exhibit in international museums and galleries. The website of Christie's Auction House discusses "the wonder from Cluj", highlighting the fact that the abandoned paintbrush factory now produces superstars of the art world, mentioning the names of Victor Man, Serban Savu, Mircea Suciu, Marius Bercea, and Ciprian Muresan (Howard 2016).

In the last years, the Romanian pavilion proved to be twice a veritable blockbuster. In 2013, the theme of the Romanian pavilion was an immaterial retrospective of the Venice Biennale, by Alexandra Pirici and Manuel Pelmus. Several art magazines and blogs wrote in superlatives about this exhibition. The magazine Art in America begins its discussion with the Romanian pavilion, characterised as "outstanding", and the famous art critic Claire Bishop commented to *Bomb Magazine* "the national exhibitions, on the whole, are not interesting. Romania and Great Britain, I enjoyed." (Russell, 2013) In Kunstkritik, Power Ekroth and Helga Just Christoffersen, who confess that they have often visited it and have spent several hours here, discuss the Romanian Pavilion. Art critic Federico Florian has published an ample account in *Klat Magazine*, and *Zero Deux* stated that it would award it a Golden Lion. Writing for Architect Magazine, Aaron Betsky discusses the Romanian pavilion as second in his top five selection, and Adrian Searle (2013) includes it in his account of the ten most important artistic events of the year in The Guardian.

In 2015, the Romanian pavilion was dedicated to the aforementioned Adrian Ghenie. The New York Times, The Financial Times, The Art Newspaper, Apollo magazine, and Artsy all covered Ghenie's exhibition at the Biennale. Les Echoes considered the Romanian pavilion by far the best pavilion at the Venice Biennale. News of the Artworld included the Romanian pavilion among "the five must-see pavilions", and The Art Newspaper considered it to be one of the seven most interesting. The magazine Dazed ranked Ghenie among the ten most important artists.

Besides Adrian Ghenie, the main curators of the Biennales have also selected other Romanian artists to the international pavilion: Stefan Bertalan, Geta Brãtescu, and Andra Ursuta in 2013, and Victor Man in 2015.

5 CONCLUSIONS

As we have seen, the political class of Romania and Hungary relate much differently to the idea of progress: while "progress" is an acceptable concept in Romania, which is included among other political slogans, the objective of progress has disappeared from political discourse in Hungary. What is more, other ideas associated with progress (rationalism, liberalism, emancipation) have become targets of the attacks launched by the dominant political force. The attitudes of political power towards art could not be further apart than they are in Romania and Hungary. In Romania, the political elite does not concern itself with the arts. Art is neither supported nor restricted by politics in Romania. As a result, artistic production largely depends upon private initiative, and, consequently, artistic production is exceptionally colourful and varied. In Hungary, the subsidising of art via the MMA is part of a clearly distinguishable cultural politics, also actively directed at establishing itself on the institutional level. State-subsidized art must meet ideological criteria: it must be "national" in character and has to promote Hungarian cultural values. If we examine the international artistic presence of the two countries, we find that Romanian artists are recognised, and Romanian artistic projects are appreciated, while Hungarian art is unable to get itself noticed.

Although the idea of progress can seem outdated, the radically opposite direction, the glorification of the past and the exclusive cultivation of national traditions leads to provincialism and a marginal position within the art world. One surely can move backwards in the art, but the art world will not appreciate this.

BIBLIOGRAPHICAL REFERENCES

Alexander, C. Jeffrey. (1990) Between Progress and Apocalypse: Social Theory and The Dream of Reason in The Twentieth Century. In: Alexander, C. Jeffrey and Piotr Sztompka (eds.) *Rethinking Progress. Movements, Forces, and Ideas at The End ofThe 20th Century*. London Sydney Wellington: UNWIN HYMAN, 15–38.

Alexander, C. Jeffrey and Piotr Sztompka. (1990) Introduction. In: Alexander, C. Jeffrey and Piotr Sztompka (eds.) *Rethinking Progress. Movements, Forces, and Ideas at The End of The 20th Century*. London Sydney Wellington: UNWIN HYMAN, 1–12.

Byrd, Antawan I. and Reid Shier. (2013) *Art Cities of the Future: 21st-Century Avant-Gardes*. London: Phaidon Press.

Eyerman, Ron. (1990). Intellectuals and Progress: The Origins, Decline, and Revival of A Critical Group. In: Alexander, C. Jeffrey and Piotr Sztompka (eds.) *Rethinking Progress. Movements, Forces, and Ideas at The End of The 20th Century*. London Sydney Wellington: UNWIN HYMAN, 91–105.

Fekete György. (2015) Az minek számít, amikor valaki szétrúgja a miniszterelnök fejét? [Online]. [Accessed 7 January 2017]. Available from: http://24.hu/belfold/2015/09/11/fekete-gyorgy/

—. (2013) Karakteresen Nemzeti. [Online]. [Accessed 7 January 2017]. Available from: http://mandiner.hu/cikk/20121203_fekete_gyorgy_karakteresen_nemzeti

Földes András. (2014) Nem lesz szappanbuborék a Mücsarnokban. [Online]. [Accessed 7 January 2017]. Available from: http://index.hu/kultur/2014/09/11/mucsarnok_a_muveszet_ne_kritizaljon_hanem_gyonyorkodtessen/

Hegyi Lóránd. (1983) *Új szenzibilitás*. Budapest: Magvetö

Howard, Maria. (2016) The abandoned paintbrush factory producing art's new superstars. [Online]. [Accessed 7 January 2017]. Available from: http://www.christies.com/features/Adrian-Ghenie-and-the-Cluj-paintbrush-factory-producing-art-superstars-7122-1.aspx

Lyotard, Jean-François. (1984) *The Postmodern Condition: A Report on Knowledge*. Manchester: Manchester University Press.

Parkányi Ráab Péter. (2014) Magyar müvészeti maffia. [Online]. [Accessed 7 January 2017]. Available from: http://magyarhirlap.hu/cikk/4224/Magyar_muveszeti_maffia.

Russell, Legacy and Deitz, Bibi. (2013) Faces of the Biennale. *Bomb Magazine*. [Online]. [Accessed 7 January 2017]. Available from: http://bombmagazine.org/article/7266/

Sárkány Mihály and Somlai Péter 2003. A haladástól a kontingenciáig. Vázlat a szocio-kulturális evolúció változó elméleteiről. Szociológiai Szemle 3, 3–26. [Online]. [Accessed 7 January 2017]. Available from: http://www.matarka.hu/cikk_list.php?fusz=9326

Searle, Adrian. (2013) Top 10 Art Shows of 2013. *The Guardian*. [Online]. [Accessed 7 January 2017]. Available from: https://www.theguardian.com/artanddesign/2013/dec/18/adrian-searle-top-10-art-shows-2013

Thompson, Kenneth. (1990). Secularization and Sacralization. In: Alexander, C. Jeffrey and Piotr Sztompka (eds.) *Rethinking Progress. Movements, Forces, and Ideas at The End of The 20th Century*. London Sydney Wellington: UNWIN HYMAN, 161–182.

Online resources:

An Immaterial Retrospective of the Venice Biennale. Facebook. [Online]. [Accessed 7 January 2017]. Available from: https://www.facebook.com/AnImmaterial RetrospectiveOfTheVeniceBiennale/

MMA website. [Online]. [Accessed 7 January 2017]. Available from: http://www.mma.hu/hu/az-akademiarol

Orbán Viktor 2015. Beszéd a Fidesz tisztújító kongresszusán. [Online]. [Accessed 7 January 2017]. Available from: https://www.youtube.com/watch?v=owK1TM0JNZQ&t=1699s

Plan B website. [Online]. [Accessed 7 January 2017]. Available from: http://www.plan-b.ro/index.php?/about/about–press/

SF MOMA 2012: Six Lines of Flight. [Online]. [Accessed 7 January 2017]. Available from: https://www.sfmoma.org/exhibition/six-lines-of-flight/

The Paintbrush Factory website. [Online]. [Accessed 7 January 2017]. Available from: http://fabricadepensule.ro/en/about/

SZIMA website. [Online]. [Accessed 7 January 2017]. Available from: http://mta.hu/szima/a-szechenyi-irodalmi-es-muveszeti-akademiarol-105824

The promotion of art on the path of socio-cultural development

Maria João Delgado & Maria Heloise Albuquerque
University of Lisbon, Faculty of Architecture, CIAUD, Lisbon, Portugal

ABSTRACT: This paper is set out to show how the different forms of dissemination and sharing of art work, regardless of the artistic discourse they contain, identify and express the social, cultural and political ideas that, at a given moment, dominate society.

We present a literature review, in a reflective dimension, on the confluences between art and society, in the different expositive trajectories that supported the artistic and intellectual conceptualizations throughout time.

From the museum's first exhibitions, independent art galleries associated with the avant-garde movements, to the present urban space intervened by Street Art, we tried to understand how plastic artists are aware of the role of art in social criticism. The representations and artefacts reinvented and transformed by the urban creators feed new paradigms of management and expositive communication.

Exhibition spaces act not only as a means of communication and artistic representation but also as a means of socio-cultural affirmation.

Keywords: Art exhibitions; museums; artistic vanguards; Street Art.

1 INTRODUCTION

Exhibitions, as a socio-cultural expression that simultaneously reveals the development of new ideas, both around art, science and technology, have assumed different forms over time, in an individual or collective demonstration of a social attitude.

Exhibitions are born from the concept of collections that exists since the Roman Era, with the idea of gathering and presenting in forums, theatres and thermal baths the rare objects of war spoils. The Greek had already exhibited surgical and astronomic objects mixed with statues from the great masters of culture or with other cult objects, satisfying the myths of those times (Newhouse, 2005).

The extensive available literature makes it possible to point out that museums, which were initially created as immutable institutions and autocrats, followed the changes of the various eras they passed through, in a constant adaptation of their mission, contents and means of action.

If for a long time museums were given the task of acquiring, preserving heritage and fostering artistic creation and diffusion, contemporary museology, observant of new social, cultural and economic dimensions, did not stop at the level of exhibitions; it also creates a broad vision of culture, associating to the exhibition practice the educational responsibility and cultural diffusion, providing services and products from the cultural market. In the field of their physical spaces and content, today museums tend to open up more and more to their public, integrating innovative structures.

As mentioned in an article, today's museum "invests in the development of cultural activities (...) guided tours, educational visits for children, adults or professionals, bars, restaurants, rest areas and recreation" (Albuquerque and Delgado, 2015, pp. 6416).

However, in many situations, cultural institutions have been replaced by spaces with a more communitarian and democratic character, such as urban space, in a direct response to the messages, wishes, and even provocations of the artists that represent the meeting of certain communities.

In 2001, UNESCO highlighted the importance to the right to equal access and participation of all people, both in the artistic process and the scientific and technological knowledge, emphasising its contribution to the economic growth of societies.

This clearly demonstrated that the interconnection of these dimensions – artistic, social, cultural, economic, and technological – contributes to social inclusion, cultural diversity, and human development.

The assertion of exhibitions and the growing artistic and socio-cultural interdependence, which was felt more profoundly since the 1960s, not only did not stay on the sidelines but also in many ways became a social phenomenon. The exhibition context is a fundamental aspect for the understanding of the artwork and it has been a key theme, since the end of the XIX century in the setting of exhibitions (Barker, 1999). However, the exhibition design also reveals the social character attributed to artwork, "a work of art is not the product of a genial individual, but the form of expression of a social state" (Nobis, 1998: 149).

However, museums have not always welcomed all artists, as happened with avant-gardes the end of the XIX and XX centuries and, nowadays, with graffiti artists. These excluded artists discover new exhibition spaces by bringing their work to the public and provoking reflections and debates around the critical problems of society.

The exhibition design always accompanied the progress and socio-cultural changes that have manifested themselves both in the public cultural institutions and in the exhibitions of individual initiative.

2 EXHIBITION HALLS AND CANONS

Along with the trajectory of the birth of museums, between XVII and XIX centuries, regular exhibitions and temporary shows emerge in different European cities. These are precursors of the "Salons", promoted by groups of artists, which arise alongside social, economic and cultural developments (Newhouse, 2005).

In Paris, the "Académie Royale de Peinture et de Sculpture" organized an exhibition in the "Salon Carré du Louvre", in 1725, attracting a large number of visitors and, in 1768 in London, the Royal Academy of Arts was created and considered at the time by many artists as the best space open to the public to exhibit works of art.

However, the significant increase in the number of works of art on display, because of the commercial pressures of buyers and artists, and the lack of exhibition space, led to the devaluation of these events. Despite several attempts to improve the way of exposing all the works, which were heaped on the walls of the various rooms and galleries, from the floor to the ceiling, the exhibition of the paintings and sculptures continued not to respect any aesthetic and expository criteria. There was no formal agreement or historical sequence in the arrangement of the works, by themes or artists, as a priority given to commercial aspects only (O'Doherty, 1999).

This author comments on the "Salons" in this way:

The Salon itself implicitly defines what a gallery is, an appropriate definition of the aesthetics of the period. A gallery is a place with a wall, which is covered with a wall of pictures. The wall itself has no intrinsic aesthetic; it is simply a necessity for an upright animal (O'Doherty, 1999; 15).

In addition to the artists' dissatisfaction with the presentation of their works, the jury of selection of works gave preference to the great masters of painting already consecrated, who had a monopoly of production for the sale of paintings and sculptures, creating a confrontation of interests with the new generations of artists.

Individual exhibitions or independent groups of artists born from Associations of Artists, Independent Associations of Artists and Secession began to appear, in a statement against the selection made by juries and in favour of a new form of organisation and presentation of the work exposed.

Newhouse (2005) attributed to the French painter Jacques-Louis David the beginning of this practice, with the opening of an individual exhibition in 1799. This exhibition, in which the artist was responsible for how works were presented, ensured successful sales for five years. The most paradigmatic case was that of Gustave Coubert when he submitted to the selection jury a painting to be presented at the first Universal Exposition in Paris in 1855, which was rejected. As a reaction, the painter organised a private show in a tent that he built near one of the doors of the building, which he calls the Pavilion of Realism, where, in an act of commercial innovation, he sold his paintings, photographic reproductions and the exhibition catalogue accompanied by a manifesto.

In 1863, following the protests of the artists and with the endorsement of Emperor Napoleon III, the works of art rejected in the official "Salon", which was intended for artists' members of the "Académie Royale de Peinture et de Sculpture", were exposed in what became known as the "Salon des Refusés". This was the first and last time that Napoleon allowed such an exposition, considered as the affirmation of exhibitive modernity. During his government, all the works that escaped the themes and techniques accepted by the Academy continued to be rejected.

Gustave Coubert, who continued to see his paintings rejected by juries, organised another private exhibition in 1867, inviting Edouard Manet to exhibit his works, which had also been rejected by the "Salons" over the years. The success of these exhibitions, with a high number of visitors and sales, encouraged other artists to follow the same path (Celant, 1996), opening doors for the introduction of new criteria and exhibition environments.

This was the case of the exhibition of the *Société anonyme des artistes-peintres, sculpteures et graveurs*, which, outside official tutelage, organised an exhibition in Paris, in 1874, in the studio of photographer Félix Nadar, whom the critics later called Impressionists. Thirty-one artists participated in this show, among which were Edgar Degas, Camille Pissarro, Auguste Renoir, Claude Monet, Alfred Sisley, Berthe Morisot, Armand Guillaumine and Paul Cézanne.

This group would organise approximately eight private exhibitions, which had great relevance in the affirmation of these avant-garde artistic movements, both by the artistic discourse used and by the configuration of the exhibitions and the promotion of new spaces of culture. The resort not only to newly opened art galleries but also to other makeshift spaces, such as the *Café Volgini*, in Paris, which in 1880 received the works of Gauguin at the same time as the Universal Exhibition (Altshuler, 1994).

The *Salon des Indépendants* in 1884 is another event marked by the initiative of the artists, Odilon Redon, Georges Seurat and Paul Signac, who try to emancipate the work from the exhibition space, defending its neutrality. As Newhouse (2005) comments, this position anticipates the White Cube that marked the XX century in what concerns the conceptual principles of the design exhibition.

The 1905 *Salon d'Automne*, which brought together works by Renoir, Picabia, Kandinsky or Cézanne, was fundamentally known for the presentation of the group of artists "Les Fauves", and for the repercussion that these artists' works, the intense chromaticism and the freedom of artistic language, had in public. The exhibition "Sonderbund 136", organised in Cologne in 1912, became paradigmatic in its composition, like the Viennese Secession in 1902, with the introduction of the concept of total work.

Although in Europe the exhibitions of the individual initiative were present and opened to the discussion of the great artistic and social issues of its time, the Armory Show – The International Exhibition of Modern Art – was the first joint exhibition of painting and sculpture of the European and American vanguards. Held in New York in 1913, The Armory Show stood out both for the presentation of modern art to a public still accustomed to realism, as well as for the form of exhibiting, which emphasised cultural and educational aspects, in addition to the contemplative act (Celant, 1996).

After the first decade of the XX century, several independent associations continued to organize exhibitions in Europe, which stood out not only for their artistic expression and in the way they exhibited and used the exhibition space, but also, and above all, for the use of critical texts, catalogues, magazines and organization of debates, highlighting their ideologies, their political, cultural and social ideas.

This is the case of the "Der Blaue Reiter" (Blue Knight) group show, organised in Munich in 1912, and the Dadaists, whose exhibitions definitely marked the exhibition scene in the beginning of the century, with provocative politicised interventions, particularly the 1st DADA International Fair, in 1920.

Also, according to Nobis (1998), the exhibition "L'Espace des Abstraits" in 1927/1928, proposed by Lissitzky, had both space and the visitor at the centre of its preoccupations. This space, which lived from the dynamics that it established with the work of art, invited the visitor to participate in a scenographic game, to discover the various works of art.

The International Surrealism Exhibition, conceived by Marcel Duchamp, which included Salvador Dali, Man Ray, George Hugnet, among others, held at the Charles Ratton Gallery in Paris, and in New York, both in 1938, did not stop at the formal presentation of works within a conventional space, but transfigured this space, presenting the works and challenging simultaneously all the senses (Altshuler, 1994). In this exhibition, the work of art places objects of daily life in the artistic plan, and according to Uwe M. Schneede (1998), the artistic work subjugated to the created exhibition environment. This environment becomes an integral part of the work of art, where walls, floor, ceiling and the works presented were part of the exhibition. The exploration of the visitor's emotion opens doors to the aesthetics of reception (Eco, 1993).

Marcel Duchamp, considered the precursor of the minimalist and conceptualist tendencies of the second half of the XX century, goes from painting to the selection of industrialized objects, decontextualizing them and elevating them to the category of artistic object, by enhancing their aesthetic capacities and incorporating them in his artistic works, which he calls ready-made.

Simultaneously, it shows the intentional relation between works of art and their placement, the so-called site-specific, which refers to the minimalism of the 1960s and 70s, and to the new conception of the exhibition design.

Works of art are an integral part of the work and the place where they are presented, requiring the physical presence of the observer to complete the artistic process.

One of the unique and highly successful exhibitions was the "When Attitudes Become Form" exhibition organised by Harald Szeemann at the Kunsthalle, in Bern, in 1969, with artists from New York and various European cities. The complexity of the exhibition rested on both objects, which were no longer seen as an autonomous system, and in events, the behaviours and gestures of artists "where the interior attitude of the artist had never been directly processed" (Szeemann, 1998, p. 372). Artists focus on conceptual art, of which the raw materials are the concepts, and on the connection between the object and the environments that provoke a reaction in the viewer.

As we will see next, in the last decades of the XX century, artists, aware of the role of art in social criticism, open new ways to share and communicate with the public, promoting urban space as a place of challenge and social and artistic representation.

3 THE CITY AS AN EXHIBITION SPACE

The visibility of art in public space has its origin in the most conceptual and social aspect of art, graffiti, which arises in Philadelphia in 1970 when some young artists begin to paint their signatures on the trains and walls of Manhattan.

The graffiti, associated with acts of transgression and irreverence of groups that wanted to give voice to the protests against the evictions of the communities that lived in the suburbs of the cities, that were being transformed to give place to the renovation of urban constructions and motorways, begins to manifest itself through codes and inscriptions left on the walls of the buildings of the cities. They wanted to spread their signature quickly, or tags, highlighting their presence in the neighbourhoods.

In 1980, they present themselves as an artistic movement, whose natural evolution contemplates, not only the place but the style, the technique, the originality and its quality, constituting itself as a form of leaving their mark and signalling possession of the territory, neighbourhood or street of each group.

Parallel to graffiti, associated with a territorial brand and having the tag as essence, Hip-Hop is born as a cultural movement among some young people from the suburbs of New York City.

With a huge socio-cultural impact, this movement assumes diverse cultural dimensions, including movements such as Street Art or Post-Graffiti – which technically resemble Graffiti. If Graffiti uses proprietary and closed codes to the graffiti community, Street Art, maintaining the same conflict between legal and illegal, opens up to society with noticeable messages. The interconnections between Graffiti and Street Art are very strong, with modes of expression in constant evolution, which makes it difficult to define the boundaries between these two artistic activities that draw on very similar techniques and intentions. However, Street Art, associated with art that occurs on the street, includes, in addition to the visual arts, other creative expressions that take place in collective spaces: theatre, human statues, performances, installations, video mappings, among others.

Schwartzman (1985, p.7) states that the work of street artists attempts "communicate with everyday people about socially relevant themes in ways that are informed by esthetic values without being imprisoned by them".

Urban art has turned out to be a cultural innovation that encourages reflection, giving voice to all communities with their own languages. Writers express themselves through symbols and critical comments they leave on the streets, conveying meaningful messages.

The urban space becomes the urban art scene, where it is the responsibility of the public to decide the intrinsic value of these artistic interventions. Groups of people who remain anonymous since personal and individual recognition are not valued perform many of these works.

We are living a period of new ways of seeing and producing images, based on the diversity of materials and media, where the works of Street Art, which uses both signatures with different calligraphy and images, represent different types of artistic language and messages directed at street people.

Over the past few years, graffiti artists have been using a wider scope of expression. Personal style is free to develop without any constraints, and stickers, posters, stencils, airbrush, oil-based chalk, all varieties of paint and even sculpture are used. Most artists have been liberated from relying solely on the spray can (Ganz, 2006: 7).

In a synthetic way, we present some of the main Street Art techniques associated with the plastic arts, which we can find in our cities, such as a mural, stencil, stickers, among others.

The mural, closely linked to architecture and executed on a large scale, is a work done directly on a wall or panel, which gives it an apparent character of greater permanence in time than a screen or paper. In general, the message of the murals is accessible, direct and shared by many people.

In turn, the Stencil technique allows the reproduction of the image in a very fast and effective way in the different supports of the city. It uses a mask on paper, cardboard or other material to limit and define the images that are executed with spray paints. These two techniques are widely used by some urban creators, to make social and political criticisms, namely Banksy. This is one of the most internationally recognised artists whose pieces, with his signature, reach very high sales values. One of his works, the 2012 Slave Labour, was removed from a building in north London where it had been painted, transferred to a canvas and sold at an auction by the owners of the wall. This raised issues of property rights and regulation of these works, as these artists remain mostly anonymous.

Banksy gives great value to the context as a message. At the height of great instability in relations between Israel and Palestine, and taking into account the symbolism of the barrier separating two populations, the artist chose this wall to leave an image of social and political criticism, incorporating images of escape, always associated with humorous elements.

Stickers are directly associated with the tags that use to spread their name. Like the Stencil, this technique developed in the 1990s in response to repression against graffiti.

Paste up, wheat pasting or flyposting, which consists of the glueing of posters, is another technique that, like the previous ones, is quick to execute and does not require any improvisation in the act of placing. However, all the work is previously prepared. One of the most important works carried out in 2007 with this technique, mainly for thematic creativity, the message and the social and political challenge that it contained, is the illegal exhibition "Face 2 Face", by the French photographer JR. He created a street exhibition from photographs taken by Palestinians and Israelis, who practised the same profession, side by side on the walls of Palestinian cities.

Maintaining the objectives of spreading names, images and messages, arises the technique of mosaic, inspired by a video game, arises in which the artist intends to transform the world into a game by glueing small squares of glass, stone, plastic or tile until an image is obtained.

Clean graffiti or reverse graffiti is a technique that exploits the creation of images from the contrast between dirty surfaces and clean surfaces with the help of a water pressure gun. It can also be seen in the transformation of abandoned spaces and artefacts, which punctuate cities, aesthetically altering them, valuing degraded areas and leaving a social message. These installations, associated with three-dimensional artistic interventions, have a more interventionist and protesting character regarding social, political or other issues.

The processes of interpenetration between materials and graphic processes, urban support, creativity in artistic activities and expression of messages, direct contact with the illegality of the activity and, often, anonymity, acquired relevance in the Portuguese case, both in the dimension it presents in the context of cities, and in the strong dynamism, debate and social intervention with which it has been affirmed in the field of urban art.

In Portugal, the mural painting gained a great projection from 1974 onwards, where the political slogan is an ally to the urban art. Plastic artists participate in the elaboration of large collective murals in public places, many of them with messages targeting directly the observers (Brito, 2013).

It was not until the mid-1980s that the art of graffiti finally settled in Portugal. The 21st century marks the development of Street Art in Portugal, with new writers and crews, which fill cities with tags and images. Currently, we find in Portugal a group of Street Art artists, internationally recognised, whose works bring art, space and urban life together, through the artistic expression that marks the beginning of our century.

4 FINAL REMARKS

From antiquity to contemporaneity, this process of approximation of artists to the community translates into several phenomena, and it has found in exhibitions in museums, galleries or even in the public space, a way to publicise their work and simultaneously relate to the public, in a socio-cultural commitment.

The emergence of new artistic expressions, in its most varied forms of representation and presentation, reflects and generates social implications, creating a strong link between art and the social and cultural development of populations.

In this sense, urban art connects to museums and art galleries to spread the new artistic trends and becomes part of the cultural fabric of cities.

The representations and artefacts reinvented and transformed by the urban creators feed new paradigms of management and exhibitive communication. The experiences of urban intervention emerge as a liberating form of artistic expression and the limitations imposed by museums and galleries, with an innovative proposal to place art on the stage of the city, ensuring the visibility of the work of all artists and the accessibility of the information conveyed by these Works to the populations.

The relevance of the approach to the history of exhibition spaces, with the presence of the socio-cultural dimension, opens up for proposals to reflect on the conceptual and technical evolution of urban art, placing it in the field of artistic trends. If the internet and the digital media are disseminating and uniting this community of artists, it is therefore important to understand what unites and separates their artistic proposals and how they enable the reading and interpretation of the various messages, bringing together communities with different cultural identities and values.

BIBLIOGRAPHICAL REFERENCES

Albuquerque, M.H. and Delgado, M.J. 2015. Sustainable Museographies – The Museum Shops. *Procedia Manufacturing, Elsevier.* [Online]. (3), 6414–6420. doi:10.1016/j.promfg.2015.07.912

Altshuler, B. 1994. *The Avant Garde in Exhibition: New Art in the 20th Century.* New York: Harry N. Abrams.

Barker, E. (ed.). 1999. *Contemporary Cultures of Display.* New Haven, London, Yale University Press: The Open University.

Brito, N. 2013. *A Contestação Social e a Criação Literária na Arte Urbana pós 25 de Abril em Portugal. Conferência Arte e Cultura Urbana.* CELE: UNAM. Available from: http://www.cd25a.uc.pt/media/pdf/Biblioteca%20digital/AContestacaoSocialea%20CriacaoLiterariana%20ArteUrbanapos25de%20Abrilem%20Portugal%20-_Nuno%20Brito.pdf.

Campos, R. 2010. *Por que Pintamos a Cidade? Uma Abordagem Etnográfica do Graffiti.* Lisboa: Fim de Século.

Celant, G. 1996. A Visual Machine. In Greenber, R., Fergunson, W., Nairne, S. ed. *Thinking about exhibitions.* London: Routledge.

Eco, U. 1993. *Leitura do texto literário – Lector in fabula: A cooperação interpretativa nos textos literários.* 2ª edição. Lisboa: Presença.

Ganz, N. 2009. *Graffiti World: Street art Form Five Continents.* New York: Harry N. Abrams Inc.

Newhouse, V. 2005. *Art and the power of placement.* New York: Monacelli Press.

Nobis, B. 1998. El Lissitzky, L'Espace des Abstraits. In: Kluser, B. and Hegewisch, K. eds. *L'art de l'exposition.* Paris: Editions du Regard, pp. 145–157.

O'Doherty, B. 1999. *Inside the White Cube: The Ideology of the Gallery Space.* 4th edition. Berkeley and Los Angeles: University of California Press.

Pastor Homs, I. 2009. *Pedagogía museística. Nuevas perspectivas y tendencias* actuales. Barcelona: Ariel Patrimonio.

Rico, C. 1994. *Museos, arquitectura, arte: Los espacios expositivos.* Espanha: Silex Ediciones.

Schneede, U. 1998. Exposition Internacional du Surréalisme, Paris 1938. In: Kluser, B. and Hegewisch, K. eds. *L'art de l'exposition.* Paris: Editions du Regard, pp. 173–187.

Schwartzman, A. 1985. *Street Art.* New York: Doubleday.

Szeemann, H. 1998. When Attitudes Became Forme, Berne 1969. In: Kluser, B. and Hegewisch, K. eds. *L'art de l'exposition.* Paris: Editions du Regard, pp. 369–382.

Part IV

Humanities

Progress versus decadence. In the pursuit of a demystification of the black legend of the Portuguese Empire

Maria Leonor García da Cruz
Centro de História da Universidade de Lisboa, Faculdade de Letras da Universidade de Lisboa, Lisboa, Portugal

ABSTRACT: Developing controversies from the historical past of Portugal supposedly divided into cycles of progress and decadence, historians, scholars, politicians and economists from 19th century Portugal resorted to testimonies of previous centuries to prove their theses. The testimonies selected from the 16th and 17th centuries helped them consolidate a black legend of the Portuguese Empire, mostly because of the fragmentation and lack of contextualization to which such discourses were subject.

There must be a revision of contemporary concepts and interpretation that requires a close rereading and objective analysis of the discourse of 16th century authors, their complete works, deepening their philosophy of History and the factors that they valued in their time as conditions of progress and decadence, comparing them with each other and with other different testimonies.

Keywords: Philosophy of history, progress, empire, laws, faith, public good

The profiles of the Portuguese Empire and the definitions of management guidelines by mainland Portugal have been the target of different critical reflections throughout the Modern Age and the Contemporary Era, be it due to the circumstances and aims in which they are conceived (result of the author's issues of contemporaneity), be it due to those persons pronounce (politicians, academics, legal experts, economists…).

How can one confront such profiles with other phenomena of a national-imperial configuration, profiles that differ according to times and to geopolitical areas, and also to cultural interconnections? What kind of perceptions can one examine through speeches that, above all, reveal different types of national awareness, greater or smaller sensitivities to political and social, economic and institutional, ideological and functional issues?

Behind every speech, one discovers a Philosophy of History, a general interpretation of the course of the historic process, which, very often, reveals how much opinions are divided when defending in the interpretation a linear course of events or a cyclical course of phenomena. The coverage of the national-imperial reality also differs between authors-witnesses. Some of them show a comprehensive overview of the historic reality and of its natural diversity and complexity, some others seem to conform to the effects of strict organic structures, albeit, who knows, admittedly regulatory pertaining to the cluster.

In particular, in Portugal in the 19th Century, at the time of continued crises, and, at the same time, of regeneration projects, one notices ideas and more, an ideological literature, which conceives global interpretations of the equilibrium of the national historic process, be it to justify stages of decadence or to justify trails leading to the future (Matos, 1998; Serrão, 1983).

In a stance which is assumedly contrary to the historiography of the imperial venture, a cultivator of archetypes based on great figures or heroes of the past in the Discoveries and in the Portuguese Overseas Expansion, stands out: Alexandre Herculano in the middle of the 19th Century, a medievalist, who, although a believer in progress and Civilization, affirms the decadence of his times, placing the entire Modern Age, from the 16th Century, under trial and rendering it a target for doom.

In addition to oeuvres written in a militant spirit of ideologies and institutions – such as his *História da Origem e Estabelecimento da Inquisição em Portugal* [History of the Origin and Establishment of the Inquisition in Portugal] (1854–59; Herculano, 1975) – marked by a strong subjective load in their treatment of documental testimonies, Herculano writes stories, rich in critical intensity of the Modern age, where he appears to want to disassemble stone by stone, a historiographic structure, erected around a stage viewed by many as a period of glory and progress in Portugal's national history.

Such devastating discourse, revealing, above all, the degradation, the shadows behind the actions and phenomena of an apparent radiance that he himself calls the "anti-lesson", is revealed in *Carta V sobre a História de Portugal* (1842) [Letter V] and in *Cogitações soltas de um Homem Obscuro* (1846) [Loose Cogitations of an Obscure Man] and also in the *Solemnia Verba. Segunda carta* (1850) (Herculano, 1985) but we anticipated it in other prior works, such as *Historiadores Portugueses. IV* (1840) [Portuguese Historians], *Arqueologia Portuguesa* (1841–43) [Portuguese Archaeology] and *Pouca Luz em Muitas Trevas* (1844) [Few Light and too much Darkness] (Herculano, 1985).

For him, the name Renaissance itself would have been a "cruel mockery" (1842; Herculano, 1985). Why? Because through the appearance of progress and civilisation embossed on the territorial and commercial peak of the empire, on the printed word and literary progress, he unveils a historic change corrosive of order, of peace and morality, which the monarchy had developed until the end of the 15th Century. He sees in D. João II the real person responsible for the beginning of the decay, continued, despite appearances, by D. Manuel, albeit decadence would only become truly visible in the later years of D. João III. These two sumptuous reigns already concealed the "ill body of Portuguese civilisation, which rushing along to perish from its greed for conquests" (1842; Herculano, 1985).

The 16th Century would indeed be a pivotal century of the death of the "inner energy of a people". Herculano puts himself in a polemic position by judging History based on such outcomes of royal action that he considers negative, in fact, absolutely rampant.

What, then, stopped the nation's progress and in which phenomena was it based on?

According to our distinguished medievalist, because there are cycles in the unfolding of national history, the peak was what had taken place from the foundation of the nation until the end of the 15th Century. The Discoveries and seaborne ventures parallel to the strengthening of royal authority, in particular with D. João II, albeit dated from that preceding period, formed, from his point of view, (in D. Afonso V's reign) an accessory to the much sought-after conquest of Morocco and, then, to the ideal that at the time was behind the conquests, that is to say, the reaction to the domination of Islam. War, until then, had been nationalistic and "for religion and freedom", therefore, logic and just.

Only after this era of boom and progress did the turning point take place with the changes to the royal policy, the new monarch wishing that the overseas Expansion, once incited, would become a source of wealth and aggrandisement of the monarchy itself. The Crown would gain but not the country in Herculano's words. This was the rupture to the historical tradition of investing in the "sap of the social tree" (1846; Herculano, 1985), the State in the person of the king. Politically and economically wise, D. Manuel would have followed that inclination.

Then followed a time of contradictions, a dominant propensity prevailing, both from a political and economic viewpoint, laden with new ideological guidelines, which imposed a commercial investment, attaching less importance to agricultural production, and from the viewpoint of the exclusiveness of the Crown (royal holder of trade) and control of the sea, this, moreover, according to Herculano, based on unsure grounds.

"National glory" (1850; Herculano, 1985) of the realm and of public welfare, therefore, disguises a one-man government and one that is solely in the royal interest, with the Discoveries being managed negatively and degenerative. Such deterioration reached the national temper, which was now carried away with ostentation with an oriental flair, with luxury, cinnamon and pepper, gold and marble, abandoning the harsh fights in Northern Africa which took place in the name of the "glory of the Cross" (Herculano, 1977). The loss of stances in Morocco would disclose a dying nation and the defeat at Alcácer Quibir in 1580, "the death of the nation" (1844; Herculano, 1985), revealing D. Sebastião's frustrated attempt to recover political ideals when the old society had perished.

The allegedly long term victorious overseas Expansion – for Herculano Discoveries and conquests "in the venerable shadow of the cross" (1846; Herculano 1985) – has a negative connotation because it meant making a tremendous effort to succeed in far-off enterprises (including the luxuries from India and the gold from Brazil, accompanied by covetousness and looting) which, being above the country's resources, exhausted it.

The State extensively characterised by this author as an entity beyond that of the Crown[1], albeit dead displayed a grandeur, which was actually only the Crown's due to its wealth. The nation's soul had died, decrepit in its own day, morally corrupted and materially depleted. Nothing had benefitted.

In an appraisal of the historical process, what would then be the definition of progress and national glory? Great deeds for sure, but which ones? It would be necessary to reassess the deeds based on their usefulness, aims and ethics: were they useful for the moral and material progress of the people? Were the reasons always noble and pure? Was it a national impetus or the accomplishment of a prince's whim and ambition? Were they haunted by huge crimes? Note that in his discourse on the Modern Age (that, one must bear in mind, is not his speciality as a historian), Alexandre Herculano points out only the negative aspects and takes them as universal, from Charles V's tyranny in the American conquests to the rapes and cruelties of the Portuguese in their own conquests.

The country only gained a fleeting moment of wealth and a fictitious prosperity, easily obtained by crime, destructive of working habits and the economy,

[1] One could almost foresee a design of a few decades later, already in the first quarter of the 20th Century, when the jurist Maurice Hauriou defined Public Law as being far beyond the State, viewing the forces and organizations underlying it, in a reflection that links legal and sociological phenomena. Of course, in this studious State it would correspond, in the case of a monarchical institution, the Crown, which, in its capacity as leading entity "institution, pillar of institutions", it would regulate general rules (disciplinary law) and would assume an interior balance of forces, in this case, rulers and the ruled.

squandered by vice. What could then reverse the situation and save the national vigour now in a state of detachment? Herculano found it in a different political regime, which was opposed to absolute monarchy, since he considered that this one had annulled, already in D. João II's day, with his unitary inclination, feudal traditions, aristocratic independence and democratic or municipal freedom (1846 and 1841–43). The conditions for progress would, subsequently, be active work, severe habits, love for moderate freedom and the wish to cultivate the art of peace.

It should be pointed out how much as been compiled and redesigned in consecutive editions of the *Opúsculos*, statements and compilations of texts by Herculano on the Modern Age, with, for example, the new title *Cartas sobre a História de Portugal* [Letters on the History of Portugal] in the posthumous 1881 compilation, or a supplement of disperse documents published in prior years, such as, *Historiadores Portugueses* (1839–1840) and *Cogitações* (1846), in 1884 (Herculano, 1985). In the 1985 edition, articles by the author from 1837 are attached to several Chroniclers and Garcia de Resende, on Damião de Góis and Diogo do Couto. *Arqueologia Portuguesa* (1841–1843; Herculano, 1985) also brings together an ensemble of three documents from 1571, 1578 and 1580, a fourth, a translation without a signature, from 1582, in the 1985 edition (Herculano, 1985). But the observations that the author makes on documents from the 16th Century do not always follow a rigorous historical analysis as he selects them according to a rather limited previously construed restricted survey and elaborates in order to demonstrate a previously designed thesis.

Like Herculano, and influenced by him, other 19th Century authors underline the existence of phases of progress and peaks as well as of decadence in Portugal's history by connecting characteristics and situations of mainland Portugal to the vicissitudes of the overseas empire, to a definition of conquest and to the "fumes" of India. Rebelo da Silva tells us about the "Asian delights" that poisoned the "old and robust Portugal" in his *História de Portugal nos Séculos XVII e XVIII* [Portuguese History in the 17th and 18th Centuries] (Silva, 1971). As Borges de Macedo pointed out when he commented on this oeuvre, the author's interpretation of opposite phases of the historic process was directly associated with the priority he gave to the political factor and by identifying absolutism and centralization[2], in a profound detachment with regard to the role of regional bodies in public life.

Scholars such as Antero de Quental or Oliveira Martins, among others, stressed the peak and the decadence of the Portuguese historic process, although the first writer pointed out the decline in a more comprehensive fashion, interconnecting the mainland state of affairs, especially in the economic field, with the overseas vicissitudes in *Causas da Decadência dos Povos Peninsulares nos Últimos Três Séculos* [Causes of the Decadence of the Peninsular Peoples in the last Three Centuries] (1871; Quental, 2010). But as Hernâni Cidade ask in that respect, did no sector escape from the decadence of the 17th, 18th and 19th Centuries, not even external conditioning factors? (Cidade, 1980) One clearly must bear in mind that the historiographical advances on Brazil, the explorations of the Bandeirantes and the resistance and expulsion of the Dutch, undeniable reflection of expansionist strength, had not yet been undertaken in Antero's day. Nevertheless, and above all, his discourse is filled with a preconceived thesis that he wished to demonstrate at all cost (Pilar Vasquez Cuesta, 1994).

According to Antero de Quental, the progress of Portugal, the fruit of intelligence and vitality but also of originality, that matched this country with great culture nations, occurred in the Middle Ages. Yet, according to him, and unlike Herculano's perception, the Discoveries led to an even greater glory, which exceeded Europe in geographical studies and in the great navigations. Creation and freedom of thought were still partially here from the 16th Century. It was only during the second half of the 16th Century that a calamitous revolution took place, globalising phenomena being at the origin of such decadence, from Politics to Spirit (thought and moral) and Economics.

What was destroyed as a result of the monarchy becoming absolute was, according to him, local freedom, including the autonomous spirit of the population, the Cortes, the rise of the middle classes and tolerance towards the Jewish people and the Moors, when on, the other hand, a sterile and pompous palatial aristocracy was used and economic life weakened. From the point of view of the Spirit, the country was closed to "moral freedom", a means of civilisation allegedly conquered by the Protestant Reformation. In Portugal, the alteration of Catholicism took place through the Council of Trent (marked by the oppressive inquisitorial action, permeated with mechanics,

[2] The concepts of Absolutism (19th Century creation) and absolute monarchy, and despotism, the Modern State, powers existing in society, solidarities and resistances, among other concepts, have been the object of profound analysis both at national and international level, the investigation going forward by way of the enrichment of the question focusing on many structural, circumstantial and conditioning factors, as well as interdisciplinary practice. The fact that historical analysis removes itself from the labels and stigmas placed by scholars such as Herculano for the Modern Age, this does not mean that nowadays there are not widely different points of view: Jorge Borges de Macedo (theory of history), Martim de Albuquerque (2012, History, Law and political views), António de Oliveira (2015, Politics, Economics and Society), Maria do Rosário Themudo Barata Cruz (1992, institutional History), António Manuel Hespanha (Law and Sociology), accompanied by debates abroad which involve Roland Mousnier and Fritz Hartung, Suarez Fernandez, Fernandez Alvarez, Vicens Vives (1984), G.R. Elton, Jean-Philippe Genet and other researchers, some of whom working in international teams.

the tool of Rome's policy). Antero opposed the Catholicism in his belief in progress, in science, in philosophy and in a free conscience. Only in some self-critical manifestations on the miseries of everyday social life could one express, according to him, a certain novelty in a framework that was unintelligent and dark.

Debauchery would equally be highlighted in the lack of a modern spirit, fundamentally prolific, developing, by contrast, enlivening conquests, a warrior spirit. He points it out as opposed to manual work and trade. In the framework of the anticlericalism of his own time, he also highlights how greatly one fought in those conquests in the name of the Catholic faith and not in the name of the Portuguese nation. He also notes demographic decrease as a consequence of the war, emigration and poverty.

Antero essentially demonstrates a thesis on the end of the progress of a time of agricultural boom, flourishing trade and a demographic growth. In the decline following it, honourable noblemen became courtiers, the men of the people, drawn by urban centres, especially Lisbon, became lackeys, there was an increase in the contingents of poor people, and the wealth of the mighty who became rich in India and Brazil served as a means to pay the imports, larger than the exports, and no investment being made on the land. With the developments of the conquests, the Portuguese-man changed from owner and farmer to soldier and adventurer, with an atrophy of the national Spirit. A trench was then drawn between conquerors and losers (due to the aristocratic and religious features of the warrior Spirit), Portugal being unable to spread its Civilisation to other peoples.

One is clearly before an ungrounded historiographical discourse, which is also strikingly caustic and polemic.

Not so very far from Antero de Quental's interpretations as regards important legacies left by Portugal to the world (1879b; Martins, 1994a), besides the concept of decadence from the 16th Century, Oliveira Martins also developed topics not always coincidental. He also defended the aristocratic nature of the Spirit of the Portuguese (1875) and his disgust at miscegenation with Brazilian Indians, also justifying the existence and survival of slavery, in his opinion (1948).

He thought that the vigour of the Portuguese people crystallised "in three acts: political imperialism, discoveries and conquests, religious absolutism", a phenomenon that incidentally was present in Europe and aggravated in the Iberia Peninsula with the Jewish and Moorish issue (1891; Martins, 1986). He shaped, however, a new military and commercial aristocracy (that would replace the military one) and by comparing the 16th Century with his own times, he considered covetousness and greed more in line with his days (1889; Martins, 1994b). Be it as it is, in the times of "pharaoh" kings, such as D. João III of Portugal and Filipe II of Spain, individualism, Jesuitism and intolerance, the corrupting gold originating from the conquests would be classified as "a perversion of the Spirit". Once again, historical inaccuracy is felt when exact times are judged, mixed with global balances of centuries.

In his overall assessment, Oliveira Martins, marked by his own experience and the context of the end of his Century, commented on economic aspects in more detail and in a different way from the thinkers previously analysed. He pointed out the predominance of forms of mercantile activity, such as banks and speculation, arising from the growth of commercial enterprises due to the discoveries, which triggered the prevalence of utilitarian inclinations. Depopulation of the fields and influx of wealth into Lisbon, monarchy's support in overseas money, in slavery and in the commercial aristocracy, but insufficient Treasury funds for public expenditure such as the reform of the University, and in India, what a "leech" that left the Portuguese impoverished and perverted by luxury (1879a; Martins, 1987). Brazil's "coloniser king", as D. João III was recognised as being (1880; Martins, 1978), was not able to put an end to "India's orgy", nor did D. Sebastião with his tyrannical regulations, could work on the moralization of society or save the nation with his African policy.

The national nature was corrupted, and the results were felt in the 16th Century as much in India as in Portugal, with special reference to the luxury and misery of the nobility, the pretension of the plebes to become aristocrats, the scorn for the mechanical arts, the sacrifice of industry and of the Jews, the bargaining of Justice. Similar phenomena to the ones created by the influx of riches from India in the 16th Century occurred again in centuries to come with Brazilian mines, with gold and diamonds, what a world of show off, of ephemeral riches, a number of reform attempts being only sterile efforts (vide Pombal), always a lack of structural reforms in Portugal. He asked, in his day, (1894; Martins, 1996), Brazil lost a long time ago, if Angola could still, similarly, save the nation…

It must be noted that for a balancing of progress and decadence in Portugal's history, these thinkers from the 19th Century, marked by the historical scenario of their time and by a decadent gaze (Medina, 1980), let themselves be influenced by testimonies from previous centuries, testimonies that, albeit, were chosen by their critical discourses. Also, note that the choice of these authors is guided or even laden with prejudice, and even their discourses, basically moralising and very often reformist, are partially chosen, cut out from the overall narrative, only existing illustrative passages for theses to be demonstrated.

Such circumstances were pointed out by Maria Leonor García da Cruz in *Os "Fumos da Índia"* (1998b), when gathering testimonies from 16th Century writers such as Gil Vicente, Sá de Miranda, Garcia de Resende, António Ferreira or Camões, taking into account in the global economy of their complete work concepts and sentences normally cited. She draws, in her studies (Cruz, 1998b), attention to the fact that such works are, incidentally, well known and publicised (not totally silenced) in the following centuries and had to intersect with other critical interventions of subsequent eras. Davison Winius also analyses the caustic but reforming arguments of Diogo do Couto (ed. 2001) and Francisco Rodrigues Silveira (ed. 1996), who witnessed the evils of India from the end of the 16th Century and remarks on the analyses that were made of such evils, mainly in the 19th Century, with consequences to this day. He concludes that A. de S.S. Costa Lobo (1987), when he published excerpts of Silveira's manuscripts in 1877 in *Memórias de um Soldado da Índia* [Memories of a Soldier in India], instead of encouraging it to be read in depth and put into perspective, the "sad service he rendered Silveira was to distort it to the point of becoming unrecognizable" (Winius, 1994).

Some of the most often cited writers in the 19th century are from the first half of the 16th century. They are the creators, albeit freely critical, of an era that those believing in the decadence considered the peak. This means that they reveal the evils of society that existed prior to the phenomena that in the 19th century the writers refer more often to as global and date them from the late 16th century (tyranny of the pharaoh kings, acting through war, laws, the closing of the Cortes, supporting the warfaring and commercial aristocracy; Catholicism with the inquisition and Trent guidelines; spirit of conquest without interconnection with non-European peoples and with the deterioration of habits for luxury and corruption).

Garcia de Resende (ed. 1991) and Gil Vicente (ed. 1984) highlight the desertification of the fields and the attraction for Lisbon and the armadas, the opulence of new products, of wealth and ambition for a greater status, but none of this would contradict the effective increase in population that continued to be felt throughout the century and fed the overseas expansion, the growth of institutional complexity not only in the Empire but also on Mainland Portugal, redistribution of revenues and privileges to aristocrats and the middle classes by the Crown for services rendered for the public good.

If in addition to a dynamic society undergoing total transformation, evident in these and other 16th century authors, the danger of a change in morality/spirituality is stressed, criticising as did Sá de Miranda (ed. 1937) the ostentation and disorder caused by covetousness and permanent dissatisfaction, in the 16th century, echoed in these and other discourses in subsequent centuries (Loureiro, 2015), permanent transformation/correction was believed to be possible.

Instead of a notion of tyranny from the King or the Church, the concept of a paternal monarch was common (Albuquerque, 2012), an essentially Christian monarch who encouraged and was supported by "healthy consciences" and by an elite of scholars (noble or not noble), who take care to act by way of justice and the enforcement of the law on the four continents, as testified by legal and administrative documents of the time and authors, such as António Ferreira (ed. 2000), Diogo Bernardes (ed. 1946) and Pero de Andrade Caminha (ed. 1791). "Industrious" Lisbon, not only nautical and commercial, works and inventories from mid-16th century reveal – João Brandão (ed. 1990) and Cristóvão Rodrigues de Oliveira (ed. 1987). The dialogue in the Cortes (Anonimous, 1538) and outside them between populations and natives of the kingdom are presented in central and local assemblies, thus once again contradicting the theses analysed above, in hearings in a number of administrative bodies, in official opinions requested by the monarch from cities and individuals of different backgrounds either directly or by letter (Cruz, 1998a; Cortesão, 1968–82).

The increase in business and dialogue with communities alien to Portuguese habits (Barreto, 1983), within and in the shadow of the empire (Bethencourt, 2010; Godinho, 1990), and creations such as "casados" and new Christians under Portuguese sovereignty, or changes in Europeans due to native influences (Xavier, 2008), fruit of the in-depth analysis of current research, require, on the one hand, a critical review of prejudiced interpretations of interconnections with other peoples.

"Progress" and "decadence" prove to be extremely relative concepts, focused by scholars on different eras of the historical process. Furthermore, there is an argument for a parallelism and simultaneity in terms of regeneration and degeneration in all sectors of public life. A discourse that considers such simultaneity without considering different paces in interconnecting phenomena proves to have no scientific basis due to lack of precision and objectivity.

BIBLIOGRAPHICAL REFERENCES

Albuquerque, Martim de. (2012) *O Poder Político no Renascimento Português*. 2nd ed. Lisbon: Babel.

Anonimous. (1538) *Capítulos de Cortes e Leis que se sobre alguns deles fizeram.*

Barreto, Luís Filipe. (1983) *Descobrimentos e Renascimento: Formas de Ser e Pensar nos séculos XV e XVI.* 2nd ed. Lisbon: IN-CM.

Bernardes, Diogo. (1946) *O Lima.* vol II. Lisbon: Sá da Costa.

Bethencourt, Francisco & Curto, Diogo Ramada (eds). (2010) *A Expansão Marítima Portuguesa, 1400–1800.* Lisbon: Eds. 70.

Brandão, João. (1990) *Grandeza e abastança de Lisboa em 1552.* Lisbon: Livros Horizonte.

Caminha, Pedro de Andrade. (1791) *Poesias.* Lisbon: Academia Real das Ciências.

Cidade, Hernâni. (1980) *Antero de Quental: A obra e o homem.* 3th ed. Lisbon: Arcádia.

Cortesão, Armando & Albuquerque, Luís de (eds). (1968–82) *Obras Completas de D. João de Castro.* vol. 4. Coimbra: AICP.

Couto, Diogo do. (2001) *O Primeiro Soldado Prático.* Lisbon: CNCDP.

Cruz, Maria do Rosário Themudo Barata. (1992) Para a História da Europa no Século XVI: Tipologia de Regime e Instituições. *Estudos em Homenagem a Jorge Borges de Macedo.* Lisbon: INIC, 199–225.

Cruz, Maria Leonor García da. (1998a) *As Controvérsias ao Tempo de D. João III sobre a Política Portuguesa no Norte de África.* Lisbon: CNPCDP/Mare Liberum.

—. (1998b) *Os "Fumos da Índia": uma leitura crítica da Expansão Portuguesa.* Lisbon: Cosmos.

Ferreira, António. (2000) *Poemas Lusitanos*. Braga: Universidade do Minho.

Godinho, Vitorino Magalhães. (1990) *Mito e Mercadoria, Utopia e Prática de Navegar: sécs.XIII a XVIII*. Lisbon: Difel.

Herculano, Alexandre. (1975) *História da Origem e Estabelecimento da Inquisição em Portugal*. 1st ed. 1854–59. Lisbon: Bertrand.

—. (1977) A perda de Arzila (1549). *Poesias*. T. 1. Lisbon: Bertrand.

—. (1985) Arqueologia Portuguesa (1841–43), Carta V sobre a História de Portugal (1842), Cogitações soltas de um Homem Obscuro (1846), Historiadores Portugueses. IV Garcia de Resende (1840), Solemnia Verba. Segunda carta ao Sr. A.L. Magessi Tavares (1850), Pouca Luz em Muitas Trevas. 1579–1580 (1844). In: Custódio, Jorge, and J.M. Garcia (ed.) *Opúsculos IV*. Lisbon: Presença.

Lobo, A. S. S. Costa. (1987) *Memórias de Um Soldado da Índia*. Lisbon: IN-CM.

Loureiro, Guilherme Maia de. (2015) *Estratificação e Mobilidade Social no Antigo Regime em Portugal (1640–1820)*. Lisbon: Guarda-Mor.

Martins, Oliveira. (1948) Os Povos Peninsulares e a Civilização Moderna. *Páginas Desconhecidas*. Lisbon: Seara Nova.

—. (1978) *O Brasil e as Colónias Portuguesas*. 7th ed., 1st ed. 1880. Lisbon: Guimarães.

—. (1986) *Camões Os Lusíadas e a Renascença em Portugal*. 4th ed., 1st ed. 1891. Lisbon: Guimarães.

—. (1987) *História de Portugal*. 19th ed., 1st ed. 1879a. Lisbon: Guimarães.

—. (1994a) *História da Civilização Ibérica*. 12th ed., 1st ed. 1879b. Lisbon: Guimarães.

—. (1994b) *Portugal nos Mares*. 1st ed. 1889. Lisbon: Guimarães.

—. (1996) Ao Leitor (Na Terceira Edição). *Portugal Contemporâneo*. 10th ed., 1st ed. 1894. vol. 1. Lisbon: Guimarães.

Matos, S. C. (1998) *Historiografia e Memória Nacional 1846–1898*. Lisbon: Colibri.

Medina, João. (1980) *Eça de Queiroz e a Geração de Setenta*. Lisbon: Moraes.

Miranda, Francisco Sá de. (1937) *Obras Completas de Francisco Sá de Miranda*. vol. 2. Lisbon: Sá da Costa.

Oliveira, António. (2015) *Capítulos de História de Portugal (1580–1668)*. vol. 3. Coimbra: Palimage.

Oliveira, Cristóvão Rodrigues de. (1987) *Lisboa em 1551: Sumário em que brevemente se contêm algumas coisas assim eclesiásticas como seculares que há na cidade de Lisboa (1551)*. Lisbon: Livros Horizonte.

Quental, Antero de. (2010) *Causas da Decadência dos Povos Peninsulares nos Últimos Três Séculos*. Lisbon: Tinta-da-China.

Resende, Garcia de. (1991) Miscelânea. *Crónica de Dom João II e Miscelânea por Garcia de Resende*. New ed. Study by J.V. Serrão. Lisbon: IN-CM.

Serrão, Joel. (1983) Esta Palavra 'Decadência'. *Temas de Cultura Portuguesa*. Lisbon: Livros Horizonte.

Silva, Rebelo da. (1971) *História de Portugal nos Séculos XVII e XVIII*. New ed. Study by J. Borges de Macedo. Lisbon: Imprensa Nacional.

Silveira, Francisco Rodrigues. (1996) *Reformação da milícia e governo do Estado da Índia Oriental*. Study by Barreto, Teensma and Winius. Lisbon: Fundação Oriente.

Vasquez Cuesta, Pilar. (1994) As causas da decadência dos povos peninsulares – um elo português numa corrente de obras portuguesas sobre o tema. *Antero de Quental e o Destino de uma Geração. Actas do Colóquio Internacional no Centenário da sua Morte*. Lisbon: Edições Asa, 107–112.

Vicens Vives, J. (1984) A Estrutura Administrativa Estadual nos Sécs. XVI e XVII. In: Hespanha, A.M (ed) *Poder e Instituições do Antigo Regime. Colectânea de Textos*. Lisbon: Fundação Calouste Gulbenkian, 201–230.

Vicente, Gil. (1984) *Copilaçam de todalas obras de Gil Vicente*. 2 vols. Lisbon: IN-CM.

Winius, George Davison. (1994) *A Lenda Negra da Índia Portuguesa*. Lisbon: Antígona.

Xavier, A. B. (2008) *A Invenção de Goa. Poder Imperial e Conversões Culturais XVI–XVII*. Lisbon: ICS.

Shadows of orientalism in Portugal – some notes on theories and practices in Macao early narrative images (16th and 17th centuries)

Ana Paula Avelar
CHAM, FCSH, Universidade NOVA de Lisboa, Univerisdade dos Açores, Lisboa, Universidade Aberta, Lisboa, Portugal

ABSTRACT: Imperial Portuguese imagery unveils three discursive tropes as primordial signs of the representation of an idea of empire – the cross, the crown and the sphere. These three signs embody the representation of the Portuguese monarchy since the remote 16th century and play a role in the building of an Orientalism in Portugal. Having in mind the notion of Orientalism as an ongoing conceptual building, I will approach the first narrative images of Macao that emerge in reports about the Portuguese presence in that space.

My focus will lie on Duarte Barbosa's and Tomés Pires' texts, and on Fernão Lopes de Castanheda's and Gaspar da Cruz's chronicles, in order to show how space summons the imagery of a historical time. In these texts, we are not yet before an idea of progress, since they only confront the different, either incorporating or rejecting it.

Keywords: Empire, Macao, representation, chronicles, imagery

It is impossible to ponder on the topic Shadows of Orientalism in Portugal – Macao early narrative images (16th and 17th centuries) without having in mind that Portuguese imperial imagery unveils primordial signs such as the crown, the cross, and the sphere.

These signs represent, at a symbolical level, the Portuguese monarchy since King Manuel's reign, from the late fifteenth-century to early sixteenth (1495–1521) and play a relevant role in the building of an orientalism in Portugal.

This orientalism must be conceived as those narrative practices where the "art" of description, communication, and representation unfold. These discursive practices reveal economic, social and political purposes while emerging under different aesthetic forms (Avelar, 2010: 83).

They encapsulate an orientalism that derives itself from the way colonialism has evolved. It is thus in a historiographical context that these orientalist discourses must be put in perspective, in order to surpass generalist approaches:

> the critique of colonial representations appears increasingly abstract and disengaged with both its own history as well as the specific colonial histories that it seeks to explain. (Burke, Prochaska, 2008: 2)

Although both the texts and the historical contexts may be autonomous, they are dialectically entwined, thus building each other.

When we analyse stories dealing with China, and more specifically with Macao, one must bear in mind that these discourses unfold from, in Delacroix's words:

> une tension indépassable entre le souci de rendre compte de ce qui s'est passé et un questionnement qui émane, pour l'essentiel, du présent de l'historien. (Delacroix et al., 2010: 18).

Orientalist signs emerge from daily life representation in the History of Portuguese presence that started to be written since the 16th century.

This aspect gains a particular strength when we ponder on certain rituals such as the so-called "tea-cult" that go along with the making of an oriental imaginary. These rituals are signalled out through the centuries in those texts that describe European presence in China and Japan. One must remind of both Wenceslau de Moraes' 19th century book on this topic and Eça de Queirós' *Mandarin*, where we find a passage about it.

In the 16th century, Friar Gaspar da Cruz's Treaty where is detailed told the things of China with their peculiarities and the kingdom of Ormuz (1569) already mentions the fact that the Chinese offered their foreign guests a red beverage, which they called tea (D'Intino, 1989: 202). In Father Matteo Ricci's Histoire de *l' Expédition Chrétienne au Royaume de la Chine 1582–1610* we find an explanation about the way both Chinese and Japanese prepared this beverage, the same thing happening in the first part chapter three of Father

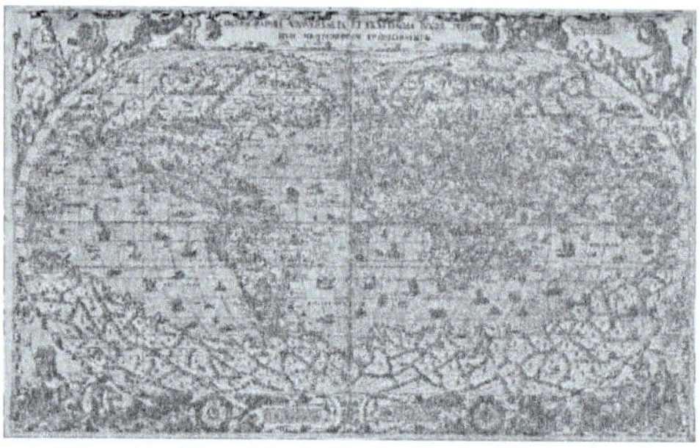

Figure 1. Giacomo Gastaldi, Planisphere http://www.wikiwand.com/it/Planisferi_del_XVI_secolo.

Álvaro Semedo's Report of the Great China. This ritual actually was a social sign of the hospitality, which should be offered to visitors.

However, all these references about the tea-cult and ritual are placed in a wider historic and cultural context. The idea of empire lies at the centre of these historiographical texts. In the 16th century, Portugal aimed to be a *Respublica Christana*, under the emperor's sceptre, as a propagator of the true faith upon the earth. Under this sign, a New Golden Age should emerge. The somehow constant contact with Eastern lands modelled a European-Portuguese identity that was getting in touch with new realities. It is no accident that those who write these texts help themselves with their own European references in order to explain the different, the Other.

Both Duarte Barbosa and Fernão Lopes de Castanheda claimed that the Chinese language reminded of German. Tomé Pires also stated that: They wear round silk net caps like the black sieves we have in Portugal. (Pires, 1944: 116)

In this vein, Gaspar Correia mentions the fact Canton is as big Évora, and João de Barros writes that the Chinese sound like Greeks. A good example of the comparative strategy between Europe and these Eastern lands is Luis de Fóis' late 16th century Treaty of brief and succinct contradictions and differences of social mores between the people of Europe and this province of Japan.

In the mid-16th century, the Casanatense Codex shows a drawing about the differences between Japanese clothes and European ones. These watercolours of peoples from the Cape of Good Hope to China had, however, a limited circulation in Europe.

Until the early 17th century the cartographic knowledge of Chinese lands lied mainly on Giacomo Gastaldi' map, which had been drawn around fifteen sixty-one at the request of the Venetian senate. Europe actually was eager of narratives of Asian lands, namely China.

The cultural world also wanted to know these new spaces. I must point out just one example, the one of Montaigne, who, although focused on South America, also read works such as Simon Goulart's *History of Portugal*. The Chinese:

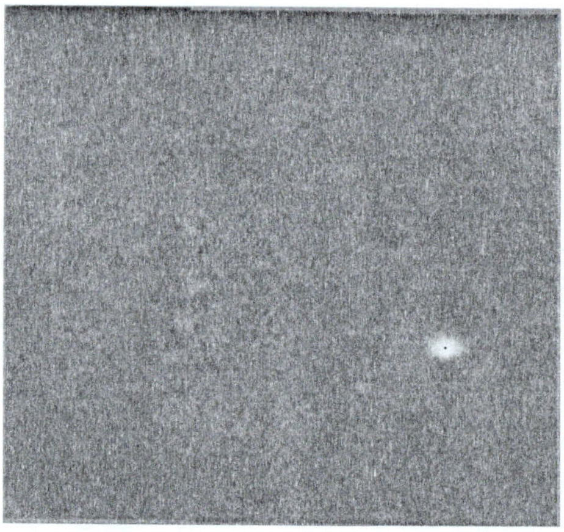

Figure 2. Left side, 3 panels of the Kunyu Wanguo Quantu, *A Map of the Myriad Countries of the World*, by Matteo Ricci https://commons.wikimedia.org/wiki/File:Kunyu_Wanguo_Quantu_by_Matteo_Ricci_Plate_1-3.jpg.

> En la plupart de leur maniere de viure ils se gouuernent à la façon des François & Alemans, prenent plaisir d'ordinaire à iouer, rire, danser, & boire dautant. (Goulart, 1588)

Montaigne also got in touch with other authors, like Juan Gonzalés de Mendonza's French printing of History of the things, rituals and mores of the great Kingdom of China (1585). D. E. Mungello informs us that Mendoza:

> (…) was sent in 1580 by way of Mexico to China, carrying a letter to the Chinese emperor from Philip II of Spain. He encountered difficulties in Mexico, however, and was forced to return to Europe. He never actually visited China, but while in Mexico obtained materials dealing with China from Spanish missionaries based in the Philippines. Later in Rome, he used these materials to write his Historia, the first European work to provide detailed information on China. (Mungello, 1989: 184)

After having spent some time in China both Matteo Ricci and Álvaro Semedo wrote their Works about this space. These encounters allowed the rising curiosity about them. Ricci even tried to map this wide realm.

It is not my aim to make an exhaustive description of the evolution of European knowledge about China. Yet one must remind the fact that the descriptions of those spaces started as early as the first Portuguese contacts with Eastern lands. The Chinese seas are mainly described in the 16th century by the texts such as Fernão Lopes de Castanheda's, João de Barros', and Gaspar Correia's Expansion chronicles about the Portuguese presence in Asia (Avelar, 2003: 15–17). Damião de Góis' and Jerónimo Osório's chronicles on King Manoel and Francisco de Andrade's on King John the third provide further information on this topic. Duarte Barbosa's and Tomé Pires' travel reports,

Afonso de Albuquerque's policies described by his son Brás de Albuquerque, António Galvão's Discoveries Treaty, Fernão Mendes Pinto's pilgrimage, and other missionary texts provide further descriptions about China (Loureiro, 2009: 181).

In these different historiographical discourses, the Chinese seas emerge as one of the multiple spaces where Portuguese domain lies. These narratives often repeat early information about this space and these peoples provided by Duarte Barbosa's and Tomé Pires' early reports (c. 1515–1516). Along these early data, each author emphasizes those dimensions he considers more relevant.

The textual matrixes for this early group of texts are those written by Tomé Pires and by Duarte Barbosa. We must bear in mind that both authors follow the same narrative structure. In *Summa Oriental*:

> The king of China is a heathen with much land and many people. The people of China are white, as white as we are. Most of them wear black cotton cloth, and they wear sayons of this in five pieces with gores, as we do, only they are very wide. In the winter they wear felt on their legs by way of socks, and on top well-made boots which do not reach above the knee, and they wear their clothes lined with lambskin and other furs. Some of them wear pelisses. (Pires, 1944: 116)

The chronicler compares the way of dressing with the European:

> They wear round silk net caps like the black sieves we have in Portugal, they are rather like Germans. They have thirty or forty hairs in their beards. They wear very well-made French shoes with square toes. (Pires, 1944: 116)

Tomé Pires also refers daily life habits, such as food, meat and beverages that are consumed by the Chinese:

> All the Chinese eat pigs, cows and all other animals. They drink a fair amount of all sorts of beverages. They praise our wine greatly. They get pretty drunk (…). Those who are to be seen in Malacca are not very truthful and steal — that is the common people. They eat with two sticks, and the earthenware or china bowl in their left hand close to their mouth, with the two sticks to suck in. This is the Chinese way. The women look like Spanish women. (Pires, 1944: 116)

On the other hand, according to Duarte Barbosa:

> The inhabitants of the country are white men, tall, well-made and gentlemen; and so likewise the women. They have got only one defect, that their eyes are very small, and on their chins, they have three or four hairs and no more; the smaller their eyes are, so much the prettier they think them; and the same as regards the women. They are very smartly dressed, clothed in silk and cotton and woollen stuff, and their costumes are like those of Germans; they are shod with soft leather boots and shoes, like the people of a cold country. They have a language of their own, and the tone of it is like that of Germans. (Dames, 1989: 213)

Barbosa provides detailed information on this Chinese daily life that Tomé Pires has already mentioned:

> They drink several kinds of wine, and many times during their meals. They also eat the flesh of dogs which they hold to be good meat. They are men of truth and good gentlemen: they are great merchants of all sorts of goods. They make much porcelain in the country, and very good, which is a great article of commerce for all parts. (Dames, 1989: 213)

Besides the porcelain, Barbosa lists a whole series of goods that delight the Europeans:

> There also grows and is produced in this country of China much very good silk, of which they make a great quantity of stuffs; that is to say, damasks of all colours, satins of several kinds, and brocade. There is much rhubarb in this country, and much musk, very fine silver, seed pearl, and pearls that are not very round. Many of these Chinese take their wives and children continually in the ships in which they live without possessing any other dwelling. (Dames, 1989: 214)

The second group of texts reveal the report of the authors' permanencies on Chinese seas. This is what happens in rather different narratives such as Cristóvão Vieira's, Vasco Calvo's and Galiote Pereira's letters and Friar Gaspar da Cruz' *Treaty of the Things of China*, which is the first printed book about China that gets European circulation.

The author points out information about spaces and peoples. After an aesthetic description of the way they look, Friar Gaspar da Cruz provides detailed information about the way they both women and men dress, and about their social behaviour. ((D'Intino, 1989: 202).

Friar Gaspar da Cruz also emphasizes the civility that, in his view, lied at the core of Chinese social behaviour.

Narratives about China keep on gathering information and details about both physical and social fields, often based on daily experience. Yet these descriptions follow the points of view that we find in earlier Portuguese narratives on Asian lands. This is a focus that emerges from the sea towards the land; a point of view that evolves in time anchored in the visual observation.

Father Manuel Teixeira's letter written on the December the first fifteenth sixty-three, reveals Macao from geographical coordinates, along with the description of the Chinese's complexion.

The permanence both in China and in Macao echoes in the way these authors tell their story. Historic time is conceived as linear and progressive, entwined with memory which somehow plays the role of discourse building. On the other hand, commemorative reminiscence is key in Fernão Mendes Pinto's Pilgrimage.

Macao emerges as space where the Portuguese feel as confident and secure as if they were in Portugal (Alves, 2010: 776).

Its importance somehow echoes in Pedro Barreto de Resende drawings of Macao's square in his *Book of Drawings of All the Fortresses, Cities and Towns in the Eastern State of India.* António Bocarro (Goa, February seventeenth sixteen thirty-five) also conceived his drawing according to Fernão Mendes Pinto's memory.

We are not yet before an idea of progress, since all these texts only confront the different, either incorporating it or rejecting it. On the other hand, as we have shown above the historical time that emerged in historiographical discourse about China and Macao followed a progressive linearity, rather than a cyclic one, which still prevailed in Eastern shores.

During the following centuries, this progressive linearity had a great impact on the way both narratives and visual representations depicted a space, the Eastern/Chinese one, and an empire, the Portuguese one. In addition to being an important element for us to understand coeval reality, those texts functioned as a matrix for other texts to come. Knowing them allow us to get a more rigorous insight about the different nuances of a topic that, even silently, stood at the core of the encounter with the Other; as Walter D. Mignolo has stated: "understanding the past [is] to speak the present." (Mignolo, 2003: 455).

BIBLIOGRAPHICAL REFERENCES

Alves, Jorge S.dir. (2010). *Fernão Mendes Pinto and the Peregrinação,* Lisboa: Fundação Oriente.

Avelar, Ana Paula (2003). Figurações da alteridade na cronística da Expansão, Lisboa: Universidade Aberta.

Avelar, Ana Paula (2010). Construindo um conceito: O Orientalismo nos primeiros escritos portugueses sobre a China e Macau. *In Macau na Escrita, Escritas de Macau,* Lisboa: Edições Húmus e CEC.

Burke, E. Prochaska, D. (2008). *Genealogies of orientalism: history, theory, politics,* Neb.: University of Nebraska Press.

Delacroix, C. et al. (2010). *Historiographies, I,* Paris, Editions Gallimard.

D'Intino, Raffaella (1989). *Enformação das cousas da China –textos do século XVI,* Lisboa: IN-CM.

Dames, Mansel (1989). *The book of Duarte Barbosa...,* New Delhi: Asia Education Services.

Goulart, Simon (1588). *Histoire de Portugal, contenant les entreprises, nauigations & gestes memorables des Portugallois... comprinse en vingt livres... traduits du latin de Ierosme Osorius..., les huit suiuans prins de Lopez de Castagnede et d'autres historiens...* Paris: Abel l'Angelier.

Loureiro, Rui Manuel (2009). *Nas partes da China,* Lisboa: CCCM.

Mignolo, W. (2003). *The Darker side of the Renaissance,* Mich: the Michigan University Press.

Mungello, D.E. (1989). *Curious Land-Jesuit Accommodation and the Origins of Sinology,* Honolulu: University of Hawaii press.

Pires, Tomé (1944). *Summa Oriental of Tome Pires and the book of Francisco Rodrigues,* London: Hackbut Society.

The idea of progress and the practice of slavery in the second half of the 18th century

Maria do Rosário Pimentel
CHAM, FCSH, Universidade NOVA de Lisboa, Universidade dos Açores, Portugal

ABSTRACT: This paper essays to confront the theory of progress with the practice of slavery and the discussion concerning its abolition. Due to the vastness and complexity of this theme, only the first French abolitionist movement, happening in the second half of the 18th century, will be considered. By then the European cultural elite assumed several principles as undisputable. Namely:

1. that all human beings are born equal;
2. that the principle of individual freedom was a right given by Nature to allow every individual to fulfil his destiny;
3. that rationality was the basic principle for all free human beings to be able to pursue her or his happiness;
4. that the concept of humanity included the notion of its unity and diversity;
5. that the progress of the human mind had its origins in the inalienable rights given to humans by Nature;
6. the general principle of progress was based on the right to freedom.

Considering all this, one is led to conclude that the practice of slavery and its traffic must be studied as social habits that abased human existence and were opposed to the new rationalist conceptions concerning the individual was taken for granted already by the end of the 18th century.

Keywords: Progress, slavery, freedom, Enlightenment, French Revolution

1 INTRODUCTION

The general notion of progress, which implies changing for better the living conditions of both individuals and communities, is an ideal that one finds in every stage of History. It is a natural inclination for human nature to project in the future, more or less distant, the desire to live better, both as individual and as a society. This natural inclination is born from the uncertainty and the actual living conditions. It is yearning, an intent carved by unpredictable paths.

However, the idea of progress implies a progressive change along a linear time period, with the aim to achieve a predetermined purpose, considering humanity's ultimate end. In these conditions, the meaning of History remains hostage from a predestination that guides human path according to its ultimate aim.

The classical mythological descriptions and the eschatological and prophetic narratives offer the idea of a course where the necessity for change is felt. These concepts are still quite far from the modern notion of progress. It is enough the presence of a providential entity that oversees the world and holds the power of justice over humankind to withdraw the plan from any possible connection with the origin of the modern concept.

In addition, it is not enough the replacement of the notion of time's circularity, implied in the theory/myth of eternal return. For instance, the mythic golden age that is already conceived in Hesiod according to a notion of linear time, is set in the past and not in the future, therefore opposing the modern concept of progress.

However, the Judaeo-Christian tradition and the classical cultural heritage are full of extraordinary and lasting progresses that Western civilisation still now admires and enjoys as forms worthy of being presented as models to future generations (Hazard, 1971: 340). The same may be said regarding the Arabic legacy and those from every civilisation that contributed to the European cultural heritage.

2 THE BIRTH OF THE MODERN CONCEPT OF PROGRESS: ORIGINS AND EVOLUTION

In Western civilisation, the idea of progress tended to be associated with the idea of History itself, following the change of values proper to the theocentric medieval societies to the Renaissance anthropocentric models. With the Renaissance and Humanist movements several changes took place:

- the importance attributed to the individual and his/her rational ability;
- the interest for the classical texts;
- simultaneously, the concerns for human daily reality;
- the awareness of a fast and overwhelming transformation of the known world, produced by secular movements;
- finally the adoption of a historical vision of reality.

All these "discoveries" led the way to a notion of progress in accordance with the changes witnessed, thus contributing to the theory of progress.

In this moment in history, one is still considering a pre-concept, the laying of the base-stones for the ideal of progress that the Enlightenment will conceptualise and popularise as a political ideology and that will be pivotal in the ideological and reformist movements of the 19th century.

At the background of the ideal of progress was western civilizational evolution, a step towards a future with better living conditions. This implied, by then, a reformist disposition shielded with knowledge and critical thinking so that changes could be instituted without objection. The idea was developed in western culture, but with the Enlightenment philosophy, the idea of civilizational evolution became applicable to every society, other than the European one. This one was chosen as the pattern for every other.

In the several theories of progress, regardless of their ideological orientation, there are three fundamental notions: the notion of the linearity of time oriented to a goal, the idea of a unified humanity which would evolve in a pre-conceived path; and finally the conception of the human being as one able to self-determination and world changing. These notions, previously inscribed in Christian tenet, due to Renaissance and Humanist values, came to be understood as secular notions.

2.1 *From the Middle Age to the 18th century*

Christianity had already taken a stand, in an eschatological movement of unique actions, towards a future inserted in a linear time, considering humanity would achieve redemption and eternal bliss. Adding to the linear notion of time, Christianity presented humanity as a whole, converging as such to the same closing. There was also the idea of the "new man" able to transform the world. But in this line of thought and conception of humanity, God is a constant. By direct emanation, it is He who, in the *Sermon of the Mountain*, announces eternal life and the conditions the believers had to follow to achieve that goal. Every human being, even those who were slaves, had the freedom to choose if they did, or did not wish to follow the Christian message, though the choice and procedures were to be judged by God.

Obviously, human will and the "freedom" of the Christian choice, that Augustine of Hippo explained using the concept of "free will", were limited. The philosopher Magalhães Vilhena points out that Augustine's theory divided more than united the believers' thinking, but from those discussions, some notions got weight and opened the way and, up to a certain point, had a direct influence in preparing the awakening of the ideal of Progress (Vilhena, 1979: 29–31). Even though, there was a step missing: the human being had to become conscious that he was, in fact, completely free to act, and that step was only taken in the Modern Age. Physical and intellectual freedom was a condition *sine qua non* for the rational performance, to become an agent of progress.

Even though Christianism had sparked a faint notion of progress, it could not be in the origin of the concept which born in opposition to Christian philosophy and social order. While Christianism conceived humans as sinners that only through God's grace would attain divine scheme, after death, in the concept of progress human being is endowed with greatness, ability and will to achieve happiness, provided by material progress, in an earthly society, regulated by nature. The concept of progress inscribes itself in the urban daily life, a secular space, and it is connected to a dynamic of growth of both individual and society, changes taking place with one possible direction: continual perfectioning.

In the passing from the Middle Ages to Modern Age a series of circumstances gave birth to the necessary conditions for the development of fundamental base concepts for the future ideal of progress. It is a cultural process that, in a slow transformation and in close relation with society, allows Renaissance culture to overcome medieval one. The reactions against scholastic and the general anxiety experienced during the late Middle Ages are, already the announcing of a new *Weltanschauung*, a new conception of the human being and of the possible knowledge, notwithstanding being still framed by a medieval culture.

To the breaking of medieval cultural unity, with the supremacy of reason above faith, the separation of philosophy from Theology, the breaking of the balance between the rational and the dogmatic, between nature and the divine grace. Also important were several social, economic and political conditions that cause a commercial increase, the development of boroughs that were the starting point of cities with great economic impact, the development of the bourgeoisie, the newest social class that prospered in the commercial activity and the cities.

The bourgeoisie had seen, for very long, its power and aspirations limited by the feudal system. This group wished to be recognised as important and, most of all had the ambition to see the formation of a cultural structure that would make valid the power it had in fact, though not by right, that would ratify its activities and would establish its specific Weltanschauung. The citizens were successful people, who made their fortune due to their own endeavours and who had confidence in their skills. Traditional culture did neither meet their claims nor contribute to the acquisition of knowledge suited to their businesses.

City laypeople found in the Renaissance premises the wished alternative for the creation of a culture of secular ballast that did not identify itself with medieval Christian thinking. The humanist pagan perspective proved essential, mainly the more advanced stages of literary but also the ideological and critical humanism. It became a true "school of dialogue" where Renaissance individuals learned that, after all, not everything was encapsulated in the old medieval knowledge and that the times they were living were different from every other period of western history.

Alternatively, to the Christian construction and explanation of the world, Renaissance culture dignified the individual as the most perfect of all creations and placed him in the highest place in the scale of values. However, this disposition did not entail any rejection of divinity. God had only been placed further away from the earthly sphere. New values became priorities: the value of work, pleasure, daily life, physical and intellectual education, the formation of the "good citizen", the *Uomo Universalis*, prepared to take an active part in society, in civic action and to rule all areas of knowledge. Nature had provided him with a rational ability and the faculty to experiment and observe. All this was aimed at the "new man", confident in his capacities, intervener and proud of being exalted and ennobled by his own successes and not by his genealogical filiation (Dias, 1969: 1–16).

The religious crises are also part of the context. They grew stronger since the 16th century and that led to the Christian Church Schism with the triumph of Reformation. Two major players in the religious dispute were Erasmus, the Christian humanist who, without ever living the Church, was acutely critic of its uses, and Luther who force the radical separation and the introduction of reformed practices and believes. The new church promoted the exaltation of labour, recognising it as a virtue, thus creating the religious basis for capitalism. According to Protestant ethics, work becomes the goal of existence and the best way to prevent against temptations. Thus, success came to be interpreted as a reliable way to identify those predestined for salvation. Furthermore, the accumulation of wealth through work was not only morally justified but also recommended (Dias, 1969: 17–36)

The formation of the new religious movements gave origin to religious wars that spread through northern and central Europe. In France, where battles were particularly bloody, the St. Bartholomew's Day massacre became a sad symbol of an insane conflict (Diefendorf, 1985; Kelley, 1972; Smither, 1991). By the 1598 Nantes Edict, the state established Catholicism as the official religion and gave the French Calvinist liberty of cult. Following this event there grew a certain tendency to tolerate practices different from those instituted by social structures.

This disposition became a ballast for the future concept of tolerance, and an unstoppable movement ever growing. In the 17th century, the notion of tolerance lost the character of political instrument and became a moral principle defended by several religions. The foundations turned out to be ethical and philosophical and answered to a commitment to freedom of conscience, and to the notion of the individual citizen who had the right to use his own mind and choose his faith.

Alongside with the defence of tolerance a new thesis gained strength: the one on the separation of Church and state, being Locke one of its most renowned pleader (1689). In the 18th century, the concept of tolerance spreads beyond the religious field. Voltaire in *Dictionnaire philosophique* [Philosophical Dictionary] presents tolerance as "l'apanage de l'humanité" [an attribute of humanity] (1878) and in the *Treatise on Toleration* the notion gains a wider meaning becoming an attribute of Nature and defining it as the respect for all different believes existing on the planet as a means to fully enjoy individual freedom and social progress (2016).

The concept appears in articles 10 and 11 of 1789 Declaration of the Rights of Man and of the Citizen:

> 10 – No one may be disturbed on account of his opinions, even religious ones, as long as the manifestation of such opinions does not interfere with the established Law and Order.
> 11 – The free communication of ideas and of opinions is one of the most precious rights of man. Any citizen may therefore speak, write and publish freely, except what is tantamount to the abuse of this liberty in the cases determined by Law. (*Declaration of the Rights of Man and of the Citizen*, 1789: 2)

In this summary of the evolution of the concept of Progress, one cannot leave unnoticed the Discoveries and the consequent period of European expansion. These are two different moments of the same movement that was the foundation of a major evolution in knowledge, questioning and obliterating much of the traditional knowledge. The discoveries and the expansion allowed a major enlargement of mental openness and one of the lengthiest commercial and cultural planetary interchange. The contact with the different realities discovered induced a diversity of perspectives and knowledge. The new proofs, the result of the assimilation of diversity, promoted a transformation of concepts, renewed relationships and processes of adaptation. In this context, one has to mention the changes in the concept of humanity and the reflexions of the School of Salamanca on the philosophy of natural law as a step for the elaboration of Modern Natural Right and the inclusion of a public international right.

The Discoveries promoted the interest for technical progress, for a knowledge that could be proved and, in this way, for the construction of instruments that, with accuracy, might have an impact on reality (Dias, 1988: 103 ff). They promoted nautical and cartographical improvements, the studies on cosmography and geometrical planning. The Discoveries developed the attention to landscape, for geography, pharmacology, medicine, and they also increased new ways for recording every new thing discovered. It became essential to observe, to experience, to test, to order, to explain and to describe.

It became necessary to study the monsoons to know which was the best time to travel, what was the best way to communicate and send news. There was the development of a quantitative way of thinking, essential for travelling, commercial exchanges, the loading of boats, to register wealth, taxes, interests, and all this in the short term had an impact on everyday life. The world unravelled a myriad of new lands, seas, new stars in the sky, new continents, and new rivers. Myths fell apart: the existence of a tenebrous sea, the existence of an uninhabited torrid area. On the other hand, there was the confirmation of the antipodes. The word discovery and the verb discover gained a new life, a potent meaning (Godinho, 1998: 55–82).

Doubts and perplexity fuelled scientific development that had the capacity to invent a powerful ally. Though the circulation of manuscripts being still intense, the press was the great divulger of the new knowledge. The Renaissance *Uomo Universalis* understood that he needed to become an expert on specific areas. Whatever the field of study he had to become also a creator of new technical instruments needed for new experiences. The Renaissance *Uomo Universalis* understood the need for specialisation in precise areas and with that understanding came another: he had to become a craftsman and inventor of the tools he needed. Necessity widened his art as invent and reformer. While some took interest in what they observed in nature, others searched in vain inside the classical texts for an explanation of what they actually observed. Some, through the telescopes and the microscopes, saw and showed, for the first time worlds never before suspected. Francis Bacon, a harbinger of philosophy and modern science, registered the impact of the great sailing expeditions in the progress of knowledge, as well as the fact that these coincided with the dawn of modern science. There is the definite fall of traditional "truths" produced by religious interpretations and the study of nature became the distinctive mark of modern science. More dedicated to experience than to speculation, joining empiricism and rationalism, inductive approaches with deductive ones, science progresses swiftly, though facing a strong opposition from traditional ways of thinking and from the Church. Even after a formal separation of science from theology, the relationship between both was always complex (Hazard, 1971: 125). One has only to remember the repercussions caused by the substitution of the classical geocentric theory, supported by the Church and justified by the Christian Aristotelianism, by Copernican heliocentric theory. Copernicus's fear was so overwhelming that the theory was only published after his death. Galileo recovered his predecessor's work, built a more powerful telescope, and confirmed Copernicus theory. In 1632, in the book entitled *Dialogue Concerning the Two Chief World Systems*, Galileo disclosed his conclusions, which were considered heretic by the Inquisition Court.

The historical conditions that have been summarised imply a change of values tending to the exaltation of human beings, the increased worth of critical reasoning, the secularisation of culture and the conception of a freedom of conscience and action. They also reveal the tendency to replace a hierarchic society sanctioned by God, based on the obedience to dogmas and on the duties to the divinity and the Prince, by another kind of society based on secular and rational principles. One can still recall that the new vision of the cosmos and the rethinking of the individual as citizen of the whole world, the feeling of greatness and power, the eagerness to dominate nature and the universal and eternal laws, this dazzlement caused by the feeling of superiority created the illusion that humans could act as gods on earth. This was a historical construction that developed itself through a long period, and that from the Renaissance reverberated through ages.

Paul Hazard, referring to the crises of European conscience writes:

> But who fed this critical thinking? Where did it gain the strength and audacity? From where did it come? From the end of time; from the Greek Antiquity; from this or that philosopher living in a heretical Middle Ages; from this or that remote source; and certainly, beyond any doubt, from the Renaissance. (1971: 340)[1]

It is in this historical state of affairs that the idea of progress has its ferment. During the Quarrel of the Ancients and the Moderns (begin in Paris, in 1687), the question of the superiority or inferiority of the present regarding the past was formulated by the Modern. They, unlike their opponents, considered themselves superior, not only because they thought of themselves as being at the front of a process of constant evolution, but also because the new reality asked for new models, due to the evident inadequacy of the old ones. The modern intellectuals were aware of the relevance of the idea of progress, of the power of science and their political value. From this moment forward a new ideology was born, founding the sovereignty of the present over the past, strengthened by the idea that knowledge and its accumulation would drive society towards a better future. It appears simultaneously with the beginning of the Enlightenment. In opposition to the clergy and the aristocracy, it was the system of ideas that suited a rising bourgeoisie during the 18th century.

[1] Whenever an English translation or edition could not be used a free translation will be provided and the original text presented in and endnote. Original text in the edition used: "Mas quem nutriu este pensamento crítico? Onde ganhou força e audácia? Donde vem, enfim? Do fundo dos tempos; da antiguidade grega; deste ou daquele doutor duma Idade Média herética; desta ou daquela fonte longínqua; mas, sem dúvida nenhuma, do Renascimento".

3 THE ENLIGHTENMENT AND THE IDEA OF PROGRESS

Not all Enlightened thinkers agreed with this idea and some accepted it with reticence. Giambattista Vico, for instance, considered one of the pioneers of modern thinking on the historical time, did not conceive progress in a linear evolution. Other thinkers, as the Abbé de Saint-Pierre, believed only in the intellectual and technological progress, but not in the moral one. Diderot believed only in the enhancement of the individual, considering the moral progress and the evolution of the masses as impossible. Jean-Jacques Rousseau was quite a critic when, according to his theory of the "good savage", that civilizational progress was the cause of chaos of modern life.

In what concerns the formation of the idea of progress there are two unavoidable characters: Turgot, with his *Tableau philosophique des progrès successifs de l'esprit humain* [*On the Historical Progress of the Human Mind*][2] and Condorcet with the work *Esquisse d'un tableau historique des progrès de l'esprit humain* (1795) [Sketch for a Historical Picture of the Progress of the Human Mind][3].

Right in the middle of the 18th century, Turgot's *Tableau philosophique des progrès successifs de l'esprit humain* presented a definitive step towards the formation of the idea of progress even though he did not offer the conception as being a global process, irreversible and continuous. According to Turgot progress would always be dependent on favourable circumstances and the existence of extraordinary people (some kind of geniuses) that would be the promoters of progress. Furthermore, still according to Turgot, progress could be stopped, or even recede significantly. He uses the Middle Ages as an example of a period when, in spite of the decadence and receding, there was also some degree of progress. In a deeper analysis, one would be able to observe both the variation and the little progress. Explaining it in a different formulation, the human mind is able to achieve some degree of progress even during the most obscure and barbarian periods. According to Turgot (1991: LIII-LVIII) human societies should be able to adapt themselves to the new state of affairs without violence, slowly, and always in a linear path.

Condorcet was Turgot disciple and followed his master's thought and, without the Christian connotations implicit in Turgot's work, he proclaimed the continuous and indefinite perfectibility of the human mind and the whole of the human species, through the transmission of a heritage in constant evolution. As Prior points out, every preceding theory prepared Condorcet's, but for the first time history is presented divided according to the progress of knowledge and not according to the political phenomena (Condorcet, 1970: XXXIV–XXXV).

In the *Sketch* (2012b)[4] Condorcet presents human evolution, since primordial times, divided into ten ages. He also defines the necessary and immutable laws that lead "the eternal chain of human destiny" (2012b: 147). And he does so by stating that the progress of the human mind:

> … is subject to the same general laws that can be observed in the development of the faculties of the individual, and it is indeed no more than the sum of that development realised in a large number of individuals joined together in society. What happens at any particular moment is the result of what has happened at all previous moments, and itself has an influence on what will happen in the future. (2012b: 2)

Condorcet's way of searching through history does not aim to find the immutable laws, rather withdraws from the past guidance on the progress that might help in future enhancements.

> Such observations upon what man has been and what he is today will instruct us about the means we should employ to make certain and rapid the further progress that his nature allows him still to hope for. (2012b: 2)

According to Condorcet progress is the constant law in human history, therefore a reflection on the past and the present might contribute to speed progress so that humanity might get closer to the future, the supreme law of human development, get closer to its indefinite perfectibility. That is the aim of Condorcet's work:

> Such is the aim of the work that I have undertaken, and its result will be to show by appeal to reason and fact that nature has set no term to the perfection of human faculties; that the perfectibility of man is truly indefinite; and that the progress of this perfectibility, from now onwards independent of any power that might wish to halt it, has no other limit than the duration of the globe upon which nature has cast us. This progress will doubtless vary in speed, but it will never be reversed as long as the earth occupies its present place in the system of the universe, and as long as the general laws of this system produce neither a general cataclysm nor such changes as will deprive the human race of its present faculties and its present resources. (2012b: 2)

Determined his goal, Condorcet will centre his attention on public education, for the ideals of the French Revolution could only be consolidated with an educated people. He even presents, in 1792, a project for public education to the Nacional Assembly, so that

---

[2] Regarding Turgot another edition was also used: (Turgot, 1991).

[3] Regarding Condorcet's works the following editions were used: (Condorcet, 2014; 2012a; 2012c; 1970; 1968b; 1822).

[4] For information concerning Condorcet life, works, introduction to the author's political thought see Steven Lukes and Nadia Urbinati introduction (Condorcet, 2012a: XV-XLII).

human mind could progress (Condorcet, 1792). He believed that schooling was a public right and that if well defined and executed it would correct the natural differences of human faculties.

The optimistic postulate of "indefinite perfectibility" agreed with the propagation of the Lights to all human beings who, with more knowledge, could use reason, in freedom. Actually, the progress of human mind was entangled with the progress of science, with the indissociable connection with Nature and with the respect of human natural rights

> to show why, in spite of the transitory successes of prejudice and the support that it receives from the corruption of governments or peoples, truth alone will obtain a lasting victory; we shall demonstrate how nature has joined together indissolubly the progress of knowledge and that of liberty, virtue and respect for the natural rights of man; and how these, the only real goods that we possess, though so often separated that they have even been held to be incompatible, must on the contrary become inseparable from the moment hen enlightenment has attained a certain level in a number of nations, and has penetrated throughout the whole mass of a great people whose language is universally known and whose commercial relations embrace the whole area of the globe. (Condorcet, 2012b: 6)

According to Condorcet, this state of affairs had already been achieved by some Anglo-Americans and those French people who, at the time, were freed from the bond of despotism, superstition and religious prejudices (2012b: 126, 133; 1970: 164, 204–205). Perfectibility will have to be achieved at two levels: by the progress of the human mind: and by the perfectibility of the human species, necessary condition for the true progress towards freedom and happiness. However, regarding the second level, says Condorcet, nothing has been done:

> There are a few circumstances from which the philosopher can take consolation, but he is still afflicted by the spectacle of the stupidity, slavery, barbarism and extravagance of mankind; and the friend of humanity can find unmixed pleasure only in tasting the sweet delights of hope for the future. (2012b: 123)

It was also important to know which changes or acquisitions would lead to a higher perfection. Which was the most urgent path humanity should tread voluntarily and consciously to attain that condition? Condorcet believed that French society was living a transition from the ninth to the tenth and last level or age in human evolution. The transition concluded humans of ages to come would only have to contribute to the "continuous and indefinite perfectibility" having three main principles to obey to:

> Our hopes for the future condition of the human race can be subsumed under three important heads: the abolition of inequality between nations; the progress of equality within a single people; and the true perfection of man. (2012b: 126)

From these conditions Condorcet believed would follow peace and happiness. Through his entire program, to be put into practice by politicians as "friends of humanity", there is an obvious ideological dimension of the idea of progress.

4 THE CLASH OF THE IDEA OF PROGRESS AND THE PRACTICE OF SLAVERY

The French Revolution would be considered the triumph of the bourgeoisie and therefore of the idea of progress. But it will be also the moment when issued from the Revolution, there would be written the first laws that, in Europe, will condemn in a radical form slave trade and slavery. Apparently, progress and slavery were on a collision course. Condorcet thought so when he defended that the "unhappy slaves" should be given back their freedom "they should never have lost". However, in 1788, in his second edition of ***Réflexions sur l'esclavage des nègres*** [Reflexions on the slavery of Negros][5] Condorcet stresses that, in spite of what might be expected the captivity of Negros has had, in the last times mighty and powerful apologists, and protestations had been unable to suspend those people progress (2014: 114)

When Condorcet wrote his reflections, when the topic was slavery, what was meant by it was a specific modality of submission known afterwards as "Black slave trade" since the majority of the slaves traded was Africans. It began in the 15th century, but it was with the colonisation of the American continent and the increasing demand of slave workforce that the transatlantic trade had an exponential increase, becoming pivotal to mercantile capitalism. Slave workforce was used in the exploration of colonial products highly valuable in European markets.

In the 18th century, slave trade achieved the highest number of slaves transported. However, the same century gave birth to the first abolitionist movements that opposed the legitimacy both of trade and of slavery itself (Pimentel, 1995: 143–160).

In England, the intense evangelism of the antislavery movement's leaders and both the political and economic interests help the adhesion and prestige of the movement that by the end of the 18th century was accepted both in political and popular circles. The British antislavery movement had a strong help from the economic changes that took place in the islands. If on one hand, the technological and industrial development could do without slaves, on the other, there was an increasing need for raw material and new markets. Under these circumstances, from an economical point of view, the abolition of slave trade seemed doubly

[5] The first edition appeared in 1781, and both editions published under the pseudonym of Mr. Schwartz, a word play with the German word for "black" or "negro".

advantageous, allowing the creation of new markets in Africa to British industrial products and on making possible the access to raw materials, especially, for textile production.

The independence of the United States deprived the British increasing industry of American cotton. This surely contributed to British activists' reinforced power for appealing for the liberation of Black slaves as a way to slow down, or at least delay the development and industrialisation of America.

The previous century the picture had been quite different. In 1715, the British, under the protection of the *aciento* granted by the Spaniards in the Utrecht Treaty, they had not only transported to Spanish colonies 4800 slaves per year, but they also profited transport as many as they wanted to any part of the world. By 1750, the British renounced the *aciento* but kept the slave trade.

However, by the end of the 18th century, the British started to look at slave trade by a different angle from the one that before ensured them the monopoly of this trade. Now, the slave trade was considered a strong setback to British economic development (Pimentel, 1995: 309–311).

In France, the movement had a different appeal. The force given to the philosophical arguments based on natural rights, the principles of justice and the aims of Enlightened intellectuals gave the anti-slavery movement a more elitist character. Furthermore, the political and economical French conditions also contributed to frail the development of the anti-slavery movement and opened to greater ambiguity the positions assumed. The French Illuminists, even though they had opened a new chapter in the controversy regarding slavery, in most cases, the arguments proved inconsistent. They contested slavery but, at the same time, they shared the conviction of their peers that considered Africans as inferior beings that could not have the same rights as French people.

Voltaire expressed this idea when he wrote on the intellectual ability of Africans:

> One can say that if their intelligence is not of a different sort than ours, it is surely quite inferior. They are not able to pau attention, they have difficulty combining thoughts and they do not seem fit neither to the advantages nor the misuse of our philosophy. (1829: vol. 1, 274)[6]

Montesquieu, in his writings, pointed out the economic importance of slavery, though assuming for a moment the position of a defender of slavery, which was not his own:

> If I had to defend the right we had of making Negroes slaves, here is what I would say: [...] Sugar would be too expensive if the plant producing it were not cultivated by slaves. [...] Those concerned are black from head to toe, and they have such flat noses that it is almost impossible to feel sorry for them. One cannot get into one's mind that god, who is a very wise being, should have put a soul, above all a good soul, in a body that was entirely black. (Montesquieu, 1989)[7]

Condorcet notes these contradictions. Using the alias "Un Ermite de la forêt de Sénart" [An eremite from the Sénart forest] he wrote in Journal de Paris, June 22, 1777, about that contradiction, stressing that while some defended the freedom of the arts, at the same time, there was the importation of Africans to whom the freedom to enjoy art was denied (1968b: vol.1, 344–348).

The colonial and economic state of affairs was the major liable factor in the practice of slavery. The colonial administrators' letters since long emphasised the growing dangers issued by racial and social unbalance felt in the colonies, asking for urgent reforms. Shortly before the French Revolution, the S. Dominique colony, the French crown's jewel in the Caribbean, the violence exploded to dramatic proportion. The echo of the Enlightenment ideas, the downfall of absolutism, and the rising of bourgeoisie triggered the feelings of colonial landlords that wanted to defend their interests. From 1791 to 1794, S. Dominique was the stage of brutal slave rebellions that ended up in a civil war.

The feeling of imminent danger, as well as the conscience of the atrocities committed, influenced philosophers' humanism driving them to express anti-slavery arguments. On August 11, 1791, the National Constituent Assembly forbade by edict slave trade. In the same year, the edict of September 28th declared the freedom, social and political right to every individual living in France, regardless of their skin colour. In March 1792, trying to control the rebellions the National Legislative Assembly conceded the civic and political right to coloured freemen living in the colonies. In February 4th 1794, after the speech made by a black deputy from S. Dominique, about the suffering and claims of slaves, France declared the abolition of slavery, enthusiastically applauded and voted unanimously, in the middle of shouts praising the Republic, the Convention and Freedom (Cochin, 1979: 11–14; Hoffmann, 1973: 99). For a moment, all lead to the conviction that the victory of anti-slavery was due to the Convention.

However, the abolition, which raised no theoretical doubts, turned out to be a practical problem. The sudden cut in the supplying and use of slave workforce, the fall of colonial productions, namely of sugar, led to the downfall of the whole economic system that was the financial backbone of France in the 18th century. Furthermore, violence continued to rule in the colonial world, partly due to the claims of equal rights, for blacks and mestizos, some of them free and rich,

[6] Original text in the edition used: "On peut dire que si leur intelligence n'est pas d'une autre espèce que notre entendement elle est fort inférieure. Ils ne sont pas capable d'une grande attention; ils combinent peu, et ne paraissent fait ni pour les avantages ni pour les abus de notre philosophie".

[7] The French edition used also in the composition of this text was the one published in 1849 (Montesquieu, 1849: 203–204).

others poor and slaves, but for white people of the lower social status. All of them got together forming one of the most diverse alliances. As Michèle Duchet summarised, the universalism of the principles gave to practice advantage over the theory, and after the illusion of the first moment, slavery and slave trade were re-established during Napoleon consulate, under the law of May 1802, replacing the regulations prior to the French Revolution (1979: 42). Then, the end of slavery was designated as a philanthropic mistake, an indiscrete measure (Duchet, 1979: 23, note 21).

The abolitionist cause, therefore, diluted itself without contestation. One has to consider relevant Chateaubriand's words written in 1802, in *Génie di Christianisme* [The Genius of Christianisme]:

> With eloquent words everything was lost, event pity faded for who would date still to make a plea in defence of the Negros after the crimes they have committed? (1975: 134)[8]

The events come to prove that, after all, ideas alone cannot act on reality and that the 1794 abolition was not exactly the outcome of a reflexion on slavery and the living conditions of slaves, but a mere coalition of reformist and humanitarian concerns. Abolition was, in those social, political and economic conditions, far from appearing beneficial. Slavery was the wave that stirred the European social crust and that after its implementation threatened to destroy the French empire.

How far could the abolition of slavery go? Would it be possible to frame it in the context of the Declaration of Human and Citizen Rights? The slaves were, in France, "extraordinary instruments" of progress, but was it possible to include them as a target for the civilizational progress? But of which civilisation? Who would be the human beings living in it?

The whole question was too complex and too controversial both in the political and economic spheres as in the philosophical discourse. Times were then diverse, moving at different speeds. Even so, reality was interconnected in a complex net of constancy and transformation, of ambiguities and contradictions, of harmonies and conflicts.

The right to freedom, that in the 18th century was an irrefutable philosophical principle, became the focus of discussions, both in those favouring abolitionism and slavery. Natural freedom was not only the right each individual had to decide his/her own present and future. It was also, as it is mentioned in *Histoire philosophique et politique des établissements et du commerce des européens dans les deux Indes* [Philosophical and political history of the Europeans establishments and commerce in both Indies], the most distinctive quality of the rational being, the principle of his vices and virtues, a precious and inalienable right (Raynal, 1782: 118, 126, vol. VI).

It was an idea of freedom that tried to put together European and non-European causes, "primitive" and "civilised" people, blacks and whites. However, in a world inhabited with extravagant superstitions, barbarian practices and outdated legislation that oppressed freedom, good luck was definitely not favouring the non-Europeans (Raynal, 1782: 105, 125–126, 133–134).

Meanwhile, Condorcet, in his *Réflexions sur l'esclavage des nègres* [Reflections on the slavery of Negros], in 1781, compares slavery to a theft with more painful consequences because the individual was simultaneously deprived of his possessions, his abilities and the means to recover them. In the bottom line, it meant to take away everything Nature had given him to preserve his life and have the necessary means to survive (2014: 23)

The famous ideals of the French revolution: freedom, equality and fraternity; are expressed in Declaration of the Rights of Man and of the Citizen, voted and approved in 1789. In the first version, one can read in article 1 that "Men are born and remain free and equal in rights. Social distinctions may be based only on considerations of the common good"; in article 2 one reads that "The aim of every political association is the preservation of the natural and imprescriptible rights of Man. These rights are Liberty, Property, Safety and Resistance to Oppression". Article 4 states that "Liberty consists in being able to do anything that does not harm others: thus, the exercise of the natural rights of every man has no bounds other than those that ensure to the other members of society the enjoyment of these same rights. These bounds may be determined only by Law" (*Declaration of the Rights of Man and of the Citizen*, 1789).

However, the second version of the Declaration, written in 1793, there is a different composition with significant changes. Article number 1 states the aim of a society: "The aim of society is the common welfare. The government is instituted in order to guarantee to man the enjoyment of his natural and imprescriptible rights"; and in article 2 the rights are explained: "These rights are equality, liberty, security, and property". Meanwhile, Article 3 simply declares: "All men are equal by nature and before the law". Article 18 is a novelty for it regards the condition of subjection, stating: "Every man can contract his services and his time, but he cannot sell himself nor be sold: his person is not an alienable property. The law knows no such thing as the status of servant; there can exist only a contract for services and compensation between the man who works and the one who employs him" (*Declaration of the Rights of Man and of the Citizen*, 1793)

Due to the form this article is written, subjection seems to identify itself with the more traditional servitude, that was still legal, or with a similar modality between free poor people that was the outcome of the huge misery that existed in France at the time.

Nothing allows us to interpret these articles as including the black slaves. The lack was not due, certainly, to their silence, because if there was a

[8] Original text: "Avec des grands mots on a tout perdu: on a éteint jusqu'à la pitié; car qui oserait encore plaider la cause des noirs après les crimes qu'ils ont commis?"

moment in history when slaves made their voices heard loud and clear this was one of the most tremendous and dramatic, as already referred. One has to find it odd there is no allusion, whatsoever to them, furthermore because the arguments on this matter were pressing. May it be that the colonial events were not important enough for a reflection on slave trade and slavery? Or is the silence the outcome of that reflection?

Reading Condorcet there is no doubt that the matter was under questioning. In his *Remarques sur les pensées de Pascal* [Comments on Pascal's thoughts], in 1774, he mentions that slavery was far from being only a bad outcome of a false principle of conscience. In the case of the Africans, it was also a distorted vision of reality, because it was allowed and justified for the Africans but condemned as an "awful barbarity" when applied to white people. Says ironically Condorcet "but they are black and that changes all our ideas" (1968a: 644–647)[9]. Regarding this issue, there should ever exist a false principle of conscience.

Therefore, Condorcet advised the readers to release all slaves because such kind of practices could not exist inside a nation that pretended to be free, support equality and achieve excellence. How could talk about the *Declaration of the Rights of Man and of the Citizen*, of inviolable liberties, citizens security if, systematically people most sacred laws were violated? Again, in an ironical tone Condorcet proposes to the *Corps électoral contre l'esclavage des noirs* [Electoral body against the slavery of Negros], in February 1789 that it should be added to the *Declaration of the Rights of Man and of the Citizen* a single word: WHITE, changing the text to express that:

> All white people are born free and equal according to a method to determine the level of necessary whiteness. (Condorcet, 1968b: 469–472)[10]

Paraphrasing Raynal (1782: vol VI, 105–106), while in Europe the flag of fraternity was unfurled, a black slave whose hard work offered those same Europeans the colonial delicatessen, was forgotten from that glorious cry.

Slavery was an unjustifiable institution, but while it was not abolished, it should be faced with realism, meaning, providing the means for it to become less rigorous and proceed to a gradual abolition. If, by any chance, "the voice of humanity" became powerless to face the landowners' interests, it was advisable to warn them that the slaves neither needed the owners' generosity or pity to liberate themselves. In this case, Raynal was calling upon the right to rebel as one of freedom's dimensions, a resort to be used when every other failed (Raynal, 1782: vol VI, 138–139).

Following Raynal's thinking, other writers, fearing the economic repercussions, advised a gradual emancipation, accompanied by reforms, instead of an immediate and complete release that might turn out to be equally prejudicial to the liberated slave.

However, there were also those who tried to demonstrate that slavery allowed the slave to benefit from better living conditions than European serfs, and admitted the idea that, though being a "chronical misfortune", slavery should be considered an advantage for blacks and whites alike, and should be extended to European working system (Biondi, 1973: 26; Malouet, 1788: 33–42)

In Diderot et D'Alembert *Encyclopédie* (Diderot et al., 2016), Le Romain's entry entitled "Nègres considérés comme esclaves dans les colonies de l'Amérique" [Negroes, *considered as slaves in the colonies of America*] (Le Romain, 2003) one finds the insistence in the idea that the development of the new found lands would only be possible by the use of slaves. Le Romain states:

> These black men, born vigorous and accustomed to coarse food, find comforts in America that make animal life much better there than in their own country. This change for the good puts them in the position to stand up to the labor and to multiply abundantly. (11: 80)

However, Condorcet stresses the need of workforce did not imply the workers' slavery. Black should be free and be paid for their work as the European workers. Liberty and property were natural oppositions. The Physiocrats, as Quesnay and Turgot, the moral and utilitarian philosophy of Francis Hutcheson and Adam Smith defended that free workers were not only more productive as also less expensive. Furthermore, it would reinforce a major interest for the development of new technologies and an environment more favourable to happiness and progress (Condorcet, 2014: 36–38; Duchet, 1979: 40; Smith, 1981: 199–200).

The physiocrats thinking used antislaveryism as an extension of the fight held in Europe to replace servitude by free workers. The Abbé Beaudeau expressed this way of thinking when, in 1766, in the book *Éphémérides du Citoyen* [Citizens' Ephemerides], he compared the situation of European serfs and black slaves, and pastor Frossard, in the book *La Cause des esclaves nègres et des habitants de la Guinée* [The Negros' slaves cause and the inhabitants of New Guinea], followed the same line of thought stressing that both situations were on the same level (Duchet, 1971: 164).

Political economy and humanism were allies in this fight that led to the conclusion that slavery was an economic mistake, not only because the slave work was not lucrative but also because it was a waste of the capital invested. Regarding the colonies, a reform was urgent to replace the old colonial policy based on slave labour by a system more suited to metropolitan bourgeoisie's interests. Only with individual freedom, said Benjamin-Sigismond Frossard, would it be possible to conciliate prosperity and national development with happiness and individual interests. Without it

---

9 Original text: "mais ils sont noirs et cela change toutes nos idées".

10 Original text: "Tous les hommes blancs naissent libres et égaux en droits; donner une méthode pour détermine le degré de blancheur nécessaire".

either America or Europe could not live in security, enjoying humanist behaviour, productive work and conditions to develop employment and industry (Frossard, 1789: 172–175).

In this state of affairs, if it was difficult to support freedom, inequality was considered a hard to ignore evidence and when it came allied with prejudices of inferiority it only gave strength to the pro-slavery arguments. The idea that there were human beings that were born to become slaves had already lost all credibility, but the inequality/inferiority was a powerful argument. Voltaire was one of the Enlightenment philosophers who clearly defended the inequality of other people compared to Europeans. In the introduction to *Essai sur les mœurs et l'esprit des nations* [Essay on the behaviour and the Spirit of Nations], he states that only a blind man would be unable to see that whites, Negros, albinos, Hottentots, Lapponians, Chinese, and Americans were entirely different races and that the Negros, regardless of the climate they lived in, produced always "animals of their own kind" and mulattos were but a "bastard race" (Voltaire, 1829: 19–20). According to Voltaire slave trade was but a confirmation of Europeans' superiority (*Appud.* Condorcet, 1968b: vol. IV, 504).

One has to bear in mind that the idea of inequality was not used exclusively in racial questions, but it was applied to the white society too. In this case, it was even considered not only necessary but also beneficial to the society's balance. Voltaire, in the entry *Égalité* [Equality] in the Dictionnaire philosophique, declared it was convenient the social division of individuals into two groups: the rich who led and the poor who served, and he conclude that evil did not reside in inequality, but on dependency (Voltaire, 1878: vol. XVIII, 475).

On the other hand, Jean-Jacques Rousseau, who did not believe in the possibility of equality, not even in a natural stage, did not accept that inequality had to lead inevitably to the superiority of ones over the others (Rousseau, 1997). Diderot, pondering on species diversity, in the entry *Humain [Human]*, states that that the original human colour is white:

> White appears therefore to be the primitive color of nature that climate, food, and custom alters, and causes the gradation from yellow and brown to black. [...] From what precedes it follows that in all of the new world that we have thus far traversed, there is only the one same race of men, more or less dark skinned. The Americans come from the same stock. The Europeans come from the same stock. From north to south one perceives the same varieties in the two hemispheres. All goes therefore toward proving that the mankind is not composed of essentially different species. The difference from whites to browns comes from food, customs, habits, climates; that from browns to blacks, from the same cause [...] There was therefore only one original race of men, which being multiplied and spread over the surface of the earth, has produced over time all of the varieties which we have just mentioned; varieties which disappear over time, if one imagines that people move all at once, and that they find themselves either necessarily or voluntarily subjected to the same causes which have acted upon those who are believed to have occupied those regions (2010: 347–348).

There is an implicit idea of hierarchy in Diderot's words, however, he clearly rejected every argument that led to the justification of slavery or any form of subjection, concluding that in Africa and America, human beings became the worst calamity to their own species (1973: 504–505 vol. 15). He called the attention to the fact that, sooner or later, the subjected people would rebel, destroying everything the colonisers had built, if their rights were not respected (1973: 496, 500–503, 507, vol 15)

European philanthropists tried to prove that black people were not in any way inferior to other human beings, and if they were in a less developed state regarding their rationality or way of being that was due solely to slavery imposed on them by Europeans that so perpetuated their ignorance and abused their weaknesses (Diderot, 1973: 501, vol. 15).

Thomas Jeffery in an exposition regarding North America stresses the abusive treatment imposed on black slaves, that instead of being considered as human creatures, with a soul, reason and feelings, were considered as "the trash of nature's beings" with no human rights or privileges (Lecointe-Marsillac, 1968: 94–95).

In Brazil the situation was identical. The Jesuit Antonil, in 1711, reported that black slaves lived in a real hell and to achieve work efficiency there was a systematic use of violence and constraint, and sometimes the practices were so cruel that had to be considered as surpassing simple bestiality, hard to be "supported either by barbarians" and even more "by Christians" (Antonil, 1967).

Victor Schoelcher, who, in 1848 proclaimed the abolition of slavery in France, almost fifty years after the failure of the first attempt, stated that there existed a considerable unevenness between Africans and Europeans abilities. However, that unevenness did not have its roots in any genetic variation, but in the conditions of life to which the slaves were subjected. He was convinced that with a similar education, the unevenness would soon disappear. Schoelcher declares that the same phenomenon could be observed among the Europeans who were subjected to servitude in Eastern Europe (1976: 146–156, 162–167).

5 THE DIFFERENT "SCIENTIFIC" THESIS USED TO DEFEND OR JUSTIFY SLAVERY

However, in what regards the differences that quickly transform themselves into inequality and inferiority, there is still one darker step to descend. What kind of being was a black? In some cases, he was considered the creature who in *the chain of being*, so popular in the 18th century, was placed between the animals and humans. The notion of a *scala naturae* was not a novelty of the 18th century. During the Renaissance, there was an increasing interest in the intermedia forms between animals and plants. It is in this period that the *scala naturae* starts to be systematised, achieving its prime with the 18th-century naturalists. The bottom line had to do with the discoveries made during the Renaissance that questioned the inflexibility of God's creation that the Church sanctioned: God had created all living beings and these, obeying the Creator's will, maintained the initial perfection (Pimentel, 2010: 52). On the other hand, the mechanical development, though being a strength inherent to nature itself led to the combination of the theory of evolution with the presupposed idea of Progress. Every organism that made part of the chain presented a unidirectional evolution and was ordained, in a progressive and continuous way, starting with the simpler and moving to the more complex.

> That body of truths which is attained by going through the enormous chain of natural entities; the successive links that leads from brute matter to the lowest degree of organic life, from organic matter to that which gives the first indications of spontaneous movement and sensitivity, and, finally, from that to man; the relations of all these entities to man, either relative to his needs or in the analogies which bring him nearer to them, or in the differences that separate him from them: this is the picture which natural history gives us today (Condorcet, 2012b: 112–113)[11].

The process was not one of adaptation, which implied transformation, and since not all connections between different species were known, it was necessary to look for the missing links of the chain of being. Lamiral was one of the authors who classified the black slaves as being part of the animals that only existed to serve and be submitting. According to Lamiral before the black people there had existed an intermediary who lived isolated in forests, in a wild state and that no one had ever seen (Lanthenas, 1968: 33–38).

The need to explain the somatic and cultural differences among human beings, and particularly the physiognomic traces of black people was felt since Antiquity. Between the 16th and 18th centuries the notion of degeneration, according to Judaic-Christian perspective of the corruption of the human spirit as consequence of the original sin, became a recurrent theme to justify the inferior condition of non-European populations. The religious interpretations spread over skin colour the mark of sin, which extended through generations. With the naturalistic theories, in the end of the 18th-century and particularly in the following century, differences came to be represented as the structural reality of the individuals themselves. Religious preconceptions, environmental and social influences, biological factors, extrinsic and intrinsic forces, evolutionary and degenerative processes were all arguments presented as causes for diversity. If Europe represented civilisation in its most advanced stage, due to an extraordinary development that went far back in time, and if evolution was positive and linear, human groups that were different could only be understood as the result of stagnation or degeneration.

The development of this kind of reasoning had to bring to the discussion the problem of the origin of humanity and this would give birth to two major anthropological positions: the monogenic theory, defending the unity of humankind, and the polygenic theory contesting a common origin. The Enlightened anthropological reasoning the monogenic theory was dominant. However, at the same time, a new idea began to grow defended that humanity did not have a unique ascendency. The polygenic thesis defended that different races had different origins and lacked any affinity or kindred. This theory, that up to the 19th century had feeble influence, has, however, its roots in the 16th century. Both theories were concerned with the differences. The monogenic defenders it was the consequence of degeneration. For the polygenic apologists, the differences were the result of Nature itself responsible for establishing among several beings, humans included, a hierarchy of physical and psychic differences. In practical terms, both theories led to a hierarchy of different types of humans. Both were in essence discriminatory.

From the mixing of the ecological and biological thesis, the outcome was an intermediary theory that deafened the existence of biological changes, caused by environment and transmittable from generation to generation. Around this group of ideas, many different interpretations were born, defending that black people should be considered as a different species from the rest of humanity. This conclusion was reinforced by Buffon and Pierre de Maupertuis theories, though both believed in the unity of the human species. Buffon considered being unacceptable the idea that African children were born to white and became black due to solar exposition, considering that skin pigmentation was transmitted from parents to offspring and only a very long period might explain the change. The Black colour should not be considered as a specific quality, but rather an acquired one in consequence of a long and constant exposition to specific environmental conditions turned into a hereditary characteristic. Blacks and whites had the same origin and were members of the same species. However, to Buffon, as well as for the majority of 18th-century thinkers, the white man represented the norm, being the black people a deviation. Because of climatic and chemical reactions the Africans had, to use Buffon's words, "degenerated" (1752: 292–295, vol. 4). Humankind was not constituted by essential different species, but by different varieties of the same species and this meant that human societies that did not have the means to initiate an evolutive process were inevitably condemned to

[11] For a more profound study of the concept of "the chain of Being" and its evolution refer to Lovejoy (1964).

degenerate. This means that different circumstances, particularly the environment, vices, illnesses, food and social habits might represent obstacles for the progress of individuals and annulled the ability to move from the stage of savages to civilised beings.

According to Cornelius de Pauw the long exposition to the tropical environment was enough to sparkle a physiological transformation that "disfigured" Africans, with the possibility of causing the destruction or change of certain organs, as the brain, that would affect their cognitive abilities (Cohen, 1981: 125)[12].

Montesquieu, before Cornelius de Pauw, had also used the climate factor theory and concluded that the vast majority of African population was either savages or barbarians, which was, he says, proved by the way they engaged in commerce. The fact that African valued more a necklace made of glass beads than gold itself proved they were of poor intelligence (1989: books 14, 15, 21).

6 CONCLUSION

When European thinkers, led by the idea of progress, turn their inquiries into the past and evaluated their evolution from Antiquity until the 18th century, they felt pride in their civilisation. But when they compared themselves with non-European people they were overwhelmed by an excessive feeling of superiority. Europeans assumed themselves as having the most developed civilisation in the whole world. The jurist Charles de Brosses, in 1760, in the book *Du Culte des dieux fétiches* [On the Cult of the fetish gods], stated that some people were still in their "infancy", meaning in a stage of development rudimentary that other have long surpassed (Cohen, 1981: 119).

Europeans had already lived through the same level of development the African societies were still experiencing, but one day every society would achieve perfection. Holbach explained that a chain of consecutive experiences would lead human savages to the stage Europeans would consider as a civilised society, paying attention to the most sublime sciences and knowledge (Cohen, 1980: 121). However, for the time being, European elite considered that African societies were still far from such an elevated stage, considering that it had taken more than a millennium for Europe to achieve it.

The faltering and the subterfuges used by most European intellectuals reveal that the economic, mental, and cultural conditions were not still ready to allow the abolition of slavery that humanitarian principles claimed- There are several and different interests in this issue and there is a clear separation between the theoretical and the practical discourses. Slave situation is a clear example of the mismatching between the humanist ideal and actual living conditions. One may consider that the route is travelled over in two different speeds: the idea of progress, seen in a global perspective was gradual, continuous and positive and the issue of slavery, a constant in human history, always oppressive and humiliating of human condition corrupted and brought to a halt lords and slave societies.

Slavery is the antithesis of freedom, however, it was considered, erroneously, as an essential factor for progress.

[12] Refer also to the original Cohen's edition (1980).

BIBLIOGRAPHICAL REFERENCES

Antonil, André João. (1967) *Cultura e Opulência do Brasil.* Texto da ed. de 1711, intr. e vocab. por A. P. Canabrava. São Paulo: Companhia Editora Nacional.

Biondi, Carminella. (1973) *Mon frère, tu es mon esclave.* Pisa: Editrice Libreria Goliardica.

Buffon, M. de. (1752) *Histoire Naturelle.* 5ª ed. vol VI. Paris: De L'Imprimerie Royale.

Chateaubriand, François-René. (1975) *Génie du Christianisme.* Nouvelle édition revue avec soin sur les éditions originales précédez d'une étude littéraire sur Chateaubriand par M. Sainte-Beuve. vol. 2. Nendeln: Kraus Reprint.

Cochin, Augustin. (1979) *L'abolition de l'esclavage.* Fort de France: Désormeaux.

Cohen, William B (ed). (1981) *Français et africains. Les noirs dans le regard des blancs, 1530 1880.* Trad. Camille Garnier. Paris: Gallimard.

— (ed). (1980) *The French Encounter with Africans: White Response to Blacks, 1530–1880.* Bloomington: Indiana University Press.

Condorcet, Jean-Antoine-Nicolas de Caritat. (2014) *Reflexões sobre a escravidão dos negros.* Trad. de João Tiago Proença. Lisboa: Antígona.

—. (2012a) *Political Writings.* Lukes, Steven and Urbinati, Nadia. New York: Cambridge University Press.

—. (2012b) The Sketch. *Political Writings.* New York: Cambridge University Press, 1–147.

—. (2012c) On slavery; Rules for the Society of the Friends of Negroes (1788). *Political Writings.* New York: Cambridge University Press, 148–155.

—. (1970) *Esquisse d'un tableau historique des progrès de l'esprit humain.* Texte revu et présenté par O. H. Prior. Paris: Librairie Philosophique J. Vrin.

—. (1968a) *Remarques sur les pensées de Pascal.* vol. III. Stuttgard: Bad Camustatt.

—. (1968b) Lettres aux auteurs du journal de Paris de 9 de Juin de 1777. Stuttgard: Bad Camustatt.

—. (1822) *Esquisse d'un tableau historique des progrès de l'esprit humain; suivie de refléxions sur l'esclavages des nègres.* Paris: Chez Masson et Fils.

—. (1792) *Rapport et projet de décret relatifs à l'organisation générale de l'instruction publique Présentation à l'Assemblée législative: 20 et 21 avril 1792.* Available at: http://www.assemblee-nationale.fr/histoire/7ed.asp. (Accessed April 2017).

Declaration of the Rights of Man and of the Citizen. (1793) Available at: http://www.columbia.edu/~iw6/docs/dec1793.html. (Accessed April 2017).

Declaration of the Rights of Man and of the Citizen. (1789) Available at: http://www.conseil-constitutionnel.fr/conseil-constitutionnel/root/bank_mm/anglais/cst2.pdf. (Accessed April 2017).

Dias, J. S. da Silva. (1988) *Os Descobrimentos e a problemática cultural do século XVI.* Lisboa: Editorial Presença.

—. (1969) *A política cultural da época de D. João III.* Coimbra: Universidade de Coimbra.

Diderot, Denis. (2010) *Humanity (Type).* Translated by Naomi J. Andrews. Originally published as "Humaine espece," Encyclopédie ou Dictionnaire raisonné des sciences, des arts et des métiers, 8:344 (Paris, 1765). 8. Ann Arbor: Michigan Publishing, University of Michigan Library. Available at: http://hdl.handle.net/2027/spo.did22 22.0000.242. (Accessed April 2017).

—. (1973) *Oeuvres complètes. Edition chronologique.* Introduction de Roger Lewinter. [Paris]: Le Club Français du Livre.

Diderot, Denis & Alembert, Jean le Rond D'. (2016) *Encyclopédie, ou dictionnaire raisonné des sciences, des arts et des métiers, etc.* Morrissey, Robert and Roe, Glenn (eds). University of Chicago: ARTFL Encyclopédie Project (Spring 2016 Edition). Available at: http://encyclopedie.uchicago.edu/. (accessed April 2017).

Diefendorf, Barbara. (1985) Prologue to a Massacre: Popular Unrest in Paris, 1557–1572. *The American Historical Review* 90 (5): 1067–1091. http://www.jstor.org/stable/18 59659.

Duchet, Michèle. (1979) Les réactions face au problème de la traite négrière: analyse historique et idéologique. *La traite négrière du XVe au XIXe siècle.* Paris: Unesco.

—. (1971) *Anthropologie et Histoire au siècle des lumières. (Buffon, Voltaire, Rousseau, Helvétius, Diderot).* Paris: François Maspero Editeur.

Frossard, M. (1789) *La cause des esclaves nègres et des habitans de la Guinée, portée au tribunal de la justice, de la religion, de la politique.* Lyon: Imprimerie d'Aimé de la Roche.

Godinho, Vitorino Magalhães. (1998) Que significa descobrir? In: Novaes, Adauto (ed) *A descoberta do homem e do mundo.* São Paulo: Companhia das Letras.

Hazard, Paul. (1971) *Crise da Consciência Europeia.* Tradução e notas de Óscar de Freitas Lopes. Edições Cosmos.

Hoffmann, Leon-François. (1973) *Le Nègre romantique. Personnage littéraire et obsession collective.* Paris: Payot.

Kelley, Donald R. (1972) Martyrs, Myths, and the Massacre: The Background of St. Bartholomew. *The American Historical Review* 77 (5): 1323–1342. http://www.jstor.org/ stable/1861309.

Lanthenas, François-Xavier. (1968) M. Lamiral réfuté par lui- même. *La Révolution Française et l'abolition de l'esclavage. Textes et Documents.* t. VII. Paris: EDHIS.

Le Romain, Jean-Baptiste-Pierre. (2003) *Negroes.* Translated by Pamela Cheek. Originally published as "Negres," Encyclopédie ou Dictionnaire raisonné des sciences, des arts et des métiers, 11:80–83 (Paris, 1765). 11. Ann Arbor: Michigan Publishing, University of Michigan Library. Available at: http://hdl.handle.net/2027/spo.did2222.0000 .029. (Accessed April 2017).

Lecointe-Marsillac. (1968) Le More-Lack, ou Essai sur les moyens les plus doux et les plus équitables d'abolir la traite et l'esclavage des nègres d'Afrique. *La Révolution Française et l'abolition de l'esclavage. Textes et Documents.* Paris: EDHIS.

Locke, John. (1689) *A Letter about Toleration.* Available at: http://www.earlymoderntexts.com/assets/pdfs/locke1689 b.pdf. (accessed April 2017).

Lovejoy, Arthur O. (1964) *The Great Chain of Being: a study of the history of an idea.* Cambridge, Mass.: Harvard University Press.

Malouet, Pierre Victor. (1788) *Mémoire sur l'esclavage des nègres.* Neufchatel.

Montesquieu, Charles de Secondat baron de. (1989) *The spirit of the laws.* Translated and edited by Anne M. Cohler, Basia Carolyn Miller, Harold Samuel Stone. Cambridge: Cambridge University Press.

—. (1849) *L'Esprit des lois.* Paris: Librairie de Firmin Didot Frères.

Pimentel, Maria do Rosário. (2010) *Chão de Sombras; Estudos sobre Escravatura.* Lisboa: Colibri.

—. (1995) *Viagem ao fundo das consciências. A escravatura na Época Moderna.* Lisboa: Edições Colibri.

Raynal, Abbé Guillaume Thomas. (1782) *Histoire philisophique et politique des etablissements et du commerce des européens dans les deux indes.* Gèneve: Chez Leonard Pellet.

Rousseau, Jean-Jacques. (1997) *The discourses and other early political writings.* Cambridge: Cambridge University Press.

Schoelcher, Victor. (1976) *Des colonies françaises. Abolition immédiate de l'eslavage.* Rep. de l'edition de 1842. Paris: Paguerre Editeur.

Smith, Adam. (1981) *Riqueza das Nações.* Lisboa: Fundação Calouste Gulbenkian.

Smither, James R. (1991) The St. Bartholomew's Day Massacre and Images of Kingship in France: 1572–1574. *The Sixteenth Century Journal* 22 (1): 27–46. http://www.jstor.org/stable/2542014.

Turgot, Anne Robert Jacques. (1991) *Discursos sobre el progresso humano. Estudio preliminar.* Trad. y notas de Gonçal Mayos Solsona. Madrid: Editorial Tecnos.

Vilhena, Vasco Magalhães. (1979) *Progresso: História breve de uma ideia.* Lisboa: Caminho.

Voltaire. (2016) *Treatise on Toleration.* Translated, with an introduction abd notes by Desmond M. Clarke. London: Penguin.

—. (1878) *Dictionnaire Philosophique.* Paris: Éd. Garnier.

—. (1829) *Essai sur les Mœurs et l'esprit des nations.* Paris: Chez Lefèvre, Libraire, Werdet & Lequien Fils.

On the "bread of Brazil" in the colonial period: Uses, habits and production

Ana Carolina De Carvalho Viotti
Department of History, São Paulo State University, Franca, São Paulo, Brazil

ABSTRACT: Being so consumed and used, the manioc, during the colonial period, received the nickname of "bread of Brazil". This text aims to highlight and historicize, amongst notes, reports and books produced between the Seventeenth and early Nineteenth centuries, the importance that the root acquired as food for the population in the tropics, especially to the enslaved and in the Atlantic trade. The scientific discourses elaborated with the ideals of agricultural progress in the late Eighteenth, to boost production of the genus, will be also presented and discussed.

Keywords: manioc; feeding; colonial Brazil; agriculture; progress.

1 INTRODUCTION

In a sort of rendering of accounts of the goods and income of the Jesuit college of Santo Antão de Lisboa in Brazil, written by Father Estevão Pereira in 1635, there were listed the annual expenses for the sustenance of the slaves. Apart from fresh meat "for their eating from time to time", or "whale, cod, sardines," there was an expressive expenditure with manioc flour. The priest says that, in addition to the flour that the mill yields, all of which was spent on blacks, at least 200 more bushels were needed, and the ordinary was worth at least 200 réis. (Descrição… 23 Aug 1635).

At the time, the flour, which had long been produced by the natives and had been tried – with a good deal of approval, by the way – by many chroniclers and travellers in the Sixteenth century, already configured the first item in the barns of those who owned slaves, and it replaced the traditional wheat flour based bread on most and varied tables. In this brief text, we will present some of the main reports about the uses and consumption of the root, highlighting the growing importance acquired in the daily life of the colonists of all kinds, especially as the ration of the slaves. Likewise, we will point out the attempts to make its production more and more efficient, to meet the domestic and foreign demands of the Portuguese seaborne empire, later becoming the State of Brazil.

2 ORDINARY TO FIND AND TO EAT

Since the earliest times, manioc had been defined as one of the "sustenance of the inhabitants of Brazil, whites, Indians and slaves of Guinea". Pero de Magalhães Gândavo, in 1576, emphasized that, among all possible products, it occupied "the first place" (Gândavo, 1930, 175–6). In the following years, the by-products of manioc, cassava or carimã, to mention some of its nicknames, seem to have lost no space in the dishes of the groups cited by the Portuguese historian and chronicler, on the contrary: the volume of news about its consumption increased – including, it's important to mention, an expressive use in feeding the poor population of the Kingdom, as shown by the continuous communications requiring the substantive increase on the despatch of flour (Ofício… 2 Nov 1800) –, notably among the captives, the main target of our observations. In 1640 Recife, Johan Nieuhof noted that it was "with its flour that Brazilians, as well as Portuguese, Dutch and blacks, make bread" and that "apart from wheat, it is the best" (Nieuhof, 1942, 289). In 1666, on one of the earliest accounts of Pernambuco after the expulsion of the Dutch (França, 2012, 133), the Lord of Mondevergue reported on the controversial practice of the acquisition of bread by the slaves themselves, since "the lords allow only a few hours of the week to be dedicated to their sustenance, which they employ by planting manioc to eat [...]" (França, 2012, 237). On the outskirts of the town of São Paulo, sometime between 1670 and 1719, Father Manoel da Fonseca claimed that the "common sustenance" was due to "the manioc, from whose roots flour is made" (Fonseca, s/d, 136). In that interval, precisely in 1695, François Froger reported, from São Sebastião do Rio de Janeiro, that the root "replaces bread in much of America", and explained the process of obtaining its flour: "the root is crushed and its juice is collected, which, being a powerful poison, is poured into a hole dug in the ground to prevent animals from drinking it" (Froger, 2008, 65–66), the domestication developed by the natives.

In the early decades of the Eighteenth century, the Jesuit André João Antonil reported the expenditure of flour for the diet of the enslaved in Minas Gerais and Bahia, listing it among the genera found for purchase and stating that "all the slaves want sustenance and clothing, medicines, nursing and nurses", and, for this purpose, "fields of many thousands of manioc are necessary" (Antonil, 1997, 75). The observations of Nicolas La Caille, in 1751, on the common combination of manioc flour and fish to nourish the captives are also known. Shortly before, the probable merchant Le Gentil La Barbinais, in a troubled passage through the Baía de Todos os Santos in 1714, claimed to know that "the master gives to his slave only 7 pounds of manioc flour a week, and it is up to the slave to arrange everything else that is necessary to live, by work or by means of petty thefts" (La Barbinais, 2012, 533). South of there, in Rio de Janeiro, the chief surgeon of the First Fleet, Arthur Phillip, in 1787 described, among all sorts of earthly gifts, the cultivation of a large quantity of manioc for the consumption of slaves and poor people. He gives a prominent place for the pestled fruit of the manioc, when he says that the manioc flour, which served as bread for the thousands of men living in the tropics, was one of the examples of the triumph of the human spirit over nature, a proof that the desire of the Providence develops all our potentialities for the benefit of our conservation. (Phillip, 2008, 240).

Some years later, by the end of the Eighteenth century, Luís dos Santos Vilhena, responsible for giving an account on daily colonial life in a series of letters, said that it was customary to distribute to the slaves "a fourth of Flour and three and a half pounds of dry and salted meat to sustain them for ten days" (Vilhena, 1922, 188), and the root being, according to him, "the fundamental basis of which we can say, supports the subsistence of Brazil and especially all the inhabitants, naturals and foreign" (Vilhena, 1922, 205). This entanglement of references, in short, corroborates the assertion that manioc has fed, with greater or lesser prominence, most of the Portuguese slaves in America, and received more and more attention from the political administration and the slave masters during the colonial period.

3 THE OBLIGATION OF PLANTING

The cultivation of manioc, indispensable for the maintenance of the bodies in the colony, then became demanded and regulated by all the governance. In Salvador in 1687, a law was published, analogous to the Dutch legislation that was current in previous years, (Schwartz, 2004, 383) determining the obligation to plant 500 root pits per slave employed in sugarcane plantations (Carta ..., 9 Aug 1687). Then, in 1701, a ban on the cultivation of cane was established on farms with less than six slaves, which should be devoted exclusively to the cultivation of manioc, a measure criticized by the city councils: the officers of Rio de Janeiro in 1703, for example, denounced the "damages that would befall on the inhabitants of this place if they integrally fulfilled the law on the plantation of manioc" (Carta ... 11 Mai 1703); in of Recife, in the following year, asked that the cultivation of the flour be carried out by the farmers with less than three slaves (Carta ... 9 Apr 1704). Those impositions had their price: also in 1704, the governor of the captaincy of Pernambuco, Francisco de Castro de Moraes, addressed a letter to King Pedro II exposing the urgency and necessity of granting to the farmers of manioc, if a substitution of part of the sugarcane production by the root was desired, the same privilege that the cane farmers had (Carta... 8 Mar 1704). The following year, apparently for not having obtained a positive return of the monarch, he wrote again to denounce the "losses" due to the application of that law (Carta... 27 Nov 1705).

The necessity of its cultivation – and of the expansion of the crops – was therefore continuous. When a "sun with excessive ardour" reached the tropics between 1722 and 1723, culminating in a severe drought and lack of supplies, the "História da América Portuguesa" registered the pains of its absence. The greatest lack among all genders, it would not be wrong to say. The effect of this inclement was, as said by the historian Sebastião da Rocha Pita, devastating: it struck all America, drying the waters, spoiling the fruits, sterilizing the crops and killing the cattle, so that besides the lack of all food, the greater lack was of manioc flour, which is the common bread of the inhabitants of this State, and for this reason its price in the provinces of Pernambuco, and Rio de Janeiro, reached three thousand two hundred and four thousand réis per bushel. The case was was better in Bahia due to the action of governance, according to Rocha Pita: the Viceroy helped to repair the calamities of the weather and ordered several villages to reduce the roots of manioc to flour, encouraging the farmers to make new plants for the times to come. For this purpose, it is known that he sent large sums of money for distinguished officers to get some flour, usually given to the Prison Infantry, to the ordinary people, and to the great factories of the Engenhos, and farms, which manioc plants usually are grown for their families, and slaves, had been sterilized by the drought. (Pita, 1730, 632–633).

Other times of famine and lack of "common sustenance" were identified everywhere throughout the colonial period. The events reported by the city councils and governors of several captaincies are common in the last decade of the Eighteenth century. The case of Pernambuco is quite illustrative. In 1792, the then governor and captain-general, Tomás José de Melo, sent a letter to Fernando José de Portugal, the governor and captain-general of Bahia, "with a great deal of sentiment", expressing the miserable state of which was reduced this captaincy with the lack of that flour, there experienced, due to a great drought in the year 1791. Considering this situation, to avoid the imminent ruin and the dire consequences that could follow from the lack of first necessity genders, Melo wrote, begging to his fellow, not only for charity, compassion, humanity and happiness of those loyal vassals, but for the survival of the population, a vessel carrying manioc flour (Ofício… 26 Mar 1792).

Bearing in mind the news from the Governor of Pernambuco, it seems that the lack of such ingredient, common to everyone's table, even at those tables whose guests could afford wheat flour bread, showed itself even more conspicuous amongst the slaves. In addition, the royal administration observed the plight of those unfortunate, demanding the governor to carry out solutions. In September 1795, he had to answer a letter from the interim Secretary of State for the Navy and Overseas, Luís Pinto de Sousa Coutinho, in which he was strongly recommended to cultivate the manioc culture, and to send to the Court all the flour that was not necessary to support the families and slaves of this captaincy. Although he should follow such recommendation, he couldn't: the drought that had devastated the land for at least three years, since a third of the population, including the troops and blacks, had suffered from starvation (Carta… 15 Sep 1796). He wrote from Recife, but the situation did not look different in Olinda. The following year, the council officers requested the intervention of D. Maria I to solve the lack of manioc flour, to prevent those people from "falling into ruin"; another five years would pass and the Governing Board of the captaincy of Pernambuco would officiate the new secretary, proposing measures to be taken in order to remedy the lack of basic necessities genders such as meat and manioc flour. (Ofício… 14 Dec 1801)

4 ON BOTH SIDES OF THE ATLANTIC

It is worth remembering that the production of this flour also served the African continent: it was extremely necessary for the slave trade (Relação… 1756), in a way that as more consignments of "Guinea slaves" were demanded, the more manioc crops had to be planted. Especially after periods of drought, as described above, the inhabitants of Brazil hard felt the intense flow of these supplies to the African coast. The Crown came to determine, with such situation in mind, the obligation of slave merchants to have captives specially dedicated to the planting of manioc, as can be seen in the correspondence, as early as July 24, 1725, between Manoel Rolim de Moura, then governor of the captaincy of Pernambuco, and the monarch: "in view of the damage caused to my vassals of the Brazilian State by the dearth of supplies […] of the many and continuous vessels that go to Costa da Mina on account of the slave business", the king claims as

> servido mandar promulgar uma lei que todas as pessoas que mandassem as tais embarcações tivessem escravos, plantando mandioca, em tal forma que houvesse farinhas para eles e para o povo (Carta… 24 Jul 1725)[1].

However, the measure, like so many others, was not utterly fulfilled by some or proved insufficient for others, varying by time and place, and requiring reiteration repeatedly.

5 INCREASING THE PRODUCTION

The increase in the production of manioc flour was a recurring theme, moreover, in communications between colony and kingdom, to promote the spread of plantations, from north to south, particularly for domestic consumption. For example, in 1760, the letter sent by Manuel Bernardo de Melo e Castro, governor and captain-general of the State of Pará and Maranhão, to the Secretary of State for the Navy and Overseas, Francisco Xavier de Mendonça Furtado, informed the strict observance of the metropolitan guidelines: there, he expressed all his concern and care with the manioc plantations, as one of the principal objects of interest of the settlements. Also, he reminded the importance, use and consequent reward that one has as a farmer, not only for the subsistence of life with abundance but still for an increase of opulence of the whole land (Ofício…13 Oct 1760).

The recommendations are disseminated and the enthusiasm with manioc plantations becomes increasingly demanded. A few years later, in a letter between the officers from Recife council and the King D. José I, the petitioners ask to impose penalties on farmers who did not plant manioc in an adequate and sufficient quantity for the "common good" and "sustenance of families and households", since many did not care to supply their homes and farms with the basic item of ordinary ration and meal (Carta… 23 Jul 1763). Later, in 1766, the governor of São Paulo, Luís António de Sousa Botelho Mourão, published an order to increase the cultivation of the root, to which some councils promptly responded: in Cananeia, the officers were committed to the "increase of the manioc plant cultivation, in the present and for the future", in Iguape, they had published a "notice to [meet the demands] with full disposition", regarding the development of the manioc crops as well as and in the "disways" of the flour, through a "diligence". A similar reaction was observed in Taubaté.

At the end of the century, back to Pernambuco, there is more news about the "fulfilment of orders to favour the cultivation of manioc flour" (Ofício… 17 Apr 1798), straight from D. Tomás José de Melo, governor of those lands. In the dawn of the Nineteenth century, Dom Rodrigo de Sousa Coutinho, the first count of Linhares, sent a royal order to José Joaquim Nabuco de Araújo, which, among other guidelines, dealt with the measures in which Nabuco encouraged the "Senhores de Engenho" to cultivate manioc at least for their slaves. In other words, such measure indicated his concern about the need to provide, on the

[1] "Appropriate to have a law enacted so that all the people who commanded such vessels would have slaves planting manioc, in such a way that there would be flour for them and for the people".

part of the captive holders, at least the basic gender to their plates. It should be noted, however, that Coutinho, while judging Nabuco's action as deserving of "royal approval", emphasized that the Prince Regent did not agree that pecuniary penalties should be imposed for this purpose, because the planting must be a result of the conviction that the owners have of its utility, and not effect of coercion (Ordem Régia, Jul 1800).

What is seen, roughly speaking, is that even with the obvious need to extend these crops; it was notorious that many masters and farmers were reluctant to use the labour force and lands they owned for this type of cultivation, generally less profitable than sugarcane or cotton. In addition, there were fluctuations in the price paid for flour, which then influenced the increase in the final values of products: many – or all – of Brazilian goods demanded the labour of enslaved people, fed by manioc flour. José Cesar de Meneses, governor of Pernambuco in 1780, denounce this, for example, (Ofício… 3 Jun 1780), Antonio José Calmon de Sousa e Eça, chief captain of the town of Jaguaripe (Bahia), on more than one occasion (Ofício… 20 May 1788), and Jerônimo José de Melo and Castro, governor of Paraíba, in 1787 (Ofício… 28 May 1787). The board was set, therefore, with the tension between the need to feed oneself and the slaves, the need to meet the demands of the Kingdom to supply the Empire's capital and the slave trade, as well as the goal to obtain maximum return on investments in agriculture.

6 IMPROVING THE PLANTATION

It is in this context of requests – often orders – to increase production that agricultural treatises with scientific rigor (Amaral, 2012; Kury, 2012; Serrão, 1998) on the cultivation of the manioc root took place. The observations of Agostinho Joaquim do Cabo, botanist-gardener of the philosophical expedition of Alexandre Rodrigues Ferreira to the extreme north of Brazil, on the management and versatility of preparing the root, date back to 1788. In his memories, he explains accurately about two types of flour, "seca e de água", and about several other preparations: "tucupi", "auarubé", "tapioca", "polvilho", "gomas", "carimãs", "beijus", porridges, "guaribas" and even beverages from the fermentation of the root, such as "pajuarus", "mocororós", "caribés" and spirits, describing the handling and qualities of each of the recipes, with the expected detail of a naturalist. In his words, the plant was used and consumed:

> [...] principalmente [pelos] os índios e os moradores brancos pobres, que não tem posses para terem o seu pescado, e ainda alguns que o tem, passam muitos dias com a mesma, por causa da estéril situação em que alguns se acham. (Cabo, 20 Feb 1788)[2]

The black and slave population was not mentioned, certainly, because they did not appear systematically as labourers on the banks of the Rio Negro (Reis; Silva, 1989). A speech in the "Academia Real de Ciências de Lisboa", in 1821, also published as a memory, refers to the products that could be obtained from the root, its species and the processes to which it could be submitted – "shaved, grated, squeezed, sifted, cooked and the flour called 'de pau"' (Barros, 1821, 65) –, also aiming to use such procedures in an American vegetable that had found a place in European dishes, the potato.

Concerning the planting techniques, since the first volume of "O fazendeiro do Brazil", published in 1798, Friar José Mariano da Conceição Veloso defended that it was necessary that they abandon the uncouth and rough rural economy of the primitive tenants of Brazil (Veloso, 1801, xxv), to plant and reap the fruits of the land by method, to take into account not chance and seasonality, but science[3]. Among the procedures exposed by him in subsequent volumes of this huge work, attention is drawn to the planting of manioc and cotton plants together, so that the latter did not entangle and hinder harvesting. The choice of manioc for such an enterprise was, as he explains, "the abundance of it" (Veloso, 1806, 236) since it was "a very versatile and very useful root" (Veloso, 1799, 44). It is not excessive to emphasize that in specialized publications and later agriculture manuals, the advice to dedicate part of the land to this crop, both for a pestle and for cooking, persisted:

> espécies [que] não são para polvilho nem farinha, mas para sobremesa, [são] para se cozinhar com carne e para os escravos comerem cozida à noite com melaço, que dá uma boa ceia. (Fonseca, 1863, 27)[4].

Such use of manioc was constantly praised, for it being very healthy, as well as for the convenience of the price and its national production (Castro, 5 Jul 1788), which is why its inclusion in the most varied menus was encouraged and, throughout the colonial period, it became an item of first necessity, shared between whites and blacks.

7 MANIOC AND PROGRESS

It is appropriate to be said that, although appreciated and praiseworthy, manioc, especially when it came to the feeding provided to the slave, should be configured as one of the items of the dish, and not the only one. It would not be rare, however, to hear that

[2] "Mainly by the Indians and the poor white inhabitants, who do not have the resources to have fish, and some who have it, spend many days on it because of the barren situation in which some find themselves".

[3] In the same sense, José Gregório de Moraes Navarro (1799) encourages the use of the plough.

[4] "Species [which] are not for 'polvilho' or flour, but for dessert, [are] to be cooked with meat and to be eaten by the slaves at night, cooked and with molasses, which gives a good supper".

the flour, without any other ingredient, constituted the slave's bread. The Jesuit Jorge Benci, in this sense, questioned:

> como é possível que o escravo ou escrava, andando em contínua lida e trabalho, sustente a vida com uma ração escassa de farinha de pau, sem outra coisa que a ajude a levar? Como se há de sustentar o miserável escravo sem outra coisa mais que uma triste ração de farinha de pau? Não vedes que isto é fazê-los comer terra? (Benci, 1977, 63)[5].

The poor settlers also had to seek other ingredients, such as local fruits and vegetables, as well as pork or beef, to supplement their food. Travellers in the colonial period also described the use of manioc by wealthy lords or merchants: in this movement, the plant remained as the main responsible for the sustenance in the tropics.

The production of the root is considered part of national progress. In several nineteenth-century publications dealing with agriculture, such as those printed by the "Sociedade Auxiliadora da Indústria Nacional", whose objectives were to acquire projects, machines, models and inventions that could contribute to the growth and prosperity of the national industry of this Empire, the dissemination and increase of planting techniques are not forgotten. Nor did the various agriculture manuals published in the first half of that century. It was stated that the more prosperous the production, the greater the progress of the nation. In addition, the progress of the cultivation of this American root, configuring a shared nutrition base throughout Portuguese America, at first, and afterwards in the State of Brazil, did not fail to contribute to the feeding of dishes and of plans, from the most particular to the most "national", of many of its inhabitants.

BIBLIOGRAPHICAL REFERENCES

Amaral, Ilídio do. (2012) *Nótulas históricas sobre os primeiros tempos da Academia das Ciências de Lisboa.* Lisboa: Edições Colibri.

Antonil, André João. (1997) *Cultura e opulência do Brasil.* Belo Horizonte: Editora Itatiaia Ltda.

Barros, José Villela de (1821) *Memória ou exposição do methodo de plantar, e colher no Brasil a Mandioca [...]. História e memórias da Academia Real das Sciências de Lisboa*, Volume VI. 1821.

Benci, Jorge. (1977) *Economia cristã dos senhores no governo dos escravos.* São Paulo: Editorial Grijalbo.

Cabo, Agostinho Joaquim do. *Memória sobre a mandioca ou pão do Brasil,* [...] 21,2,008 – Manuscritos Biblioteca Nacional do Brasil(BN). Barcelos, 20 Fev 1788.

Carta ao governador Luís Antônio de Sousa Botelho Mourão comunicando publicação de ordem [...] I-30,22,002 n°001 – Manuscritos BN. 9 Mar 1766.

[5] "How is it possible that the slave, living in continuous work and labour, sustains life with a meagre ration of wood flour, with nothing else to help him carry on? How can one sustain the miserable slave with nothing but flour? Don't you see that this is to make them eat dirt?"

Carta ao governador Luís Antônio de Sousa Botelho Mourão tratando de ordem recebida [...] I-30,22,005 n° 001 – Manuscritos BN. 16 Mar 1766.

Carta de D. Álvaro da Silveira e Albuquerque, ao rei sobre o prejuízo que teriam os moradores. [...] Arquivo Histórico Ultramarino (AHU), ACL, CU, 017, Cx. 7, doc. 790 11 Mai 1703.

Carta de Francisco de Castro de Moraes, ao rei [D. Pedro II], sobre a lei editada para [...]. AHU, ACL, CU, 015, Cx. 22, doc. 2004. 24 Nov 1705.

Carta de Francisco de Castro de Moraes, ao rei sobre a necessidade [...] AHU, ACL, CU 015, Cx. 21, doc. 1957. 8 Mar 1704.

Carta do Governador da capitania de Pernambuco, D. Manoel Rolim de Moura, ao rei D. João V [...] AHU, ACL, CU, 015, Cx. 31, doc. 2864, Recife, 24 Jul 1725.

Carta do governador do Brasil Matias da Cunha para S. Magde., sôbre a informação [...]. AHU_ACL_CU_017, Cx. 28, D. 3421_3422. 9 Ago 1687.

Carta dos oficiais da Câmara de Olinda à rainha [D. Maria I], sobre a falta da farinha de mandioca [...]AHU, ACL, CU, 015, Cx. 193, doc. 13277, 14 Mai 1796.

Carta dos oficiais da Câmara de Taubaté ao governador Luís Antônio de Sousa Botelho Mourão [...] I-30,22,018 n° 006 – Manuscritos BN, 06 Abr 1766.

Carta dos oficiais da Câmara do Recife ao rei, sobre a falta de plantações de mandiocas [...]AHU, ACL, CU, 015, Cx. 100, doc. 7793. 23 Jul 1763.

Carta dos oficiais da Câmara do Recife ao rei sobre a proibição aos [...]. AHU, ACL, CU, 015, Cx. 21, doc. 1967. 9 Abr 1704.

Castro, Martinho de Melo e. Ofício a Fernando José de Portugal sobre a ordem de Sua Majestade a respeito do [...] I-33,28,034 – Manuscritos BN. 05 Jul. 1788

Descrição dos bens e rendimentos do colégio de Santo Antão de Lisboa [...] Coimbra. ANTT – Cartório dos Jesuítas, maço 13, n. 20, fl. 9. 23 Aug 1635.

Fonseca, Antonio Caetano da. (1863) *Manual do agricultor dos generos alimenticios, [...].* Rio de Janeiro: Eduardo e Henrique Laemmert.

Fonseca, Manoel da (s.d). *Vida do Venerável Padre Belchior de Pontes, [...]* (1752). São Paulo: Companhia Melhoramentos.

Froger, François. (1999) In: FRANÇA, Jean Marcel Carvalho. *Visões do Rio de Janeiro colonial – antologia de textos, 1531–1808.* Rio de Janeiro: José Olympio.

França, Jean M.C. (2012) *Construção do Brasil na literatura de viagem dos séculos XVI, XVII e XVIII.* Rio de Janeiro: José Olympio.

Gandavo, Pero de Magalhães. (1930) *Diálogos das Grandezas do Brasil.* Rio de Janeiro: Oficina Industrial Gráfica.

Kury, Lorelai B.; Gesteira, Heloisa (org).(2012) *Ensaios de história das ciências no Brasil: das Luzes à nação independente.* Rio de Janeiro: EdUERJ.

La Barbinais, Le Gentil. (2012) In: FRANÇA, Jean Marcel Carvalho. *A construção [...]* Rio de Janeiro: José Olympio.

Navarro, José Gregório de Moraes. (1799) *Discurso sobre o melhoramento da economia rustica do Brazil [...].* Lisboa : Na Of. de Simão Thaddeo Ferreira.

Nieuhof, Johann. (1942) *Memorável Viagem Marítima e Terrestre ao Brasil (1640–1649).* São Paulo: Martins Editora.

Ofício ao governador da Bahia sobre o preço da farinha de mandioca. I-31,27,029 – Manuscritos BN, 1770–1781.

Ofício ao governador e capitão general da Bahia sobre a impossibilidade de praticar [...] II-33,20,015 – Manuscritos BN, 20 Mai 1788.

Ofício da Junta Governativa da capitania de Pernambuco ao visconde de Anadia, [...] AHU, ACL, CU, 015, Cx. 230, doc. 15543. 14 Dez 1801.

Ofício de D. Tomás José de Melo, a Rodrigo de Sousa Coutinho,[...]. AHU, ACL, CU, 015, Cx. 201, doc. 13760. 17 Abr 1798.

Ofício de Luís Pinto de Sousa, sobre as providências [...] AHU, ACL, CU, 015, Cx. 190, doc. 13117, 15 Set 1795.

Ofício do coronel Jerônimo José de Melo e Castro, a Martinho de Melo e Castro, informando da preferência [...]. AHU_ACL_CU_014, Cx. 30, D. 2178. 28 Mai 1787.

Ofício do governador de Pernambuco a Fernando José de [...] II-32,33,017 – Manuscritos BN. – 26 Mar 1792.

Ofício do governador e capitão-general do Estado do Pará e Maranhão [...] AHU, ACL, CU, 013, Cx. 47, doc. 4298. 13 Out 1760.

Ofício do intendente-geral da Polícia, Diogo Inácio de Pina Manique, [...] AHU_CU_009, Cx. 113, D. 8816. 2 Nov 1800.

Ordem Régia. II-32,25,015 – Manuscritos BN. Jul 1800.

Pita, Sebastião da Rocha. (1730) *Historia da America Portugueza [...]* Lisboa Occidental: Na Officina de Joseph Antonio da Sylva.

Reis, João J.; Silva, Eduardo. (1989) *Negociação e conflito: a resistência negra no Brasil escravista*. São Paulo: Companhia das Letras.

Relação das despesas com tabaco, aguardente, praças e direitos [...] AHU, ACL, CU, 015, Cx. 82, doc. 6867. ca. 1756.

Schwartz, Stuart B. (2004) O Brasil colonial, c. 1580–1750: as grandes lavouras e as periferias. In: Bethell, Leslie (org). *História da América Latina: América Latina Colonial*. São Paulo: Edusp.

Serrão, José Vicente. (1998) O pensamento agrário setecentista (pré-'fisiocrático'): diagnósticos e soluções propostas. In: CARDOSO, J. L. (org). *Contribuições para a história do pensamento económico de Portugal*. Lisboa: D. Quixote.

Vilhena, Luiz dos Santos.(1922) *Cartas de Vilhena. Notícias Soteropolitanas e Brasílicas (1802)*. vol. I, Salvador, Imprensa Oficial da Bahia.

Progress in the European periphery: An impracticable theory in 19th century Azores?

Susana Goulart Costa
CHAM, FCSH, Universidade NOVA de Lisboa, Universidade dos Açores, Portugal

ABSTRACT: During the second half of the 19th century, the Azores experienced a progress bustle, similar to what was happening in Continental Portugal. Located in the European periphery, the project was based on European paradigms that Azorean elite absorbed while travelling to France, England, and Germany. Literary, philosophical, and political books bought in major foreign urban centres, including North American cities, also played a major role. The project covered in several areas: in the economic area, privileging an agriculture of exoticism, with the experimentation of what was considered, at the time, exotic botanical species: tea, hydrangeas, camellias, etc. In the social field, developing numerous systems to favour the underprivileged. In education by encouraging the teaching of children and adults. In the technical and industrial areas by designing the construction of several factories, and investing in maritime and land transport systems.

However, tension quickly triggered between the dream of progress and the political, social, and cultural network that supported it. In the burgeoning periodical press, criticism sustained that, due to the power centred in the country's capital, Lisbon, the Azores islands did not receive enough means to promote progress. The latent tension between the projects of popular "happiness" and the effective opposition ended up mirrored on the hard and limited conquest of administrative autonomy that the archipelago achieved in 1895.

The theories of progress became impracticable and 19th-century changes were only attained in the following century.

Keywords: The Azores, 19th century, Progress, Autonomy

1 INTRODUCTION

The liberal movement in Portugal had in the Azorean archipelago one of its main stages. The war between Absolutists and Liberals divided the country during the 1820s.

The conflict riches one of its most important moments from 1830 to 1832, when King Pedro IV settled in Terceira Island, where he organised the liberal movement, its army and planned to conquer mainland Portugal, dominated by the absolutists supporting Prince Miguel.

The liberal victory, in 1834, became crucial to the history of the archipelago and led to the extension of the General Captaincy, an administrative structure of absolutist nature established in 1766.

The Islands were divided into three administrative districts: Ponta Delgada that included the Islands of São Miguel and Santa Maria; Angra, governing The Islands of Terceira, São Jorge and Graciosa; and Horta that included the Islands of Faial, Pico, Flores, and Corvo. This new political organisation, established in 1835, lasted more than 100 years.

2 LIBERAL PROGRESS IN THE AZORES: A THEORY TO PUT TO PRACTICE

The liberal movement, based on the principles equality, freedom and fraternity, proclaimed by the French Revolution in 1789, underlined dynamically the life in the Azores archipelago from the 1830s on. A holistic development was dreamt, which should encompass the various social groups and trigger the economy in the primary, secondary, and tertiary sectors.

Insular elite coordinated the leadership of this development. With broad political and social power, supported by an agricultural economy, the main Azorean landowners of the liberal generation sought to develop the islands according to European standards.

Travelling to France, England, and Germany and educating their offspring according to "European fashion", the insular elite conceived a plan for progress with two concrete areas in mind: the economical and the socio-cultural (Barreira, 1997).

2.1 *The economic focus*

In economy, the main objective was aimed at the development of the agricultural sector. In this context, the founding of Sociedade Promotora da Agricultura Micaelense (The Micaelense Farmer Promotion Association), in 1843, set the paradigm for innovating agrarian economy, which would give positive signs as early as 1848, with the opening of the First Agricultural-Industrial Exhibition, in São Miguel.

The agricultural focus reflected the principle of an articulated relationship between material and spiritual progress. António Feliciano de Castilho's book *Happiness by Agriculture* (1849), published in Ponta Delgada, is an interesting example of this philosophy that proclaimed the social welfare of the people was dependent on the investment in agriculture (Castilho, 1849).

However, obviously, the investment in the production of cereals was continued, for this was the base of the insular food system, the so-called Mediterranean diet, introduced from the beginning of Portuguese settlement, in the 15th century.

On the other hand, mainly since the early 30s, there started the plantation of orange trees. The investment was justified in view of the interest that the British market showed regarding the importation of oranges. The export of these fruits became crucial to the economic dynamics in the archipelago, especially to São Miguel and Terceira Islands. Due to the external pressure, local authorities have implemented a network of services adjusted to the designated "orange economy". Jobs were created for the plantation and work in the orange gardens, the harvesting, and the construction of proper containers for gathering, in the construction of containers, inland transport of crates and, finally, the transfer of oranges from small boats to large ships with British flags sailing to Bristol.

The commercial dynamics based on oranges production and exportation assumed such proportions that, at the end of the 50s, the main beneficiaries pressured for the construction of an artificial port in Ponta Delgada (São Miguel Island). The aim was to provide the city with an efficient infrastructure, able to respond to the growing mercantile needs of the island. The construction of the artificial port of Ponta Delgada began in 1861, fully subsidised by private funds belonging to the Island oligarchy.

The focus on orange production was influential to the primary sector mainly in São Miguel and Terceira: between 1861 and 1870, Ponta Delgada District shipped 2,098,540 boxes, Angra 298,793 and Horta 85,041 boxes (Dias, 1995: 196).

All this dynamism, perceptible from the 1820s to 1860, placed the Azores on the map of non-European economic interests. The Sephardic Jewish communities based both in Morocco and in London took charge in controlling the export business.

Figure 1. The Dabney's Family in Faial (http://blogorigens3.blogspot.pt/2015_01_01_archive.html).

Regarding Jewish presence in the Azores, there is evidence of their activity since 1819. Groups of Jews previously settled in Morocco arrive at San Miguel, drawn in by the island's economic development.

From the 1830s, the number of Sephardic Jews living in the archipelago increased, centring their economic and financial activities mainly in San Miguel, but also in Terceira and Faial Islands. Traders *par excellence* and with a renowned ability to sustain efficient economic networks throughout the world, their presence was very important for the transformation of consumption the paradigms. They were fluent in the English language and moved skilfully in British mercantile routes. Therefore, they seized the advantage of Azorean growing development to implement a business structure that greatly worried local shopkeepers.

Regarding the American presence, there is the example of the Dabney in Faial Island. In 1806, the American President Thomas Jefferson appointed John Bass Dabney, a Boston native, to the position of General Consul of the United States to the Azores. In Faial, John Dabney spurred the maritime trade, not only through orange exportation but also mainly from the sale of wine. This was produced in Pico, Island which featured (and still does) excellent characteristics for the wine production. However, given the profile of the Pico coastline, very irregular and rugged, it was impossible to berth vessels of large or medium tonnage. However, Horta port, in Faial Island, had good natural conditions and guaranteed security to American ships that berthed there. This relationship between Pico and Faial proved crucial for the economic survival of the two islands, supported by the trade of wine.

John Dabney founded the "Dabney Empire", concentrating in Horta a diverse range of maritime services. Among these, there was the supply and repair of American whaling ships that crossed the seas in the Azores; the commercial links between the USA and the Azorean Islands, allowing the entry of American goods in European ports; and the supply of coal, essential to fuel the ships that crossed the ocean (Costa, 1996–1997).

The cosmopolitanism of the archipelago was determinant for the amendment of part of the harsh island landscape that during the second half of the 19th century was oriented to the agricultural experimentation. This way, the Islands became a natural laboratory, as they had been in the early period of settlement. One of the most interesting features that mirror this purpose is the construction of botanical gardens, clearly of English influence, where the cultivation of exotic flowers, trees, and shrubs was experienced. The first experimentations with hydrangeas, camellias or cater started in this period, with the importation of plants from Australia, New Zealand, Brazil, Japan, etc.

2.2 *The socio-cultural aspect*

Central to liberal views of progress was the understanding of an essential development transversal to society. It was understood that welfare, substitute to progress, required material and cultural prosperity.

In the Azores, similarly to what was happening in other places, the pulsing elite assumed the promotion of Arts and Sciences. The beneficiaries of this elitist benevolence would be the other social groups. The emphasis was focused on education, a crucial tool for a harmonious growth of the Islands rural society. The aim was to promote theoretical and traditional education but also with an emphasis on practical knowledge, fitting the effective needs of the archipelago. Schools were built for children and young adults, but they opened their doors to the adult population through the sponsorship of evening classes, trying to tackle the high rates of illiteracy in the Azores, then reaching 80% of the population.

The whole project was founded by the Sociedade dos Amigos das Letras e Arte (Society of Friends of Arts and letters) founded in Ponta Delgada in 1848 (1849).

This partnership not only invested in the promotion of the arts, especially theatre and music, as devoted itself to the creation of primary schools, rehearsing a peculiar teaching method of reading, which would be the forerunner of the "Portuguese Method of Castilho" (Dias, 2000).

This Society mirrors the consistent importance of the 19th-century associative structure that exceeded the enlightenment individualism. Consequently, in the second half of the 19th century, there is an increase of associations of diverse nature (Silva, Susana Serpa, 2000).

One of the consequences of this associative movement may be seen in the proliferation of music bands that grew up in many parishes. In 1831–1832, two bands landed in Angra do Heroísmo and Ponta Delgada, as part of the liberal army that came from France and England. They stayed for a short period, accompanying the military expedition of Mindelo, but their influence in military groupings remained strong. An 1831 law determined that all the regiments had to have a musical band. Shortly after civilian groups formed their own band so developing a strong musical culture still present in Azorean society (Cabral, 1985).

Figure 2. Church of Our Lady of Victory (Furnas, S. Miguel). It was built by José do Canto (1820–1898), an Azorean intellectual and garden builder. https://www.facebook.com/HistoriaDosAcores/photos/a.228753753896151.41228.197544470350413/560108087427381/?type=3&theater.

3 WAS THE LIBERAL PROGRESS IN THE AZORES AN IMPRACTICAL THEORY?

The liberal optimism of the first half of the 19th century began to suffer severe setbacks from the 1860s onward. Indeed, the second half of the 19th century brought many problems to Azores social and economic improvement and general progress.

The main difficulties were felt in the primary sector. On the one hand, the orange trees began to show signs of disease. The production decay was a fact since the 1870s, and with it came a sharp decrease in revenue to local islanders. On the other hand, the strength of the Dabney economic empire in central islands began to dwindling too. The causes were mainly two: the vineyards were infested by oidium, and by the 50s, a plague of phylloxera contributed to a severe decrease in wine production (Sousa, 2004).

Concerned with this reality, the insular cultural and economic elite continued to draw on European models, trying to apply in the Islands a draft agricultural conversion, inspired by English Industrial Revolution. Thus, the introduction and promotion of agro-industrial plants, namely tea, tobacco, bulrush, and sweet potato, was meant to replace the orange groves.

These new products became the pivots of the primary sector restructuration in the Azores, in an attempt to continue to ensure the "Happiness for Agriculture".

However, Portugal was living a difficult political period, disoriented due to the alternation system between Progressives and Regenerators. In the international context, the country was plunged into a deep international crisis, facing the constraints caused by decisions taken at the Berlin Conference. The economic and financial crisis of the Kingdom was increasingly serious and, in order to control it, the Government increased taxes and protected the metropolitan productions.

It is in this critical context that the Portuguese Government determines a set of measures that was received with dissatisfaction by the Azoreans. The standardisation of currency value affected the Islands economy where the coin had a lower value than in mainland Portugal. Taxes increased, including those focused on the production of alcohol extracted from sweet potato that had become the key production to replace the decrease of orange groves. The extinction of judicial entities in the Azores, forcing an increasing expense in solving legal disputes, was a further blow that increased islanders' dissatisfaction (Leite, 1995).

The growing difficulties of the primary sector, especially in agriculture, which employed most of the workforce, led to increased emigration. Most immigrated to Brazil, strengthening the industrial entrepreneurs' difficulties in obtaining cheap labour workforce.

To the process of deterioration of economic conditions, the Islanders felt increased dissatisfaction with the political decisions applied to the archipelago. Recognising the international strategic relevance of the Azores, attested by the establishment of submarine cables on the island of Faial by England, Germany, and United States, and the growing interest of foreign naturalists in the Islands, the Azoreans rose their dissenting (TELO, 1993).

This political hubbub is mirrored in the increase of political parties that could control the regional periodical press known at that time. In the newspapers, there was a call for civic intervention, encouraging citizens to act to force change the way the Islands were governed. For the regional oligarchy, there was a need for a new political scenario was. It seemed the only way to defend the economic development of the archipelago, consolidating the diversity of agricultural experiments, reinforcing the structures of land and sea access and decreasing the State taxation, in order to experience a situation of economic and social comfort (Santos, 1996).

They asked for administrative autonomy. In late 19th century, the idea that the Azoreans should manage the archipelago was consensual. The local elite echoed the autonomist voices. Aristides Moreira da Mota, the main leader, wrote the first autonomist project, dating from 1892, which would be enacted, albeit mitigated, on March 2, 1895. This way, each of the Azorean districts could apply, by a majority of two-thirds of eligible citizens, the "free administration of the Azores by Azoreans" (João, 1991).

4 CONCLUSION

Azores administrative arrangement regulated by the 1895 Decree reflects the decline of the progressive paradigm that characterised most of the 19th century. Despite considerable efforts to invigorate the Islands' economy in their various sectors, trying to introduce European models, the archipelago was engulfed in numerous difficulties (Ferreira, 1996). This tension between the needs of the Azorean periphery and **political** centrality constraints limited the reach of Portuguese benefits and investments that were effectively achieved in the archipelago. (Cordeiro, 1992).

In fact, progress was a utopia that a minority elite tried to put to practice. Realistically, the development has benefited few landowners' families, leaving few crumbs for the majority of the population, forced to resort to emigration to survive or contenting themselves with a harsh everyday frugality. Similarly, the raised wealth by commercial dynamics from 1830 to 1870 was only effective in the island of São Miguel, extending very slightly to Terceira and Faial.

All the other islands were on the fringes of the optimistic investment and found themselves with economic and social problems that would last until the 20th century.

The 1895 administrative autonomy, only effectively applied in the districts of Ponta Delgada and Angra do Heroísmo, eventually became unable to resolve the underlying problems of the Azores. Although they could solve administration problems, district revenues were incorporated in the country's financial core. Consequently, local authorities identified the problems, had the respective solutions, but they did not have the funds required for achieving them (AAVV, 1995).

Therefore, by the end of the 19th century, progress in this Azorean periphery seemed but an impractical theory.

BIBLIOGRAPHICAL REFERENCES

(1849) *Estatutos da Sociedade dos Amigos das Letras e das Artes em São Miguel.* Ponta Delgada: Tipografia da Rua das Artes.

Barreira, Cecília. (1997) *Onde está a Felicidade. O conceito de Progresso Técnico no Século XIX.* Lisboa: Universitária Editora.

Cabral, J. M. (1985) *Filarmónicas da ilha de São Miguel.* Angra do Heroísmo: Instituto Açoriano de Cultura.

Castilho, António Feliciano de. (1849) *Felicidade pela Agricultura.* Ponta Delgada: Typ. da Rua das Artes.

Cordeiro, C. (1992) *Autonomia e Identidade Nacional: Os Açores na 2.ª Metade do Século XIX*. Coimbra: Faculdade de Letras, 263–279.

Costa, Ricardo Manuel Madruga da. (1996-1997) "O Século Dabney no Fayal. Subsídios para a compreensão de uma estratégia". *Boletim do Núcleo Cultural da Horta*, Vol. XII: 143–150.

Dias, Fátima Sequeira. (1995) "A importância da economia da laranja no arquipélago dos Açores durante o século XIX". *Arquipélago-História*, 2ª série, I, nº 2: 189–240.

Dias, José Teixeira. (2000) "Castilho – Leitura Repentina. Método original?" *Arquipélago-História*, IV (2):465–480.

Ferreira, José Medeiros. (1996), "A autonomia dos Açores e os regimes políticos em Portugal" *Ler História*, vol. 31: 157–167.

João, Maria Isabel. (1991) *Os Açores no século XIX: economia, sociedade e movimentos autonomistas*. Lisboa, Edições Cosmos.

Leite, José Guilherme Reis. (1995) *Política e administração nos Açores de 1890 a 1910: o 1° movimento autonomista*, Ponta Delgada: Jornal de Cultura.

Santos, João Marinho dos. (1996) "Para a história da autonomia dos Açores: a introdução forçada da fisiocracia no Arquipélago". *Revista Portuguesa de História,* vol. 1, t. 31: 575–593.

Silva, Susana Serpa. (2000) "Aspetos da vida social e cultural na segunda metade do século XIX". *Arquipélago-História*, 2ª série, IV, n° 2: 299–358.

Sousa, Paulo S. (2004) "Para uma História da vinha e do vinho nos Açores (1750–1950)". *Boletim do Instituto Histórico da Ilha Terceira*, 62: 57–159.

Telo, António José. (1993), *Os Açores e o controlo do Atlântico*. Lisboa: Asa.

Torres J. (1849) *Relatório dos trabalhos da Sociedade dos Amigos das Letras e das Artes em S. Miguel.* Ponta Delgada: Typ. do Correio.

Progress and remembrance of the Finnish Lutheran work in the Ondonga Kingdom in South West Africa

Kim Stefan Groop
Department of Church History, Faculty of Arts, Psychology and Theology, Åbo Akademi University, Turku, Finland

ABSTRACT: The concept of progress has always been pivotal in the Protestant missionary movement. At the same time, this movement has had an ambivalent relationship with progress. On the one hand, progress has been needed in terms of advancement of the missionary and development of his or her work. On the other hand, the concept of progress has been questioned, as something worldly, self-seeking, and potentially immoral.

Under scrutiny in this article is the work of the Lutheran Finnish Missionary Society in the late 19th and early 20th centuries in Ovamboland in South West Africa, in what is today northern Namibia. Both among Finnish Lutherans and among Ovambo traditionalists there was a desire for progress. The encounter between Lutheran Pietism and Ovambo culture was at times uneasy. Yet, at the end of the day, it gave rise to other, sometimes unexpected, kinds of progress.

In this article, also the legacy of the Finnish Missionary Society in Namibia is studied. In the cultural memory among the Ovambo, the relative tension between faith and culture is visible. Nonetheless, the progressive dimensions in, and discernible achievements through, the Finnish-Ovambo relationship assume a far more prominent position in northern Namibian remembrance.

Keywords: Ondonga, Namibia, Lutheran mission, initiation rites, cultural memory

1 KING SHIKONGO SHAKALULU AND THE DAWN OF MISSION WORK IN OVAMBOLAND

The mission history in Ovamboland[1] in what is today northern Namibia and southern Angola dates back to the year 1870 when the first missionaries and carpenters from the Lutheran Finnish Missionary Society (since 1985 the Finnish Evangelical Lutheran Mission) arrived at Omandongo. In one of the first meetings with Ondonga King Shikongo shaKalulu, Pietari Kurvinen, speaking in the Herero language, expressed his joy to adhere to the wish of the king whom the missionaries had heard "wanted teachers and the word of God". The missionaries, according to Kurvinen, did "not seek worldly goods, but that your souls may be saved" (Peltola, 1958: 41).

Kurvinen's speech to the Ondonga king demonstrates two typical features in the missionary encounter with local communities in the late 19th century in Africa. Firstly, the missionaries were "called" by local rulers who viewed foreigners as a key to advancement. Kurvinen and his colleagues had travelled from Finland to Ovamboland because Shikongo shaKalulu had asked German Rhenish missionary Hugo Hahn for missionaries. The Ndonga and other Ovambo ethnic groups were interested in progress. They wanted "European" goods and skills. Blacksmiths, weapons, and alcohol were high on the kings' wish lists. There is little evidence to suggest that the Ovambo were discontent with their customs, or living conditions, but they were not uninformed about life outside the tribal boundaries either. The Ovambo ethnic groups had raided each other, traded, and sometimes even protected each other from mutual enemies. Many had experienced Portuguese trade, in goods and slaves, and others had escaped to neighbouring kingdoms and chieftains to avoid punishment or death in their own quarters. (Tönjes, 1996: 84–89, 116–122). King Shikongo shaKalulu hoped that the Finnish missionaries would help him achieve superiority over the other kings and chiefs, in terms of goods and military ability, but he was also open to other for him unknown areas where he could advance his wealth and influence.

Secondly, Kurvinen's speech to the king demonstrates that the Finnish missionaries had their suspicions of what the king wanted. They wished to

[1] The name Ovamboland or Amboland was used by early travellers, but also by the Finnish missionaries, to describe an area in what is today northern Namibia and southern Angola. The Ovambo people have nine tribes or sub tribes with different dialects. Still today, the name Ovamboland is often used, although it is considered more politically correct to refer to the four regions, which the Ovambo inhabit: Omusati, Ohanguena, Oshana and Oshikoto.

demonstrate that they had something else in mind. Kurvinen was careful in his choice of words. Emphasising that "having heard of this [the king's calling]" their "hearts caught fire out of love to you" (Peltola, 1958: 41), the missionaries positioned themselves not as servants to the king, but as heavenly messengers with a different agenda; to make sure that souls are saved. From this start in 1870, it would take 42 years before the aspirations of the two parties – the Finnish missionaries and the Ovambo kings amalgamated, when the first Ondonga king, Kambonde kaNgula was baptised on his death bed in 1912 (Notkola and Siiskonen, 2000: 37).

2 MISSIONARY CHRISTIANITY AND PROGRESS

Finnish missionary presence among the Ovambo ethnic groups would lead to a number of fascinating encounters, clashes and developments. In this section, I will discuss three aspects pertaining to progress in and of the missionaries work. Particular focus is on the work in the Ondonga kingdom.

Firstly, contrary to what the missionaries themselves often suggest, the missionary vocation – or "calling" – indeed entailed a progress or career. Many of the early missionaries came from humble backgrounds. Martti Rautanen is a good example. He barely made it through missionary training due to his limited schooling prior to being admitted to the seminary of the Finnish Missionary Society in Helsinki. Nonetheless, after graduation, he, and his peers were ordained as a pastor for the mission field, and Rautanen would achieve relative success as a missionary leader.

In some aspects life as a missionary was cumbersome. Many missionaries and family members died in malaria (Peltola, 1996: 106–108), and the relation to the leadership in Helsinki was at times complicated. The missionaries were expected to answer to the directions given by the leadership at home – a leadership, which had little understanding of the situation in Ovamboland. Nonetheless, for most of the 19th century Finnish missionaries, serving for the Finnish Missionary Society was a great career opportunity. These young men were given considerable responsibility and power among local inhabitants and colonial personnel alike. Martti Rautanen may have struggled in missionary seminary, but already before turning 40, he had become the leader of the work in Ovamboland (Peltola, 1996: 109). With time, he would also become "an Ovambo" folk hero with his own nickname – Nakambale, the man with the hat. In 1925, he was also awarded an honorary doctorate for his life-long achievements (Peltola, 1996: 346).

Secondly, the missionaries conveyed a message of progress but it was, predominantly, a spiritual one. When Finnish missionaries wrote or talked about progress, it was a progress from a fallen world, or, when considering the Ovambo, a progress from a state of paganism, to salvation. It was a progress comparable to the journey, which the character "Christian" makes in John Bunyan's classic "the Pilgrim's Progress" (Bunyan, 1678). Between start and goal, there were many challenges and these challenges could bring you down. While the Ovambo people did not separate between culture, religion, and life at large, the missionaries did. In missionary thinking, life could, and should, indeed, involve faith dimensions. Religion should thus not be separated from life at large. However, culture, in missionary thinking, was a tricky business. Most Finnish missionaries shared a Pietistic worldview, and to them, both Ovambo and Finnish culture had a great number of elements that were of little benefit to a Christian who was caring for his or her salvation (Groop, 2014: 87–89).

On the Ovambo mission field, there was one cultural rite, which the missionaries did not like at all: the female initiation. This rite-de-passage looked somewhat different in different tribes, but it had the same purpose everywhere: to transform girls into women. In Ondonga, it was called ohango, in Oukwanyama, it was called efundula, and in Ombalantu, it was known as olufuko (Nampala and Shigwedha, 2006, 27–48). Girls who were found pregnant before or during ohango, efundula or olufuko would pull dishonour over their families and could be killed. After initiation, however, the girl was regarded as a woman and considered ready for motherhood; either in a monogamous or in a polygamous relationship. If a girl wanted to take the leap and become a Christian, the missionaries demanded that she renounced this female initiation or pagan wedding as it was commonly called in missionary circles. However, for this girl to do that – at least for the first 40 years of missionary presence in Ovamboland – it meant that she renounced the very heart of her own culture. In the first written church rules in the emerging Ovambo church, in 1924, participation in ohango (here termed in the Ondonga language) was viewed as the greatest of offences and would lead to excommunication from the church (Groop, 2016: 56). Today it may be easy to criticise the stiff missionary resistance to female initiation as extreme, or wrong. Nonetheless, at the time, the Finns and Ovambo were too far apart in their understandings to be interested in negotiations on the matter. Due, at least in part, to the missionary resistance to female initiation, the Christian parishes grew slowly during the first decades of missionary presence in Ovamboland. Thus, the focus on spiritual progress hampered mission growth. Only after 20 years of work, the Finnish field started to grow. In September 1890, at least 14 newspapers in Finland reported that 52 people had been baptised at Olukonda and that there were 280 Christians on the Finnish missionary field (see for instance Satakunta, 1890: 18). Still, in the year 2017, the Evangelical Lutheran Church in Namibia has an uneasy relationship with female initiation – which has re-emerged through the annual "Olufuko festival" in Outapi (Nambala, 2012).

Thirdly, the Ovambo in the 19th century were pragmatic in their relation to the Finnish mission. They were interested in progress, but, before making any commitments, they wanted to know that they would win more than they would lose. One of the key reasons why King Shikongo shaKalulu had wanted missionaries was his belief that they would bring traders and other treats. The missionaries were largely unwilling to provide the kings with weapons. They frequently gave tobacco to them, but, under no circumstances, they would provide them with alcohol. Learning that they had less use of the missionaries than they had wished, all rulers with Finnish presence but Shikongo shaKalulu made the missionaries leave their domains. Like shaKalulu, most subsequent Ondonga kings considered the missionaries necessary in a time of rapid change. They made good use of their Finnish guests in their aspirations to keep abreast with change, and ahead of their rivals. They gained from Western health care, and they made good use of the missionaries' political expertise. Some even took an interest in building construction. In the 19th century, the most durable houses in this part of today's Namibia were built out of sun-dried tiles and had grass roofs. The missionaries' houses had windows and wooden doors with locks. Already in 1883, King Kambonde kaNankwaya and some other members of the royal family had started building houses the "European" way (Rautanen, Missionstidning för Finland, 1883: 182). As for the ordinary men, women, and children, there was always an element of opportunism, and a wish to advance in life. Once they knew that their participation in church and school activities had the king's blessing it became a matter of personal choice. As the Finns became more appreciated from the end of the 19th century onwards, becoming a Christian turned into a favourable option for more and more Ovambo. The ever-improving health care at the Lutheran stations was highly appreciated. Furthermore, schools provided opportunities to advance in life, and Christianity itself attracted ever more people with its promise of a better life and ultimate salvation.

There is little doubt that the key to the hearts of the Ovambo people – and to missionary progress – was health care (Saarisalo, 1971: 97). Already the first day when Martti Rautanen arrived at Olukonda in 1880 he walked to King Kambonde kaNankwaya to treat his feet (Rautanen in Missionstidning för Finland 1881: 4). This was a task, which he had inherited from his recently deceased predecessor Gustav Skoglund. He also treated the successor, Kambonde kaMpingana, for nearly thirty years until his death in 1909. If the female initiation had hampered the early mission, helping the royal family medically had the opposite effect. Within short, Rautanen had also become the king's teacher and adviser. Though no kings of the 19th century opted for baptism, their trust in the Finnish missionaries grew. In addition, as the relation between kings and missionaries improved the Christian parishes grew. By 1900, there were around 1,000 Christians in the Ondonga kingdom in Ovamboland, and about ten years later the work started to expand rapidly (Peltola 1958: 163). Shortly after the turn of the century, the Finns resumed their work also in other Ovambo polities. In 1912, King Ķambonde Mpingana's successor Kambonde kaNgula was baptised on his deathbed. His successor Nambala yaKadhikwa was also baptised later the same year. This marked a definitive milestone in the Finnish work (Peltola 1996: 246–248).

3 THE MEMORY OF PROGRESS

Let us in this last section of the article look at the Finnish mission history in the collective memory of the Ovambo people. With remembrance and memory in this article, I am referring to a collective or cultural memory. Thus I am reaching beyond a personal memory and consider a "memory" which has been orally communicated from one generation to the next, but also mediated and remediated through other channels; not least through literature. (See for instance Assmann, 2011 and Erll, 2011).

In the many discussions and interviews, which I have had with Ovambo people over at least ten years, I have been surprised by how positively they recollect the Finnish past in Ovamboland. Logically, I should suspect that people are only telling me – a Finnish citizen – what they believe that I want to hear. Therefore, I have sometimes challenged them, by pointing at negative encounters in "their" mission history, but mostly my informants have not changed their attitude. The same message seems to echo also in other studies of Finnish-Namibian heritage. For instance, Ian Fairweather, who has studied the old missionary station at Olukonda, signals that he would have expected a more critical view of his informants' view of Olukonda. He writes that this mission station is celebrated as an important place for many things, including both the preservation of Ovambo tradition and the, somewhat opposing, transition of Ovambo communities to modernity (Fairweather, 2006: 163, 164 and 166). What Fairweather highlights is the memory of progress. This is a central aspect of the long encounter between the Ovambo (culture) and Finnish (Lutheranism), which is frequently brought up in discussions and interviews. Fairweather's informants in the Ovambo village of Olukonda did not remember the colonial era in terms of the imposition of cultural values, but indeed "in terms of a cleavage between those who developed a close identification with Finnish mission Christianity and those who did not." (2006, 166) This tally with my experience of the Ovambo memory of their missionary past. In interviews and discussions, informants have sometimes talked about missionary strictness, and about a division into those who joined the mission and those who remained traditionalists, the Ovambo do not generally seem to hold the opinion that the Finnish missionaries imposed their values upon them. What many have brought up, however, are issues of progress and development. Those who have defended mission Christianity (such as Shivute, 2017 and Nelumbu, 2010) have spoken about the extraordinary Ovambo progress from a "primitive" state, wearing animal hides, to modernity or "civilisation".

In December 2016 and January 2017 I discussed with and interviewed several individuals from the Ovambo ethnic group in order better to understand their memory of their past. The topic of these discussions did not have much with progress to do as such, but rather with missionary heritage at large. Nonetheless, the common denominator in all these talks was the aspect of progress and development. In the following, I will discuss two of these interviews, with people whom I expected would be rather critical of the Finnish missionary past and the handling of Ovambo culture within the Lutheran church.

Professor and surgeon Filemon Amaambo is in his 60s and works at the Onandjokwe hospital in Oniipa. This hospital was built by Finnish missionaries, under the supervision of Medical Doctor Selma Rainio, who is also buried near the hospital. Amaambo does not talk much about the spiritual work of the Finns. Instead, he prefers to view the Finnish mission past in Namibia through some of the changes that it gave rise to, and their benefits in society. He gives at least four examples of beneficial progress. Firstly, the early missionaries, through their presence, brought postal services to Ovamboland, where in particular Olukonda was an early hub. Secondly, Amaambo emphasises that missionary presence in Ovamboland brought about a development of the road network. Thirdly, the missionaries introduced new ways of building houses. Fourthly, Amaambo mentions health care. When being asked about the meaning of Olukonda, Amaambo instead draws attention to the Onandjokwe hospital as a prime example of a Finnish legacy in northern Namibia, which is being cherished today. Amaambo maintains the fact that it all started at the doorsteps of the local authorities of the time since the missionaries settled close to the local rulers. He maintains that the critique, as regards missionary influence on local culture, had to do with a wish for speedy progress. The Finns were

> looking for certain outcomes too early, too quickly. They wanted to see certain outcomes within their short lifetime. They did not leave everything up to natural evolution to take place over decades or centuries. (Amaambo 2016)

Nonetheless, Amaambo has a largely pragmatic view of the development. In his recollection of the Finnish missionary past, change and progress is pivotal and should be viewed over longer periods. He does not, as he says, "spill blood" over opinions that missionaries "forced change" upon the people.

Nangola Mbumba is the General Secretary of Swapo (South West Africa People's Organisation); the leading political party in Namibia, and he is in his 70s. Like Amaambo he also brings up the tension between Christianity and culture and mentions as examples female initiation (ohango and olufuko), and the change of dress code among Christians in the late 19th century. Whereas Mbumba regrets that female initiation was not allowed within the church, he seems to respect the decision of the missionaries of the time. He simply states that it was "impossible to adapt the two things together". The two systems could not be mixed. However, Mbumba also signals that he is not entirely content with the outcome, but the issue of female initiation has to be "discussed and checked in the future". As for the change of dress code, Mbumba expresses his understandings for the outcome, pointing at hygiene issues and other realities as a reason why western clothes were introduced. However, above all he highlights the Finnish mission – the missionaries and the education system – in relation to the struggle for freedom;

> The contribution of the Finnish missionary is no longer just measured in terms of the church work or the health or education work. It is also [measured] in terms of nationalism in Namibia [...] The missionaries were proud of us, that we did what we did, sacrificed a lot and we are a part and parcel of a new Namibia. (Mbumba, 2016)

Mbumba views the Finnish mission through political spectacles and views political freedom as a definite milestone. He emphasises the whole period from 1870 and the arrival of missionaries such as Martti Rautanen through to his own education under Finnish teachers such as Lahja Lehtonen in the 1960s. Also the role of UN Special Representative (and later President) Martti Ahtisaari, in the transition to independence, is mentioned. Mbumba highlights the Lutheran schools and in particular Oshigambo High School, and its significance for the training of future leaders. While the missionaries were worried about the keen interest in politics among many of the students, and their involvement in the fight for independence, the link between the teachers and students were never broken. Also as regards the somewhat later development shortly before independence in 1990, as Mbumba describes, the socialistic Swapo party realised that it would benefit in the upcoming elections if it had good relations with the Lutheran church – the largest church in Namibia. Thus, the history of unbroken relations between the (mission) church and the political movement prevailed, and socialist principles were put aside for the sake of progress and winning the election.

To both Amaambo and Mbumba the Finnish Missionary Society, and the coming of Christianity to northern Namibia, is connected with progress. It is not particularly the Christianisation of the Ovambo North, which they highlight. This does not mean that they do not respect the spiritual aspect of the Finnish work. But what they "remember" is development and they highlight examples from their own history as well as social and scientific contexts; the Olukonda mission station and national heritage, the Onandjokwe hospital, and the achievement of independence in Namibia.

4 CONCLUSIONS

When trying to track and understand progress in the history of Africa involving European missionaries, it is important to realise that the concept of progress has always had different meanings for different people and groups of people.

Both the Ovambo and the Finns wanted to make progress in life. The Ovambo rulers wanted access to European skills and goods in order to survive in the stiff geopolitical competition. In general, the Ovambo were open to solutions, which they felt that could help them advance in life, and this included Lutheran schools, health stations and churches. Many of the Finnish missionaries had humble backgrounds and becoming missionaries can be viewed as a "blessed" career opportunity. On the missionary field, local inhabitants and colonial administration personnel welcomed their expertise alike, and some of the missionaries achieved relative success.

To the Finnish missionaries in the Ovamboland, in the late 19th century and early 20th century, progress in life was, above all, understood as a spiritual journey from heathendom to salvation. As in the 17th century classic "the Pilgrim's progress" by John Bunyan, this journey as a Christian involved many stops and challenges. Between spiritual and material progress, priority was given to the former. Nonetheless, as a bi-product of the spiritual journey, which the missionaries introduced, Ovambo Christians obtained education and, in the second half of the 20th century, many were engaged in politics.

To the Ovambo, it was important to advance in life according to long traditions, and one of the cornerstones in Ovambo society was the female initiation into womanhood, which in Ondonga was called ohango.

Of course, at least initially, progress according to the Ovambo view of life tended to clash with missionary understandings of life and eternity. Those Ovambo who chose to become Christians could no longer participate in all Ovambo festivities. This relative tension between "missionary" Christianity and Ovambo culture would lessen with time, but still, in 2017, the Evangelical Lutheran Church in Namibia struggled to accept female initiation. While this tension has recently been reactivated in the cultural memory in northern Namibia, through the return of organised female initiation in the annual "Olufuko festival" in Outapi, it is not tensions that people are generally talking about when remembering their missionary past, but progress. In discussions and interviews, the missionaries are often projected against the Ovambo past, when people were dressed in hides and smeared their skin with red ochre. This part of history is often brought up as a kind of distant and vague origin, which few people would want to return to. The interviewees, mentioned in this article, highlighted the missionaries' role in the development of health care, of infrastructure as well as in relation to the process leading to liberation and Namibian independence.

BIBLIOGRAPHICAL REFERENCES

Assmann, Astrid. (2011) *Cultural Memory and Western Civilization: Arts of Memory*. New York: Cambridge University Press.

Bunyan, John. (1678) *The Pilgrim's Progress From This World to That Which Is to Come*. London: Nath. Ponder.

Erll, Astrid. (2011) *Memory in Culture*. New York: Palgrave Macmillan.

Groop, Kim. (2014) "Between Pietistic Heritage and Olufuko Revival: Female Initiation and the Evangelical Lutheran Church in Namibia". *Journal of Theology for Southern Africa* vol. 150.

Groop, Kim. (2016) "Divine Law Enforcement and Mission Transculturality: The Finnish Missionary Society and the emergence of the first Church Rules in the Ovambo mission field in South West Africa". In: I. Dahlbacka, K. Groop and J. Dahlbacka, ed. *The Shifting Boundaries of Tolerance: Inclusion, Exclusion, and Religious Communities of Memory*. Åbo: Studies on Religion and Memory.

Groop, Kim and Amaambo, Filemon. (2016). Interview regarding Lutheran heritage in Namibia.

Groop, Kim and Mbumba, Nangolo. (2016). Interview regarding Lutheran heritage in Namibia.

Groop, Kim and Nelumbu, Martin. (2010). Interview regarding national reconciliation in Namibia.

Groop, Kim and Shivute, Tomas. (2017). Interview regarding Lutheran heritage in Namibia.

Missionstidning för Finland (1881, 1883).

Nambala, S. V. V. (2012) *The Evangelical Lutheran Church in Namibia*. Pastoral Letter for Press Release.

Nampala, Lovisa and Shigwedha, Vilho. (2006) *Aawambo Kingdoms, History and Cultural Change: Perspectives from Northern Namibia*. Basel: P. Schlettwein Publishing.

Notkola, Veijo and Siiskonen, Harri. (2000) Fertility, Mortality and Migration in SubSaharan Africa: The Case of Ovamboland in North Namibia, 1925-1990. London: Palgrave Macmillan.

Peltola, Matti. (1958) *Sata vuotta suomalaista lähetystyötä 1959–1959*. Helsinki: Suomen lähetysseura.

Peltola, Matti. (1996) *Martti Rautanen: Mies ja kaksi isänmaata*. Helsinki: Suomen lähetysseura.

Saarisalo, Aapeli. (1971) Etelän ristin miet: Martti Rautasen elämä. Porvoo, Helsinki: Werner Söderström osakeyhtiö.

(1890) Satakunta: Sanomia Länsi-Suomesta.

Tönjes, Hermann. (1996) *Ovamboland: Country, People, Mission*. Windhoek: Namibia Scientific Society.

No past – no progress: Uses of history as a prerequisite of progress in the reinauguration of The Luther Church Helsinki

Jakob Dahlbacka
Department of Church history, Faculty of Arts, Psychology and Theology, Åbo Akademi University, Turku, Finland

ABSTRACT: This quotation, ascribed to the architect Zaha Hadid: "There cannot be progress without facing the unknown" – which also serves as the caption for this conference – depicts progress almost as a (foolhardy) step into the unfamiliar, a leap of faith that, with any luck, might result in an improvement or development of a previous condition.

Although this statement may be true to some extent, however, it does fail to acknowledge another equally or perhaps an even more important aspect and prerequisite of progress, namely knowledge of and reference to the past. This reverse side of the coin is perhaps most poignantly expressed in common phrases such as "learning from the past" or "learning from history". Whether we actually learn something from the past might well be disputed, but what stands unchallenged is that the past has a great deal to say with regard to the present and the future.

In this article, I will try to illustrate and discuss the significance of using the past when aiming at some sort of progress. I will do this by looking at how the 'relaunch' of an old church building in Helsinki, Finland, was staged, narrated, and subsequently promoted to the general public. As I will demonstrate, the history of the church came to play a crucial role in this story and, I would suggest, the uses of history probably supported the church's progress.

Keywords: Uses of history, memory politics, church building, historical narrative, progress.

1 FROM NIGHTCLUB TO CHURCH – AN UNLIKELY SUCCESS STORY

In worldly terms, the history of the 'Luther Church Helsinki' resembles a classical 'comeback story' by typical Hollywood standards. As such, it almost serves as an ideal image for progress itself. Using a more religious vocabulary, the story of the church could be portrayed as one where good conquers evil, where words like 'miracle', 'homecoming' and 'resurrection' are used frequently, and where allusions to the biblical narrative are common. Such was the message given by the church congregation before the reopening, by those speaking at the opening ceremonies, and by the media at the time of the inauguration (Filén, 2016; Takaisinkotiin, 2016).

The church, which was originally erected in 1894 at the very heart of the Finnish capital by The Lutheran Evangelical Association of Finland (Koskenniemi, 1967: 214), was reinaugurated in the summer of 2016; thus, bringing to a close a long and unlikely series of events. Consequently, it was an occasion of great interest, highlighted not only by the Christian press but also by the secular media.

To some extent, the media coverage was part of the plan of the Evangelical Association, in their aim to promote the relaunch of the church. However, this does not fully explain the multiplicity of reports in newspapers, podcasts and blogs. More likely, the initial interest was fuelled by the fact that the occurrence itself served as a much longed-for, and unexpected exception to the norm, in which churches around the world, on a daily basis, are desacralized or otherwise lost to activities that are more mundane.

As I will try to demonstrate, what made this specific transformation or conversion special, was the fact that the history of the church formed such a glaring contrast to its future, sacral plans. For instance, practically all the news-articles emphasized the fact that the church had served for many years as a nightclub. Consequently and in comparison, the opening ceremonies themselves, consisting of a service and an inaugural ceremony, did not receive any publicity or reporting worth mentioning. On the other hand, what did attract attention was firstly the initial announcement concerning the projected conversion, and, secondly, when it was later discovered that the original altar cross had been found and was to be carried back to the church. This, I will return to.

The Lutheran Evangelical Association of Finland also lay behind the relaunch; this association is a revivalist movement within the 'folk church' of Finland. The association has its origins in the great revivals that spread across the nation in the 19th century. Many of these movements started out as being critical of the majority church, but very few of them actually left the church. This is also the case with the Evangelical Association, which is one of the largest revivalist movements in the country with a membership of just less than 4% of the population (Niemelä & Salomäki, 2006).

Opinions about the revivalist movements have always been divided. Similar to the majority church (The Evangelical Lutheran Church of Finland), the revivalist movements have had their ups and downs as regards membership. Even though the future looks relatively bright for the Lutheran Evangelical Association of Finland (Niemelä & Salomäki, 2006: 369), for a long time, a gradual fall into obscurity seemed to be the fate of the Luther Church Helsinki. Its time of prosperity having long since been gone. In the 1960s, the number of church activities started decreasing – something which can probably be related to the decrease of activities in the association at large (Huotari, 1981, 34). In 1989, for economic reasons, the church was sold to an insurance company, which, in its turn, rented out the property (Dahlbacka, I, 2016:177–178). By the mid-1990s, the former church was functioning both as a restaurant and as a nightclub (The Luther Church Helsinki, 2017).

If you include in this colourful history, the bombing that the church was subjected to during the second world war, and the squatting it fell victim to during the 1980s (The Lutheran Evangelical Association, 2014), then the high spirits among the members of The Lutheran Evangelical Association of Finland are understandable when it was announced that the property would once again serve as a church building. This chain of events was labelled a 'new beginning', a 'return home', and even an 'awakening from the dead'. Consequently, considering everything that the church has endured, it is fair to say that the story contains all the elements necessary for creating a strong success, comeback, or retribution story.

In a sense, the journey that the church has made is in itself an example of progress. Naturally, this becomes most evident in the material alterations that have been made to the building. This originally gothic style prayer house is now a modern church in what could be labelled a functional style. In addition, on a visionary level progress can also be noticed. Having originally been a rendezvous mainly for the members of the Evangelical movement to which it belonged, the congregation's manifesto today is to be an "open, homely and praying community for the people of Helsinki". This includes, among other things, services in English for the international churchgoers. The target audience seems to be much more diversified than previously (Huhtinen, 2016: 12–13), and inspiration for church activities is being imported from such faraway places as Redeemer City to City in New York (Takala, 2016: 6–11).

Considering the length of time that has passed since the last regular services were held in the church, and taking into account the relatively ambitious future plans that the congregation holds for the church, this venture could almost be described as an example of Zahid Hadid's "step unto the unknown". Will present day members of the association, who have no immediate and personal connection to the church building, feel a sense of belonging? Will people from outside the association find their way to the church and manage to make it their own? Will the church, as the vision states, become a community for the people of Helsinki?

Naturally, only time will tell. However, interestingly enough, when promoting the church, the emphasis has mainly been placed on telling the story – or rather history – of the church, rather than on boosting its future activities. In a sense, it seems as though the vision of becoming a church for the people of Helsinki is being realized with the aid of the historical narrative that has been created around it. The past is used to pave the way for the future.

2 MEANING-MAKING THROUGH HISTORICAL NARRATIVES

Looking at what is currently happening from the perspective of historical studies – and more precisely from the perspective of history didactics and studies on uses of history – might clarify matters. In historical studies, the idea of the past forming the basis of the present and of the future is no novelty, but rather a matter of course that elucidates why attention has been placed on the church's past rather than on its future.

A basic line of thought is that the past, the present, and the future all influence each other and interact with one another: How we look at history affects the way we look at the present and the future, and, vice versa, our interpretation of the past is dependent on who we are today and tomorrow (Johansson, 2012: 40–41; Wiklund, 2006: 46).

For instance, the German historian Reinhart Koselleck expresses this by stating that the past and the future are interwoven in the present: "experience is present past" and "in the today; is the future made present" (2004: 259). On the other hand, if one prefers St. Augustine:

> What is now plain and clear is that neither future nor past things are in existence and that it is not correct to say there are three periods of time: past, present and future. Perhaps it would be proper to say there are three periods of time: the present of things past, the present of things present, the present of things future" (1953: 350).

This being the case, i.e. that the three time dimensions are so tightly intertwined and affect each other, and that our perceptions of the past influence our plans for the future, it is clear that 'possession' of the past is a decisive force in the present. History and memory are very often used to achieve certain objectives. This is what is termed 'Uses of history' or 'Memory politics'.

To put it bluntly: the past can be, and is often, used to manoeuvre and manipulate the future. The famous Czech-born French writer Milan Kundera, in his novel *The book of laughter and forgetting*, points this out unequivocally:

> The struggle of man against power is the struggle of memory against forgetting [...] They shout that they want to shape a better future, but it's not true. The future is only an indifferent void no one cares about, but the past is filled with life, and its countenance is irritating, repellent, wounding, to the point that we want to destroy or repaint it. We want to be masters of the future only for the power to change the past (1996: 4, 30).

However, things need not be driven to such extremes, nor is it exclusively politicians or leaders who use the past in a desire to become masters of the future. It is common to every human being and group of people.

According to the German history theoretician Jörn Rüsen, people use historical narratives – i.e. the past – in order to formulate their identity, to orientate themselves, and to

> articulate those stories which are needed to give one's own life a senseful temporal direction in the changes and alterations of time (1988: 28).

To summarize, people use the past to form their identity and to orientate themselves. This being so, narratives about the past become important and a usable means of mediating, influencing and legitimizing values and actions for those who want to direct the future (Nordgren, 2006: 36).

Narratives or stories that are easy to relate to, that bind together the past, the present, and the future, and that help people to orientate themselves in a larger context, are especially useful. Such narratives Rüsen labels 'historical narratives' (1988: 25–26). In other words, in order to justify, legitimize, or simply explain progress it has to be anchored in the past.

3 FACING THE ALREADY KNOWN AS A PREREQUISITE OF PROGRESS

Thus far, the line of argument has been rather theoretical. In this concluding chapter of the article, my object is to anchor theory to practice by looking at a few aspects of the relaunch of the Luther Church Helsinki, in which the past played an important role.

As I have already indicated, those in charge of marketing the church have never concealed the multifarious history of the church, even though, for instance, its intermezzo as a nightclub might be considered as being quite far from its original, sacred purpose. Quite the reverse, these 'lost' years have become the cornerstones of the narrative, and they are highlighted whenever the relaunch of the church is mentioned.

To some extent, this course of action serves to frame the story of the church as a typical success, or comeback story: when all hope seemed gone, a miracle happened and the underdog won ... This, I would suggest, is a pre-structured, conventional, familiar, and easily accessible trope (Zarzycka, 2017: xiii), which appeals to the spectator and is capable of conveying the plot even to those who have no personal relationship with the church itself. As such, it is of course of great value, since most people can relate to this metaphor – if not on a personal level – then at least on an affiliative level.

Evidently, it has also been easy to draw a parallel between the events that took place while the church was not a church and well-known biblical stories. For instance, figures of speech such as 'welcome home', 'the church that rose from the dead", 'no place in the inn', and the 'Parable of the prodigal son' are examples of allusions to biblical stories that have been used to portray and frame the progress.

Naturally, biblical stories play a pivotal role within Christian groups, as they are familiar to all and as they are regarded as cultural texts around which memory communities build their identity (Assmann 2006, 104). With reference to Rüsen's characterization of well-functioning historical narratives, these allusions and metaphors, mentioned above, make it easy for the presumptive (Christian) audience to relate to the story and to the progress. Consequently, applying these biblical stories in a use of history directed at a Christian group makes the narrative especially powerful. This is because the target audience is familiar with the wording and because the cultural text gives the argument a legitimization (Hervieu-Léger, 2000: 86–88, 97). I have suggested that this kind of uses of history – i.e. where cultural texts are referred to in order to strengthen the argument – could be labelled 'Prophetical use of history'. A Prophetical use of history "not only legitimizes predictions and statements about the present time and the future with examples from history but also guarantees them an authority that is based on a cultural text" (Dahlbacka, J, 2016: 25).

Another event that, with the benefit of hindsight, transpired to be both the key element and the culmination of the whole narrative, was a single object from the past. When the church was sold, its religious interior was either dismantled or destroyed. However, in the wake of the announcement concerning the reopening, news also spread about the discovery of the church's altar cross.

The cross – a simple handcrafted cross of wood, believed to have been lost forever, miraculously survived both the emptying and the eviction. A young theologian and future priest, Jukka Hildén, who by chance had happened to pass by the construction site, picked the cross up from a rubbish skip, transported it to the city of Turku, and kept it in a basement storeroom. There it lay untouched for more than twenty years.

Prior to the inauguration, a website called 'takaisin kotiin' (back home) was launched, announcing the return of the cross to its 'home' in the church. It was to be carried by volunteers, all the way from Turku to Helsinki – a distance of more than 150 kilometres. Around 150 people volunteered to carry the cross for some part of the distance.

According to the web page, previously mentioned, the purpose of the retrieved cross and the 'crusade' was to call attention to the fact that it is never too late to start all over again. Here as well, the allusion to the Christian doctrine of grace is obvious. However, it is hard not to notice the resemblance to both the original 'finding of the cross-myth' (*Inventio Crucis*) that served, among other things, to legitimize Jerusalem as a Christian city (Limor, 2014: 37), and to the medieval crusades, which aimed at reclaiming the city from the Muslims, who, allegedly, had unrightfully occupied it.

In the light of this, reuniting the cross with the church can be said to have had a legitimizing purpose. In a sense, bringing it 'back home' demonstrates the long-established right of The Lutheran Evangelical Association of Finland to possess the building. In addition, it affirms that the building was a church from the very beginning, and what happened during those few years in between was only a minor interruption. Using the argument of Jörn Rüsen, this narrative consequently binds together the past, the present, and the future.

By way of conclusion, I have tried in this article to suggest and exemplify how the past is a prerequisite, sometimes even a driving force, for progress. The article began with a quotation of Zaha Hadid stating that there can be no progress without facing the unknown. Thus, I assume it is suitable to end with another statement, which summarizes the argument of this article: There cannot be progress without facing the already known.

BIBLIOGRAPHICAL REFERENCES

Assmann, Jan. (2006) *Religion and Cultural Memory: Ten Studies*. Stanford, CA: Stanford University Press.

Dahlbacka, Ingvar. (2016) Lutherkyrkans altarkors – en märklig historia in *Hemåt – Evangelisk årsbok* 2016. Vasa, Slef-media: 177–183.

Dahlbacka, Jakob. (2016) Religious Uses of History as Inclusion and Exclusion in Dahlbacka I, Groop K, Dahlbacka J (Eds.) *The Shifting Boundaries of Tolerance. Inclusion, Exclusion, and Religious Communities of Memory*. Åbo: Studies on Religion and Memory.

Hervieu-Léger, Danièle. (2000) *Religion as a Chain of Memory*. Cambridge UK, Malden USA: Polity Press.

Huhtinen, Pekka. (2016) Kansainvälinen kohtaamispaikka etsijöille in *Sanansaattaja*, 1/2016: 12–13.

Huotari, Voitto. (1981) *Kirkkomme herätysliikkeet tänään*. Pieksämäki: Sisälähetysseuran kirjapaino Raamattutalo.

Johansson, Maria. (2012) *Historieundervisning och interkulturell kompetens*. Karlstad: Karlstad University Studies.

Karin Filén. (2016) website of Finnish Broadcating company YLE. Available at: https://svenska.yle.fi/artikel/2016/05/12/helsingforskrog-blir-kyrka-igen-vid-fredriksgatan-42 (Accessed 29.1.2017).

Koselleck, Reinhart. (2004) *On the Semantics of Historical Time*. New York: Columbia University Press.

Koskenniemi, Lauri. (1967) *Suomen evankelinen liike*. Helsinki: Suomen luterilainen evankeliumiyhdistys.

Kundera, Milan. (1996) *The book of laughter and forgetting*. London: Faber and Faber limited.

Limor, Ora. (2014) Conversion of Space in Katznelson I & Rubin M (Eds.) *Religious Conversion. History, Experience and Meaning*. Farnham, Surrey, England & Burlington, USA: Ashgate: 31–60.

Niemelä, Kati & Salomäki, Hanna. (2006) Herätysliikkeet 2000-luvun kirkollisessa kentässä in *Finnish Journal of Theology* Vol. 111. 4/2006: 359–371.

Nordgren, Kenneth. (2006) *Vems är historien? Historien som medvetande, kultur och handling i det mångkulturella Sverige*. Diss. Umeå: Karlstads universitet, Estetisk-filosofiska fakulteten.

Rüsen, Jörn. (1988) Functions of historical narration: proposals for a strategy of legitimating history in school in Angvik M. et al. (Eds) *Historiedidaktik i Norden 3*. Malmö: Lärarhögskolan i Malmö, 19–40.

St. Augustine. (2014) The Confessions. Translated by Vernon J. Bourke in *The Fathers of the Church Vol. 21*. Washington, D. C. The Catholic University of America Press.

Takaisinkotiin (2016) website for carrying back the cross. Available at: http://takaisinkotiin.fi (Accessed 29.1.2017).

Takala, Elina. (2016) Tehtävänä rakastaa stadilaisia in Sanansaattaja 1/2016: 6–11.

The Lutheran Evangelical Association. (2014) Available at: https://www.sley.fi/node/24694 (Accessed 29.1.2017).

The Luther Church Helsinki. (2016) Available at: http://www.lutherkirkko.fi (Accessed 29.1.2017).

Wiklund, Martin. (2006) *I det modernas landskap. Historisk orientering och kritiska berättelser om det moderna Sverige mellan 1960 och 1990*. Diss. Stockholm/Stehag: Brutus Östlings Bokförlag Symposion.

Zarzycka, Marta. (2017) *Gendered Tropes in War Photography. Mothers, Mourners, Soldiers*. New York, Ny & Abingdon, Oxon: Routledge.

Art and progress; Portuguese colonial representations in the great world exhibitions

Maria João Castro
CHAM FCSH, Universidade NOVA de Lisboa, Universidade dos Açores, Portugal

ABSTRACT: The great exhibitions – world, universal, international, or colonial – promoted in the second half of the 19th century and the 20th century in the light of western development were one of the most important factors in spreading the idea that progress, an outcome of the Industrial Revolution, was essential. In effect, right from the start, these events were privileged spaces for presenting and promoting the socio-economic development of an imperial Europe, which had mapped out the world. The colonial idea was not forgotten, and its representation increasingly dominated the great exhibitions where it had been present ever since the first exhibition of 1851. As a result, the configuration of this faraway reality was brought to the metropoles of the Old Continent, and the hegemony of western civilisation was highlighted, as was its role as a civilising agent in a combination capable of bringing progress to the overseas possessions.

In this context, the great exhibitions of the 19th and 20th centuries were used as vehicles to legitimate and disseminate the colonial policies of the European empires, and to spread propaganda about them. However, they were used also, and primarily, as vehicles for an artistic culture that reflected the worldview of a society that saw itself on a global scale and was the fruit of progress, a progress it felt an urgent need to disseminate.

Keywords: Empire, Colonial Art, Exhibitions, Contemporary Art

1 GREAT EXHIBITIONS

In the mid-19th century, an ambition for universality, a desire to spread knowledge of the benefits brought by industrialisation, and faith in progress meant that a need arose to create an exhibition structure that could bring together the new urban experience – and so the great exhibitions were born. The 'Universal Exhibition' became a platform where the results of development on various levels, and in different registers, could be exhibited in a compact fashion. It brought under one roof the most glorious aspects of the 19th century. Exhibiting everything from agriculture to mining, from machinery to art, it brought the public a view of the human cosmos immersed in a new movement, finding its raison d'être in one of the main identifiers of the age: the birth of the phenomenon of the masses.

Held in the modern metropoles, these events reproduced in miniature a new model of urban life, imprinting a new rhythm on the city, which was adapting to the novelty of crowds that were avid with curiosity and interest to witness and experience the recent fruits of progress. In addition, these exhibition centres functioned in two ways: they showed the balance of the work already done and provided a foretaste of what still remained to be achieved. Visited by a huge public who were extremely eager to see and experience a new reality, the great exhibitions became excellent experimental laboratories in technological terms, turning themselves into places of the future yet at the same time spaces for revisiting the past.

If in their genesis, these exhibitions proposed to display progress, they themselves were the result of that very same progress as they were built in areas redeveloped for the purpose and were backed up by a series of (support) infrastructures, which were also the result of progress. They were a manifestation of prestige and ostentation and the stage on which nations affirmed their economic, technological, and even cultural power, thereby revealing their own aspirations to embrace progress. In other words, these showplaces encapsulated what the 19th century understood as modernity, turning themselves into epicentres of cosmopolitanism that urgently wished to demonstrate and teach the virtues of the present age and confirm predictions for an exceptional future. Besides this, the exhibitions provided unique opportunities to highlight new landscapes, especially those that exhibited the local indigenous artistic culture of the distant possessions belonging to the European colonial empires.

In the context of this reflection, what is of interest is to focus on a tripartite strand consisting of progress – empire – art, all brought together and shown in these unique spaces. Consequently, we opted to refer only to those exhibitions whose typology fits the triangulation mentioned above. The success of the first exhibition in London in 1851 set the calendar in motion for the international exhibitions held in other European capitals. These expanded and widened their scope while restructuring themselves according to a genealogy with a strong colonial bent. To highlight this perspective, large colonial representations were increasingly installed, the objective of which was to reproduce in miniature all the resources of the empire.

The very first exhibition on a truly international scale – the Great Exhibition of the Works of Industry of All Nations (London, 1851) – included a small colonial wing limited to India, the jewel in the British crown. As a result, this prompted a fascination with colonial possessions and the exotic. Without any doubt, one of the objectives of this exhibition envisaged right from the start was to convince the public that the empire was an extension of the metropole and was capable, even at a distance, of mirroring and enhancing the source of progress.

At the second international exhibition (Paris, 1855), the colonial section's area was increased thanks to the growing interest in the distant exotic possessions that were part of all the overseas empires.[1] However, it was in the 1867 exhibition (also in the City of Light) that national pavilions, supposedly designed according to the respective architectonic typologies of each nation, made their first appearance.

The next event in 1878 saw the birth of the 'rue des Nations'. This highlighted the essence of this international fair through an eclectic montage of façades all lined up and designed in accordance with the aesthetic image of each country. In addition, an autonomous area for the colonies was created.

The 1889 exhibition commemorated the centenary of the French Revolution. Its symbol was the Eiffel Tower, which became not only a very popular panoramic observatory but also an icon of industrial progress. The colonial section was displayed in a large 'tableau vivant' divided into four zones: Arabia, Oceania, Africa, and Asia. "A maior sensação foram os nativos trazidos das colónias asiáticas e africanas"[2] (SOUTO 2011:211). This form of presentation meant that the backward nature of the colonial area contrasted dramatically with the European development evident in the rest of the exhibition.

The 1900 Paris Exhibition, the last and the most grandiose ever mounted, offered an overview of the century now gone, and was supported by a series of infrastructures that made an apologia for progress. If initially, these were essentially industrial; from 1900 on, they became linked to an aesthetic of illusion and magic: the recreational component became requisite. Within this context, and with Art Nouveau at its apogee, railway stations, railway bridges, and the first Metro lines were inaugurated with electricity acting as the great fairy godmother (Guerreiro, 1995: 48–57). Film projections, the Ferris Wheel ['la Grande Roue'], the escalator, or a 'journey' in a Trans-Siberian Railway carriage became the attractions that allowed illusion to triumph and exalted progress. At each new universal exhibition, the shows became progressively more mechanised with state-of-the-art technologies occupying the place of honour.

In relation to art, "O lugar privilegiado concedido à arte neste programa não tem equivalente na história das Exposições Universais do século XIX"[3] (GUERREIRO 1995:70). The exhibition claimed its status as the beacon of the universe in matters of art (showing the triumph of Impressionism), but an art that must be measured by technique; a technique in its turn had to respond to the calls of art. The event was renowned for its prestige and became a place of pilgrimage. It included an important foreign and colonial section dominated by panoramic views and a 'Tour du Monde' pavilion inside which the visitor went on a journey around the world that evoked and simulated such different realities as a Chinese pagoda, an Indian temple, or elephants from Cambodia. Once again, but this time integrated into the foreign landscape brought to the exhibitions, came the human landscape. Exotic peoples coming straight from the colonies helped the tableaux presented to come to life, turning them into authentic 'Human Zoos' that quickly became famous.

Parallel to these industrial, commercial, technological and artistic exhibitions on a European scale, the growing importance of the imperial colonies meant that in 1883 the first Colonial Exhibition (the International and Colonial Exhibition) was inaugurated in Amsterdam.[4] However, it was only in the 20th century that this new type of exhibition would catch on and enjoy a new profile that aimed to show the inhabitants of the metropole the different facets of the colonies. This resulted in spectacular reconstructions of the natural environments and monuments of Africa, Asia, and Oceania together with the obligatory staging of performances by inhabitants of the colonies. Once they were emancipated, the colonies reflected not only the imperial policy of the western powers but also regarded themselves as an idealisation of those exotic territories, inviting people from the metropole to 'travel' through the overseas colonies without ever leaving home. Hence the importance of the presence of indigenous people, the 'Other', whose purpose was to certify the 'authenticity' of the representation. By bringing back 'living pieces of the

[1] For example, the third Exhibition (London, 1862) witnessed the rise of the fashion of Japonism (the influence of Japanese arts, especially wood engraving, clothing and decoration), with Japan becoming known in Europe for the first time.

[2] "The biggest sensation was the indigenous peoples brought from the African and Asian colonies."

[3] "The privileged place given to art in the programme found no equivalent in the history of the Universal Exhibitions of the 19th century."

[4] Portugal was not represented. It should be remembered that imperial Europe's interest was focused on the 'Scramble for Africa', a fact that would give rise to the Berlin Conference two years later (1884–1885).

empire', the colonial exhibitions became hybrid spaces for convergence that sought to emphasise the benefits of European colonialism.

Numerous exhibitions were held between the mid-19th and mid-20th century, but, given the scope of this paper, we shall mention only the following: the Colonial and Indian Exhibition in London (1886), the Universal, International and Colonial Exhibition of Lyon (1894), the Portuguese Insular and Colonial Exhibition of Porto (1894, see below), the International and Colonial Exhibition held in Rochefort-sur-Mer (1898), the Colonial Exhibition of Paris (1907), the National and Colonial Exhibition of Marseille (1922), and, of prime importance, the International Colonial Exhibition of Paris (1931). This last event marked the end of an era – the 'Age of Empire' (Hobsbawm 1990) – that would never return.

All these events sought to encourage intercolonial relations, highlighting the achievements of human progress idealised in an age of imperial expansion. The fact that France invited the great imperial nations to be present served a dual purpose: to demonstrate that several countries followed the same colonial approach as France herself did, and in parallel to prove that France was better suited than any other country to celebrate the future of colonialism.

However, the Second World War altered and rescaled world geopolitics, and the end of the conflict coincided with (and led to) the beginning of the breakup of the European colonial empires.

2 PORTUGUESE REPRESENTATIONS

Portugal was present at the great European exhibitions right from the beginning and gradually gave increased importance to her representation.

This cycle of Portuguese participation began with the first exhibition held in London in 1851 where a Portuguese section, albeit mostly limited to a show of agricultural products, was erected. At the following exhibition in the English capital in 1862, the Portuguese commission was presided over by the king himself, Dom Fernando II (1816–1885), showing the growing importance given to these displays of national products. However, it was in Paris in 1867 that the Portuguese pavilion (built according to a design by Rampin Mayor) took pains to link the Portuguese representation to its empire by imparting an image of sumptuous magnificence and exoticism. The neo-Manueline architecture included oriental-style images of an Anglo-Indian nature, thereby showing the influence of the fashion for Orientalism that was so much in vogue at the time.

At the 1878 Paris Exhibition, Portugal was represented by two pavilions. One was in the 'rue des Nations' and reproduced the south portal of the Jerónimos Monastery (which together with the Tower of Belém was one of the two most famous monuments from the time of the Portuguese Discoveries). There was a carefully studied relationship between the interior and the exterior, and internally the pavilion alluded to the cloisters of Belém and the Monastery of Batalha in acknowledgement of the Manueline art inspired by the Discoveries. The second pavilion, the Pavilion of the Colonies designed by Jean-Louis Pascal (1837–1920), was eclectic in that it combined Manueline elements with polychrome neo-Mudéjar 'azulejo' tiles in a conventional and static arrangement with a vaguely Orientalist flavour. The job of filling this space was given to the commissioner Luís Andrade Corvo (1824-1890), who focused on the work he had been doing at the Colonial Museum of Lisbon during the previous two years.

Figure 1. Portuguese Pavilion in the 1867 Universal Exhibition in Paris. Archivo Pittoresco. Semanario illustrado, Vol X, N.° 7, 1867, p. 49.

At the 1889 Paris Exhibition, the Portuguese pavilion displayed features that referred to the Tower of Belém, the monument mentioned above from the time of the Portuguese Discoveries. Inside the pavilion, eleven of the sixteen rooms were devoted to the Colonial Exhibition. They again were organised by Andrade Corvo and contained objects that had been lent by the Lisbon Geographical Society (created in 1875).

At the 1900 Exhibition, Ventura Terra (1866–1919) designed the Pavilion of the Portuguese Colonies set up in the Trocadéro. It reflected not only the formal influence of French architecture (Ventura Terra had been a student at the École des Beaux-Arts in Paris) but also looked towards the Portuguese maritime discoveries (COFFIGNON 1900:289). It is worth mentioning here that in order to remind visitors that the Portuguese had always been a nation of seafarers, the façade of the Trocadéro pavilion dedicated to the colonies showed a group of women symbolising the colonies who held a shield bearing the arms of Portugal.

As for Portugal itself, the phenomenon spreading throughout the Old Continent of hosting exhibitions did not pass unnoticed. As a result, the country that laid claim to possessing the first global empire (1415)

Figure 2. Universal Exhibition of Paris, 1878. Façade of the Pavilion of the Colonies. O Occidente, Vol. I, N.° 23, 1878, p. 177.

Figure 3. Colonies Portugaises. Carte réclame Sicard (éditeur). 1900. Paris, Musée des Civilisations de l'Europe et de la Méditerranée.

as well as the last European empire (1974) mounted several exhibitions: the Portuguese Insular and Colonial Exhibition in 1894, the Colonial Exhibition in 1934, both in Porto, and finally in 1940, but this time in Lisbon, the Great Exhibition of the Portuguese World. This one was to commemorate the double centenary of the Foundation of the Nation (1140) and the Restoration of Independence (1640).

The 1894 Exhibition in Porto commemorated the birth of Infante D. Henrique (1394-1460) but the 1934 event was the first that was truly used by politics as propaganda for the empire. The 'Estado Novo' regime of Oliveira Salazar (1889–1970), which had been consolidated by the 1933 Constitution, was inspired by the 1931 Colonial Exhibition of Paris to initiate a new colonial policy (based on the 1930 Colonial Act)[5]. The justification for such an event came from the idea that most Portuguese people had not the slightest idea of the reality surrounding their greatness, be it from the material point of view of the extent of Portugal's territory, or from the moral, political, and spiritual point of view. These were the result of the historical fact of five hundred years of colonisation, and the contemporary fact that Portugal still retained the majority of her overseas territories. The exhibition paid particular attention to the fact that Portugal, "the country that through its discoveries gave new worlds to the World",[6] was the oldest of all the colonising countries. The whole exhibition area was centred around the Crystal Palace, re-baptised the Palace of the Colonies, and was organised in such a way as to allow 338 institutional and private stands to be installed, as pointed out in the 'Guia Oficial dos Visitantes' [Official Visitors' Guide] and in 'O Império Português, Álbum-Catálogo Oficial' [The Portuguese Empire: Official Album-Catalogue], both published at the time. The whole exhibition sought to show mainland Portugal the 'life' of people from the colonies in a similar way as had been done in Paris. Thus indigenous people from Cape Verde, Angola, Mozambique, India, and Macau were all brought to Porto, and 'villages' from Timor and Guinea were set up next to a lake, thus bringing the recreated scenes 'to life'. In addition to what was designed and built (monuments, pavilions, and kiosks) and a huge variety of exhibited objects (notably of indigenous and colonial art), a series of movable structures were erected (for example, an overhead cable car and a tourist train). As well as all this, there were colonial troops, theatre and cinema with a colonial influence, diplomas for attendance, and prizes. The event also promoted conferences and publications, parades and processions.

However, it was perhaps the 'Double Centenary' Exhibition held in 1940 that became the great imperial parade of contemporary Portugal, commemorating

[5] Decreto n.° 18 570 de 8 de julho de 1930, published in the Diário do Governo N.° 156, I Série. Online at https://dre.pt/application/dir/pdfgratis/1930/07/15600.pdf (accessed on 16.1.2017).

[6] A phrase that has now become iconic and appeared in bold type in the first Portuguese Colonial Exhibition's leaflet, Porto, Lito Nacional, 1934 [Graphic visual].

as it did eight hundred years of history and imperial expansion. Belém, in Lisbon, was chosen as the location for the event, which was organised around the newly created Praça do Império. The choice was intentional as it validated the historicist discourse that was anchored in important moments of the nation's past and was so dear to the dictatorship. In addition, the square was surrounded by elements alluding to the Discoveries – Jerónimos Monastery, the River Tagus, Praça Afonso de Albuquerque, and the Tower of Belém, all symbols that referred to the golden days of the nation, the metropole of a vast empire. The choice of Belém was therefore not gratuitous given that it was fundamental to legitimating the greatness of Portugal's overseas history – it was from Belém that the ships set sail on their voyages of discovery and to which they later returned. Indeed, the event did not forget to include a historical section or to mount a colonial section.

With regard to the former, the following structures are worth noting: the Formation and Conquest Pavilion, the Independence Pavilion, the Pavilion of the Discoveries, the Sphere of the Discoveries, the Colonisation Pavilion, the Pavilion of Portuguese People in the World, and the Metropolitan Ethnography Section with its Regional Centre containing representations from the organisation of 'Portuguese Villages' and its Popular Art Pavilions.

In the Colonial Section installed in the Tropical Garden, there were indigenous peoples from the Portuguese colonies who acted as performers. This Section consisted of five pavilions (representing the overseas territories) and two streets (from Macau and India) which gave the space a 'picturesque' and exotic feel in a clearly fictionalised appropriation of the overseas possessions. This staging of the overseas experience led shortly afterwards to Diogo de Macedo (1889–1959) publicly defending the creation of a national Museum of Colonial Art. The idea behind this was to use the museum not only for conservation purposes but also as a preventive measure as it was believed that through educating the public, their sensibility could be nourished, and a taste for art encouraged. However, this wish was never fulfilled.

It should be mentioned here that the timing of the exhibition, although planned before the outbreak of the Second World War and adversely affected by the Spanish Civil War dragging on, meant that its success did not impact on the rest of Europe, which was suffering numerous crises at the time from which Portugal officially remained apart. Consequently, the only foreign pavilion in Belém was that representing Brazil.

The organisation of the Porto Colonial Exhibition in 1934 had served as a rehearsal for this much larger event, and Portugal's recent participation in the universal exhibitions in Paris in 1937, and New York and San Francisco in 1939, provided an experience that helped build an international perspective on exhibitions. Overall, the programme was extensive and ambitious. The exhibition was seen as a way to demonstrate progress besides also being a space to display the new Portuguese paradigm of developing infrastructures in her overseas colonies, which attested to the country's ability to achieve its goals. As to the exhibition area and surroundings, the expropriation of land adjoining the future exhibition area and the eviction of local residents must be mentioned as should the demolition of a considerable number of buildings that, in just eleven months, gave way to a historical city made of stucco, wood, plaster, and paper. The building of access roads and infrastructure (the Lisbon-National Stadium motorway, the Marginal coast road, and the Belém ferry terminal), the connection of the national radio station, Emissora Nacional, to the colonies together with a variety of works carried out to improve, remodel and renovate existing structures all served to enhance the entrepreneurial role of the 'Estado Novo' and consolidate it in the eyes of the public. As for the overseas territories, the event highlighted Portugal's entrepreneurial skills by exhibiting models and pictures of the pioneering works built in the colonies – bridges, railways, dams, art centres, hospitals, and schools. It was a whole building programme following the metropolitan pattern and undertaken in the image of other European empires.

3 FINAL CONSIDERATIONS

From the end of the 19th century, the Great Exhibitions cultivated the people's fascination with the exotic and colonial possessions. This reflected the growing importance of the imperial colonies and the growing emphasis given to the arts and to (extra-)European culture. The authorities who presided over their organisation sought to convince the public that the metropole provided the source from which the modern era and progress flowed, building itself up as the matrix of a 'superior' civilizational culture. To highlight this perspective, extensive scenic sets were built reproducing the overseas territories. Their purpose was to display a miniature version of the resources of the empire. In fact, the 19th century exhibitions encapsulated what the century understood as modernity while the 20th century events showed it as contemporaneity. Progress built upon science and industry, and cosmopolitanism based on the idea that human knowledge and production were transnational, objective and without limits, organised as the expression of supranational progress, ended up providing symbolic material for the cult of the nation, for the construction of nationalisms, and for the propagation of the importance of the imperial colonies.

From 19th-century display cabinet to the current 'exhibition-spectacle', these exhibitions have proved to be factors in development and (trans)national affirmation. In addition to mobilising extraordinary resources, they have contributed to the urban and architectonic development and have collaborated to spread knowledge of science and technology, ideas and principles, culture and civilisation. Even though

Figure 4. EXPO'98. The Oceans, a Heritage for the Future". Lisbon, 1998.

the first exhibitions tended to veer towards the hegemony of the host nation, since the second half of the 20th century the emphasis has been on the dialogue between peoples and nations, thus greatly contributing to 'institutionalised international cooperation'. The future tendency increasingly seems to be one whereby permanent structures and infrastructures that can later be used for different purposes or even designed with this in mind, should be an outcome of the exhibition, thereby increasing the scope of the importance of culture and art in relation to progress on a global scale. In this respect, the growing self-affirmation of countries that were once territories belonging to European empires does not prevent them from referring to the colonial period of their history but now focuses on it in the light of the post-colonial dynamic.

The last major international exhibition of the 20th century, Expo '98, was held in Lisbon in 1998, and it dealt with this universe. Its theme was 'The Oceans, a heritage for the future', and it celebrated the 500th anniversary of the arrival of Vasco da Gama (c. 1460–1524) in India in 1498. This is an indication of how even at the close of the 20th century colonial echoes and spin-offs still resonated and had a place in the world: a different place to be sure, a place of inheritance and testimony to a unique age, and one from which the new autonomous nations were carved.

In the final analysis, it is important to be aware of the major role played by art and progress in the building of imperial representations, a role that allowed the metropolitan population to discover and be seduced by faraway exotic places with growing success. This contributed to an increase in travelling and colonial tourism as well as to the construction of a coloniser-colonised image set with a strong nationalist flavour. As exponents of a European ascendancy that sought to serve as one model for colonial globalisation out of all the disparate theories and practices, the Great Exhibitions of the 19th and 20th centuries showed how progress engendered by the Industrial Revolution was the right path to take to build the so-called 'world of tomorrow'.

This paper had the support of CHAM (FCSH/NOVA-UAc), through the strategic projected sponsored by FCT (UID/HIS/04666/2013).

BIBLIOGRAPHICAL REFERENCES

Ageorges, Sylvian. (2006) Sur les traces des expositions universelles – Paris 1855–1937, Paris, Parigramme.

Aimone, Linda; Olmo, Carlo. (1993) Les Expositions Universelles 1851–1900, Paris, Éditions Belin.

Allwood, John. (1977) THe Great Exhibitions, Londres, Studio Vista.

Coffignon, A. (1900) Le Pavillon du Portugal. L'Exposition de Paris. Encyclopédie du siècle. Paris, Libr. Illustré.

Findling, John. (ed.) (1990) Historical Dictionary of World's Fairs and Expositions, 1851–1988, New York, Greenwood Press.

Greenhalgh, Paul. (1988) The Expositions Universelles, Great Exhibitions and World's Fairs, 1851–1939, Manchester, Manchester University Press.

Guerreiro, António. (1995) Exposições Universais: Paris 1900, Lisboa Expo 98.

Hobsbawm, E. J. (1990) A Era do Império 1875–1914, Lisboa, Editorial Presença.

Martins, João. (2000) Arte Efémera em Portugal. Pavilhões Portugueses nas Exposições Universais do Século XIX, Lisboa, Fundação Calouste Gulbenkian.

Souto, Maria Helena. (2011) Portugal nas Exposições Universais 1851–1900, Lisboa, Colibri.

Progress on display: Universal Exhibitions in the second half of the 19th century

Ana Cardoso de Matos
CIDEHUS, University of Évora, Évora, Portugal

ABSTRACT: Industrial exhibitions, going back to the second half of the 18th century, were always considered an important means of promoting the industrial development of countries and diffusing technical and industrial progress. After London organized the first Universal Exhibition, in 1851, this event category gained an international character. These exhibitions were also an opportunity to present new materials and building techniques. The Universal Exhibition was visited by a very diverse public from politicians and engineers to workers.

Keywords: Universal Exhibition, Science, Technique, Progress, Study visits

1 INTRODUCTION

In recent years, Universal and International Exhibitions have enjoyed growing interest among historians from various fields. Such interest may in part explained by the diversity of ways in which these events can be approached: while they were important in the evolution of the economy and the internationalization of trade, they also played a decisive role in the diffusion of technical and scientific advances and in the layman's contact with those advances. Other aspects still must be considered when looking at Universal Exhibitions, such as the case of political mediation or the links between these events and the development of architectural styles or museums (Matos et al., 2010).

Figure 1. View from the Knightsbridge Road of the Crystal Palace in Hyde Park for Grand International Exhibition of 1851. Dedicated to the Royal Commissioners, London: Read & Co. Engravers & Printers, 1851.

2 FROM INDUSTRIAL EXHIBITIONS TO UNIVERSAL EXHIBITIONS

Industrial exhibitions, going back to the second half of the 18th century, were always considered an important means of promoting the industrial development of countries and diffusing technical and industrial progress. After London organized the first Universal Exhibition, in 1851, this event category gained an international character.

When we look at the chronology of universal exhibitions we see that, up until the 1870s, they were held in either London or Paris, the capitals of those countries which retained the economic power of Europe.

Only in 1873, after the Franco-Prussian Empire was constituted, did Vienna hold its first Universal Exhibition. Paris was again the venue for subsequent universal exhibitions, in 1878, in 1889 and in 1900, but London only saw, in 1870, an international workers' exhibition.

When we look at the chronology of universal exhibitions we see that, up until the 1870s, they were held in either London or Paris, the capitals of those countries which retained the economic power of Europe. Only in 1873, after the Franco-Prussian Empire was constituted, did Vienna hold its first Universal Exhibition. Paris was again the venue for subsequent universal exhibitions, in 1878, in 1889 and in 1900, but London only saw, in 1870, an international workers' exhibition.

Outside the European continent, the first universal exhibition took place in New York, in 1853. In 1876, the celebration of the American Independence Centennial provided the pretext for organizing an exhibition in Philadelphia.

Table 1. Source Schroeder-Gudehus and Rasmussen.

Table 1 -Universal and Industrial Exhibitions held in the second half of the 19th century

Year	Universal International	Place	N° of visitors
1851	Universal	London	6.039.195
1853	Universal	New lork	
1855	Universal	Paris	5.162.330
1862	International	London	6.096.617
1865	International	Porto	
1867	Universal	Paris	11.000.000 to 15.000.000
1873	Universal	Vienna	7.255.000
1876	International	Philadelphia	10.000.000
1878	Universal	Paris	16.156.626
1880	International	Melbourne	1.330.000
1888	International	Barcelona	2.000.000 to 2.300.000
1889	Universal	Paris	32.250.297
1893	Universal	Chicago	27.500.000
1897	International	Brussels	6.000.000
1900	Universal International	Paris	50.860.801

Source: (Schroeder-Gudehus and Rasmussen, 1992)

Figure 2. New York Crystal Palace designed by Karl Gildemeister. The image is an "oil-colour" plate by George Baxter, London, dated 1 September 1853.

The differences between universal and international exhibitions were not always clear. Thus, the exhibitions held in Philadelphia in 1876, in Melbourne in 1880 or in Barcelona in 1888, although they were called "international", are usually counted among universal exhibitions due to the number of visitors and the impact they had on international trade and on their host cities.

The second half of the 19th century also saw various theme exhibitions, as was the case of the one on electricity held in Paris in 1881. These exhibitions were in line with the advances made in several fields of knowledge, of which electricity was an important example.

3 UNIVERSAL EXHIBITIONS AND THE DIFFUSION OF SCIENCE AND TECHNIQUE

Organized at a steady rhythm throughout the 19th century, universal exhibitions (table 1) were a place where scientific, technical, and industrial advances were publicized – especially those which might find applications in industry or contribute to the advancement of urban spaces and everyday life. Donald Cardwell considers that universal exhibitions are important for the history of technology and, in his own words, "the exhibitions, taken as a whole, give some impression of the 'state of the art' in the various technologies" (Cardwell, 1994: 304, 285).

Universal exhibitions also served as a means of disseminating and globalizing science and technique, a fact which was stressed by Prince Napoleon at the time of the 1855 Universal Exhibition in Paris: "o problema do futuro consiste em fazer partilhar à universalidade o que até aqui tem sido reservado para o pequeno número"[1] (Diário do Governo, 1861: 1163).

In this dissemination, the role of the press became increasingly important, and for this reason, the editorial capacities of the various countries assumed a great weight in the 1855 Exhibition. In that event, books were considered the expression of national cultures and provided the standard for appraising the different countries' degrees of "illumination" (Lisboa, 1924: 138–139). King Pedro V himself, in the journal of his travel to that exhibition, mentions this fact when he praises French typography which, thanks to the inclusion of pictures among the texts, had contributed to "the spreading of knowledge in mechanics, mathematics, and natural sciences" (Lisboa, 1924: 149). In those days, however, the great novelty was photography (Souto et al., 2012). Considered one of the symbols of the "intimate link between science and the arts" (Damásio, 1854: 250–251), photography appeared in the Paris Universal Exhibition of 1855 as the representational art of the future. And in the Paris Universal Exhibition of 1900, the Lumière brothers' animatography was seen as a way of surpassing distances and bringing the peoples of the world closer together.

Universal exhibitions permitted countries' economic powers to be compared. The awards collected, along with the evaluation of each country's participation, worked as a way of placing them in a hierarchy regarding technical, scientific, and industrial matters. In the Universal Exhibition held in London in 1851, Great Britain, which obtained 78 gold medals, was considered the most developed nation in industrial and technical terms, followed by France with 52 medals. Nevertheless, as mentioned by Adolphe Blanqui, who represented the Paris Royal Academy of Science during the London Exhibition, for the French, the main result of the Exhibition is the universal, absolute and undisputed recognition of their superiority in terms of art and taste (Bergeron, 1998: 21).

Throughout the second half of the 19th century many international exhibitions were held, several of them thematic in nature like the exhibition on electricity hosted by Paris in 1881. This stemmed naturally

[1] "the challenge of the future consists of sharing universally that which has, so far, been kept to a few"

Figure 3. Paris Cristal Palace 1867. Photo of anonymous author. Library of Congress: Prints and Photographs Division.

from the scientific advances occurring in certain areas, and their subsequent technological applications.

The recreational character associated with universal exhibitions assumed growing importance in the second half of the 19th century. In the 1889 Exhibition, various voices criticized an excess of recreation, which tended to relegate to a secondary position the goals which had initially guided these events. The 1900 exhibition confirmed this trend, leading someone to complain that "De plus en plus les expositions universelles perdent leur premier caractère (…). L'Intérêt de l'industrie et du commerce n'est plus que le prétexte, l'amusement son but "[2] (Mattelart, 1994: 150).

4 A "STAGE" FOR SHOWING ADVANCES IN CONSTRUCTION TECHNIQUES AND MATERIALS: FROM CRYSTAL PALACES TO THE EIFFEL TOWER

By demanding the construction of a series of pavilions for showing their machines and products, exhibitions brought to light the development and use of new materials and building techniques - a case in point being the Crystal Palaces, which showed the association of glass and iron.

Some of these constructions, e.g. the Palace of the Paris 1867 Universal Exhibition, were ephemeral, while others resorted to the idea of prefabrication, allowing them to be disassembled and later reassembled somewhere else, e.g. the Crystal Palace of the London 1851 Universal Exhibition, which was subsequently moved to Sydenham, on the city's outskirts. Still, others led to the development of prefabrication techniques, as was the case with the Eiffel Tower. Some of the buildings that stood the test of time became symbols of the cities in which they were raised: the Eiffel Tower, built for the Paris 1898 Universal Exhibition, and both the Grand and the Petit Palais, made for the 1900 exhibition held in the same city, which became, respectively, an important exhibition centre and a museum of the decorative arts.

[2] "Increasingly, universal exhibitions have been losing their primitive characteristics (…). Interest in industry and trade is merely a pretext, the main goal being recreation…. "

Figure 4. Petit Palais, Published in N.D. Exposition Universelle de 1900, an album of 50 fine photographic views. Paris: 1900. This image is available from the Brown University Library under the digital ID 1303845244984376.

The Eiffel Tower and the Halle des Machines, built for the Paris 1898 Universal Exhibition, were crucial in affirming the use of iron and steel in modern construction, thanks to the demonstration of their multiple constructive potentials. The command held by engineers over new materials and their skills in mechanics and strength of materials were evident in these buildings and, consequently, the triumph of iron architecture in this exhibition resulted in a stronger affirmation of engineers.

Universal exhibitions also served to make known other building materials. The 1878 exhibition helped demonstrate the growing importance of Vicat and Portland cement in public works, while the 1900 exhibition definitively asserted the use of concrete.

5 MEETING AND CONFRONTATION OF DIFFERENT COUNTRIES

Universal Exhibitions were often seen as places of gathering for peoples from different parts of the globe, with the coexistence of countries assuming a peaceful character. Nevertheless, these exhibitions were also, as mentioned by Armand Mattelard, " un endroit autour duquel se cristallise la peur de l'autre"[3] (1994: 148) – not only because the coming together of people brought along fears of epidemic contamination, but also because the show-off of each country's military and industrial power was a way of measuring the respective strengths of the various countries attending these events.

In 1878, in the aftermath of the Franco-Prussian war, the Universal Exhibition held in Paris was considered by the newspaper *Tages Press* to be a great moral victory for France. The newspaper *Nouvelle Presse Liberé*, for its part, saw this exhibition as a truly political event, a revenge more splendid and lasting than any political victory. The *Gazette d'Augsbourg*,

[3] "A place where fear of the other crystallises".

in turn, stressed the fact that, in the space of just seven years, France had managed to recover from the war and to give through this exhibition a good picture of its productive capacities. From London, Eça de Queirós wrote: "desde que a Exposição se abriu, e que a França celebra em Paris a sua grande festa de ressurreição, toda a Europa tem um tom mais calmo; (...). Exala-se da Exposição, parece, uma emanação de concórdia, de trabalho, de civilização que enche os espíritos de um salutar desejo de fraternidade e de paz"[4] (Queirós, s/d, p. 340).

At the end of the 19th century, when the conquest of colonial markets increased the tension among the most industrialized European countries, the presentation of their warfare-related industries assumed special relevance.

Table 2. (Matos, 2012: 304).

Committee appointed to study the Paris Universal Exhibition of 1855

Name	Training	Activity pursued at the time of the appointment	Studies required at the exhibitions
Júlio Máximo de Oliveira Pimentel	Chemist	Professor at Instituto Industrial de Lisboa	Chemical industries
José Vitorino Damásio	Engineer	Interim Director of Instituto Industrial de Lisboa	Steam engines and railways
Sebastião Betâmio de Almeida	Chemist	Escola Industrial do Porto	Chemistry
João de Andrade Corvo	Engineer	Professor at Instituto Agrícola de Lisboa	Agricultural machinery
José Maria da Ponte e Horta	Engineer	Professor at Escola Politécnica de Lisboa	Steam engines
Sebastião José Ribeiro de Sá	Engineer	Head of the 'Repartição das Manufaturas' at the Ministry of Public Works, Trade and Industry	Agricultural products, machines, and tools

6 STUDY VISITS AND RECREATION VISITS

Interest in these exhibitions arose among the populations of several countries, leading to visits by a numerous and diversified public. Thus, alongside engineers and entrepreneurs looking for technical novelties, and intellectuals concerned with the material expression of the current culture, the space of the exhibitions also received members of various countries' elites, in addition to a multitude of people full of curiosity, for whom the manifestations of science and technique were both marvellous and incomprehensible. The London Universal Exhibition of 1851 was visited, during its 141-day span, by 60.000 people. Some 50 years later, the Paris Universal Exhibition of 1900 received 8 times as many visitors.

Since exhibitions were spaces of diffusion for technical, industrial, and agricultural progress, national governments appointed committees charged with studying the advances made in the various sectors of the economy. In the committees appointed by the Portuguese government, chemists, engineers, and agronomists had an outstanding role, as they possessed the skills required to evaluate the advantages of the technologies presented, and to indicate which of these were best suited to satisfy the country's needs (Matos, 2004).

A decree dated March 29th, 1855 appointed a committee that should study, at the Paris Universal Exhibition of the same year, "progressos e melhoramentos das diferentes artes e ofícios, a fim de que os esclarecimentos e indicações que se colhessem servissem para guiar a indústria nacional, encaminhando-a no seu desenvolvimento e aperfeiçoamento sucessivos"[5].

It was made up of five scholars and an administrative official, all of whom were trained in chemistry and/or engineering.

The committee appointed to study the London Universal Exhibition of 1862 was required to produce reports on the advances concerning industrial mechanics, with a special focus on engines, spinning and weaving machines, railroad materials and the manufacture of wagons and means of transportation for ordinary roads.

In 1878, one of the main issues faced by the railroad network was the installation of narrow-gauge railways. So the engineer Cândido Xavier Cordeiro, who was intensively involved with Portuguese railroads – namely the one linking Porto and Braga – was appointed to study that issue at the Paris Universal Exhibition (Carvalho, 1906: 363).

The fact that exhibitions were seen as an efficient way of instructing workers, based on direct observation of machines and industrial objects, led several governments to send representatives of the various industries to universal exhibitions (Matos et al., 2010). With that aim, in 1855 the Portuguese government required the director of the Instituto Industrial de Lisboa to choose 10 workers – 5 from Lisboa and 5 from Porto – to visit the Paris Exhibition at the government's expense. In Porto, the representatives of the various industries were selected by the Associação Industrial Portuense.

In the wake of these exhibitions, some museums were created which extended in time the diffusion of each country's scientific, technical, and industrial advances, and whose goal was to support the instruction given by establishments of technical education. The Universal Exhibition of 1851 gave rise to the first science museum – the South Kensington Museum. When the exhibitions closed, the conservatories of arts and trades and the museums of technology and industry requested the offer of many of the objects which had been on display, thus contributing to the transfer and spread of technology.

[4] "Since the Exhibition opened, with France celebrating in Paris its great feast of resurrection, the whole of Europe enjoys a calmer tone; (...). It seems that the Exhibition exudes an emanation of concord, of work, of civilization, which fills the spirits with a healthy desire of fraternity and peace".

[5] "The progress and betterment of the different arts and trades, so that the clarifications and hints thus obtained might guide the national industry, directing it in its continued development and improvement".

7 EXHIBITIONS AND CONGRESSES: FROM PROGRESS IN NEW SCIENTIFIC FIELDS TO THE IMPLEMENTATION OF INTERNATIONAL PROCEDURES

Universal exhibitions thus acted to attract science congresses which discussed many of the main issues faced by civil societies and national economies. As the 19th century wore on, the number of congresses held simultaneously with exhibitions increased. Focused on areas such as rail transportation and electricity, they contributed to defining international norms and procedures in fields such as communications, transportation, and energy. For instance, the congress on industrial property held at the time of the 1873 Vienna Exhibition proposed the first international convention on brevets. In 1878 was organized, simultaneously with the Paris Exhibition, a congress on literary property, presided over by Victor Hugo, which resulted in the creation, in 1881, of the Berne International Union for the protection of literary and artistic works.

These congresses gave visibility to the world of science and stimulated the appearance of emerging sciences (Rasmussen, 1989: 23–44) such as statistics and anthropology (Benedict, 1983; Corbey, 1993).

8 CONCLUSION

The Universal and International Exhibitions held throughout the second half of the 19th century were important means for the diffusion of scientific, technical, and industrial advances and, being open to the general public, they contributed to the globalization of science and technique (Lafuente et al., 1998: 31–38). The buildings required by the exhibitions helped affirm new architectural styles, advances in building materials and techniques, and the role of engineers as a professional category. The congresses which accompanied these exhibitions contributed to the affirmation of new scientific areas, and to the establishment of international norms.

These events, which were decisive in defining a new relationship between science and the public (Bensaude-Vicent, 1993) contributed to a wider acceptance, by people who lacked technical training, of innovations linked to both the urban and the domestic spaces.

BIBLIOGRAPHICAL REFERENCES

Benedict, Burton. (1983) *The anthropology of world's fairs. San Francisco's Panama Pacific International Exposition of 1915.* Berkeley: The Lowie Museum Anthropology.

Bensaude-Vicent, Bernadette. (1993) Un public pour la science: l'essor de la vulgarisation au XIXe siècle. *Réseaux* (58): 47–66.

Bergeron, Louis. (1998) *Les industries du luxe en France.* Paris Ed. Odile Jacob.

Cardwell, Donald. (1994) *The Fontana History ofTechnology.* Londres: Fontana Press.

Carvalho, Antonio Luciano de. (1906) Elogio historico de Candido Xavier Cordeiro. *Revista de Obras Públicas e Minas* Tomo XXXVII (442–444).

Corbey, Raymond. (1993) Ethnographic Showcases. *Cultural Anthropology* 8 (3): 338–369.

Damásio, José Vitorino. (1854) Abertura das aulas no Instituto Industrial de Lisboa; Relatório do Director interino do dito Instituto. *Boletim do Ministério das Obras Publicas, Comercio e Industria* 10.

Diário do Governo. (1855).

Diário do Governo. (1861) n° 102.

Lafuente, Antonio & Saraiva, Tiago. (1998) Ciência, Técnica e Cultura de Massas. In: Mourão, J. A, Matos, A. C and Guedes, M. E (eds). *O Mundo Ibero Americano nas Grandes Exposições.* Lisboa: Ed. Vega, 31–38.

Lisboa, Academia das Ciências de. (1924) *Escritos d'El-Rei D. Pedro V. Coligidos e publicados pela..* Coimbra: ACL.

Matos, Ana Cardoso de. (2004) World exhibitions of the second half of the 19th century: a means of updating engineering and highlighting its importance. *Quaderns d'Història de l'Enginyeria* VI: 225–235.

—. (2012) À mi-chemin entre études et " plaisir " : les visites des Portugais aux expositions universelles de Paris (seconde moitié du XIXe siècle) ". In Carré, A.L., Corcy, MS, Demeulenaere-DouYère, C. et PÉREZ, L. (dir.), *Les expositions universelles à Paris au XIXe siècle. Techniques. Publics. Patrimoines*, Paris, CNRS, 299–314.

Matos, Ana Cardoso de, Gouzevitch, Irina & Lourenço, Marta Catarino. (2010) *Expositions universelles, musées techniques et société Industrielle/World Exhibitions, Technical Museums and Industrial Society.* Lisboa: Ed Colibri.

Mattelart, Armand. (1994) *L'invention de la communication.* Découverte. Paris.

Queirós, Eça de, *Cartas de Inglaterra e Crónicas de Londres*, Lisboa, Edições Livros do Brasil, s/d.

Rasmussen, Anne. (1989) Les Congrès internationaux liés aux Expositions universelles de Paris (1867-1900). *Mil Neuf Cent. Cachiers Georges Sorel. Revue d'histoire intelectuelle* (7): 23–44.

Schroeder-Gudehus, Brigitte & Rasmussen, Annae. (1992) *Les Fastes du Progrès. Les guides des Exhibitions universelles. 1851–1992.* Paris: Flammarion.

Souto, Maria Helena & Matos, Ana Cardoso de. (2012) The 19th century World Exhibitions and their Photographic Memories. Between Historicism, Exoticism and Innovation in Architecture. *Quaderns d'Història de l'Enginyeria* 12 (57–80).

Does acculturation mean progress?

Renata Seredyńska-Abou Eid
University of Nottingham, UK

ABSTRACT: The unprecedented post-accession migration from Poland to the United Kingdom highlighted inter-cultural issues, including acculturation as an element of adaptation to a new life. This paper presents selected findings of a qualitative research study 'Translating Cultures, Adapting Lives' in an attempt to substantiate the discussion whether acculturation aimed at adaptation can be perceived as progress. The concluding remarks emphasize that the processes of acculturation in the context of migration are powerful, enriching and transformative for individuals; therefore, the experienced change and the acquisition of culture correlate with progress.

Keywords: Progress, acculturation, Polish migration

1 INTRODUCTION

Since Poland's accession to the European Union in May 2004, hundreds of thousands of Polish people decided to migrate to other EU countries in search for a better life. The largest group chose the United Kingdom (UK) as their destination. The 2011 Census revealed that Polish is the second most spoken language in England and third in Wales (Office for National Statistics 2013, BBC 2013).The unprecedented in its scale migration to Britain had a significant impact on the socio-cultural aspects of both migrant and local communities.

This article attempts to establish to what extent acculturation as an inevitable element of migration and the transition it brings to the lives of relocating individuals can be understood as progress. If the term is defined as a 'develop[ment] towards an improved or more advanced condition' (OED 2017), an immediate question arises how subjective entities such as cultural values and traditions can fit with a measurable concept.

The discussion is based on a doctoral study project 'Translating Cultures, Adapting Lives,' conducted in the East Midlands between 1st May 2013 and 31st March 2014. The mixed-methods study targeting Polish post-accession migrants was designed to collect qualitative data through semi-structured interviews, a focus group, and an online and face-to-face questionnaire. All collected data was also supplemented by ethnographic observations and a study of online communities[1] carried out between 2008 and 2014. Apart from individual interviews, a set of institutional interviews (e.g. with the Polish Consul) were conducted in order to establish the response of the authorities and adaptation support options that migrants could possibly consider in constructing their lives in the East Midlands.

2 THE CONCEPT OF ACCULTURATION

The term 'acculturation' denotes social, behavioural and psychological changes induced by inter- and cross-cultural encounters, which in the context of migration are unavoidable and may eventually lead to cultural adaptation of a migrant group. The term was initially defined by Robert Redfield, Ralph Linton and Melville Herskovits (1936: 149) as a concept that 'comprehends those phenomena which result when groups of individuals having different cultures come into continuous first-hand contact, with subsequent changes in the original cultural patterns of either or both groups.' Redfield (et al. 1936) also emphasised the distinction between 'acculturation' and 'cultural change,' as the former being an element of the latter. Nonetheless, the most popular definition of 'acculturation' in academic literature seems to be Berry's et al. (2007: 349), according to which 'acculturation' is 'a process that individuals undergo in response to a changing cultural context.' It can further result in observable changes in behaviour.

In addition to those terms, in the literature of the subject there exist other terms such as 'enculturation' and 'deculturation,' which are not just opposites as

[1] Results of the desktop research on Polish Online Communities are available in my article 'The Role of Online Communities for Polish Migrants in the UK' in Isański, J. and Luczys, P. (eds.) Selling One's Favourite Piano to Emigrate. Newcastle: Cambridge Scholars, 2011.

the use of prefixes would suggest. Edward T. Hall (1989: 229) defined enculturation as 'a process of learning culture,' which '[links] individuals to their primary cultural contexts' (Berry et al. 2007: 349) and which should not be confused with 'acculturation' – a process involving a whole group. These ideas could be lead to understanding 'enculturation' as a process of learning the first culture and situating 'acculturation' within the context of second and other cultures. Nonetheless, 'deculturation,' another related term was coined by Daniel Kunene (1968: 19) for whom this term was a neologism describing the dominance of one culture and its denial of the right of existence towards the other culture 'a necessary precedent to the process of acculturation [...] unless a cultural vacuum [was] to be presumed.'

Therefore, it can be assumed that upon arrival in a new society, migrants' 'acculturation' takes place. Berry (2005: 698) describes it as a 'dual process of cultural and psychological change that takes place as a result of contact between two or more cultural groups and their individual members.' Hence, cultural adaptation, which is an ultimate goal of relocation, can be viewed as a process of socio-cultural adjustment of migrants to the new environment. Stevens et al. (2004) also stress the psychological elements of the acculturation processes, which underpin migrants' emotional ties with both cultures, the target one and the home one. Hence, an element of transcendence is noticeable in the process of 'acculturation.' Kim (2001: 197) localises the latter within Erikson's concept of 'self-actualising people' and Ricoeur's 'transcendental ego.'

3 ACCULTURATION OF POLISH MIGRANTS IN THE UK

Although the Polish and British cultures are similar in some points and divergent in others, many of Polish migrants in the UK need to adjust to the modes of life in the destination culture. Under the condition of willingness, they can achieve those adaptational cultural objectives through opening to and accepting cultural transmission (Berry et al. 2007), which requires highly individualized and subjective processes of enculturation – referring to primary cultural contexts (Hall 1989, Berry et al. 2007), socialization – 'deliberate shaping' (Berry et al. 2007: 30) and adaptation – individuals' internal processes (Berry et al. 2007: 352). The pursued outcome is a culture change, of which one frequent form is acculturation.

3.1 *Initial cultural challenges*

In the process of acculturation, with cultural adaptation in view, initial challenges faced by migrants are a frequent phenomenon. They can clearly demonstrate the spectrum of differences between the home and host cultures. The TCAL participants categorised those issues in three groups: (1) everyday aspects that needed adjustment with hardly any option to change them (e.g. traffic, separate taps, ethnic diversity, accents or tea with milk), (2) attitudes and behaviours (e.g. mock politeness, indirectness, confusing question 'How are you?' or smile to everyone) and (3) issues that required learning and/or improvement (e.g. language, the habit of queueing or idioms and metaphors). Some participants, however, could not remember the initial difficulties (e.g. Janka) and some survey respondents admit vaguely that 'there were a few things' ('było kilka takich rzeczy').

Despite reported initial challenges and varied fulfilment of pre-arrival expectations, a majority of the TCAL participants believe that it is important to learn local customs of the host community, in line with the saying 'In Rome do as the Romans do.' This can indicate their awareness and acceptance of the potential horizontal transmission of cultural elements (Berry et al. 2007) as a method adjusting to the destination community. TCAL interviewees shared the methods they used to achieve the objective of adapting to life in the UK. Dorota summarized her learning curve by saying that '[o]n average, it was me trying to understand them' while in her further, she elaborates how curiosity and her inquisitive nature led her to gradual adjustment to the local culture. Hania, however, described the British as 'a positive and helpful nation' ('ten naród [...] jest bardzo pozytywnie i przychylnie nastawiony'). While Iwona emphasized that individuals need to 'socialise' and 'learn through experience,' Katarzyna added that one 'need[s] to take things as they have a positive attitude.' These comments explicate the results of a close contact with the host culture as potentially altering individuals' own culture present at the point of their arrival in the UK. To what extent this could be seen as progress is debatable and directly related to individual objectives of migrants.

3.2 *The significance of the English language*

A vast majority of the TCAL participants agreed that 'proficiency in English is important for [their] live[s] in the UK' (Q18) and described it as 'fundamental' ('fundamentalna') although, in their comments, many respondents admitted that they did not speak English upon arrival in the UK. The survey participants further explained that handling work without language would be challenging, e.g. 'I wouldn't be able to do my job if I was not fluent in the language' ('bez biegłej znajomości języka nie byłabym w stanie wykonywać swojej pracy') or 'especially at work' ('zwłaszcza w pracy' R...). Similarly, individual interviewees admit that English language proficiency is significant in the UK and many indicate a language barrier and communication problems as their initial challenges (Edward, Bogdan and Hania). Furthermore, Patryk and Edward would strongly recommend potential newcomers to have a certain proficiency in English; however, it needs to be noted that there is also some expectation towards the local community to learn Polish to communicate with the migrants. This view is supported by a comparison of the English and other migrant ethnogroups, such as Kurds and Arabs, who 'learn Polish to communicate with Poles in shops and in business while the English are unchangeable' ('Kurdowie, Arabowie uczą się polskiego, żeby dogadać się z Polakami w sklepie, biznesie. Angllicy są niezmienni'). This could be perceived as linguistic imperialism and anti-acculturation, which would be in line with a generalized view that 'compared to other nations Poles integrate more slowly into new societies, and in many only partially' (Dyczewski 2002: 19) and further evidence that for centuries the Polish language and culture were dominant within the Polish-Lithuanian Commonwealth (Kłoskowska 2012). Nonetheless, among the responses, such an attitude was in minority.

It is noteworthy that almost all participants declared that their proficiency in English improved since they had relocated to the UK, regardless of their length of stay in the country, though initially due to their low language proficiency some participants had to apply certain strategies, e.g. using translation, speaking a third language or learning English intensively, which clearly shows that in their migrated lives the individuals put a certain degree of effort to communicate within the host culture.

This approach is also confirmed by the Polish Saturday School in Skegness, which offers English support classes for those children who have language problems in English schools. As stated by a school committee member and a mother of one of the students, the school organization that is based on the age group and proficiency in both languages, Polish and English, seems to be highly effective from the students' point of view (SPEA). Furthermore, Moja Wyspa (My Isle), a Polish community platform, offered the English Zone, a section where language issues and linguistic intricacies can be discussed for the benefit of users. Although the Polish language is a strong and uniting cultural element of the Polish community, the eagerness to adjust through learning and improving the host language depicts a mostly integrative strategy of acculturation, which encompasses the retention of the culture of origin and exchanges with the host culture(s) (Berry et al. 2007).

3.3 *Culture acquisition through engagement with the host culture*

The TCAL participants reported that they obtain information about British culture and lifestyle from a variety of sources, including mass media, social media and social contacts (both Polish and British). While Polish platforms offer an abundance of information on Polish UK life and aspects of significance to newcomers, many websites provide information about translation services, advice on various application forms such as the national insurance number, social benefits, residence and citizenship. All information has been available in the Polish language, which clearly indicates and restrains the audience. It needs to be noted that apart from newspapers, television and radio in English, the rrsponsentd enumerated also radio broadcast in Polish, which can refer to PRL (Polskie Radio Londyn – Polish Radio London), not necessarily a radio station based in Poland. In this case, it could be seen as acculturative as the broadcast focuses on life and work of Polish people in London, though it is available in the whole country.

In order to establish further the ways acculturation occurs among Polish migrants, the TCAL participants were prompted to indicate what they have learnt in the UK, apart from the language, in terms of behaviour, customs, or habits. Some participants struggled with specifying what they have acquired in the UK, which can discern from direct responses such as 'don't know' ('nie wiem'), 'hard to say' ('ciężko powiedzieć'), or 'I don't know, honestly' (Janka). Some other participants questioned whether there is anything you can learn from the English ('Nie wiem, czego się można od Anglików nauczyć?' Edward) or made statements such as 'I don't think there is a lot to learn' (Daniel) and declared that they 'did not learn anything from them [the British]' ('Ja się od nich niczego nie nauczyłem' Edward). Nevertheless, most participants enumerated the elements of life in which they experienced changes due to their residence in the UK. In her interview, Iwona made an explicit link between the learning curve and the purpose of migrants' decision to settle in the UK by saying that '[i]f they came to improve themselves, they probably learnt the language, maybe new qualifications, attitude … a different attitude to things.' Iwona's explanation emphasizes the possible enhancement of skills and change of attitudes, which points towards the integrative and assimilative types of acculturation (Berry et al. 2007). This is further illustrated by respondent R191 who describes how they learnt manners and politeness:

> I say 'sorry' all the time, even if it is not my fault (I mean situations in shops, queues at tills or pushing on the bus and so on). I thank drivers for letting me cross the road. What else can I say … I have learnt better manners and learnt to be polite to other people.
>
> Przepraszam za wszystko, nawet jeśli to nie moja wina (chodzi mi o sytuacje w sklepach, kolejki do kas, pchnięcia w autobusach itd.) Dziękuję kierowcom za przepuszczenie mnie jako pieszej przes pasy. Co tu dużo mówić … nauczyłam się większej kultury oraz uprzejmości dla drugiego człowieka.

Moreover, the general picture of habits acquired by other TCAL respondents involves contradictions, such as (everyday) politeness ('(codzienna) uprzejmość'), 'courtesy' (Dorota) and 'smiling at strangers' ('uśmiech do obcych ludzi') as opposed to 'fake smile and mock politeness' ('sztuczny uśmiech i udawana uprzejmość'). Another example of collected contradictory data can be 'good manners' ('dobre maniery') versus 'lack of manners among the British' (Edward). In addition, while some participants indicated that a family-oriented approach is missing in the British culture, others stated that in the UK they acquired a habit of Sunday family trips ('niedzielne wycieczki rodzinne'), which are clearly contradictory. Such discordant responses emphasise the individual, subjective and background-related nature of perception as well as potential internal, from the UK perspective, differences in migrants' culture of origin. Nonetheless, those of the quoted examples that refer to acquired behaviours or habits demonstrate acculturation; however, all cited examples provide supporting evidence for the significance of reflection within the inter-cultural context of Polish migration to the UK migration.

In order to further exemplify their acculturation pathway, participants mentioned elements that in the inter-cultural comparison were classified as more British features, i.e. 'assertiveness', 'calmness (in solving problems),' 'going to a pub after work' ('wychodzenie do pubu prosto po pracy na drinka'), or small talk ('rozmowa o niczym'). In addition, many responses specify typically British habits, such as 'queuing' or its variation in the form of 'not pushing the (bus) queue' ('nie wpychać się w kolejkę do autobusu'), a very frequent use of phrases such as 'thank you,' 'sorry' and 'please' or greeting people with the question 'How are you?' Aspects such as cuisine and celebrations were also exemplified among the responses. The participants itemized traditions and celebrations such as Easter egg hunt, Red Nose Day, Pancake Day, Christmas parties or 'Indian celebrations' while cuisine elements included English breakfast, 'tea with milk,' 'afternoon tea' or 'pudding for Xmas.' A few participants also reported that they rather celebrate Halloween or 'walk on the streets and watch vampires' than follow the traditions of All Saints Day (1st November). The last example shows a more conscious approach to acculturation as the respondents compared their cultures, i.e. the culture of origin and destination while all the specified aspects of culture that Polish migrants acquired in the UK show acculturation by active engagement in the host culture and do not necessarily testify to the denial of the culture of origin.

Furthermore, some participants declared that they 'learnt to distinguish among races, nationalities and ethnic groups' ('rozróżniać rasy, narodowości i grupy etniczne') while others learnt openness ('otwartość' Hania). Further examples relate to very individual experiences. While 'buying your own drinks at weddings' ('kupowanie sobie drinków na weselach') can be seen as a clear example of unfulfilled expected similarity (Jandt 2016), '[s]nort[ing] cocaine' could be categorised as a modern international social problem.

Moreover, acculturation through engagement with the target culture was also tested by prompting the participants to share their memories of their participation in the Diamond Jubilee of Queen Elisabeth. The celebrations of 2012 did not arouse much interest among survey respondents, some of the interviewees reported that they joined in the celebrations. Although Bogdan emphasized only the fact that there were a few days of public holiday, which for him and his family resulted in 'a trip to York' ('pojechaliśmy na wycieczkę […] do Yorku' Bogdan), Hania in more detail described how they celebrated at work. She reported that they had 'cakes' ('ciasta upieczone'), which 'raised money for charity' ('pieniądze poszły z tego na charity'), and that 'the whole ward was watching the television coverage till midnight' ('[transmisja] tak, oczywiście. Cała, cała. Na oddziale wszyscy chyba do północy').

Another example, testing acculturation by participants' willingness to introduce British Christmas traditions into their Polish celebrations, expectedly generated varied responses that ranged from 'none' ('żadnych') through more tentative 'rather none' ('raczej żadne') and 'I don't know' ('nie wiem') to particular examples and comments such as 'all' ('wszystkie') or 'already done.' Those eager to incorporate new customs exemplified such traditions as 'Boxing Day,' 'minced pies,' 'Christmas pudding,' 'Christmas/Xmas crackers,' or '(roast) turkey' ('pieczenie indyka' or 'zrobić indyka w piekarniku') On the other hand, limited acculturation, or possibly its extended process, can be inferred from comments that explicate lack of engagement, e.g. 'I don't know their traditions' ('nie znam ich tradycji' or 'as for me, nothing fits: turkey???, boxing day???' ('nic mi nie pasuje: indyk???, boxing day???').

3.4 *Impact of acculturation on Polish migrants*

The most noticeable impact of acculturation is observable in terms of the Polish language. Linguistic borrowing and contextual adaptation of terms seem to be the most noticeable signs of attrition. The first example depicts how the simple language of an everyday conversation can pose a challenge to the user. In the first example observed in 2008, in an instruction a father was giving his young daughter, the Polish word sok (juice) was replaced by its English equivalent, i.e.

> Father: […], no pij juice.
> ([…], drink your juice)

Such a replacement was permissible due to the simple structure of the sentence and the availability of a single word equivalent. It may have happened because it was convenient for the speaker in that particular situation or because at that moment he was incapable of producing the Polish word. While such interference may not affect the speaker, it may have a lasting impact on the language development of the young daughter, the recipient of the message.

Another example of lexical borrowing can have more lasting communication impact as it represents the formation of neologisms. Namely, one of the cultural adaptation questions tackled the issue of separate taps in the UK. The responses included the option of a mixer tap (bateria z mieszaczem). One of the participants in his comment on the issue stated:

> Nie przyzwyczaiłem się [do oddzielnych kranów]. Montuję miksery.

(I have not got used to [separate taps]. I fit in mixers.)

By using the word 'miksery' ('mixers') the speaker created a neologism as in Polish the word 'mikser' refers to a 'kitchen blender' rather than a mixer tap. The latter is simply called a tap. Such a situation could have happened as the speaker wanted to take a linguistic shortcut in explaining his approach to the issue. Nonetheless, such a use of Polish in Poland would likely result in some misunderstanding while in the UK situation, it seems to be justified by the context.

Contextualised adaptation of terms that do not exist in one of the migrants' languages seems to be a natural process of cultural adjustment in the host country. The participants of the TCAL study were asked to evaluate the use of two terms that are commonly used by Poles in the UK: (1) w UKeju (in the UK) and (2) Jeżdżę dablem (I drive a double-decker).

The first example, w UKeju (in the UK), represents the adaptation of an English term to reflect the specificity and precision of the name of the country. In Polish, similarly to many other European languages, the names England or Great Britain are frequently used with reference to the UK, which can be offensive through its exclusionary effects. Hence, Polish migrants started using the original term in a partly polonised form, i.e. often the pronunciation of the letters remains English /ju: keɪ/ while a relevant Polish inflection for a locative is added, hence w UKeju /wju:keju/. It needs to be noted, though, that a reverse process does not take place, i.e. when speaking English, Poles do not tend to say in Polska (Poland).

The other example also depicts the practicality of adapting work-specific terms by Polish bus drivers in the UK. Since there is no equivalent in Polish as double-decker buses are not common, one would have to use a complex descriptive term, such as 'autobus piętrowy' (a multi-storey bus) or 'autobus dwupoziomowy' (= a two-storey bus). Therefore, Polish drivers in the UK adapted the local term and polonised the spelling (i.e. 'dabl' or 'dabel') and the sounds. It seems that such borrowings are unavoidable in migrants' adapted lives, as through their context-specificity they tend to improve communication with the host community and within the prescribed community of practice. These examples, however, may be categorized as the linguistic habitus, which Thompson (in Bourdieu 1991: 17) defines as 'a sub-set of dispositions acquired in the course of learning to speak in particular contexts.' Nonetheless, the context-specific adaptation of terms faced less understanding among many TCAL respondents who, with two exceptions of Katarzyna and Patryk, were rather disapproving and emphasised an apparent lack of possibility for such lexical adoptions.

4 PROGRESS?

The results of the TCAL study clearly show that the English language is important for migrants in the lives in the UK while their acculturation occurs subjectively and at many levels. To what extent these indicate progress depends on set objectives and the scale of the fulfilment of pre-arrival expectations. Although most participants emphasised the significance of the English language to their migrated lives and declared that their proficiency in English improved, neither in this nor in other studies on Polish migrants did the respondents recognize the English language as their socio-cultural capital, as defined by Bourdieu and Coleman (Halpern, 2005), that could define their future prospects. The language has not been perceived, perhaps not yet, as a potential door opener in terms of migrants' future professional and personal opportunities. This does not mean that there is no progress for the concept includes re/construction, not only discovery. Hence, improving or acquiring another language is progressive for the individuals involved as it transforms their lives, enriches their existence and opens future opportunities. Although individuals may not realise the potential, future developments in their lives may prove the progressive nature of their path.

Regarding language interference and the influence of the target language on the native one, lexical borrowing and adaptation are well known processes in the historical development of languages. It can be argued that context-specific lexical borrowing refers to the concept of identity reconfiguration and the diachronicity and and multi-layeredness of culture in the intercultural context of migration. Hence, similarly to acquiring customs and traditions, linguistic changes may constitute progress in that identities evolve rather than new forms supersede previous identities.

5 CONCLUSION

The acculturation and adaptation processes among Polish migrants in the UK occur at multiple levels and are therefore individual and subjective. While motives for relocation, expectations towards the destination culture and identity issues play a certain role in shaping the attitudes of migrating individuals, migrants' knowledge of the host culture, their sources of information about local lifestyle and reflection upon initial challenges faced in the target culture are crucial elements of their acculturation and pursued adaptation. Although most participants feel they have adapted, their engagement in the destination culture and involvement in matters of the Polish community vary. Nonetheless, the process of acculturation cannot be prevented and therefore evolutionary changes in the cultures of individuals take place while their identities are reconfigured and reconstructed. Since acculturation has a transformative power and leads to an enriched life, it can be assumed that, a broader cultural range developed by individuals equips them to handle their lives more successfully. Thus, if acculturation means empowerment, transition and enrichment, it can be perceived as progress.

BIBLIOGRAPHICAL REFERENCES

BBC (2013) 130,000 *Speak No English* – Census. [online] 30 January. available from <http://www.bbc.co.uk/news/uk-21259401>

Berry, J.W. (2005) 'Acculturation: Living Successfully in *Two Cultures'. International Journal of Intercultural Relations* [online] 29, 697–712. available from <http://www.sciencedirect.com/science>

Berry, J.W., Poortinga, Y.H., Segall, M.H., and Dasen, P.R. (2007) *Cross-Cultural Psychology*. Cambridge: Cambridge University Press.

Bourdieu, P. (1991) *Language and Symbolic Power*. Cambridge, MA: Harvard University Press.

Dyczewski, L. (2002) 'Values and Polish Cultural Identity'. in *Values in the Polish Cultural Tradition: Polish Philosophical Studies*, III. ed. by Dyczewski, L. Washington: The Council for Research in Values and Philosophy.

Hall, E.T. (1989) *Beyond Culture*. New York: Anchor Books.

Kim, Y.Y. (2001) *Becoming Intercultural: An Integrative Theory of Communication and Cross-Cultural Adaptation.* Thousand Oaks and London: SAGE.

Kłoskowska, A. (2012) *Kultury Narodowe U Korzeni*. 2nd edn. Warszawa: Wydawnictwo Naukowe PWN.

OED (2017) 'Progress'. in Oxford English Dictionaries [online] available from <https://en.oxforddictionaries.com/definition/progress>

Office for National Statistics (2013) *Language in England and Wales*, 2011.

Redfield, R., Linton, R., and Herskovits, M.J. (1936) 'Memorandum for the Study of Acculturation'. *American Anthropologist* [online] 38 (1), 149–152. available from <http://onlinelibrary.wiley.com/doi/10.1525/aa.1936.38.1.02a00330/abstract>

Birth of Cape Verdean man in the Writing of Jorge Barbosa

Hilarino da Luz
CHAM, FCSH, Universidade NOVA de Lisboa, Lisboa, Portugal
Universidade dos Açores, Portugal

ABSTRACT: Jorge Barbosa described Cape Verdean culture as being Luso-African. According to his approach, and according to history, the archipelago of Cape Verde has a distinctly local culture, in whose roots African and European elements can be found, with equal importance. In Cape Verde, there was thus the fusion of African and European (mainly Portuguese) cultural traditions. Owing to its history, geographical location and some sociological constraints, the archipelago is one the most wonderful experiences of the "fusion of opposite souls", which resulted in a unique and harmonious people.

In order for this to take place, there was the will of the Portuguese who, while seeking to solve their domestic problems, disembarked at new horizons, adapting to different situations and playing a part in the birth of new peoples, as can be seen in the author's unpublished column "Comemorações de Camões" (Celebrations of Camões).

Jorge Barbosa addressed the initial reality of the islands, since the arrival of the settlers to the birth of the Creole "soul". In his view, everything sprang from the bold Portuguese spirit which, even though it faced huge storms, increased its kingdom in an epic way, as can be read in the poem "Expansão" (Expansion), in which the author praised Portugal for its expansionist vision.

Keywords: Cape Verde; Jorge Barbosa; social experience.

This article intends to deal with some of the ways the history of Cape Verde is reflected in the writing of Jorge Vera-Cruz Barbosa, also known as Nhô Jorge. Barbosa was born in the island of Santiago on 22 May 1902 and died in Portugal, in Cova da Piedade, on 6 January 1971, from a heart condition. His literary work includes poetry, Crónicas de S. Vicente, letters, some articles, short stories and an unpublished novel called Bia Graça, which he left unfinished.

According to his approach, and according to the history of Cape Verde, this archipelago has a distinctly local culture, in whose roots African and European elements are found at pretty much the same level. In fact, Cape Verde, a small insular state composed of ten islands and some islets, was found uninhabited by navigators at the service of the Kingdom of Portugal; this happened during two journeys and the islands were found in two groups, between 1460 and 1462. The first group – made up of the Eastern islands (Santiago, Fogo, Maio, Boa Vista and Sal) – was discovered in 1460 by António de Noli, while Infante D. Henrique was still living. Therefore, although the name of its discoverer, Diogo Afonso, was only found in the royal charter of donation of 29 October 1462, the second group – made up of the Western islands and some islets (Brava, S. Nicolau, S. Vicente, Santa Luzia, Santo Antão and the islets Raso and Branco) – was discovered at the end of 1461 and beginning of 1462, by order of D. Fernando, the brother of D. Afonso V.

Therefore, there was in Cape Verde the fusion of African and European (mainly Portuguese) cultural traditions. This process began with the island of Santiago in 1462, owing to its excellent geographical location and abundance of water. The second island to be peopled was Fogo (1480-1493), as it was geographically close to the island of Santiago. It offered good conditions for the production of cotton, one of the main products sold along the coast of Guinea-Bissau. In 1490 the island of Boa Vista began to be peopled with shepherds who were made responsible for the livestock that existed there, just as the island of Maio when Rodrigo Afonso received it as captain-donee of his uncle Diogo Afonso. In the beginning, all that existed was the cultivation of cotton and the breeding of livestock by some slaves. In 1545 João da Fonseca began to people the island of Brava. The island of Santo Antão began to be peopled around 1548, at which time it was granted to Pêro da Fonseca by D. João III.

That peopling was not effective, which is why at the end of the sixteenth century there were few people on the islands belonging to the Barlavento group. The island of S. Vicente appears as a special case within this context, as it did not offer conditions favourable to the establishment of a permanent population. Only the transatlantic crossings of the end of the eighteenth century and beginning of the nineteenth century would awaken the interest of international navigation in the island. The first proposal to people that island was drawn up in 1734, after several failed attempts, including a plan by Queen D. Maria II in 1838 that intended its settlement to be made with white people from the Azores and Madeira, owing to the difficulties felt in agriculture there. Only in the mid-nineteenth century was its settlement effective and definitive, with Cape Verdeans from the islands of Santo Antão, São Nicolau, Boa Vista and Maio. Although it was peopled late, it quickly achieved great importance in the economic and cultural life of the archipelago, because of its Porto Grande (Big Port).

In that way, owing to its history, geographical location on the Western African coast (at about 500 km of the promontory that gave it its name) and sociological constraints, Jorge Barbosa considers the archipelago to be an experience of the "fusion of opposite souls", which resulted in a specific human people (the Cape Verdeans). For that to happen, there was the will of the Portuguese who, in seeking to solve certain problems, moored at new horizons, adapting to different realities and playing a part in the birth of new peoples, as can be read in the column "Celebrations of Camões", when the author mentions that: "Every Portuguese man held within him a latent sense of national expansion, of Lusitanian projection, because Portugal, a poor country, needed riches to establish its kingdom and defend its nationality" (Barbosa, 1952, 11).

From that point of view, Jorge Barbosa approached the very initial reality of the islands, since the arrival of the settlers to the birth of the Creole "soul". According to him, everything sprang from the bold Portuguese spirit which, even though it faced huge storms, increased its kingdom in an epic way, as can be read in the poem "Expansão" (Expansion), in which the author praised Portugal for its expansionist vision. That praising view is found in the refrain: "Small and heroic/Lusitanian Kingdom"; and the author describes that kingdom as being "heroic", even though it has been crushed by Castile and the "dreadful sea": "It would seem to have been crushed/between brave/crafty Castile/and the rampart of waves/of the dreadful sea" (Barbosa, 2002, 311). Therefore, the small size of the Portuguese territory limited the lives of the Portuguese people, which is what had led them to expand it: "Scarce ploughing land/little land/so little/the Kingdom no longer had/more land/suitable to plough" (Barbosa, 2002, 311).

The same way he highlighted the Portuguese expansion, he also pointed out the consequences of the Portuguese reasons for setting out to sea, thus managing to expand their empire, when he mentioned that: "Thus the Lusitanians/set out to sea/ploughing and overcoming distances afar. //Thus they expanded/the little land/of the small and heroic/Lusitanian Kingdom/into a scattered/and colourful empire" (Barbosa, 2002, 311).

As this was a text praising the expansionist history of Portugal, which increased its empire, we wonder about what might have led a "claridoso" realist, who would prove to be a "pamphleteer", a man ahead of his time, to write a poem of this nature, with no reference to independence. What that approach shows is that he felt he was an integral part of that "Lusitanian Kingdom", even if, along with the other "claridosos", he fought for Cape Verdean sociocultural affirmation as a unity of the "scattered/and colourful empire". The poem discloses a dubious and unknown side of the author, presumably a consequence of particular features of the history Cape Verde, found by sailors at the service of D. Afonso V who, using "bulgy caravels", moored there to the satisfaction of the King and "misfortune" of the Cape Verdeans; this gives the impression that he would prefer the islands to be discovered by other powers, as can be read in the poem "Descoberta" (Discovery): "Thus one day/the islands were discovered/for the glory of King/Afonso, the African/and our misfortune" (Barbosa, 2002, 310).

According to the author, that discovery, which occurred "millennia" after the geological formation of the islands took place, was made by strong and brave men, scorched "by the sun and south wind". They left from the "Promontório de Sagres" (Sagres Promontory), the core of Portuguese maritime expansion; those men were still rough, had long beards and traces of salt on their faces, as can be read in this poem: "Then they spun/millennia and stars/exact in space/until the bulgy caravels/arrived. // They were seamen who came/from the Sagres Promontory/men with wide chests/brave burnt/by the sun and south wind. // People of a rough tongue/and long beards in the wind/bearing invisible traces/still humid/of the ocean foam" (Barbosa, 2002, 309).

In seeking to understand the several aspects concerning the existence of the Cape Verdean man, it is important to mention that his destiny began with the departure of a ship from the aforementioned promontory. This ship carried those Europeans who contributed to the settlement of the archipelago, which was uninhabited when the settlers arrived. The latter only found carnivorous "birds of prey" with curved, sharp beaks, strong claws and long-distance vision. They also found vegetation brought by the birds when there were strong winds, as can be seen in the poem "Prelúdio" (Prelude), dedicated to his friend António Aurélio Gonçalves: "When the discoverer arrived at the first island/there were no/naked men/or naked women/peeping/innocent and fearful/from behind the vegetation. // Nor poisonous arrows crossing the air/nor cries of alarm and of war/echoing across the hills. // There were only/the birds of prey/with sharp claws/the seabirds [...] // When the discoverer arrived/ [...] // at that/at that initial hour/this destiny still of us all/began to be fulfilled" (Barbosa, 2002, 99-100).

Jorge Barbosa wished to say that "Cape Verdeanity, as a result of the arrival or departure, exists; or rather, is found in the existence, in the continual becoming, in the centuries-old sedimentation of the parts in presence, both from the ethnic, cultural and geographical points of view" (Veiga, 1989, 1). The poem "Panorama" (Panorama) suggests that people from all walks of life took part in the sociocultural formation of the country, such as: "burnt sailors/pirates, slaves, /adventurers, /convicts,/noblemen, slave traders, /donees of the Islands,/Chief Captains" (Barbosa, 2002, 36).

All those people will have been taken by a "ship" that, because of the influence of slave trade, the author called "slave ship". They were carried in large numbers and squashed onto the holds so that they could be sold and expand the Portuguese empire. That project devastated many families and societies, as slaves were taken away from their country, family and friends, having no freedom of choice. Some stayed in the archipelago and took part in the establishment of its society, by constantly socializing, as can be noticed in a long poem, "Relato da Nau" (Report from the Ship), about a ship that was driven by the wind: "It used to be/the first slave ship/heading towards the Archipelago/being swiftly driven/by the trade winds. // It had an ample paunch/just as befitted/the transport of hundreds of captives/squashed on the holds./The ship traded slaves. However, it had/something grand about its high masts/with its crossed yards. //The white sails puffed up/by the salty blows of the breeze/gave it the arrogant/and round outline/of the breasts of pigeons. // Unsinkable/and unbreakable/the ship held in its sway/the weight of gales and of waves/ [...]" (Barbosa, 2002, 420).

The slaves were carried in appalling conditions; they met their physiological needs on the holds of ships, where they also had a lot of trouble breathing: "The ship had the odour/of genitals/faeces/and micturitions/that crept up nauseously/from the hold/where black men held captive along the African coast/were carried" (Barbosa, 2002, 421).

Those journeys were marked by unforeseen events, due to the obstacles presented by nature. The ships in which people travelled faced cyclones and storms of the "equinox" that sometimes made them sink. Those weather setbacks would dismantle the "ship", by means of a harsh blow, leaving it "adrift" for "three days and three nights". During that time, its experienced captain sought to defend it from the avalanches and whirlpools of the waves of the sea until the wind calmed down, which would only happen at sunset.

Many of the slaves died of anguish, thirst, hunger and suffocation. Those were thrown into the sea, vertically and without receiving "spiritual protection". The procedure for getting rid of the bodies was very arduous, which made the captain distribute a "pint of rum" and a "naked female slave" to his crew, as prizes. After handing them their prize, he would go to the "deck", having also lighted "seven candles" corresponding to the "seven days" of the week to "Our Lady of Navigators", a very common image in Portugal during the fifteenth century, whom navigators would call upon to ask for protection in order to return to their homeland alive: "He would only stay on deck/to look out. // In front of/Our Lady of Navigators/in the cabin, he lit/seven candles in a circle/the first of the seven that would then be/seven times/seven days of the week" (Barbosa, 2002, 423).

The dismantling of boats was very common while carrying slaves to the archipelago. That gave rise to thousands of deaths; those who survived went to America and Europe and were also used in the settlement of that small insular territory, namely through intercourse with white Europeans, leading the author to state that: "Thus moored/the first batch/that came captive/for the settlement of the islands" (Barbosa, 2002, 423). The first Cape Verdean man was born in the city of Ribeira Grande (currently, Cidade Velha), on the island of Santiago. Then a later expansion to the remaining islands took place, as previously mentioned. That city, considered the birthplace of the Cape Verdean Nation, started to decline because of frequent pirate attacks. The inability of Cape Verdeans to hit the "agile" thieves, who were well prepared, made that important trading centre start to decline economically: "At the time of the pillages/there were the noisy/war weapons/loaded from the muzzle/used to shoot from the top/of the parapet of the port/at the pirate ships/that would come to attack. // [...] // They did not have the precision/nor the movement/essential/to take aim/neither did Sir/the First Captain know/the complex secrets/of ballistics. // [...] // After all, the mad/agile pirates/would jump along the unguarded beach/with whistles and screams/the clinking of swords/and the plunder would quickly/take place" (Barbosa, 2002, 307–308).

Having that city as his place of origin, the Cape Verdean man resulted from a centuries-old union of "two opposite souls", fruit of a long and enduring process, which generated a single being, descending from the "blood of the African slave" and from the "adventurous" European, as Jorge Barbosa mentions in the poem "Povo" (People): "A clash within a single soul/of two opposite souls/searching for each other, blending into each other/into a centuries-old fusion; // a clash within a single blood/of the strong African blood/with the adventurous blood/of the men of the Expansion; // a clash within a single being/of two poles in contact/in the insistent projection/of many generations [...]" (Barbosa, 2002, 46).

That fusion triggered in the Cape Verdean the indecision about whether he is African or European, giving the impression of the apparent lack of a cultural identity, although Jorge Barbosa highlighted the latter when he referred that "in the soul of the people remained/ this deep anxiety/ – something undecided/between the tropical climate/and the mirror of Portugal" (Barbosa, 2002, 46).

Gabriel Mariano believes this indecision (Portuguese or African) was created by Cape Verdean students living in the Metropole. According to him, to approach the question that way means to reject the specificity of the process of cultural miscegenation that occurred in the archipelago, as the following passage from his article "Negritude and Cape Verdeanity" attests to: "Cape Verdean students in the Metropole have been divided over the regional character of the Creole people: Portuguese or African? [...] In fact, once the question has been formulated, the answer it immediately triggers is the following: either Cape Verdean or Portuguese or African. However, this would only be exact if human peoples would repeat themselves as identical copies. If they would repeat themselves in all their details when transplanted from their environment of origin to different environments. Furthermore: to put the problem in those dilemmatic terms (either one thing or the other…) is to reject out of hand, and without any scientific grounding, the appearance of culturally mixed expressions or the opportunity for them to appear.

It must be noted that, even though the author expresses that indecision, in the poem "África" (Africa), written in 1935, he shows us his awareness of being African in the most varied aspects of his everyday life and history. Therefore, he also speaks of the different wounds inflicted upon the Africans by the major colonial powers that, besides enjoying the local riches, conditioned the lives of the population with the imposition of their power. The author also deals with some cultural aspects, namely art in which artists use wood to carve "strange animals and divinities": "AFRICA!/the wounds in your body have not yet healed/from the hard-fought/battles of rapacious attacks/made for the glory of the conquests/and for the prestige of the empires! // Threats still hover/from the dominating/masters/of your destiny! // […] // AFRICA!/in your fauna there is every marvel of creation/every concert, every sound, – enraged howls/[…] // AFRICA!// from your drums/in whose war-like and barbarian rhythm/I picture agile, tattooed black men/dancing and fighting. // […] On the banks of your rivers,/under the shadow of your jungles/under the heat of your sun,/in your fertile womb,/the energies of your race sleep,/until the panting hour comes,/the bugle hour of your triumphant morning!" (Barbosa, 2002, 334–336).

Therefore, besides presenting us with Cape Verdean issues, the poet also shows us cases as those mentioned in the poem "Ocorrência em Birmingham" (Incident in Birmingham), where he describes a racist episode that happened in a tobacconist: "John/from Birmingham/Alabama/USA // entered the tobacconist. // He was insulted/punched/and kicked out. // Out in the street/the policeman/beat up/ knocked down/spat on/and arrested the troublemaker.// Rogue black!" (Barbosa, 2002, 316).

The main character of the text has a single name and had a tragic end because he crossed the border between the black man and the white man in the South of the United States. How that action unfolded resulted from the fact that John was considered "black", therefore "rogue". That approach to racism reproduces a serious problem that black people faced in the United States and was echoed by the reflections made by the author around questions of identity, acculturation and social justice. All in all, Jorge Barbosa takes the long acculturation process at the origin of the Creole man as a point of reference, while simultaneously revisiting the idea of a dominated and imperialist Africa.

BIBLIOGRAPHICAL REFERENCES

Amaral, Ilídio de. (1964) *Santiago de Cabo Verde: a terra e os homens*. Lisboa: Junta de Investigação do Ultramar.

Barbosa, Jorge. (1952) Comemorações de Camões in Cabo Verde, 34, 11.

—. (2002) *Obra Poética*. Lisboa: INCM.

Barcelos, Cristiano José de Sena. (1899) *Subsídios para a História de Cabo Verde e Guiné*, parte I. Lisboa: Tipografia da Academia Real das Ciências de Lisboa.

Luz, Hilarino da. (2008), *Jorge Barbosa: o claridoso realista*, Dissertação de Mestrado apresentada à FCSH, UNL, 2008.

—. (2013), *O Imaginário e o Quotidiano Cabo-verdianos na Produção Literária de Jorge Barbosa*, Tese de Doutoramento apresentada à FCSH, UNL.

Mariano, Gabriel, Negritude e Cabo-verdianidade in Cabo Verde, 104, 6–8.

Pereira, Daniel A. (2005) *Estudos da História de Cabo Verde*. Praia: Alfa Comunicações.

Santos, Elsa Rodrigues. (1989), *As Máscaras Poéticas de Jorge Barbosa e a Mundividência Cabo-verdiana*. Lisboa: Caminho.

Veiga, Manuel. (1989) *Signos e Símbolos em Jorge Barbosa: uma tentativa de análise semiológica*. Praia: Instituto Cabo-verdiano do Livro e do Disco.

Community empowerment and progress in Africa – notes from the field

Ana Maria Martinho Gale
CHAM, FCSH, Universidade NOVA de Lisboa, Lisboa, Portugal
Universidade dos Açores, Portugal

ABSTRACT: This paper presents information on how progress and development in Africa have been discussed at the national, regional and international levels in recent years.

It aims as well at presenting examples of good practices in terms of localized community experiences. The cases chosen highlight the importance of cultural and political heritage, transmission of knowledge, and generational interactions.

We focus on the case of Guiné-Bissau, and on the role of culture and community building as means to encourage the discussion and implementation of new solutions for old and recent problems.

Keywords: Africa, development, Guiné-Bissau, culture.

1 PROGRESS IN AFRICA

When discussing how African countries deal with the regulation of progress, it is very common to simplify this issue by looking at the international indexes on human development and crisis management. This view often hides endogenous readings on reality, mainly in two ways: how localized practices of progress and control of responses to conflict operate; how recent events, mainly since the late 80s, determine the relation between international commitments and national realities.

Some case studies as presented by other researchers and by myself prove that African realities, when considered through a perspective based on community practices, have found through time responses to principles of governance and to national challenges.

Recent cycles and levels of higher education training in and outside the continent show that capacity building is a reality and that development occurs often against all odds even in countries struggling with democratic institutionalization.

In a discussion paper presented to the DPADM (Division for Public Administration and Development Management United Nations Department of Economic and Social Affairs) in November 2005, Alphonse Mekolo and Valentina Resta, its authors, remind us that

> issues of governance are not new in Africa and that the application of governance principles was often disturbed by wars as well as socio-political or economic crises affecting some countries on the continent. (. . .) this paper underlines efforts made by African countries in implementing principles of good governance while at the same time facing the challenges of development which they experience with acuity since the 1990s.
> (unpan1.un.org/intradoc/groups/public/documents/un/unpan021509.pdf, p. 2)

As the UN's Millennium Development Goals (http://www.unfoundation.org/what-we-do/issues/mdgs.html) launched a path towards the pursuit of common grounds on core issues such as the eradication of extreme poverty and the pursuit of literacy in its many fronts, Africa responded to this challenge through the involvement of institutions such as the African Union (AU) in programs addressed specifically to comply with these criteria.

Today, the AU is working towards the 2063 agenda combining past, present and future actions. One might say that the basic challenges have yet to be met in many countries and regions. Still, this shows how the internal discussion is being held in the continent and how the AU members are working together. The general plan for this agenda reads as follows:

> It is a strategic framework for the socio-economic transformation of the continent over the next 50 years. It builds on, and seeks to accelerate the implementation of past and existing continental initiatives for growth and sustainable development.
>
> Some of the past and current initiatives it builds on include: the Lagos Plan of Action, The Abuja Treaty, The Minimum Integration Programme, the Programme for Infrastructural Development in Africa (PIDA), the Comprehensive Africa Agricultural Development

> Programme (CAADP), The New partnership for Africa's Development (NEPAD), Regional Plans and Programmes and National Plans. It is also built on national, regional, continental best practices in its formulation.
> (https://au.int/agenda2063)

Though its Flagship Projects seem to be more focused on trade and economic development, the project of free circulation in the continent and the E-networks to be created, as well the expected "Silencing of the guns" may bring hope to cultural and educational development with impact on regional and local dynamics.

Africa is moving and it is doing so at a very fast pace. In July 2017 the Mo Ibrahim Foundation (MIF), an institution that has been working towards Good Governance in Africa, is set to sign a Memorandum of Understanding (MoU) with the African Peer Review Mechanism (APRRM) on the margins of the Twenty-Ninth Ordinary Session of the Assembly of the African Union in Addis Ababa, Ethiopia, on 4 July 2017. The Mo Ibrahim Foundation becomes this way the third Strategic Partner to sign the MoU besides the African Capacity Building Foundation (ACBF) and the United Nations Economic Commission for Africa (UNECA). This Foundation epitomizes the role of African structures that focus on the support to good practices at the highest level. It encourages responsible leadership through its Prize which

> is awarded to a former Executive Head of State or Government by an independent Prize Committee composed of eminent figures, including two Nobel Peace Prize Laureates
> (http://mo.ibrahim.foundation/prize/)

This Prize has been awarded to leaders such as Pedro Pires (Cabo Verde), Nelson Mandela (South Africa), Joaquim Chissano (Moçambique) and rewards contributions to the consolidation of democracy, social and economic stability. It sends in fact a very powerful message towards change and critical analysis of the continent's reality. At the same time, it lists the identification of opportunities and threats Africa is facing.

One of the main groups that it chooses to favour is youth. The Report "Africa at a Tipping Point" lists the many risks for the continent mostly due to the lack of opportunities for young people.

Considering that 60% of the population is under 25, the threats are as challenging as the opportunities.

> Today, 60% of the continent population is already under 25 years of age. By 2050, Africa will be home to 452 million people under the age of 25. Their drive, ambition and potential provide African countries with an extraordinary asset. But this demographic dividend is at risk of being squandered.
> (http://mo.ibrahim.foundation/news/2017/creating-economic-opportunities-young-africans-continents-urgent-challenge)

Prone to the impact of violence and terrorism, young people face a difficult future even though they show growing levels of literacy and access to higher education, often outside of the continent.

The extensive list of joint declarations, partnerships and agreements seem to cover all that is needed as far as the identification of strategic steps. Some case studies also help understand what can actually work through such programmes. As pointed out in the report quoted above under the DPADM, The Africa Governance Forum (AGF), The Africa Governance Inventory (AGI), The African Peer Review Mechanism (APRM) and the The Brussels Programme of Action (BPOA), just to name a few, show how much has been done through time (1997–2015) in order to cover convergence, monitoring and accountability in governance.

Though extensively ratified, other conventions, on Human Traffic, Environment and Cultural Heritage, are for different sets of reasons facing challenges that need more answers at the regional and continental level. (http://www.africaprogresspanel.org/publications/policy-papers/2015-africa-progress-report/)

Guiné-Bissau is, as I see it, and considering my experience of fieldwork in the country, a potential case study that can bring some light on the critical combination of challenges, threats and achievements in African countries today.

2 GUINÉ-BISSAU: CHALLENGES AND CURRENT SITUATION

It is a fact that many countries are more obviously exposed to the many threats pointed out above. Guiné-Bissau has been dealing with extreme political unrest, electoral problems due to ethnically and economically induced violence, corruption.

Seen as a failed state in western media and many international institutions, namely due to the exposition to the drug traffic, Guiné-Bissau represents a somehow radical case study concerning the premises exposed above.

If in fact the country has been struggling with recurrent conflicts, mostly politically based, it is also true that the response to this reality exists and is coming from a variety of sources. Such reaction, mostly cultural, is connected with activities pertaining community empowerment and youth movements.

These movements are present in particular in the reactions of bloggers, rappers, NGO leaders, film directors and musicians.

Bloggers and journalists such as Didinho, Adulai Djaló "Lai", Suleimane Djaló and Umaro Djau (see http://bambaramdipadida.blogspot.pt) have been pointing out the many dangers the country is facing and have also brought their claims to international attention. As Djau refers in a recent text

> Thus, the challenge of finding all the root causes of our problems is on us, all of us. But, will you

> dare to look around and challenge yourself as a true patriot on behalf of every Bissau-Guinean, regardless of their social, cultural, and religious background or will you be willing to flush down the river everything for which all those people who came before us stood for, fought for, and lost their lives for? And finding all the root causes has to start with a true understanding of our social fabric – including values and assumptions – which gives meaning to our history and existence as a pluralistic nation.
> (http://www.gbissau.com/2016/09/14/guinea-bissau-reflecting-on-behalf-of-future-generations/)

In the words of Adulai Djaló the political independence has fallen short in its results due to the lack of commitment to the future generations.

> A independência política sem a procura para a autogestão dos recursos (humanos e materiais) o país não se alavanca rumo a independência econômica, e muito menos sonha com a prosperidade. Triste é de constatar o futuro nebuloso para a sociedade guineense presente e para a geração vindoura. [Political Independence without the search for autonomous management of human and material resources will not drive the country towards economic independence and it will not allow prosperity. It is sad to see how the future is gloomy to the present Guinean society and to the generations to come] (http://www.gbissau.com/2016/12/06/opiniao-desunidos-para-nos-autodestruirmos-e-adiar-o-futuro-da-geracao-vindoura/)

Responding to a recent speech from the President, José Mário Vaz, in which he suggested that young people should go back to rural activities and the fields, to prevent foreigners from taking over, Djau challenged the timing of this idea, considering that what the country needs is in fact that opportunities are equally open to all regardless of their ethnic or socioeconomic origin.

This is a very relevant claim, since the cycles of poverty in Guiné-Bissau are directly related to the access to school, technology and resources, which are not usually as functional in rural areas. But in the President's mind, the occupation of the land seemed to be a key factor. Are the peoples of Guiné-Bissau going to forget where they came from as they move to the cities? It is very unlikely, so far the sense of belonging and cultural heritage is undisputed in this country.

Lineage is a fundamental matter and oral history still keeps clear track of family and collective memory. (see Christoph Kohl: 2010, 92)

Still, the presence of foreigners does raise a lot of concerns in rural and fishing communities but that has a different expression as I expect to show ahead.

The "Lei de terra", concerning the regulation over its uses and property, seems to be in fact a strong expectation on the part of the population. (see http://www.radiojovem.info/2017/07/26/sobre-os-jovens-o-regresso-ao-campo-a-lavoura-e-a-terra/)

The political leadership is contested in what relates to the evolution of the country in the last 43 years (see Santos Fernandes: 2017) mainly by the new generations, educated in higher education institutions and often abroad. There are in fact more and more vocal accusations of a lack of democratic responsibility and values on the part of the successive governments.

Fernandes' work is based on interviews conducted by academics of Universidade Colinas de Boé to political leaders. Their conclusions: the struggle for power after the independence has been confused with the struggle for protagonism at all costs and these premises are stated as a warning to politicians.

3 NOTES FROM THE FIELD

The importance of NGOs in Guiné-Bissau is very public and unquestionable. There are around 12 national organizations, and they cover the different areas that have been forgotten by the state: Education, Microcredit, Communication, Women's Empowerment, Handcraft, Environment, Land Ownership, Technologies.

One of the most successful areas of their intervention are the Rádio and TV Comunitárias. Their success is easy to explain. In fact, as listed in AD/Intervenção Comunitária (http://www.adbissau.org/intervencao/comunicacao-comunitaria) there was a progressive appropriation and training on the part of the local communities of the equipment and its technical usage, with no calendar pre-established for the process. This allowed a natural order of management and it also left space for the suggestion of innovations and proposals.

These radios were not planned as an external resource but as a means for the communities to express themselves. Also, the presence of cultural values associated to these projects, conquered the "homens grandes" (traditional leaders) and determined their localized importance thereof.

The fact that young men and girls also got progressively involved in the management and control of the local radios was instrumental for the sustainability of these resources. (see https://www.youtube.com/watch?v=EZOq_rC3jkM)

The success of the show of Rádio Kasumai born under this reality, "Bedjas na Manti", for instance, became a national reference.

Due to the enthusiastic reception of the radios, the NGO AD (Acção para o Desenvolvimento) launched in 2001 the first Televisão Comunitária, TV Klélé.

As the local media pursue their role in guaranteeing social cohesion and sustainability (see Patrícia M. Paula: 2011) there are two other cases that should be pointed out as examples of how Guiné-Bissau is showing to be a dynamic country at the local level regardless of the difficult political situation and of the slow application of African Development Programmes.

The first case is Neram N'Dok, a community project in the Bijagó islands. It is an example of participatory governance set in the islands of Urok, a marine area protected but threatened by illegal fishing.

Conservation and traditional cultural practices ensure a sustainable model under the control of the locals. Neram N'Dok is also about gender equity and of how tradition and modernity come together to determine the best solutions for all and for the future generations.

Also concerning the Bijagós archipelago, the movie Kadjike (Sacred Forest), by film director Sana Na N'Hada, from 2013, though being a fictional work, is deeply rooted in collective beliefs and practices of the region. As Fernando Arenas points out the film focuses

> on the clash between traditional Bijagó culture and the global drug trafficking network, the film is set in an atemporal place in harmony, according to Bijagó cosmogony, with God Nindo's creation. The natural landscape and the environment play a key role in the defense of its symbiotic relationship with traditional society and culture as the destructive forces of modernity intruded upon this pristine corner of the planet. According to the Bijagó creation myth—which is presented in voice-over by the Kriol-speaking narrator (Sana himself)—Acapacama, the original female inhabitant, founded the first village of Nocau. Before it was time for her to leave the world, she charged her four daughters with the care of tradition, the ocean, the forest, and the wind. Acapacama's heritage was to be protected by her descendants. The Bijagó society at the center of *Kadjike* not only lives in harmony with its natural ecosystem, but is also portrayed as an egalitarian society in terms of gender roles. (https://muse.jhu.edu/article/614578).

In the words of the director himself "I didn't come into cinema because of talent but because I felt obligated to tell certain stories. There has always been a question of necessity."

(http://africasacountry.com/2015/03/sana-na-nhada-guinea-bissaus-revolutionary-filmmaker-is-one-of-africas-most-important-directors-today/).

> Despite all the uncertainty facing his country today N'Hada remains hopeful about the future. As we speak, he is already turning his mind to his next feature, a film documenting the positive effects of independence in his homeland.
>
> With *Luta Ca Caba Inda* (The Struggle is Not Over Yet), an ongoing project first shown in 2012, N'Hada may yet bequeath his most profound legacy to Guinean cinema. Along with Gomes [film director Flora Gomes] he has set out to find and make accessible the remains of raw film material made in the country after independence but either lost or damaged in the era of political upheaval. (Idem)

Sana Na N'Handa and his work are a very good example to illustrate how Africa's intellectuals and communities are taking in their hands the solutions to problems, through self-determination and their own definition and interpretation of progress.

"Who will not admire the advances of this century?" (Eça de Queirós, *Civilização*)

José Carlos Vasconcelos e Sá
ULHT, ECATI, Lisbon, Portugal

ABSTRACT: The text starts by discussing the concept of progress by anchoring it to the concept of linear, chronological and historical time, due to alphabetic writing, as opposed to circular, smooth-running, repetitive time, characteristic of illiterate societies. In a second phase, it analyses the moment when linear time goes into crisis and, with it the idea of progress. This situation, that late 19th century and early 20th century European modernity faces, evokes, to illustrate it, two texts by some other Portuguese writers: Eça de Queirós' and its tale "Civilização" and Fernando Pessoa and its poem "Ode Triunfal". At last, it examines the still persisting effects of that crisis on contemporaneity, that speed and acceleration are the most visible manifestations.

Keywords: Linear time, crisis, progress, modernity, speed

1 TIME CONTROL

Progress, from Latin *progressus*, which means advancement. Human societies have always progressed, in other words, they changed. However, such changes were not always perceived the same way. Such understanding relates to the way in which each society conceives time[1], i.e., in turn, time, a condition immanent to the human being insofar as it inscribes us in history, ties directly to the cognitive technologies available in human communities. In illiterate societies – with no alphabetic writing – these changes were widely imperceptible. Ruled by the concept of a cyclical and circular time, where repetition structured life, illiterate societies used as a prevailing form of communication, the oral expression. Having a memory of its culture the myth that periodical repetition demanded in order to enable its permanence as a collective reference, embodying the community and by those means ensuring its survival. The traceless development makes change became almost imperceptible, but it still slowly happens.

After the introduction of alphabetic writing, the concept of time was profoundly changed[2]. The undefined circularity led to a straight-line representation, in which particular occurrences were written on, in chronological order. Where once there was a stretchy present of some kind, a blurry mismatch between then, now and after – since it establishes them externally – the writing demands a new order, clearly distinguishing past, present and future – in the sense that it set them externally- imposing the vision of sequence and recurrence. This order brings the creation of prose, straight-line narrative, which registers, in a historical context, while describing, in the context of the history, the most noticeable changes of the course of time. Regarding the substantially different temporal relationships which both oral expression and writing[3], intellectual technologies have with ways of life, influenced on the one hand by nomadism and recollection and on the other by agriculture, P. Lévy writes:

[1] The human conceptions about time are innumerable and come not only from philosophy, but also from the sciences and even from literature. From these varied analyses, we retain, in this context, those that relate time to movement and succession, as, for example, Aristotle's. Now the idea of succession allows the quantitative measurement of time (hours, days, years ...), also giving rise to its "qualitative measurement" through writing and history.

[2] Here I follow the argument developed by Pierre Lévy in "*As Tecnologias da Inteligência*", cap. '*A Escrita e a História*'.

[3] About the effects of writing in culture, Lévi-Strauss comes closer to McLuhan's analyses rather than Pierre Levy's, who we have been quoting. Lévi-Strauss, reacting to the idea according to which the knowledge of alphabetic writing would be determinant for the constitution of any "superior" society, underlining very negatively such accomplishments, agreeing with McLuhan on this topic. In fact, in "Tristes Trópicos", he writes: "The only phenomenon that has faithfully accompanied it (the writing) is the formation of cities and empires, in other words, the integration of a considerable number of individuals in a political system and its hierarchical position in castes and classes. That is, in any case, the typical evolution that we have been witnessing from Egypt to China, when the writing emerges: it seems to favour men exploitation more than their enlightenment (…). If my hypothesis is correct, it's necessary to admit that the printed publication primary function was to facilitate servitude" (Lévi-Strauss, 318, 1957).

The writing establishes a new relationship with time: when hunting or reaping we have immediate possession of the capture or wild fruit; defeat or success is decided in the field. Agriculture, by contrast, implies a thoughtful organisation of time, a whole system of backwardness, a speculation on the seasons". (Lévy, 112, 1994).

Through History, the effect of writing has also been to mobilise time, subjecting it to the human need of contemplation of achievements. With it, time is captured inside human logic allowing, to some extent, its manipulation; with it, the perceptions of change might become clearer because significant events that impact it not only rely on human choosing, but they also remained fixed on their chronology. Time is no longer external to man, it is no longer out of man's reach as it was the case with circular time to which he was subjected to, by contrast, now he commands it, controls it. The writing establishes a form of "time sense" that merges with a "history sense" reality particular to men. Time and history walk together, controlled by men towards a bright future, also part of the human[4] reason, Enlightenment's fundamental belief. Progress, humanity's driving force, is not only controlled but also captures. It will be what individuals make of it[5].

2 TIME CRISIS, EÇA AND PESSOA: PROGRESS IN PORTUGUESE

The enlightened architecture of time was conceived with wide, sober and seemingly firm facades; if it would be a building it would have open interiors, simple but beautiful. It would virtually welcome all. It would invite us in, enjoying an environment infused with happiness. However, soon we begin to hear Enlightenment's adverse forces. We now notice two voices of Portuguese literature – the novelist Eça de Queirós (1845–1900) and the poet Fernando Pessoa (1888 – 1935) – who tackled very differently the crises, striking with ever more intensity, the bases of the positivity and the building of the Enlightenment. As representatives of Portuguese modernity, both writers express in their characters the feelings of European modernity: Eça with the apathetic boredom towards "civilization" divided between a measured refusal of progress, eagerly accepting rural life, reaction of a romantic character to reason exaggeration. Álvaro de Campos, one of the four heteronymous of Pessoa, a full acceptance of the electric universe demanding the futuristic way, the interface between man and machine.

Regarding Eça de Queirós we will refer to the tale "Civilização", which was a sketch of the novel *A Cidade e as Serras*; regarding Fernando Pessoa, we will quote his heteronymous Álvaro de Campos and the poem "Ode Triunfal".

2.1 *Eça and the decadence of the western man*

The episode we will mention is part of the tale "Civilização". Jacinto, the main character, is a rich Portuguese noble, who lives surrounded by everything created by progress and bequeathed by civilization because he believes this is the only way that a civilized man can experience the modern culture. Zé Fernandes, Jacinto's intimate friend, the tale's narrator, describes the literature available at Jacinto's small palace in the following manner: "An afternoon I wanted to copy a dictum by the economist Adam Smith and I searched along 8 meters worth of political economics bookshelves! My friend Jacinto thought he was brilliantly provided with all of the essential intellectual works – and also of stupidity" (Queirós, 3, 1892). Later on, realizing the environment surrounding Jacinto's work office, the novelist writes:

> O que, porém, mais completamente imprimia àquele gabinete um portentoso carácter de civilização eram, sobre as suas peanhas de carvalho, os grandes aparelhos, facilitadores do pensamento, – a máquina de escrever, os autocopistas, o telégrafo Morse, o fonógrafo, o telefone, o teatrofone, outros ainda, todos com metais luzidios, todos com longos fios. Constantemente sons curtos e secos retiniam no ar morno daquele santuário. Tique, tique, tique! Dlim, dlim, dlim! Craque, craque, craque! Trrre, trrre, trrre!... Era o meu amigo comunicando. Todos esses fios mergulhados em forças universais transmitiam forças universais. E elas nem sempre, desgraçadamente, se conservavam domadas e disciplinadas! (ibid, 3–4)
>
> [What truly gave Jacinto's work office a portentous civilisation nature was the oak candleholders, the large appliances, thought facilitators – the typewriter, the reproducer, the Morse telegraph, the phonograph, and others, all with shiny metals and long threads. Dry and short sounds were constantly ringing through the sanctuary's warm air. Tick, tick, tick! Dlin, dlin, dlin! Crack, crack, crack! Trrre, Trrre, Trrre! It was my friend communicating. All these fibbers immersed in universal forces sent

[4] The relationship between writing and reason is analysed by Pierre Lévy who considers the phonetic writing as not being only a mere mean of expression of knowledge, sciences, philosophy and law but ontologically constitutes it. It creates, in opposition to the oral expression, a new form of knowledge of reality through the theoretical exposition, whose truth is exterior to the subject, valuing for its inherent logic.

[5] Christianity reinforced the linear conception of time, that writing has imposed, from the revelation theme, criterion that, relied on the concept of divine providence, creates the theology of history. From that moment on, a transcendental entity would own the time course that the humans would be object to. With the industrial revolution, the revelation theme is replaced by the reason theme and the linear guardianship of the history changes again: the transcendent reason gives place to the immanent. It is the Enlightenment and later on the positivism, with its belief in the technique and in Men that found the theory of progress to which we have been referring.

universal forces. Unfortunately, they did not always remain tamed and disciplined!][6]

The last sentence referring to unruly technology is used by our writer to describe the episode involving Jacinto, his friend Zé Fernandes and the Gouveias sisters, invited to a soirée, during which the host wanted to show the immense qualities of the phonograph, which, in addition, received the "round and ocular" voice of their close relative. The voice questioned rhetorically and emphatically:

> Quem não admirará os progressos deste século?" Furiosos, enterramos uma almofada na boca do fonógrafo, atirámos por cima mantas, cobertores espessos, para sufocar a voz abominável. Em vão! sob a mordaça, sob as grossas lãs, a voz rouquejava, surda mas oracular: – "Quem não admirará os progressos deste século? (ibid, 4)
>
> [Who will not admire the advances of this century?" The increasingly thunderous and frightening voice became immune to the several attempts of muting it: "Angry, we tucked a pillow in the phonograph's mouth, threw blankets over it, thick covers to smother the abominable voice. To no avail! Beneath the gag, the thick wools, the voice was raspy, deaf, but oracular.
>
> – Who will not admire the advances of this century?][7]

This briefly quoted episode, apart from its grotesque scene, reflects a subtle irony, which does justice to the idea of Queirós' text being outlined as a survey, as a research on the European men's culture. The irony of a mechanical anomaly leading to the repetition of a question which is simultaneously an answer, kind of "forces" to reflect on the place of progress and its main strength, the skill, in the culture.

The topic of machine's "indiscipline" and "disobedience", which Eça explicitly stated, refers to distrust, repudiation and even outrage that much of the nineteenth century displayed towards the industrialization and its effects. Such events, frequently violent, by both the labour movement, such as the ones carried out by luddites throughout Europe and the nineteenth century's conservative elite – which one of the leading representatives was undoubtedly Ortega y Gasset – who ferociously criticised the consequences of Industrialization which created the mass man, whose spirit, he characterised as ignorant, selfish and weak, became increasingly dominant, contaminating the vital roots of western culture[8].

However, the motto with this episode is the repetitive question: "Who will not admire the advances of this century?" The rhetorical interrogation is a clear reference to the inquiry suggested by the text: What is the value of progress in western civilization? Such reflection can be made – we think – from two myths of our culture: The myth of Sisyphus on the one hand and the myth of the Fall on the other. Quite briefly, the first one tells the story of Sisyphus who had disobeyed the gods due to the passion he had for earthly life. In addition, for that reason, he was condemned to make a stone roll uphill, but when the stone reached the top it fell back to the foothills, making Sisyphus' job permanently unfinished. Like Sisyphus, humans lived a repetitive, equal, redundant, useless, hopeless, therefore meaningless life. The quote insistently repeated in the tale, due to a machine's apparent anomaly, refers to this meaningless human life, now realised through the objects produced by progress. Such malfunctioning would be its truth, as what is mischievously generated by progress is the same beneath a new cover. Its truth refers to that flaw, to a lack of meaning[9].

---

[6] Author's free translation

[7] Author's free translation.

[8] The progress criticism, based on the industrialization and the type of society it engendered, had a lasting continuity through all the 20th century. In sociology, Max Horkheimer and Max Weber analysed the domain of the instrumental reason in modern societies and Weber, deeply disenchanted with modernity, concluded that we live in as "iron prison"; later on, Marcuse – together with Horkheimer, Adorno and Benjamin, representatives of the critical theory and Frankfurt school – would mention the production of a unidimensional men. In philosophy, the technique manifest itself in a relevant theme when Heidegger publishes, in 1949, the article "La Question de la Technique", a fundamental text in which the technique is considered as a form of *aletheia,* in other words, of truth. The philosopher demonises the modern technology, he analysed as a truly independent force of the human will: "In their ways of operation, (the technique) provokes nature, demanding from her the liberation of energies that can be explored and accumulated. The essence of modern technology should be understood as a process of nature disclosing with a special characteristic of incitement in relation to it". (Vasconcelos e Sá, 2001, 126). The objects produced by it "without real individuality" – in reality, degraded objects of the human – constitute the proof that the German philosopher exhibits to justify his Conception of the essence of technology. According to him, it cannot be conceived as a simple result of the individual or collective human decision, but as something autonomous in relation to the human, to which he calls *Gestell*, that means the transcendental pre-condition of technology. Then, *Gestell* means the dimension of the modern technology that orders or commands the particular way of this disclosure. This disposition is understood as an impersonal cognitive structure or an impersonal will that not only provokes the world but also and essentially, incites the human being, in a systematic, precise and constant way, to provoke the world. The action of the impersonal wills always disclosure the nature in the same way". (ibid, 127-28).To conclude this topic, the mass culture, itself a result of the world of machines, namely film, still mute, also censored the emergent industrialization, or using Chaplin's accurate irony, like on "Modern Times", or dramas like "Metropolis" from Fritz Lang.

[9] It was Albert Camus, who contemporaneously analysed, with his recognised lucidity, The Myth of Sisyphus. With only 29 years old, Camus published in 1942 "Le mythe de Sisyphe", a text about the absurdity of contemporaneous life, its absence of sense, questioning, in this context, suicide's morality.

On the other hand, referring to the Myth of the Fall, which, as we know, narrates the story of the original sin, is directly related to the progress topic. Adam and Eve, disobeying God, ate the forbidden fruit from the tree of knowledge and consequently were, expelled from paradise.

However, the guilt from then on carried by humans fades away over time as men's achievements provide evidence of his true potential. The wording of Eça's sentence is substantial to the effect of excluding anyone from not admiring the significant human capacity to create the "advances of this century". That is the context in which G. Steiner in "La Barbarie de l'ignorance" wonders whether Adam and Eve's retreat from Eden was a separation. Conflicting, without a doubt, but a separation, which Nietzsche recognized when he proclaimed "the death of God". The hypothesis is that this event has lodged a will expressed by a man who, in ownership of his writ of emancipation – to which Greeks gave the name of logos, a word that simultaneously meant knowledge, truth, rationality, order, language – will be allowed to seek his autonomy.

If that would be the case, there would not be as much a punishment but a separation. This is the idea implied in the expression used by Picasso, when he refers to the artistic creation work and relates it to God, stating that "He is the real competitor" (Picasso, apud Steiner, 2004), expression that desacralizes the maker, putting him at the similar plan as the human being. That proximity effect between both paradoxically constitutes that which allows the separation idea. Progress, proof of human autonomy, would then be at the foundation of the secession between God and Humans. Jacinto's boredom felt in front of the progress works (the Parisian ennui, that contaminated the intellectuals from the time) to whom everything was already "dull", may be understood as remorse from that forgotten guilt, remorse that makes him come back to his small Eden, his home in the mountains.

2.2 *Álvaro de Campos and the Glorification of Progress*

One of the heteronyms of Fernando Pessoa, Álvaro de Campos (along with Ricardo Reis, Alberto Caeiro, and Bernardo Soares), had several creative phases: decadent at first, and futurist/sensationist after – it is at this latter moment that the "Triumphal Ode", to which we shall refer, belongs – and, finally, its pessimistic phase. His Triumphal Ode also responds to the boredom that stifled a part of the nineteenth-century and the beginning of the twentieth-century. It belongs to the positions of the Italian Futurist Movement, whose proposals, utopian for its time, but today very real, started from the glorification of technology and progress, including the desire for fusion between man and technology, thus anticipating a thematic universe that brings together topics that today are central to contemporary thinking, such as artificial intelligence, biotechnologies or post-humanism.

Futurism, as an organic movement, in spite of its relatively short life, politically and culturally influenced Mussolini's fascism and Hitler's Nazism. Safeguarding the futurists' political topics, such as the defence of authoritarianism, the call to violence, the glorification of force and aggression, which do not belong strictly to Pessoa's vocabulary, the "Triumphal Ode", nevertheless, shares with them this desire for the transcendence of the human, offered by the technological progress, and thus making possible the overcoming of time and space. Such affinity is well illustrated by comparing the following excerpts from the "Triumphal Ode", written in 1914, to Marinetti's First Futurist Manifesto of 1909. In the "Triumphal Ode", Campos, at one point writes:

> Ah, poder exprimir-me todo como um motor se exprime!
> Ser completo como uma máquina!
> Poder ir na vida triunfante como um automóvel último-modelo!
> Poder ao menos penetrar-me fisicamente de tudo isto,
> Rasgar-me todo, abrir-me completamente, tornar-me passento
> A todos os perfumes de óleos e calores e carvões
> Desta flora estupenda, negra, artificial e insaciável!
>
> [If I could express my whole being like an engine!
> If I could be complete like a machine!
> If I could go triumphantly through life like the latest model car!
> If at least I could inject all this into my physical being,
> Rip myself wide open, and become pervious
> To all the perfumes from the oils and hot coals
> Of this stupendous, artificial and insatiable black flora!][10]

In turn, Marinetti proclaimed:

> (…) we shall sing the nocturnal fervour of arsenals and shipyards gleaming under violent electric moons; railroad stations that devour smoke-feathered serpents; plants suspended in the clouds by the curled ropes of their smoke; clouds that ride the rivers like giant gymnasts, shining in the sun with a twinkle of knives (…) (Marinetti, apud, Berman, 1986).

With Eça de Queirós and Fernando Pessoa, we have seen two attitudes that are inscribed in the idea of a crisis our Western culture experiences. Jacinto turns his back on progress, to the city, taking refuge in the past, slow, and rural, extensive time. On the other hand, the engineer Álvaro de Campos (such was the profession imagined in the biography written by Pessoa for this heteronym), despises tradition, the past, longing for

[10] Translated by Richard Zenith.

the electric future to take over the vital forces of man, in order to rebuild his out-of-date humanity.

However, both Pessoa and Marinetti, in conjuring up their futurist desire for the creation of their model of a New Man who would be "a nonhuman species" (ibid, 20), could not have assumed, possibly, that their aspiration would begin to take shape one hundred years later.

3 TOTAL MOBILISATION: SPEED AND ACCELERATION

The exaltation of technologies and, especially, the will to relate organic and artificial, that at the beginning of the 20th century was considered utopic, is part of the time acceleration process that Ernst Jünger translated, summoning the volcano image which represents modern societies, the association between war and progress, to which he named total mobilisation. The concept, described in the texts "Total Mobilisation" (1930) and "The worker" (1932) constitutes an analysis of modern technique through Nietzsche's concept of will to power.

> Total Mobilisation – says Jünger – is consummated by itself much more than by ourselves; it is, in war and in peace, the expression of mysterious and compulsory claim to which that period of masses and machines submits us. (Jünger, 2002: 198)

All modern societies, be it capitalist or socialists, suffer, according to Jünger, from this total mobilisation, in which the modern technique is the driving force and the image of the worker, that encompasses the technique and the human, without, however, limited to it. Communities celebrate progress, but what they are really doing without knowing, is submitting to the logic of total mobilisation, both technological, corporate, military and social; what the communities are doing without knowing it is to accept those destructive and auto destructive devices chat follow only the logic of their own movement, their own advancement.

The transformation of life into energy, about which Jünger talks, has, since the 1950s – with the arrival of Cybernetics and the concept of information, and later on with computing and computer science, artificial intelligence, information societies and globalisation – its translation into the idea of speed. In fact, information technologies radically modified the human experience by modifying their two essential axes[11]: the time compressing it through technological acceleration until it reduces it to the instant, in a way that, paradoxically, we can say that we live today in a time without time and space – creating another space, new, virtual, in which we increasingly relate. Today, more than ever, Marx's sentence from the "Communist Manifesto" from 1848: "All that is solid melts into air" (Marx, apud, Bergman, 1986) is no longer metaphorical but it assumes a literal meaning, in a way that specific atoms from things in the world transform itself in digital bits.

The time crisis we have been referring to, and through which culture questions progress, manifests in our contemporaneity through speed and acceleration, symbols under which modernity has grown. Paul Virilio recognizes them such relevance to his comprehension that he founded Dromoly, the sciences that study the speed from an anthropological point of view, as a cultural and political value. To the author, in contemporaneous societies the dizziness of the speed and acceleration destroys the duration of time; if progress is just like the Latins understood it, 'advancement', what we have ahead, today, is much more the speed's inebriation and light-headedness rather than any human sense of the purpose of that 'advancement'. Politics, in this global, interdependent space, increasingly volatile, losses its control capacity, allowing imponderability, the growth of insecurity, exclusion, making us live in an ever more "state emergency".

The strong criticism that Virilio draws against the effects of the acceleration to which contemporary societies are subjected by the techno-capitalist logic, has its reverse in the positions stated by two professors of the University of London, Alex Williams (sociology) and Nick Srnicek (philosophy) expressed in the Accelerationist Manifesto[12] published in June 2013. This leftist tendency of accelerationism seeks to overcome neoliberalism that has captured science towards the interests of capitalism, drastically reducing its creative capacities. Neoliberal capitalism, unlike nineteenth-century bourgeois capitalism, does not only generates poverty, injustice and oppression but also blocks progress, constantly threatened by the logic of desire of the market, which is increasingly besieged by a

[11] The changes produced by NIT (New Information Technologies) in the human experience, namely, with the alteration of time by continuously accelerating it, supressing, therefore, the duration, causes a psychological dysfunction difficult to overcome on the human being. Probably this maladaptation of the human time to mechanical time, a much faster one, will be at the basis of the incorporation, in our body, of non-organic elements created by biotechnologies, which will allow us, eventually, to become compatible with the environment created by the machines. It is in this context, which is currently ours, that we gradually leave the human universe to enter the post humanism.

[12] Accelerationism is a movement of ideas, which proposes to deepen the capitalism's "contradictions", radicalize it, so that its dissolution capacity is accelerated in a way that it self-destroys. Nick Land, English philosopher rooted in China, considered one of the major representatives of accelerationism, defends that neo liberalism radicalization for not agreeing with that bourgeois politics in itself, but for exclusively nihilist reasons. In other words, accelerationism will have, at least, two different trends: one of Marxist traditions originating in France, very influenced by philosophers like Deleuze, Guattari and Derrida, whose proposals are substantiated in a political program, as seen with Alex Williams and Nick Srnicek; another one more ontological and less political, represented by Nick Land, of anarcho-capitalist trend, interested in the technological improvement of mankind, endowing him of a more advanced cognitive architecture.

saturated and exhausting production of objects. The manifesto advocates a policy that frees science from the interests of capital, and that rescues and develops the productive forces of cognitive workers engaged in intellectual work, considered the nerve centre of capitalist exploitation.

Progress, change, the "advancement" to which Latin makes reference, has, in our contemporaneity, those who distrust it – all those who wrote about the implosion of space-time, the derealisation of the real, the end of History – but also, as we have just seen and among many others, those who believe in it. The crisis that modernity shows is more a crisis of human beings who knew how to use the anthropotechnics that they themselves created to become human, perhaps now, too human; Perhaps today we are faced with more radical choices that frighten and threaten us in such a way that we think that everything will end. However, it is at times like this that it is worth remembering the verse of Hölderlin, which Heidegger quotes in his text on the technique: "Mais, là où il y a danger, là aussi / Croît ce qui sauve "(Hölderlin apud Heidegger, 38, 1954).

BIBLIOGRAPHICAL REFERENCES

Queirós, E., (2013) [1902], *Civilização*, LusoLivros, in: https://www.luso-livros.net/Livro/contos-de-eca-de-queiros/, accessed on 20 February 2017.

Berman, M., (1986), *Tudo o que é Sólido Desmancha no Ar – A Aventura da Modernidade.* S. Paulo: Editora Schwarcz.

Cordeiro, E., (s/d) Técnica, mobilização e figura – a técnica segundo Ernst Jünger, www.bocc.ubi.pt, accessedon 23 January 2017.

Heidegger, M. (s/d) [1954] *La question de la Technique*, [PDF] La question de la technique – Monoskop https://monoskop.org/.../Heidegger_Martin_1958, accessed on 25 february 2017.

Jünger, E. (2002) "A Mobilização Total". *Natureza Humana – Revista Internacional de Filosofia e Psicanálise*, S. Paulo.

Lévi-Strauss. (1957) *Tristes Trópicos*, S. Paulo: Ed. Anhembi.

Lévy, P. (1994) *As Tecnologias da Inteligência.* Lisboa: Piaget.

Pessoa, F., (1994), *Poesias de Álvaro de Campos.* Lisboa: Ática.

Steiner G. Spire A. (2004) *Barbárie e Ignorância.* Lisboa: Fim de Século.

Vasconcelos e Sá, J. C. (2001) "A Crítica da Técnica e da Modernidade em Heidegger e McLuhan", *Interações*, 1: 124-137.

Williams, A. e Srnicek N. (2013) ACELERAR MANIFESTO: por uma política aceleracionista (Tradutor: Bruno Stehling – UFRJ. url: https://syntheticedifice.files.wordpress.com/2013/08/manifiesto-aceleracionista1.pdf accessed on 23 february 2017

Humanism and technology: Comparative analysis of More's *Utopia*, Bacon's *New Atlantis*, and Miguel Real's *O Último Europeu 2284*

Maria do Rosário Monteiro
CHAM, FCSH, Universidade NOVA de Lisboa, Portugal,
Universidade dos Açores, Portugal

ABSTRACT: Being a creation based on Christian Humanism, Utopia went through several changes, many of which are already present in More's narrative. Bacon's New Atlantis brought one of the first radical transformations about. The closed, agrarian, and immutable Utopian island opened itself to the world and promoted change through controlled scientific research. However much of what More conceived was maintained: New Atlantis is still an island lost somewhere in the middle of the ocean, protecting its privacy and secrecy, evolving through the inclusion of selected external information, but avoiding exchanged. In Miguel Real's *O Último Europeu 2284* [The Last European 2284] the narrative starts in part of the European continent, with diffused borders, but unquestionably defined as a perfect utopian society, technological and scientifically advanced, and actively engaged in providing its citizens with the necessary *otium*, in the sense of the Greek word *skole*, meaning intellectual activity. After the destruction of this utopian and pacifist Europe, there is the creation of a new one, in the almost desert island of Pico, in the Azores. Once again, this utopia is crashed, and what endures as hope for future generations is a handwritten manuscript describing the two shattered future Utopias. The thread to explore in the paper is that of how humanism evolved through the last 500 years, taking these three paradigmatic literary utopias as base stones.

Keywords: Utopia, Thomas More, Francis Bacon, Miguel Real, Humanism, Technology, Progress.

1 MORE'S *UTOPIA*

Much has already been said on More's *Utopia*[1] (1965), and whole libraries are still waiting to be filled with a myriad of still unwritten texts, in different languages and times, presenting new analysis triggered by this short narrative. This is a rare hypothesis only bestowed upon the most excellent literary achievements, upon those that overcame the hard test of time and culture and, regardless of readers' particular circumstances, will keep on challenging new interpretations and triggering new emotions. *Utopia* is a never-ending spring of inspiration and the best and more thorough criticism remains incomplete. Even the outstanding André Prévost's analysis, probably one of the best ever written, in my opinion, cannot give all the answers (1978).

In this paper, More's *Utopia* will be the foundation stone to study some aspects of the relationship between humanism and science choosing to focus on three different but complementary periods. According to Pina Martins, an outstanding Portuguese scholar on Renaissance and humanism, the *studia Humanitatis* are so designated because they are the best way to form the whole human being. Cultural education gives form to the free citizen, the one who is fully aware of taking part in the life of his city (Pina Martins, 1969: 5). According to Pina Martins, this movement is not a rupture with the past, but rather an integration; the ancient being renewed in its grafting of the new; in these terms, the lesson seems to have persisted until the Enlightenment (1969: 8).

It is now consensual the idea that Erasmus and More worked together in an educational project that would be concluded with the publication of a diptych constituted by *In Praise of Folly* (1509) and *Utopia* (1516) or "in praise of *sapientia*", the discourse of the *Morosophos* (Pina Martins, 2009: 82; Prévost, 1978: 66).

For *Utopia* to achieve its goal in Erasmus/More project, the praise of knowledge had to be grounded in the hard reality of everyday life. Therefore, *Utopia* could not be confined to the second book, the first to be written. It had to confront what might be possible someday, somewhere, with what actually was. Moreover, the harsh reality of the present is the only thing that can be changed (the past is gone; the future is a mere potentiality). Therefore, a Lucian's *Morosophos* as Erasmus saw Thomas More had to show the two sides of the coin. One cannot change what one does not know or the result will be mere wishful thinking or to use a semantically more appropriate word "*rêverie*".

[1] There are several editions of Thomas More's *Utopia*. The one I chose to use is the standard Yale edition, therefore all citations will refer to this edition, but whenever necessary, other editions will be referred in the endnotes.

Though the Humanists had a weird sense of humour, they were all deeply committed to the task of being, primarily, educators of citizens.

They also knew that, as Pico so eloquently put it, human beings have the ability to be either similar to angels or to the beasts according to one's choices. The role of the educator is to deal with both kinds. They found the best and the worst in all places, at every social level.

Yes, the humanists were elitists; they formed a closed group to which one had access by merit. They were quite a small group in European society, in spite of the important transformations, for better and for worse, that came out of their written texts, and their "games" of influences and their letters.

Humanists were all connected with the various political spheres, within the church but mainly within governments and courts and their survival depended on these political ties. However, they were also fiercely critical and experts on what we might call "semantic entanglements". They were always on the verge of the abyss.

This powerful small elite was unanimous in the concern and search for the true meaning of words, on their destiny as educators, on their belief that the ways of the world could be changed through human endeavour, on their urge to restore ancient wisdom and use it to make a better society. They no longer believed their fate depended on spiritual or divine intervention, though they were all religious minds. However, one unsolved question divided them: how could they change society? Working within the political spheres, or stepping out and defending their independence as thinkers and scholars. This is one of the debates expressed in *Utopia*, alongside with the discussion of the best way to govern a city.

In the space allotted to me, I will only be able to scratch surfaces. The search for science in Utopia hardly goes further than the humanities. That is the ground the first humanists' new better, the one that should be at the basis of all other knowledge. There are several humanists who were doctors, mathematicians, and so forth but what we now call science would spring only in the second period of the Renaissance.

More's Utopians have a natural inability for invention. Foreigners bring the press and books from antiquity. Nevertheless, utopians lack the sparkle for invention precisely because they do not engage in exchanges with the rest of the world. However, as Lyman Sargent says, if we were poor peasants in 1516 we would find life in Utopia to be extremely appealing. On the other hand, maybe not, because as peasants:

> [We] would not have heard of it at all, and even in the unlikely event that [we] could read, [the book] was only available in Latin. [...] Secondly, if [we] did come to hear of it, [we] would have filtered it through [our] understanding of the world, an understanding in which reason played a very small part. (Sargent, 2004: 2–3)

Maybe this was one of the things More would like to discuss further with Raphael. Or again maybe not, because:

> When Raphael had finished his story, many things came to my mind which seemed very absurdly established in the customs and laws of the people described – not only in their method of waging war, their ceremonies and religion, as well as their other institutions, but most of all in that feature which is the principal foundation of their whole structure. **I mean their common life and subsistence – without any exchange of money**. This latter alone utterly overthrows all the nobility, magnificence, splendour, and majesty which are, in the estimation of the common people, the true glories and ornaments of the commonwealth.
>
> I knew, however, that he was wearied with his tale, **and I was not quite certain that he could brook any opposition to his views, particularly when I recalled his censure of others on account of their fear that they might not appear to be wise enough, unless they found some fault to criticize in other men's discoveries**. (More, 1965: 245)

So definitely, Raphael had a bad temper! However, hidden in the constant irony that flows through *Utopia*, the problem of the lack of exchange goes further than simply the absence of money. The problem with the utopians is their lack of individuality, a value too great for going unnoticed by any humanist.

This "praise of sapientia" shares with the *Praise of Folly*, or Lucian's *True Story* the subtle irony, the same sense of humour, the same taste for the half said and the half implied, for veiling the true meanings under layers of misguided hints.

Probably that is why we keep on scrutinising this complex tapestry that requires intelligence and also the knowledge and the ability to place it in it proper ambience: inside an elite, writing and laughing and taking pleasure in shared knowledge, books and ideas, but acutely aware of the reality and its ugliness.

2 BACON'S *NEW ATLANTIS*

Thomas More and Francis Bacon share many things. Both were politicians, both fell in disgrace – one lost his head the other his honour. Both were eager for knowledge, and both were humanists, though in different periods. I confess my complete agreement with Brian Vickers regarding a certain lightness that leads some critics to classify Bacon as anti-humanist (2000). This can only be defended by someone forgot to look at the sources, the way humanists taught us to do. Or as Vickers puts it: one should always "trust the author rather than his interpreters" (2012: 204).

Bacon had a humanistic upbringing, and the disciplines of the trivium were fundamental for his varied activities, either public, political or scientific. As many humanist, among them Erasmus, Bacon had a strong reaction to Ciceronianism, to the overvaluation of rhetorical devices over the matter discussed. However, this by no means allows us to consider Bacon an anti-humanist. As Vickers proves:

> 'The duty and office of Rhetoric', [Bacon] wrote, 'is to apply Reason to Imagination for the better moving of the will.' Well aware that rational processes can be disturbed, Bacon declares that 'the end of Rhetoric is to fill the Imagination to second Reason, and not to oppress it'. Its role is corrective, supportive, protecting reason from disturbance and forming a channel for the passage of virtue. (2000: 151)

This said Bacon's *New Atlantis* is a literary narrative that follows the path first scoured by More. Both use Plato as their starting point, both are concerned with the transformation of human conditions, both imagine an ambiguous narrative though by different means: More using irony and semantical contradictions, Bacon by leaving his text "incomplete". I do believe that Bacon wanted the narrative to be as it is, opened to the future, using rhetoric to trigger Reason.

While More is concerned with changing living conditions through a lengthy process of individual civic and moral education, Bacon believes science will be the answer to:

> The prolongation of life. The restitution of youth to some degree. The retardation of age. The curing of diseases counted incurable. [. . .] The increase and exaltations of the intellectual parts. (Bacon, 2002a: 488–489)

More would certainly agree with all this, but while utopians try to follow an ecological natural life to achieve a longer and healthier lifespan, Bensalamites believe this will be achieved through science.

For both authors, only an educated mind can accomplish the discovery of better ways of living. However, Bacon discovered the major weakness in the utopian society: it cannot stay isolated; it cannot become frozen in time. Bacon found the answer in Plato, though not in the same dialogue More uses. Bacon's philosophical base stone in The *Laws*, Plato's late revised version of the ideal city[2].

There are, I think, many reasons for Bacon choosing of the *Laws*, instead of the *Republic*, as More did. To mention briefly one: the political and religious situation in Britain. Bacon was a defender of monarchy as the best political system. However, Parliament and Court were in turmoil almost since the beginning of Stuart dynasty, and Puritans were increasingly pressing for more Parliamentarian power and enhanced control of kingship.

Bacon needed his utopia to be an aristocratic one, with different social and economic levels, but one that had the possibility to evolve in time. That is impossible in Hythlodaeus' Utopia. Bacon saw society as a living organism, and to preserve it from degeneration Plato offered two basic solutions – prospective travels abroad and a ruling council. Says Plato in the long and painstaking (and painful) dialogue:

> **No state will ever be able to live at a properly advanced level of civilization if it keeps itself to itself and never b comes into contact with all the vices and virtues of mankind**; nor will it be able to preserve its laws intact if it just gets used to them without grasping their raison d'être. In the mass of mankind you'll invariably find a number—though only a small number—of geniuses with whom it is worth anything to associate, and they crop up just as often in badly-ruled c states as in the well-ruled. So **the citizen of a well-run state, provided he's incorruptible, should go out and range over land and sea to track them down, so that he can see to the strengthening of the customs of his country that are soundly based, and the refurbishing of any that are defective. Without this observation and research a state will never stay at the peak of perfection; nor will it if the observers are incompetent.** (Plato, 1997: 951a–d)[3]
>
> **This council**, which should consist partly of young men and partly of old, must have a strict rule to meet daily from dawn until the sun is well up in the sky. Its membership is to be: (1) those Priests who have won high distinction, (2) the ten Guardians of the Laws e who are currently the most senior, (3) the Minister of Education for the time being, together with his predecessors in office. No member should attend alone: each is to bring a young man of his own choice, aged between thirty and forty. **The discussion at their meetings must always center round their own state, the problems of legislation, and any other important point relevant to such topics that they may discover from external sources. They must be particularly concerned with those studies which promise, if pursued, to further their researches by throwing light on legislative problems that would otherwise remain difficult and obscure. Whichever of these studies are sanctioned by the older members should be pursued with all diligence by the younger.** (1997: 951d–952a)

Here Bacon found the key to justify the importance of the House of Salomon or the House of the six days. Evolution is guaranteed through the exchange of ideas, artefacts, and knowledge. There is no reason to impose communitarianism, considered inadequate,

[2] For recent detailed and thorough analysis of Plato's *Laws* see Christopher Bobonich (2010; 2002).

[3] Emphasis added.

and the King, though a central figure in the social hierarchy, may become none existent in political terms, since:

> [The Council Members] have consultations, which of the inventions and experiences which we have discovered shall be published, and which not: and take all an oath of secrecy, for the concealing of those which we think fit to keep secret: though some of those we do reveal sometimes to the state, and some not. (Bacon, 2002b: 487)

We all know that who controls knowledge does indeed also hold the power. Therefore, the members of the House of Salomon, who choose those fit to carry on the research, ensuring the continued survival and legacy of this body of knowledge, and the constant perfecting of that political and social community, govern New Atlantis, in fact.

3 REAL'S *O ÚLTIMO EUROPEU 2284* [THE LAST EUROPEAN 2284]

Miguel Real's novel (2015), published last year in celebration of *Utopia* 500 years, follows More and Bacon humanist tradition transposing it to modern Europe, or even more accurately to the Modern European Union. Today we look at this Union and it seems to be on the verge of dissolution. In *The Last European* the European utopia is utterly destroyed leaving one single last prove of its existence – the novel itself, handwritten by the last living European citizen. With his death, it is not only Europe that disappears from earth; it is our whole culture that is destroyed.

There are several important innovations regarding this novel that, as More's and Bacon's, is a mixture of utopia and dystopia from the point of view of the narrator/protagonist known as Headmaster.

All history comes from his perspective, in a mixture of times, memories and present actions journalistic report The Headmaster is the last heir of humanist culture and through his long life and profession as keeper of European Museums; he is simultaneously witnessing and founder of utopias and their subsequent destruction/evolution to dystopias.

The book he commissions himself to handwrite is to become the proof utopia is possible, but also easy prey for dictatorial governments.

Reading *the Last European* we learn how a small conclave in the European continent managed to create and endure, for a century, a kind of technological utopia. Baconian science attained its peak with the production of endless ecological energy, taken from earth's core, and the creation of a supercomputer that controls everything: food, energy, provided unbreakable security, and controlled citizens' freedom, not by constraining them but by giving them all they wish. The inhabitants did not work, simply enjoyed the virtual reality that allowed them to experience skydiving, mountain hiking or a day at the beach, or whatever they wished.

A board of scholars supervised urban life, decided on behaviour penalties, and protected European cultural heritage.

This utopia is menaced by a dictatorial imperial China and ignored by an American utilitarian empire. Until the day the Chinese managed to temper with the source of energy and by cutting its supply bring New Europe to a dramatic, bloodshed invasion and utter destruction.

The Headmaster escapes with a few hundred citizens and lands on the Island of Pico, in the Azores. There, in the small island inhabited by one Portuguese family, The New Europeans rebuild a small ecological utopia from scratch. Everything goes according to plan, in utter secrecy, until the use of the internet, recovered after the construction of some computers using parts and bits of old ones brought from nearby islands, calls the attention of the American empire. What triggers their attention to the Island was not the amount of information exchanged through the net, but the quality of the searches only needed and used by well-informed experts.

As they discovered the island was inhabited they also discovered a rare and highly important natural matter whose scarcity made it more important than gold, and definitely more important than life itself.

Military troops are sent to the archipelago America claims as its property. The Europeans are tricked with promises of living as a free and united community within American soil, but in fact, they are too educated and inquisitive and would disrupt American monotony. So adults and adolescents are exterminated in secrecy, and young children delivered to American families to be raised as common ordinary indifferent people.

Summing up Miguel Real creates a utopia based on science and knowledge and humanist values that are destroyed by European hubris, by their absolute faith in the technology and knowledge and by their confinement. The second utopia is a modern ecological one, not perfect, rather seeking constant improvement, both intellectual and scientific. Moreover, this too is destroyed because it exposed itself and was powerless when confronted with utilitarianism and stupidity

Utopia will always be transitory, as life itself.

BIBLIOGRAPHICAL REFERENCES

Bacon, Francis. (2002a) *The Major Works*. Edited with an introduction and notes by Brian Vickers. Oxford: Oxford University Press.

—. (2002b) New Atlantis. *The Major Works*. Oxford: Oxford University Press, p. 457–489.

Bobonich, Christopher. (2010) *Plato's Laws: a critical guide*. Cambridge: Cambridge University Press.

—. (2002) *Plato's Utopia Recast: His Later Ethics and Politics*. Oxford; Nova Iorque: Clarendon Press; Oxford University Press.

More, Thomas. (1965) *Utopia*. Surtz, Edward and Hexter, J. H. 4. New Haven and London: Yale University Press.

Pina Martins, José V. de. (2009) *Estudo Introdutório à Utopia Moriana*. 2ª ed. Tradução, prefácio e notas de comentário de Aires do Nascimento. Estudo Introdutório de José V de Pina Martins. Lisboa: Fundação Calouste Gulbenkian.

—. (1969) Humanismo e Renascimento. *Revista da Faculdade de Letras, série III* (11): 5–54.

Plato. (1997) Laws. In: Cooper, John M and Hutchinson, D. S (eds). *Complete works*. Indianapolis: Hackett Pub, p. 1318–1616.

Prévost, André. (1978) Introduction. In: Prévost, André (ed) *L'Utopie de Thomas More*. Paris: Mame, 33–275.

Real, Miguel. (2015) *O Último Europeu 2284*. Alfragide: Dom Quixote.

Sargent, Lyman Tower. (2004) The Necessity of Utopian Thinking: A Cross-National Perspective. In: Rüsen, Jörn, Fehr, Michael and Rieger, Thomas (eds). *Thinking Utopia: Steps into Other Worlds*. Nova Iorque: Berghahn Books, 1–16.

Vickers, Brian. (2012) Bacon and rhetoric. In: Peltonen, Markku (ed) *The Cambridge Companion to Bacon*. Cambridge: Cambridge University Press, 200–231.

—. (2000) The myth of Francis Bacon's 'anti-humanism'. In: Kraye, Jill and Stone, Martin W. F (eds). *Humanism and early modern philosophy*. New York: Routledge, 135–158.

Imagining the future: A view of progress in H.G. Wells' science fiction

Leonor Sampaio da Silva
CHAM, FCSH, Universidade NOVA de Lisboa, Portugal
Universidade dos Açores, Portugal

ABSTRACT: Progress is a crucial concept in all science fiction. This literary genre resorts to literature in order to test a scientific hypothesis, which is supposed to bring the positive emotions ensuing from progress. Four books by H. G. Wells provide the literary corpus for this paper. Three are fictional narratives — *The Island of Doctor Moreau* (1896), *The Invisible Man* (1897) and *The Country of the Blind* (1904). The first one revolves around a scientist devoted to control evolution and speed its pace in order to create a wholly rational new set of beings; the other two bring us into contact with two male characters whose main trait is to be invisible to the eyes of the communities they arrive at (an English Sussex village and a South-American valley, respectively). A third narrative, *A Modern Utopia* (1905), mixing essay and fiction, is set on a planet exactly like ours, with the same continents and islands, seas and the moon, botanic species and animals, but differently organised. Although diversely constituted, all these books conclude that progress must always have a collective scope instead of fuelling personal ambitions.

Keywords: Progress, H.G. Wells, future, science fiction, happiness

1 THE PROGRESS PRINCIPLE AND SCIENCE FICTION

The starting point of this paper is an understanding of progress such as discussed by Amabile and Kramer (2011) who identify it as an important element of society. According to the authors, it makes work meaningful, triggers the positive emotions of satisfaction, gladness and joy, leads to a sense of self-worth, boosts creativity and productivity, feeds motivation, deepens engagement, and changes the perception of human life as it creates breakthrough opportunities and fosters a sense of accomplishment. However, progress also has a dark side. Its function as a powerful stimulant of happiness is often threatened by a number of setbacks, such as "hitting dead ends while trying to solve a vexing problem" or "being blocked in attempts to meet a goal" (Amabile and Kramer, 2001: 74). Failing to find relevant information and being unable to learn from previous failures are further obstacles to progress.

The idea of progress as a catalyst of happiness underlies Wellsian science fiction. His scientific narratives invite the reader to confront historical reality, namely the belief in science and knowledge as a tool to success, and to meet either probable or unsuspected outcomes of scientific advancement. As Philmus argues, science fiction can be interpreted as a literary strategy of "interpreting sectors of historical actuality" (1970: 30) in which outstanding inventions or discoveries, at least initially, fuel feelings of self-satisfaction as well as create bright anticipations of future times.

Progress is an underlying motive in the four books selected for the present analysis. The assumption that the world can benefit from science is never dismissed, even if it requires respect for nature and education for the general purpose of collective well-being.

2 THREE FAILED HYPOTHESES

2.1 *Controlled evolution*

In *The Island of Doctor Moreau* (1906), progress is rooted in the aspects previously mentioned. Wells puts into play the possibility of speeding up the process of natural evolution. Drawing upon Darwin's theory, Moreau embodies the scientific ambition of playing God in a universe in which science had destroyed the myth of divine creation. In addition to this, Moreau is also inspired by the discovery that all living forms are plastic and, therefore, susceptible of being transformed both in physical and in moral terms. The latter form of plasticity leads the scientist to inculcate in the humanised animals a deification of himself as Creator, thus completing the image of progress as leading to such an extreme degree of motivation and self-worth that it transcends the natural limits.

The scientist's devotion to his work is self-absorbing and blinds him to the setbacks he constantly meets. Unable to learn from his failure and to subject his personal ambition to the rule of Nature, he ends up dead at the hands of one of his creatures, thus showing how progress is totally hindered when humankind proves unable of solving a vexing problem which in this case is that of the stubborn return of the beast no matter how deep it is plunged into the bath of physical pain.

Wells manages to present to the reader a parody of civilisation. The reversion of the Beast Folk into the state of nature suggests the uneasy balance and drama opposing the natural and the artificial at the core of human condition. More importantly, the absence of a purpose to the pain inflicted to the creatures is what causes the final catastrophe. We are led to the belief by the narrator that had there been a humane scheme apart from the "bloodless intellectualism" (McConell, 1981: 5) of Moreau, an emotional attachment to living forms and a moral response to suffering, the final outcome might have been different. It was the utter absence of a collective design that aborted the scientific project cherished by Moreau.

As often happens in Wellsian fiction, widespread destruction ensues from oblivion to the common good. The punishment of scientific activity when is disrespectful of common good leads to the liberation of the creatures from artificial constraints, to the death of the team making the experiences and to the rescue of the narrator. The final disorder suggests the possibility of a future reordering. The final words of Prendick are an eloquent rendering of the conscientious mind: "And so, in hope and fear, my story ends".

2.2 *Invisibility*

In *The Invisible Man*, the hypothesis under study (invisibility) is based on advancements in the field of molecular physics, namely in what concerns the refraction of light and the optical density of matter. In a nutshell, the initial thesis is that the visibility of a body is determined by its behaviour in face of light; the body may absorb it, reflect it, and refract it or a combination of the three. If none occurs, it is not visible.

The core of this narrative is also the individual ambition of a scientist, Griffin. What drives Griffin to discover the invisibility formula is not the will to share with his fellow human beings the advantages of such knowledge, but the predictable power he would come into. As is stated in chapter IX, "An invisible man is a man of power". Hundred and ten years before (1787), Jeremy Bentham had published an essay in which the same conviction is clearly stated. "Panopticon or the Inspection-House" presented a solution for controlling the behaviour and the productivity of individuals and groups for political, clinical, economic, teaching and penal purposes, by resorting not to force but to the adoption of an architectural model that allowed a viewer to see without being seen. The panoptic, so would the model be called, consisted of a circular structure in which an observer placed on a central tower would monitor what was happening in the sun-lit rooms in all the perimeter of the building. In these cells, there would be students, workers, patients, or prisoners, according to the social function of the construction. The good behaviour of the observed depended on the uncertainty of being or not being watched, that is, the suspicion of being permanently surveyed, even when there was nobody in the tower, would be a deterrent to bad behaviour; the power of the observer grew in the exact proportion of his/her invisibility. Bentham, in 1787, very clearly stated that power does not derive from what you see or know, inasmuch as from what the others do not see and do not know.

Motivated by the lust for power, scientist Griffin finds out and uses on himself the invisibility formula. The exaltation mood blinds him to the disastrous consequences of his act: to become invisible, he needs to free himself from clothes, from footwear and from all kind of accessories. However, his nudity makes him vulnerable to Nature. Cold is an insuperable opponent (and an invisible one as well); the snow that piles up on his head and his shoulders, the rain that turns him into a shiny surface and the fog that makes him a humidity bubble with a human shape give away his presence; his feet are printed on the soil. Moreover, loneliness is absolute, creating fear and distrust in all people.

The first time we meet this character is upon arrival to Iping, a quiet Sussex village, on a cold February day. Completely covered by a coat, overcoat, hat, two pairs of gloves, scarf, glasses and bandages around his head, his looks raise awkwardness in the inhabitants, who weave several bizarre theories about his origin and intentions. Has he escaped justice, is he a dangerous anarchist, a harmless lunatic, a half-breed with white and black patches on the skin, as horses do? Or is he a supernatural being? The mystery shrouding the enigmatic visitor poisons the thoughts of a once unstressed population. A great discomfort finds a way to their hearts, evolving to terror once his secret is known.

Indeed, the combination of irascibility and egocentricity in Griffin's temperament increases. As the siege closes in on him, he trespasses on private property, violates everyone's privacy, steals, assaults, and even kills. And because an invisible man needs a visible sidekick to carry his baggage and watch over his books, he ends up choosing a society pariah as his partner. After this poor man escapes, he searches the help of an old university colleague. However, Kemp, in spite of his scientific education, does not approve either the ends or the means used by Griffin. At the end of the book, he who believed that invisibility would make him the one-eyed man in the country of the blind, ends up dead and in a state of complete demise: chased by all, without his beloved books and clothes, the final image of the invisible man is of utter social, physical and intellectual loss.

Both Moreau, the biologist, and Griffin, the physicist, are extremely intelligent men who erect plans for a new order of things where progress favours first and foremost their private ambitions. Griffin's death not only makes visible the naked body of one who aimed at absolute power and who thought could win over nature; it is equally visible that overwhelming ambition causes total vulnerability.

2.3 *The one-eyed man as King*

There is a second invisible man in a work from 1904, called *The Country of the Blind*; yet, in this case, the invisibility is not the result of a scientific discovery but of a natural evolution. In *The Country of the Blind*, a series of natural catastrophes — flooding, landslides, etc. — resulted in severing the ties that bound that land to the rest of the world. Protected from outer influence by high mountains, the valley not only reflects the insularity (literal and symbolically) conventionally associated with a utopian universe reminiscent of Paradise: the mild climate, nature's bountiful flow of food, the tameness of the beasts and the leniency of habits of the people living there come together to mirror the notion of utopia as a reward granted to the best men and women. Except for one odd circumstance that in the past had blotted the happiness of the inhabitants: an unknown disease had spread, without containment, and had gradually blinded those living there. One day, someone from the community had decided to venture into the world seeking God's help; took a bar of silver and never returned. Fifteen generations later, only blind children were born and not even the memory of sight remained in that paradisiacal valley. Likewise, the memory of a world beyond the rocky boundaries had been lost.

The arrival of the protagonist follows the pattern of a troubled entrance into the utopic world. Nunez, coming from a neighbouring land, a man well-travelled, well-read and enterprising, was climbing the Andes with a group of English people when disappeared. Sliding and falling, he happens upon the valley, where he finds a community that sleeps during the day and toils at night, walks in single line, builds houses without windows and paints them in the most improbable colours, even by today's standards (dark grey, dark brown, olive-green).

When he realises that he is in a place where everyone is blind, Nunez remembers immediately the popular proverb that had excited the main character of the previous narrative — "in the country of the blind, the one-eyed man is king" — and imagines the immense power he will have by using such advantage. Nevertheless, the result is quite different from the imagined one. Instead of a triumphal arrival, a sequence of misfortunes places him in a frail spot before the community. He is constantly tripping because he cannot see in the darkness (for it is at night that all the inhabitants are awake and active); he uses words that are not understood by the blind speakers (as seeing, sight, to see); due to a less accurate hearing, he is not capable of pursuing instructions as plain as "follow the sound of my feet"; his communication skills using the valley's code are basic, and he fails at producing and interpreting almost inaudible sounds, subtle intonations and different degrees of touch — summarising, he's quickly declared as an inferior human being, physically and mentally, a savage to whom everything must be taught from the beginning and from whom a slower and more difficult learning process than that of the local children is expected. As with other savages inhabiting pages of European literature, the culturally predominant community even changes his name, using the designation of the place he claims to be of his origin. Nunez is, hence, the Bogota of the blind. Moreover, in the name of truth and its ethical and intellectual edification, he is forced to swear to a creed that includes denying the existence of sight and the properties of the physical space as the mountains, the clouds, the moon, the stars and the world outside the valley.

He cannot even hide, for the keen ear of the inhabitants and the layout of the town do not make an escape easy. Because he cannot count on no one's help, he is a weak opponent and easily defeated in a physical confrontation. The need to survive ends up breaking him into respecting the Law that regulates that community. As time goes by, he falls in love with the only pretty woman, universally rejected due to the tone of her voice, deemed unpleasant. When the amorous relationship becomes public, the youngsters persecute them for fear of degeneracy of the race, and the elders decide that only by complying with a certain condition could the intruder become a full citizen. In order to attack the source of Nunez's mental weakness, a drastic measure is proposed: to surgically extract his eyes. To them, the surgery would irrefutably eliminate the cause of the permanently demented state of his mind, given to delusions and hallucinations with non-existing entities, as clouds and stars. Thanks to science, there was a solution to the problem, but Wells' utopia draws back in face of Nature's power. Nunez leaves during the sleep of the blind to meet the morning light on the other side of the mountain. By refusing the science that would blind him, he surrenders to dawn and to a world of shapes and colours.

3 A FOURTH HYPOTHESIS: USING SIGHT TO SEE THE POSSIBLE

A Modern Utopia ends with a return home as well. There, as well (or especially there), the notion of utopia reigns as Double and Image of the real. Clear in the hybrid quality of the writing, which combines elements of the essay and from fiction, the half-breed nature of this text is carried out by using several unfolding that reinforces the Duplication process from which the Images are created: the utopian place is a planet that replicates planet Earth exactly. To this planet, two travellers enter: the narrator and a companion with whom he is always in discord. Not only has each one of them their respective doubles in the planet of Utopia, as each animal, plant, place, object and inhabitant with an earthly existence (including the readers of the book, and even the heavenly bodies) has another (and better) version in the utopian planet.

It is not irrelevant that the event founding the duplication of the worlds is the moment of reading, for this event in itself reveals several unfolding: reading is a consequence of the previous act of writing; it's the point of convergence between two minds (the mind that communicates and the mind that interprets) and constitutes an image of the possible at the level of consciousness and will.

Besides, this Modern Utopia expresses without hesitations the utopian legacy at the same time that it takes on a difference concerning the past. The modern element that tells it apart brings novelty compared with the status quo model for the genre: Wells' utopia invests in building up verisimilitude. In the past, the perfect symmetry of the streets, the uncharacteristic uniformity of the buildings and the clothes went as far as the indistinctiveness of the human element. The new utopia praises the importance of freedom (considered more important than virtue and happiness), respects privacy and values aesthetics. For no one acknowledges as valid or likeable a social order that voids personal identity, the society of the future welcomes diversity of opinion (there is even a vegetarian ecologist), money and socioeconomic differences (here based on talent and not by birth), wealth and private property, critical thinking and free transit of all inhabitants.

Travelling is, by the way, a requirement for personal education and inevitable in a world peopled by migrant populations that frequently resort to an efficient transportation network in order to know the world where they live. In 1905, H. G. Wells predicts globalisation associated to the shortening of distances, and multi and intercultural thoughtfulness: the modern utopia, unlikely its predecessors, is friendly to visitors, even if they tell implausible tales of an interplanetary voyage driven by an act of textual imagination. In addition, because it's fundamental that meetings in this new world be peaceful, political unification around the World State produces a linguistic reflex — only one language is spoken on the entire planet. This language of synthesis is a sort of English, but more spacious due to the greater attraction power that it exerts over the other linguistic systems.

In spite of the political and linguistic unification, the modern utopia is not able to present itself uniformly, as the existence of small sub-worlds contained in the interior of its world serves to show. Given that perfection is a goal to attain, but not attainable, the double of Earth did not manage to eradicate the indolent, the invalid, the void of character or of intelligence, the alcoholic, the useless, the criminal, the incompetent and the social parasites. By valuing life and freedom, the juridical system only puts into force sentences that vary between the extremes of poverty and of exile to islands that function as penal colonies. In the words of the author, it is necessary to resort to 'social surgery' to stop a handful of toxic individuals from destroying the world. Each island has its criminal community and its members live in freedom amongst equals and in obeisance to their own laws. The penal colonies are so distinct from each other that although the island for alcoholics is much more joyful, it is far less organised than the island for incurable swindlers, where the establishment of order is taken seriously so that the pleasure of infraction can be real. The ports and seas are kept under watch so that no prisoner may leave the island of exile.

If the penal system mirrors the difficulty in amending socially nefarious behaviours, other cases reveal the success of the innovative measures used in that space. Amongst those, fighting discrimination against women is of importance. Wells believes that the power of men over women is based on the fact that all differences between the genders represent an economic disadvantage to women. As the main difference lies in the reproductive system, modern utopia made motherhood a professional choice so that women may emancipate. The woman that pursues the career of motherhood with professionalism and competence is handsomely remunerated, for this is considered a crucial occupation for the future of the State and the entire community. Professionalisation of motherhood frees women from mercenary marriages and contemplates supplementary rewards for each child who becomes a successful adult.

4 CONCLUSION

We can understand each one of the four Wellsian narratives here analysed as an important stepping-stone for the conception of Progress. H. G. Wells saw progress and happiness as a real possibility requiring careful planning. That it is possible may be proven by the image of Utopia as a Double of Earth. However, even the first three narratives are more than seemingly failed attempts at picturing progress. If *The Island of Doctor Moreau* offers a convincing argument against acceptance of limitations imposed by a cold and self-centred authority, it also provides a clear image of the hero as one who can balance knowledge and unselfish love of humanity. In *The Invisible Man*, the possibility of improvement loses ground in face of the stronger intention of raising awareness to be blinded by power, but this narrative reinforces, nonetheless, the message of the last work chosen, where the vision of a perfected society is more vivid. Likewise, in *The Country of the Blind*, the choice of returning to the world beyond the mountains is a clear sign of abandonment of unchecked ambition.

In all we encounter the portrait of social convulsion caused by the uneven distribution of power, leading the reader to conclude that a world peopled by blind and invisible people is of no advantage to the former or the latter. As Borrello (1972) stresses, Wells was aware of the many perils surrounding humankind in the age of progress: an all-pervading sense of estrangement, alienation and loneliness jeopardising effective communication, doubts about the future, the effect of dehumanising technology and a haunting suspicion of impending catastrophe are felt alongside an unfaltering optimism.

Another conclusion from the conjoint reading of these narratives is evident: not only everything that exists cannot be seen, as not everything that cannot be seen is inexistent. *A Modern Utopia* is the proof of that. The same principle of analogy that tells the difference between the Image and its Object will tell the difference between the utopian planet and the Earth. The same process of making the absent present that shares the definition of Image explains why textual imagination can represent what has not yet had historic existence. The same outlook that deciphers the invisible by dominating visual syntax observes the double of the planet from Earth. As much as the Image is not mixed up with the reasons of the reality, it copies, evokes, or recreates, progress in H. G. Wells leans on the principle of verisimilitude to dialogue with historic reality without becoming same with it. Modern Utopia is a Double offered to the Earth inhabitants so that by contemplating the replica of the known world, they can stare at themselves in the mirror, and see themselves, not as they really are, but as they could be.

What is Earth lacking to become the better image of itself as it sees itself in the mirror? The Double of the Earth is in everything the same, except in the mental content of life. What tells the two planets apart is "a different literature, a different philosophy and a different history". By declaring invalid the ambitions of the blind and invisible protagonists of the other narratives, A Modern Utopia invites us to develop a new habit of thought. As stated at the end of the work, "It spoils the world of everyday to let your mind run on impossible perfections".

The places we know may in the future welcome a world state inhabited by happy citizens. For that to happen, humankind must reject the type of blindness that encourages ambitions of power. Thus, *A Modern Utopia* presents us more than a mere duplication of the Earth. It declares that progress must always have a collective scope. Its positive influence must nourish the dream of universal freedom, accomplishment and happiness; it must fight private ambitions and outworn social conventions. It must be aware of science but reject the posture of invincibility. It must embrace the lesson that only through cooperation can we are given the chance to move forward, conducting our lives in line with the principle of meaningful accomplishment.

BIBLIOGRAPHICAL REFERENCES

Amabile, Teresa, Kramer, Steven. (2011) *The Progress Principle*. Boston, Mass.: Harvard Business Review Press.

Bentham, Jeremy. (1995) *The Panopticon Writings*. Ed. Miran Bozovic. London: Verso.

Borrello, Alfred. (1972) *H. G. Wells: Author in Agony*. Illinois: Souther Illinois University Press.

McConnell, Frank. (1981) *The Science Fiction of H.G. Wells*. New York: Oxford University Press.

Philmus, Robert M. (1970) *Into the Unknown: The Evolution of Science Fiction from Francis Godwin to H. G. Wells*. Berkely & Los Angeles, California: University of California Press.

Wells, H. G. (1968) "The Country of the Blind", in: *Selected Short Stories*. Harmondsworth: Penguin Books.

—. (1968) "The Invisible Man", in *Selected Short Stories*. Harmondsworth: Penguin Books.

—. (1975) *The Island of Doctor Moreau*. London: Pan Books, Ltd, in association with William Heinemann, Ltd. 1975 (Heinemann, 1896)

—. (1930) *Tono Bungay and a Modern Utopia*, London: ODHAMS PRESS LTD.

Two dystopic visions on the relationship of humans and progress – Emile Souvestre and Cordwainer Smith

Maria do Rosário Monteiro
CHAM, FCSH, Universidade NOVA de Lisboa, Portugal
Universidade dos Açores, Portugal

ABSTRACT: Zaha Hadid's statement that is used as an epigraph to this book is also the cornerstone of this essay: "Non puo esserci progresso senza affrontare l'ignoto". This sentence has an ambiguous meaning and can be interpreted in at least two different ways: either as a natural challenge or as an aggressive defiance. This ambiguity encompasses the relationship between individuals, communities and progress, reminding the image of Janus, each facing a different side, both forming the same and a different entity. It is a natural and complementary ambivalence. Every endeavour undertook by promoters of progress has a degree of uncertainty, a pending threat of failure, and the outcome always produces positive and negative consequences frequently in uneven ways.

Culturally, in a consistent way at least from the beginning of modernity, Western civilisation has regarded progress as a natural unstoppable endeavour. Sometimes even as a duty of every rational educated person – to pass (or trespass) the frontier of the known, to act, to evolve, to transform, to change, and to discover the "God given world". This almost linear way to understand progress, to view reality from a dominant, sometimes exclusive point of view, tends to erase the notion and effects of negative consequences both on individuals and on communities, assumed by the dominant culture as acceptable collateral damages in the name/notion of rational evolution.

In this essay, the main goal is to defy this dominant trend of evolution as based mainly on material, objective, rational progress, and defy the unique view of a future based on quantifiable and technological evolution. This challenge will be done with the comparative analysis of two literary texts: Emile Souvestre's *Le monde tel qu'il sera* and Cordwainer Smith's "Alpha Ralpha Boulevard".

Keywords: Progress, Emile Souvestre, Cordwainer Smith, unique versus multi-visions of reality, utopia, dystopia

1 FROM ENLIGHTENMENT TO THE RATIONAL, POSITIVISTIC POINT OF VIEW OF PROGRESS

1.1 *Enlightened philosophical optimism*

The evolution of Western civilisation, from Antiquity to the 20th century, has followed a pattern based on the assumption that there is one world, one reality, and to understand it meant to use an exclusive rational point of view. There were exceptions to this dominant perspective, variations on emphasis, periods of radical defence and sceptical perspective. The history of the idea of progress is not linear, for human history did not evolve this way through time. But the idea of progress is profoundly connected with another fundamental idea analysed by Arthur Lovejoy in his exemplary study *The Great Chain of Being* (1964).

The idea of progress and its connection with science and technology has its origin in the Enlightenment. It is an 18th-century creation, though their "inventors" seemed incapable of realising the full extension of such a connection. As Lovejoy states:

> To many eighteenth-century minds this conception of a world in which, from the beginning, no emergence of novelty had been or would hereafter be possible seems to have been wholly satisfying" (1964: 243)[1].

Among others, Leibniz had opened the hypotheses of creation being a multitude of worlds (that implying a multitude of different forms of reality), but this world, the one we live in, is seen as the best

[1] Abbé Pluche wrote in a book that became almost a best-seller considering the time of publishing, that: "Nothing more, therefore, will be produced in all the ages to follow. All the philosophers have deliberated and come to agreement upon this point. Consult the evidence of experience; elements always the same, species that never vary, seeds and germs prepared in advance for the perpetuation of everything, ... so that one can say, Nothing new under the sun, no new production, no species which has not been since the beginning." (Apud. Lovejoy, 1964: 243)

possible world, in accordance with Western religious and cultural *Weltanschauung*.

Following some 17th-century "giants", Leibniz, Diderot, Voltaire, Rousseau, Newton and so many other philosophers and scientists opened the door to the multitude, diversity and evolution but did not cross its threshold and break off from the idea of an accomplished, finished universe.

> WHEN the principle of plenitude was construed either religiously, as an expression of the faith in the divine goodness, or philosophically, as an implicate of the principle of sufficient reason, it was, as usually understood, inconsistent with any belief in progress, or, indeed, in any sort of significant change in the universe as a whole. (Lovejoy, 1964: 242)

Commonly, the optimism implied, for instance, that Leibniz's theory suited well the 18th-century intellectual elite (Leibniz, 1934: 52–87; 1902: 65–248). Optimism is the fundamental feeling shared by scientist and philosophers during Enlightenment: the absolute faith in Reason facing an infinite universe. The enlightened reason will free itself from constraints of religion and will focus its attention on the intellectual improvement of beings, not through the use of science and technology, but through knowledge, understood as cumulative and progressive (Condorcet, 1822). As Raymond Trousson states even in the first utopia build on the idea of infinite progress, *L'An 2440. Rêve s'il en fut jamais* (Mercier, 1801: first ed. 1771):

> L'essentiel demeure, en 2440, la vie simple et austère fondée sur la morale et la vertu, et l'étude de la nature a toujours pour fin première de faire comprendre et admirer la création selon les principes, non plus des Églises oppressives et dogmatique, mais d'un pieu déisme. Astronomes et savants sont donc devenus, non seulement des hommes de science respectés, mais "les hérauts de la Divinité". (2003: 79)

It was all a matter of time and effort to educate the individuals so they would see the light.

Some years after Mercier wrote his *Utopia*, Condorcet, in *Esquisse d'un tableau historique des progrès de l'esprit humain* (first edition 1795) expresses the optimism that will stay as the peculiar mark of the 18th-century elite:

> Si l'homme peut prédire avec une assurance presque entière, les phénomènes dont il connoît les lois; si lors même qu'elles lui sont inconnues, il peut, d'après l'expérience du passé, prévoir avec une grande probabilité les événements de l'avenir; pourquoi regarderoit-on comme une entreprise chimérique celle de tracer avec quelque vraisemblance, le tableau des destinées futures de l'espèce humaine,

Figure 1. Frontispiece de *L'Encyclopédie*, 1751.

> d'après les résultats de son histoire. (1822: vol. 1, 327)[2]

The same mark we can find previously in the American Declaration of Independence, in the final version revised by Benjamin Franklin (1776):

> We hold these truths to be self-evident, that all men are created equal, that they are endowed by their Creator with certain unalienable Rights, that among these are Life, Liberty and the Pursuit of Happiness. That to secure these rights, Governments are instituted among Men, deriving their just powers from the consent of the governed.

In addition, in the French Revolution motto (1789) "Liberté, égalité, fraternité.

1.2 *Rational positivist optimism and the other side of the coin*

By the end of the 18th-century, a profound transformation occurred in Western Weltanschauung. The romantic ideals swept through European countries: rules, unanimity, classical ideals were replaced by a

[2] Quoted in modernised French by Trousson (2003: 74).

Promethean view of humanity[3]. From philosophy to the arts, from economy to politics, human endeavour promised an age of bliss, transformation and progress. Progress was the key word: from the wealth of the nations to the transformation of social and political realities, everything seemed possible.

The progress of science gained a purpose: knowledge was to be used for humanity sake in concrete, objective creations. Science and technology were to be united. The utopian ideal, expressed until then mainly through philosophy and literature gave way to utopianism (Trousson, 2003: 85–91).

The last decades of the 18th and the first decades of the 19th centuries had one goal: progress. The Industrial Revolution fuelled the optimism, inspired philosophers and entrepreneurs who would promise the end of poverty, wealth and prosperity. A golden age for all. Its advocates were Saint-Simon (1760–1825), François Marie Charles Fourier (1772–1837), Étienne Cabet (1788–1856), Robert Owen (1771–1858), among others. In all of them, we can read the promises of the new age.

> L'imagination des poètes a placé l'âge d'or au berceau de l'espèce humaine, parmi l'ignorance et la grossièreté des premiers temps; [...]. L'âge d'or du genre humain n'est point derrière nous, il est au devant, il est dans la perfection de l'ordre social; nos pères ne l'ont point vu, nos enfants y arriveront un jour; c'est à nous de leur en frayer la route. (Saint-Simon, 1859: 328)

As Trousson explains the dream became more comfortable, there should be a general moral improvement, but at the same time, material wealth, in a renewed world where industry and technology spread a fair abundance (2003: 85).

However, Romanticism, even after being surpassed by Realism, left a permanent footprint in Western civilisation: the end of unanimity, the importance of the individual, the liberation from pre-existing codes, the creation of hybrid new genres, the appeal of emotions and their valuation, the importance of the unconscious impulses, etc.

The positivist euphoria never quite silenced pessimism. It laid dormant, almost inaudible, but it was there and grew from the 30s on to explode finally in the creation of dystopias, the dark side of western society. When one looks at the history of utopia, it becomes evident that the 19th century marks the beginning of the end of belief in ever achieving a future state of eutopia[4]. Most utopias written during the 19th and 20th centuries reflect a total belief in science and technology in a naïve way. With the exception of Cabet's *Voyage en Icarie*, published in 1839, there is no true innovation in utopia form or philosophy[5]. Comparing with the previous centuries it is quite a poor site.

The motto that governed the 19th-century utopias may be epitomised in Cabet's words:

> Et remarquez-le bien ! les grandes découvertes dans les sciences et dans l'industrie ne font pas seulement des Révolutions *scientifique* et *industrielles*, mais aussi des Révolutions *sociales* et *politiques*; car tout se tient, tout se lie, ou plutôt tout se confond dans la Nature, qui n'est qu'une immense UNITÉ, dans aquelles nous ne distinguons des éléments divers que pour aider la faiblesse de notre intelligence. (1845: 468)

The problem arose precisely on the social and political spheres. The objective evolution on science, technology, in industry and manufacturing, and in communications was undeniable. However, far from this scientific and technological paradise stood millions of underprivileged, the victims of economic crises, of dismissal, those who had no voice, no education, and no future.

The ground became, little by little, fertile for a different form to express the hope of a better future. Until then, the wastelands of industrialisation fuelled disbelieve. Using the hybrid form of the novel, another romantic gift, dystopia gave what utopia could no longer offer: a voice to express the doubts, the mistrust and even the desolation with humanity's present

[3] Trousson (2003: 88) quotes a short piece written by Alfred Desessarts, in 1833, entitled "Prométhée; scène psicologique", and published in *Revue Universelle: Bibliothèque de l'homme du monde et de l'homme politique*. This short texts ends with Prometheus looking at the future of humanity that, finally, promises to justify the actions that led to his punishment: "Ne savais-tu pas aussi qu'en symbolisant ma douleur, cette douleur trouverait de la pitié et soulèverait d'éternelles plaintes; que sur mon rocher je représenterais l'humanité étroitement enchainée? Va! l'exemple de mon châtiment se brisera contre l'insouciante ardeur de ceux qui se précipiteront sur mes traces ver la science. La science, je la vois empreinte du caractère des siècles, grave ou folle, pieuse ou athée, je la vois aspirer à saisir et lever le voile divin; je la vois, à pas lent lents et mesurés, ou impétueux et mal calculés, pénétrer dans le labyrinthe inconnu pour arriver à un sanctuaire, imaginaire peut-être, du moins tant qu'on n'en aura pas trouvé les portes. Bien! science humaine ! courage ! fille de la terre ! Je la vois déjà, je la vois toujours ! elle creuse de ses larges mains, elle efface, creuse encore, grave profondément, elle sue à édifier; puis, obéissant à sa loi d'instabilité, elle laisse une époque railler, détruire le travail de l'époque qui l'a précédée, engendrée ... Mais l'œuvre se lègue, se continue, et la science rajeunie, raffermie par son espoir, franchit les dernières limites et va sur le trône d'éternité s'assoir, toute palpitante, auprès du Dieu immobile! ...." (Dessessarts, 1833: 189).

[4] Comparing the 19th to the 18th century there is no doubt that there was a huge decay in the writing of utopias that persists until now and the creation dystopias and their progressive increase in the number from the mid 19th century and through the entire following one.

[5] Cabet brings no real innovation of form, but uses the text to expose his political philosophy where science and technology function as the cornerstone of political, social and economic development that he considers inseparable.

and future, not only in the collective sphere, but particularly taking into account the individual in conflict with dominant society's values and practices, or, putting it more clearly, the dominant social class of entrepreneurs and capitalists, the bourgeois elite. Far from bringing happiness to all, the industrial transformations have deepened social differences, factories quite different from the pleasant and healthy working stations described by socialist utopians, where silent machines are operated effortlessly by few watchers (Trousson, 2003: 129).

The reality the dreamers of progress could not, or would not see, was the one described by Dickens in *Hard Times*, in 1854, the same described by Pierre-Joseph Proudhon in 1846 (1867: 135–173; Trousson, 2003: 131). The weight and oppression call to mind Piranesi's Prisons with the tiny little human figures overwhelmed by the dark, gothic and gigantic prisons walls (Ficacci, 2001).

Piranesi picture

2 A FIRST CRY OUT AGAINST POSITIVIST UTOPIA OR A CALL FOR HELP

In 1846, just one year after Cabet's quoted words above, a French novelist published the first known dystopia. Emile Souvestre's *Le monde tel qu'il sera* [*The World as it shall be*] (Souvestre, 1846a)[6]. Ironically, Souvestre had been an adept of technological progress, a follower of Saint-Simon and other French philosophers of scientific and technological progress.

This novel is Souvestre only adventure in science fiction, or what French critics call *roman de l'avenir*. If Thomas More with his foundational text has determined the future persistence of both utopia and dystopia as a new literary genre, as Versins refers, Souvestre's work became "the model for everything that will be written in the genre during the 19th and 20th centuries" (1984: 824).

Of course, Versins is thinking in terms of dystopia's major thematic: the danger of technology when used for particular or unclear motives, the possibility of creating human-made hells, etc. However, when we compare Souvestre's text with Wells, Huxley, or Orwell, surely much was anticipated by Souvestre. However, there is still some naïveté in this first dystopia. The futuristic universe still bears the marks of romanticism and is still very far away from Orwell or Huxley's realism, a cruel disappointment, even some sense of a claustrophobic universe.

Souvestre's "enchantment" with Saint-Simon and Fourier theories of progress led him into believing that art should have a useful purpose, should be a way to educate women and the working class (Plötner-Le Lay, 2006: 29). By 1845, he distanced himself from social utopianism, though remaining a Republican and adopting a Christian social and moral philosophy, as *Un philosophe sous les toits – journal d'un homme heureux* testifies[7].

Unlike *The Attic Philosopher*, *The World as it shall be* had a short literary fortune: three French editions in the 19th century and only two in the 21st century ... (Souvestre, 2013; 2002; 1871; 1859; 1846a), a Portuguese adaptation in 1859[8], reissued in revised edition with a critical introduction by Fátima Vieira in 2006 (Souvestre, 2006; 1860) and an astonishing number of six Spanish translations, all printed in the 19th century ... (1890; 1883; 1868; 1857; 1852; 1846b)[9]. The first English translation is only published in 2004.

As Utopias, dystopias are based on the actual state of affairs the author lives in and, in 1846, the social and political situation in France was characterised by unsteadiness.

> The Second Restoration, the failures of July's Monarchy and the Second Republic had vast consequences that went beyond the political sphere, affecting the whole society. By the time Souvestre was writing *Le monde tel qu'il sera*, France was facing a violent social and economic crisis: famine, unemployment. Lack of social protection drove many people to despair; the industrialists cut their losses by closing factories, ignoring the social consequences. ... (Monteiro, 2007: 287)

The main characters are a French couple who visited for three days a supposed Utopia set in the year 3000. What they discover, however, is an unfair and unequal society where both rich and poor human beings are deprived of their humanity for the sake of profit. In that dystopian future, the political structures are perfect because they are inexistent and/or ineffective! Stupidity seems to have become the most effective human ability, action and interaction are dominated by faked scientific principles, morality is absent, and there is no social interaction.

> The three days Maurice and Marthe visit the world in the year 3000 lead them to a conclusion: the future promised by industrialization

[6] For English readers of Souvestre's novel I suggest the excellent Clarke's English translation (Souvestre, 2004).

Part of the analysis of Souvestre's and Cordwainer's texts were previously published, in a more thorough version, in "Two Technological Dystopias: *Le monde tel qu'il sera* and *Alpha Ralpha Boulevard*" (Monteiro, 2007). The perspective adopted eight years later will be, necessarily somewhat different, in a way I hope will be complementary.

[7] The first dated edition of this work is 1850. One knows that in 1851 the novel was acclaimed by the French Academy, in the same year, Souvestre's editors published a third edition and the first English translation appeared in 1853 (1853; 1851; 1850).

[8] The Portuguese edition includes twenty illustrations drawn by the illustrator Vidal Junior who copies the original ones, making few changes, mostly minor details that do not alter the overall design.

[9] These are the Spanish translations that I was able to identify so far. In these six editions, apparently only the 1846 does include illustrations.

> and technology will enslave humanity in a vortex of spiritual degradation. (Monteiro, 2007: 290)

The Enlightenment ambition and faith on a rational humanity, one that has seen the Lights of reason, will not happen in the following centuries. As Kumar states most 19th and 20th centuries writers of Utopias/Dystopias were "overwhelmed by what seemed to them the invincible folly and stupidity of mankind" (Kumar, 1991: 91). Souvestre's dystopia set the mood to face utopias from the mid 19th century onward.

> … the realization of utopia was bringing in a world of unprecedented servility and sterility, a world where old forms of tyranny were returning in the new guise of mass democratic politics and benevolent state planning. (Kumar, 1991: 93)

However, probably worst than tyranny and collective stupidity, the major menace that utopia presented was the eminent annulment of free will, the complete destruction of individual ego, producing people completely and hopelessly incapable of remonstration (Trousson, 2000: 183).

3 THE CONFIRMATION OF THE "UTOPIAN MENACE" – "ALPHA RALPHA BOULEVARD"

In Thomas More's *Utopia*, that is, from the beginning of the genre, Utopia and its twin Dystopia have functioned as Kumar says, as two faces of the same coin, and I add, the obverse being positive (utopia) and the reverse being the negative (dystopia). However, looking at the history of Utopia in the past five hundred years one gets sometimes the feeling that we were given a "loaded coin", tending to flip systematically to the obverse side, as if this would inevitably be humanity's destiny.

This feeling has its source in the very nature of utopia, a literary genre deeply connected with the dominant *Weltanschauung*. Born in a period of discovery but also of political, social and religious controversy, More's *Utopia* places side by side both twins, as if in equilibrium: neither dystopia was completely negative nor Utopia was faultless.

However, Modernity was characterised by a general feeling of euphoria and belief in the supreme abilities of the human being. It is the period of Enlightenment, followed by the French, the agrarian and the industrial revolutions, all triggered by intellectual, scientific and technological progress. During this period, there is no place for doubts: Utopia reigns and the reverse hides in the shadows, waiting the right moment to challenge the *status quo*.

Souvestre brought the negative twin to daylight and gave dystopia the possibility of avenging centuries of silence. The revenge was intense: Utopia became the enemy, the hell on earth, the future "Big Brother" that would control individual egos, silencing them, annihilating individuals and turning them into herds.

> In the name of a very near and threatening future, utopia warned and recalled people to reason rejecting the myth of indefinite progress and extreme industrialization that Nodier and Souvestre had denounced. (Trousson, 2000: 185)

If there is still some romantic and naïveté in Souvestre's dystopia, the 20th century struck a deadly blow to the naïve imagination. Now, the situation was reversed: dystopia warned against the evils of static utopia. Evolution is the key to understanding life, and human life is governed by the same natural law: live, adapt and survive. So are human endeavours. After a long period of apparent stagnation in a dystopian vision, the 20th century witnessed several changes in the way Western society perceived the future. This had its reflection in the history of utopia that also had to change and adapt in order to survive in a more realistic and less naïve culture. Since the second half of the 20th century, the announced death of utopia proved to be unrealistic, for the drive for living and imagining a better future is an unconscious one, shared by all humanity. Vita Fortunati summarised this evolution:

> Utopia […] is no longer static and is no longer a system that has been planned one time for all, but it is a Utopia as a continuous battle to achieve a better world. (Fortunati, 2000: 642)

Utopia and dystopia had the help of science fiction to make their point. In addition, if dystopia used this genre from Souvestre onward, utopia profited from a change in science fiction writing, which occurred in the 60s. The era of faith in science and technology ended and science fiction, from the 60s onward, gave voice to this change.

> [The new science fiction] is characterized by literary experimentation and by a growing interest in human sciences, such as psychology, and sociology. These new themes introduced in science fiction's canon were accompanied by a lack of faith in human intelligence, in the human ability to achieve perfection, and a lesser concern for the scientific accuracy of its extrapolation. These alterations made possible the creation of "ambiguous utopias" such as Cordwainer Smith's "Alpha Ralpha Boulevard".[10] (Monteiro, 2007: 291)

The short story "Alpha Ralpha Boulevard" was first published in 1961 (Smith, 1993) and it presents several transformations regarding utopian tradition. The most

[10] For more developed information and analysis of Cordwainer Smith's works please see Burn (1975), Elms (1984), Lewis (2000), Hellekson (2001), Monteiro … (Monteiro, 2014; 2012; 2007).

evident is the abandonment of the narrator functioning as an outsider who is guided through the utopian place and describes afterwards what he saw. In this short story, the main character is not an observer but a participant first person narrator, who constantly reformulates his judgements, who is biased and changes his expectations.

However, there are several similitudes between *The World as it shall be*, and "Alpha Ralpha Boulevard". For instance, the narrative action starts in the Utopian world that will turn out to be dystopic. The cause of the dystopian state of affairs is not inequality, or moral flaws, or profit, or the extreme liberal code of "laissez-faire laissez-passé" among human beings. It has a cost that is paid by the Underpeople, the genetically modified beings who sustain, literally, the utopian bliss of "genetically pure" human beings.

> Through a long period of evolution, human beings achieved Utopia by the 141st century. However, two millennia later, perfection is destroying humanity because "happiness can kill people as softly as shadows seen in dreams" (Smith, 1994: 290). The world government, called the Instrumentality of Man, had sworn to preserve humanity pure and happy, taking advantage of all the benefits of space colonialism, technology and science. After two millennia of perfection, they are forced to admit that they were wrong, that perfection does not fit human nature. (Monteiro, 2007: 292)

Perfection is now seen and felt as unnatural. Flawlessness is adverse to evolution, for what is perfect simply *is*, never *becomes*. Life does not function in accordance with the rule of perfection, but with the rule of evolution. Perfection is killing humanity: the lack of surprise, of problems, of challenges. The lack of uncertainty that gives life its flavour and its intrinsic value.

> We are sworn to uphold the dignity of man. Yet we are killing mankind with a bland hopeless happiness which has prohibited news, which has suppressed religion, which has made all history an official secret. I say that the evidence is that we are failing and that mankind, whom we've sworn to cherish, is failing too. Failing in vitality, strength, numbers, energy. (Smith 1994, 290)

The short story starts in the moment utopia begins to be dismantled: the cholera virus is spread and the news report the first fatalities, the first people being freed from perfection, from a predetermined 400 years lifespan. The old cultures and reintroduced, people are implanted with memories, different languages, different nationalities. The main characters will play the role of Adam and Eve of the new imperfect world.

This transformation is gradual and once again, it is decided by a supreme government and applied by the Underpeople. These, as Souvestre's workers are the anonymous and ignored victims of human utopia. However, they are not victims, as in Souvestre's texts. They are creatures of their own destiny: they had no political or social relevance, lived underground, away from the humans who they serve, but even so, they built their own society, their own religion, amusements and economic structures, their moral values and their political aspirations.

Moreover, they watch the new humanity giving the first step in an unpredictable life. They watch the new Adam and Eve (the characters Paul and Virginia) destroy paradise in a similar way that led the primaeval humans to be expelled from the Garden of Eden: expressing doubt regarding authority's actions, the reality they are experiencing, that was imposed upon them. It is up to Eve/Virginia to "bite the apple":

> Paul, why does it all happen so fast? This is our first day, and we both feel that we may spend the rest of our lives together. There's something about marriage, whatever that is, we're supposed to find a priest, and I don't understand that, either. Paul, Paul, Paul, why does it happen so fast? I want to love you. I do love you. *But I don't want to be made to love you. I want it to be the real me.* (1993: 382)[11]

Virginia has the active role, as in the myth of Eden, and Paul plays the passive role of Adam, bringing to mind the character described in *Paradise Lost* (Milton, 2005). Their quest will be the quest for truth concerning their most intimate reality. They do achieve it and it is up to a member of the Underpeople to express that knowledge concerning humanity's inmost nature, to create it by expressing it aloud:

> All of us have been worried about what true people would do to us when you were free. We found out. Some of you are bad and kill other kinds of life. Others of you are good and protect life. (1993: 398)

4 CONCLUSION

The Enlightenment idea of progress as unstoppable was correct. However, the same does not apply to the absolute faith in considering it the way to perfection. History has proved them wrong. Evolution does not follow a linear route; neither does it have only a positive side. Perfection is not a natural law, though it may and should be a goal. Only in aspiring to improve, in doubt, in facing problems and discovering new ways to solve them can society achieve progress, but it cannot become a goal in itself, only an instrument for human evolution. Utopia will always be where we set our eyes on, but it will not be the final goal. Progress can only be beneficial if it is subjected to the law of uncertainty, the same that governs life. Absolute truths, definitive decisions, final assessments are doomed to be proved wrong. Every action or decision, every discovery or new technology will have to be considered

[11] Underline added.

as having two complementary consequences: positive and negative. Human kind has to learn to balance their diverging inclinations. The future progress may offer humanity will have to be "an ambiguous imperfect Utopia where people have to battle constantly to make it a better world" (Monteiro, 2007: 295)

BIBLIOGRAPHICAL REFERENCES

Burn, A. (1975) Paul Linebarger. In: Porter, Andrew (ed) *Exploring Cordwainer Smith.* New York: Algol, 5–11.

Cabet, Etienne. (1845) *Voyage en Icarie.* 2ª ed. Paris: Bureau du Populaire.

Condorcet, Jean-Antoine-Nicolas de Caritat. (1822) *Esquisse d'un tableau historique des progrès de l'esprit humain; suivie de refléxions sur l'esclavages des nègres.* Paris: Chez Masson et Fils.

Dessessarts, Alfred. (1833) Prométhée. *Revue Universelle: Bibliothèque de l'homme du monde et de l'homme politique.* VI. Bruxelles: Louis Hauman et C°, 184–189.

Dickens, Charles. (1920) *Hard times.* Londres: J.M. Dent.

Elms, Alan C. (1984) The Creation of Cordwainer Smith. *Science Fiction Studies* (34): 264–283.

Ficacci, Luigi. (2001) *Piranesi. Catalogo completo delle acqueforti. Catálogo completo de grabados. Catálogo completo das águas-fortes.* Koln Taschen.

Fortunati, Vita. (2000) Utopia as a Literary Genre. In: Fortunati, Vita and Trousson, Raymond (eds). *Dictionary of Literary Utopias.* Paris: Honoré Champion.

Hellekson, K. L. (2001) *The Science Fiction of Cordwainer Smith.* Jefferson: McFarland & Company, Inc.

Kumar, Krishan. (1991) *Utopianism.* Buckingham: Open University Press.

Leibniz, Gottfried Wilhelm Freiherr von. (1934) *The Philosophical Writings of Leibniz. Selected and translated by Mary Morris.* Londres: J. M. Dent & Sons.

—. (1902) *Discourse on metaphysics, correspondence with Arnauld, and Monadology.* With an introd. by Paul Janet; translated from the originals by George R. Montgomery. Chicago: Open Court Pub. Co.

Lewis, A. R. (2000) *Concordance to Cordwainer Smith.* Framingham: NESFA Press.

Lovejoy, Arthur O. (1964) *The Great Chain of Being: a study of the history of an idea.* Cambridge, Mass.: Harvard University Press.

Mercier, Louis-Sébastien. (1801) *L'an deux mille quatre cent quarante, rêve s'il en fût jamais; suivi de L'homme de fer, songe.* nova ed. Paris: Lepetit jeune et Gerard.

Milton, John. (2005) *Paradise lost.* ed. ilustrada. Introduced by Philip Pullman. Oxford: Oxford University Press.

Monteiro, Maria do Rosário. (2014) From More's natural Utopia to Cordwainer Smith's artificial pseudo-utopia. *Spaces of Utopia: An Electronic Journal* 3: 27–40. http://hdl.handle.net/10362/14927.

—. (2012) People e Underpeople ou as involuções e evoluções da humanidade. In: Rosa, Jorge Martins (ed) *Cibercultura e Ficção.* Lisboa: Documenta, pp. 331–340.

—. (2007) Two Technological Dystopias: *Le monde tel que'il sera* and *Alpha Ralpha Boulevard. Morus – Utopia e Renascimento* (4): 285–296. http://www.revistamorus.com.br/index.php/morus/article/view/190; http://hdl.handle.net/10362/14303.

Plötner-Le Lay, Bärbel. (2006) *Émile Souvestre: Écrivain Breton et Saint-Simonien 1806–1854.* Morlaix: Skol Vreizh.

Proudhon, Pierre-Joseph. (1867) *Système des contradictions économiques ou philosophie de la misère.* 3ª ed. vol. IV. Paris: Librairie International.

Saint-Simon, Claude Henri de. (1859) De la réorganisation de la sociéte européene. *Oeuvres Choisies; précédées d'un essai sur sa doctrine.* 1st ed. 1814. vol. 2. Bruxelles: FR van Meenen et Cie, Imprimeurs, 251–328.

Smith, Cordwainer. (1994) *Norstrilia.* Framingham, MA: NESFA Press.

—. (1993) Alpha Ralpha Boulevard. In: Mann, James A. (ed) *The Rediscovery of Man. The Complete Short Science Fiction of Cordwainer Smith.* Framingham: The NESFA Press, 375–399.

Souvestre, Émile. (2013) *Le monde tel qu'il sera en l'an 3000.* Illustrations et texte d'après l'édition originale de 1846; préface de Claire Barel-Moisan; illustré par MM. Bertall, O. Penguilly et St Germain. Morlaix (Finistère): Skol Vreizh.

—. (2006) *O que Há-de Ser: O Mundo no Ano Três Mil.* Edição de texto, introdução e notas de Fátima Vieira. Acomodado ao gosto português por Sebastião José Ribeiro de Sá. Famalicão: Quasi.

—. (2004) *The World as it Shall Be.* Edition and Introduction by I. F. Clarke. Translation by Margaret Clarke. Middleton, Conn.: Wesleyan University Press.

—. (2002) *Le monde tel qu'il será.* Edição facsimilada. Apex.

—. (1890) *El mundo tal y como será en el año 3000.* Valencia: Pascual Aguilar.

—. (1883) *El mundo tal y como será en el año 3000.* Traductor desconocido. Imprenta de Aznar.

—. (1871) *Le monde tel qu'il sera.* Poissy: Typ. S. Lejay et Cie.

—. (1868) *Lo que será el mundo en el año 3000.* Traducción de F. N.: Librería Cuesta.

—. (1860) *O que há ser o mundo no anno tres mil.* 1ª ed. Acomodado ao gosto português por Sebastião José Ribeiro de Sá. Lisboa: J.M. Correia Seabra & T. Quintino Antunes.

—. (1859) *Le monde tel qu'il sera.* Paris: Impr. de Ch. Jouaust.

—. (1857) *El mundo tal cual será el año 3000.* Traductor desconocido. Madrid.

—. (1853) *The Attic Philosopher in Paris: or, a Peep at the World from a Garret ... From the French of E. Souvestre.*

—. (1852) *El mundo tal cual será en el año 3000.* Madrid: Ediciones Populares.

—. (1851) *Un philosophe sous les toits, journal d'un homme heureux.* 3rd ed. Paris: Michel-Lévy frères.

—. (1850) *Un philosophe sous les toits, journal d'un homme heureux.* Paris: Michel-Lévy frères.

—. (1846a) *Le monde tel qu'il sera.* Ilustré par MM Bertall, O. Pengilly et S' Germain. Paris: W. Coquebert.

—. (1846b) *El mundo tal cual será por Emilio Souvestre traducido por un español del año 3000.* Barcelona: Imprenta de José Matas, calle de la Platería.

Trousson, Raymond. (2003) *Sciences, techniques et utopies: Du paradis à l'enfer.* Paris: Harmattan.

—. (2000) Dystopia. In: Fortunati, Vita and Trousson, Raymond (eds). *Dictionary of Literary Utopias.* Paris: Honoré Champion, 180-185.

Versins, Pierre. (1984) *Encyclopédie de l'utopie, des voyages extraordinaires et de la science-fiction.* 2ª ed. Lausanne: l'Âge d'homme.

Emil Cioran and Bruno Taut: Utopia as a flight from progress?

Andrea Franceschini & Paolo Vanini
Dipartimento di Lettere e Filosofia, Università degli Studi di Trento, Trento, Italy

ABSTRACT: Emil Cioran (1911–1995) and Bruno Taut (1880–1938) are two very different intellectual figures not only as for their primary occupation but also as for their worldview. However, in their theoretical reflections, both the authors addressed the question of utopia. In fact, according to them, the most important contradictions and challenges of modernity converge around this thorny issue. On the one hand, Taut was a German expressionist architect who, in the first half of the twentieth century, imagined the utopian city as a meeting place of the requirements of the industrialized society (functionality, productivity, efficiency) and the inescapable anthropological need for beauty and harmony. On the other hand, Cioran refused to work his whole life and settled in Paris in 1937 as a stateless person of Romanian origin. Such rejection of any formal employment accounts for his sharp criticism of utopia, which he conceives as the place where the diabolical ideal of a humanity constrained by the symmetries of work and the ugliness of fatigue is realized. This paper aims to show that Cioran's and Taut's conclusions are only apparently in contrast: indeed, in dealing with utopian hopes and impossibilities, both authors are forced to distance themselves from the modern idea of *progress* – conceived as a symbol of a "dogmatic" and "enlightened" historical inertia towards the moral, political and scientific perfection of humanity.

Keywords: Cioran; Taut; Utopia; Glass architecture.

1 A PRISON OF GLASS

In *Histoire et utopie* (1960) Cioran examines the "mechanisms", which characterize utopian thought and likens the effort to design a *perfect future* to the philosophical search for "a *timeless history*. A contradiction in terms, inherent in the hope of a new kingdom, of a victory of the unsolvable at the heart of becoming. Our dreams of a better world are based on a theoretical impossibility. Hardly surprising if, in order to justify them, we must resort to *solid* paradoxes" (Cioran, 1960: 170).

In this passage, Cioran alludes to the Renaissance debate over the ideal city, originally developed in the framework of architectural thought between the fifteenth and sixteenth centuries. In particular, Leon Battista Alberti, in *De re aedificatoria*, establishes the theoretical principles, which underpin the design of the ideal city, whose structure must reflect the perfect rationality of the Cosmos. The aim of his treatise is to reach a perfect complementarity between the location of the building and its public function, with the implicit assumption that social harmony cannot be achieved within a city whose space of coexistence is structurally disharmonious.

According to this "humanistic" model, people of different professions and social groups are assigned a dedicated area where they can practice their job. This idea brings about two important consequences, which will play a central role in the following utopian literature. On the one hand, if citizens are guaranteed a place where they can fulfil the *function* assigned to them, the ideal city has no place for the *outcasts*, since everyone, from the rich to the poor, occupies a certain position and responds to the social duties, which this role entails. On the other hand, the integration of marginalized people within the social body is not a sign of an egalitarian society, but rather of a society whose hierarchical structure is so defined *as not to admit unproductive exceptions*: just like the sewers, even the poor can be useful (Camporesi, 1973).

Cioran believes that this principle materialized in Cabet and Fourier's grotesque utopian socialism and, in particular, in the concept of Phalanstery (i.e., the ideal building designed to host the utopian socialist community) and its architectural as well as literary variations (Cioran, 1960: 163). The arithmetic and flawless division of the Phalanstery reminds Cioran of the atrocious harmony of a perfectly planned factory, where the right to work is *deified* in the alienating obsession with performance. Ironically, he recommends "the description of the phalanstery as the most effective vomitive I know" (Cioran, 1960: 495).

Unfortunately, Cioran detects reverberations of the Phalanstery even in the crystal symmetries of the contemporary metropolis, which he criticizes following some suggestions offered by Georg Simmel (Cioran, 1995: 947). Simmel compares the city to an "acoustic cave", which essentially consists of two different elements: the *transparency* of the glass buildings and the mechanical *noise* of the engines and the public transport (Blumenberg, 2009: 59). In this city, where everything can be seen because it is made of glass, nothing can be understood and there is no one to talk to because the noise surrounds the individuals and isolates them from each other. In *La Chute dans le temps* (1964) Cioran describes this solitude through the caricature of the car drivers: which are prisoners of their vehicles and "cut off from every root" of the ground.

This lack of "contact with the ground" symbolizes the lack of contact with life, which characterizes the alleged progress of modern knowledge (Guerrini, 2002). By encouraging "our dreams of power and negation", this knowledge complies with the anthropological attitude of dilating in a false and unreal world, "for once we know ourselves, once we feel ourselves to be man, we tend to gigantism, we want to seem *larger than life*" (Cioran, 1964: 530). According to Cioran, the desire to distort our real size, pretending to be someone we are not, depends on the dread of accepting our true nature. He claims that the human being is "the great *deserter of being*", "an *abnormal* animal" desperately seeking for his own essence: "A shadow grappling with images, a somnambulist who *sees himself* walking, who contemplates his movements without discerning either their direction or their cause" (Cioran, 1964: 528–529).

Through these metaphors of the "shadow" and the "sleepwalker", Cioran uses the concept of transparency to highlight the hiatus between the rational ability to see and the ethical necessity of living. In an early work published in 1937, he depicts this paradoxical fracture: "While walking on the street, somehow or other the world seems to exist. However, by looking at it through a window, everything becomes unreal. How can the transparency of a glass detach us from life to this extent? Actually, a window detaches us from the world more than a prison wall. By looking at life so much, we consign it to *oblivion*" (Cioran, 1937: 323).

2 NUANCES OF A HOPE

The history of twentieth-century architecture seems to sum up, in its basic outline, the escape from being which Cioran exemplifies with utopian illusions. Up until now, architecture seemed unwilling to abandon the path of a progressive "estrangement from the world", which was already undertaken at the dawn of the modern movement. The triumph of the optical perspective over the haptic and multi-sensory dimension of human experience in the contemporary city is coupled with the production of increasingly impenetrable (though more and more transparent) architectures (Pallasmaa, 1996). In this respect, the history of glass architecture seems to coincide with the growing gap between human beings and the world. Nevertheless, in the hopes of its earliest supporters, glass architecture was entrusted with a diametrically opposite meaning: the thinning of the architectural membrane – enabled by the large windows and the concrete structures – seemed to respond to the urgency of community rapprochement among human beings and a reconciliation between human beings and the world (Whyte & Schneider, 1996).

The glass architecture revolution begins around mid-nineteenth century with the construction of the Crystal Palace (1851) and reaches its full maturity around 1914. In this date, the pamphlet *Glasarchitektur* was published by the visionary expressionist writer Paul Scheerbart, and the glass Pavilion was designed and built by Bruno Taut for the Deutscher Werkbund Exhibition in Cologne (Biraghi, 2008: 200–208). Subsequently, the glass revolution found its dialectical fulfilment in Mies van der Rohe's skyscrapers. Bruno Taut's Pavilion was symbolically dedicated to his friend Paul Scheerbart, regarded as the theoretical pioneer of a new architecture and hence also of a new civilization. Located in a peripheral position of the Exhibition, the Pavilion is an eccentric building, a dome-shaped structure made of reinforced concrete with a circular plan, on whose surfaces coloured glass plates are set like jewels. By virtue of its sculptural and pictorial qualities, the Glaspavillon shows the total disjunction between Taut's poetics and the classicist principles prevailing among the Werkbund at that time.

Inspired by Scheerbart's ideas – whose quotes decorate the lintel – the Pavilion seems to give voice to a deep desire of architectural emancipation from the bonds of matter and history: a detachment from the existent, in the name of a deeper rapprochement between human beings and the world. According to Scheerbart, glass architecture should have facilitated the transition from a culture of closed spaces and segregation to a civilization allowing sunlight, moonlight and starlight to penetrate into human dwellings: his prospect was highly focused on the symbolic and aesthetic qualities of architecture, guided by a deep conviction that a bright architecture would have brightened humanity itself. Therefore, the urgency of glass architecture is not advocated by Scheerbart and Taut in the name of the progress or on the basis of functional principles, but rather from a utopian perspective tinged with marked religious and even mystical traits, particularly as for Scheerbart's writing. Scheerbart imagines a humanity purified by the constant vision of beauty created through a joint use of glass, colour and light. A cosmic dimension is what he aims to recreate by imagining a house as a window on the world, communicating with the sky and the surrounding nature. Hence, the architecture theorized by Scheerbart represents the antithesis of the Prussian *Mietskaserne*, which was interpreted by Benjamin himself as the architectural expression of a society – the German one – increasingly hierarchical and militarized (Benjamin, 1930). The crystalline and chromatic architecture made of glass thus represents a positive emancipation of human spirit from an oppressive and repressive materiality. The smooth and transparent surfaces cherished by Scheerbart were intended to help overcome human separation from the world, by facilitating the contemplation of beauty and lessening the line between humanity and nature. According to Scheerbart, "the man who sees the splendours of glass every day cannot have ignoble hands" (Scheerbart, 2004: 96).

This cult of beauty is fully accepted by Bruno Taut. An emotional and aesthetic tension pervades the vitreous space of the Pavilion. Nothing is left of the disenchantment and of the manly acceptance of the modern-metropolitan life, so widespread among the majority of modern architects: according to Taut, architecture should not be a reflection of the contemporary *Zeitgeist*, but rather its critical and utopian expression. As the supreme representative of the "*phantastische Architektur*" (Conrads & Sperlich, 1960), Taut is the pioneer of an expressionistic and organic modernity that does not accept easy compromises with the *status quo*: especially, Taut does not accept the ideal of the technological progress as the necessary *status quo* of our future society. Even his sketches for an *Alpine Architektur* (2004) are inspired by Scheerbart's theories and reflect a similar need to escape from a world torn apart by utilitarianism, hatred and war. From the bottom of the bleak and muddy trenches where millions of bodies were trapped, Taut's spirit rises to the crystal purity of mountain peaks, shortly envisaged as a possible home for humanity.

The desire to escape from a world at war is the root of his idea of *Stadtkrone* (city crown), developed in an eponymous work in 1919, symbolically dedicated *dem Friedfertigen*, or, literally, to the one who is ready for peace. The "city crown" gives voice to an idea, which is ancient and revolutionary at once. Throughout all times and cultures, the "city crown" is the symbolic centre of the community, whose fundamental aspirations seem to "crystallize" within it. The Strasbourg Cathedral, Mont Saint Michel, the Yavuz Selim Mosque, the Acropolis in Athens, Angkor Wat are only a few examples cited by Taut to support his idea that a "crystallized religious conception" is the goal of every great architecture (Taut, 2002: 51–52). The *Stadtkrone* illuminates the surrounding architectures, provides an organic sense to the city and a symbolic meaning to the community. As Taut claims, "the greatest buildings were derived from the highest thoughts" (Taut, 2002: 53). Such architectures express the deep need for meaning rooted in humankind. According to Taut, the architect's task is to convey these puzzling tendencies and materialize them in an architecture, which is not a simple functional construction, but rather a concrete thought or perhaps a "solid paradox" as Cioran would argue.

Therefore, architecture responds to a need that is both new and deeply rooted in human nature, a need for meaning and happiness: "A deep desire directs us all: according to Aristotle, we want cities in which we can live not only safely and healthily, but also happily" (Taut, 2002: 55). Against the architecture of isolated functions – which results in a city that looks like a "body without a head" – Taut opposes a socialist architecture characterized by religious traits and based on "*sozialer Gedanke*", a feature deep-seated in European culture between the nineteenth and twentieth centuries. The socialist ideal in its over political (*überpolitisch*) sense, i.e., as a moral and utopian inspiration, would then represent the possible root of a new *Stadtkrone*, which could shed new light and bring forth a new meaning to the architecture of the contemporary city. Taut highlights two fundamental trends of modernity: the aforementioned *sozialer Gedanke*, and the yearning for happiness through aesthetic pleasure. In fact, Taut conceives this apparently hedonistic drive as a noble desire for an "elevation above everyday existence". For this reason, the centre of his ideal city is entirely devoted to the arts. Arts should be no longer confined to the museums, but rather living in a phantasmagorical architecture geared towards strengthening its ties with sculpture and painting, thus forming an almost Wagnerian *Gesamtkunstwerk*. The *Stadtkrone* – a pure architecture liberated from any function, a *Kristallhaus* revealing its magnificence especially at dawn and dusk – towers above the centre with its cruciform plan reflecting the trajectory of the Sun: "Infused by the light of the sun this crystal house reigns above the entire city like a sparkling diamond, a sign of the highest serenity and peace of mind. In its space, a lonely wanderer discovers the pure bliss of architecture. While ascending the stairs to the upper platform, he looks to the city at his feet and beyond to the sun rising and setting, towards which this city and its heart are so strongly directed" (Taut, 2002: 69).

The idea of transparency and total organization in Taut's phantasmagoria – glass, the ever-present reference to the Sun, the city crown – seems to follow the logic of the "utopia of form" analysed by Manfredo Tafuri (Tafuri, 1973: 47–48) and marks a strong difference from the genuine "plan ideologies" subsequently emerged during the 1920s with Le Corbusier and Ludwig Hilberseimer. As a utopia of form, Taut's ideal city is also exempt from any desire for total control over urban space based on labour division and economic principles. Conversely, it embodies a deeply aesthetic model, which has little or nothing in common with principles of economic rationalization. The ideal of beauty is therefore opposed by Taut to the prevailing materialistic and utilitarian culture of his time.

Taut is a Romantic. He aims at an organic *Gesamtkunstwerk*, i.e., a city conceived as an organism, not as a machine. His logic is radically utopian and, as such, unrealizable, doomed to bear no relation to the progress of history. Being timeless and spaceless, it also avoids the Dialectic of Enlightenment, which strongly affected modern urban planning. Unconcerned about problem solving, Taut's utopia still shines like a crystal, like a dream from which we are destined to wake, but in which we found ourselves at ease while sleeping, just as in Scheerbart's novels. What other meaning can be ascribed to the improbable alpine architecture or the crystal crown in a city where nothing (or little) is defined but the role of art and beauty? From the 1920s onwards, Taut himself awakes from that dream, by designing social housing for a "coloured Magdeburg" as well as public housing in Berlin for the GEHAG (*Gemeinnützige Heimstätten-, Spar- und Bau-Aktiengesellschaft*). Through these concrete and discrete projects, his utopia raises to an undeniable ethical stance: in the *Hufeisensiedlung*, for instance, the city crown becomes the community centre around which the houses are arranged; colour plays a fundamental symbolic and psychological role in all the *Siedlungen* he designed. Taut becomes the architect of the real working class. His utopia as an escape from hatred never became a flight from humankind.

Glass surfaces, as stressed by Cioran, isolate us from life: we become orphans of the world insofar as we simply observe it. Taut's prisms and coloured architectural surfaces seem to be related not so much to the curtain walls of Mies' skyscrapers, but rather to the stained glass windows of Gothic cathedrals. They express the fundamental will to allow the world to penetrate into human houses. Their goal is the experience of light, a secular and humanistic experience. Yet, they bear no relation to the panoptic vision of a human eye dominating a world reduced to the landscape.

Bruno Taut's life and work, thus, runs along two separate but parallel paths: on the one hand, the dream narrative of utopia; on the other hand, the pragmatic work of the people's architect. Both of them represent a confutation against the tyranny of progress conceived as a modern paradigm, which denies the ethical need for happiness, and the aesthetic need for beauty. As for the latter, utopia seems to play the role of a memorandum, i.e., an implicit and everlasting reference: utopia lives in Taut's work as, by recalling one of Benjamin's images, the little door through which the Messiah may appear (Benjamin, 1995). Beauty has to be nourished as a promise of happiness: this is, perhaps, the deepest meaning of Taut's criticism of progress.

3 CONCLUSION: LOOKING FOR A BETTER PLACE

In Modern Western philosophy, the ideal of scientific progress is implicitly related to the image of a utopian society, in so far as "the good state of Utopia" is a political consequence of a "rational planning" for the future of humankind. We suggest that Progress has been conceived as the utopian direction human Reason was supposed to follow in order to reach a perfect condition. However, this positivist image of Progress was destabilized between 19th and 20th century: in the very moment in which the technological development of Western society showed its inability to provide freedom, justice, and equality.

From this point of view, the apology of beauty proposed by Taut is a Romantic reaction against a world, ruled by the terrifying principle of utility. The same world romantically and ironically denied by Cioran. Nevertheless, Cioran is different from Taut, because he is dramatically torn between the mystic desire for sublime beauty, and the sceptical conviction that reality was doomed to the worst (van Itterbeek, 1999: 83–105; Id., 1998: 69–79; Mutin, 2004; Valcan, 2008). His scepticism forces him to distance himself from utopia, in particular from political utopia, regarded not as a hope for progress, but rather as the revelation of the "bloody farces", which condemn both the present and the future (Cioran, 1949: 3).

Anyway, this is the very point of contact between Cioran and Taut. Faced with a humanity in which everybody seems to die of ideals – in which the need for truth and for "scientific assurances" easily turns into ideological dogmatism –, the ability to doubt oneself becomes a utopian ideal to strive for. In this regard, Cioran proposes to consider "futility" an ideal to press for; he aims to "to put [us] on guard against the Serious, against that sin which nothing redeems" (Cioran, 1956: 338–339). We shall bear in mind that these words were written in the existentialist Paris of the late 1950s, a period in which too many philosophers confused the ideology of freedom with the ability to be free. Against them, Cioran maintains the need to be *dégagés* towards everything, to learn to "be able to commit yourself to anything without adhering to it" (Cioran, 1956: 338). According to Cioran, this exercise of non-adherence to one's own certainties implies the aesthetic value of scepticism, because it is nice to doubt. The one who doubts can see reality from an upside down perspective, just as Taut did through the "chromatic lenses" of his glass palaces and colourful houses.

Regarding modernity, Cioran writes that the cathedrals, "having lost the impulse that supported their mass, turned back into stone, they shrink and slip [...] Nothing gothic is left in the substance, nothing gothic left in ourselves" (Cioran, 1956: 309–310). Taut, in this respect, never surrendered to the evidence of a stone. The colours of his windows certainly had an ethical value, not just an aesthetic one, as they had to be mirrored by the soul of the beholder. As Cioran claims, certain colours teach us to recognize "nuances of trumpery" (Cioran, 1960: 439) and, by getting away from them, to look for a better place.

BIBLIOGRAPHICAL REFERENCES

Benjamin, Walter. (2002) *Il casermone* [1930]. It. tr. by G. Schiavoni. OCWB. Vol IV (Scritti 1930-1931). Ed. by R. Tiedemann & H. Schweppenhäuser. It. ed. by E. Ganni, with the collaboration of H. Riediger. Torino: Einaudi: 196–202.

—. (1995) *Tesi di filosofia della storia* [1940]. Angelus Novus. Saggi e frammenti, ed. by R. Solmi, with an essay by F. Desideri, Torino: Einaudi: 75–86.

Biraghi, Marco. (2008) *Storia dell'architettura contemporanea I. 1750–1945*. Torino Einaudi.

Blumenberg, Hans. (2009) *Uscite dalla caverna*. It. tr., Milano Ed. Medusa.

Camporesi, Piero. (1973) *Il libro dei vagabondi*. Torino: Einaudi, introductory essay.

Cioran, Emil. (1995) *Des larmes et des saints*. 1st ed. 1949. tit. or. Lacrimi şi sfinţi), Fr. tr. by S. Stolojan, in *Œuvres*. Paris: Gallimard.
—. (2011) Précis de décompositon, in *Œuvres*, "Bibliothéque de la Pléiade". 1st ed. 1949. Paris: Gallimard, 2011.
—. (1956) Tentation d'exister, in *Œuvres*. 1st ed. 1949. "Bibliothéque de la Pléiade". Paris: Gallimard.
—. (1960) Histoire et utopie, in *Œuvres*, "Bibliothéque de la Pléiade". Paris: Gallimard.
—. (1964) La Chute dans le temps, in *Œuvres*, "Bibliothéque de la Pléiade". Paris: Gallimard.
—. (1995) *Entretiens*. Paris: Gallimard.
—. (1997) *Cahiers*. Paris: Gallimard.
Conrads, Ulrich. Sperlich, Hans G. (1960) *Phantastische Architektur*. Stuttgart: Hatje.
Gagnon, Remy. (2010) L'animal vague d'Emil Cioran. Réflexion sur l'humanisme, *Alkemie*, n. 6, 2010: 59–65.
Guerrini, J.C. (2002) Vision du temps. Cioran, analyste de la réaction, de l'utopie et du progrès, *Mots*, n° 68.
Hacen, Aymen. (2005) Réflexions sur la pensée tragique de Cioran, in Van Itterbeek, Eugène (ed.). *Cahiers Emil Cioran. Approches critiques*, vol. VI, Sibiu-Leuven: Éditions de l'Université Lucian Blaga / Les sept dormants: 9–18.
Mutin, Rachel. (2004) *Cioran. La passion de l'absolu dans une âme sceptique*, Thèse de doctorat, Lyon: Université Jean Moulin III.
Pallasmaa, Juhani. (1996) *The Eyes of the Skin. Architecture and the Senses*. Hoboken: John Wiley & Sons.
Quarta, Cosimo. (1993) *L'utopia platonica. Il progetto politico di un grande filosofo*. Bari: Dedalo.
Scheerbart, Paul. (1982) *Architettura di vetro*, It. tr. by M. Fabbri, with an introductory essay by G. Schiavoni. 1st ed. 1914. Milano: Adelphi.
Sciolla, Gian Carlo. (1975) (ed. by), L*a città ideale nel Rinascimento*. Torino: UTET.
Tafuri, Manfredo. (1973) *Progetto e utopia*. Architettura e sviluppo capitalistico, with an introduction by F. Purini, Laterza. Roma: Bari.
Taut, Bruno. (2004) *Alpine Architektur. A Utopia*. Eine Utopie [1914], ed. by M. Schirren. Munich: Prestel.
—. *Die Stadtkrone* [1919], ed. by M. Speidel. Berlin: Gebr. Mann Verlag.
—. (2007) Nippon. Mit europäischen Augen gesehen [1933], ed. by M. Seidel, Gebr. Mann Verlag, Berlin.
—. (2007) Japans Kunst. Mit europäischen Augen gesehen [1933–36], ed. by M. Seidel, Berlin: Gebr. Mann Verlag.
Valcan, Ciprian. (2008) *La concurrance des influences culturelles françaises et allemandes dans l'œuvre de Cioran*. Bucarest: Institutul Cultural Român.
Van Itterbeek, Eugène. (1999) Cioran et le romantisme allemand, in *Cahiers Emil Cioran. Approches critiques*, vol. II: 83–105.
—. (1998) Expérience extatique et connaissance de soi chez Emil Cioran, in *Cahiers Emil Cioran. Approches critiques*, vol. I: 69–79.
Whyte, Iain Boyd & Schneider, Romana. (1996) *Die gläserne Kette. Eine espressionistische Korrespondez über die Architektur der Zukunft*. Ed. by I.B. Whyte e R. Schneider. Ostfildern: Hatje.

Inner and outer reality in film storytelling: Wandering between vertexes of human reality foundations

Carlos Figueiredo
University of Lisbon, Faculty of Architecture, CIAUD, Lisbon, Portugal

Inês Coimbra
University of Lisbon, Faculty of Architecture, Lisbon, Portugal

ABSTRACT: We do have memories, concepts, identity and experiences, building a self-inner world that gives us the notion of reality about ourselves and the world we live in. Each of us has a differentiated reading of what surrounds us, of our environment, a self-consciousness that is unique, an "intuitive" sense of what is true or not, reality or fiction, before our eyes.

What is the foundation of our concepts and notions we use that allow our discernment capacity to tell apart reality from dream/illusion/hallucination? How are we aware and understand the features and concepts of different alternate worlds and their meanings?

Different films convey all range of contrasting times, concepts and realities, about our world and alternate ones that can coexist in the same film fiction. In a film, we are immersed in the plot and telling, the characters' inner lives, their feelings and in the fictional world, they live in. But the tools we use to decide if we – and the characters – are experiencing reality, "dream time", "mind time" or an "out of this world" reality are the same as in our "real" life.

What are the foundation types where our concepts and notions are rooted? The ones that allow us to decide what reality is and what is not? This inner capacity of judgement about reality also makes alternative universes presented understandable and readable.

We consider three major families of these foundations: Dream and Mind, Scientific and Metaphysical based concepts types.

Keywords: inner self, dream and reality, film storytelling, alternate universes, reality foundations

1 REALITY, MEANING AND THE SELF

The way we see start from ourselves

The Human Being must give consistency and meaning to his surroundings so that he can understand it. We can only deal with what makes sense to us:

> The way that we interpret the events that shape our lives (…) is dependent to a large extent upon processes of fantasizing: of imagining a portrait of our world that lends coherence to what is essentially the chaos of existence. (Walters 2011a)

Man and the Universe in which he lives are only reconcilable and relational if Man establishes in himself a relation of meaning with what surrounds him: "The schism between the self and the world is absolute and irreparable." (Kravanja, 2011:120)

The dream is a fiction lived by Man as a reality, a construction that is based on himself, on his realities and experiences, in his universe: "Awakening from the dream means to awaken from what we thought was reality but which is not." (Botz-Bornstein, 2008:98)

Even when we dream, the centre of dream fictions is always ourselves:

> The one orientational anchor that is never lost in dreams is the self, the 'I' who dreams. (…) It is (…) the organizational unit of consciousness: I dream, therefore I am. (Hobson, 2003: 142)

Realities apprehend and understood as a phenomenon

The perception of an event and its understanding as such depends on being not only looked at but also seen, identified and interpreted as a phenomenon. An event understanding is not assured, only with their detection and admission as reality.

This understanding depends on being able to give meaning and coherence to this phenomenon, according to our view of the reality of the world presented and of ourselves. This capacity is based on the logical and emotional structure that we built with the experience of all our living and information acquired in our life, with its synthesis.

When we live a cinematic fiction, the same principle applies to us as spectators:

> Directors create reality and illusion within each film; we interpret what is real and what is dreamed by interpreting filmic cues and applying our own memories, perceptions, points of view, and self-awareness. (Halpern, 2003: 161)

The Need to give meaning and coherence to all that surrounds us

When an event exists in our daily reality, we have to obtain its meaning and coherence through our notion of reality. We have the imperative need to make it compatible with our world and reality – our universe – giving it meaning, even if for that we must broaden our horizons and definitions: our notion of reality.

An event that we must assume as a reality, but which we cannot include in the sphere of what we can give sense and relate becomes an enormous source of conflict and anguish, creating a tension that only can be solved with its integration:

> Unlike ordinary anxiety that may be triggered by specific events (...) existential anxiety refers to uncertainties relating to the human conditional, existence itself (Popovic, 2001: 32)

Differently, we do not read the realm of fantasy as reality: we read it as speculation, as something that could exist. We realize it, but we do not think it exists in our reality.

Fantasy has a coherence in itself, it does not have to be integrated with the reality of our world and to have in it a sense and coherence nor being accepted as reality:

> When faced with fantasy in films, however, I am conscious of being shown a series of events that, according to the rules of reasonable logic, could never occur within my reality (Walters, 2011b)

Yet, Fantasy refers to the universe of our reality, the only one we know and can understand:

> "Fantasy recombines and inverts the real, but it does not escape it: it exists in a parasitical or symbiotic relation to it. (Jackson, 1981)

Man, as a protagonist in a film must create a connection between himself and the world that surrounds him, in which he lives:

> Protagonists build a bridge between their selves and the diegetic world they inhabit (...). The protagonists' 'inner world' of thoughts and feelings. (Kravanja, 2011: 119)

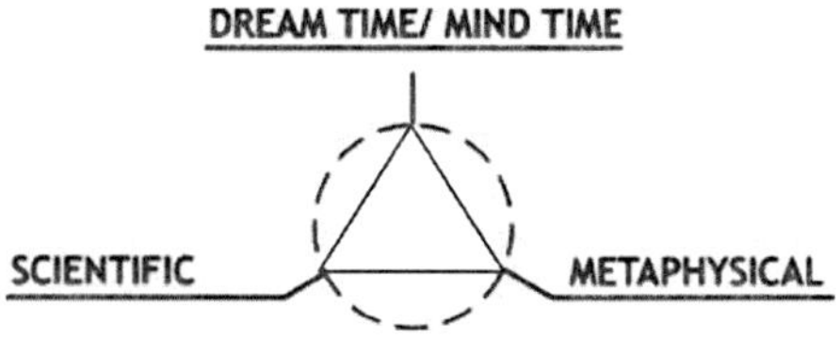

Figure 1. Placing of several Cinematographic Fictions in the Origin Root Type Foundation Triangle, along with pairs of the triangle's vertex.

2 WANDERING BETWEEN REALITIES FOUNDATION VERTEXES

Different Films convey all range of contrasting times, concepts and realities, about our world and alternate ones that coexist in the same film fiction, plot and telling. The tools we use to decide if we – and the characters – are experiencing reality, "dream time", "mind time" or an "out of this world" reality are the same.

What could be the foundations where our concepts and notions are rooted? The ones that allow us to decide what is reality or not? This inner self-reality capacity of judgement also makes alternative universe fiction presented to be understandable and readable.

In the following three films, three major families of these foundations are considered: Dream and Mind, Scientific and Metaphysical, based on concepts types, represented and related according to an Origin Root Type Foundation Triangle:

3 BETWEEN SCIENCE AND METAPHYSICS

Science or Metaphysics justifying realities: the two sides of the same coin?

In *Metropolis* (Lang, 1927) a new futuristic and technological city is proposed as well as Good and Evil confront: there is the Lords surface world as there is the slave's underground world that keeps everything working. Freder, the son of the dictator Joh Fredersen, transport the viewer through this entire social, technological, metaphysical and even romantic opposed sides of the film world.

In this film, there is no portrait or view revealing the mind state or inner life of characters as the viewer is never placed in the mind time of characters; a distinctive aspect of the German Expressionist Film. Otherwise, the inner emotions and despair of characters reflect themselves on the outer world, exaggerating, deforming, bending, and symbolizing the space they inhabit.

There are two moments where a parallel reality world coexists with the Metropolis one. Those ones are already alternative reality proposed words, in relation to ours:

– Maria is a preacher that in her underground Proto-Christian Chapel tries to call the slaves to the

Figure 2. The Energy Central Machine, where the underground slaves work, metamorphosed in a monster machine meaning it takes the slaves lives.

have hope, not to react with violence, but to free themselves with positive actions. Rotwand, an evil person, used a sophisticated laboratory to give live to a Robot that he managed to give the appearance of the real Maria: the Maria Robot. His laboratory his full of metaphysical signs, as is Maria Chapel, of opposed values.

- Another scene where when we can see this overlap is in the huge energy power machine where dozens of slaves work to exhaustion point, and that suddenly metamorphoses in an evil monster that eat the slave's workers, in a symbolic picture of this machine roll in their lives. (Fig. 2)

Along the film and storytelling, we have the Science vertex, the Metaphysical one deeply interlaced, creating, and showing a panel of different alternative realities in this fantasy film about the reality Metropolis is.

4 BETWEEN MIND AND SCIENCE

Dream within a dream and the unclear line between realities

In "Inception" (Nolan, 2010a), Cobb and his partner Arthur work in the area of espionage, whose specialty is stealing secrets from people's minds during

Figure 3. Cobb and Ariadne find themselves in a dream state in which she is the architecture of this dream state.

their dream state. We're constantly confronted with a scenario where – through a machine – the character can share a lucid dream state and act individually over it: "Inception (...), refracts current conceptions about dreams – particularly lucid dreaming" (Greenberg 2010), like in the scene where Cobb and Ariadne, the new apprentice, submerge together in a lucid dream state to test her ability to shape the space limits.

Walking through the city, Cobb explains to her that in this dream stage it is our mind that builds the space and interferes with it. The machine needs two elements: the architect who envisions and plans the space, and a second element, who will populate it with his unconscious.

In this case, where Adriana is the architect, she tests her ability and begins by bending the city itself. Challenge to test this infinitive ability to create new worlds, she starts to bend the city itself upside down, bending the physical laws, by walking over this upside-down city level and elevate stairs and bridges, with only the power of her mind.

This start to get out of hand when the people of this world start attacking and killing them. Since the city transformation where so out of our daily reality, Cobb unconscious detect that someone else was creating that world and, by so, they forced who were dreaming to waking up, by killing them in the dream.

Although the characters tried to have a verification method, we never completely understand if they are in a dream state or not. The only elements that are presented in the crossover between realities are the close-up camera shot of their open and close eyes movements.

Later, after invading the mind of the entrepreneur Saito, they are contracted to implement an idea in the mind of Fischer. After getting into Fisher's mind, this men's unconscious attack starts creating a war zone within the dream state. Without being able to wake up they must go into the deepest part of Fisher mind's, to place an idea – what they call of inception.

They enter in a dream within a dream (nesting), in a succession of parallel realities where time perception goes slower as they go into deeper levels: "Some dreams reflect events occurring in the dreamer's environment as he or she is sleeping." (Hobson, 2003:148).

In this scene, where they find themselves in three different dream levels, we can understand how they communicate between those levels: while they run in a fast speed car that crashes, in another level the van movement defines the orientation of the hotel room where Arthur is, making that dream the space rotating in the same direction.

Although everything that we see is the product of the character's mind, the scientific explanation of the machine and their rules, allows the viewer to classify this reality as being founded between the vertex of the Dream/Mind Time and the scientific vertex.

5 BETWEEN MIND AND METAPHYSIC

Emotional communication established through objects between alternative universes

In the film "What Dreams May Come" (Ward, 1998), Chris and Annie, a happy couple, are irreversibly shaken by the loss of their two children's. Four years later, Chris dies in an accident, leaving her wife in a middle state, between earth and the after world. After Chris realized that his presence is destroying and consuming Annie, he accepts his own death and passes the light at the end of the tunnel to do the crossover; a common sense visual reference to the passage to afterlife world.

When he arrives in heaven, he faces a painted world, surrounded by mountain ranges, flower valleys and lakes. This space is parallel to the one that Annie painted on a canvas and offered him as a gift for their 20th wedding anniversary.

Annie painted a three-canvas scenarios composition of this world that allows a connection with the metaphysical world of Chris.

- The first canvas showed a blurred woman in pain. Not only represents Annie interior felling but also makes a visual connection with the first appearance of Albert on earth, guiding and explaining to Chris the rules of the new reality that he is living now.
- The second frame shows the place where Chris and Annie first meet, with the additional element of the house that she idealizes to live on – she mixed elements from their memories with the dreams of the couple (fig 04, 1st frame);
- The third frame contained a new painted element, a tree that was created after Chris death. She only introduces this new element because she believed that he would be able to see it too (fig 04, 2nd frame). In Chris heaven, the tree appears at the same time as Annie painted it on our material world (fig 04, 3rd frame). This is an element that crosses the border between the two universes, at the same time, not only an element in Chris landscape but as an element that allows establishing an emotional communication between them.

By the time Annie let go this fantasy, believing that all these toughs only happen in her mind, she drops water over the painted leaves of the tree, ending up having a direct reaction on Chris's tree by losing his leaves. Both characters fell the pain of the distance caused by the worlds that tear them apart, trespassing the borders of the worlds and their logic.

Figure 4. Interaction between two characters through universes apart: ours and an after this life Metaphysical one.

The three sequence of frames shows different moments in time of the same background landscape used as a symbol of communication between the different universes and their reality's: the first frame shows the moment in which Chris and Annie first meet in the mountain's (a memory reference); the second one presents a suffering Annie on earth, when she is next to the painted canvas, and the third frame we can see Chris in the actual paint made heaven landscape.

To assign a symbolic meaning to this place, the director used a series of cinematic tolls as the visual repetition of the place – between memories, the paint and heaven, as well as values added by the character and context – by this place where they first met, all with visual effects of the dominant colours in the painted sky.

We can interpret this metaphysical world as series of more elaborate projections of our terrestrial memories, mixed with our dreams and imagination. This peculiar feature allows the viewer to classify this foundation as a segment between the Dream Mind Time vertex and the Metaphysical one:

> The protagonists are metaphysical subjects that do or do not structure the course of their lives according to a homogenous flow of meaning: can be disturbed by external contingencies. It is a matter of the permeability and the flexibility

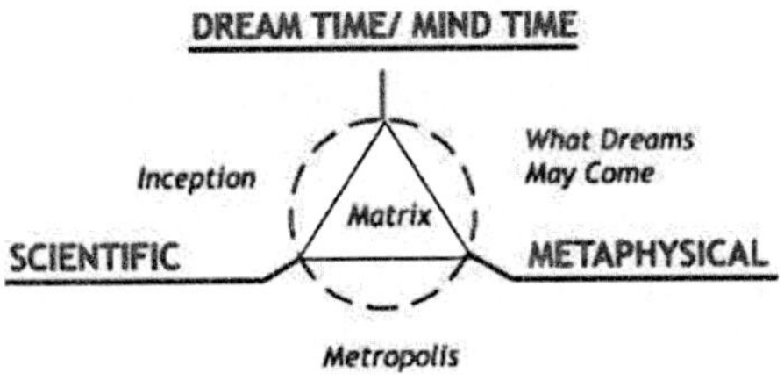

Figure 5. Origin Root Type Foundation Triangle in Films: our concepts and notions, which build each, own existing world of reality.

> of the boundary between the self and the world. (Kravanja, 2011: 119)

Just like when Albert explained to Chris when he enters the painted sky: "Annie gave you a start, like holding onto handrails. Now you're creating an entire world from your imagination, from paintings you love anything you want." Even more, after they arrive at the painted house, Albert opens a hole in the wall of paint, opening a window to another more realistic landscape. Albert explains to him: "(…) this is your world now. The thought is real physical is the illusion. Ironic, ah?" (Ward, 1998).

6 CONCLUSION

As the Universe in which we live is only recognizable and reconcilable having a relation of meaning to ourselves, there we must be able to give meaning and coherence to any event there or inside us. This evaluation is made according to our view and understanding of our reality system, the one of our world and of ourselves; we have no other.

Any event that exists in our daily reality must be given meaning: it is imperative for us the need to make any event detected compatible with our own universe – our world and reality notion, so we can place that event in our reality system and so be able to understand it.

Films convey all range of contrasting times, concepts and realities, about our world and alternate ones that coexist in the same film fiction, plot and telling. This cinematic fiction tells us about reality and fantasy worlds, but these are understood according to the very same principles by us as spectators.

Even if we do not read Fantasy as a reality, it must have coherence. Fantasy Worlds does not have to be integrated with our reality and world, but they also must anchor in our universe and reality, whatever any parallel reality Fantasy proposes.

In life as in film, the experience of reality can be of our "daily world", "dream time", "mind time" or a "out of this world" one: in any of these cases what are the origin types where our concepts and notions of reality judgement are rooted?

Considering that, films are emulations of ourselves and our inner and outer world and realities, from the filmic fictions analysed and their telling, a triangle of three major vertexes or families of foundations seems to emerge an Inner/Mind vertex, a scientific vertex, and Metaphysical one.

These could constitute concepts families or types – foundation vertexes – around witch our concepts, that allows giving events an identity and status of reality, seems to be based or rooted.

The diagram below, labelled "Origin Root Type Foundation Triangle", illustrates this idea. Could be that Man reality concepts and justifications are formed from one or several of these sources? Only from these?

As explained, in the three films analysed the nature of the proposed words and their understanding seem to be rooted in pairs of this triangle vertexes:

- Inner based (mind or dream state) with Science based roots in "Inception" (Nolan, 2010a).
- Metaphysical with Mind based roots in "What Dreams May Come" (Ward, 1998).
- Science with Metaphysical based roots in "Metropolis" (Lang, 1927).

Therefore, these films wander in the each of the sides of the proposed triangle.

But with other previous film analysis that do not fit in the size of this article, one can say that some films roots their fictional alternative realities around only one vertex: the Science one, as in "Interstellar" (Nolan, 2010b), the Mind one, as in "A Beautiful Mind" (Howard, 2001) and the Metaphysical one, as in "Dragonfly" (Shadyac, 2002).

We can assume that there are Film Fictions that manage to base their alternate realities in all three vertexes, as for example "Matrix" (Wachowski, 1999), where we have a world ruled by machines and A.I. (Science validation base), we have a prophet lady that says to Neo the future and things transcend the known rules (Metaphysical validation base) and also where it is the inner life of Neo, his Mind life, that subverts the Science imposed rules of the virtual constructed world.

In all these films an alternate world of reality clashes with our daily one, that we take for granted! However, can Man grasp any universe or reality beyond this ancestral Triangle?

BIBLIOGRAPHICAL REFERENCES

Botz-Bornstein, Thorsten. 2008. *Films and Dreams: Tarkovsky, Bergman, Sokurov, Kubrick, and Wong Kar-Wai*. 1. ed. Lanham, MD: Rowman & Littlefield.

Greenberg, Harvey. 2010. *Inception and the Cinematic Depiction of Dreams*. University of Florida – USA: PSYART.

Halpern, Leslie. 2003. *Dreams on Film: The Cinematic Struggle between Art and Science*. Jefferson, N.C: McFarland & Co.

Hobson, J. Allan. 2003. *Dreaming: An Introduction to the Science of Sleep*. 1. ed. in paperback. Oxford: Oxford University Press.

Jackson, Rosemary. 1981. *Fantasy: The Literature of Subversion*. London: Methuen.

Kravanja, Peter. 2011. 'Cinema, Contingencies, Metaphysics'. *Image & Narrative* 12 (2): 118–24.

Popovic, Nash. 2001. 'Existential Anxiety and Existential Joy'. *Practical Philosophy*, September.

Walters, James. 2011a. *Fantasy Film: A Critical Introduction*. English ed. Oxford; New York: Berg.

—. 2011b. Fantasy Film: A Critical Introduction. English ed. Oxford; New York: Berg.

Filmography:

Wachowski Brothers. 1999. *The Matrix*. DVD. Science Fiction. USA, Warner Bros. Pictures.

Howard, Ron. 2001. *A Beautiful Mind*. DVD. USA. Warner Bros. Pictures.

Lang, Fritz. 1927. *Metropolis*. DVD. Sience Fiction. Germany, Universum Film (UFA).

Nolan, Christopher. 2010a. *Inception*. DVD. Fantasy. USA, Warner Bros. Pictures.

—. 2010b. Interstellar. DVD. Science Fiction. USA, Warner Bros. Pictures.

Shadyac, Tom. 2002. *Dragonfly*. DVD. Fantasy. USA, Universal Pictures.

Ward, Vincent. 1998. *What Dreams May Come*. DVD. USA, Polygram.

Image References:

Fig 1 – Origin Root Type Foundation Triangle (from authors)

Fig 2 – Metropolis:

Wifi qwriter Blog. Dawsomn, Steven. 2011. Metropolis Cometh. [Accessed 4 April 2017]. Available from: http://hifi-writer.com/wpblog/?p=3071.

Fig 3 – Inception: IMDb. 2010. *A Origem*. [Accessed 3 April 2017]. Available from: http://www.imdb.com/title/tt1375666/mediaviewer/rm1069515520.

Fig 4 – What Dreams May Come:

Fig 4 – (Image 1):

Nielsen, J. C. 2014. *What Dreams May Come*. [Accessed 3 April 2017]. Available from: http://www.robinwilliams.net/whatdreamsmaycome.php

Fig 4 (Image 2): Smith, W. C. 2015. *What Dreams May Come*. [Accessed 3 April 2017]. Available from: https://beausmithsblog.wordpress.com/author/williamcsmith2015/

Fig 4 (Image 3): IlPoetaLibero. 2016. Realtà immaginarie. [Accessed 3 April 2017]. Available from: https://ilpoetalibero.wordpress.com/

Fig 5 – Origin Root Type Foundation Triangle in Films (from authors).

David Byrne's *True Stories* – Progress, narrative, fragment, and collage

Mário Avelar
Universidade Aberta, CEAUL-ULICES, Lisboa, Portugal

ABSTRACT: This article analyses David Byrne manipulation of the road movie genre in *True Stories*. The road movie genre, with its own specific codes and themes, claimed a recurrent presence in 20th century American cinema characterized by the protagonist's progress and the process of unveiling the Other both in himself and in the spaces and peoples he is confronted with. I show how *True Stories* starts by simulating a revision of this genre while building an intense dialogue with pop culture, and the video-clip aesthetic, and culminates in a deep encounter with an American mythical structure anchored in an idea of social, economic, and personal progress.

Keywords: Fragment, Myth, Other, Progress, Road movie.

> "... au paradis il n'y a pas de musique. Au paradis, il n'y a pas de tristesse, pas d'ennuis, pas de problèmes: ils n'ont pas tellement besoin de musique.
> La musique, c'est pour émotions, la tristesse, le désir, la nostalgie."
> Ryuchi Sakamoto, "Les images ne mentent pas"
> "We move along so quickly on roads going to nowhere."
> Lee Ranaldo, *Road Movies*
> "Gigantesque paysage d'usines, de viaducs et de voies ferrées. Et puis tout d'un coup East Orange et la campagne la plus carte postale qui soit, avec des milliers de cottages propres et nets comme des jouets au milieu de grands peupliers et de magnolias" Albert Camus, *Journaux de voyages*

The road movie genre, with its own specific codes and themes, claimed a recurrent presence in 20th century American cinema characterized by the protagonist's progress and the process of unveiling the Other both in himself and in the spaces and peoples he is confronted with. In this article, I will show how David Byrne's *True Stories* puts forward a revision of this genre's euphoric progress ethos through an intense dialogue with pop culture, and the video-clip aesthetic.

In *True Stories,* the road *topos* becomes both a spatial interval and a leitmotif for the emergence of a narrative anchored in a collage of newspaper articles, which Byrne articulates in order to build his singular approach to American emblems and myths. Since the uncanny stands at the centre of each story/fragment, Byrne's assemblage oscillates between pop realism and surrealism, thus aesthetically revising the genre's rationalist and analytical background. Rhetoric tropes (metonymy and synecdoche) and strategies (allegory and anagogy) that underlie the movie subtext in its dialogue with visual emblems and cultural myths help us to decipher Byrne's debunking of the genre. They must thus stand at the centre of my analysis. Eventually, I also ponder on the function of music as a cultural subtext that takes part of these rhetoric strategies.

The initial frame unveils a road cutting out the desert plane, stretching in the horizon until it converges to a single vanishing point where earth and sky meet. The road and the wide, open desert spaces are obvious visual American clichés. For us, Europeans, they are easily identifiable due to their permanent presence in our collective memory woven by cinema, and mostly by two genres: The Western and the road movie. Both genres mapped out narratives, signs, colours, and recurrent spaces that helped building an image, and a mythical idea of America. Independently of the generation, one belongs to, of time and space; they keep on appealing to each personal conscious or subconscious idiosyncrasy. Simone de Beauvoir, for instance, stressed this subconscious appeal: "... ces grands horizons sont fascinants. Comme j'ai hâte de partir en liberté sur ces routes..." (Beauvoir, 1997: 120)

When it emerged in the 1950s Pop art reinforced America's visual recognition when it summoned an ensemble of cultural signs identified with economic progress, like the Coca-Cola bottle, and many other commercial products and brands. Pop art represented graphically signs, which were supposed to be artistically insignificant, artistically irrelevant: publicity illustration, magazines and newspapers, furniture, fast food, movie stars, pin-ups, comics. Fernand Léger, Marcel Duchamp, and Giorgio De Chirico stand out among those who helped to forge pop art aesthetic background (Lippard, 1967: 91, 16). De Chirico was one of the first artists to insert a commercial brand in his 1916 and 1917 paintings.

In line with Surrealist and Dadaist traditions, the American artists who came of age in the 1950s somehow denied the notion of the sacred. From Cubism, pop art assimilated the collage aesthetic strategy, with its multiple layers of cultural strata. Collage also allowed the artists to incorporate in a single object the cultural and visual diversity of American landscape. American artist William C. Seitz, who entitled his exhibition in the Museum of Modern Art (Fall 1961) "Assemblage Art", put forward the notion of collage environment – commercial signs, lights, advertisement, junkyards, slums, and 'new antiques', which had appealed to the visual artists since Apollinaire. (Lippard, 1967: 79). In its visual impact, American landscape should be read as collage environment. David Byrne incorporated this dimension both in his songs and in his video-clips; e.g., "Standing tall by the side of the road, /I would pretend that I was a bill-board." (Byrne, 1988: 1) Among these signs, the voyage and the road are recurrent archetypal presences; movement and fluidity stand at the core of his mythical perception of America.

In her anthology *Pop Art*, Lucy R. Lippard reminds that pop art emerged from visual clichés (Lippard, 1967: 7). In the same book, English art critic and senior curator of the Solomon R. Guggenheim Museum Lawrence Alloway links the origins of pop art to America as cliché (Lippard, 1967: 9). Visual dimension, cliché, and America somehow become a triad at the core of pop art. He goes on stating that pop art is mainly a product of a society of wide horizons, always mutating, such as America; always looking towards the future (Lippard, 1967: 11). A society, I should add, where optimism stands at the centre of a national ethos radically connected with an idea of both social and personal progress. In its spontaneity and in its positive attitude towards life pop art somehow embraces an American ethos: an ethos diving deep in American founding narratives and myths.

> For an ordinary, innocent beholder the axis visual dimension-cliché is mainly anchored on unconditional representation. Noël Carroll develops this concept in his book *Philosophy of Art* in straight connection with three other concepts: lexical representation, conditional specific representation, and conditional generic representation. Unconditional representation:
>
> is the sort of representation that obtains by triggering the audience's innate recognitional capacities – capacities that enable viewers to recognize that the referent of Mona Lisa is a woman simply by looking at the picture. (Carroll, 2000: 50)
>
> In this line of thought:
>
> much representation in the arts – for example, in mass market movies, TV, theatre, painting and sculpture – proceeds by triggering innate recognitional capacities and, in that sense, is immediate (i.e., not mediated by the manipulation of an arbitrary or conventional code). (Carroll, 2000: 50)

As I have shown elsewhere (Avelar, 2004: 13–22) the signs that frame the American scenes dive deep in mythical founding narratives and *topoi*. This engagement in the past, this architextuality, unveils, however, political discourses. In the early 1980s I remember being in a lift in Moscow, with my American blue jeans. At one floor the lift stopped, the doors opened and a cleaning lady entered. She pointed at my jeans and exclaimed "America! Coca Cola! Freedom!" Some of us in this room may remember that in the Portuguese 1970s Coca Cola triggered a different political analogy: "Coca-Cola, Imperialism's dirty water!" As Carroll wisely remarks:

> [t]hough representational painting, drawing, and sculpture depend primarily on unconditional representation – as theater, film and TV do – they also depend on conditional representation, deploying titles, captions, and backstories (e.g., knowledge of the myths, historical events, religious parables and so on that the images depict) in order to be fully intelligible. (Carroll, 2000: 50)

Intelligibility may be conditioned when our knowledge of the past, of History, relies mainly on visual information; when this occupies the space of the real thing.

We are aware of the fact that the knowledge of the past provided by Historians has a narrative dimension, and depends on the point of view of who writes, of who structures it. These are supposed to be however, narratives anchored on rigorous factual analysis; they are supposed to provide readings close to bygone realities. The fact is that the generation who grew up in the 1950s, who grew up during the early days of pop art, relied mainly on TV to get information and knowledge in order to build their own personal insights of History, of the past. TV narratives were, for the innocent beholder, the real thing. TV narratives were History.

David Byrne, himself a European and a pop artist too, who grew up in the 1950s, welcomes and incorporates the visual cliché in his 1986 movie *True Stories*. His is an attitude of hospitality towards the powerful visual signs that have become part of the contemporary Western culture and of the perception we have of America. Among these stand the open road and the wide-open spaces I mentioned above with which the film begins. They are part of American utopia, of American mythical origins. The viewer's perception of these signs, of the sounds of nature, and of the voice of the little girl dressed in white (a sign of purity and innocence) singing is interrupted by the voice of the narrator who says: "This is where the town begins", and starts tracing a historical timeline, a timeline of progress. The first information he provides is that this was the place where everything began. So private History – the narrative of a mythical town of Virgil in Texas – and general History, private myth and the myth of origins coincide. This really is a special place in the heart of a mythical desert. We could bring to mind Beauvoir's words: "Nous étions au cœur de la ville et cependant dans un désert." (Beauvoir, 1997: 195) However, a question remains: is this factual truth?

In its elliptical structure, this timeline of civilizational progress over-simplifies History, change and evolution. The shots, fragments of old movies, TV series and documentaries show how History has been made available, knowable to us viewers; how History and TV coincide. Though a historical narrative may be simulated, these are not facts but a narrative built upon narratives. These may be picked up either on cinema and on TV or in newspapers and in magazines. Here the author actually got his true stories; the true stories that helped him to create "a film about a bunch of people in Virgil Texas", as the casual information on the black screen goes.

Underscored by an anaphoric discourse ("and then…and then …and then"), a torrent of frames revealing dinosaurs, Indians, Spaniards, Mexicans, pioneers, settlers, the gold rush, industries – cotton, cattle, oil, electronics, the 150th anniversary of Texas, flows on screen picked up in drawings, old movies, TV series, icons of pop culture, mixed with private confessions, like the narrator's love for dinosaurs. Clips from visual culture stand out as creators of a discourse on reality. Emerging from the distant, remote, mythical, imaginary past the elliptical enunciation thus flows into a mythical real present: The Celebration of Specialness of Virgil, Texas.

The narrator so far had been just a voice. In his *Sun Times* review Roger Ebert writes that:

> [h]e is a thin, quiet, withdrawn figure with a voice so flat that you have to listen to the pauses to figure out when the sentences end. He drives a new red convertible and wears Saturday night cowboy clothes. (Ebert, 1986)

The narrator emerges then on the screen and, in a family reunion simulacrum, he projects a series of slides of this town. The casual pose is mirrored in the way he dresses, since "to be something is to be in relation to something other" (Desmond, 2004: 4). His Stetson, his cowboy boots, his self-imposing obvious Texan look turns him into a visual cliché, that is far from being an ironic parody of Texan culture; far from being an allegory, and above all a political allegory: he just wants to dive deep into a culture, to be part of the gang, to be welcomed as another sign among recognizable, ordinary, identifiable signs. Hospitality towards the other and desire to dive deep in otherness thus coincide. Albert Camus had identified this trace of American character on he wrote in his Journal: "Leur hospitalité, leur cordialité est du même goût, immédiate et sans apprêt. Ce qu'il y a de meilleur en eux." (Camus, 1978: 29)

The slides' projection ends up when mimicking Buster Keaton, he enters the screen, when he cuts through the screen because unlike Keaton who is assimilated by film, or Tom Baxter in Woody Allen's *The Purple Rose of Cairo* who frees himself from the film, the narrator just moves into another scenic dimension, the curtain behind the screen. The fourth wall is thus unveiled, and performance becomes self-evident. This is the first time that Byrne breaks the fourth wall. He will often recede while he is narrating in his car in a documentary-like casual pose.

The simulation of the genre, the documentary, is crucial here, because, although uncanny, these are supposed to be true stories. The artist collected some of them from tabloid newspapers and magazines: the woman who never leaves her bed, whose gadgets remind Chaplin's in Modern Times; the couple who for many years communicates only by way of their children; the odd exotic fashion pageant; the karaoke-like contest in a shopping mall where, among others, people impersonate Prince, Meatloaf and Billy Idol. Byrne admitted having invented some of these stories. Anyway, as a line goes in the end of the film: "If you can think of it, it exists somewhere." To these stories he adds characters like the lying-woman – with whom Burt Reynolds had been in love, who met the real Rambo in the jungle in Vietnam when she was a nurse there, who also was born with a tail, and who wrote half of Elvis' songs-, the voodoo practitioner, the conspiracy theorist preacher inspired in the Church of SubGenius (Byrne is a SubGenius himself), and Louis Fyne, the man obsessed to find someone to love, who keeps a "Wife Wanted" sign on his front lawn, and who provides the main narrative thread. His surname mirrors the casualness, the lightness of tone that pervades through the film.

There is a moment close to the beginning when the narrator is walking around in the shopping mall and meets two young boys close to a magazine stand. They are reading some story in a magazine and laughing like fools. We are led to assume that they must be reading and sharing some crazy, uncanny stuff. The narrator looks at them and smiles. This episode reveals what is supposed to be the film's relation to reality and the viewer's attitude towards it: sometimes life is stranger than fiction and we just have to take it as it is. This attitude somehow echoes Nancy Marner's approach to American western pop art in "Pop Art in California", when she underlies its innocence, its laudatory vision of the American scene (Lippard, 1967: 151). There is, however, no political statement, no partisan-like attitude, and no strategy of the seduction, of imposing one's agenda upon the viewer. This tone is very clear in the narrator's words at an early stage of the film:

> I have something to say about the difference between American and European cities, but I forgot what it is. I wrote it somewhere. (Byrne, 1986)

In the early eighties, deep cultural changes were taking place in America. Among these stood out two signs, two fragments – metonymies – of America's allegory of progress, of economic and social achievement: ITEC, which brought along with them more fluid business interactions, and shopping malls, new places of social relations. Significantly, the first focus on Virgil's identity is the ITEC's facility, the Varicorp

Corporation, the new city icon, an icon of social and economic progress – the new technologies utopia.

There the narrator meets Louis Fyne, who, along with the town, is the movie central character. They will meet again in the shopping mall, which has replaced the downtown as the centre of American activities. There, he adds, people are inventing their own system of beliefs. After all, shopping is a feeling, reminds this section subtitle.

The narrator walks around the shopping mall the same way he drives through town – "Freeways are the cathedrals of our time, someone said, not me", he comments while driving his red convertible and looking at the eye of the camera in an ongoing dialogue with the viewer. There is no predetermined target, no structured discourse; only the flowing of time eventually leading to the celebration (al) show. Meanwhile, he welcomes every new sign, either a homogeneous ensemble of houses or Fyne's tennis shoes, with an approving smile. The ordinariness of these signs requires the viewer's unconditional representation: they belong to our collective American imaginary; like the baseball cap, they function as metonymies of America's culture. They represent the American peaceful scary middle-class imaginary. We have seen them in *Poltergeist*, in *Edward's Scissors Hands* and in *Desperate Housewives*. Cinema has specific instruments in order to raise them to a mythical order. Janet Maslin stresses the importance of Byrne's statement on perception:

> When I see a place for the first time, I notice everything – the colour of the paper, the sky, the way people walk, doorknobs, every detail.' (Byrne, 1986)

And she adds: "True Stories makes its audience notice things in that way, too." (Byrne, 1986)

In *True Stories,* photography is the main provider of an uncanny mythical dimension. Roger Ebert reminds that:

> The movie was shot by Ed Lachman, who has become a brand name for people interested in offbeat directors. He was the guy who followed Werner Herzog to the slopes of a volcano that was about to erupt to film "La Soufriere," and he has worked for Wim Wenders, Shirley Clarke, Bernardo Bertolucci, Jean-Luc Godard... This time, he finds a new look: His landscapes and city scenes are like those old postcards in which everything seems slightly skewed. His buildings look like parodies of buildings. (Ebert, 1986)

The houses – the signs – are not depicted in their own materiality, but as fictional signs inspired by previous perceptions, the postcards, the light of a movie, a video-clip. In their fictive dimension, one recognizes a myth anchored in economic, and, eventually, social progress.

Simulacrum and performance thus prevail in *True Stories*. The narrative actually begins with the building of a stage – a nuclear sign of the performative dimension – where the specialness that is about to celebrated will take place. Emphasising the progress that lies at the core of this narrative, the viewer is led to witness the ongoing process of building the stage. At the same time, the viewer watches Louis in the process of creating the sad song he will sing later on Celebration Day. His song, his performance of "People Like Us" coincides with the narrative's climax.

Song and stage, art and performance, therefore, go hand in hand. This climatic song had been preceded by several tunes that belonged to different genres – from American pop to Mexican pop, from gospel to country, from rock to children tones. Through these harmonies, through musical genre diversity, Byrne provides a cultural metonymical approach to American cultural, social and ethnic diversity; they summon and echo a cultural melting pot. The genres do not evoke a hierarchy; they are part of a continuum. Byrne will use this metonymical musical strategy again in a later CD, *Rei Momo*, where he will revisit several musical genres – cumbia, orisa, merengue, mapeyé, bomba, salsa, reggae, samba, charanga, pagode, bolero – that mirror a world of coexisting diversity.

In the end of the movie, the white dressed girl who had left the initial frame re-enters the screen. Circularity is somehow accomplished when, still dancing along the road, she almost fades into the distance. A new sound provides the background for her performance, "City of Dreams", a song by the *Talking Heads*, and a periphrastic naming of the ultimate achievement: paradise. Yet, as paradise lies ahead, somewhere in the future, music is still required to remind us how important our imagination, our dreams are for an individual and collective progress. In addition, this progress, as the road movie keeps on reminding, is inseparable of *this road to nowhere* fluid soil.

BIBLIOGRAPHICAL REFERENCES

Avelar, Mário (2004). *História(s) da Literatura Americana,* Lisboa: Universidade Aberta.

Beauvoir, Simone de (1997). *L'Amérique au jour le jour – 1947*, Paris: Folio.

Blasing, Mutlu Konuk (1987). *American Poetry – The Rhetoric of its Forms,* New Haven: Yale University Press.

Byrne, David (1986). *True Stories* London: Faber.

__________ dir. (1986). *True Stories*. USA: True Stories Venture – Warner Bros.

__________ (1988). *Naked*. CD – USA: EMI

Camus, Albert (1978). *Journaux de voyages*. Paris: Gallimard.

Carroll, Noël (2000). *Philosophy of Art,* London and New York: Routledge.

Deleuze, Gilles (1983). *L'Image-Mouvement*, Paris: Les éditions de Minuit.

Desmond, William (2003). *Art, Origins, Otherness – Between Philosophy and Art*. New York: State University of New York Press.

Ebert, Roger (1986). *"True Stories"*. *Sun Times*. October 31. (http://rogerebert.suntimes.com/apps/pbcs.dll/article?AID=/19861031/REVIEWS/610310304/1023

Lippard, Lucy R. (1967). *Pop Art (World of Art)*. London: Thames and Hudson.
Maslin, Janet (1986). "David Byrne in *True Stories*" (New York Times) (http://www.davidbyrne.com/film/true_stories/press/truestories_newyorktimes.php)
Ranaldo, Lee (2004). *Road Movies*. New York: Soft Skull Press.
Sakamoto, Ryuchi (2017). "Les images ne mentent pas". *Cahiers du Cinéma,* 731, Mars, 14-17
Thomas, Julia ed. (2000). *Reading Images* New York: Palgrave.

Lisbon through the eyes of a writer: Literature, architecture and city memories in the novel *O Secreto Adeus*

Miguel Baptista-Bastos
University of Lisbon, Faculty of Architecture, Lisbon, Portugal

ABSTRACT: Baptista-Bastos (1933/2017) can be considered the last writer in Portugal, whose work of fiction has usually been circumscribed thematically to the city of Lisbon. Spanning several generations, his work undergoes a constant change, depending on how the city ages and on the urban transformation that follows: writer and city converge in a unique and singular process, where both protagonists walk side by side through time and space: always together and never apart.

Using language, the reader grasps an idea of the city, throughout the ages and different periods.

Is this an architecture that exists as a theme for the existence of a narrative, or is this a narrative that absorbs architecture in order to conduct its own fiction? The memory of a place and its spatial composition are joined and transformed into a written record.

In our investigation, we attempt understanding his written work within the limits of an urban context. It will include an interview with the author himself. The interview will focus mainly on his first novel, *O Secreto Adeus* [The Secret Goodbye], written in 1963, alongside the various places in Lisbon where the action of the novel takes place. We will attempt a diachronic comparison of these same places over time: from the time narrated in the novel until the present day.

Keywords: Lisbon, Literature, Architecture, Representation, Reinterpretation

1 PRE-INTRODUCTION

The strength and novelty of Baptista-Bastos' work, initiated in 1962, has matured in the past decades, without losing its focus, because all his writings are set in the same place, which acts as its "secret soul". His work has been studied by various academic investigations in the Portuguese-speaking world. The narrative elements of his work are simultaneously visual and active. This could be attributed to him being considered a dean of national journalism. In his novels, he uses a literary technique borrowed from journalism, with its effective visual and body components.

Something is shared between the reader and the books: the relationship between the subject (the characters) and the spatial setting (Lisbon) is permanently in an unstable equilibrium, that one could compare to a dance, where one partner's step depends on his partner's body, and the narrative arises from this dialogue. The vast knowledge the author possesses of his native language is extended with the knowledge of his native city, in a specific movement: streets, squares, avenues and people emerge from the dark ink of the written words, connected by a story. Using language, the reader grasps an idea of the city, throughout the ages and different historical periods; from the early twentieth century until the present day. Baptista-Bastos' literary career can be divided into two phases corresponding to two periods in Portuguese modern history: the period during Salazar's dictatorship, characterised by censorship, and the period starting in the 1974 revolution, until today. There is a literature imprisoned in a censored city and a liberated literature in a free city.

Is this an architecture that exists as a theme for the existence of a narrative, or is this a narrative that absorbs architecture in order to conduct its own fiction? The memory of a place and its spatial composition transformed into a written record. In our investigation, we attempt understanding his work in the urban context, including an interview with the author himself concerning the places where the novels take place in different times. We aim to observe the city as presented in his first book and now, that is an architectural construction, which moves from the narrative to the space of architectural representation.

Something is shared between the reader and the author: the relationship between the subject (together with the characters) and the spatial setting (Lisbon) is permanently in an unstable equilibrium, that one could compare to a dance, where one partner's step depends on his partner's body, and the narrative arises from this dialogue.

The plot of the novel revolves around different memories of the same places in the city. Its reading may be useful in understanding many facts occurred in Lisbon during the last fifty years. We will consequently try to dissect an architectural construction expressed as a written narrative, in an investigation into the representation of space in architecture.

2 THE WRITER AND THE CITY (AN INTRODUCTION)

We propose to observe a city reinvented by a writer. In this way, we shall attempt to investigate an alternative way of thinking and clarifying architecture and urbanism. Narrative techniques have at their disposal a social archaeology, which allows for an analysis of the city in its various contexts, its habits and lifestyles in a specific time and place, through the memories transmitted by literature. The city in question is Lisbon and its narrator is Baptista-Bastos, the writer. We are in the year of 1963 – when he wrote his first novel *O Secreto Adeus* [The Secret Goodbye] (1963). Today, we are far from those dark times, when Portugal lived in the shadow of a fascist regime referred to as "New State"[1]. However, there are still traces in the city and in a society that remain, and the reading of this book will help us understand this obscure period, in a city filled with light, but at the same time, filled with sadness. Fiction is a filter that helps us understand what went on and how it spread. These sombre times are described in this novel, where the plot and its physical setting are inextricably linked.

Describing a melancholic country, introverted by Salazar's regime, and demanded an accurate and analytic study, resulting in a highly critical reflexion made by its literature. This analysis has the originality of being a representation of Lisbon that, in spite of exposing the existing political censorship, revealed a modernity that avoided the "typical and folkloric" image of Lisbon imposed by the regime. This book was innovative in the context of Portuguese literature because it introduces a new style, marked by a permanent confrontation with the establishment and the use of very short sentences of quick speech, which suggest images, with Lisbon permanently on the background.

With a journalistic career that began in the early 1950s, Baptista-Bastos' first book draws on his personal experience to narrate an episode of a journalist who tells us about various situations that occurred in Lisbon. In terms of a narrative style, *The Secret Goodbye* changes from Portuguese Neo-Realism aesthetics, so much in vogue in the 1950s, to a social representation, due to its urban character, presenting a portrait of Lisbon, which we can still recognise today. We can say that it is the first Portuguese book of fiction that is not attached to the neo-realistic Portuguese literary movement, because it is much more expressive, straightforward and totally attached to the city, as opposed to what was done at the time. A journalist that uses a journalist's technique whilst writing a book that tells the story of a reporter. In an autobiographical tone, the book reveals a highly controlled society, a city saddened by the impossibility of self-expression. Moreover, the old city and the new one are intermingled through fiction. The author uses the description of the city as a means to reveal an enormous counterpower, opposing himself to the political and cultural dominion held over the country. Strangely, this book went unnoticed to the eyes of the national commission of censorship, but, after its publication, the author, an opponent of the regime, suffered its consequences, and was temporarily forced to leave the country, since he was at imminent risk of being imprisoned.

3 THE WRITER AND THE CITY (LITERARY AND URBAN CONTEXT)

The writer has written more than a dozen books, including fiction, short stories and interviews, where you will always find the urban setting of Lisbon, making this an unusual body of work in the national scene: no other writer has his work built around a single place.

Lisbon is the filter through which he can see things and the world. Literature and architecture merge into an exposed urban contemporary reality, based upon a dynamic narrative, where the journalistic style is used like a sequence of representations where the action takes place.

The Secret Goodbye is a story set in the same epoch of its writing, where the main character observes and narrates a sequence of events in a contemporary Lisbon, even if his will to narrate these events is always conditioned by the censorship of the government. The architecture of the city is gradually revealed, always circumscribed to an urban perimeter defined by buildings, streets, squares and neighbourhoods. Places that emerge in a special way in a literary invocation, using textual language placed at the service of human memory, which gives structure to an event in a specific location.

The mere fact that the plot is set in a specific city and a specific time, suggests unusual representations that cannot be understood with a simple photographic image. The characters in the book appear in situations that are linked to Lisbon during this unique period of Portuguese recent history: an archaeology of the city discovered in a text of fiction.

In this fable, the characters are dependent not only on the plot but also on the buildings which are converted into fiction, that is: the places described by the author are easily recognized by any inhabitant of Lisbon, but the book changes them by moving them to a different age – the outside world converted into an inner world.

The places mentioned in literature are presented in forms infinitely different from the real architectural

[1] "New State" is the name given to the authoritarian, autocratic and fascist political regime that ruled Portugal for 41 years without interruption, since the approval of the 1933 Constitution until it was overthrown by the Revolution of the 25th of April 1974. Salazar governed the "New State" between 1933 until his death in 1968, and Marcello Caetano from 1968 until 1974. This was the authoritarian regime with the longest duration in Western Europe during the 20th Century.

object. The architectural and topographical reality is a pretext in the construction of a plot for the understanding of a specific time and place.

However, this book does not intend to be the literary guide for a touristic view. It aims to expose a metropolis in a simple manner, in a specific time: an urban memory reinvented by fiction. This passion for a certain place does not force the reader to embark on a pilgrimage; instead, it forces him to rethink another urban reality of the metropolis; a kind of literary tour without the need of going to the places where the action takes happens.

The literary tour of *The Secret Goodbye*, is very visual, using narrative techniques borrowed directly from journalism[2], however, it is also ambiguous, for the architectural setting which can appear to have a very strong image, can shortly afterwards acquire a faltering atmosphere. The characters are in permanent dialogue between themselves, always circumscribed to a time and place exclusive to Lisbon, and always immediately recognised by a local inhabitant; however, this city is not shown or put on display. The background is not described; the narrative technique takes us to the architecture of the city, as opposed to other writers, as is the case of Eça de Queirós[3], where a detailed description of the architecture was an inseparable part of the story – as if his work of fiction totally depended on this description.

Throughout the book, the architectural narrative, accomplished by the description and the placing of the actors in the different places, becomes useful in understanding the work of literature – aspects like the climate, seasons, trees, aromas and the political context. These different moments are immediately seen and understood by the reader.

The social context commands the order of events, together with the different places, because it clarifies the nature of the different characters. This metaphorical fusion between the city and its persona, made in such a way that one can describe the other, is the core of this novel's fiction.

This book put Portuguese literature in a new contemporary standpoint, inspired by a literary journalistic style, with the use of short sentences that, despite being very descriptive, are treated like images: narrated images. This style was also largely inspired by cinema, since the author was a film critic and participated in the script and the organization of some films, which laid the foundations for the New Portuguese Cinema, as is the case with the film *Belarmino* (Fernando Lopes, 1964), where he plays an active role as the interviewer.

Literature was perhaps its most important instigator, for this form of expression was permanently under the attention of the censors. (The censors were not very concerned with the other arts, which had a secondary role, for these were heavily dependent on Salazar's political apparatus.) Literature was the Portuguese artistic expression with the greatest degree of freedom, because it was not dependent on a patron, as opposed to painting, architecture, or sculpture, and it could easily be published and divulged at a minimum cost. This obviously puts the writers under a greater risk, since their artistic expression has a much more intervening and critical role in society.

The development of a different style in Portuguese society arose out of an urgent need. The sixties and a group of Portuguese intellectuals were the instigators of this new ethics and aesthetics.

The sixties in Lisbon was one of the most important, influent and culturally active periods in Portugal, for they marked the birth of a new modernity in Portugal, clearly influenced by the Neo-Realism of the fifties. However, this generation did not exactly owe obedience to this movement, in contrast to what happened in the rest of Europe.

4 LISBON 1960s (URBAN CONTEXT)

Alongside the traditional historical city, another city was emerging, much more visible in the urban topography, but less recognisable, for it was not much promoted by Salazar's regime: the beginning of modern architecture in Lisbon. New neighbourhoods were planned, like the Olivais[4] that was the largest example in Portugal of urban planning in accordance with the principles of the "Charter of Athens", still known today as the "garden-neighbourhood".

A new bridge crossing the Tagus was built between 1962 and 1966, uniting the margins of the cities of Lisbon and Almada. It was named after Salazar (after the April Revolution it was renamed "25th of April Bridge" – which is its current name). This was a huge transformation in the social context; it originated a much closer relationship between both cities and accelerated urban growth.

The Metro of Lisbon was inaugurated in 1959. The first underground transportation network that opened in Portugal. It united the old and the new parts of the city. It was an extraordinary innovation, for the first time inhabitants of Lisbon travelled in a fast and silent manner.

The "Avenidas Novas" was the name given to a large area of urban development, which expanded the city of Lisbon to the north, at the end of the 19th century and in

[2] Before becoming a writer, Baptista-Bastos was a journalist, essayist, and film critic.

[3] José Maria de Eça de Queirós (1845–1900) is one of the most important Portuguese writers. He was the author of highly acclaimed novels: *The Maias* (2002a), and *The Crime of Father Amaro* (Queirós, 2002b). He is considered by many to be the best Portuguese realist novelist of the 19 century.

[4] The Urban Plan of the Olivais was developed between 1955 and 1962. In 1960, the Municipality of Lisbon defined the urban layouts and a set of rules under which housing would be designed. The detailed plans for these urban layouts began in the same year. In total, the three chosen layouts – North Olivais, South Olivais and Chelas – occupied an area of 710 hectares – which for Lisbon was an enormous area. The principles of the "Charter of Athens" (neighbourhood units, green spaces, high-rise construction) were widely applied in this part of the city.

the first half of the 20th century. This growth stabilised in the 50s and 60s when the latest modernist constructions exposed a new way of life. The city's intelligentsia who, since the 119th century, had regularly met in the old cafés of Lisbon, such as the "Brasileira", the "Nicola" or the "Café Gelo", also occupied these new areas. This is the case of the group of the "Vá-Vá" café, who gave birth to the "New Portuguese Cinema".

5 LISBON 1960s. (INTELLECTUAL/ARTISTIC CONTEXT – BREAK)

The beginning of the 1960s is marked by social unrest (University strikes), important left-wing clandestine movements and by a cultural and editorial agitation, persecuted by the state police PIDE[5]. This atmosphere thrived in the universities, where intellectuals passionate for literature, painting, architecture and films, established themselves in the cultural life of their country.

Immersed in this context *The Secret Goodbye* clearly mirrors how this suffering but a rebellious generation saw things. The book was first edited in the collection "NOVOS" [New] published by Portugália. This collection had an enormous impact in the cultural circles of the time, for its authors were all young with a single book published and disliked by the ruling power. Besides Baptista-Bastos, another author also initiated his career in this collection; considered one of the most important books published in Portugal in the second half of the 20th century. We are talking about Herberto Helder and *Os Passos em Volta* (1963) [The Surrounding Footsteps].

It was also at this time that a new "Brutalist" architecture appeared in Portugal, through the work of Nuno Teotónio Pereira and his building "Franjinhas", which moved away from modern and neo-realistic aesthetics.

As mentioned before, the "New Portuguese Cinema" was an Avant-guard movement that in spite of the dictatorship in the 1960s, managed to break with the current ideology initiating a movement that would establish itself in the following years. It was inspired in the French Nouvelle Vague and in Italian Neo-Realism. Young educated university students, seduced by an alternative type of cinema, which was available in small film-clubs, joined this movement. The two most famous films from this period are *Os Verdes Anos* (Paulo Rocha, 1963) and *Belarmino* (Fernando Lopes, 1964). Baptista-Bastos had an active participation as the interviewer/actor in the film *Belarmino*.

6 INTERVIEW WITH THE WRITER

Miguel Baptista-Bastos (MBB) – What was Lisbon like in the sixties?

Baptista-Bastos (BB) – The atmosphere felt in Portugal, particularly in Lisbon, was very unhappy, people lived in tragic times [. . .] absolutely tragic! And this moved me deeply. My first book of fiction, *The Secret Goodbye*, is born of this discontentment [. . .]. I understood that the world was changing, because, at the time, I was very aware of things, and I started to see the uprising in the so-called third world, particularly in Indonesia, with an international meeting, called the "Conference of Bandungue", in 1955, which I followed through the international newspapers. I read, amongst others, *Le Monde*, whenever it was possible and was very well informed. I quickly understood that the world was going to change, do not ask me why, because I do not know. But I understood that the world, the world where I lived, was changing [. . .] and it did! Dictatorships changed and even religion's vision of the world changed, by abdicating from its responsibilities. The church abdicated and that was truly disastrous. However, it was also then that the church was challenged from within, and this was very important, for its own evolution [. . .] for everything!

(MBB) – You mentioned that the world was changing and we can see the changes, but this thought was common to several people and forms of expression, for example, there is a breakthrough in Portuguese cinematography with two films that initiate the movement of the "New Portuguese Cinema", with *Os Verdes Anos* and *Belarmino*. You also made this happen in literature, in poetry with Herberto Helder, in architecture with Nuno Teotónio Pereira [. . .] so at that time we can see the appearance of a series of...

BB – Things were in fermentation [. . .]. I mean, because both *Belarmino* and *Os Verdes Anos* were important for Portuguese cinema (they still are today), but they are two opposite visions. One more rooted in the working class, *Belarmino*, where in an interview I conduct and which occupies most of the film, he says that he could have been a champion, but he was not, because people would not allow it, the surrounding atmosphere would not allow it, the world was closed and draught-proof. The vision in the film *Os Verdes Anos* was more bourgeois, more, how can I say this: more moderate. It was more cautious. After all, it is a love story, the courtship of a girl by a boy who came from the countryside, and apparently, there is more to it than that. Everything is set in Avenida de Roma, when it was very different from what it is today, filmed by a bunch of young men who have all given up now, they are all dead now (pause) by a group of young men, in this case, the group of Paulo Rocha the film director, a very good friend of mine, who lived in the same building of the café "Vá-Vá", which was a very important meeting point for intellectuals in Avenida de Roma, which was the new part of town. [. . .] And the story was filmed there. He used to say I was a kind of fetish-actor for him and I took part in several films, in small parts, or as an extra [. . .]. Paulo would call me and say: "I'm going to make another film and, once again, you have to take part! (laughs)". He had studied in Paris, in the IDHEC, the institute

[5] The International Police for the Defence of the State (PIDE) was a police force that existed in Portugal during the period of the "Estado Novo". It was the political police and responsible for the security of the state.

of advanced studies in cinematography, and culturally France was important then. Not so much today, but then, Paris had everything; it had philosophy, painting, poetry, the biggest film library in the world, directed by a man called Henri Lanoe, everybody and everything supported these cultural movements, and, as a result, it gave birth to the French *Nouvelle Vague* [. . .]. Obviously, it can all be disputed (laughs).

MBB – It was called *New Wave*, *Nouvelle Vague*. And it is readapted to the Portuguese context. But there are common points; there is a new vision of modernity.

BB – I would not just call it modernity, I would call it a great indisposition and despair felt by this generation, I mean . . . we can define it as a tormented reaction, because we all had more or less the same concerns, with different ways of solving them. Different aesthetics leading to a common ethics: a tormented modernity. (Baptista-Bastos et al., 2014)

7 CONCLUSIONS

This work is based on research that does not exist in Portugal. Literature, interpreted as a platform for an architectural inspiration, is something enriching for the imagination of the architect. A multidisciplinary research is imperative, especially today, in an increasingly homogeneous and equal world, which often does not accept diversity, as a factor of progress, for an aesthetic and an ethics in art: this case, literature and architecture. Baptista-Bastos, recognised as one of the great Portuguese writers of the second half of the twentieth century and the last great chronicler of Lisbon, uses architecture as inspiration. Why not use urban literature as an inspiration for architecture?

Finally, this research is being developed for a future book, where with these premises, the progress and multidisciplinarity between the Arts and Letters are promoted.

BIBLIOGRAPHICAL REFERENCES

Baptista-Bastos, Miguel & Gomes, Diana. (2014) Interview with Baptista-Bastos recorded on October 2014.

Baptista-Bastos, Armando. (1963) *O Secreto Adeus: Romance.* Lisboa: Portugália.

Belarmino. (1964) Lisboa: Fernando Lopes.

Helder, Herberto. (1963) *Os Passos em Volta: Contos.* Lisboa: Portugália.

Os Verdes Anos. (1963) Lisboa: Paulo Rocha.

Queirós, Eça de. (2002a) *The Maias.* Trans. Patricia McGowan and Ann Stevens. Manchester: Carcanet Press.

—. (2002b) *The Crime of Father Amaro.* Translator Margaret Jull Costa; Cover illustration: Paula Rego. Sawtry: Dedalus Ltd.

Part V

Science/technology

The electric motor in Portugal: Technological progress and industrialization

Ana Cardoso de Matos
CIDEHUS, University of Evora, Evora, Portugal

Maria da Luz Sampaio
CIDEHUS, University of Evora, Evora, Portugal
IHC, Faculdade de Ciencias Sociais e Humanas, Universidade NOVA de Lisboa, Lisboa, Portugal

ABSTRACT: The introduction of the electric motor in Portugal is due to foreigner companies such as Siemens, AEG, ASEA, and ACEC. They were important levers to move the electrical sector, the electricity network, and the technological progress. Portugal only started the manufacture electric motors in large quantities in 1946 when ACEC, EFA and CUF united to set up to develop a business project and the introduction of the electric motor in Portugal has common aspects with others European countries.

In this paper we add different views and through the analysis of photograph we describe the entry of the electric motor into the factory and the impact of electrical motor in the manufacturing space, which allowed, not only, to rationalize the flows between the machine but also and operating a new circuit between the different machines or operations.

Keywords: motor, factory, electricity, technological progress

1 THE PROGRESS OF ELECTRICAL NETWORK OF PORTUGAL

In Portugal, the introduction of electricity dates back to the end of the 19th century. The transference of technology was possible due to the foreign banks capital, electro-technical engineers, and entrepreneurs. This became evident in the case study of CRGE- Companhia Reunidas de Gás e Electricidade, founded in 1891 or SOFINA, Société Financière de Transports et d' Entreprises Industrielles, that since 1905, began to invest in several countries, such as Spain and in 1913 it became the majority stockholder of CRGE, by subscribing 96,000 shares. SOFINA's presence in Portuguese companies extended to the Sociedade Energia Eléctrica, founded in Porto in 1908. Another example of electricity companies acting in Portugal is given by AEG, which by 1910 had installed in the country several hydro-electrical and thermal power stations, e.g. the stations in Évora, Chaves, Régua, Lamego and Angra do Heroísmo, which produced energy for lighting and industrial consumption. Their action extended to the construction of electrical power stations linked to tramways in the country's main cities – Lisbon, Porto and Coimbra.

In 1917, the country had 39 electrical power stations – 12 hydro-electrical, 21 thermal and 6 mixed-type. Their combined power output of 13,136 kW, with an average of 336.8 kW per station and an average consumption of 1.9 kW per 100 inhabitants. The consumption data in Portugal, in the first quarter of the 20th century, shows a pronounced distance between the Portuguese market (covering only 15% of the country's municipalities) and the consumption levels in the main European countries. Interest in hydro-electricity, which had begun at the end of the 19th century, saw an increase during the First World War due to the shortage of imported coal, which would be reflected in the activity of thermal power stations – especially those belonging to the big companies in Lisbon and Porto, namely CRGE and Sociedade de Energia Eléctrica (Matos, Sampaio 2016).

The presence of electrical enterprises such as Siemens in Portugal dates back to 1876 when the company installed a furnace with heat recovery system in the Real Fábrica de Vidros da Marinha (a glass factory). In the ensuing years, Siemens supplied machines and other equipment to several Portuguese companies, including those needed for the installation of the tramways of Oporto, which began working in 1895, and by the tramways of Lisbon, which started operating in 1901. Initially, acquisitions were made through Madrid, where the German group Siemens founded, in 1899, the Compañía Anónima Española de Electricidad, but in 1905, a new branch was created in Lisbon, under the name Companhia Portuguesa de Electricidade Siemens-Schuckert Werke, Lda, with an office in the city of Oporto (Alves 2005: 55). The opening of this branch was a sign of the development, which the Portuguese electricity market had undergone since the end of the 19th century.

Figure 1. 3-phase Motor Type MK - B, de cage, or squirrel. Source: *ASEA – Journal*, 1924, p. 12.

Another enterprise with interest in the Portuguese market ASEA - Allmänna Svenska Elektriska Aktiebolaget, a Swedish enterprise responsible for the introduction of the electrical motor. Jonas Wenströn conceived a new motor, after visiting the International Exhibition in 1881 (Barnevick 1983: 12), and in 1894, he patented his invention and in the letter of presentation he wrote:

> My invention relates to dynamo electric machines of that particular cases in which the armature, field magnet and the conductors thereon are all stationary, the rotary part being a body of iron called the "Keeper" [...]. My Invention consists in the construction whereby I am enable to operate a machine of this description without eddy currents with construction, consist of an armature having a continuous surface facing the keeper. (United States Patent Office, 1894: 1)

ASEA launched, in 1924, a new engine that incorporates the characteristics of the induction motors produced until then by the Swedish company. The new engine was more appropriate for textile industry, carpentry, mechanical metalworking and had the advantage of shortening the length of the free shaft between bearings, thus reducing the risk of fouling the area between the stator and the rotor (ASEA Journal, 1924: 12). These motors were designed to be used in the mines and had a housing, specially built for the collector rings. They could have different forms of joining to the machine tool: either direct or by transmission belt and with or without third bearing. When this engine was conceived and designed, it was intended to maximize the use of the motor, exploring options that had been extensively studied at the time introducing two important innovations: issues associated with bearings and the introduction of a new ventilation system. In Portugal, Jayme da Costa enterprise will be the branch of ASEA until 1960, only by trading its products.

The technological evolution and improvements introduced in electric motors were constant, driven by the growth of cities, their lighting networks and the pressure for technological solutions. The emerging sectors were requesting, namely, more power and more speed for their machine. The introduction and improvement of the engines are associated with the requirements of higher mechanical yields, which industrial development needed at each stage. Portugal started the manufacture of electric motors in large quantities in 1946 when ACEC, EFA and CUF joined to develop a business project that in 1960 took the name of EFACEC. The technologic input of the Belgium partner ACEC was crucial to the development of the business project and made the technological transfer possible.

EFACEC origins go back to the year 1915, with the setting up of a small factory called *A Moderna – Sociedade de Serração Mecânica*, which would later become Electro-Moderna Lda, located in the city of Porto. Born in a context of a shortage of raw materials and highly expensive imported equipment, such as electrical engines, this small, workshop based enterprise sought to cater to the market as far as its technical capacities could go. To that end, the company set up a Secção Metalúrgica (Metallurgical Division) for repairing electrical machinery. Also, making good use of its managers' and technicians' knowledge, it sought to find solutions for the lack of imported equipment – by repairing machines, recycling materials and finally manufacturing electrical engines (EFACEC 2009).

2 THE ELECTRIC MOTOR AND THE TECHNOLOGICAL PROGRESS

The term electric motor is attributed to the object that produces mechanical energy regardless of the type of energy that powers it: thermal, hydraulic, gas, wind or solar. (Larousse 1971) The electrical current activates other machine tool, transforming electric energy into mechanic energy.

Electric motors are used in a variety of industrial and domestic situations, and they can be found in the interior of ventilators, extracting bombs, home appliances, or attached to tool-machines, etc. They can be fed directly by a portable battery, by continuous current or alternated current, and this can be produced locally or originate from a central located several miles away. (Herléa 1985).

The installation of a motor depends on its effective power that will be proportional to the set of operating machines and devices that it is going to activate. That has not always been taken into consideration by investors because the relation between power and the installation of machinery is not the main concern when deciding to invest. This is a reason why, in the Portuguese industry, hydraulic motors became less important. Thermal motors were better integrated and that was a crucial factor for the establishment of the steam machine as the main industrial motor (Santos 2000). In the specific case of the district of Oporto,

Figure 2. Machines-tools using shafts and belts. Use of hydro or thermal energy. Source: Centro Português de Fotografia /DGLAB/MC, PT/CPF/ALV/003914.

one can verify an increase in the number of steam machines between 1890 and 1910: starting with 37 machines installed with a power of 1.627 hp rising to 104 steam machines with a power of 13 043 hp. Most of them were bought in England and France, the main exporters to the Portuguese market (Santos 2000, p. 278).

Initially, the use of thermal or hydraulic motors demanded that shafts and belts connecting the motor to the transmission that in turn was connected to each one of the operating machines made mechanical transmissions. This process demanded an initial preparation of the motor (the process of heating), as well as an energy transmission path that extended over several sections and spaces, in which importantly losses were registered, especially, if the belts were poorly installed or were deteriorated, forcing high maintenance costs. On the other hand, with electric motors, even if there were losses in the dynamo (generator), we have less energetic waste because the electricity has a regular supply; however, in the early years it was normal to have failures, in particular at summer, when the Portuguese rivers have a low flow rate. The electric motor had a regular mechanic activation, mainly behind its wheel, what produce also an inert energy use in the industrial process. Its introduction, joined with the introduction of tool machines allowed for each one to be autonomous, while a group of operating machines made the transmission of power with steam or hydraulic motor. The changing of the transmission networks to an individual transmission optimized the space inside the factory and the productive capacity without the need for additional investments. It even reduced costs, not requiring lifting mechanisms and heavy components because the individual transmission allowed the establishment of a direct connection between the tool and the piece. Furthermore, the installation of an electric motor per machine had the advantage of reducing accidents, giving the operator the possibility to directly regulate and interact with the machine. Besides, it was no longer subject to the rotation of axle lines or transmission belts that extended through the workspace. Progressively, the tendency was for each machine to be conjoined with a motor,

Figure 3 and 4 . Portuguese Industrial units with electrical motors in different phases of its evolution. Source: Enterprise Archive: Jayme da Costa SA.

that was turned on and off according to the needs of the operator. In regarding this, the engineer responsible for the 1ª Circunscrição Industrial (1st Industrial circumscription) services stated that,

> Os motores eléctricos pelo pouco espaço que ocupam, facilidade de montagem e funcionamento com reduzida mão-de-obra para a condução, comodidade e economia resultantes do emprego de motores individuais para cada máquina ferramenta ou grupo de máquinas ferramenta, etc., estão a ter uma crescente aplicação". (Viegas, 1924: 17–18)[1]

[1] Translation: "the small space occupied, the facility of assembly, the operation of reduced labour, convenience and economy resultant from the use of individual motors to each machine-tool or group of machine-tools are reasons why electric motors are having a growing application" (Viegas, 1924: 17–18).

In Oporto and Lisbon, in the decade of 1920, electricity was already used in several factories associated with tool-machines, mostly in electric motors of small power. For a traditionally disperse industry, set in workshops and domestic units, the introduction of electric motors must have greatly contributed to the progress of big industrial units, as well as to the increase of productivity and to give a breath of fresh air to the domestic industry (Matos, Mendes and Cruz 2004: 216).

There are several types of electric motors, its manufacture and characteristics must be analysed from distinct processes: the scientific and technological progress of the 19th and 20th centuries that allowed for industrial development with the use new energy sources (first the steam and later electricity), and improved the routes of technological exchange between northern and southern Europe. Also important was the search for new materials and new metals that modified production procedures, expanded markets and means of consuming.

Electric motors started being manufactured in the 19th century, but big scale production depended on electric generators and electricity distribution networks, structures that would be developed in the beginning of the 20th century (Costa 1945: 5).

Through the 20th century, the electric equipment companies dedicated themselves to manufacturing industrial motors. The most common was a cage or squirrel motor. It was used in various machines, both in industrial and domestic sectors as, for instance, the first washing machines.

From the first motors until the current ones, scientific and technological development, namely electrical and mechanical calculation, the model and production of new construction materials and the evolution of manufacturing processes, allowed for the perfecting of the characteristics of the electric motor. Therefore, since the beginning of the 20th century until the 1980s these changes, among others, contributed to a reduction of motors weight and size to less than half. Gradually, a more resistant and armoured frame, other metallic alloys and different ventilation systems were developed as applications became more diversified. This technological evolution is essentially due to the production of lighter metals, the replacement of forged iron by aluminium in the armour, the development of new isolating materials that can stand higher temperatures and a progressive adaption of the motor's power and dimension.

Hydraulic and thermal motors were under the control of Circunscrições Industriais (Industrial Circumscriptions), but the inspection of the electric motors belonged to the Repartição das Indústrias Eléctricas (Division of Electrical Industries). However, the surveillance was reduced because they were considered "secondary" motors for (re)transforming steam or hydraulic energy. This way of interpreting the electric motor caused a lack of information regarding its operation, and consequently, one has now less knowledge of the installed power and the progressive functioning entry. Its installation and maintenance became a business for small companies that placed new steering wheels, performed its rewinding, did cleanings, changed oils and replaced parts.

The data collected until now suggests an increase in the entry of foreign electric machines in Portugal, after the First World War. This led the government to change the importing rates, raising tax deductions on this type of material. Its systematic entry, in the 1920s, is due to the start of operation of the first hydroelectric plants allowing the production of cheaper energy in larger quantities and making possible the progressive entry of electricity in public lighting and in industrial activities.

3 THE IMPACT OF ELECTRIC MOTOR

The electric motor brought several changes to the manufactures' space, which gained more linearity, new places for the rheostats, the feeding panels, and the energy cut appears, as well as for the switches, light bulbs, and outlets. Like other countries, in Portugal, since 1920–30 we found in the photographs of the industrial units that motors were on all the devices and the workspace became more rationalized and gave manufacturers flexibility in the design of the buildings and in the arrangement of the machinery (Devine 1983: 364).

The process of electrification and the introduction of the electric motor into the factory brought changes that permitted not only to rationalize the flux between worker and machine tool but also the operational circuit between different machines and operations. The machinery aligned in the operational chain is now reinforced by the electric motor that shapes the entire production process and each operator gains prominence by controlling its own machine and electric connection, obtaining a larger responsibility and articulating himself with an indispensable new member of the factory: the electrician. Gradually he became more qualified after receiving specific education in Industrial Schools. In these, since 1925, there were taught specific courses and starting in 1948 Portuguese Industrial School started teaching courses for electrician assembler. The diploma was a real innovation, because it introduced a preparatory cycle in technical education, extending the scope of training, vocational guidance, and thus functioning as a means to the rise of the popular classes (Stoer, Stoleroff and Correia 1998: 9).

The motors were initially made of forged iron, had big dimensions and were heavy. Its components were joined by screws of different sizes and had electric wires covered in cotton, different openings, handles, and bases. The use of aluminium, bakelite and other plastics improve the object and its use. The production of motors altered the manufacturing processes and the technical characteristics of the motors. These became lighter, with simpler armours, and joints between its components. Initially, the ventilation systems were not diversified; the most used one being the axle that had a direct impact in the heating motor, which forced the operator to place his hand on the armour to check if it could work longer. With technical progress, different ventilation systems were implemented and the motor was able to stand higher temperatures and work for longer hours. The copper wire that used to be layered in cotton was now covered in a plastic sleeve.

4 CONCLUSION

Today, the motor is still used in the same contexts as initially: in agriculture, activating bomb-motors and other agriculture machines, in the industry, it is still joined with many machines or as a generator of mechanic force. Its characteristics are now more clear and versatile; its manufacture was standardized as result of the organization and regulation of the markets.

BIBLIOGRAPHICAL REFERENCES

ASEA Journal. Anos 1924–1928. Västerås – Sweden.

Barvenik, Percy. (1983) *Electrifying experience: A brief account of the first century of the ASEA group of Sweden 1883–1983, with a special postscript on the future*. Sweden: ASEA.

Centro Português de Fotografia/DGLAB/SEC, PT/CPF/ALV/003914.

Costa, Jayme da. (1945). *Jayme da Costa*. Porto.

Devine, W.D (1983). From shafts to wires: Historical perspective on electrification. *Journal of Economic History*. Cambridge: Cambridge University Press.

Efacec. (2009) *EFACEC 60 anos: 1948–2008*. S. Mamede de Infesta: EFACEC.

Herléa, Alexandre. (1985) *Les Moteurs*. Paris: Presses Universitaires de France.

Madureira, Nuno Luís and Matos, Ana Cardoso de. (2004) A tecnologia. In Pedro Lains and Álvaro Ferreira da Silva (Orgs.), *História Económica de Portugal 1700-2000*, vol. II – o século XIX (pp. 189–218). Lisboa: Imprensa de Ciências Sociais.

Madureira, Nuno Luís, and Matos, Ana Cardoso de. (2004) A tecnologia. In Pedro Lains and Álvaro Ferreira da Silva (Orgs.), *História Económica de Portugal 1700-2000*, vol. II – o século XIX (pp. 189–218). Lisboa: Imprensa de Ciências Sociais.

Matos, Ana Cardoso and Sampaio, ML. (2016) The Branches of large electricity Companies in Portugal. From trade to the transfer and adaptation of technology (20th century). in *Mondes Electriques*. Bern: Peter Lang.

Matos, Ana Cardoso de. (2010) Electricidade, produção, distribuição e consumo. In *Dicionário da História da I república e do republicanismo*. Vol. I: A-E (pp. 1104–1111). Lisboa: Assembleia da República.

Matos, Ana Cardoso de (Coord.). Mendes, Fátima et al (2003) *O Porto e a electricidade*. Lisboa: EDP/ Museu da Electricidade.

Santos, Maciel. (2000) *Os capitais metalúrgicos em Portugal 1840–1930*. Tese de Doutoramento, vol. I e II. Faculdade de Letras da Universidade do Porto.

Stoer, Stephen R.; Stoleroff, Alan; Correia; José. (1990) O novo Vocacionalismo na Política Educativa em Portugal e a Reconstrução da Lógica da Acumulação. In *Revista Crítica de Ciências Sociais*, 29, Fev. de 1990, 11–53.

United States Patent Office. (1894) Jonas Westron, of Örebro, Sweden, Dynamo-Electric Machine, p. 1, Feb. Signification Forming part of Letters Patent nº 515,386 – Letter Jonas Wenström, Dynamo Electric. Machine, patented February 27, 1894 http://patentimages.storage.googleapis.com/pdfs/US515386.pdf

Viegas, José dos Santos Salvador. (1926) *Relatório dos serviços da 1.ª Circunscrição industrial*, ano 1923. Boletim do Trabalho Industrial, 128.

Wenström, Jonas. (1894) *Dynamo Electric Machine*. Patented 27th February. Specification forming part of letters patent nº 515.386. Available: http://patentimages.storage.googleapis.com/pdfs/US515386.pdf. 2015-06-02

Morphizm – Systemic digital graphics

Aleksander Józef Olszewski
Faculty of Art, Kazimierz Pułaski University of Technology and Humanities, Radom, Poland

ABSTRACT: The article presents the author's memories of the beginnings of his interest in computer art and its applications of the design works. In the further part, the author presents the method of the morphological analysis and its application in the area of digital graphics and education as well.

Please, do not change page numbers, do not change headers, and do not put footnotes. In substitution, if necessary, use endnotes.

Keywords: Systemic art, systems, analysis and synthesis, homolog, morphological product

1 INTRODUCTION

My first encounter with a computer goes back to the 80's when my first attempts to use it were in the implementation of a master's thesis in the area of design that I supervised.

In the early 90s I bought my first computer, a colour dot-matrix printer, a digital camera and a manual scanner. With this equipment, my struggle with the world of digital art started and it still fascinates me to this very day. For me, the most important part of the creative process is the ability to quickly verify your idea. None of the media I know provides such a possibility. By using digital technology, I am looking for an art form that does not imitate traditional techniques. It is, therefore, important not to imitate and duplicate, but to look for the features that are possible to obtain using digital technology. The projects that I am currently working on seek to reduce the effects arising from the use of programs and focus on the most important thing – the idea. I gave it a working name, morphism.

During the period from 1994 to 2015, I created on a computer a few cycles of painting, drawing and graphics using digital technology, which were inspired by the relativity of space and saved in binary code, among others.

The world of lines created from a sequence of points shaped as a square began a compositional adventure with my imagination. These points, their linear collections in various configurations, are the backbone of an image built by me. Most of my compositions have a square shape or its multiplications; the divisions occurring within the image are the result of the division of the plane into "n" squares. The final image is thus the result of intuition and the internal structural division. An example is the series of "3 × 3 Względność" ("3 × 3 Relativity"), or the last cycle "3 × 3 × 3" relating to the graphic interpretations implemented systemically. Computer painting and drawing still fight for a place in the area of classical techniques. In my opinion, the more important issue is the idea, and what leaves a trace of a line, either a finger on the sand, or a trace of carbon, graphite, coloured pencils, ink, stylus etc. is of secondary importance. A picture or a drawing even made by the use of a computer technique if it exists in a single copy, still remains a picture or a drawing.

I think that every era expanded a range of technical capabilities aimed at creative expression through the subsequent discovery of previously unknown media. That is how I treat my work with a computer. All works, to a greater or lesser extent, are implemented by the language of geometry. It would seem that this language has exhausted its possibilities and yet, you can still discover new meanings with its help. My thoughts on systemic operations in digital art are related to this geometric language.

2 AROUND THE CONCEPT OF DIGITAL ART

I have followed discussions on digital art in various forums and would like to share a few reflections. Digital Art is very young because not much time has passed since the establishment of the first "PC", giving developers a wide access to this tool (Pamuła, 2004). However, the use of a computer for artistic purposes happened much earlier. In connection with the use of computer tools in the arts, there are all kinds of opinions. Due to the nature of perceiving the problem, I will quote some of them. We must remember that the process of creating digital art brings together two fundamental aspects of the work, the semantic one illustrating the content and the formal one, built by the use of artistic language. The carrier of these elements is the digital record. I do not think that the problem was in the "digit".

Basic elements from which a language, like musical notation, could be created, have been sought in the art for a long time. It is sufficient to recall W. Kandinsky and his primer of basic forms. The digit for such a way of thinking seems to be ideal. But among "Orthodox" art creators, there is a view that digital art is "suspicious". The description of the world has an analogue dimension, which is able to present all artistic strategies. I will try to present views on the subject.

A few years ago on the internet forum MTG I read Alicia Candiani's text on digital art with its perverse message of a digital umbrella as protecting evil practices in contemporary art (. . .) that in the history of art of the Western world never before were artistic practices determined so clearly through the prism of used technical means. In the end, we do not specify, writes Candiani "architecture as the art of the brick", painting as "the art of the brush" or cinema as "the art of celluloid" (. . .). Then the author says that "digital art" is an expression of dominance and limitations both in relation to the same art and artists. Thoughts kept returning to this statement which annoyed me and I wondered how it really is – whether Candiani was right or not. I must admit that the description of the situation was outlined impressively, even very effectively, but is it correct? The basis for consideration of the impact of technology on artistic realisations is glaringly visible in graphics art. In artistic graphics, the type of technique used allows the specialist to identify the technology. Steel engraving carries the technique, material and form in its name that cannot be accomplished by the woodcut technique. I believe that the critical attitude of the author to digital imaging stems from the wrong understanding of digital art.

It seems you need to look at digital technology from the side of its multiple applications. And so, if we draw on paper and then using the scanner transform it into a digital image, simply by doing this, it does not automatically become digital graphics. What is the photograph of this drawing? Only a photograph and reproduction. Similarly, in the case of a drawing scanned in a scanner. At an exhibition abroad, I saw a digital graphic piece which had the area of several square meters. It looked like a street billboard; it dominated and subordinated other exhibits on display. This strategy used by the artist seems ambiguous in terms of art because, in this way, each work reproduced by digital technology would have the right to be called digital graphics. One could ask the next question then, is a woodcut of dimensions 10 × 15 cm scanned, enlarged to dim. 10 × 15 m, and printed in digital technology a digital graphics piece, a woodcut or just a reproduction? I guess we have no doubt that we are dealing with the most ordinary reproduction so that it is signed. The technique of digital duplication has nothing to do with digital graphics. If we approach this technique in an orthodox way, a lot of, so called, "computer graphics" works would lose their right to exist. I do not intend to take away the right to exist from this form of expression, but I think that it should bear the name of the intermedia graphics. However, with one caveat – there must be an interference in the material previously introduced to "a computer", through the use of software tools.

The fact that both the computer and the Internet have become a commonly accessible technology opened an area that previously was only available to selected artists who had access to classic digital graphics technology. Currently, there is no limitation on access to a computer and programs for generating a digital image or, as others say, for creating digital graphics. This availability, however, causes some kind of a mess in the naming and identification of the technique.

For me, an important element of creative work in this technology is the ability to verify quickly the idea. None of the media I know provides this opportunity. There is also a possibility of a very deep penetration into the structure of the work, and many other possibilities. In digital technology, I am looking for an art form that does not imitate traditional techniques, nor duplicate material made in the classical technique. Thus, I look for those features that are only possible thanks to software and implemented in digital printing technology.

A radical solution is the way a Japanese artist creates; he writes an algorithm of a graphics piece, without imagining it, and he simply accepts the result of his calculations generated this way, an image. One could suggest that this algorithm is simply a matrix layout. It is interesting that the author does not interfere with the work created in such a way. The first images of digital artwork were created in a similar manner (Pamuła, 2004: 73). We thus could call this way of creating digital artwork purist – it was the art of rigorous formal discipline [1916–1925]. There is also another clue leading to the use of digital technology – it is the designing of painting, drawing and graphics, by transforming, mixing, and the collaging of photographs, the result of which is a project implemented in a classical medium. It can also remain in such a form, but then a more appropriate name would be mix media or intermedia graphics. Currently, many artists successfully use the former and the latter activity.

The source of identification of classic graphics is a matrix that determines the technique and it is a very important issue for the people involved in printmaking.

Grzegorz Banaszkiewicz, in his discussion on the matrix, formulated the following view, which could be described briefly: everything is a matrix, as long as you can get a picture using this matrix, i quote:

> It is worth noting that, in a broad sense, the matrix can arise and exist independently of the will and human intervention, as a record of a physical process or phenomenon. The ability to copy, which is the essence of the matrix, seems to stem directly from the nature of the world, the nature of phenomena, which formed the matrix; Man is only an explorer and user of methods of using this property of the matrix. According to Janusz Kaczorowski, created by artists themselves or only used by them, a graphics

> matrix, in accordance with its potential, at the right time and using the necessary technical procedures, in a process called matrixing, creates the "second life" of graphics i.e. an image or an object, being the direct footsteps of the previous existence, which was the source and condition necessary for the final form of the graphics. (Banaszkiewicz, 2006: 46)

The matrix is associated with the word "memory" and the possibility of duplication. It is also one of the main features of the matrix; its materiality in the case of numbers is debatable because the virtual matrix is tempered by its physical aspect. We realise that the materiality of the matrix is something secondary in relation to the idea (algorithm), the source of which is the brain, whose more modest imitation is a hard disk of a computer. Is a brain the matrix of art? In 1998, in one of my university studies, I put forward questions regarding the identification of specific ways to create a digital image. These questions are still valid today.

The easiest way to answer these questions is from a technical point of view. And so: the picture is a trace of a tool, and it exists in a single copy.

> To say that a computer image is a drawing or a painting, you must destroy its record (matrix) in computer memory; then the prints exist in one copy, which is a feature of a drawing and a painting (uniqueness). As for other prints, you can specify its edition number, e.g. of computer graphics, and only then destroy or archive a computer record. (Hoffman et al., 1998: 104)

Writing back then about computer graphics I used the notion of recording and computer memory, which is, in fact, a matrix of digital graphics. You can wonder what a "virtual matrix" has got to do with a mathematical algorithm, which is the basis of the virtual existence of this matrix. Therefore, it is possible that it is the algorithm, which is a matrix that holds the image. So the question is, "Is it possible that mathematical models "rule" digital images?" Looking at the model of the work of a computer, you can say that there are images, the source of which is information recorded with use of the algorithm and there are algorithms describing the introduced images.

Quoting Piotr Zawojski's opinion in his book 'Elektroniczne obrazoświaty. Między sztuką a technologią' ("Electronic Picture words. Between Art and Technology"): "digital image is an image reduced to a digital matrix (. . .) there is no need to refer to any model or a real object . . ." I think it is a contribution to the further defining of the dependence of a matrix and a mathematical model or algorithm which, in its essence, is the original source of an image, i.e. the actual matrix. The figure below shows a record of my name in binary code and raster. Creating a record of the notion we do not know what a graphic image will look like. The interesting thing is that any modification of that image changes the original meaning of the record.

Figure 1. 1001111100110010100111011010100010110101111010011100101110010 01.

> The name of the author of this text in binary code and raster, and the question whether it is a matrix or image – or both.

I have come across a view, which questioned the material being of digital graphics, i.e. paper, believing that this physical material environment has no meaning. The statement, however, drew attention to a very important issue of the materiality of digital graphics. Does digital graphics need paper?

I think a very good thing happened when artists using a computer decided to describe their work as digital graphics (in Polish cyfrografia, a term introduced by R. Małoszowski in his Internet publication) (http://www.wsp.krakow.pl/konspekt/konspekt9/digital.html) Computer graphics is a term used to define the design work of a computer engineer, who provides us with a specific product, which is a computer program. Only the latter, in conjunction with a computer, creates an environment for creative work in digital art.

Graphic techniques and associated matrices, which are worth recalling, at the time of its birth were intended to serve a purely practical function, the function to duplicate information. For the purposes of multiplication, printing and various forms of media image were invented. All these inventions were designed to reach the largest audience of the image. When having reached its technological peak, different types of printing techniques were being replaced by a new invention, and with time passed, they were moving to the area of art of unique or limited editions prints losing its original, purely utilitarian function. I think the same will be with the computer (digital printing), which revolutionised printing and the design process in different areas of life.

For some artists, but also collectors, the important value is the issue of manual labour. It should be noted that the creator operates a tool, not their hand; a trace a pencil, brush or a graphic board are an intermediate trail of their hand. Summing up these considerations, I think that these issues are of secondary importance. In art, you have to have an idea, and how it will be implemented is secondary. It is the technique that must determine the right feature for itself, and in any case, it cannot mimic or imitate traditional techniques, it must be different from them, it must free itself from them, as it will decide the future of its role in art. D. J. Bolter aptly emphasises the role of this invention.

> Computer – is a technology that defines the 20th century, and so fundamentally defining the image, the shape and dynamics of our century, so it is impossible not to use it in the process of artistic creation. (Zawojski, 2000: 36).

It should be noted here that it has been more than half a century since the invention of the first computer.

At this point, I would like to once again refer to my own experiences, which I presented in the magazine "Arteria" 4. During the period from 1994 to 2006, I created a series of graphic works in digital technology, inspired by, among others, the space, and its relativity recorded in binary code. The world of lines created from a sequence of pixel points having the shape of a square began the adventure of my imagination. These points, their linear collections, in various configurations, are the backbone of an image built by me. Most of my compositions are the shape of a square or their multiplications, and divisions that exist within the image come from a plane divided into "n" number of squares and are related to the permutations, which I carried out in the 1970s. The number "n" of squares comes from the raster binary image recording. The final result is the image and is thus the result of intuition, sensitivity to a colour and the structure of the plane (Olszewski 2004: p. 53). An example of this is the series SLH – Car Skyline – relating to the digital graphic interpretation of time, place and to the space observed during the driving of a car. In the titles, the works are described with their own numbers, the place of a route, country, town, the number of a digital photography and date. Through such describing and recording, I physically refer to the real space in which I was personally. Each of these works has a real picture of the place, time and space encoded with a digit (Olszewski 2004: 55). The method of registration of each digital image provides an authentic, true source for digital graphics. After all, you have to remember about the beginning of each creative being. A bridge between the technology of machines and "the technology of the human mind" is a universal space of memory. However, the digital memory is just a prosthesis of human possibilities. 100 or 200 MB does not mean anything until the picture remains a permanent part of our consciousness (Białostocki, 1983:13).

Małoszowski's opinion on graphics cloning corresponds well with my view on this issue:

> For the creator, it is more important that cloning allows for deep penetration into the genetics of an image, for transformation and improvement within the file version of the matrix (. . .) also the author sees its very important benefit – its mobility and susceptibility to cooperation with different technologies. (. . .) Digital graphics, therefore, is not an imitation of the classical disciplines of art, but an autonomous technology with its own means of expression, tools and exhibition spaces. (. . .) It retains most of the artistic advantages which have been equated with classic media till nowadays, however it introduces its own means of expression. This means that by using digital graphics one can pursue artistic works referring to the traditional workshop through the skills acquired in the classical disciplines (the ability to draw or paint). However, you can create separate aesthetic phenomena on the basis of the information file. Through the ability to clone (each piece is identical, the matrix is not destroyed), in a sense, it is indestructible)
>
> [http://www.wsp.krakow.pl/konspekt/konspekt9/digital.html]

As for the durability of magnetic media, opinions are divided, but we think that progress in this area will eliminate this drawback.

At this point, I would like to supplement the text with my remarks on the systemic activities, which are significantly associated with digital graphics.

2.1 *Systemic activities*

The concept of the system is one of the most basic and most general concepts in modern science. Therefore, it is very differently defined by various authors, although there is a common school of thought among researchers regarding its content.

The term "system" comes from the ancient Greek noun "systematic" meaning connect, connection, whole. Those "whole" and "part" are concepts being considered by the ancient Greeks. A part is smaller and belongs to the whole. The whole comprises parts and is full (a complete set of parts). In the whole, no parts are missing and outside the whole, there is no part that does not belong to it. The whole arranged and intersecting with each other are called a system, and parts of such a system are its elements.

The systemic description of phenomena in design, or as it is called systeming (Dorosinski et al., 1981: 89), which consists of a systemic analysis and synthesis, is to identify the basic action in the practical situation. The structure, in the systemic terms, is a relational network between the system components. In our understanding, the structure is the organising and distribution of elements in the system of the form. The form and is an expression of the internal structure of things. The analysis and synthesis of the properties of these elements lead to the construction of the system.

2.2 *Morphological analysis – fundamental problems*

The study of morphology is described as the search for a sub-set of "n-connections" in the morphological product. The term "connection" is defined as an object of considerations, built in a specific manner and obtained from finite sets. If a connection has "n" components or elements, it is known as the "n-connection" where:

A = {a1, a2, a3, ...} aα; B = {b1, b2, ... bz}; .... N = {n1, n2, n3 ... n}.

Sets A, B .... N are called the forming sets (in "systeming" they stand for sub-systems), and forming a set of forming sets {A, B, C .... N} is called morphology (in "systeming" it is the tested system).

By taking one and only one element from each forming set, and placing them in the same order in which forming sets are given, we organise the "n-connection", for example (a2, b1, c3 ... n); (a4, b5, c1 ... n); etc. The "n-connection" set is called Zwicky's "product of morphology."

The search for solutions using morphological analysis should be carried out in five stages:

Stage 1 – includes the accurate (precise) formulation of the problem.

Stage 2 – is related to the isolation and defining of forming sets, being part of solving the problem A, B, C, ... N and homologs aα, bz, ... nw (homologs in the morphological analysis are the equivalents of elements in "systeming").

Stage 3 – involves sorting all the solutions in the n-dimensional "morphological box" as well as a two-dimensional array in which the forming sets are recorded in rows. Such an array is called the "morphological matrix". A solution is defined in the form of a chain of connected components.

Stage 4 – involves assessing solutions, e.g. by the point method.

Stage 5 – concerns the selection and implementation of the solution, which has been rated as the best.

2.3 *Implementation of graphics project '3 × 3 × 3' using a morphological analysis method*

Large squares – where a1, a2 a3 ... an

Medium squares, B – b1, b1, b1, ... bn

Small squares, C – c2, c2, c2, ... cn

Colour D – d1, d2, ...n

The thickness of the line E – e1, e2, ...n

Variable location G, -g1 asymmetry.

The morphological product for a set of 3 × 3 × 3

L = 3 × 3 × 3 × 2 × 2 = 108

2.4 *Exemplary solutions of my realisation of a series of graphics "3 × 3 × 3"*

I named the method presented above – morphism, and below I attempt to define it.

Morphizm – This term describes the activities and results in the area of art, which use the morphological analysis method, specially developed for technical design. I distinguished morphism from the area of systemic design, using some of the elements. The feature identifying this trend is the morphological matrix, which is a set of components with specific characteristics, composed on a plane and in a space in a certain way specified by the artist. Each element constituting the set is determined by the creator. The search for a set of elements forming the work can be done in various ways, both objective, e.g. calculated and subjective, e.g. at random. The selection and the method of connecting belong to the creator of a composition.

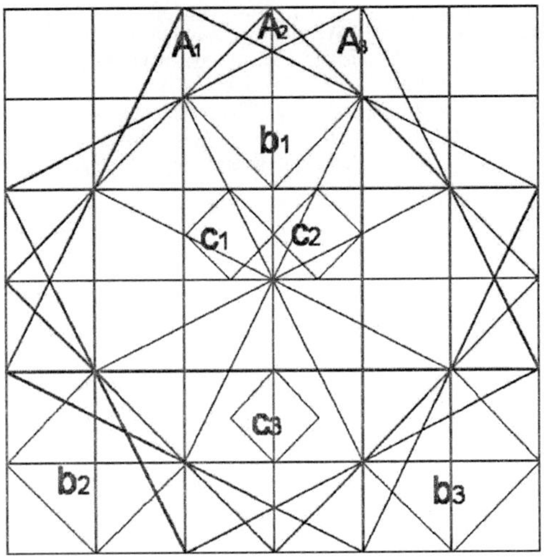

Figure 2. Morphological analysis of the squares A, B, C.

Macierz morfologiczna

Zbiory form	składowe (homologi)							
	1	2	3	4	5	6	7	8
A	a1	a2	a3	a4	a5	a6	a7	
B	b1	b2	b3	b4	b5	b6		
C	c1	c2	c3	c4	c5			
D	d1	d2	d3	d4				
E	e1	e2	e3	e4				
F	f1	f2	f3					
G	g1	g2						
H	h1							

Figure 3. Permanent location F f1 symmetry.

BIBLIOGRAPHICAL REFERENCES

Banaszkiewicz G. (2006), The definition of graphics, *Arteria periodical* No. 4.

Białostocki, J. (1982), *Symbols and Images*, Warsaw.
Dorosinski, W. C.; Gasparski W.; Wrona, S. (1998), *Outline of Design Methodology*, Warsaw.
Hoffman, H.; Łukaszewski, B.; Olszewski, A. (1998), *Fundamentals of Visual Art – selected issues*, Radom.
Pamuła J. (2004) An Outline of the History of Computer Art', *Arteria periodical* No. 2.
Olszewski A. (2004) A Car Line of the Horizon, [in] Aleksander Olszewski (catalog prepared on the occasion of 30 years of creative work), ed. R. K. Bochyński, Radom.
Zawojski, P. (2000), *Electronic Pictureworlds. Between Art and Technology*, Kielce.

Optical fibre or a critical social history of light

Szymon Wróbel
Institute of Philosophy and Sociology of the Polish Academy of Sciences, Warsaw, Poland
Faculty of "Artes Librales" at the University of Warsaw, Warsaw, Poland

ABSTRACT: The author reflects on the modern history of light from Descartes (natural light) through Walter Benjamin (light of lanterns and lamps) until Eben Moglen (modern technologies based on optical fibre). The main purpose of this text is to give outline of a modern, critical social history of light. The author's primary concern is the relationship between light and power and knowledge alongside the process of lighting understood as observation (surveillance) and information (awareness). The metaphor of light allows to trace the dialectic of Enlightenment mired in the figure of hunter, panoptic apparatus, and modern wearable technologies. The fundamental idea is that the beam of light is as enlightening as it is blinding. This perspective allows the author to reconstruct genealogy of progress in which light is as much the emancipator as it is a police interrogator. Finally, reflections on the concept of phenomenon and semblance attempt to transcend limitations of the two major 20th century paradigms in thinking; phenomenology and psychoanalysis. The former deals with what is displayed or illuminated (phenomena) and the latter with what is overexposed (traumas). The author seeks to transcend the two paradigms of speculative philosophy on matter and memory in Bergson. Conclusions are of theoretical and political nature. The latter challenge the idea of the public sphere being the realm of the visible while the former question established belief that knowledge is but seeing formatted by concepts.

Keywords: light, panopticism, optical fibre, hunter, phenomenon

1 HOUSE OR STREET

It is in the early work of René Descartes entitled "Le Monde ou Traité de la Lumiére" written in the years 1629–1633 and published only posthumously, where the Enlightenment utopia of light begins. In the famous opening Descartes proposes "to treat here of light" which provides evidence that the understanding of the light is, from the very outset, central to his project (Descartes, 1629/1998: 3). The utopia of light endeavours to disperse all the darkness and to diffuse brightness over the world. It is a new world phenomenological vision in which the once proclaimed visibility of all phenomena must never again fall into darkness. The lecture, wherein the principles of the physical world are presented altogether with the full description of the structure of the imaginary "new world" illuminated by the eternal light, should read "as if my intention were simply to tell you a fable [une Fable]" (Descartes, 1629/1998: 32).

This is the world where the sun never dies. While a bedtime story creates the path to sleep and into the darkness, Descartes spins his fable in the same fashion as Scheherazade does, that is, to make the night pass by and to prevent the reader from falling into sleep. While nowadays the fable of world-light has become a reality and is as common as a newspaper story, an inherent part of our radiant everyday world, it was at the time the centre of heated debates on the nature and the source of light. Let us recall only in "Dioptrique" from 1637 where Descartes formulated the wave theory of light, and was the first to describe the phenomenon of refraction of light necessary for the explanation of rainbow formation. This discovery was, indeed, an important part of his considerations in "Les Meteores" in the same 1637.

The interest in light professed by Enlightenment philosophers did not merely stem from its scientific prominence at the time. Of course, Kepler's theory of the retinal image and his inventions such as the refracting telescope or eyeglass lenses on the one hand and optical telescope and crystal lenses of Christiaan Huygens on the other were the most exemplary scientific breakthroughs. At the same time, however, light became the main utopia of Enlightenment, providing it with a tool for illuminating the present and the future. Just as utopia belongs to the era of light, so does eternal light belong to the new world. Today, when both our "day" and "light" are artificial we are doomed to live by "false utopias". Utopia of Enlightenment was the utopia of transparency.

In contrast, the postmodern-world "utopia" of the fully illuminated and light-polluted globe is the utopia of shadow i.e., the possibility of hiding away from the excess of "bad visibility" in the world illuminated by a planetary scale electric glow. Utopias were once full of "prophetic power" and exercised the "power of prophecy". Such utopias were akin to a lamp, lantern or torch, in that they were the mechanism of lightning up the darkness. Today, utopia escapes into the darkness and desires fulfilment beyond electricity, the visibility of network, the prevailing wiring, the optical fibres and away from the omnipresent electronic tracking of our corporal presence. Let me only recall that Thomas More's island if Utopia was, before the conquest by Utopus, a "city-in-the-dark".

The same Benjamin who so carefully diagnoses the subject concealed in the pouch of apartment calls Paris the capital of the nineteenth century, and it is mainly because of the ubiquity of light. Passage ways of Paris – according to Benjamin – were to be "the scene of the first gas lighting" while in fact becoming infinite passages for ever unsatisfied eye of a stroller [flaneur] a exploring the illuminated displays and goods illuminating their exchange value – price (Benjamin, 1938/2006: 18-23). The end of light is the end stage of commodity fetishism and the exchange of goods, but it is also the end of the civilization of the city. For a bourgeois nothing is as fearful as the return of the darkness being equivalent to return to what is prenatal, unorganized, and barbaric. Was it not the reason why, as newspapers reported recently, following the attack on "Charlie Hebdo" on the 7th of January 2014 and then again after a the attacks of November 13, 2015, the lights in Paris suddenly went out? Lights out in Paris mark the end of civilization.

2 HUNTER OR ECLIPSE

Let us begin from the paradigm of hunter seemingly immersed in the Palaeolithic darkness, where the nomads under the cover of night engaged in big game hunting. God was, of course, the first hunter. "And God said, Let there be light: and there was light. And God saw the light, that [it was] good: and God divided the light from the darkness. And God called the light Day, and the darkness he called Night. And the evening and the morning were the first day" (Genesis 1, 3–5). It is not by coincidence that both our tormentors and ordinary police officers form a boundary of light and darkness between us and themselves. Police officers do it professionally by throwing at us a beam of blinding light while they remain in the shadows. Perhaps the hunter is Imago, the last stage in the development, the stage in which the individual attains maturity and becomes complete, the stage in which it is non-moulting and capable of reproduction.

Nimrod, mentioned in "Genesis" as the great-grandson of Noah, grandson of Ham and son of Cush, was "a mighty hunter" whose kingdom was Babel (Genesis 10, 9-10). The authors of "Dialectic of Enlightenment" – Theodor Adorno and Max Horkheimer make Nimrod simply a human. They also claim that the reign bears the marks of the transformation of nature and society into a "hunting ground" (Adorno, Horkheimer, 1944/2002). The world was originally God's hunting ground, and only later Nimrod's. After God's withdrawal into darkness, and then following Nimrod's death under the rubble of the tower of Babel, nature has become man's hunting ground. The reign over nature is the principle on which the very notion of the subject is founded. The end of this reign can only occur when humanity rediscovers its naturalness. "Naturalness" can only mean darkness; i.e., the lack of artificial lighting. Being the subject means, therefore, being in control of the hunting ground, which is a place of light.

Adorno and Horkheimer identify their work to be that of a detective tracking down crimes committed after Enlightenment. Detective is a modern hunter. "What we had set out to do – write Horkeimer and Adorno in the preface to Dialectic of Enlightenment – was nothing less than to explain why humanity, instead of entering a truly human state, is sinking into a new kind of barbarism" (Adorno, Horkheimer, 1944/2002: XIV). For the authors reason is a tool for calculation and planning which in relation to the objectives always remains neutral. The element of reason is coordinating, tracking and hunting. "Culture has evolved under the shadow of the executioner; Genesis, which tells of the expulsion from Paradise, and the Soirees de Petersbourg are in agreement on this. [...] One cannot abolish terror and retain civilization. Even to relax the former means the beginning of disintegration" (Adorno, Horkheimer, 1944/2002: 180). Joseph de Maistre's "Les Soirees de Petersbourg" raise the question of light and civilization, sunset and sunrise. However, the philosophy of Joseph de Maistre is not the philosophy of light as it is the philosophy of the guillotine.

Light in Enlightenment takes many forms. This beam of light is by no means uniform. Enlightenment invents institutions such as hospitals and schools hand in hand and in the same spirit as it invents prisons and leprosaria. It is of importance to acknowledge the fact that within the dialectic of Enlightenment liberation goes hand in hand with enslavement and light with a shadow. Horkheimer and Adorno add that man's self-control virtually entails destruction of the subject because even if controlled, oppressed and defeated nature (darkness) is still the only all-determining reference system – it is life itself: "[t]he history of civilization is the history of the introversion of sacrifice – in other words, the history of renunciation" (Adorno, Horkheimer, 1944/2002: 43). Society-light arises when man denies nature-darkness in himself (in the body), and does so in order to control the nature outside (dark forest).

3 PANOPTICON OR A TRAP

Adorno makes, therefore, the paradigm of hunter a paradigm of Enlightenment. Today, however, the hunter's place is occupied by advanced technology with such symbols of its reign as currently high-profile drones, nuclear submarines or satellites. With this in mind it is fair to ask how to understand the paradigm of research on techniques as instruments of control, supervision and constant surveillance. What is, in fact, demonstrated in various analyses of the relationship between control and "visibility", in reconstructions of "optics" and "the figure of hunter and victim", in the new technologies and what might be called "phenomenology of drones", is that the panoptical paradigm is still present (Chamayou, 2015; Chamayou, 2012).

Horkheimer and Adorno in "Notes and Sketches" ending a rather unusual structure of "Dialectic of Enlightenment" note that incarceration was a bourgeois affair. In the Middle Ages incarceration was reserved solely for the offspring of princes who in themselves symbolized an inconvenient hereditary claim. Common criminals, on the other hand, were tortured to death, often publicly, to instil in the masses a respect for order and law, since – the authors observe – "the example of severity and cruelty reaches the severe and cruel to love" (Adorno, Horkheimer, 1944/2002: 187). Although the martyrs of bourgeois regime do not die subjected to long days and nights of breaking wheel torture, they perish mentally in the great prison buildings being unable to set example for anyone, just like the insane forced into lunatic asylums. Ever since Thomas Hobbes' times the great dilemma of power-sovereign was weather to execute a culprit in the daylight spectacle of the law, or eliminate one silently under the cover of darkness, or, so to speak, in nature – as if it were redundant species (Hobbes, 2005).

Prison, from the point of view of "Dialectic of Enlightenment", reflects the bourgeois mode of life. The rows of cells represent monads in the true Leibnizian sense. The prison is the abode inside-out. The Monads have no windows through which they could separate themselves from the substance, nor they have direct influence on one another. Lives of the imprisoned are coordinated by God/Engineer, the Great Supervisor. "The human being in jail is the virtual image of the bourgeois type he has yet to make himself in reality" (Adorno, Horkheimer, 1944/2002: 188). Prison is the image of the bourgeois working world, an emblem of hatred of human beings for what they are yet to make themselves become. Prison may raise hatred to the world of lights. Seclusion imposed from outside permeates individuals and invades their bloodstream. Their well-trained souls and happiness are as bleak as the prison cells which the supervisors can already do without, since the entire humanity has fallen to them as spoils.

These final comments appended to "Dialectic of Enlightenment" outline the essential content of "Discipline and Punish" by Michel Foucault. The main difference, however, is that the former see prison as dark and anonymous while the latter as illuminated and individualized. Blood always belongs to someone, just as the fingerprints.

In the modern world a hunter is a detective. Previously, however, a detective (illuminator) partially played the role of a victim (illuminated). Biographies of modern detectives follow the biography of Eugene-François Vidocq – a French criminal who became the founder and first director of the crime-detection Sûreté Nationale. Formerly a thief, Vidocq offered his services as an informant to the police and was soon appointed the leader of informally organized a plainclothes unit, the "Security Brigade" formed of ex-convicts whose new role was confined to infiltrate the criminal world. In 1827 Vidocq handed in his resignation, founded a paper factory, and took to writing Diaries. It was after Vidocq that the character of Balzac's Vautrin was modelled. Following the bankruptcy of the factory, Vidocq founded "Office of Information" considered to be the first known detective agency. Vautrin/Vidocq earns a nickname Trompe-la-Mort (Dodgedeath or Cheats-Death) as the one who cheated death by managing to avoid the death sentence repeatedly. Death can be fooled only by falsification of death or birth certificate. Vautrin/Vidocq is born again in a dark dungeon and then again he dies in the street light when betraying his friends of the night (criminals) for the sake of the fake glow of work for police.

Both in "Discipline and Punish" and "The Birth of the Clinic" Foucault associates vision and power when making numerous references to commanding eye, tracking eye, or supervising eye (Foucault, 1993; Foucault, 1999). Next to the major technology – the telescope, the lens, or optical fibre – accompanying the advancements in physics and cosmology, modernity procured quite a number of much more modest techniques and intersecting observations (insights) that "must see without being seen"; "using techniques of subjection and methods of exploitation, an obscure art of light and the visible (…) secretly preparing a new knowledge of man" (Foucault, 1975/1978: 171). Here we enter a new world, in which vision means is more than a vision of human and more than visibility of things. Today, light penetrates intercellular spaces and its vision now includes the molecular structure of things.

In this world, a new microscope is being invented, deployed, and tested, a telescope of moral purity and decency. The obsession of being watched and the privilege of observing – these are the two major trends of modernity, overlapping and intertwined. "Full lighting and the eye of a supervisor capture better than darkness, which ultimately protected. Visibility is a trap" (Foucault, 1975/1978: 200). Panopticon, so accurately described by Foucault, is another optical device. It is a machine for watching, a kind of camera obscura, where man is neither spied on nor man does the spying, but instead resides permanently in a transparent building where the exercise of power is under the control of the whole of society. "Panopticon is a royal menagerie – writes Foucault – the animal is replaced by man, individual distribution by specific grouping and the king by the machinery of a furtive power" (Foucault, 1975/1978: 203). The basic unit of architecture is no longer a theatre nor a stadium but a Panopticon

– a cylindrical device invented by Jeremy Bentham (Bentham, 1995). Panopticon Prison Design consists of a circular structure with an inspection house at its center with large windows overlooking the internal façade of the structure. In this new technology of power the eye is a single cell equipped with power. The tool of this power is viewing through the periscope.

4 CLOUD OR WEARABLE TECHNOLOGIES

In "Treatise on Light" Descartes observes that rays of light are different from material corpuscles in that "they are nothing material in the medium through which they pass", and that they designate only in what direction and according to what determination "the luminous body acts on the body it is illuminating" (Descartes, 1629/1998: 3). The nature of light allows it to penetrate matter "in depth" and "into the distance". Light illuminates not only our faces and geometry of our bodies, it can also illuminate e.g., the neuronal structure of the brain. Descartes sensed that refraction of light differs from one medium to another, he also assumed that the force of light can be increased according to the quantity of the rays that come together and by diverse dispositions of the bodies through which it passes. I do not know whether Descartes foresaw the invention of optical fibre – transparent fibre used as a means to transmit light as information by propagating light signals.

Although in "Discipline and Punish" Foucault gives great insights into microphysics of power, Foucault did not foresee the invasion of light on the atomic structure of the body. Let me ask, therefore, whether the intervention of optical systems in the sphere of social and organic life has reached such a level that we may already pronounce that we live in the age of massive population control exercised by infinitely deep penetrating light. If so, what forms of agency in political and public debate can still be professed with such a strict supervision? Is this the brain itself that is the dwelling of man, and the public space today are websites illuminated by the lantern's light? Has the network society restored is us a sense of the natural contact with the light – be it the entire length of the fibre, or on one of the server clouds? Just as there is no atlas of computing clouds, and neither can man foretell the future from these clouds nor predict their behaviour, so it seems impossible to understand the anatomy of clouds, a subject pertinent to the philosophy from Aristotle, vital for both William Shakespeare and Richard Dawkins. Should we now be deluded into believing that computing clouds are the foundation of our freedom? Is freedom equal to access to the server? Is the gate to freedom now hoovering almost like providence, equipped with default settings in the cloud ready to perform any configuration in any given place on the earth and at any time of the day or night? Is this new freedom a part of the new world of fibre optics and globalized anarcho-capitalism hidden behind a chaotic cloud of stock market operations (Moglen, 2010)?

Perhaps, we are now facing the need to rethink the problem of light and obsessive lighting of everything. Perhaps, the world has accelerated and we have already entered a new wave of techno-panoptic modernity where lighting and tracking is performed by devices equipped with sensors continuously monitoring the amount consumed and burned calories, distance travelled, visited locations and the amount of time spent in them. The so-called trackers included in wearable technologies often take a form similar to a hospital wristband or can be placed in a pocket or yet clipped to clothing. As opposed to old pedometers, today's trackers process, visualize and send our data to the cloud, as well as learn our behaviour and get to know us better. Our data is stored in the cloud and is accessible from anywhere, anytime, be it on your computer, tablet or smartphone. It allows us to easily track our progress over time and compare results with users' community (Horvath, Hoge, Cameron, 2016; James, Petrone 2016).

Here is the modern version of investigative journal: our track data laboriously recording the history of our lives, record of the victim's moves. Trackers are perfect hunters whose final hunting ground has become our bodies. The tracks and traces recorded by these trackers are like traces of a citizen imprinted in his apartment – the way Benjamin saw it – these are the traces of our presence in the world. Not only does the Basis tracker measure temperature, heart rate and number of sleepless hours, it automatically detects the difference between walking, running or cycling. Tracker Shine modestly measures our activity, monitors the amount of calories burned and hours slept. Pulse saves steps, total distance ran, calories burned, and on top of that total duration and quality of sleep. Fitbit Force motivates the user to keep moving, saves steps, distance, calories burned, stairs climbed, and every active minute during the day. At night it tracks the stages of sleep and can wake the user at the right time with a vibrating alarm. So does the list continue.

What is the meaning of all this? Should we construe that light entered into a new accelerated phase, phase velocity in which it lightens up not only streets and our homes, but our brains, chemical parameters of our blood, the information stored in a cell, in our DNA, daily cycle of our key organs, and our body movements? In such an optical fibre-world we would no longer be ordinary citizens or subjects sensitive to light, but rather bio-citizens, brain-subjects, self-neural beings (Rose, 2013, 2008, 2007; Malabou, 2013, 2012, 2009). Is there, in this new optical fibre-world, any place for the source of light other than that of technical lighting (surgery)? Is there in this world of neuro-consumers, neuro-strollers, and neuro-citizens a space for a spectacle in sunlight? After all, there is still space for "cloud" and so there is space for the information meteorology.

In "Treatise on Light" Descartes admits that he knows of "only two sorts of bodies in the world in which light is found, to wit, the stars and flame, or fire" (Descartes, 1629/1998: 6). Neither the society of the spectacle nor the society of "optical fibre cloud" knows the light of the fire nor the light of stars. In this society all we know is the light of monitor screens. In the same treatise Descartes commits to present evidence that the light which is in the objects is of the same nature as the light in our eyes (Descartes, 1629/1998: 3, 6). In his fable Descartes adduces an experiment demonstrating that the inhabitants of another planet and another galaxy can see objects quite like us – Earthlings, but without the help of sunlight (Descartes, 1629/1998). The term "the society of the spectacle" was first used by Guy Debord in 1967 to describe the contemporary times – unlike in Foucault where the term is employed to describe the Antiquity. The contemporary times for Debord is the world that has passed through the focus of the optic and is its own inverse. Illusion has become a reality, "[t]he real consumer becomes a consumer of illusions. The commodity is this factually real illusion, and the spectacle is its general manifestation" (Debord, 1967/1995: 47). The category of semblance is central here and it shall be dealt with at the end.

5 VIDOCQ OR SEMBLANCE

Two great philosophies of the twentieth century, namely phenomenology and psychoanalysis, still live in the shadow of light and breathe light. What psychoanalysis fears is waiting for moments when light goes out. Josef Breuer compared mental apparatus to a malfunctioning lightening machine. „The brain however – concludes Breuer in his and Freud famous work "Studies on Hysteria" – behaves like one of those electrical systems of restricted capability which are unable to produce both a large amount of light and of mechanical work at the same time. If it is transmitting power, only a little energy is available for lighting, and vice versa" (Breuer, 2008: 195). The brain should not be likened to a telephone wire which functions only when it is excited but to a telephone line through which there is a constant flow of current so that it produces excess excitation and may become energetically exhausted. What is requested from such a system is that simple establishment of a contact (plugging into socket) sets it in operation. Yet, what must be assumed is constant operation voltage and constant stand-by mode which expends energy.

Phenomenology, contrary to psychoanalysis, fears turning on another light source. Phenomena constitute the totality of what can be brought to light and what lies in the light of day. When a being shows itself as something which in itself it is not, it signifies "semblance". Semblance (Schein) concerns what reality which is not what it gives itself out to be, where "not" is used in privative sense. In contrast, "appearing" is a not showing itself. In this instance, "not" must by no means be confused with the privative not which determines the structure of semblance (Heidegger, 1927/1962: 25 onwards). Privative (privatio – deprivation) character of semblance entails lack, loss, and absence. Semblance should not, therefore, be understood as "illusion" or "chimera" but instead as self-showing itself, and as what allows itself to be looked at. Semblance in "in plain sight" as it is "object of seeing". This is what Kant understood when employing a complex, two-fold meaning of the term "appearance" (Erscheinung).

Towards the end of this paper I ask once again: what is the status of light and what is the effect of light? Light affects eye in that images are produced. The question, however, is, whether our contemporary world illuminated by artificial light of numerous screens "embodies semblance" or neutralizes it? Or, rather, should we say the two eventualities are in fact spiritual twins (doppelgangers)? After all, conjuring up mysteries – this game of appearances – and resolving them is in itself a complementary activity; the former being directed inversely to the work of removing mysteries, i.e., abolishing of appearance.

In "Matter and Memory" Henri Bergson conflates images with things and thoughts with movement (Bergson, 1896/1990). For Bergson, however, matter is the aggregate of "images" (Bergson, 2012: 3), and image itself is a being which in itself is more than an idea (idealism) but less than a thing (realism). In effect, image is something halfway between a "thing" and an "idea". Bergson further adds that thinking about images in space, or the interval between things and ideas is a matter of common sense. Finally, it is from Bergson that we learn that such an approach to the problem, i.e., placing things in images, allows us to avoid the pitfalls of phenomenology, i.e., thinking about the world and existence in terms of appearance and reality. Images exist before the division between appearance and reality. Perhaps, we should only be afraid of images created by devices which themselves emit light and which themselves are oblivious of light understood as the Other; be it a computer, a TV screen, or yet a tracker. Perhaps, we should be least afraid of light, be it optical fibre, or even light without the source of light. After all, optical fibre is only a transparent medium for transmission made to a diameter slightly thicker than that of a human hair. Light travels down a fibre-optic cable by repeatedly bouncing off the walls as a result of the continuous decrease in the refractive index, a marvel probably not foreseen by Descartes in his "Treatise on Light".

Above all, therefore, we should be afraid of a double informant, a police spy disguised in the form of wearable technologies. We should be afraid of Vidocq. It is the one with a knack of spinning bedtime stories. Perhaps the true problem is neither light nor darkness, but the great divide in between, i.e., the distinction between day and night, and the divider of day and night. In Leibniz's metaphysics God is a substance devoid of point of view, God is the whole made up of all parts, and as such is devoid of prospects.

BIBLIOGRAPHICAL REFERENCES

Adorno, W. Theodor. (1966/2004) Negative Dialectics. Trans. E. B. Ashton. London and New York: Routledge.

Adorno, W. Theodor, Horkheimer, Max. (1944/2002) Dialectic of Enlightenment. Trans. Edmund Jephcott. Stanford: Stanford University Press.

Benjamin, Walter. (1982/2005) The Arcades Project. Trans. Howard Eiland & Kevin McLaughlin. Cambridge, MA. & London: Belknap Press.

Benjamin, Walter. (1938/2006) The Paris of the Second Empire in Baudelaire, [in:] Walter Benjamin: Selected Writings, Volume 4: 1938–1940, edited by Howard Eiland, Michael W. Jennings. Cambridge, Mass.: United States.

Bentham, Jeremy. (1995) The Panopticon Writings. Edited and introduced by Miran Bozovic. London: Verso.

Bergson, Henri. (1896/1990) Matter and Memory. Trans. N.M. Paul and W.S. Palmer. Zone Books.

Breuer, Josef. (1895/1955) Studies on Hysteria. Trans. James Strachey. [in:] The Standard Edition of the Complete Psychological Works of Sigmund Freud, Vol. II. Hogarth Press: London.

Chamayou, Grégoire. (2012) Manhunts. A Philosophical History. Trans. Steven Rendall, Princeton University Press.

Chamayou, Grégoire. (2015) A Theory of the Drone. Trans. Janet Lloyd. The New Press.

Debord, Guy. (1967/1995) The Society of the Spectacle. Trans. Donald Nicholson-Smith. Zone Books.

Descartes, René. (1629/1998) The World or Treatise on the Light, [in:] Descartes, René. The World and Other Writings. trans. Stephen Gaukroger, New York: Cambridge University Press.

Foucault, Michel. (1975/1978) Discipline and Punish: The Birth of the Prison. Trans. Alan Sheridan, New York: Random House.

Foucault, Michel. (1963/1973) The Birth of the Clinic: An Archaeology of Medical Perception. Trans. Alan Sheridan, New York: Random House.

Hobbes, Thomas. (1651/2010) Leviathan: Or the Matter, Form, and Power of a Common-Wealth Ecclesiasticall and Civill. Edited by Ian Shapiro. Yale University Press.

Horvath, Joan, Hoge, Lyn, Cameron, Rich. (2016) Practical Fashion Tech: Wearable Technologies for Costuming, Cosplay, and Everyday. Apress.

James, Daniel A., Petrone, Nicola. (2016) Sensors and Wearable Technologies in Sport: Technologies, Trends and Approaches for Implementation, Springer Singapore: Springer Briefs in Applied Sciences and Technology.

Malabou, Catherine. (2013) Self and Emotional Life: Merging Philosophy, Psychoanalysis, and Neuroscience, New York: Columbia University Press.

Malabou, Catherine. (2009) What Should We Do With Our Brain? Trans. Sebastian Rand. New York: Fordham University Press.

Malabou, Catherine. (2012) The New Wounded: From Neurosis to Brain Damage (New York: Fordham University Press.

Moglen, Eben. (2010) Freedom In the Cloud: Software Freedom, Privacy, and Security for Web 2.0 and Cloud Computing, https://www.softwarefreedom.org/events/2010/isocny/FreedomInTheCloud-transcript.html

More, Thomas. (1516/1967) Utopia. Trans. John P. Dolan. [in:] GREENE, James J. and Dolan John P., [eds.]. The Essential Thomas More. New York: New American Library.

Rose, Nikolas. (2007) The Politics of Life Itself: Biomedicine, Power, and Subjectivity in the Twenty-First Century, PUP.

Rose, Nikolas. (2008) Governing the Present: Administering Economic, Social and Personal Life, with Peter Miller.

Rose, Nikolas. (2013) Neuro: The New Brain Sciences and the Management of the Mind, with Joelle M. Abi-Rached PUP.

The Holy Bible, an Exact Reprint Page for Page of the Authorized Version Published in the Year MDCXI. Oxford: Oxford University Press, 1833.

Automatic workflow for 4D-BIM based modelling

Carmine Cavalliere, Guido Raffaele Dell'osso & Maria Angela Leogrande
Department of Civil, Environmental, Land, Construction Engineering and Chemistry, Polytechnic University of Bari, Bari, Italy

ABSTRACT: Four-dimensional (4D) simulation could reduce the uncertainty relating to the building processes increasing the visibility of the construction workflows and supporting the detection of the potential conflicts and safety hazards at the construction site. The aim of the research is to provide an automatic approach for the real-time scheduling through the Building Information Modelling (BIM) methodology in order to select the best scheduling alternative. The availability of information relating to the duration of the work tasks execution, within a unique digital building model, allows performing 4D modelling in the early building design stages in order to enable stakeholders to assist decision-making and to optimise the site layout plans and the work tasks themselves. The object identification codes (ID) and the implementation of the temporal parameters linked to the building information model elements, within the 3D model, make possible the automatic execution of the 4D BIM. This methodological approach could decrease the very time-consuming activities of data entry that are usually conducted at the end of the construction (building) stages. Furthermore, this procedure allows increased the coherence of information among different design activities, such as between modelling and scheduling that are often found to be asynchronous and affected by inconsistencies.

Keywords: Construction management, 4D-simulation, BIM.

1 INTRODUCTION

The increasing complexity of buildings and the dynamic nature of the construction workflows make it difficult to complete a project in accordance with the specifications of time, costs, quality and safety.

One of the major challenges, especially for large construction projects, is to visualise mentally the on-site work task activities, which often lead to misinterpretations. Four-dimensional (4D) simulations could improve the quality of the project execution planning increasing the availability of information, the integration and communication between stakeholders and, at the same time, it could minimise the occurrence of potential conflicts (Hartmann et al., 2008; Staub-French et al., 2008; Chau et al., 2004).

The 4D BIM-based modelling application is also suitable for construction quality management to help project participants to understand better the requirements acceptance and to communicate in a visualised manner (Chen and Luo, 2014). 4D simulations based on the Building Information Modelling approach make possible the continuous optimisation of the project in the early design phases supporting the decision-making process.

In order to guarantee a wider use of this application, some researches developed automatic procedures capable of readily analysing the different scheduling alternatives (Haiati et al., 2016; Montaser and Moselhi, 2015; Bhatla and Leite, 2012) and methods to perform quantitative analyses for space management (Jongeling et al., 2008).

This paper proposes an automatic and iterative procedure to conduct a 4D BIM developed in a case study.

2 BACKGROUND

2.1 *Building information modelling*

BIM is the necessary evolution of the construction industry paradigm related to the growing complexity of the building process.

BIM allows stakeholders to manage the digital information of the building throughout its entire lifecycle providing a knowledge domain of the construction processes coming from policies, technologies and processes interaction (Succar, 2009). This methodological approach proposes a re-engineering of the design practices moving the building design from two-dimensional (2D) based documentation to n-dimensional (nD) modelling in which the parameters related to the whole lifecycle of the building are connected into the digital building prototype (Mihindu and Arayici, 2008).

2.2 *Temporal parameters into building information modelling*

The n-dimensional design approach guarantees to govern time as a manageable parameter within the fourth dimension (4D).

According to Brito and Ferreira (2015, p. 379) 4D simulations, when applied to construction management, make possible a number of benefits:

1. Effort reduction in visualisation and mental interpretation of the schedule;
2. Capacity of identification potential conflicts of space and time during construction;
3. Support in the insertion of equipment, construction site elements and resources in space for planning analysis;
4. Support in the sequencing and ideal suitability of the schedule work rate;
5. Visualisation of possible impacts caused by changes in planning and
6. Easiness of integration and communication between all project stakeholders.

Hence, 4D modelling provides a higher quality project meeting the predicted cost, duration, and quality.

In this scenario, some studies developed methods for the optimisation and the implementation of 4D simulations, starting from the innovation of 4D BIM applications in construction and engineering projects (Lopez et al., 2015). Montaser and Moselhi (2015) depicted an automated methodology for constructing a 4D BIM by use of specific software and the development of an Application Programming Interface (API) setting a newly added attribute to each 3D BIM object. Other researches presented some developments for automatic workspace conflict verification system (Moon et al., 2014; Su and Cai, 2014). Kim et al. (2013) developed an automatic method for the construction progress measurement by comparing the as-built model with the as-planned parameters through the use of a 4D BIM and 3D data resulting from remote-sensing technology.

The combination of various colours is often used to characterise the different construction states (Chang et al., 2009). Indeed, the use of colours provides the possibility to increase the integration and communication among the stakeholders and to identify the potential conflicts more easily.

It should be emphasised that 4D simulations, as well as the other n-dimensional modelling, require an accurate digital building model related to the aim of the model itself. Thus, particular attention should be paid to the information required to perform a 4D model. In this connection, Boton et al. (2015) described the level of graphical and temporal details required for the 4D models according to the definition of the different Levels of Development (LOD).

The opportunities provided by the automatic mechanism of the BIM processes, and especially the possibility to monitor over time the evolution of the construction states, are also improving how safety aspects can be approached. Dynamic conditions in terms of risk management at the construction site can be tracked in the early stages of the building design using methods for automatically detecting hazards. Zhang et al. (2013) developed algorithms based on an automated safety-rule model checker that can automatically analyse the digital building model to detect fall-related hazards and suggest the relating preventive measures to users.

Scientific and technological developments make possible faster management and sharing of information even playing a significant role in reducing incident rates (Cheng et al., 2011; Teizer et al., 2010). Hadikusumo and Rowlinson (2004), for instance, implemented an innovative method for safety hazards identification at the construction site by using a design-for-safety-process (DFSP) tool.

3 FRAMEWORK AND METHODOLOGY

Moving from the scenarios analysed in the previous sections, the research proposes an automatic method for the 4D modelling using specific software. The proposed methodology makes possible the automatic achievement of a 4D model starting from the 3D modelling by including new temporal parameters. Furthermore, the procedure allows the iterative data updating based on needed design changes without requiring manual adjustments.

The methodology shown in figure 1 (Fig.1) is cyclically developed by consequential activities starting from a 3D model made by Autodesk Revit 2017.

The automated procedure was tested on a BIM model of a single-family house with two floors, having an internal surface area of 177.25 square meters. The structure is made of beams and pillars realised with cast-in-situ reinforced concrete, ground beam foundations, brick walls (30 cm) and floors (25 cm).

Figure 2 shows the temporal progress as displayed in the 4D simulation software, while the workflow used for its processing is depicted in Figure 1.

The 3D model has the geometrical information and the temporal attributes in order to govern effectively the 4D simulations. The following parameters are set for each BIM object:

Process: necessary for the definition of the element type to be realised;

Unique Code (UC): necessary for the unambiguous characterization of the specific element and for the correct connection of data coming from 1.1 and 1.3 activities within the activities 1.4 (Fig. 1);

- Priority: activities to be completed before starting the construction of the element;
- Time: days required for the construction of the element;
- Planned to Start (PtoS): starting date of work (yyyy/mm/dd HH:mm);
- Planned to Finish (PtoF): completion date of work (yyyy/mm/dd HH:mm);

- Activity: needed to be able to visualise 4D simulation in Autodesk Navisworks Manage 2017.

After the 3D modelling phase, an IFC file is exported for the subsequent import into the 4D simulation software.

The second activity of the workflow (activity 1.2, Fig. 1) concerns the development of an algorithm for the management of data in Autodesk Dynamo Studio, before the subsequent export to Microsoft Excel. The implementation of data from Revit into Dynamo Studio is necessary in order to create a platform that interacts with Revit and Excel. The script made in Dynamo Studio, by using the ExcelWriteToFile function, allows to automatically creating a list of all the 3D model elements with the related previously set temporal parameters necessary for the creation of 4D simulation in Navisworks Manage.

The subsequent activity 1.3 concerns the export of the previously Dynamo Studio data set into a spreadsheet (CSV file). This step allows the generation of a list of activities (Cartel 1, Fig. 1) organised in chronological order and highly manageable by all the participants in the project.

The 4D modelling phase requires a prior import of the IFC file from 3D model made by Revit and the CSV file from Excel (activity 1.4, Fig. 1). Thereby it is possible to obtain the visual simulation of the on-site work task activities.

The automation of the process is made possible by setting the Unique Code attribute (UC). In particular, the UC attribute is assigned to each element of the model and therefore characterises both the information of the 3D model and that contained within the Excel data (since they come directly from the same Revit model).

The automatic association by rules in Navisworks allows the linking of each construction task related to the construction of the model elements (resulting from the CSV file) to each 3D model element (resulting from the IFC file). This connection is made possible based on the UC attribute shared between the geometrical data from Revit and the temporal parameters from CSV file. After making the association, it is possible to conduct automatically the 4D simulation. It should be emphasised that the direct export from the 3D model to the 4D simulator software still allows the reading of the temporal attributes but it does not ensure a complete process automation. In this case, the manual procedure requires a high computational charge and a lack of the automation of the workflow that would lead to a low level of temporal parameters management also resulting in the increase of related errors.

When traced backwards, the methodological approach depicted in figure 1 allows making any temporal parameters changes (e.g. in connection with different design choices) and then automatically updating the data into the 3D model, during all stages of the building process. The changes are directly performed in the 4D simulator acting on the temporal parameters or graphically modifying the temporal bars into the timeliner section (activity 2.1, Fig. 1). This procedure

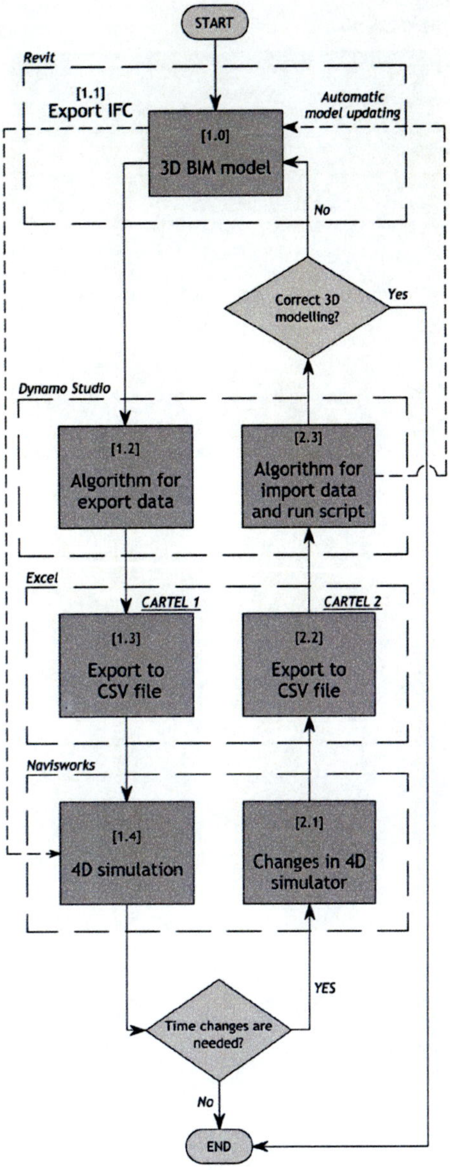

Figure 1. Automatic workflow for 4D modelling.

allows stakeholders to visualise changes on a real-time basis.

After the changes, the subsequent phase (activity 2.2, Fig. 1) consists of exporting a new CSV file (Cartel 2, Fig. 1) in order to create an additional worksheet in which it is possible to display only the information of interest to be re-imported into the 3D Revit model. The ExcelReadFromFile function in Dynamo makes it possible to associate the new temporal parameters as modified in Navisworks (resulting from Cartel 2, activity 2.2, Fig. 1) to the related Revit model elements, again thanks to the Unique Code sharing. The

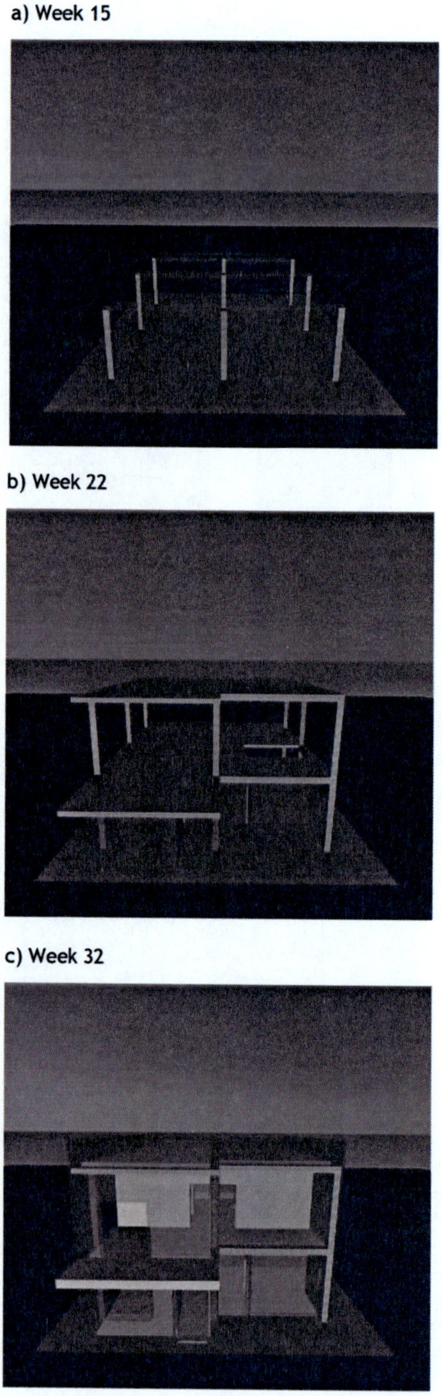

Figure 2. 4D simulation along schedule.

UC attribute allows attaching the modified temporal parameters to the corresponding elements of the model. This way, the changes done in Navisworks Manage are directly visible in the Revit 3D model and the temporal attributes of the objects are automatically updated only by running the script in Dynamo Studio. The workflow appears to be concluded if there are no more changes to be made either to the model or to the time schedule. In the case of further necessary design changes related to the temporal parameters, it is possible to start from the activity 2.1 according to the analysed procedures and in accordance with the flow chart illustrated in figure 1. Obviously, in the case of changes related to the modelling of the 3D model (e.g. geometrical changes) it is needed to start from activity 1.0.

4 CONCLUSIONS

Very often, the lack of adequate information about the temporal aspects of the building process causes high computational costs. This traditional design approach leads to inconsistent information and is often affected by errors. In the traditional practices, the project schedule activities are conducted through the development of two-dimensional (2D) documents, such as a Gantt chart or PERT chart. Such 2D-based support does not allow project teams to visualise the temporal progress of the work tasks execution but this activity is assigned to the mental visualisation and the experience of the designers. Moreover, these types of documents require manual adjustments in correspondence with changes in the design choices resulting in a very time-consuming activity.

The BIM approach allows the connection of information from a 3D model to a simulator able to generate automatically a 4D simulation. The potentiality of this approach makes it possible to identify potential conflicts and on-site spatial or temporal interferences, identify risk situations and supports the activities for the reallocation of resources. Furthermore, 4D simulations can constitute the basis of coordination meetings ensuring greater collaboration and communication between all project participants.

This study has depicted a workflow for applying an automated procedure for the 4D simulations. After implementing the algorithms for importing and exporting data in specific software, any changes of temporal parameters necessary to execute the 4D simulation are directly visible on the 3D model and vice versa. Such automated workflow allows project teams to have always-consistent data and to analyse the work task execution at any phase of the building design process.

BIBLIOGRAPHICAL REFERENCES

Bhatla, A. and Leite, F. (2012) Integration Framework of BIM with the Last Planner System. In *20th Annual Conference of the International Group for Lean Construction*. San Diego State University, San Diego.

Boton, C., Kubicki, S. and Halin, G. (2015) The Challenge of Level of Development in 4D/BIM Simulation Across AEC Project Lifecyle. A Case Study. *Procedia Engineering*, 123, pp. 59–67.

Brito, D. M. and Ferreira, E. A. M. (2015) Strategies for Representation and Analyses of 4D Modeling Applied to Construction Project Management. *Procedia Economics and Finance*, 21, pp. 374–382.

Chau, K. W., Anson, M. and Zhang, J. P. (2004) Four-Dimensional Visualization of Construction Scheduling and Site Utilization. *Journal of Construction Engineering and Management*, 130, pp. 598–606.

Chang, H. S., Kang, S. C. and Chen, P. H. (2009) Systematic procedure of determining an ideal color scheme on 4D models. *Advanced Engineering Informatics*, 23, pp. 463–473.

Chen, L. and Luo, H. (2014) A BIM-based construction quality management model and its applications. *Automation in Construction*, 46, pp. 64–73.

Cheng, T., Venugopal, M., Teizer, J. and Vela, P. A. (2011) Performance evaluation of ultra wideband technology for construction resource location tracking in harsh environments. *Automation in Construction*, 20, pp. 1173–1184.

Hadikusumo, B. H. W. and Rowlinson, S. (2004) Capturing safety knowledge using design-for-safety- process tool. *Journal of Construction Engineering and Management*, 130, pp. 281–289.

Haiati, O., von Heyl, J. and Schmalz, S. (2016) Bim and Sequence Simulation in Structural Work – Development of a Procedure for Automation. In *Proc. 24th Ann. Conf. of the Int'l. Group for Lean Construction*. Boston, MA, USA, pp. 73–82.

Hartmann, T., Gao, J. and Fischer, M. (2008) Areas of Application for 3D and 4D Models. *Journal of Construction Engineering and Management*, 134, pp. 776–785.

Jongeling, R., Kim, J., Fischer, M., Mourgues, C. and Olofsson, T. (2008) Quantitative analysis of workflow, temporary structure usage, and productivity using 4D models. *Automation in Construction*, 17, pp. 780–791.

Kim, C., Son, H. and Kim, C. (2013) Automated construction progress measurement using a 4D building information model and 3D data. *Automation in Construction*, 31, pp. 75–82.

Lopez, R., Chong, H.-Y., Wang, X. and Graham, J. (2016) Technical Review: Analysis and Appraisal of Four-Dimensional Building Information Modeling Usability in Construction and Engineering Projects. *Journal of Construction Engineering and Management*, 142(5), pp. 1–6.

Mihindu, S. and Arayici, Y. (2008) Digital construction through BIM systems will drive the Re-engineering of construction business practices. In *12th International Conference Information Visualisation*, pp. 29–34.

Montaser, A. and Moselhi, O. (2015) Methodology for Automated Generation of 4D Bim. In *5th International/11th Construction Specialty Conference*. Vancouver, British Columbia, pp. 1–10.

Moon, H., Dawood, N. and Kang, L. (2014) Development of workspace conflict visualization system using 4D object of work schedule. *Advanced Engineering Informatics*, 28, pp. 50–65.

Staub-French, S., Russell, A. and Tran, N. (2008) Linear Scheduling and 4D Visualization. *Journal of Computing in Civil Engineering*, 22, pp. 192–205.

Succar, B. (2009) Building information modelling framework: A research and delivery foundation for industry stakeholders. *Automation in Construction*, 18, pp. 357–375.

Su, X. and Cai, H. (2014) Life Cycle Approach to Construction Workspace Modeling and Planning. *Journal of Construction Engineering and Management*, 140, pp. 1–12.

Teizer, J., Allread, B. S., Fullerton, C. E. and Hinze, J. (2010) Autonomous pro-active real-time construction worker and equipment operator proximity safety alert system. *Automation in Construction*, 19, pp. 630–640.

Zhang, S., Teizer, J., Lee, J. K., Eastman, C. M. and Venugopal, M. (2013) Building Information Modeling (BIM) and Safety: Automatic Safety Checking of Construction Models and Schedules. *Automation in Construction*, 29, pp. 183–195.

Progress of what civilisation?

Teresa V. Sá
Section of Social and Territorial Sciences, Faculdade de Arquitectura, Universidade de Lisboa, Lisboa, Portugal

ABSTRACT: This article analyses some transformations in the meaning of the word progress over time, from the 19th century, when it was in fashion, to the present day. It will briefly look at how progress has become "progresses" and has undergone a series of metamorphoses that have shaped it into growth, development and sustainability, creating new versions. The central idea of progress, however, is always one of technological and scientific expansion, and prioritised investment in it is never truly questioned.

It is precisely this aspect that will be critically examined in the article, finding within it a naturalisation of progress. Finally, based on an interpretation of a tale by Eça de Queirós, Civilização, the need for a debate is suggested, regarding the forms of progress and civilisation that are able to encourage the political creation of conditions for a "good life".

Keywords: Progress, science and technology, nature, civilisation

1 FROM PROGRESS TO "PROGRESSES"

Words have a time and a place. They can come into fashion, go out of fashion, have one meaning and then another. They have a physical, social and cultural context. All this can be plainly seen in everyday life. Words dress us, i.e. they place us within the social hierarchy, in a physical place and at a particular historical time. The word progress is a good example of the importance of words' position in time and space.

But some words also cause "feelings" (Bauman, 2003). Community is one of those words: it inspires trust and security in us; we can count on one another; there are no strangers, no dangers; we feel safe. Community indicates something good, calm and peaceful. But this feeling of peace drives away the word's true meaning. Bauman states that there is a difference between the "community of our dreams" and the "really existing community". "Do you want security? Give up your freedom, or at least a good chunk of it. Do you want confidence? Do not trust anybody outside your community. Do you want mutual understanding? Don't speak to foreigners nor speak foreign languages." (Bauman, 2003:10).

The word progress, so often used by social philosophers in the 18th and 19th centuries, also created a good feeling. Society was marching in the right direction, and it just needed to follow that natural direction: – the direction of progress. We were all going to live better, with more freedom and more security. A word full of hope, which calmed people down, which reassured them about the future.

In Europe, the 19th century was the century of progress: "the 19th century lived in the conviction that acceleration of material progresses would produce unprecedented changes for human societies." (Julliard, 1997:159). Carl Becker, in 1921, stated that belief in progress was the compensation mankind had found to replace belief in Providence (in Lasch, 2002) – belief in religion is replaced by belief in science and technology. Both Condorcet and Comte argued very clearly that development in scientific knowledge that would allow for greater and greater domination of nature involved the idea of perfectibility, in which scientific knowledge (truth), virtue and joy come together.

Comte sought to answer the questions: "Where are we going? Where is the development of humanity leading us? Is it led in the 'right direction' and is that the direction of my goals and ideals?" (Elias, 1999:39). He wished to answer these questions using scientific knowledge.

The word progress appears in the Littré dictionary of French (1859–1872) with two meanings: 1) progressive movement of civilisation, political institutions; 2) forward movement – the action of advancing (in Taguieff, 2001).

The first definition contains the idea of value, which to a certain extent corresponds to the thought of those who defended progress in the 19th century; the second definition, which means an advance without a positive or negative connotation, corresponds to the first meaning of the word progress. It was Rousseau, in 1755, who introduced the idea of perfectibility to public opinion and debate. In his view, it did not mean "perfecting" but rather that which would allow man to leave his "primitive state", that which leads him to be transformed. It is "this distinctive and almost unlimited faculty (. . .) [that] causing over the centuries his acumen and his errors, his vices and his virtues to flourish, which eventually makes man a tyrant over himself and nature." (Rousseau, in Taguieff,2001:61).

This second definition of progress, of moving forwards, without any objective other than to advance further, advance without stopping, appears to match the use of technology and science by contemporary society in the Western world. It keenly brings to mind the use of NIT (new information technology) in today's society, the applications of scientific knowledge in the field of biology and chemistry, in which new discoveries join more new discoveries, transforming all of our lives and making it possible to achieve more and more innovations without anyone pondering enough about what drives us to transform in general and transform ourselves in such a way. Innovations become progresses in themselves and for their own sake. Progress is transformed into progresses.

In fact, the idea of progress today increasingly corresponds to a march forward without a goal, bringing ourselves closer to Rousseau's perfectibility than Comte and Condorcet's ideas. This trend is based on two fallacies decried by Castoriadis about social representations of science decried by Castoriadis (Castoriadis, 1990). One is dealing with science as a pragmatic truth, i.e. it can be checked through facts. This is therefore a thought that addresses the how and not the why. The other comes from the scientific notions of the 19th century and argues that scientific knowledge gets closer and closer to reality, i.e. previous thought is less true than new thought. These two fallacies, to follow Castoriadis' thinking, are based on the idea that scientific thought understands reality better and better, and that, for that to happen, the philosophical aspect related to the why no longer makes sense. The problem of truth, the question raised by Auguste Comte, whatever the answer found to it, the problem of knowing whether or not the path followed is the right path, no longer makes any sense. Science serves to provide quick answers to the problems affecting society. And one of the problems facing society today is enabling each and every person to consume more, so that he/she can move more and communicate more. As a significant portion of people in the most developed countries can consume more, move more, communicate more, a very small group at the top – an economic elite, which dominates the entire process – gets stronger. Society increasingly performs technical-pragmatic actions, the major goal of which is solving concrete problems that make it possible, simultaneously or later on, for them to be appropriated by big capital." (Castoriadis, 1990).

2 PROGRESS: METAMORPHOSES OF THE WORD

Pierre-André Taguieff presents six traits of the modern view of progress, from Bacon to Condorcet: "Progress is a necessary, continuous, linear, cumulative, irreversible and unlimited process." (Taguieff, 2001:58). It seems that any of these would fit with our society. The idea of progress is based above all on scientific development allied with technology. Science and technology will make it possible to make more and more items, more and more cheaply, and these items can be used by a larger and larger number of people.

The religious dimension has in fact been lost over time, shifting from progress to representing a myth, the "myth of progress" as von Wright calls it (in Bouveresse, 2017). That myth is founded in the idea of considering progress like any other "natural and necessary" thing (Bouveresse, 2017:72). What is new in this idea, is not that humanity makes different types of progress, but the conviction that progress is part of human beings' nature as a species and has no limits (Bouveresse, 2017).

However, the word progress has stopped being the great sacred word of our time – the promise of a better world associated with it in the 19th century was quickly denied at the start of the 20th century with the Great War – and so the word has gradually been replaced by other less ambitious words like growth, development, sustainability[1].

After the Second World War, economic growth emerged as a keyword for the economic reconstruction of Europe, as well as for the third world. A much less ambitious word, growth embodied the goal of managing to achieve more and more with fewer and fewer resources. Technology, numbers, quantification and rationality directed the process. The greater the technological development, the faster objectives were achieved. Human, social and ecological aspects were not taken into account, or they were underestimated or subordinated to the obligation to "grow".

However, during the period of economic growth that took place in post-war Europe, the expansion of the welfare state and workers' struggles in more developed countries led to a bigger and broader middle class. This process was accompanied by the attainment of new social rights, more purchasing power, the demand for and acknowledgement of equal rights for women, growing guarantees of security and a multiplication of new freedoms. As Bauman states (2001), a pact or commitment was established between workers and capital, the terms of which were used by capitalists to impose the rules. Capitalists were nonetheless aware that they could not survive without the cooperation of the working class – both parties recognising, through various negotiation practices, the need for or legitimacy of the other.

In the meantime, a section of the hard left had begun to criticise the ideology of growth because, in its economic rationale, it tended to impose cost-benefit assessments as a fundamental factor; in other words, it underestimated or subordinated anything that could not be to some extent quantified. This criticism of growth found a wider audience, to the point of becoming commonplace in official discourse when

[1] On this point, I shall follow Castoriadis' thought in his article "Reflections on 'rationality' and 'development'".

the negative aspects arising from the industrialisation process became more extensively noticeable. These aspects were reflected in threats connected to the depletion of resources and the degradation of nature.

The term growth is therefore gradually being replaced by development. The idea of "quality of life" has emerged and has been rapidly quantified, with indicators sought that are able to measure this new variable. "Economic indicators" have been joined by "social indicators" or "well-being indicators" (Castoriadis, 1978:83). Development is a much more reassuring word than growth because it is intended to take into account social aspects related to people's well-being.

Nonetheless, as Castoriadis (1978) states, nobody wonders what development is. The word appears to be enough in itself, dispensing with all other justifications. Development is a term that leads us directly to the transformations undergone by of a living being in nature. Societies grow and develop just like animals and plants. And, taking nature as a model, everything seems simple. But it is easy to see where the problem lies in a concept that is apparently as simple as this: "it is not possible to have development without a reference point, a defined state which must be attained; and nature furnishes, for any being, such a 'final' state." (Castoriadis, 1978:87). It is precisely here that the problem can be found. Nature furnishes this final state, but in society we are the ones who furnish it. If we do not discuss that "final" state, it is because we accept that development is the ability to grow endlessly. Historical development becomes the passage from one defined state to another undefined state, i.e. undefined growth supported by the virtual "omnipotence" of technology.

In the 1980s, new problems were identified, connected to environmental degradation and degradation of the very conditions for human survival: global warming, destruction of the ozone layer, desertification (Queirós, 2003). In 1987, the Brundtland Report, Our Common Future, was published. It includes the words sustainable development, expressing the idea that the environment and development are inseparable problems. The report discusses environmental concerns but also social concerns, bearing in mind the social imbalances that development can cause: increased social inequalities, increased poverty in developed countries, famine in poor countries, etc. The fundamental idea is not allowing development to create negative consequences for future generations (Queirós, 2003: 332).

Sustainable development introduced a human side to issues that were usually thought of using economic rationality. It is a term that drives us to think about society beyond the short term and makes us return to ancient wisdoms: in Greece at the start of the 20th century, and even in later years, the oldest members of society "continued to plant olive trees and cypresses, without considering costs or returns. They knew that they should leave the Earth in good order for those who would come after them." (Castoriadis, 1978:96).

What is sustainability? This is a question similar to some that children ask adults and that, because they are so straightforward, they are hard to answer. Reflection on sustainability should lead us to pause for a moment. But, instead, the system is reorganised, as Lefebvre (2014) would say, it adapts itself, assimilates the new word and makes it a new banner. Everything goes on, just the same, and henceforth we will seek new solutions for the evils of progress, still entrusting to technology the mission of controlling technology itself.

3 PROGRESS OF WHAT CIVILISATION?

I now suggest a brief reading of Eça de Queirós' exemplary tale A Civilização (Civilisation), published for the first time posthumously in 1902. It helps us think not only about the idea of progress but about a range of questions that continue to be posed in today's society.

In his tale, Eça describes a character to us: Jacinto, a rich, cultured, civilised bourgeois man, who appears in two opposing settings: the civilised urban world of progresses, information, and useful, functional objects; and the rural world, , a slow-paced world of tradition and nature slow pace.

These two worlds are shown to us through two houses that the author describes exquisitely: the Jasmineiro palace, embodying civilisation, where a profoundly bored, tired and melancholic Jacinto lives; and the Torges manor, embodying the union between man and nature, where a happy, relaxed Jacinto can be found.

Let us take a brief look at the two houses. The form-function relationship for of the palace was perfect. Innovation, rationality and functionality were in charge. Everything was thought out, down to the finest details: "The library, (...) – contained twenty-five thousand books, fitted in ebony, magnificently covered in scarlet morocco. (...) And the only downside to this monumental warehouse of knowledge was that all those who went there inevitably fell asleep, because of the armchairs, which were furnished with thin, moveable boards for holding a book, a cigar, a pencil for taking notes, a cup of coffee, and also offered a shifting, yielding combination of cushions, where the body found, to the spirit's disadvantage, the kindness, depth and elongated peace of a bed." (Queirós:3).

At the Torges manor, the indoor and the outdoor come together: "The grand staircase led to an entirely covered veranda, following the façade of the large house, and it was adorned between its thick granite pillars with boxes full of earth, in which carnations bloomed." (ibid:13). In the kitchen, just like the old village kitchens, the objects were always the same and around them the murmur of the women who worked there could be heard: "The kitchen was a thick mass of dark tones and shapes, the colour of soot, where, on the dirt floor, a red fire glowed at the back, licking thick iron pots, (...) There, a gang of hustling, chattering women plucked chickens, whisked eggs, threshed rice, with righteous zeal..." (ibid:15).

From these two worlds, the civilised and the ancient, which Eça de Queirós brings to life and allows us to experience using domestic settings, I shall present a brief episode connected to our word: progress.

At the Jasmineiro palace, there is an area at the back of the library which was Jacinto's office. It was there that civilisation was powerfully present. This area included the most innovative items from the world of technology, which provided rapid contact with the world and the other. Eça de Queirós called these "the great instruments, enablers of thought": typewriter, duplicating machines, Morse telegraph, phonograph, telephone, théâtrophone, and others, all using shiny metals, all with long cords.". Of these shiny, brilliant devices, "constant short, dry sounds clattered in the warm air of the sanctuary." Tick, tick, tick! Ring, ring, ring! Clack, clack, clack! Whirr, whirr!... It was my friend communicating. All those cords dived into universal forces." (ibid:5).

And now for the brief event: one night, Jacinto wished to let his cousins, the Gouveias, hear the voice of your their relative, the counsellor Pinto Porto, which was recorded on the gramophone. He switched it on, and the counsellor's voice emanated from it: "Marvellous invention! Who can fail to admire the progresses of this century?" But, due to an inexplicable technical malfunction, the phonograph kept on repeating the same sentence: "Who can fail to admire the progress of this century?" They went to a different room, left the house and, when they returned in the early hours of the morning, the phonograph could still be heard: "admire....progress....century?...". It was only in the afternoon that an electrician was able to silence the phonograph. (ibid:5).

In this brief episode the user is supplanted by the technical object: there is no way of stopping the gramophone, something which seemed bizarre and surreal in the second half of the 19th century but suggests something all too real in the 21st century. The speed with which everything transpires or, to put it another way, the rapid transformations in technology that change our everyday lives do not let us switch off the "gramophone". Turning it off means not being like everyone else. It means being "excommunicated", which implies remaining outside the standards of communication, movement and relationships that modulate and regulate (or deregulate) our dealings with the world.

It is hard to dispute that the reunion with nature and the earth that Jacinto adopts as a personal solution conceals, within the criticism of "civilisation" that inspires it, something that we might call the "social question", despite its decisive importance at the time. I am talking about inequalities, relationships of power, domination and exploitation that Jacinto's solution not only does not challenge but implicitly legitimates. Nonetheless, despite its shortcomings and ambiguities, it is interesting to note that complaints about the helplessness of technical and material progress when faced with "malaise" in civilisation do not exclude, for Jacinto, a demand for perfection in the search for the "good life", which represents another idea of progress. Jacinto, in fact, does not abandon civilisation, nor does he seek to exchange it for something which could be the "state of nature" or another "natural state", given naturally: his reunion with nature does not exclude books or many other natural or mental creations from the criticised civilisation criticised, while, on the other hand, the nature that his eyes discover and admire is already a deeply historical, humanised one. So the protagonist's perspective is rather one of a civilised human rediscovering contact with nature as a cultural, civilising alternative – in the same way that his criticism of the idea that the common sense of his time made of progress, that criticism asserts the legitimacy and rationality of a progress sensed to come from another civilisation.

So it is possible to say that in its reunion with Jacinto, nature intervenes against the naturalisation of progress, a characteristic of its mythical representation that was criticised, as we have already seen, by von Wright (in Bouveresse, 2017). In brief, refusing to naturalise progress also means rejecting evolution, social order, uses and customs and the course of history. In brief, rejecting the naturalisation of progress means also rejecting the naturalisation of evolution, social order, habits and customs or the course of history. For that reason, for us to question (and question ourselves about) progress and civilisation, Eça's tale is a good starting point.

BIBLIOGRAPHICAL REFERENCES

Baumn, Z. 2003. Comunidad. En busca de seguridad en un mundo hostil, Madrid, Siglo Veintiuno.

Baumn, Z. 2001. Modernidade Liquida, Rio de Janeiro, Jorge Zahar Editor.

Bouveresse, J. 2017. Le mythe moderne du progres, Marseille, éditions Agone.

Castoriadis, C. 1978. "Reflexões sobre o "desenvolvimento" e a "racionalidade"", Histórias e Ideias nº2, Porto, Afrontamento.

Castoriadis, C. 1990. Le Monde Morcelé. Les carrefours du labyrinthe III (pp. 71–103), Paris, Seuil.

Elias, N. 1999. Introdução à Sociologia, Lisboa, Edições 70.

Julliard, J. 1997. La faute aux élites, Paris, Gallimard.

Lasch, C. 2002. Le Seul et vrai Paradis, Climats.

Lasch, C. and Castoriadis, C. 2012. La Culture de L'égoisme, Paris, Climats.

Lefebvre H. 2014, Critique de la vie quotidienne III, Paris, L'Arche editeur, (first ed. 1981).

Queirós, Eça (1902), A Civilização [Online] Available: http://bibliotecadigital.puc-campinas.edu.br/services/e-books/Eca%20de%20Queiroz%20-%20Civilizacao.pdf, [Accessed 3 January, 2017].

Queirós, M. 2003. "Questões para uma agenda contemporânea do desenvolvimento sustentável", Revista da Faculdade de Letras, Geografia I série. Vol. XIX, Porto (pp. 331–343).

[Online] Avalable: http://ler.letras.up.pt/uploads/ficheiros/3929.pdf, [Accessed 10 January 2017].

Taguieff, P.-A, 2001, Du progres, Paris, Librio.

Author index